The Routledge Handbook of African Development

This handbook presents an extensive new overview of African development – past, present and future. It addresses key core themes and topics that are pertinent to the continent's development – including sections on history, health and food, politics, economics, rural and urban development, and development policy and practice.

The volume draws on the expertise of over 60 of the world's leading scholars to provide a detailed and up-to-date analysis of the key opportunities and challenges that confront Africa, and how such issues are being addressed. Arranged by key themes, the handbook provides not only a historical understanding of the past, but also political perspectives of the future. The chapters provide critically informed analyses of their topics by drawing upon the latest conceptual viewpoints and applied experiences in Africa in a form of case studies to offer a comprehensive examination of the opportunities, challenges, key debates and future prospects.

This handbook is an invaluable state-of-the-art overview and reference concerning many different aspects of Africa's development, which should be of interest to academics in all fields of African studies, and also academics and students working in cognate disciplines such as development studies, geography, history, politics and economics.

Tony Binns is Ron Lister Professor of Geography at the University of Otago, Dunedin, New Zealand, and Visiting Professorial Fellow in the School of Global Studies at the University of Sussex, Brighton, UK.

Kenneth Lynch is a Reader in Geography in the School of Natural and Social Sciences at the University of Gloucestershire, UK.

Etienne Nel is a Professor of Geography at the University of Otago in Dunedin, New Zealand, and Visiting Research Fellow, College of Business and Economics, University of Johannesburg, South Africa.

The Routledge Handbook of African Development

Edited by Tony Binns, Kenneth Lynch and Etienne Nel

Routledge
Taylor & Francis Group
LONDON AND NEW YORK

First published 2018 by Routledge

2 Park Square, Milton Park, Abingdon, Oxon, OX14 4RN
605 Third Avenue, New York, NY 10017

Routledge is an imprint of the Taylor & Francis Group, an informa business

First issued in paperback 2020

British Library Cataloguing-in-Publication Data
A catalogue record for this book is available from the British Library

Library of Congress Cataloging-in-Publication Data
A catalog record for this book has been requested

ISBN: 978-1-138-89029-9 (hbk)
ISBN: 978-0-367-73483-1 (pbk)

Typeset in Bembo
by Swales & Willis Ltd, Exeter, Devon, UK

Contents

Figures

Tables

Boxes

Contributors

Afia A. Adaboh is a PhD student in the Global Development and Sustainability concentration at the Heller School for Social Policy and Management, at Brandeis University, Massachusetts, USA. Her research focuses on the evaluation of policies and programmes aimed at improving the quality of life of vulnerable and marginalised groups in West Africa and the Caribbean.

Clive Agnew is Professor of Physical Geography and Vice President for Teaching Learning and Students at the University of Manchester. He researches and teaches in the fields of water resources development and applied hydrology. His first degree was in Physical Geography (BSc 1976, Newcastle) and his PhD (UEA, 1980) was on water resources and drought in West Africa. Since then, he has worked extensively in both the drylands and the wetlands of Africa, Europe and the Middle East. He is currently engaged in research on the provision of water and ecosystem services in urban areas of Bangladesh and Tanzania.

Nicola Ansell is Professor of Human Geography at Brunel University, London. Her research focuses on social and cultural change in the lives of young people, mainly in southern Africa. Her current research projects are exploring how southern African cash transfer schemes are shaping generational relations, and the links between education and aspiration in remote rural areas of Lesotho, India and Laos. She is the author of *Children, Youth and Development* (Routledge, 2nd edition, 2016).

Joseph K. Assan is currently an Assistant Professor of International Political Economy of Sustainable Development at the Heller School for Social Policy and Management, Brandeis University, Massachusetts, USA. Joseph teaches on the Sustainable International Development Programme. He is the Lead Investigator on Poverty Reduction and Development Programme at the Center for Global Development and Sustainability at the Heller School.

Hazel Barrett is a social scientist who undertook her first degree at the University of Sussex in the School of African and Asian Studies. She then moved to the University of Birmingham to undertake her MA and PhD in West African Studies. Hazel has worked at Coventry University since 1992, and in 2006 she was conferred a Chair in Development Geography. She is currently a Research Professor in the Centre for Trust, Peace and Social Relations. Her main areas of research are the socio-economic aspects of development, in particular gender, health and rural development in sub-Saharan Africa. Over the last two decades, her research has been directed at the social and economic aspects of the HIV/AIDS epidemic in sub-Saharan Africa. More recently, her research has focused on the traditional harmful practice of female genital mutilation (FGM) in Africa and among the African diaspora in the EU.

Simon Batterbury is Professor of Political Ecology, Lancaster Environment Centre, Lancaster University, UK, and Principal Fellow in the School of Geography at the University of Melbourne, Australia. He has worked on rural development and soil and water conservation in rural Burkina Faso and Niger since the early 1990s, and on land tenure and mining in East Timor and New Caledonia.

Tony Binns is Ron Lister Professor of Geography at the University of Otago, Dunedin, New Zealand. Before coming to New Zealand, he worked for 30 years in the School of Global Studies at the University of Sussex, UK. He is a Visiting Professorial Fellow at the University of Sussex. Tony has been involved in teaching and research on Africa for over 40 years, and has taught at universities in Kenya, Nigeria, Sierra Leone and South Africa. He has written or edited some 20 books, 35 book chapters and over 100 peer-reviewed articles. Among his most recent books is *Africa: Diversity and Development* (with Alan Dixon and Etienne Nel, Routledge, 2012). Tony is Chair of the International Editorial Advisory Board of the *Australasian Review of African Studies*, and is a member of the Steering Committee of the International Geographical Union's Africa Commission.

Patrick Bond is Professor of Political Economy at the University of the Witwatersrand School of Governance, Johannesburg, South Africa. He works with social movements, labour and community organisations on economic and environmental justice. Recent books are (with John Saul) *South Africa: The Present as History* (James Currey, 2016) and (co-edited with Ana Garcia) *BRICS: An Anti-Capitalist Critique* (Aakar Books, 2016).

David Booth is a Senior Research Fellow at the Overseas Development Institute in London. He is interested in the political economy of development and in smart approaches to development assistance. He co-authored *Governance for Development in Africa: Solving Collective Action Problems* (Zed Books, 2013).

John Briggs is Professor of Geography at the University of Glasgow and has previously been on the staff in the University of Botswana, Lesotho and Swaziland (UBLS) and the University of Dar es Salaam. He has lived and worked in a number of African countries, most notably Tanzania, Egypt, Sudan, Rwanda, Malawi, Botswana and South Africa. His interests are in rural development issues and particularly in the relationship between indigenous environmental knowledge and agricultural practices. Currently, John is Vice Principal at the University of Glasgow with 'responsibility for the governance of academic policy and practice across the institution'.

Deborah Fahy Bryceson is Honorary Research Fellow and Professor at the Centre of African Studies at the University of Edinburgh. As an economic geographer, her research interests span: (1) livelihood, labour, mobility, deagrarianisation and urban growth in Africa; (2) the social dynamics of transnational families, creole societies, drinking patterns, and age and gender differentiation; and (3) agrarian studies, including food marketing, agricultural policy, urban food supply, and rural and urban transport patterns. She has published *Mining and Social Transformation in Africa* (Routledge 2014), *African Urban Economies* (Macmillan, 2006), *Disappearing Peasantries? Rural Labour in Africa, Asia and Latin America* (Intermediate Technology Publications, 2000) and *Women Wielding the Hoe* (Berg Publishers, 1995).

Hazel Cameron is a Lecturer in the School of Inernational Relations at the University of St Andrews. Her research focuses on issues of state crime, state responsibility and global elite bystanders to genocide, and mass violence. Her doctoral thesis and a recent monograph focus on the Rwandan genocide, and she also has a research focus on state-sponsored political violence in Zimbabwe.

Stephen Devereux is a Research Fellow at the Institute of Development Studies at the University of Sussex, UK, where he is a Co-Director of the Centre for Social Protection. He holds the SA-UK Bilateral Research Chair in Social Protection and Food Security, affiliated to the Centre of Excellence in Food Security and the Institute for Social Development at the University of the Western Cape, South Africa. His edited books include *Food Security in Sub-Saharan Africa* (with Simon Maxwell, ITDG Publishing, 2001), *The New Famines* (Routledge, 2007), and *Seasonality, Rural Livelihoods and Development* (with Rachel Sabates-Wheeler and Richard Longhurst, Routledge, 2012).

Alan Dixon is a Principal Lecturer in Geography at the University of Worcester, and a Research Associate of Wetland Action. His research has focused on human–environment interactions in wetland socioecological systems in Africa, particularly the role of local knowledge and institutional arrangements in supporting sustainable livelihoods.

David Fielding is Donald Reid Professor of Economics at the University of Otago. He studied at Oxford University and has a range of research interests in development economics and political economy, including institutional reform in sub-Saharan Africa and the economics of ethnic and religious differences.

Alexandre Apsan Frediani is a Lecturer and Co-director of MSc in Social Development Practice at the Bartlett Development Planning Unit of University College London and a founder member of the Sierra Leone Urban Research Centre. His research has focused on the application of Amartya Sen's capability approach in development practice, as well as on the thinking and practice of participatory planning and design for socially just urban development.

Erik Green is an Associate Professor in Economic History at Lund University, Sweden, and a Research Fellow in the Department of Economics at Stellenbosch University, South Africa. Green's research focuses mainly on issues related to agrarian change, rural labour and inequality in eastern and southern colonial Africa. He has published extensively on African farming and labour in colonial Malawi and colonial Zimbabwe, and the impact of settler expansion in colonial Zimbabwe, colonial Malawi and Cape Colony. He has also co-authored a textbook on African economic history and published conceptual papers on African economic history. Green is the co-founder of the international African Economic History Network (www.aehnetwork.org) and a member of the organising committee of the Annual Meeting of the African Economic History Network.

Jonathan Hobson is Head of Social Sciences at the University of Gloucestershire, UK, where he teaches sociology and criminology. A political sociologist by training, his current research includes the geopolitics of genocide recognition and transnational analysis of restorative justice approaches used for rebuilding communities.

Lauren A. Johnston is a Research Fellow in the Faculty of Business and Economics, University of Melbourne, Australia. She is an expert in development and international economics applied to Chinese and African economies, and their interaction. Lauren has worked for the Ministry of Finance of Sierra Leone, the World Economic Forum and World Bank. She holds a PhD in Economics (Peking), an MSc in Development Economics (SOAS) and a BA/BCom (Melbourne).

David Jolley is a Lecturer in International Relations at Flinders University, Adelaide, Australia, where he specialises in teaching development studies, African political history and Latin American political economy (since 2005). He has co-authored articles for the *Journal of Iberian and Latin American Research* (with Dr Geralyn Pye).

Peris Jones is Associate Professor at the Norwegian Centre for Human Rights, University of Oslo, Norway. He is a human geographer who has worked on different aspects of human rights, not least their transformative potential. His more recent work seeks to conjoin these interests as geographies of human rights, particularly in cities. Until recently, Peris was Associate Research Fellow at the British Institute in Eastern Africa, Nairobi.

Sam Kareithi is a leading development practitioner in eastern and southern Africa, with considerable experience in facilitating agricultural policies, agricultural growth and access to markets; catalysing private sector investment for market sustainability; and formulating and supporting better implementation of policy. He has taught at universities in the UK and South Africa, and has held several positions in the international development sector, including Regional Manager for the Swiss Agency for Development Cooperation (SDC), Deputy Team Leader of the DFID-financed Foodtrade Eastern and Southern Africa Programme and Team Leader of the ADF Facility in the DFID-financed VUNA programme in southern Africa.

Rebecca Kariuki is a Marie Curie Early Stage Researcher at the York Institute of Tropical Ecosystems, Environment Department, University of York, UK. Her research interests lie in the areas of ecosystem services, plant ecology, environmental modelling and people–environment interactions in African savannas.

Braima Koroma is a Lecturer at the Institute of Geography and Development Studies (IGDS) Njala University, Sierra Leone, and a Co-Director at the Sierra Leone Urban Research Centre. His research interests include urban livelihoods and the city economy, disaster risk, urban planning and development, climate change, and development impact evaluation. Braima has over 11 years' experience of teaching, research, training and facilitation, and consultancy on a broad spectrum of interdisciplinary research to examine complex development problems with various national and international organisations and institutions, including the World Bank, UN, ADB and German Technical Cooperation (GIZ). Braima has authored a number of reports and publications.

Dinar D. Kharisma is a PhD student in Global Development and Sustainability Concentration at the Heller School for Social Policy and Management, Brandeis University, Massachusetts, USA. Prior to studying at Heller, he worked with the Government of Indonesia, mostly on the development and evaluation of various national social protection programmes.

Kenneth Lynch is a Reader in Geography in the School of Natural and Social Sciences at the University of Gloucestershire, UK, where he teaches Geography and also contributes to programmes in Sociology and Criminology. Kenneth's research has focused on the relationships between the city and countryside in the UK and Africa, where he has undertaken field-based research in Tanzania, Nigeria, Sierra Leone and South Africa. His most recent projects have focused on urban and peri-urban food supply in Freetown, Sierra Leone, zoonotic disease transmission in

Nairobi, and ecosystems services for water and food security in South Africa and Egypt. He also worked for five years as Deputy Director of the University of Gloucestershire's Pedagogic Research and Scholarship Institute, where he led programmes developing research capacity and investigated undergraduate research, student–community engagement and technology-enhanced learning. In 2004, he was awarded a UK National Teaching Fellowship.

Tanya Lyons is the former President of the African Studies Association of Australasia and the Pacific (2012–2017), the Editor of the *Australasian Review of African Studies* (since 2009), and is a Senior Lecturer in International Relations at Flinders University, Adelaide, Australia, where she specialises in teaching African political history. She has co-edited the following books: *African Frontiers: Insurgency, Governance and Peacebuilding in Postcolonial States* (with John Idriss Lahai, 2015), *New Engagement: Contemporary Australian Foreign Policy Toward Africa* (with David Mickler, 2013), *South Sudanese Diaspora in Australia and New Zealand: Reconciling the Past with the Present* (with Jay Marlowe and Anne Harris, 2013) and *Africa on a Global Stage* (with Gerry Pye, 2006). Tanya has also written chapters in other edited collections, including *Australian Foreign Policy Toward Africa* (2012) and *The State of African Studies in Australia* (with Elizabeth Dimock, 2007). She is the author of *Guns and Guerrilla Girls: Women in the Zimbabwean Liberation Struggle* (2004).

Joseph M. Macarthy is an urban planner with specialty in urban planning and climate change adaptation. He lectures in the Institute of Geography and Development Studies, Njala University, Sierra Leone. Joseph's research interest centres on urban planning and management, climate change adaptation and urban development. He is also one of the co-directors of the Sierra Leone Urban Research Centre in Freetown, Sierra Leone.

Roy Maconachie is Reader in International Development in the Department of Social and Policy Sciences at Bath University, UK. A human geographer by training, his research is concerned with the political economy of natural resource management, with a specific interest in the socio-economic dimensions of the extractive industries in West Africa. Over the past decade, he has carried out fieldwork in Nigeria, Sierra Leone, Liberia and Ghana. This work has been funded by a range of organisations, including the Leverhulme Trust, the British Council, the British Academy and the Nuffield Foundation. His recent research, funded by Humanity United, the International Growth Centre and the Canadian International Resources and Development Institute, seeks to deepen understanding of the governance of artisanal and small-scale mining, and the challenges of formalising the sector. He has published more than 50 peer-reviewed journal articles and book chapters based on his research, and has provided consultancy services for a wide range of international development organisations, including the World Bank, the United Nations, the Natural Resource Governance Institute and GIZ.

Robert Marchant is a Reader in the Environment Department at the University of York, UK. His research and teaching interests focus on the theme of vegetation dynamics and ecosystem change. His PhD research at the University of Hull involved producing a 50,000-year record of forest history from the Bwindi Impenetrable Forest, Uganda, to understand forcing mechanisms (climatic, human and ecological) responsible for sedimentary and vegetation change. His current research is based in Colombia, Amazonia, Kenya, Tanzania and Uganda. Robert's research has been funded by research councils (e.g. the UK's National Environmental Research Council, the Netherlands Foundation), international agencies (e.g. WWF, the EU) and national governments (e.g. the UK and the Netherlands).

Claire Mercer is Associate Professor of Human Geography in the Department of Geography and Environment at the London School of Economics, UK. She is the author, of *Development and the African Diaspora: Place and the Politics of Home* (with Ben Page and Martin Evans, Zed Books, 2008), and she has published widely on diaspora, civil society and NGOs in Africa. Her current research is looking at the relationship between the middle classes, domestic architecture and suburban space in Tanzania's largest city, Dar es Salaam. Claire is a member of the Editorial Working Group of *Review of African Political Economy*, a member of the International Advisory Board of *Antipode*, a member of the Advisory Board of *Critical African Studies*, and an Editorial Board Member of the Development Geography Section for *Geography COMPASS*.

Simon Milligan is a freelance design, review and evaluation specialist who has undertaken assignments for DFAT/AusAID, DFID, IFAD, SDC and the EC in sub-Saharan Africa, South Asia and South East Asia. He is also a Visiting Fellow at the University of Bath, UK. Simon has held long-term positions on various aid programmes, including Community Development Specialist on an EC-financed rehabilitation and reintegration programme in post-war Liberia, State Programme Manager of a DFID-financed governance programme in Nigeria and Education Research and Analysis Specialist on the Performance Oversight and Monitoring unit of the DFAT-financed Education Partnership in Indonesia. His most recent assignments have focused on measures to promote access to markets and services, and the performance assessment of large donor-financed programmes. His PhD research at Sussex University, UK, was on livelihoods and conflict dynamics in northern Nigeria. He has published several papers in peer-reviewed journals.

Garth Myers is Professor in Urban and International Studies at Trinity College, Hartford, Connecticut. He is a geographer with expertise in urban planning and African area studies. He has conducted research in Kenya, Tanzania, Malawi, Zambia, Senegal, South Africa, Finland and the UK over the past 20 years. His research interests include African urban geography and urban planning, comparative urbanism and comparative urban land politics, urban political ecology and environmental justice, as well as urban environmental governance. He has published various books, including *Verandahs of Power* (Syracuse, 2003), *Disposable Cities* (Ashgate, 2005), *African Cities* (Zed Books, 2011) and the latest, *Urban Environments in Africa* (Policy Press of University of Bristol, 2016).

Frankline Ndi is a PhD student in the School of Geography at the University of Melbourne, Australia, studying the effects of oil palm plantation establishment and the politics of land tenure in south-west Cameroon.

Sabelo J. Ndlovu-Gatsheni is Professor and Director for Scholarship in the Change Management Unit (CMU) in the Vice Chancellor's Office at the University of South Africa, Pretoria (UNISA). He was formerly the Founding Director of the Archie Mafeje Research Institute for Applied Social Policy (AMRI) at UNISA. He has published 12 books, over 50 journal articles and over 50 book chapters. His research covers decolonial theory, African development, African history and African politics. His latest publications include *The Decolonial Mandela: Peace, Justice and the Politics of Life* (Berghahn Books, 2016) and *Decolonizing the University, Knowledge Systems and Disciplines in Africa* (Caroline Academic Press, 2016).

Etienne Nel is a Professor of Geography at the University of Otago in Dunedin, New Zealand. His research focuses on issues of local, regional and economic development in southern Africa and in Australasia. He is the current Managing Editor of the *New Zealand Geographer* and Commissioning Editor for Australasia for the journal *Local Economy*. He has published books on development issues in Africa and more generally, including *Africa: Diversity and Development* (Routledge, 2012), *South Africa: Past, Present and Future* (Prentice Hall, 2000), *Local Economic Development in the Developing World* (Transactions Press, 2005) and *Geographies of Development* (Routledge, 2017). He is currently Chair of the CROP/International Social Sciences Committee on Comparative Research on Poverty.

Philip Nel is Professor in the Department of Politics, University of Otago, Dunedin, New Zealand, and a Research Associate of the Arnold Bergstraesser Institute, Freiburg, Germany. He writes on poverty, inequality, corruption and the politics of development, and also teaches on ethics and international politics. He has a DPhil from Stellenbosch University, South Africa, and serves on the editorial boards of *Review of International Studies* and *Global Society*.

Fiona Nunan is a Senior Lecturer in the International Development Department, University of Birmingham, UK. She carries out research into the governance of natural resources, and is the author of *Understanding Poverty and the Environment: Analytical Frameworks and Approaches*.

Ben Page is a Reader in Human Geography and African Studies at University College London, UK. His research interests are broadly located within the field of development geography, bringing together interests in migration, development and the environment. His principal concern at the moment is exploring the role of the African diaspora in the UK in bringing development to Africa. Much of Ben's early work focused on water supply in West Africa as a way of linking different histories and places to broader development questions about communities, the state, infrastructure, services, participatory governance, deliberative democracy and the transformation of the landscape.

Aarti Patel is currently the General Manager of the Canterbury Community Pharmacy Group (CCPG) in Christchurch, New Zealand, which focuses on enabling inspired and successful pharmacy practice. Gaining her MPharm in the 1990s as research intern with Health Systems Trust was instrumental in developing Dr Patel's wider public health and systems thinking to investigating medicines and access. Her PhD examined the issue of quality of medicines from the perspective of consumers, policymakers, manufacturers, health providers, wholesalers and distributors. Aarti has worked as a Chief Pharmacist in the Province of KwaZulu-Natal overseeing the essential medicines programme, and in 2010 was invited to become Technical Advisor for the DFID-funded Southern African Regional Programme on Access to Essential Medicines (SARPAM). Her present role as GM of CCPG allows her to support community pharmacists to develop their wider population health focus as part of developing and delivering enhanced clinical services in primary care.

Kamna Patel is a Lecturer at University College London, UK, within the Development Planning Unit where she teaches development theory and critical insights on development practice. Her research focus is on the social impact and governance implications of state-led low-income housing schemes in South Africa and India.

Gina Porter is a Professor in the Department of Anthropology at Durham University, UK. Her research combines ethnographic approaches with a strong interest in spatial perspectives. Uneven power relationships and associated issues of exclusion are linking themes through her work. The majority of her field research has been in sub-Saharan Africa, where her current research focuses principally on daily mobilities and the impact of mobile phones on young people's lives. Her latest publications include *Young People's Daily Mobilities in Sub-Saharan Africa: Moving Young Lives* (Palgrave Macmillan, 2017).

Andrea Rigon is a Lecturer at the Bartlett Development Planning Unit of University College London, UK, and a founder of the Sierra Leone Urban Research Centre. His research focuses on how power relations affect the participation of different people and social groups in decision-making processes. He is particularly interested in how residents' participation is managed within urban development projects, particularly in informal settlements, and what are the effects on in/equality and social exclusion.

Christian M. Rogerson is Research Professor in the School of Tourism and Hospitality, College of Business and Economics, University of Johannesburg, South Africa. His major research areas are small enterprise development, local economic development and tourism studies.

Jayne M. Rogerson is Associate Professor in the Department of Geography, Environmental Management and Energy Studies at the University of Johannesburg (South Africa). Her research interests are in tourism geography, greening of the built environment and sustainability issues, with respect to cities and tourism.

John E. Serieux is an Associate Professor in the Department of Economics, University of Manitoba, Canada. His research interests are in the areas of international financial flows, debt, income inequality, institutions and structural factors in development. He is a graduate of the University of Toronto, and has done extensive work with the United Nations, particularly relating to development challenges in sub-Saharan Africa.

Danny Simatele is Professor of Environmental Management and Sustainability Science in the School of Geography, Archaeology and Environmental Studies at the University of the Witwatersrand in Johannesburg, South Africa. His research revolves around a number of thematic issues such as environmental impact assessment, urban agriculture, solid waste management, climate change adaptation, water resource management and community-based natural resource management. He has a wide experience of working in sub-Saharan Africa and has written extensively on urban and peri-urban transformation, urban environmental management, climate change adaptation, food security, and environmental justice.

David Stephens is Professor of International Education at the University of Brighton, UK. During a 40-year career, he has held senior posts at a number of universities, including Oslo, London and Sussex. He has extensive experience of researching in West, East and southern Africa, and is currently leading a funded research project in Kenya examining the impact of mentoring upon teacher development and learning. He is particularly interested in issues of culture and quality within education development. His next book will be *International Education and Development: A Narrative Approach* (Routledge, 2017).

Ian Taylor is Professor in International Relations and African Political Economy at the University of St Andrews, UK, and also Professor in the School of International Studies at Renmin University of China. He is also Professor Extraordinary in Political Science at the University of Stellenbosch, South Africa, and a Visiting Professor at the University of Addis Ababa, Ethiopia. Ian is interested in sub-Saharan Africa's political economy and its international relations, the history of Afro-Asian diplomacy, the notion of 'rising powers', and the implications for global governance and development (and for Africa specifically).

Alec Thornton is a Senior Lecturer in Development Geography at the University of New South Wales (Canberra campus), Australia, and Research Associate at the School of Tourism and Hospitality at the University of Johannesburg, South Africa. He is Vice President of the African Studies Association of Australasia and the Pacific (AFSAAP) and Co-Editor of the *Australasian Review of African Studies.*

Emma Tomalin is Professor of Religion and Public Life at the University of Leeds, UK. Her research interests include a focus on religion and global development, and religion and gender. Related to these areas, her most recent publications are *Religions and Development* (Routledge, 2013) and *The Routledge Handbook of Religions and Global Development* (2015).

Jon D. Unruh is a Professor of Geography at McGill University in Montreal, Canada. His research, applied and policy work over the past 25 years has focused on war-affected human geographies, with a particular focus on the re-establishment of land and property rights after armed conflict. His past work has dealt with Islamic, customary, statutory and war-related approaches to land rights, and how these interact with displaced populations, ethnic cleansing, restitution, legal pluralism and agriculture in post-war and peace-building scenarios.

Adriaan van Klinken is Associate Professor of Religion and African Studies at the University of Leeds, UK. His research focuses on the public roles of religion in Africa, specifically in relation to issues of gender and sexuality. His major publications are *Transforming Masculinities in African Christianity: Gender Controversies in Times of AIDS* (Ashgate, 2013), *Public Religion and the Politics of Homosexuality in Africa* (with Ezra Chitando, Routledge, 2016) and *Christianity and Controversies over Homosexuality in Contemporary Africa* (with Ezra Chitando, Routledge, 2016).

Richard Vokes is Associate Professor in the Anthropology of Development at the University of Western Australia. He has extensive research interests in the African Great Lakes region, where he has been conducting ethnographic research since 2000. He has published extensively on this region, including on new religious movements, the HIV/AIDS epidemic, the history of photography, media and social change, and the anthropology of development (education, governance and infrastructure). He is author of *Ghosts of Kanungu: Fertility, Secrecy and Exchange in the Great Lakes of East Africa* (2009), *Photography in Africa: Ethnographic Perspectives* (2012), *Media on the Move: Ethnographies of Communication and Connectivity in Sub-Saharan Africa* (with Katrien Pype, 2016), *The NRM Regime and the 2016 Ugandan Elections* (with Sam Wilkins, 2016) and *Media and Development* (2017). He is currently working on a book on the history of the developmental state in Uganda.

Olivier Walther is an Associate Professor at the Department of Political Science at the University of Southern Denmark and a Visiting Associate Professor at the University of Florida, USA. His research focuses on regional integration, cross-border trade and terrorism in West Africa. Dr Walther currently coordinates the Sahel and West Africa Club's programme on cross-border cooperation at the OECD. He is the Africa Editor of the *Journal of Borderlands Studies* and is on the executive committee of the African Borderlands Research Network (ABORNE).

Simon Willcock is a Lecturer in Environmental Geography in the School of Environment, Natural Resources and Geography at Bangor University, UK. His research focuses on issues of ecosystem services, sustainable development and migration.

Section 1
Introduction and context

1 Introduction

Tony Binns, Etienne Nel and Kenneth Lynch

Africa is a rapidly changing continent. But for many, it is still an unknown place, shrouded by myths and distorted perceptions. Despite having over 20 per cent of the world's land area and 14 per cent of the world's population, the 54 countries only accounted for 2.2 per cent of the world's trade by value in 2016, compared with 38.5 per cent from Europe and 40.7 per cent from Asia (UNCTAD, 2016). Africa is in many ways still a marginalised continent. But things are certainly changing, and are going to change much more in the near future – it is a case of 'watch this space'! Economic growth in China and India has dominated the early years of the twenty-first century and has led to the emergence of a large middle class in Asia. Despite often unreliable statistical evidence, there are signs that this may be happening in some African countries, and Africa will undoubtedly be the continent to watch in the next 20 years and beyond. Although frequently starting from a low base, economic growth rates in some African countries are currently among the highest in the world, with 13 African countries recording average annual growth rates in GDP above 6 per cent over the period 2007–2015 (AfDB, 2016).

While in some countries there have been significant achievements in democratic governance, education and healthcare, sadly Africa is still the world's poorest and least developed continent. Nineteen of the twenty lowest ranking countries in the United Nations' 2016 *Human Development Report* are in Africa – the only exception in the 'bottom 20' list is Afghanistan. With a high incidence of HIV/AIDS, Swaziland, a small country in southern Africa, has the dubious distinction of having the world's lowest life expectancy at birth of just 48.9 years, a far cry from the 83-plus years of places such as Japan, Singapore and Switzerland (UNDP, 2016). While Norway in 2015 recorded 2.6 children in every 1,000 born dying before the age of 5, the comparable figure for Angola was a staggering 156.9, the world's highest.

But it is not just economic growth and development that should concern us. Culturally, Africa is rich and vibrant, now, as in the past. The birthplace of humankind, the rich heritage of ancient civilisations, cities that long predate European contact, and a colourful tapestry of art, music and literature. But so many people are unaware of these attributes. Most people's limited knowledge of Africa largely comes from the world's media, with often sensationalised stories that frequently display a poor understanding of African countries and people, and concentrate on 'newsworthy' issues such as natural and human disasters, including drought, desertification, civil war, terrorism and famine.

In light of Africa's rapid change and ongoing challenges, it seems appropriate at this point in time to take stock of where things are in relation to a number of important issues that face the continent and its people. In a sense, this is the main motivation for this handbook. We hope it will break new ground in helping to better understand Africa, and shed light on a range of issues relating to Africa's people, environments and development. We have brought together here some 56 international experts whose contributions are closely related to their research interests on Africa. Wherever possible, we have included African writers, but many of the non-African writers who have contributed to the handbook would certainly regard themselves as 'Africanists', in some cases having devoted a lifetime's research and teaching to topics relating to Africa. We have attempted to achieve a good coverage of what might be regarded as the key issues relating to Africa. While we believe that the handbook brings together a good range of authors and important topics, we frankly admit that this is by no means an exhaustive collection. In some cases, experts we approached were busy at the time with other research and writing projects, while in other cases it was actually difficult to find individuals who could write on specific topics that the editors had identified. However, despite these inadequacies, we believe that this handbook probably represents the most comprehensive collection of contributions on Africa that has been published in recent years.

In addition to the introduction and conclusion, the handbook comprises 48 contributions, which we have grouped into 10 themed sections, as follows: 'Introduction and Context', 'History', 'Social Perspectives', 'Health', 'Environment', 'Development Issues', 'Rural Africa'; 'Urban Africa', 'Economic Perspectives' and 'Political Perspectives on Africa's Development'.

Section 1 contains three introductory chapters. Following this introduction, Richard Vokes (Chapter 2) overviews development theory and attempts by development agencies to understand how specifically African knowledge and perspectives might support development agendas and goals. Vokes argues that scholars' understandings of what the notion of 'African perspectives' on development should include have also come to recognise that in many parts of the continent, ordinary people have been living with development interventions and projects for several decades, which, for many people, actually means their whole lives.

In Chapter 3, Sabelo J. Ndlovu-Gatsheni poses a thought-provoking question: 'What does development mean for a people struggling to emerge and free themselves from the inimical legacies of enslavement, colonialism, imperialism, apartheid, neocolonialism and underdevelopment, as well as impositions of the Washington Consensus and neo-liberalism?' Ndlovu-Gatsheni interrogates elements of this question, concluding that only a combination of genuine de-imperialisation on a world scale and decolonisation is likely to result in global democracy in which Africa could be afforded space to develop.

Section 2 of the handbook includes two chapters examining perspectives on different phases of Africa's history. Erik Green (Chapter 4) poses the question: 'Did colonialism cause Africa to be underdeveloped, or did the European occupation of the continent, although morally and politically questionable, facilitate economic development by further integrating Africa into the world economy?' Green concludes that we simply know too little about the patterns, processes and legacies of colonialism, and in any case the impacts of colonialism varied significantly between regions and between different social classes. In Chapter 5, Tanya Lyons and David Jolley explore the challenges and opportunities for Africa and Africans in the post-independence period and, with reference to specific case studies, they highlight various contrasts and contradictions.

In Section 3, the focus is on a number of social aspects relating to African livelihoods. In Chapter 6, Joseph K. Assan, Dinar D. Kharisma and Afia A. Adaboh examine aspects of population growth, migration and employment with particular reference to youth in Ghana.

Hazel Barrett (Chapter 7) recognises that Ester Boserup's 1970 book *Women's Role in Economic Development* is widely acknowledged as a catalyst for the inclusion of feminist and gender perspectives in development thinking, policy and programming in Africa. Barrett charts the progress in empowering women, and shows how in recent years the Sustainable Development Goals have mainstreamed gender, acknowledging the importance of 'gender justice' in achieving meaningful sustainable development. While growth and development are occurring more in some African countries than others, there is no doubt that there is very significant social and spatial inequality. Philip Nel (Chapter 8) shows that wealth in Africa is highly concentrated, and *within-country* inequality has increased across high-, middle- and low-income states, with dire social, economic and political consequences. Nel suggests that the roots of inequality lie in the colonial past, and have been reinforced by the limited-access institutions that the colonisers established, and which generations of African leaders since then have maintained. Many would argue that education can and should play a key role in reducing equality and improving livelihoods. David Stephens (Chapter 9) identifies and then explores seven key trends in African education since independence in the 1960s, including access and enrolment, gender parity, aid and education, and the role of education in conflict, reconciliation and reconstruction. Nicola Ansell (Chapter 10) shows that Africa is the youngest continent on earth with children under 15 constituting 41 per cent of its population in 2015. However, children and childhood were the focus of very few academic studies until the late 1990s. Ansell considers children in Africa through a relational lens, examining how their lives are strongly shaped by relationships with families and other adults in their communities, by national policies and cultural practices, and increasingly by international organisations and the operations of the global marketplace. Finally in Section 3, Ben Page and Clare Mercer (Chapter 11) consider the developmental impact of the African diaspora, examining the fact that there are people of African origin living outside the continent, irrespective of their citizenship and nationality, who are willing to contribute to the development of the continent and the building of the African Union.

Section 4 includes four contributions concerning health issues. In Chapter 12, Kenneth Lynch considers some of the relationships between health and development, and examines a number of strategic challenges that need to be overcome to improve health outcomes among Africans. Increasing national investment in the health sector is vital, together with a need for more well-trained health professionals. It is a depressing fact that southern African countries have the world's highest incidence of HIV/AIDS, and in Chapter 13 Hazel Barrett traces the history and response to HIV/AIDS, and most notably illuminates the encouraging progress made in combating the epidemic since 2000. In Chapter 14, Stephen Devereux examines the important issues of food security and famine. Proposed solutions to food insecurity are presented, ranging from increased food production to social protection, nutrition interventions and improved governance. Devereux concludes that the key factor explaining the persistence of hunger and famine in Africa is probably not a lack of financial resources and technical capabilities, but rather how available resources and capacities are allocated and managed, which is largely a political and governance issue. Aarti Patel (Chapter 15) considers the important issue of access to essential medicines in Africa, and emphasises the need for changing outdated pharmaceutical laws and developing sustainable strategies for the procurement and distribution of essential medicines.

The focus of Section 5 of the handbook is the environment, and it includes four contributions. In Chapter 16, Clive Agnew considers the vital issue of water resources in a continent where there are extensive water resources that have often not been exploited. Agnew presents a chronology of water resources development, and concludes that the recent history of water resources development in Africa is dominated by engineering solutions and state-driven development for national priorities. He advocates the introduction of water exploitation strategies that place more emphasis on environmental and social costs, efficiency gains and sustainable funding. The significance of wetlands in Africa has often been underestimated, and in Chapter 17 Alan Dixon argues that rather than constituting benign places where rare ecosystems exist and humans avoid contact, wetlands are, and have increasingly become, contested socioecological spaces at the interface of people–environment interactions. While this continues to be a source of tension between conservation and development policymakers, there has been some acknowledgement in recent years of the need to develop more holistic ways of thinking about people's interactions with wetlands. In Chapter 18, Fiona Nunan considers the issue of conservation, showing that the approach taken to conservation and the nature of participation of local people differs over time and space, and that there is no universal panacea approach that should be adopted. In the final contribution to this section Rebecca Kariuki, Rob Marchant and Simon Willcock (Chapter 19) critically examine the ecosystem services (ES) approach to understanding the biophysical environment and its relevance to many rural African communities.

Section 6 of the handbook comprises six contributions on topics related to the broad theme of 'development'. Tony Binns and Etienne Nel (Chapter 20) consider the current position of Africa, and whether such terms as 'Hopeless Africa' and 'Africa Rising' are actually appropriate. The extent to which such perceptions and realities are reflected in African countries' achievement of the United Nations Millennium Development Goals (MDGs) is also examined. It is often said that regional integration would bring greater strength and bargaining power to African countries. But as Olivier Walther shows in Chapter 21, progress in this area has been relatively limited, and Africa is still one of the most expensive places in the world to do business. In Chapter 22, John Briggs picks up the theme of the MDGs, evaluating the level of their success and associated outcomes, and looking ahead to the achievement of the post-2015 Sustainable Development Goals, which he believes will provide African countries with an even greater challenge. Ian Taylor (Chapter 23) focuses on relationships between the BRICS countries (Brazil, Russia, India, China and South Africa) and African countries in providing relationship options beyond the traditional North–South axes that were developed through colonialism. Taylor examines commodity structures and trading patterns, and concludes that many African countries are trapped into low value-added production structures. The issue of aid and debt is examined by John Serieux in Chapter 24. Africa remains the world's highest aid-receiving region, and in recent years aid-giving has shifted from the exigencies of Cold War politics to financing requirements for the MDGs and SDGs, and fighting the war on terror. Although debt levels have been brought down, the greater involvement of some African governments in international bond markets is exposing these countries to the vagaries and competition of the global market. Chapter 25 focuses on the question of faith and religion in Africa, a continent where many individuals and communities are deeply religious. Reflecting this reality, Emma Tomalin and Adriaan van Klinken identify a significant recent shift in international development policy towards recognising the significance of religion in development interventions.

The significance of rural development in many everyday livelihoods in Africa is recognised in Section 7, which includes five contributions. Many would argue that consideration of the issue of land tenure should be fundamental in any rural development programme. In Chapter 26, Jon D. Unruh considers the importance of tenure security, the various types of land tenure and how these might evolve over time. Deborah Fahy Bryceson shows in Chapter 27 how 'deagrarianisation' was a key process in Africa during the 1980s when adverse terms of trade impacted on smallholder peasant exports. The search for non-agriculture income-generating activities, according to Bryceson, led to much economic stress among rural households, but also revealed significant creative ingenuity. In Chapter 28, Simon Milligan and Sam Kareithi examine the incidence of poverty in rural Africa. They examine different approaches to rural development since the 1960s, and look ahead to the future of farming and rural non-farm economies, where they believe the state will need to play a strong role together with the private sector. Gina Porter (Chapter 29) considers the significance of rural marketing, and shows how significant changes are occurring with the expansion of contract farming and supplying supermarkets, and most particularly with new technologies, notably mobile phones and the introduction of mobile banking. The relationships between rural and urban areas are often remarkably strong in African countries, and in Chapter 30 Kenneth Lynch examines the ways in which these relationships have been perceived. In recent years, it seems that international NGOs have been giving more attention to these relationships, and while rural–urban relations can be exploitative, they can also be mutually beneficial.

Section 8 of the handbook considers a range of issues relating to Africa's towns and cities. There are five contributions in this section. Chapter 31 by Andrea Rigon, Joseph M. Macarthy, Braima Koroma and Alexandre Apsan Frediani considers the politics of urban management and planning in African cities. Africa is rapidly urbanising, and the chapter addresses key issues in management and planning that concern African cities, suggesting that reforms will be needed, including decentralisation of power and fiscal autonomy, in order to cope with continuing urban growth. Kamna Patel focuses on housing issues in Chapter 32, which she regards as a political issue of great social importance across Africa. The chapter overviews the history of housing policy and the key housing challenges in the face of rapid urbanisation. Looking ahead, it is suggested that the provision of housing can be an effective way to enable everyone's social and moral claim to a right to the city, or to deny it to certain groups, which makes housing an intently political activity. In Chapter 33, Danny Simatele focuses on the important issues of unemployment and solid waste management, with particular reference to Kinshasa, capital of the Democratic Republic of Congo. He suggests that improved governance is necessary to better appreciate the urban informal sector's role in issues such as waste management, and to increase local community involvement and participation, which could generate a greater sense of commitment and responsibility by the community. Chapter 34 is concerned with the issue of food security in African cities. Alec Thornton specifically focuses on urban agriculture, which is nowadays a common feature in many African towns and cities. Thornton calls for more inclusive, resilient and liveable cities, with an integration of urban agriculture into city planning, as part of a broader policy approach to sustainable urban development. Garth Myers (Chapter 35) considers the future of African cities, which he believes are facing uncertain and varied prospects. Predicting urban futures is difficult in light of diverse trajectories of growth and demographic and economic trends. He examines the challenges and opportunities for African cities, and identifies the potential that marginal communities have for 'transformational energies and dynamism' that could produce what he calls 'emancipated African urbanism'.

Nine contributions on diverse economic themes constitute Section 9. In Chapter 36, Patrick Bond examines the issue of extreme uneven development. He critiques the concept of 'Africa Rising', and suggests that although there have been slight improvements in poverty levels, education, mortality and morbidity rates, 'the conditions for reproduction of daily life in Africa have not improved as a result of the continent's integration into world capitalism since the 1980s'. David Fielding considers financial sector development, foreign direct investment and economic growth in Chapter 37. While there is evidence of financial sector growth in sub-Saharan Africa, he suggests there is much scope for central banks and other regulatory bodies to improve the conditions under which African banks lend. Chapter 38 focuses on agriculture, and Roy Maconachie shows how it is still the focus of rural life in many communities across the continent, employing 65 per cent of Africa's labour force. The chapter examines small-scale indigenous farming systems and then large-scale commercial agriculture, before considering the increasing production of bioenergy crops for global markets. Maconachie then turns his attention to mining (Chapter 39), and the continuing dependence of many African countries on exporting primary commodities, and their vulnerability to fluctuating world commodity prices. The concept of 'resource curse' is critiqued, and how this might lead to community conflict. The significance of artisanal and small-scale mining is reviewed and the involvement of countries such as China in the 'new scramble for resources'. In Chapter 40, Christian M. Rogerson focuses on the small, medium and micro-enterprise economy (SMME), which, according to some observers, accounts for 95 per cent of all firms in sub-Saharan Africa, which are responsible for at least 80 per cent of the region's employment. Rogerson argues that business incubators could play a key role in growing the entrepreneurial base, but the incubator model will need to be adapted to suit local needs. Tourism is a growing element in the economies of many African countries, but progress in developing the tourism sector is uneven geographically. Christian M. and Jayne M. Rogerson (Chapter 41) examine different types of tourism and consider the development challenges, concluding that for successful tourism to develop, there will be required extensive innovation on the part of tourism entrepreneurs, policymakers, stakeholders and, most notably, international development agencies. Gina Porter considers the significance of transport in Chapter 42, pointing to the high transport costs in Africa due to factors such as limited markets, remoteness and accessibility. But there are some positive recent developments, such as new connectivities associated with the increasing use of mobile phones to aid transport organisation, and in some cases to actually reduce the need for mobility. The issue of 'land-grabbing', or large-scale land acquisition, has become an issue of much debate in the last decade, and in Chapter 43 Simon Batterbury and Frankline Ndi examine the nature and occurrence of this process. There is great uncertainty about the extent of land-grabbing, but research seems to suggest that rural Africans are frequently unable to participate in decision-making processes for projects that can profoundly affect their livelihoods. Lauren A. Johnston (Chapter 44) traces the development of China–Africa links, and aspects of Chinese aid, trade and investment. Johnston identifies the need for further research on China's present and potential future economic impact on African countries.

The five chapters comprising Section 10 focus on various political aspects related to development in Africa. In Chapter 45, David Booth examines the debate surrounding the issue of 'good governance' and whether it is actually good for development. He considers the concept of neopatrimonialism and compares governance and development aspects in Africa with those in selected Asian countries. Interestingly, he concludes that comprehensive governance improvement is not necessarily a precondition for economic progress in developing Africa. Hazel Cameron sheds some useful light on conflict and post-conflict issues in Chapter 46. The chapter explores African development through the lens of violent conflict, highlighting the role of development in conflict, as well as the effects of conflict on development. Cameron concludes that the poor understanding of the complex roots of Africa's violent conflicts ensures that the continent is likely to continue to experience and suffer from violent intra-state wars. Closely related to conflict and post-conflict scenarios is the question of human rights, which are explored by Peris Jones in Chapter 47. The chapter examines the social construction of rights and associated critiques and then overviews human rights discourse. Jones concludes that, ultimately, how rights are used in practice will tell us about how we are to enhance rights-based approaches for ending impunity and encouraging accountability in Africa. Very much linked with the issue of human rights is the role of the International Criminal Court (ICC) in Africa, which is discussed by Jonathan Hobson in Chapter 48. Hobson examines the role of the ICC, its relationships with the African Union, the ICC's impact and its likely future, outlining some potential solutions to the difficulties it faces. The impact that corruption has had on the continent is explored by Philip Nel in Chapter 49. The chapter discuss the scale of corruption, its effect on sustainable development prospects, and possible responses.

References

AfDB (African Development Bank) (2016) *African Economic Outlook, 2016*, Abidjan: AfDB.
UNCTAD (United Nations Conference on Trade and Development) (2016) *Economic Development in Africa Report, 2016*, New York: UNCTAD.
UNDP (United Nations Development Programme) (2016) *Human Development Report, 2016*, New York: UNDP.

2

African perspectives on development

Richard Vokes

Introduction

The notion of 'African perspectives' on development refers to three things. First, it refers to the attempts by a number of Marxist-influenced African leaders, from the 1960s onwards, to pursue models for the development of their nations that were qualitatively different from those that were being promoted by the international community. Second, it refers to the arguments of a number of 'postcolonial' scholars who argued that historical processes of European extraction from Africa were the cause of the continent's 'underdevelopment' in the first place. Finally, it refers to more recent attempts by development agencies to understand how specifically African knowledge and perspectives might forward development agendas and goals. To understand how these three elements emerged at the times they did, it is necessary to locate them within the broader trends in development theory and practice from which, and into which, they emerged.

It is often suggested that the modern history of international development in Africa began with US President Harry S. Truman's Inaugural Address of 20 January 1949. The now famous Point 4 of that speech promised to extend the benefits of America's scientific and industrial progress to *all* of the poorer nations of the world (US overseas assistance had previously focused upon only the countries of Latin America). In so doing, Truman's address enshrined international development as a key priority for US foreign policy – and by implication, as a priority for all of its international allies as well – and it soon led to the creation of America's Technical Cooperation Programme Unit (TCPU), the forerunner to the contemporary US Agency for International Development (USAID); the United Nations Development Programme (UNDP); and the World Bank's International Development Association (IDA), the part of the bank that provides concessionary loans to the world's poorest nations. For these reasons, Truman's speech is often seen as ushering in the 'Development Age' (Leys, 1996).

Truman's speech defined development in a particular kind of way. Specifically, the address effectively reiterated what some would call a 'colonial world view' by describing the problems people faced in the Global South – for example, food shortages, disease and low levels of economic development (Rist, 2014) – as stemming from their relative *lack* vis-à-vis populations in the North. Indeed, the transcript of Truman's speech became one of the first major documents to use the term '*under*development' in the sense that we now commonly understand that word.

The speech also emphasised, as a corollary to this notion of 'lack', that the solutions for Southern problems inevitably lay, therefore, in the transfer of capital and knowledge – especially money in the form of overseas development assistance (ODA), usually called just 'aid', and technical and scientific knowledge – from the North to the South. It was only by way of these flows that the countries of Africa, and elsewhere, could move from 'underdeveloped' to 'developed'; or in other words, could 'catch up' with the rich world, or 'modernise' (Lerner, 1958). In these ways, the speech also emphasised a particular meaning of the word 'development' as something that was essentially *transitive* – i.e. something that referred to an action that is done *by* one entity *to* another, by a wealthier country to a poorer one.

Truman's understandings of what the 'problem' for development was, how development practice should be organised to address this problem, and what 'development' ultimately therefore *meant*, dominated the field of development studies for several decades following. Indeed, they became *the* defining features of what is itself often referred to as the 'era of modernization' (Peet and Hartwick, 2015). Academics debate as to when, precisely, this era ended. However, one commonly held view is that it only declined following the emergence of 'neo-liberal' approaches to development, from the 1980s onwards (Sahn *et al.*, 1999). However, just because the ideas that Truman articulated *did* become the orthodoxy throughout this period, this is not to say that they remained unchallenged throughout this period, nor that alternative approaches to development were not tried during this time. On the contrary, almost immediately following the famous address, scholars, developing world leaders and others began to offer alternative ideas about, and for, development, which argued: that the countries of the Global South were *not* 'backwards', or 'lacking', compared to those of the North; that the solutions for underdevelopment did *not* lie in transfers of capital and knowledge from the North to the South, but instead in a more effective use of the resources, and of the knowledge and skills, that were already there in developing countries; and therefore that the societies of the Global South were perfectly capable of taking control of their own destinies, or in other words, of following their *own pathways for development*. Moreover, many of these ideas were made by theorists, policymakers and others who either spoke *about* Africa and/or who articulated concerns *for* Africa, or who were themselves African. Thus, these debates frequently congealed around discussions of development in Africa in particular.

Dependency theory

These arguments emerged most forcefully in what became known as 'dependency theory', which began to form just a few months after Truman's speech. In late 1949, an economist working in the Department of Economic Affairs at the United Nations in New York, Hans Singer, published what was to become one of the foundational texts for the theory, 'Economic Progress in Under-Developed Countries', in *Social Research*. A short time later, an economist at the Economic Commission for Latin America, Raoul Prebisch, published another key report, *The Economic Development of Latin America and Its Principal Problems*, which was published by Singer's department at the UN in 1950. Taken together, these two texts – which became jointly known as the 'Prebisch–Singer hypothesis' – argued through technical economic analysis that underdevelopment was not an outcome of any intrinsic quality of Southern societies themselves (i.e. it was not an outcome of the fact that they inevitably 'lacked' something). Rather, their relative impoverishment, vis-à-vis wealthier nations, was an outcome of the way in which they had been integrated into the global economic system, in a manner that had disadvantaged them from the outset, and had resulted in their becoming ever *more* disadvantaged over time. This stemmed from the fact that the global system had developed in such a way that Southern countries had always been reliant upon the export of their primary resources to wealthier nations to support the latter's large manufacturing bases, while developed nations had always been reliant on exports of manufactured goods, including to poorer nations. Yet from the start, this engendered a comparative disadvantage, given that in these markets, the relative cost of raw materials tends to always remain static, while the equivalent cost of manufactured goods will always rise more quickly (Prebisch, 1950). In other words, with each passing year, poorer countries would inevitably need to export more and more of their primary commodities in order to purchase the same volume of manufactured imports. Simply put, this system will result in their becoming relatively poorer over time.

Although the Prebisch–Singer hypothesis had some early impact among academic economists, it initially gained little traction within wider development circles (Leys, 1996: 110–118). This is partly because it was developed from a highly technical set of analyses, but more so because its logical conclusion was that the Global South needed to adopt a completely different approach to development, one that might involve them at least partly *withdrawing* from the international economic system, by isolating – or 'protecting' – their national economies, and in fact allowing *less* inflows of capital and technology from the Global North. This flew in the face of the fundamental logics of modernisation theory, as I have described it above, and ran counter to the kind of development practice that these logics informed. However, by the late 1950s, a group of Marxist scholars had taken up the key economic ideas of the Prebisch–Singer hypothesis, and had expanded these into a more wide-ranging theory: what became known as *dependency theory* (Baran, 1957; Frank, 1966; Cardoso, 1977). And a new generation of African leaders were then exposed to this theory during their university education, at tertiary institutions both in Africa and in America and Europe. When some of these leaders later went on to achieve independence for their nations, and to become the first independent-era presidents of those countries, they often brought dependency thinking with them into their new offices.

The two African leaders who were perhaps most influenced by dependency theory, and who therefore made the most sustained attempts to apply at least some of its logic within their own national development policies, were Kwame Nkrumah, President of Ghana from 1960 to 1966, and Julius Nyerere, President of Tanganyika – later Tanzania – between 1962 and 1985. Nkrumah had been exposed to Marxist thought long before dependency theory *per se* emerged, especially during his years at university in the USA, first at Lincoln College, and then at the University of Pennsylvania, from the mid-1930s onwards. During this time, he was particularly influenced by the famous Trinidadian Marxist writer C.L.R. James and the Chinese-American Marxist philosopher Grace Lee Boggs, and their thinking informed his own ideas about pan-Africanism. Given this background, it came as little surprise that Nkrumah later became sympathetic to dependency theory and, following Ghana's independence in 1957, tried to apply it in practice. In particular, Nkrumah's 1964 'Seven Year Development Plan for National Reconstruction and Development' took seriously the idea of using protectionism measures as a means for encouraging development from within. The plan heavily taxed manufacturing imports, and used the revenues raised to invest in a broad programme of technical education and training which, it was hoped, would help to drive Ghana's own industrial development (Birmingham, 1998). Meanwhile, Nyerere, who had been similarly exposed to Marxist ideas during his time in Uganda's Makerere University, and later at the University of Edinburgh, also tried to implement a version of protectionism combined with 'development from within' in Tanzania following his country's independence. In Nyerere's model for national development, which was laid out most explicitly in his famous 'Arusha Declaration' of 1967, both manufacturing imports *and* inward aid were severely restricted, and national growth and development were to instead stem from a collectivised system of agriculture called *ujamaa* – which it was hoped would in turn generate a more broad-based manufacturing sector (Nyerere, 1979; Sahle, 2002). Elsewhere, other African leaders of newly independent states also implemented at least some elements of these approaches, which collectively constituted a distinctive African kind of socialism. For example, Leopold Senghor, President of Senegal from 1960 to 1980, another key figure in the development of pan-Africanism, also implemented protectionist measures. However, for Senghor, these were most effective when used to protect not national markets, but regional ones – which were bigger and therefore had a greater chance of achieving effective development 'from within' (Hymans, 1971). This was one of the reasons why Senghor favoured plans for those West African nations that had formerly been colonised by the French – and which were therefore all French-speaking – to form a postcolonial regional federation. However, successful though all of these experiments were in certain respects, none was ultimately able to overcome the powerful negative effects of the 1970s Global Financial Crisis, to which we will turn shortly. However, before doing so, it is necessary to explore another strand of development thinking that also emerged in the 1960s.

Postcolonial theory

From the early 1960s onwards, the argument that developing nations' relative impoverishment, vis-à-vis wealthier countries, was not a reflection of any internal shortcomings on their part, but was instead an outcome of the external pressures that had been historically placed upon them, was also lent further weight by the emergence of a second – albeit in many ways overlapping – body of scholarship called 'postcolonial development theory' (Gupta, 1988; Sachs, 1992; Rahnema and Bawtree, 1997; Saunders, 2002). This broad and amorphous body of work – which eventually came to include contributions from historians, literary theorists, philosophers, anthropologists, geographers and others – can be broadly characterised as an attempt to understand the total effects and legacies of European imperialism upon the societies of the Global South. Importantly, it argued that these effects – which throughout Africa, at least, were taken to have been generally detrimental – had been not *only* economic in nature, but also psychological, social, political, cultural and demographic (the best introduction to this broad body of scholarship is Young, 2016). In other words, for postcolonial theorists, underdevelopment in Africa was not only an outcome of the kinds of unfair terms of trade – as had been described by Singer and Prebisch – but was also a reflection of the kinds of collective trauma, social disruption, cultural rupture and demographic upheavals that colonialism had also generated on the continent. From this, it followed that any attempts to address underdevelopment in Africa would need to involve not only alternative economic strategies, but also plans for improving collective well-being, for strengthening social cohesion and for promoting cultural pride.

One of the first postcolonial scholars to draw attention to the role that colonialism had played in creating complex forms of underdevelopment in Africa, and to the need to find specifically African responses to these, was Franz Fanon. Born in Martinique, in the French Caribbean, to the descendants of African slaves and Indian indentured labourers, Fanon was the protégé of another famous Caribbean Marxist and pan-Africanist, Aime Cesaire. Throughout his professional life, Fanon was greatly influenced by Cesaire's ideas, and eventually made his own name as a scholar by applying these same ideas within the context of his own profession: clinical psychiatry. Working as a psychiatrist in mental hospitals, first in Brittany, in France, and later in Algiers, in Algeria (at the height of the Algerian War of Independence from France), Fanon started to speculate that Marxist thinking might help to explain the patterns of mental illness that he observed in his postings, whereby in France, a disproportionately high number of patients in mental institutions were from colonies, while in Algeria the kinds of mental illnesses patients suffered from were particularly acute. Fanon eventually concluded that the structural inequalities inherent in colonialism, and the language through which these were enacted, could explain these patterns in both places. In short, these inequalities produced collective trauma and ruptures in pre-existing forms of social and cultural life, while the language of colonialism – which everywhere cast the colonisers as essentially superior to the colonised – served to establish collective identities in which African people came to internalise a sense of collective inferiority. For Fanon, it was this collective outlook that lay at the roots of the continent's underdevelopment. It followed, therefore, that the complete overthrow of colonial regimes, combined with the forging of a new 'class consciousness' (in which Africans saw themselves as not dependent upon their former colonial masters), were necessary prerequisites for the emergence of any sort of meaningful development on the continent. In other words, Africa needed to undergo both a political and an intellectual revolution if it was to ever address its underdevelopment. And these ideas – which Fanon articulated most fully in his seminal *The Wretched of the Earth* (1961) – later led him to throw his support behind the separatist National Liberation Front (FLN) in the Algerian War.

However, the scholar who is today most closely associated with the arguments that colonialism was primarily responsible for African underdevelopment, and that the continent therefore needed to forge its own pathways ahead, is Walter Rodney. Like Fanon, Rodney was born in the Caribbean – in his case, in the then British Guiana, now Guyana – to the descendants of African slaves, and like Fanon, it was in the Caribbean that he gained wide exposure to the ideas of Marxism and pan-Africanism. In Rodney's case, this occurred especially during his undergraduate years at the University College of the West Indies in Jamaica in the early 1960s. In 1966, Rodney completed a PhD on the history of the transatlantic slave trade on the Upper Guinea Coast at London's School of Oriental and African Studies, and this research later formed part of the evidential base for what became his seminal work: *How Europe Underdeveloped Africa* (1972). This book examined the history of Europe's contact with Africa over the *longue durée* from the fifteenth century to the 1960s, and as such in fact examined more than just the period of 'direct' European colonialism over the continent. Nevertheless, the arguments that it forwarded fell very much within the framework of postcolonial studies by suggesting that prior to their initial contacts with European sailors, traders, etc. African societies had developed in a perfectly normal and unimpeded way – in terms of productive growth (i.e. economics) individual well-being and social and moral order. However, as European contact increased, and especially as this evolved into what became the transatlantic slave trade, these normal features of development all became distorted. For example, the slave trade retarded African productivity by removing a large part of its productive population, i.e. its healthy young men; it negatively impacted individual well-being by casting Africans as slaves, and therefore as inferior – and in this regard, Rodney's arguments partly mirrored Fanon's; and it corrupted African moral orders by effectively militarising the regions in which the trade operated, as societies fought with each other either to procure slaves to be sold to European traders, or to defend themselves from such advances. Furthermore, Rodney argued that although it was the slave trade that first established all of these effects, the later period of direct European colonialism in a sense extended and routinised them – through the systematic extraction of raw materials from Africa, through the establishment of formal education systems in which Africans always underperformed vis-à-vis European standards, and through the denigration of, and at times demonisation of, African 'traditions' (Rodney, 1972). Once again, then, it followed that the only way in which Africa might ever escape from its current situation of underdevelopment would be for it to break its ties with the Euro-American world altogether, and to return to an earlier situation in which its economic, social and moral development occurred on its own terms.

At the time Rodney's book came out, it was both heralded as a masterpiece of African history – especially because of the way in which it integrated quite different kinds of data sets to develop a narrative that spanned several different historical epochs – yet also condemned for its controversial arguments about the generally corrupting effects of external interventions into Africa. Yet controversial though these arguments were, they have continued to resonate with some African scholars and politicians up until the present time – and they have even been periodically updated to reflect current international relations and realities. In recent years, the most famous example of a scholar of African descent forwarding a 'Rodney-esque' type of argument is the Zambian-born economist Dambisa Moyo. Moyo's hugely influential book *Dead Aid: Why Aid Is Not Working and How There Is a Better Way for Africa* (2009) – which even made it onto the *New York Times* Best Seller list – argues that even today, external interventions into Africa slow economic growth, negatively affect African well-being and disrupt moral orders. And most significantly of all, Moyo suggested that even *development aid itself* produces these effects. Moyo argued, based on a review of a large body of economic studies, that aid hampers economic growth in African countries by making their exports uncompetitive, increasing inflation and reducing the value of their savings. In addition, the entire aid regime, by continuously casting Africans as 'helpless', as 'needy' and as 'childlike' – think of the images used on the promotional posters of practically every humanitarian campaign in Africa over the past 50 years – effectively patronised Africans, and thereby reduced their collective self-esteem. Finally, Moyo argued that by providing African governments with a major funding stream over which their citizens effectively have no influence, aid also at best reduces accountability, and at worst props up corrupt governments (and even, in some cases, violent regimes). Either way, it erodes democracy everywhere across the continent. However, although Moyo's diagnoses of the inherent dangers of external intervention largely mirror those of Rodney, her suggested response to them is nevertheless quite different. Whereas Rodney – as the dependency theorists before him – argued that in light of these issues, African countries needed to find ways to withdraw from international markets (or at least protect themselves from the worst effects of these), for Moyo the solution lies in Africa becoming *further* integrated into free markets. Thus, *Dead Aid* argues that contemporary aid regimes should simply be dismantled, and that African economies should simply trade their way out of underdevelopment instead. Although this may carry economic risks in the short term, in the long run, Moyo argues, it will lead to greater self-esteem, and to more accountable and democratic nation states.

Living with development

By the 1980s, it had become increasingly clear that neither the kind of 'orthodox' approach to development that had been introduced by Truman, i.e. modernisation, nor the kind of alternative more protectionist approaches that had been adopted by countries such as Ghana, Tanzania or Senegal, had been greatly successful, either in significantly boosting economic growth in Africa, or in 'closing the gap' between the Western world and the African continent. One major factor here had been the oil crisis of the 1970s, and the subsequent global economic downturn, which, for a number of complex reasons, had left African countries with major foreign debt (AfDB and AU, 2009). Indeed, by the 1980s, many African countries' commitments on foreign debt repayments exceeded their ability to pay them. In other words, *they were bankrupt*. In response, both donor countries and multilateral agencies such as the World Bank and the International Monetary Fund (IMF) began to themselves move away from some of the key tenets of modernisation – especially the notion that African countries had to develop through a series of carefully controlled stages, whereby they moved from being primarily agricultural-based societies, to first urban ones, and then industrial ones – and to instead 'open up' these countries to market forces. It was hoped that this would enable market forces to simply drive economic development in their own direction. Later, it was recognised that African states also needed to provide a 'safety net' for the poorest members of their societies, in the form of free basic healthcare, free primary education and basic levels of social security (Serra and Stiglitz, 2008). From the mid-1980s onwards, this new model for African development was rolled out across Africa – on a country-by-country basis – in the form of what became generally known as 'structural adjustment programmes', or SAPs (Sahn et al., 1999).

The shift away from modernisation, and towards what has become widely known as 'neoliberal' approaches to international development, had two effects on the way in which both development practitioners and scholars engaged with African people's own perspectives on the development process. The first is that it resulted in development programmes and projects taking indigenous environmental/ecological, or traditional, knowledge (IEK or ITK) much more seriously as a resource useful for achieving development goals. The key point here is that the earlier modernisation theory had cast all forms of traditional knowledge as essentially 'backwards', and therefore a *barrier* to progress, and something to be actively replaced with more 'modern' forms of knowledge instead (Sillitoe, 2009). However, with the shift away from modernisation, IEK now came to be seen as a useful resource, and as something worth preserving and protecting in its own right, and as something that could be incorporated into natural resource and agricultural programmes as a means for reducing their risks and for increasing their productivity. Thus, from around the mid-1980s onwards, a growing number of development agencies began to explore, for example, how knowledge about local environments, and environmental risks, as might be contained in things such as traditional stories, myths and rituals, could be used to improve environmental management (Pottier *et al.*, 2003). They also examined how ownership of certain elements of IEK – for example, knowledge about the healing properties of some or other plant – might be protected using legal instruments such as international intellectual property (IP) laws (Drahos, 2014).

Second, given that modernisation had been so heavily criticised for being at best an imposition of Western-biased markets upon Africa, and at worst a continuation of European domination of the continent, its demise also encouraged development agencies to begin to explore new ways in which the development enterprise could be made more collaborative, or 'participatory' (Melkote, 1991; Friedmann, 1992). At first, these efforts focused upon how individual projects could be made more effective through the inclusion of local perspectives – especially those of certain marginalised groups, such as women – in all stages of the project management cycle, which, as a minimum list, always include the four stages of: (1) background research; (2) project design; (3) project implementation; and (4) monitoring and evaluation of results. However, from the outset, there was never one agreed way, or even set of ways, for how to do this. Instead, over the next decade or so, an extremely wide range of different kinds of tools and techniques for incorporating what became known as *participatory action research* (PAR) into development work was developed by practitioners. Indeed, so diverse was the possible range of approaches here that one early attempt to define the approach, Robert Chambers' *Rural Development: Putting the Last First* (1983), could do little more than to group them into four general categories of methods. These categories were: (1) *group dynamics techniques*: all of which focused upon improving empathy and communication between development subjects and development agencies; (2) *sampling methods*: all of which involved collecting and recording 'typical' elements of a given social world in order to understand its totality; (3) *interview techniques*: all of which involved group interviewing methods of one kind or another; and (4) *visualisation tools*: all of which, as the name suggests, attempted to get development subjects to 'picture' their issues and problems in certain kinds of ways (for a good overview of participatory methods, see Slocum, 2003). I would also note here that from the mid-1980s onwards, the methods for participatory development have been alternatively referred to as PAR, and also 'participatory rural appraisal' (PRA) and 'rapid rural appraisal' (RRA). Over the years, scholars of international development have sometimes tried to forge fine-grained distinctions between these various methodologies. However, because the very same sets of methods are referred by all of these terms, in practice, it is probably easier to just think of them as in fact one and the same thing. To start with, these kinds of participatory methods tended to be applied within only small-scale development projects – for example, those that might be carried out at only a 'village level' – and were therefore favoured by smaller NGOs. However, over time, the notion of participation expanded further to also include participation with national ministries, and even entire national governments – as a means for making even country-wide programmes better suited to the local conditions. And in this respect, they were even taken up by major multinational organisations such as the World Bank and the IMF (Lewis and Kanji, 2009).

Conclusion

In recent years, scholars' understanding of what the notion of 'African perspectives' on development should include have also come to recognise that in many parts of the continent, ordinary people have now been living with development interventions, and projects, for many decades – which, for many people, means in fact their whole lives. As such, scholars have increasingly realised that development has now, in fact, become part of everyday life in Africa (Ferguson, 1994). In other words, for any ordinary African person, even someone living in the most remote of rural villages, listening to development planning sessions, including participatory ones, attendance at development-related workshops and training sessions, and engaging with development literature, posters and other media have all become part and parcel of everyday life. In consequence, development itself is no longer 'external' to African societies – i.e. it is no longer something that is 'done to' these societies by outsiders – but it is now internal to, or embedded within, these polities (Green, 2014). Indeed, it may even be possible to start to think about how different African 'cultures of development' have now emerged, whereby people have incorporated the activities and practices that are integral to 'doing development' into their daily routines, their ritual practices and other areas of their social lives. Development is now part of everyday African life and – at least while so many African countries remain in the lowest cohort of countries on the Human Development Index (HDI) – will remain so for a long time to come.

References

AfDB (African Development Bank) and AU (African Union) (2009) *Oil and Gas in Africa*, Oxford: Oxford University Press.

Baran, P. (1957) *The Political Economy of Growth*, New York: Monthly Review Press.

Birmingham, D. (1998) *Kwame Nkrumah: The Father of African Nationalism*, Athens, OH: Ohio University Press.

Cardoso, F. (1977) 'The consumption of dependency theory in the United States', *Latin American Research Review*, 12(3): 7–24.

Chambers, R. (1983) *Rural Development: Putting the Last First*, London: Longman.

Drahos, P. (2014) *Intellectual Property, Indigenous People and Their Knowledge*, Cambridge: Cambridge University Press.

Fanon, F. (1961) *The Wretched of the Earth*, London: Grove Press.

Ferguson, J. (1994) *The Anti-Politics Machine: Development, Depoliticization, and Bureaucratic Power in Lesotho*, Minneapolis, MN: University of Minnesota Press.

Frank, A. (1966) *The Development of Underdevelopment*, Boston, MA: New England Free Press.
Friedmann, J. (1992) *Empowerment: The Politics of Alternative Development*, Oxford: Blackwell.
Green, M. (2014) *The Development State: Aid, Culture and Civil Society in Tanzania*, Oxford: James Currey.
Gupta, A. (1998) *Postcolonial Developments: Agriculture in the Making of Modern India*, Durham, NC: Duke University Press.
Hymans, J. (1971) *Leopold Sedar Senghor: An Intellectual Biography*, Edinburgh: Edinburgh University Press.
Lerner, D. (1958) *The Passing of Traditional Society: Modernizing the Middle East*, Glencoe, IL: Free Press.
Lewis, D. and Kanji, N. (2009) *Non-Governmental Organizations and Development*, London: Routledge.
Leys, C. (1996) *The Rise and Fall of Development Theory*, Bloomington, IN: Indiana University Press.
Melkote, S. (1991) *Communication for Development in the Third World*, New Delhi: Sage.
Moyo, D. (2009) *Dead Aid: Why Aid Is Not Working and How There Is a Better Way for Africa*, New York: Farrar, Straus & Giroux.
Nyerere, J. (1979) *Ujamaa: Essays on Socialism*, Oxford: Oxford University Press.
Peet, R. and Hartwick, E. (2015) *Theories of Development: Contentions, Arguments, Alternatives*, 3rd edition, New York: Guilford Press.
Pottier, J., Bicker, A. and Sillitoe, P. (2003) *Negotiating Local Knowledge: Power and Identity in Development*, London: Pluto Press.
Prebisch, R. (1950) *The Economic Development of Latin America and Its Principal Problems*, Lake Success, NY: United Nations Department of Economic Affairs.
Rahnema, M. and Bawtree, V. (1997) *The Post-Development Reader*, London: Zed Books.
Rist, G. (2014) *The History of Development: From Western Origins to Global Faith*, 4th edition, London: Zed Books.
Rodney, W. (1972) *How Europe Underdeveloped Africa*, London: Bogle-L'Ouverture Publications.
Sachs, W. (1992) *The Development Dictionary: A Guide to Knowledge as Power*, London: Zed Books.
Sahle, E. (2002) *The Legacies of Julius Nyerere: Influences on Development Discourse and Practice in Africa*, Trenton, NJ: Africa World Press.
Sahn, D., Dorosh, P. and Younger, S. (1999) *Structural Adjustment Reconsidered: Economic Policy and Poverty in Africa*, Cambridge: Cambridge University Press.
Saunders, K. (ed.) (2002) *Feminist Post-Development Thought: Rethinking Modernity, Post-Colonialism, and Representation*, London: Zed Books.
Serra, N. and Stiglitz, J. (2008) *The Washington Consensus Reconsidered: Towards a New Global Governance*, Oxford: Initiative for Policy Dialogue.
Sillitoe, P. (2009) *Local Science vs Global Science: Approaches to Indigenous Knowledge in International Development*, Oxford: Berghahn Books.
Singer, H. (1949) 'Economic progress in under-developed countries', *Social Research*, 16(1): 1–11.
Slocum, N. (2003) *Participatory Methods Toolkit: A Practitioner's Guide*, Maastricht: United Nations University.
Young, R. (2016) *Postcolonialism: An Historical Introduction*, Oxford: Wiley-Blackwell.

3

The African idea of development

Sabelo J. Ndlovu-Gatsheni

Introduction

African development can be defined from Marcus Garvey's perspective as constant and consistent African search for self-improvement. Indeed, prior to colonialism, African people, like all other human beings, constantly and consistently improved their lives and life chances. They were inventive and innovative in many ways. One can read Molefi Kete Asante and Abus S. Abarry's *African Intellectual Heritage: A Book of Sources* (1996), which contains African works of imagination, invention, cultural dynamism, political engineering, religious and economic sophistication, and originality. In this work, Africans are captured as active and independent domesticators of plants and animals. They improved their technologies from stone to iron tools. They migrated in search of better environments that enhanced life chances. They formed sociopolitical systems: lineages, clans, chiefdoms, kingdoms, states, and nations of varying sizes and complexities. Just like other human beings, Africans even meditated on the meaning of life itself, on their origins, as well as on what being human meant. This is why Asante and Abarry's (1996) work opens with a focus on African narratives of 'the creation of the universe'. Myths of foundation and legends captured the meditative African practices, for example. Like all human beings, Africans developed complex religious ideas as they constantly and consistently made sense of their lives and their environment.

But the historical experience of Africa in the past 500 years invokes a fundamental question: What does development mean for a people struggling to emerge and free themselves from the inimical legacies of enslavement, colonialism, imperialism, apartheid, neocolonialism and underdevelopment, as well as impositions of the Washington Consensus and neo-liberalism? Taken together, these inimical processes constitute what the Latin American scholars such as Anibal Quijano (2000), Walter D. Mignolo (2000), Ramon Grosfoguel (2007), Nelson Maldonado-Torres (2007) and many others have termed 'coloniality'. Coloniality names the silenced constitutive part of Euro-North American-centric modernity that captures the negative impact of modernity outside Europe and North America.

African scholars such as Peter Ekeh (1983) prefer to use the common term 'colonialism' to highlight the structural straitjacket within which Africa is entrapped and the paradigm of difference it inaugurated. Ekeh argues that 'colonialism' cannot be simplistically understood as an event (colonisation) of conquest and rule over Africa by Europeans. Rather, it must be viewed as a systemic social movement of epochal dimensions whose enduring consequences outlive the end of direct colonialism (Ekeh, 1983). Kwame Nkrumah (1965) captured the long-term consequences of colonialism and coined the term 'neocolonialism', and Walter Rodney (1972) articulated it in terms of 'how Europe underdeveloped Africa'. Concurring with this epic school of colonialism, Ali Mazrui (1986: 13) understood colonialism as 'a revolution of epic proportions', concluding that 'what Africa knows about itself, what different parts of Africa know about each other, has been profoundly influenced by the West'.

Samir Amin (1990: 58) highlights the epistemic implications of Eurocentrism and colonialism on African conceptions of development, arguing that there is no 'implementation of the principle of autocentric development' in Africa that is free from 'economic rationality observed by conventional economics'. Conventional economics, just like all other understandings of the world cascading from Europe and North America, are basically local ideas that have been universalised. They carry Eurocentrism. Samir Amin (1990) described Eurocentrism as one of the greatest 'ideological deformations' of our time as it falsely invented Europe as the centre of the world and as the inventor of all positive human values. Ngugi wa Thiong'o (1986) defined what was unleashed by Euro-North American-centric modernity in its colonial and imperial phase as a 'cultural bomb', and elaborated on the long-term consequences in this eloquent manner:

> The effect of a cultural bomb is to annihilate a people's belief in their names, in their language, in their environment, in their heritage of struggle, in their unity, in their capacities and ultimately in themselves. It makes them see their past as one wasteland of non-achievement and it makes them want to identify with that which is furthest removed from themselves; for instance, with other people's languages rather than their own. It makes them identify with that which is decadent and reactionary, all those forces which would stop their own springs of life. It even plants serious doubts about the moral rightness of struggle. Possibilities of triumph or victory are seen as remote, ridiculous dreams. The intended results are despair, despondency and a collective death-wish.
>
> *(Ngugi wa Thiong'o, 1986: 3)*

Because of enduring legacies and realities of enslavement, colonialism, imperialism, apartheid, neocolonialism, underdevelopment and structural adjustment programmes, as well as the new scramble for African natural resources, African conceptions of development continue to be a struggle for what Ngugi wa Thiong'o (2009a, 2009b) understood as 're-membering,' that is 'a quest for wholeness after over five hundred years of "dismemberment".' Development as understood from a decolonial 're-membering' perspective is basically a restorative and recovery project (Ndlovu-Gatsheni, 2015a, 2015b). After 1945, Africans and other (ex)-colonised peoples increasingly demanded that development be made one of the important human rights – the right to development. This made sense in a global terrain within which it was known that a combination of enslavement, colonisation, apartheid and neocolonialism actively denied development to those regions of the world that experienced colonial difference.

When Rodney wrote *How Europe Underdeveloped Africa* (1972), other scholars such as Andre Gunder Frank were already articulating the lack of development in ex-colonised regions as caused by peripherisation that bred dependency. Whatever critics of dependency scholarship can say about its limits, it correctly identified constraining structural conditions born out of identifiable historical processes cascading from the encounter between Africa and Europe that produced dependency. That dependency is still a central part of the asymmetrical relations of Africa and Europe, 60 years after 'decolonisation'. This means that ideas of development in Africa must transcend the 'myth of decolonisation' (Grosfoguel, 2007; Ndlovu-Gatsheni, 2013b).

In this case, development is partly a struggle to end dependency and broadly a re-humanising process after long years of denied humanity. This denial of being is well documented by Maldonado-Torres (2007) as 'coloniality of being', which captures in minute details of what I have called 'technologies of dismemberment' underpinning 'genealogies of dismemberment' (Ndlovu-Gatsheni, 2015b). The technologies of dismemberment include genocides, epistemicides, linguicides, various processes of alienation and dispossessions (Ndlovu-Gatsheni, 2015b: 32). Consequently, at the centre of the African idea of development is the struggle for restoration of denied, if not lost, ontological density. One can even evoke Frantz Fanon's (1968) ideas of the 'zone of being' and the 'zone of non-being' to elaborate on the African idea of development as part of long-standing struggles to rise from the 'zone of non-being' to the 'zone of being'. The zone of non-being is a colonially invented space to which African people were forced to inhabit. It is characterised by depravity and suffering.

At the global level:

> 'development' in our time is such a central value. Wars are fought and coups launched in its name. Entire systems of government and philosophy are evaluated according to their ability to promote it. Indeed, it seems difficult to find any way to talk about large parts of the world except in these terms.
>
> *(Ferguson, 1990: xiii)*

Inevitably, African leaders were so enchanted by the post-1945 developmental promise to the extent that they strongly believed that decolonisation would not only result in taking over of the colonially crafted colonial state, but that they would achieve full sovereignty, which would enable them to launch autonomous developmental trajectories aimed at not only reducing the inequality gap between economies of ex-colonies and those of erstwhile colonial powers, but also improving the socio-economic conditions of the formerly colonised peoples. However, 60 years since the dawn of decolonisation, 'the promised land of development remained an illusion' (Grosfoguel and Cervantes-Rodriguez, 2002: xxvi).

Today, development can be understood from two vantage points. The first is an ahistorical and apolitical technicist-liberal approach. In this perspective, development is understood in three ways. First, development is understood as a vision, description or measure of the state of being of a desirable society. Second, development is understood as a process of social change in which societies are transformed over long periods. Finally, development is conceived to be consisting of deliberate efforts aimed at improvement of human life (Thomas, 2000). The problem with the ahistorical and apolitical approach is that it does not take into account the realities of world power and global interests, as well as colonial matrices of power that constitute the discursive context within which the notions of development emerge, are defined and articulated, contested and questioned (Ndlovu-Gatsheni, 2013a, 2013b). For Africa, the long-standing challenge has been how to decolonise and democratise the racially hierarchised and Euro-North American-centric world system that is informed by power politics and coercion (Grosfoguel and Cervantes-Rodriguez, 2002: xi–xxix).

This chapter favours a historical and political perspective where development is understood as a broad process of 're-membering' of a formerly 'dismembered' people and a highly contested idea caught up within 'a terrain of hegemony and counter-hegemony' (Pieterse, 2010: 9). This approach enables one to capture modern conceptions of development as an idea and discourse while at the same time speaking to African conceptions of development informed by decolonial liberatory thought. At the same time, such an approach remains consistently sensitive to the popularisation and universalisation of development as a human aspiration that is itself entrapped within post-Enlightenment modernist conceptions of progress (Escobar, 2012). Below is a discussion of the Truman and Bandung versions of development as part of enhancing the distinctiveness of the African idea of development.

The Truman versus Bandung versions of development

The apogee of post Enlightenment conceptions of development culminated into the post-1945 Truman version of development whereby Europe and North America claimed to have been entrusted by modern history with the task of developing the Third World in its image (Truman, 1964 [1949]). At the centre of the Truman version of development is what Fantu Cheru termed 'the imperial project' that is informed by geopolitical considerations and the Global North's power calculations, as well as the consistent rhetoric of humanitarianism that conceals coloniality (Cheru, 2009).

The Truman version of development is deeply situated in the long history of the making of the modern world system and the invention of Africa that was accompanied by what Valentin Y. Mudimbe terms the 'paradigm of difference' (Mudimbe, 1988, 1994). It is traceable to what became known as 'voyages of discovery' that opened the way for mercantilism, the slave trade, imperialism and colonialism. These processes were articulated from a colonial perspective as entailing the spreading of civilisation, modernity, commerce and Christianity to those places such as Africa that were described as the 'heart of darkness' (Conrad, 1902). Consequently, development as an offspring of Enlightenment and modernity became entangled in a complex and long history mediated by the logics of coloniality informed by domination, oppression and exploitation, on the one hand, and the rhetoric of modernity that promised a brave modern society where salvation, civilisation, progress, economic growth, freedom and democracy would be the order of life, on the other (Mignolo, 2005).

The Truman version of development became contested by the Bandung decolonial version of development cascading from the Bandung Conference of 1955. The Bandung version articulated development as liberation and a human right that has to be fought for (Mkandawire, 2011). But African struggles for decolonisation and development have remained trapped within the hegemonic Truman version of development, which is backed up by what Adebayo Adedeji termed the 'development merchant system' (DMS) driven by the Breton Woods Institutions (BWI), which finance the implementation of an exogenous development agenda (Adedeji, 2002: 4). At the centre of the DMS is what David Slater termed 'imperiality of knowledge' constituted by 'interweaving of geopolitical power, knowledge and subordinating representation of the other' (Slater, 2004: 223). The DMS maintains coloniality long after the dismantlement of administrative colonialism. It still approaches Africa as a space inhabited by a people 'shorn of the legitimate symbols of independent identity and authority', as well as a 'space ready to be penetrated, worked over, restructured and transformed' from outside (Slater, 2004: 223). The DMS exists as a consortium of the International Monetary Fund (IMF), the World Bank (WB), the World Trade Organization (WTO), intergovernmental organisations (IGOs) and multinational corporations (MCs). They advance a 'Bretton Woods paradigm' of development as part of a continuation of the Truman version (Therien, 1999).

At another level, the Bandung decolonial version of development continues to contest the Truman imperial version of development through enabling the crafting of counter-hegemonic development initiatives and strategies that included: the formation of the Organisation of African Unity (OAU); the demand for a New Economic International Order (NIEO); the crafting of the African Declaration on Cooperation, Development and Economic Independence of 1973; the Revised Framework of the Principles for the Implementation of New International

Economic Order in Africa of 1976; the Lagos Plan of Action for the Economic Development of Africa of 1980; the African Priority Programme for Economic Recovery of 1986; the African Alternative Framework to Structural Adjustment Programme for Social Economic Recovery and Transformation of 1989; the African Charter for Popular Participation for Development of 1990; the United Nations New Agenda for Development of Africa of 1991; and the current New Partnership for African Development (NEPAD) of 2001 (Adesina *et al.*, 2006).

The Bandung version shared some commonalities with the 'United Nations paradigm' and its emphasis on human development. The Bandung Conference was informed by the spirit of decolonisation, South–South solidarity and the principle of non-alignment. The conference laid the basis for the struggle to decolonise and democratise the international society so as to achieve equitable representation in global decision-making bodies as it called for a New International Economic Order (NIEO) (Kahin, 1956). Development was articulated in terms of transcending colonialism and coloniality, as well as in terms of freedom, self-determination and self-reliance under state mediation and direction (Berger, 2004). The Bandung Conference also endorsed the right of governments to freely choose their own political and economic systems (Abdulgani, 1964). For all the states that emerged from colonialism, nationalism and development were articulated as inseparable processes simply because colonialism denied Africans both development and freedom. Therefore, in Africa and other ex-colonised parts of the world, nationalism and developmentalism constituted efforts at 're-membering' those dismembered societies and peoples so as to regain their lost ontological density.

Development as 're-membering' Africa and the traps of coloniality

In the language of Ngugi wa Thiong'o, African development initiatives constituted efforts aimed at 're-membering' Africans after centuries of being 'dismembered from the land, from labour, from power, and from memory' (Ngugi wa Thiong'o, 2009a: 22). Thus, development conceived in decolonial historical and political Africa-centred terms can be defined as a long-term struggle for the 're-membering of Africa' that encapsulates such initiatives as Ethiopianism, Garveyism, Negritude, pan-Africanism(s), African personality, Afrocentricity, consciencism, African humanism, African socialism(s), the Black Consciousness Movement and the African Renaissance. This point was well captured by Nigerian novelist Chinua Achebe when he said:

> You have all heard of African personality; of African democracy; of African way to socialism; of Negritude, and so on. They are all props we have fashioned at different times to help us get on our feet again. Once we are up we shall not need any of them anymore. But for the moment it is in the nature of things that we need to counter racism with what Jean-Paul Sartre has called an anti-racist racism, to announce not just that we are good as the next man but that we are better.
>
> *(cited in Moore-Gilbert, 1997: 179)*

'Re-membering' Africa entailed building unity inside individual states, aspiring for pan-African unity at the continental level, and fighting for deimperialisation, deracialisation, decolonisation and democratisation of the racially hierarchised and Western-centric modern world system that came into being in the fifteenth century. Coloniality is underpinned by theft of the history of the colonised. Inevitably, the discourse of development in Africa commenced as a struggle over Hegelian denial of African history. Since the encounter between Africa and Europe, African history experienced usurpation, if not its displacement and silencing (Depelchin, 2005). This is why Paul Tiyambe Zeleza posed the question: 'Is autonomy of African history possible, can this history be written without European referents, is it possible to liberate African history from the epistemological traps of Eurocentrism, the traps of the "colonial library"?' (Zeleza, 2005: 1).

This is also why Edward Wilmot Blyden and later Kwame Nkrumah emphasised the need to rewrite the history of Africa as part of 're-membering' the continent and the centring of 'African personality' (Nkrumah, 1964). Nkrumah's concept of philosophical consciencism articulated African development as embracing traditional Africa, Islamic Africa and Euro-Christian Africa as constituent elements of a new emergent ideology for the harmonious growth and development of society. This is how he put it:

> The philosophy that must stand behind this social revolution is that which I have once referred to as philosophical consciencism: consciencism is the map in intellectual terms of the disposition of forces which will enable African society to digest the Western and the Islamic and the Euro-Christian elements in Africa, and develop them in such a way that they fit into the African personality. The African personality is itself defined by the cluster of humanist principles which underlie the traditional African society.
>
> *(Nkrumah, 1964: 79)*

If, to African nationalist thinkers and theorists, development meant moving from one state of being that is characterised by depravity and suffering towards another that is more desirable, then that state of being was that of slavery, imperialism, colonialism, apartheid and capitalism. Such African leaders and theorists as Frantz Fanon, Leopold Sedar Senghor, Kenneth Kaunda, Julius Nyerere and Nelson Mandela emphasised the importance of humanism that was denied by the slave trade, imperialism, colonialism, apartheid and neocolonialism. Humanism in African thought is a central leitmotif of 're-membering' for a people whose very being was questioned and denied. Among Africans, there was and is a strong feeling that slavery, imperialism, colonialism, capitalism, apartheid and neocolonialism were the antithesis of humanism and resulted in a denial of humanity to black people. Independence and development had to culminate in restoration of lost African ontological density as part of the 're-membering' process. Kenneth Kaunda defined humanism as a 'philosophy of life' and noted that colonialism 'devalued man' (Kaunda and Morris, 1966: 19–21). Kaunda posited that:

> It was nationalism, of course, which restored our self-confidence, for it taught us what we could do together as men [and women], and only as men [and women] – at no stage in the freedom struggle had we the material power or military might of colonialists. It was humanity in revolt that won us freedom. [. . .] It was the triumph of a Man-centred society over a Power-centred society. This intense belief in the possibility of Man is a discovery which Africa appears to be making long after the West has discarded it.
>
> *(Kaunda and Morris, 1966: 21)*

Kaunda noted that when Europe embarked on enslavement and colonisation of other spaces and peoples, it was actualising its long-standing abandonment of humanist ideas that envisioned a humane society and world where every human species had space to live comfortably. In his 'meditations on man', Kaunda expressed his conviction that 'only the recovery of a sense of the centrality of Man will get politics back on the right track', and he posed the question: 'How can we humanise our politics in Zambia so that the humblest and least endowed of our citizens occupies a central place in Government's concerns?' (Kaunda and Morris, 1966: 41). The dominance of humanist nationalism resulted in various experimentations with African socialism, the most well-known example being that of 'ujamaa' (familyhood) of Tanzania under Julius Nyerere, which sought to provide an escape route from capitalism's idea of happiness based on 'exploitation of man by man' (Nyerere, 1967).

The search for a humane society by African nationalists is further exemplified by such charters and declarations as the Freedom Charter (South Africa), the Arusha Declaration (Tanzania), the Common Man's Charter (Uganda) and the Mulugushi Declaration (Zambia). These charters and declarations were part of nationalist humanist imaginations of a better world, free from racism, exploitation and oppression. That these declarations and charters did not eventually reflect the practical political practices of African leaders partly reflects how difficult it was to decolonise the Western conceptions of politics in terms of 'will to power' rather than 'will to live' (Ndlovu-Gatsheni, 2016).

But at another level, nationalist developmentalism was informed by a realisation that colonial domination was imposed on Africa successfully because of 'guns and steel' (Mkandawire, 2005: 13). Therefore, for Africa to survive domination and humiliation, there was a need to develop and modernise. Another important element of nationalist developmentalism was the desire to eradicate what Thandika Mkandawire termed the 'unholy trinity of ignorance, poverty and disease' (Mkandawire, 2005: 13). It is therefore important to take stock of African nationalist-inspired development initiatives since the dawn of decolonisation, beginning with the implications of the Bandung decolonial spirit and the legitimate decolonial demand for a New Economic International Order (NEIO) amenable to the possibilities of African development that is unencumbered by global colonial matrices of power.

The Bandung spirit and struggles: New Economic International Order

The Asian-African Conference that was held in April 1955 in Bandung in Indonesia constituted the highest point of the Afro-Asian solidarity movement that was informed by Afro-Asian decolonial nationalism. When the Afro-Asian movement was formed, its specific aim was to hasten the process of decolonisation (Kimche, 1973). But by 1955, the focus of the Bandung Conference was how to create conditions for global peaceful coexistence within a post-1945 dispensation that was troubled by ideological rivalry between the West and the East, which became known as the Cold War.

The Cold War was another layer of coloniality. Both the Soviet Union and the United States of America's rise to superpower status inaugurated other forms of global coloniality and imperialism. We can call it ideological coloniality. Those who met at Bandung had a clear understanding of this ideological coloniality. In the first place, the Marxist idea of unity of workers was substituted with the slogan of the unity of all ex-colonised peoples of the world at the Bandung Conference. What was emphasised was solidarity and cooperation among newly independent states in the face of a 'mixture of wooing, bullying, flattery, threats and "presents" coming from superpowers' (Jansen, 1966: 309). Richard Wright's book *The Colour Curtain* (1956) captured the decolonial essence of the Bandung Conference in these revealing words:

> The despised, the insulted, the hurt, the dispossessed – in short, the underdogs of the human race were meeting. Here were class and racial and religious consciousness on a global scale. Who had thought of organizing such a meeting? And what had these nations in common? Nothing, it seemed to me, but what their past relationship to the Western world had made them feel. This meeting of the rejected was itself a kind of judgement upon the Western world!
>
> *(Wright, 1956: 12)*

At another level, a 12-point economic cooperation agenda of the Bandung Conference identified development as an urgent priority for Afro-Asian states. It is clear that at Bandung, the African and Asian leaders realised that the development they urgently wanted could not be attained within a modern world system that was racially hierarchised and Euro-North American-centric. Decolonisation needed to be expanded to grapple with a global system that remained asymmetrical in its power configuration – a world where the newly independent African states had no voice in global decision-making. This is why the participants at the Bandung Conference were deeply troubled by their lack of participation in the international institutions that had been established to govern world affairs (Abdulgani, 1964).

One can safely argue that the Bandung Conference correctly identified 'global coloniality' as a major hindrance to the achievement of development in Africa and Asia (Ndlovu-Gatsheni, 2013b). This argument is amplified by the fact that three objectives drove the Bandung spirit, namely non-alignment in the age of the Cold War, elimination of all forms of colonialism and racism, and finally modernisation and economic development (Mayall, 1990). The idea of non-alignment spoke to an aspiration for a new international norm that gave states the right to develop and orchestrate an autonomous, freely chosen trajectory in the international system without being compelled to follow the imperial-hegemonic dictates of the superpowers. The issue of equality of states was stressed, including raising concerns about under-representation of African and Asian states on the United Nations Security Council. The Bandung spirit also articulated the problem of international economic injustices (Krasner, 1981).

What is also important about the Bandung world view was to introduce new and alternative principles for international society informed by international engagement based on dialogue and cooperation, justice, fairness and peaceful coexistence. These principles indicated the desire of those states emerging from colonialism to make a paradigm shift from imperial and colonial principles that were informed by the spirit of domination and coercion as an international mode of interaction that has been in place since the fifteenth century (Ndlovu-Gatsheni, 2013a).

The Bandung spirit laid a strong basis for the crystallisation of a Third World coalition in the United Nations, which in 1964 constituted the United Nations Conference on Trade and Development (UNCTAD) and informed the ideas that produced the Non-Alignment Movement (NAM), which was formerly constituted at a summit in Algiers in 1971. NAM included Latin American states that had gained political independence far earlier than Asian and African colonies. The Latin American states had been active within the United Nations in pushing the agenda of reforming the international economic and political system. At the Cairo Economic Conference of 1962, the Afro-Asian and Latin American states had already pointed out that the more dangerous division of the world was the North–South divide, which was taking the form of the 'haves' and the 'have-nots', rather than the ideologically informed East–West divide that preoccupied the Western powers (Nesadurai, 2005: 12).

The resolutions of the first NAM Summit held in Belgrade in Yugoslavia in 1961 and the Cairo Summit of 1962 are credited for putting pressure on the United Nations to establish UNCTAD as an agency to address Third World development issues. The problem is that UNCTAD played a marginal role in global economic governance as the United States of America and other industrial powers rejected any role for the body in trade negotiations (Taylor, 2003). At the time of the launch of UNCTAD, the Latin American economist Raul Prebisch was pushing forward the dependency thesis to explain the problem of development in the Third World. He was appointed UNCTAD's first Secretary General. Informed by dependency ideas, UNCTAD's demands included greater access to industrial countries' markets, greater self-reliance among Third World countries, the right to nationalise assets, and democratisation of all binding international decision-making based on the principle of 'one-nation, one vote' (Gosovic, 1972). These demands were informed by Prebisch's decolonial diagnosis of Third World underdevelopment in terms of its structural dependency on a capitalist core that controlled all levers of international decision-making and profitable economic activity. The proposals were rejected by industrial powers.

Third World leaders did not give up the fight for a reformed international system. The period from 1973 to 1980 were dominated by the demand for the NIEO. It was following the oil crises of 1971 and 1973 that the Third World coalition united to push further for the NIEO. They pushed for this through the G77 – a coalition of developing countries. The NIEO was informed by the dependency ideas that emphasised decolonisation of global coloniality that favoured the industrial powers and disadvantaged those countries that emerged from colonialism. The NIEO called for the restructuring of global structural regimes informing unequal trade and other economic interactions (Cox, 1979: 257). The more radical members of the G77 demanded restitution based on the notion that the industrialised powers owed something to the South as compensation for slavery and colonialism (Cox, 1979). In the spirit of the NIEO, the African leaders adopted the African Declaration on Cooperation, Development and Economic Independence in 1973, which articulated Africa's strategy for gradual disengagement from the world economy through escalation of national and continental self-reliance. This was followed by the production of the Revised Framework of the Principles for the Implementation of New International Economic Order in Africa of 1976. It was produced by the United Nations Economic Commission for Africa (UNECA) and became the intellectual and theoretical foundation for the drafting of the Monrovia Strategy for the Economic Development of Africa of 1979. The Monrovia Declaration emphasised collective self-reliance and economic integration of African economies, investment in science and technology as the backbone of Africa's development process, ensuring Africa's self-reliance in food production, and a commitment to achieve modern African economies by the year 2000 (Baah, 2003).

The African demand for the NIEO was resisted by the industrial powers as part of a communist conspiracy, as irrational and as too revolutionary. Consequently, very little headway was made simply because the powerful and dominant wanted to stay powerful and dominant. The little concession made was the adoption in 1975 of the Charter of Economic Rights and Duties of States at the United Nations General Assembly, which to a minor extent included the rights and duties of all countries to help other countries develop economically in terms of their chosen path of development (Murphy, 2005). But the overall picture is that proposals and demands from the Global South suffered rejection by the powerful industrialised nations of Europe and North America. The lack of strong unity within the Third World coalition also contributed to the failure and collapse of the NIEO. Samir Amin had this to say about the demise of the NIEO:

> So, in the end, the battle for the NIEO was lost. As well as the failure being noted, the causes have to be studied. Are they purely circumstantial (in the economic crisis)? Can they be attributed to 'tactical errors' by the Third World (its own divisions and weaknesses)? Or do these circumstances and weaknesses show the impossibility of autocentric development at the periphery of capitalist system?
>
> *(Amin, 1990: 56–57)*

The NIEO was soon eclipsed by the era of hegemonic neo-liberalism that was ushered in by the Anglo-American leaders Margaret Thatcher and Ronald Reagan in the late 1970s. But African leaders continued to try to forge ahead with African development initiatives even within a context dominated by a development merchant system that carried imperiality and coloniality.

The search for autonomous development trajectory

Over the years since the dawn of decolonisation, it became increasingly clear to many Africans that African development could not be merely a philanthropic gift from the industrialised powers of the West. It had to be planned for and fought for as part of deepening decolonisation. This is why the OAU working closely with UNECA produced a consciously inward-looking pan-African development plan that sought to articulate a comprehensive and unified strategy to reduce dependence on external powers and to replace it with self-reliance. This initiative became known as the Lagos Plan of Action (LPA) of 1980. At the centre of the plan was not complete delinking, but strategic disengagement from those features of the international economic system that were keeping Africa dependent, underdeveloped, weak and poor (Ikome, 2007).

The LPA was a comprehensive initiative consisting of 13 chapters covering all sectors of the African economy and informed by detailed objectives that included alleviation of poverty among Africans, diversification of productive economic capabilities, internalisation of forces of supply and demand, and mobilisation of Africans for production. Just like the Bandung Conference, the LPA identified Africa's development as structural and exacerbated by a hostile external environment. The LPA was informed by both dependency ideas and the spirit of pan-Africanism.

The LPA of action has been criticised for articulating a one-sided cause of the development crisis in Africa. It identified hostile external environment without paying equal attention to internal problems that are equally important in understanding the problems of African development. The emphasis on a hostile external environment led the LPA to focus too much on trade and ignore such internal constraints as a lack of serious commitment of African leadership to the development of the African continent and its people (Onimonde *et al.*, 2004). Corruption and authoritarianism were not clearly identified as internal constraints to development. The LPA is said to have been conceived as a top-down project that ignored other important sectors capable of driving African development. More importantly, the LPA identified the problems and offered some solutions, but without committing financial resources to the implementation of the initiative (Ikome, 2007). But more importantly, the LPA document 'demonstrated both a disturbing lack of imagination and a low level of consciousness of the character of the option of self-reliance' (Amin, 1999: 59).

What must be emphasised is that if the LPA was an authentically African development initiative, it was destined to suffer a lack of financial support and delegitimisation by forces of coloniality. Thus, when one assesses the failure of African development initiatives, it is important to broaden the critique to the level of global discursive terrain within which Africans fought for development. What Adedeji identified as the development merchant system worked actively to destabilise any African development initiative so as to push in an exogenous initiative that does not threaten the hegemony of the West. The LPA, despite its declared self-reliance paradigm and its robust criticism of the colonial and neocolonial heritage, 'could not escape the conventional methodology closely associated with the conventional strategy of peripheral capitalist development' (Amin, 1990: 59). Amin posed the question:

> Should development be conceived in accordance with the demand of the international order, or conversely, is it necessarily in conflict with it. Can the international order be transformed and 'adjusted' to the priority demands for Third World development, or conversely can the latter only be the result of the reverse 'adjustment'?
>
> *(Amin, 1990: 60)*

It is important to realise that the LPA was launched in the midst of the rise of neo-liberalism and the Washington Consensus.

African entrapment in the development merchant system

According to Adebayo Adedeji, 'the overarching objective of the DMS is for the African canoe to be firmly tied to the North's neo-liberal ship on the waters of globalization' (Adedeji, 2002: 4). The DMS actively works to make sure African development initiatives do not operate outside the grip of Euro-North American development orthodoxy and coloniality. If we understand this reality, it becomes clear why the powerful industrial powers of the North have consistently rejected demands for reforming the modern world system and actively made sure that no alternative development paradigms, alternatives and agendas are entertained. It is within this context that the Lagos Plan of Action (1980–2000), the African Priority for Economic Recovery (1986–1990); the African Alternative Framework to Structural Adjustment Programmes for Socio-Economic Recovery and Transformation, the African Charter for Popular Participation for Development (1990) and the United Nations New Agenda for the Development of Africa in the 1990s were not well received by the donor community and the Breton Woods Institutions.

The 1980s and 1990s witnessed the ascendance and consolidation of the neo-liberal model of development informed by post-Keynesianism. This post-Keynesianism became known as the Washington Consensus, which carried anti-state philosophies and a strong belief in the free reign of markets, privatisation and deregulation. The Berg Report of 1981 introduced the philosophy and practice of structural adjustment programmes (SAPs). As a result of the implementation of SAPs, African states lost the little that was remaining of their control over development policy. African development became driven from the outside, with devastating consequences on the African people and their leaders. Thandika Mkandawire argued that 'For Africa the 1980s and 1990s was a period of wanton destruction of institutions and untrammelled experimentation with half-baked institutional ideas. The result was "unconstructive destruction" in its most institutionally debilitating form' (Mkandawire, 2003: 10).

The consequences of SAPs included removal of the state from driving development, exacerbation of internal inequalities and worsening poverty, rehabilitation of coloniality, and loss of policy space by African leaders. Instead of the Bretton Woods Institutions accepting responsibility for drawing Africa deeper into crisis through the introduction of SAPs, the World Bank produced a document entitled *From Crisis to Sustainable Growth* in 1989, in which they blamed lack of good governance and policy reform as the cause of economic crisis and lack of development in Africa in the 1980s. In 1994, the World Bank produced another report entitled *Adjustment in Africa*, in which the issue of sound macroeconomic and structural management was defended as a prerequisite for growth and poverty reduction in Africa. Lack of development and economic growth in Africa was explained in terms of poor policy choices by African leaders, inefficiency and corruption. Structural barriers in the international political economy were not identified as a cause of underdevelopment. The African postcolonial state was identified as the major culprit inhibiting development in Africa (Fukuyama, 2004).

An African consensus emerged that SAPs were a disaster for the development of the continent. By the late 1980s, through UNECA, an alternative to SAPs was being sought, and these efforts culminated in the production and adoption of the African Alternative Framework to Structural Adjustment Programmes for Socio-Economic Recovery and Transformation of 1989. Through this document, Africans rejected SAPs and offered well-reasoned arguments for the rejection (Tomori and Tomori, 2004). This African initiative was not taken seriously and its recommendations fell on deaf ears.

The dawn of the third millennium witnessed the rise of the so-called new African leaders who were considered to be less corrupt and fully committed to the economic renewal of the continent. The new leaders committed themselves to enabling Africa to claim the twenty-first century as the African century for development. The philosophy of African Renaissance that was increasingly associated with President Thabo Mbeki of South Africa provided the discursive framework for relaunching African development. The flagship project for the new leaders of Africa became the New Partnership for African Development (NEPAD) that was adopted in 2002. NEPAD was informed by positive millenarian thinking. There was an optimistic mood that the twenty-first century would indeed be an African century where Africa would be allowed to drive its own development with the positive support of the developed countries of the North. This optimistic mood made some Africans oblivious of coloniality as an obstacle to African development.

Inevitably, unlike the LPA, which focused mainly on external structural barriers as responsible for African development failures, NEPAD highlighted such factors as bad governance, corruption and conflicts as responsible for African underdevelopment. What is distinctive about NEPAD has been its strong conviction on the possibilities of mutual beneficial partnership between the poor African countries and the industrialised and rich countries of the North. Coloniality as a global power structure that sustains asymmetrical economic power relations between the underdeveloped South and the developed North remains a destabilising and inhibiting force.

The political rhetoric of the leaders of the industrialised countries of the North misleads one to believe that there was a paradigm shift and a new international order that was favourable to African development was emerging. The G8 pledged to fully support NEPAD. The question that developed in the minds of many progressive African scholars was why the G8 was openly supporting NEPAD when throughout the postcolonial period the industrialised countries from the North opposed African development initiatives. Was NEPAD really an African initiative? Is NEPAD not part of those spurious neocolonial partnerships that hides the realities of structural inequalities? Is NEPAD part of Africa's indigenous/endogenous development agenda, or is a project cascading from the DMS but masquerading as an African development initiative? It would seem that NEPAD falls neatly within the discourse of partnerships that commenced with such initiatives as the Lome Conventions rather than the Bandung decolonial version of development. At the Lome Conventions, the industrialised European countries countered the African development initiatives by putting on the table fake partnership proposals that were not only informed by the imperial logic of donor–recipient relationships, but were basically part of the sustenance of coloniality long after the end of direct colonial administrations. The Lome Conventions were a precursor to the hypocritical discourse of 'our common interest' contained in Tony Blair's Commission for Africa (Hurt, 2007).

Conclusions: where is Africa heading?

Today, many Africans are toying with the notion of 'developmental states' in an attempt to follow the developmental path of the Asian Tigers. But the global colonial and imperial matrices of power are not enabling them compared to the time of the rise of Asian Tigers. Today, Africa is experiencing what has come to be known as the second scramble for African natural resources. The scramblers are no longer from the West only; they also come from the East. China is leading the scramblers from the East. It is within this context that there is also the discourse of 'Africa Rising' predicated on increased demands for African primary products. This time, development is narrowly defined as economic growth driven from outside by outside demand for African natural resources. African ideas must drive the development trajectory.

But the reality about the African struggle for development is that it is taking place within a modern world system that is resistant to decolonisation and an international order that consistently works towards disciplining anti-systemic movements and forces so as to give the world system a new lease of life. The world system and its shifting international orders continue to subject African development initiatives and plans to the logic of classical economic thought and the dominant logic of capitalist accumulation that created Africa as a periphery. This is why the Bandung spirit was consistently frustrated and the African demand for a new international economic order was vehemently resisted.

The acceptable model of development for Africa in the eyes of those in the Global North is one whereby African economies are adjusted and aligned to the imperatives and demands of the international division of labour. Africa has never run short of development initiatives and plans, but what has been lacking is imagination of autocentric development outside the template cascading from classical economic thought and capitalist imperatives. The DMS works actively to prevent and destroy any development initiative that threatens to subvert the modern world system and its international order.

As argued by Samir Amin (1990), Africa has been fighting for a space to originate new economic ideas and imagine new development alternatives suitable for the African reality and experience. Such a space is denied. But Africa continues to work in cohort with other developing countries for space and economic justice. To achieve the objective of building a new international economic order, there is need for Europe and North America to genuinely accept decolonisation and deimperialisation of the world system as a form of 're-membering'. Only a combination of genuine deimperialisation on a world scale and decolonisation would result in a global democracy in which Africa would be afforded space to develop.

References

Abdulgani, R. (1964) *Bandung Spirit: Moving on the Tide of History*, Jakarta: Prapanga.

Adedeji, A. (2002) 'From the Lagos Plan of Action to the New Partnership for African Development and from the Final Act of Lagos to the Constitutive Act: wither Africa?', unpublished keynote address delivered at the African Forum for Envisioning Africa, Nairobi, Kenya, 26–29 April.

Adesina, J.O., Graham, Y. and Olukoshi, A. (eds) (2006) *Africa and Development Challenges in the New Millennium: The NEPAD Debate*, Dakar and London: CODESRIA Books and Zed Books.

Amin, S. (1990) *Maldevelopment: Anatomy of a Global Failure*, Tokyo and London: United Nations University Press and Zed Books.

Amin, S. (1999) *Eurocentrism: Modernity, Religion, and Democracy – A Critique of Eurocentrism and Culturalism*, New York: Monthly Review Press.

Asante, M.K. and Abarry, A.S. (1996) *African Intellectual Heritage: A Book of Sources*, Philadelphia, PA: Temple University Press.

Baah, A. (2003) *History of African Development Initiatives*, paper presented at the African Labour Research Network, Johannesburg, South Africa.

Berger, M.T. (2004) *The Battle for Asia: From Decolonization to Globalization*, London and New York: Routledge.

Cheru, F. (2009) 'Development in Africa: the imperial project versus the national project and the need for policy space', *Review of African Political Economy*, 120: 275–278.

Conrad, J. (1902) *Heart of Darkness and Other Stories*, London: Wordsworth Editions.

Cox, R. (1979) 'Ideologies and the New International Economic Order: reflections on some recent literature', *International Organization*, 33(2): 257–308.

Depelchin, J. (2005) *Silences in African History: Between the Syndromes of Discovery and Abolition*, Dar es Salaam: Mkuki Na Nyota.

Ekeh, P. (1983) *Colonialism and Social Structure: University of Ibadan Inaugural Lecture, 1980*, Ibadan: Ibadan University Press.

Escobar, A. (2012) *Encountering Development: The Making and Unmaking of the Third World – With a New Preface by the Author*, Princeton, NJ: Princeton University Press.

Fanon, F. (1968) *The Wretched of the Earth*, New York: Grove Press.

Ferguson, J. 1990. *The Anti-Politics Machine: 'Development', Depoliticization, and Bureaucratic Power in Lesotho*, Cambridge: Cambridge University Press.

Fukuyama, F. (2004) *State Building: Governance and World Order in the Twenty-First Century*, London: Profile Books.

Gosovic, B. (1972) *UNCTAD: Conflict and Compromise*, Leiden: A.W. Sifthoff.

Grosfoguel, R. (2007) 'The epistemic decolonial turn: beyond political-economy paradigms', *Cultural Studies*, 21(2/3): 203–246.

Grosfoguel, R. and Cervantes-Rodriguez, A.M. (2002) 'Introduction: unthinking twentieth-century Eurocentric mythologies – universalist knowledges, decolonization, and developmentalism', in R. Grosfoguel and A.M. Cervantes-Rodriguez (eds), *The Modern/Colonial/Capitalist World-System in the Twentieth Century: Global Processes, Antisystemic Movements, and Geopolitics of Knowledge*, Westport, CT: Praeger Publishers, pp. xi–xxix.

Hurt, S.R. (2007) 'Mission impossible: a critique of the Commission for Africa', *Journal of Contemporary African Studies*, 25(3): 355–368.

Ikome, F.N. (2007) *From LPA to NEPAD: The Political Economy of African Regional Initiatives*, Midrand: Institute for Global Dialogue.

Jansen, G.H. (1966) *Afro-Asia and Non-Alignment*, London: Faber & Faber.

Kahin, G.M. (1956) *The Asia-African Conference: Bandung, Indonesia, April 1955*, Ithaca, NY: Cornell University Press.

Kaunda, K.D. and Morris, C.M. (1966) *A Humanist in Africa: Letters to Colin M. Morris from Kenneth D. Kaunda President of Zambia*, Nashville, TN: Abingdon Press.

Kimche, D. (1973) *The Afro-Asian Movement: Ideology and Foreign Policy of the Third World*, Jerusalem: Israel University Press.

Krasner, S. (1981) 'Transforming international relations: what the Third World wants and why', *International Studies Quarterly*, 25(2): 119–148.

Maldonado-Torres, N. (2007) 'On the coloniality of being: contributions to the development of a concept', *Cultural Studies*, 21(2/3): 240–270.

Mayall, J. (1990) *Nationalism and International Society*, Cambridge: Cambridge University Press.

Mazrui, A.A. (1986) *The Africans: A Triple Heritage*, London: BBC Publications.

Mignolo, W.D. (2000) *Local Histories/Global Designs: Essays on the Coloniality of Power, Subaltern Knowledges and Border Thinking*, Princeton, NJ: Princeton University Press.

Mignolo, W.D. (2005) 'Prophets facing sidewise: the geopolitics of knowledge and the colonial difference', *Social Epistemology*, 19(1), January–March: 111–127.

Mkandawire, T. (2003) *Institutions and Development in Africa*, paper submitted to the Cambridge Journal of Economics Conference on Economics for the Future, 17–21 September.

Mkandawire, T. (2005) 'African intellectuals and nationalism', in T. Mkandawire (ed.), *African Intellectuals: Rethinking Politics, Language, Gender and Development*. Dakar and London: CODESRIA Books and Zed Books.

Mkandawire, T. (2011) 'Running while others walk: knowledge and the challenge of Africa's development', *Africa Development*, 26(2): 1–36.

Moore-Gilbert, B. (1997) *Postcolonial Theory: Contexts, Practices, Politics*, London: Verso.

Mudimbe, V.Y. (1988) *The Invention of Africa: Gnosis, Philosophy, and the Order of Knowledge*, Bloomington, IN: Indiana University Press.

Mudimbe, V.Y. (1994) *The Idea of Africa*, Bloomington, IN: Indiana University Press.

Murphy, C.N. (2005) *Global Institutions, Marginalization and Development*, London and New York: Routledge.

Ndlovu-Gatsheni, S.J. (2013a) *Coloniality of Power in Postcolonial Africa: Myths of Decolonization*, Dakar: CODESRIA Books.

Ndlovu-Gatsheni, S.J. (2013b) *Empire, Global Coloniality and African Subjectivity*, New York and Oxford: Berghahn Books.

Ndlovu-Gatsheni, S.J. (2015a) 'Decoloniality as the future of Africa', *History Compass*, 13(10): 485–496.

Ndlovu-Gatsheni, S.J. (2015b) 'Decoloniality in Africa: a continuing search for a new world order', *Australasian Review of African Studies*, 36(2): 22–50.

Ndlovu-Gatsheni, S.J. (2016) *The Decolonial Mandela: Justice, Peace and the Politics of Life*, New York and Oxford: Berghahn Books.

Nesadurai, H.E.S. (2005) *Bandung and the Political Economy of North–South Relations: Sowing the Seeds for Re-Visioning International Society*, Working Paper No. 95, December, Singapore: Institute of Defence and Strategic Studies.

Ngugi wa Thiong'o (1986) *Decolonizing the Mind: The Politics of Language in African Literature*, Oxford: James Currey.

Ngugi wa Thiong'o (2009a) *Re-Membering Africa*, Nairobi: East African Educational Publishers.

Ngugi wa Thiong'o (2009b) *Something Torn and New: An African Renaissance*, New York: Basic Civitas Books.

Nkrumah, K. (1964) *Consciencism: Philosophy and Ideology for De-Colonization*, New York: Monthly Review Press.

Nkrumah, K. (1965) *Neo-Colonialism: The Highest Stage of Imperialism*, New York: International Publishers.

Nyerere, J.K. (1967) *Freedom and Unity*, Dar es Salaam: Oxford University Press.

Onimode, B. *et al.* (2004) *African Development and Governance Strategies in the 21st Century*, London and New York: Zed Books.

Pieterse, J.N. (2010) *Development Theory: Deconstructions/Reconstructions*, 2nd edition, Los Angeles. CA: Sage.

Quijano, A. (2000) 'Coloniality of power and social classification', *Journal of World Systems*, 6(2): 342–386.

Rodney, W. (1972) *How Europe Underdeveloped Africa*, Dar es Salaam: Bogle L'Ouverture and Tanzania Publishing House.

Slater, D. (2004) *Geopolitics and the Post-Colonial: Rethinking North–South Relations*, Malden, MA: Blackwell.

Taylor, I. (2003) 'The United Nations Conference in Trade and Development', *New Political Economy*, 8(3): 409–418.

Therien, J.P. (1999) 'Beyond the North–South divide: two tales of world poverty', *Third World Quarterly*, 20(4): 723–742.

Thomas, A. (2000) 'Development as practice in a liberal capitalist world', *Journal of International Development*, 12(6): 773–787.

Tomori, S.O. and Tomori, O.W. (2004) 'Revisiting African Alternative Framework to Structural Adjustment Programme for Socio-Economic Recovery and Transformation in contemporary Nigeria', in B. Onimode *et al.* (eds), *African Development and Governance Strategies in the 21st Century*, London: Zed Books.

Truman, H. (1964 [1949]) *Public Papers of the Presidents of the United States: Harry S. Truman*, Washington, DC: United States Government Printing Office.

World Bank (1989) *From Crisis to Sustainable Growth*, Washington, DC: World Bank.

World Bank (1994) *Adjustment in Africa*, Washington, DC: World Bank.

Wright, R. (1956) *The Colour Curtain*, New York: World Publishing.

Zeleza, P.T. (2005) 'Banishing the silences: towards the globalization of African history', unpublished paper presented at the 11th CODESRIA General Assembly Conference, Maputo, Mozambique, 6–10 December.

Section 2
History

4
Colonial Africa

Erik Green

Introduction

The immediate and long-term impacts of European colonialism on African economies and societies is a field of research that has interested historians and social scientists for a long time. European colonialism in Africa was a project that ultimately depended upon a patriarchal and racist world view. Having said that, repressive ideologies do not always cause repressive outcomes. Did colonialism cause Africa to be underdeveloped, or did the European occupation of the continent, although morally and politically questionable, facilitate economic development by further integrating Africa into the world economy? These are difficult questions and, as we hope the chapter will show, there are no straightforward answers.

The immediate impact of the European colonisation of Africa varied between different types of colonies as well as between regions. The aim of this chapter is to provide a brief summary of the economic, political and social developments in colonial Africa. Throughout the chapter, we will emphasise the variety of experiences between colonies. However, there are some common features shared by the great majority of colonies. First, colonial authorities were generally administratively and politically weak, given that colonialism seldom implied a fundamental restructuring of the African societies and economies (Austin, 2008). Second, there was no grand plan behind the actions of the colonial authorities. Navigating by trial and error is a better way of describing the actions of the colonial powers and their agencies (Cooper, 2002; Andersson and Green, 2015; Frankema *et al.*, 2016). Lastly, the outcome of colonialism cannot be grasped without taking African agency into account, and the actions of Africans can hardly be understood in simple dichotomies of resistance or collaboration (Cooper, 1981; Isaacman, 1990). With this in mind, we will begin our summary by outlining what we know about trends in economic development, human welfare and inequality in colonial Africa. We then continue by putting these trends in the context of political change and African agency.

Patterns of long-term economic development in colonial Africa

Did colonialism foster or obstruct economic development in Africa? There is no straightforward answer to this question. In the debate, there is sometimes a tendency to treat Africa as a homogenous entity that shared historical experiences. This is far from true, making generalisations difficult to defend. In addition, we lack sufficient data on economic performance. Estimates of economic growth for African countries are only available from the 1950s, and even then most economic activities took place under the radar of the colonial governments and are not reflected in any official reports or statistics (Jerven, 2012). Population data are equally scattered (Frankema and Jerven, 2014). Lack of such data prevents us from calculating the size of the colonial economies and average income per capita, as well as distribution of income among individuals and/or groups. In other words, we lack fundamental information to provide an empirically grounded analysis of the political economy of colonialism. What this means is that we actually know very little about the impact colonialism had on economic development. Having said that, the scattered data that exist indicate that most African economies were growing in the first half of the twentieth century, most significantly in the early and late colonial period, namely 1900–1914 and 1945–1960 (Prados de la Escosura, 2012). But was this growth a result of colonialism? The answer is both yes and no.

First, the impact of colonialism on African economies varied between regions, which becomes apparent when we look at the early colonial period. In parts of West Africa, the economies were already growing significantly in the precolonial nineteenth century. This growth was based on an expansion of peasant export agriculture. When the region came under European control, the colonial powers – rather than transforming the organisation of production – remained dependent upon further expansion of peasant cultivation of export crops. In that sense, one could argue that colonialism did not cause underdevelopment, but was equally not the root cause of economic growth. For eastern and southern Africa, the precolonial legacy is less clear. Differently from West Africa, regional and external trade in these areas was in decline, and colonial occupation most likely spurred economic growth in the short run through investments in infrastructure and provision of extended market opportunities that benefited European businesses, but also African commercial elites (Oliver and Atmore, 1994).

Meanwhile, early investments in infrastructure in colonial Africa as a whole were made with the purpose of boosting exports, and not to facilitate domestic market integration. Artificial borders were established and the colonies were transformed into closed economic systems. The latter implied that regional trade – of crucial importance for people and societies in precolonial times – was now officially closed (although not always in practice) as trade had to go through the colonial powers in Europe (Austen, 1987). Regional trade and market integration was sacrificed as the colonial powers either turned the colonies into producers of tropical commodities for European markets or reinforced already existing processes of export-led growth. This made African economies vulnerable to changes in the world market, and at the same time they faced falling terms of trade (Frankema *et al.*, 2015).

As we shall see below, colonial economic policies changed over time, but the role of colonies as producers of primary products was never questioned or challenged in any serious way. At the end of the colonial period, Africa was the least industrialised region in the world. The few countries that had experienced significant growth in modern manufacturing in Africa were either independent (South Africa) or semi-independent (Southern Rhodesia) or had the capacity to use repressive measures to ensure a supply of cheap indigenous labour to be employed in the manufacturing sector (the Belgian Congo). In the former two cases, white nationalism and white labour together laid the ground for an efficient industrial lobby that did not exist in most of Africa (Austin *et al.*, 2015).

The lack of a modern manufacturing sector can be explained by both structural and political factors. Prevailing factor endowments (i.e. land abundance and labour scarcity) in most parts of Africa made labour expensive and difficult to control. This made it difficult for local or international firms to establish profitable businesses. Investments in infrastructure lowered transportation costs, but instead of facilitating the emergence of local industries, these investments spurred importation of processed goods that could be produced cheaper outside Africa. Meanwhile, colonial policies did little to ease the situation. In some colonies, the local authorities began to argue in the 1930s, and more notably after the Second World War, in favour of investing in the colonies to promote diversification. Such efforts were prevented by lack of capital and a general unwillingness in Europe to allow African manufacturers to grow, as they feared that these would compete with European firms (Austin *et al.*, 2015). Colonialism facilitated world market integration, but at the same time prevented structural change and industrialisation. In that regard, one could argue that colonialism promoted short-term economic growth but hampered structural change and sustained economic development.

Living standards

If we know little about the long-term patterns of economic growth in twentieth-century Africa, we know even less about the human development and welfare of the African countries. The vast majority of Africans secured their livelihoods under the radar of colonial authorities, either by working in informal sectors or subsistence agriculture, or both. There is anthropological work of varied quality to draw on, but this is difficult to use to trace changes over time. The situation becomes even more challenging by the fact that the data available provide us with partly contradictory information about living standards in Africa. In general terms, scattered data of welfare of those engaged in the formal sectors reveal that Africans experienced increased living standards over time. Still poor, they were not trapped in structural poverty, although some of them most likely faced conjectural poverty (i.e. temporary periods of hunger crises or even famines) (Iliffe, 1995). However, bits and pieces of information from rural areas suggest that structural poverty was widespread.

One indicator of living standards that has recently been determined are long-term trends of urban real wages. Frankema and van Waijenburg (2012) compare urban real wages for unskilled male workers in nine African colonies in 1880–1965. Their estimates show that in all cases, except colonial Malawi, an unskilled worker earned a sufficient amount to support himself and his family. Since real wages increased over time in all cases, they argue that it is not accurate to claim that unskilled labour was trapped in structural poverty, though most of them remained relatively poor throughout the colonial period. This may be true, but it might also be that Frankema and van Waijenburg (2012) underestimated expenses. African urban workers usually did not only have to support their families, but relatives in their home villages.

Information from rural areas is far more limited. There is scattered evidence that farmers engaged in commercial agriculture experienced increased living standards throughout the colonial period in West Africa, and more notably in the post-war period in central and southern Africa (Austin, 2005; Green, 2005; de Haas, 2014). Meanwhile, colonial authorities 'discovered' in the 1920s and 1930s that levels of malnutrition in rural villages were far higher than they had expected (Culwick, 1944; Moore and Vaughan, 1994). In Malawi, for example, it was concluded that latent malnutrition was a common feature in rural areas (Green, 2005). These pieces of information, although incomplete, suggest that a significant proportion of the rural population was caught in structural poverty.

Levels of poverty not only differed between rural and urban areas, but between regions. Frankema and van Waijenburg's estimates reveal a significant gap between West and East Africa. Urban real wages were nearly twice as high in West Africa up to 1940. The differences can be explained by the type of colony, which in turn was largely an effect of the precolonial environment that the Europeans faced when colonising Africa. West Africa had finished the transition from slave to legitimate trade when colonised by Europe, while the same had not yet begun in East Africa at that time. The transition to legitimate trade increased the demand for both rural and urban labourers, which drove wages up. At the same time, as West Africa was already heavily involved in export trade, the colonial authorities there could survive by taxing trade while keeping income taxes low. In Nigeria, for example, a native income tax was first introduced in the 1920s, while in Ghana a similar tax was discussed but never effected (Frankema and van Waijenburg, 2014).

In central and East Africa, taxing trade was not a viable option in the early colonial period as the level of indigenous engagement in export trade was limited. This made the colonial authorities engage in policies that would attract European settlers, which they strongly believed would more efficiently exploit natural resources and thus boost exports. The colonial authorities and European settlers faced the same structural constraints as precolonial states had had to deal with for centuries, namely that in most parts of Africa land was abundant but labour was scarce. To ensure the European settlers' access to indigenous labour, the colonial authorities used different measures. In some cases, particularly in French and Portuguese colonies, but also the Belgian Congo, the use of forced labour played a central role in the early colonial period. The most common strategy, however, was to use indirect measures, most notably by introducing a differential tax system, establishing native reserves or prohibiting Africans to grow the most lucrative cash crops. The aim of these policies was to create incentives that indirectly forced Africans to offer their labour to the Europeans (Frankema and van Waijenburg, 2014). By increasing the supply of Africans in the labour market, wages were kept comparatively low while the high tax levels reduced real wages even further. Data on rural wages are very scattered, but recent research suggests that rural wages remained stable at a low level up to the 1930s in colonies where settlers played an important role in the early colonial period (i.e. Kenya, Malawi and Zimbabwe in East and central Africa) (Bowden *et al.*, 2008; Bolt and Green, 2015). In concession colonies, the situation was even worse. In the Belgian Congo, for example, the use of forced labour was widespread and investments in the welfare of the indigenous populations low, even by the standards of colonial Africa (Mkandawire, 2010). Meanwhile, various indicators suggest a convergence over time as rural real wages increased after the Second World War and colonial policies became less extractive in the settler colonies (Arrighi, 1966; Mosley, 1983).

In short, a few Africans clearly gained from the opportunities provided by colonial authorities, either by engaging in profitable commercial agriculture or finding employment in the administration and the urban sectors. For the majority, the impact of colonialism is less clear. Some may have been trapped in structural poverty while others experienced modest welfare gains. Still, one could have expected more. This was a period in which living standards of Europe's population increased significantly and equality and welfare become a priority, eventually leading to the establishment of welfare states. The idea that Africans, on average much poorer than Europeans, had the right to public support or comprehensive welfare programmes never gained any acceptance in the colonial offices in Europe or with the local colonial authorities.

Inequality

We have limited quantitative information to empirically investigate the impact of colonialism on inequality levels, although previous research on the formal economy in Kenya (Bigsten, 1987) and recent research on colonial Botswana (Bolt and Hillbom, 2016) show that, at least in these cases, overall inequality levels increased in the first half of the twentieth century.

Not surprisingly, there are indicators that levels of inequality were far higher in elite settler and concession colonies than in the peasant economies. This is partly explained by extractive colonial policies that ensured that African wages remained fairly low in the settler and concession colonies in the early colonial period (Phimister, 1988; Bowden *et al.*, 2008). Even though African rural wages increased in the post-war period, they did so at a slower pace than the incomes of the Europeans (Bolt and Green 2015; Bolt and Hillbom 2016). It is therefore safe to conclude that levels of inequality were significantly higher in settler and concession colonies at the end of the colonial period than in the late precolonial period.

The gap between European and African incomes was significant in peasant colonies as well. Yet in these cases, the opportunities for Africans to climb the income ladder were better, at least in theory. In practice, inequality between Africans most likely increased, although the income gap remained smaller than between Africans and Europeans. The groups that gained in terms of increased incomes during the colonial period were those who had the economic means and capacities to exploit the opportunities that came with the expanding market economies, those who received education from missionary schools and those with employment in the colonial administration. Those who lost – in relative terms – were those who lived far away from the commercial centres and/or lacked the means to capture new opportunities. A significant regional dimension of inequality developed during the colonial period. While commercial centres gained from investment in infrastructure, other regions were left behind and developed into labour reserves (e.g. Vail, 1975). People living in the labour reserves still had to generate cash incomes, evidence of agency and resistance in order to pay for taxes. One common strategy was to migrate to the commercial areas either within the colony or in neighbouring colonies. Both African and European commercial activities were to a large extent dependent on this inflow of African labourers from other regions. In the latter case, farmers commonly migrated with the aim of establishing their own farms, while Africans migrating to seek employment in European-controlled businesses did so on a temporary basis to generate cash that they sent or brought back home. In Ghana in West Africa, for example, people migrated to the cocoa belt and bought land in the early colonial period (Hill, 1963). Later on, as it became increasingly difficult for 'strangers' to purchase land, migrants moved into the cocoa belt as workers, but many of them stayed and eventually managed to gain access to land to cultivate cocoa (Clarence-Smith and Ruf, 1996; Crook, 2001). In southern Africa, temporary labour migration across colonial boarders was more common. For example, nearly half of the unskilled labourers employed on the European settler farms in Southern Rhodesia and in the mining complex in South Africa consisted of migrants from Nyasaland, Northern Rhodesia and Portuguese East Africa (Feinstein, 2005; Green, 2015). Successful commercial activities were clearly dependent upon regional inequalities and the mobility of Africans.

Drivers of inequality within commercial centres in Africa did not seem to differ much from what we have seen in other parts of the world, with one very important exception: social stratification in rural Africa did not lead to the rise of a group of landless labourers (Cooper, 2002). Thus, social historian Sara Berry concludes that increased levels of social stratification were characterised by 'exploitation without dispossession' (Berry, 1993: 135–158). The lack of a group of landless people can partly be explained by the relative land abundance, which ensured that even the poorest remained with some land. Colonial policies also played a role. In most parts of colonial Africa, the indigenous people, who lacked political rights, were granted the right to cultivate land as long as they belonged to the community (see below).

A group that is strikingly invisible in the colonial accounts is women. The colonial projects were deeply rooted in patriarchal ideologies. The colonial authorities not only perceived men as breadwinners, but also presumed that they were responsible for all major economic decisions made in the household. This view clashed with the realities of many agrarian societies in Africa in which women had played a significant productive role, not only as tillers of the land, but also as independent economic agents operating their own farm businesses. Precolonial societies were not characterised by gender equality, but with colonialism women were no longer recognised as economic actors, even in cases where they played an active role and made money from commercial farming (Cooper, 2002). Ironically, this became increasingly obvious in the so-called post-war development period when the colonial authorities actively began to invest in various development schemes (see below). Women were not allowed to participate in most of these programmes, which in effect meant that the modest development efforts made most likely increased gender disparity in terms of economic opportunities (Moore and Vaughan, 1994).

Colonial policies and politics

A significant feature of European colonialism in Africa was that it was a project in which 'private finance lagged behind political control' (Cain and Hopkins, 2002: 566). But political control was limited as well. Colonial authorities were financially weak, with limited capacities to administrate the territories under their control. The number of appointed European administrators provides an illustrative example. In Kenya, there were 164 European administrators governing a population of more than 3 million in 1939. In Nigeria, which had an estimated population of over 20 million, 386 administrators were employed in 1939. The small numbers led Kirk-Greene (1980) to refer to the European administrators as 'the thin white line' in colonial Africa.

Within this constraining framework, the colonial authorities could, however, make a number of deliberate choices. The most strategic of them all was whether they would try to maximise revenues, minimise expenditure or pursue a strategy of doing both. Up to the Second World War, colonial authorities tried to do both. By comparing the level and type of public investments and per capita state revenue, Frankema (2011) concludes that most African colonial states were minimalistic up to the 1940s, spending mainly on law and order, and little on schooling and healthcare. At the same time, investments in the administration, although often the largest expenditure, were kept as low as possible. For that purpose, the colonial authorities developed administrative systems that were supposed to run efficiently at a low cost.

The British system of indirect rule is the most apparent example of this. It was a system based on institutional segregation whereby Europeans and Africans were supposed to live under different jurisdictions. Africans were to be governed under identified principles of customary (then known as 'traditional') laws while Europeans were under European law. In the interwar period, the British colonial authorities identified and appointed chiefs who would ensure the maintenance of law and order, and at the same time be in charge of land distribution and sometimes tax collection in their colonies. In some cases, they successfully managed to identify people who had been members of local political elites in the late precolonial period, while in other cases it was obvious that the appointed chiefs lacked such heritage (Spear, 2003). This system solved two problems for the British colonial authorities. First, it kept the running cost of the administration low. Second, the chiefs could potentially become important allies. At the same time, the system created its own tensions among the few educated Africans who demanded equal rights and did not recognise chiefs as legitimate leaders, as well as among successful commercial farmers who wanted to limit the chiefs' power over land distribution.

Indirect rule is often contrasted with direct rule, which was the system of governance in the French and to some extent the Portuguese and Belgian colonies. In practice, however, these three systems show a great deal of similarities with the British system of indirect rule (Keese, 2007). Generally, chiefs were responsible for maintaining law and order, and were in charge of distribution of land and collection of taxes. The differences between the colonies lay in the degree of freedom that the chiefs were given. In British colonies, they were more independent and could more easily play multiple roles. In Ghana, for example, chiefs were not only representatives of the colonial administration, but also successful cocoa growers (Austin, 1987).

While it is accurate to define most colonial states as minimalistic up to the Second World War, the level at which they extracted resources from local populations varied between colonies. A common feature of most colonies, especially in early colonial times, was the use of forced labour. The European occupiers, who had officially claimed that colonisation of Africa was needed to put domestic slavery to an end, did not hesitate to use coercive measures to allocate labour for their own needs. The use of forced labour was, however, more prominent in Portuguese Africa, the Belgian Congo and French Africa, where it played a significant role as a source of revenue (van Waijenburg, 2015). Forced labour was abandoned in French Africa in 1946 but continued to be practised in the Belgian Congo and Portuguese Africa up to independence. More important than the origin of the colonisers to explain the variety of extraction, however, was the type of colonial economy.

In peasant economies, income tax levels were generally lower and the colonial authorities either actively tried to facilitate African commercial agriculture or at least did not try to prevent it. In contrast, extractive measures were more common in settler and concession colonies, including not only regressive taxation systems, but also the prohibition of Africans from growing the most lucrative cash crops (Kenya), the use of forced labour (the Belgian Congo) and/or the establishment of an institutional order based on racial segregation (Southern Rhodesia) (Berman and Londsdale, 1979; Phimister, 1988; Frankema *et al.*, 2016).

Colonial authorities began to question the minimalistic strategy in the 1930s. Just as in the rest of the world, the impact of the Great Depression had led politicians and economists to question the efficiency of markets and call for more interventionist states that actively used fiscal policies to navigate economic crises and promote long-term economic development. Europe was still not prepared to supply any additional capital to Africa to enable more interventionist colonial policies, but this changed with the Second World War. After the war, Europe faced dollar shortages and was in need of boosting production in their colonies to guarantee sufficient supplies in their domestic markets. The outcome was the establishment of various development funds from which colonial authorities could apply for loans and grants-in-aid (Austen, 1987; Manning, 1998).

With fresh capital entering the colonies after the Second World War, the local colonial authorities in the peasant economies began to initiate major development programmes with the aim of boosting African agriculture. Similar programmes were also found in settler and concession colonies, yet the programmes were not allowed to threaten the policies of racial segregation (Andersson and Green, 2015). Consequently, most of the resources were devoted to different agriculture and rural development, while investments in education and health continued to lag behind. It is popular to highlight the failed programmes, such as the groundnut scheme in Tanganyika and the master farmers and village schemes in Malawi (Kalinga, 1993; Rizzo, 2006). However, many of the projects did have a lasting impact on the proceeding developments, many of them reappearing in new forms in the postcolonial period, now supported and financed by the international donor community. The Lilongwe land development programme in Malawi (1968–1981), known as the most ambitious rural development scheme in postcolonial Africa, shared several similarities with the master farmers and village schemes, with a strong focus on supporting relatively wealthy commercial farmers (Green, 2010). Numerous projects to prevent soil erosion and deforestation in postcolonial Africa have their roots in the colonial period (Beinart, 1984).

Most Africans did not participate in the major development schemes established in the postwar period, but for the majority of Africans the colonial state became visible in a sense that it had not been before. National strategies to improve agricultural production led to significant investments in agricultural extension services, and by the mid-1950s almost all Africans in rural areas had directly or indirectly been in contact with and affected by the new demands from the colonial authorities regarding agricultural methods. Starting off as recommendations, they soon became rules that farmers had to follow in order to avoid punishment. Cooper (1981: 127) argues that in the 1950s, 'the agricultural extension agent replaced the district officer as the embodiment of colonial authoritarianism'. Few among the colonial authorities could at this time foresee that this new form of intervention would facilitate the struggle for independence.

Agency and resistance

While Europe began to supply Africa with capital to gain economically from increased exports from Africa, the colonial authorities used the fresh capital to not only strengthen the colonial economies, but also to convince the local populations that colonial rule was to their benefit, through the various development programmes. At the time, few if any colonial officials foresaw that they had just entered the beginning of serious protests against colonial occupation that in a decade would lead to the first declaration of independence (Ghana, 1957). Two decades later, the two major empires, France and Britain, had lost nearly all colonies. Even more difficult for the colonial authorities to foresee was that the post-war development efforts themselves facilitated the opposition to colonial rule. In order to understand this paradox, we must say something about organised and everyday resistance in colonial Africa.

For the majority of Africans, making their daily living mattered more than anything else. People occasionally resisted their colonial rulers or European economic interests, but resistance was not organised and was quite limited as long as most Africans lacked regular direct contact with the foreign occupants. Tax evasion – one of the most common forms of resistance – was seldom part of political struggle, but stemmed from failures to generate sufficient incomes. Africans working in mines and on European farms were far from always following the agreements of the work contracts by not showing up or not finishing work on time (Green, 2013). Just as with tax evasion, this was not driven by a political agenda. Africans violated contracts when they were needed for farm work at home in order to secure the livelihood of the family.

There was, however, one group that became organised quite early on, namely the so-called 'modern men'. These had been educated by the Christian missionaries, and in a few cases even managed to get into European universities. Initially, these movements did not seek independence, but equal political and sometimes economic rights, while at the same time they wanted to distance themselves from rural Africa in general and the 'traditional' leadership in particular. In some cases, the movements and the chiefs officially fought each other for periods (Shepperson and Price, 1958; Keese, 2007).

In this regard, the 'modern men' potentially posed a double threat to the colonial project as they questioned both the colonial authorities and the local African authorities (i.e. the chiefs). The colonial empires dealt with this group in different ways. In Portuguese Africa, it was mainly met with oppression. France tried to integrate them into the colonial project by dividing the Africans into two categories: *évolués* (Western-educated Africans) and *paysans* (peasants). A few selected in the former group were given the same status as French citizens and in the post-war period were even allowed to vote for their own representatives to the French parliament. The other category would be kept satisfied by lowering taxes and ending forced labour. In Britain, which had put a lot of effort into the development of indirect rule, the group remained largely neglected by the colonial authorities (Cooper, 2002).

With the end of the Second World War and the beginning of increased state intervention in colonial Africa, the rules of the game were about to change. Hundreds of thousands of Africans had participated in the war as soldiers fighting Nazi Germany. When they returned home, they discovered they were still lacking the political rights that they had fought for on behalf of the Europeans. Some political reforms were implemented in the immediate post-war period, especially in French Africa, where Africans were allowed to elect their own representatives to the French parliament. These representatives pushed for the abolition of forced labour, which came into effect in 1946 (Manning, 1998). However, on the whole, Africans found themselves in the same position as before the war, experiencing political discrimination and huge economic inequalities. The tipping point had been reached. African urban workers increasingly manifested their dissatisfaction at being paid less than the Europeans, and urban strikes became increasingly common (Manning, 1998; Cooper, 2002). In the rural areas, the everyday resistance continued, but it slowly merged with the political struggle for independence.

The introduction of numerous development programmes in rural Africa in the 1950s, coupled with the introduction of legally enforced farming practices, meant that the rural population – which had remained partially independent during the early colonial period – now experienced tightened control from the colonial authorities. This, together with the dramatic increase of strikes in the urban areas and increased colonial intervention, turned the social movements into regular political parties fighting for political independence. For the first time, that identified a common cause with the Africans in the rural areas who were resisting the newly introduced stipulations on agricultural methods (Isaacman, 1990). The urban-based 'elite' efficiently used this increased rural resistance to strengthen and intensify the political struggles for political independence. The end of colonialism was about to come.

Conclusion

While it is impossible to defend colonialism on moral and political grounds, any analysis of its social and economic impact requires objectivity. What is made clear from the short review above is how difficult it becomes to draw any definite conclusions regarding the impact of European colonialism in Africa. This is for two reasons. First, we simply know too little. Second, the impact of colonialism varied significantly between regions and between different social classes. Having said that, it is possible to identify some general patterns.

The colonial authorities were, in general terms, financially weak and lacked the capacity to establish efficient administrative systems. The paradox is that this sometimes gave substantial space of manoeuvre for various local economic and political elites, while in other places it can explain the existence of extractive institutions and practices. In Ghana, for example, indigenous commercial farming – with its roots in the precolonial nineteenth century – was not only allowed to expand during the colonial period, but was protected from European competition. In colonies such as the Belgian Congo, Kenya and Southern Rhodesia, the colonial authorities took a different path and protected European interests from African competition, and implemented repressive policies to ensure that European economic interests had access to fairly cheap labour. These contrasting strategies shared the same purpose: increase the revenues to the colonial state.

The relatively weak colonial states also make it impossible to understand the development trajectories of colonial Africa without taking African agency into consideration. Africans were not passive victims, but neither were they a homogenous group with common interests. Colonialism increased the social and economic divisions within and between African communities. Some missionary-educated Africans distanced themselves from their rural background, while others tried to organise rural communities against European economic and political interests. Most Africans, however, were occupied making sure that they had enough food on the table, and if that meant collaboration, resistance or neglect of the colonial authorities mattered less.

Meanwhile, European colonialism was fundamentally a conservative project. Africans were often portrayed as significantly different from Europeans. Africans were perceived as belonging to rural areas, tilling the land and governed by so-called chiefs. Africans that were not primarily farmers were either ignored or actively opposed. African men – it was believed – were the breadwinners, while women were considered to be responsible for domestic work. African women were therefore never consulted when it came to matters related to economic activity.

The conservative roots of colonialism – independent of the colonising country in origin – are most visible in the long-term legacy of the project. European colonisation meant that Africa's role as a supplier of low-value commodities on the global market was reinforced at the same time as regional trade and integration were prevented. This created traps of slow economic development and high levels of poverty, which have proven difficult to break free from. In short, colonialism prevented Africa from freely endeavouring for the benefits that arose from the spectacular economic development that signified twentieth-century global history. Only the future will let us know if recent economic developments in Africa signify that the continent is finally breaking free from its colonial legacies.

References

Andersson, M. and Green, E. (2015) 'Development under the surface: the rise of indigenous agriculture as an unintended consequence of settler institutions in Southern Rhodesia, 1896–1962', *Journal of International Development*, DOI: 10.1002/jid.3102.

Arrighi, G. (1966) 'The political economy of Rhodesia', *New Left Review*, 39: 237–265.

Austen, R. (1987) *African Economic History: Internal Development and External Dependency*, London: James Currey.

Austin, G. (1987) 'The emergence of capitalist relations in South Asante cocoa-farming, *c.*1916–33', *Journal of African History*, 28(2): 259–279.

Austin, G. (2005) *Labour, Land and Capital in Ghana: From Slavery to Free Labour in Asante, 1807–1956*, Rochester: Rochester University Press.

Austin, G. (2008) 'The "reversal of fortune" thesis and the compression of history: perspectives from African and comparative economic history', *Journal of International Development*, 20: 996–1027.

Austin, G., Frankema, E. and Jerven, M. (2015) *Patterns of Manufacturing Growth in Sub-Saharan Africa: From Colonization to the Present*, Working Paper 17, Utrecht: Centre for Global Economic History.

Beinart, W. (1984) 'Soil erosion, conservation and ideas about development: a southern African exploration', *Journal of Southern African Studies*, 11(1): 52–83.

Berman, J. and Lonsdale, J. (1979) 'Coping with the contradictions: the development of the colonial state in Kenya, 1895–1914', *The Journal of African History*, 20(4): 487–505.

Berry, S. (1993) *No Condition Is Permanent: The Social Dynamics of Agrarian Change in Sub-Saharan Africa*, Madison, WI: Wisconsin University Press.

Bigsten, A. (1987) *Income Distribution and Growth in a Dual Economy: Kenya 1914–1967*, Gothenburg: University of Gothenburg.

Bolt, J. and Green, E. (2015) 'Was the wage burden too heavy? Settler farming, wages and the profitability of settler agriculture in colonial Malawi, *c.*1900–1960', *Journal of African History*, 50(2): 217–238.

Bolt, J. and Hillbom, E. (2016) 'Long-term trends in economic inequality: lessons from colonial Botswana, 1921–74', *Economic History Review*, DOI: 10.1111/ehr.12326.

Bowden, S., Chiripanhura, S. and Mosley, P. (2008) 'Measuring and explaining poverty in six African countries: a long-term approach', *Journal of International Development*, 20: 1049–1079.

Cain, P.J. and Hopkins, A.G. (2002) *British Imperialism, 1688–2000*, London: Routledge.

Clarence-Smith, W.G. and Ruf, F. (1996) 'Cocoa pioneer fronts: the historical determinants', in W.G. Clarence-Smith (ed.), *Cocoa Pioneer Fronts Since 1800*, London: Macmillan, pp. 1–22.

Cooper, F. (1981) 'Africa in the world economy', *African Studies Review*, 24(2/3): 1–86.
Cooper, F. (2002) *Africa Since 1940: The Past of the Present*, Cambridge: Cambridge University Press.
Crook, R.C. (2001) 'Cocoa booms, the legislation of land relations and politics in Côte d'Ivoire and Ghana: explaining farmers' responses', *IDS Bulletin*, 32(1): 35–45.
Culwick, G.M. (1944) 'Nutrition in East Africa', *Africa: Journal of the International African Institute*, 14(7): 401–410.
de Haas, M. (2014) *Measuring Rural Welfare in Colonial Uganda: Why Farmers Would Not Work for Wages*, African Economic History Network Working Paper No. 18.
Feinstein, C.H. (2005) *An Economic History of South Africa: Conquest, Discrimination and Development*, Cambridge: Cambridge University Press.
Frankema, E. (2011) 'Colonial taxation and government spending in British Africa, 1880–1940: maximizing revenue or minimizing effort?', *Explorations in Economic History*, 48(1): 136–149.
Frankema, E. and Jerven, M. (2014) 'Writing history backwards or sideways: towards a consensus on African population, 1850–2010', *Economic History Review*, 67(4): 907–931.
Frankema, E. and van Waijenburg, M. (2012) 'Structural impediments to African growth? New evidence from real wages in British Africa, 1880–1965', *Journal of Economic History*, 72(4): 895–926.
Frankema, E. and van Waijenburg, M. (2014) 'Metropolitan blueprints of colonial taxation? Lessons from fiscal capacity building in British and French Africa, 1880–1940', *Journal of African History*, 55(3): 371–400.
Frankema, E., Williamson, J. and Woljter, P. (2015) *An Economic Rationale for the African Scramble: The Commercial Transition and the Commodity Price Boom of 1845–1885*, NBER Working Paper No. 21213.
Frankema, E., Green, E. and Hillbom, E. (2016) 'Endogenous processes of colonial settlement: the success and failures of European settler farming in sub-Saharan Africa', *Revista de Historia Económica / Journal of Iberian and Latin American Economic History*, DOI: https://doi.org/10.1017/S0212610915000397.
Green, E. (2005) *Peasant Production and Limits to Labour: Thyolo and Mzimba Districts in Malawi, Mid-1930s to Late-1970s*, Stockholm: Almqvist & Wicksell International.
Green, E. (2010) 'State-led agricultural intensification and labour relations: the case of Lilongwe land development program in Malawi, 1968–1981', *International Review of Social History*, 55(3): 413–446.
Green, E. (2013) 'Land concentration, institutional control and African agency: growth and stagnation of European tobacco farming in Shire Highlands, *c.*1900–1940', in E. Hillbom and P. Svensson (eds), *Agricultural Transformations in a Global History Perspective*, London: Routledge, pp. 229–252.
Green, E. (2015) *The Development of Settler Agriculture in British Africa Revisited: Estimating the Role of Tenant Labour in Southern Rhodesia, c.1920–60*, paper presented at the World Economic History Congress, Kyoto, 3–7 August.
Hill, P. (1963) *The Migrant Cocoa Farmers of Southern Ghana: A Study in Rural Capitalism*, Cambridge: Cambridge University Press.
Isaacman, A. (1990) 'Peasants and rural social protest in Africa', *African Studies Review*, 33(2): 1–120.
Iliffe, J. (1995) *Africans: The History of a Continent*, Cambridge: Cambridge University Press.
Jerven, M. (2012) 'An unlevel playing field: national income estimates and reciprocal comparison in global economic history', *Journal of Global History*, 7(1): 107–128.
Kalinga, O. (1993) 'The master farmers scheme in Nyasaland, 1950–1962: a study of a failed attempt to create a "yeoman" class', *African Affairs*, 92(368): 367–388.
Keese, A. (2007) *Living with Ambiguity: Integrating an African Elite in French and Portuguese Africa, 1930–61*, Stuttgart: Franz Steiner Verlag.
Kirk-Greene, A.H.M. (1980) 'The thin white line: the size of the British colonial service in Africa', *African Affairs*, 79(314): 25–44.
Manning, P. (1998) *Francophone Sub-Saharan Africa, 1880–1995*, Cambridge: Cambridge University Press.
Mkandawire, T. (2010) 'Thinking about developmental states in Africa', *Cambridge Journal of Economics*, 25(3): 289–314.
Moore, H.L. and Vaughan, M. (1994) *Cutting Down Trees: Gender, Nutrition, and Agricultural Change in the Northern Province of Zambia, 1890–1990*, London: James Currey.
Mosley, P. (1983) *The Settler Economies*, Cambridge: Cambridge University Press.
Oliver, R. and Atmore, A. (1994) *Africa since 1800*, Cambridge: Cambridge University Press.
Phimister, I. (1988) *An Economic and Social History of Zimbabwe, 1890–1948: Capital, Accumulation and Class Struggle*, London: Longman.
Prados de la Escosura, L. (2012) 'Output per head in pre-independence Africa: quantitative conjectures', *Economic History of Developing Regions*, 27(2): 1–36.

Rizzo, M. (2006) 'What was left of the groundnut scheme? Development disaster and labour market in Southern Tanganyika 1946–1952', *Journal of Agrarian Change*, 6(2): 205–238.

Shepperson, G. and Price, T. (2000) *Independent African: John Chilembwe and the Nyasaland Rising of 1915*, Blantyre: CLAIM.

Spear, T. (2003) 'Neo-traditionalism and the limits of intervention in British colonial Africa', *Journal of African History*, 44(1): 3–27.

Vail, L. (1975) 'The making of an imperial slum: Nyasaland and its railways, 1895–1935', *Journal of African History*, 16(1): 89–112.

van Waijenburg, M. (2015) *Financing the African Colonial State: The Revenue Imperative and Forced Labour*, African Economic History Network Working Paper No. 20.

5

Post-independence Africa[1]

Challenges and opportunities

Tanya Lyons and David Jolley

Introduction: perspectives on post-independence Africa – colonialism and decolonisation

Africa has long been imagined by the West as the 'heart of darkness' ever since Joseph Conrad depicted this bleak view in his 1899 novel of the same name. Conrad's character Kurtz's depiction of 'the horror, the horror' has thus remained etched in the international psyche as of undeniably African origin. These images of 'darkness' and 'horror' are regularly reinforced by the media's fascination with representations of the regular 'tribal' violence that occurs during civil wars and conflicts, all too commonly reported 'out of Africa' (Ofuho, 2003). Yet relying upon these Western historical and contemporary 'images of Africa', as sourced via popular media or classic literature, results in a very limited understanding of what contemporary Africa really *is*.

What is Africa? This is not an unwarranted question in the West, because there are many assumptions that Africa is just 'one place' – and there is very little awareness that it is a continent with 55 different countries, each with its own history and political intricacies. There are also assumptions that 'Africa – as one place' is shaped by the same social, political and economic forces, with the same social, political and economic outcomes. Furthermore, any examination of contemporary 'Africa' must acknowledge and include all of the associated regions – southern Africa, East Africa, West Africa, North Africa and the Horn of Africa. While this mix of nation and region is clearly challenging to some international relations theorists (Engel and Olsen, 2005), it is necessary to acknowledge because of the multiple differences and interconnections that span the continent.

The 'modern' history and politics of Africa *have* been well documented and analysed by historians and social scientists (e.g. see Chazan *et al.*, 1999; Thomson, 2000; Gordon and Gordon, 2001; Reid, 2012). Thematically, their texts have fitted neatly into the discourse of African studies. With its origins in the 'colonial project', African studies has since gone through a myriad of changes and transformations, influenced by postcolonial, feminist and postmodern theories. More recently, the 'decolonial' perspective (see Ndlovu-Gatsheni, 2012, 2013, 2015) has called for a new 'decolonisation' to occur across Africa, encompassing the desire for Africa and other postcolonial states to be rid of all colonial 'baggage' inherited at independence. This perspective also emphasises the need for Africans to be at the centre of their own history, including the writing of it.

It is important to acknowledge the legacies of colonisation, because these legacies have shaped African societies today (e.g. see Chazan *et al.*, 1999; Thomson, 2000; Gordon and Gordon, 2001; Cooper, 2002; Edie, 2003; Schraeder, 2004; Reid, 2012). However, as will be demonstrated below, this factor alone does not explain all of Africa's contemporary woes. These woes are highlighted by the contrasts and contradictions that dominate Africa's post-independence developments, and will be discussed below in relation to the Rwandan genocide and the end of apartheid, and the rise of Boko Haram and the Nigerian film industry. This chapter will then explore the challenges and opportunities for Africa and Africans in the post-independence period. In order to achieve this, a discussion of the definitions of terms below will first enable us to clarify what we mean by post-independence Africa.

Post-independence Africa: contrasts and contradictions

In 1994, two separate events in two different post-independence African nations illustrated the differences and contradictions across the continent, confirming how difficult it is to comprehend Africa 'as one place'. First, in April 1994, South Africans celebrated the end of apartheid and the beginning of a new democracy, a 'rainbow nation' led by one of the most respected African leaders, the late Nelson Mandela. However, at the same time, the small central African nation of Rwanda erupted into a genocidal civil war, one that the international community failed to both comprehend and prevent. This 'heart of darkness' overshadowed the 'rainbow nation', and reinforced 'pessimism' and long-standing bewilderment about Africa (Lorch, 1994; Wallace and Gordon, 1994; Ayittey, 1998; Conrad, 2013 [1902]).

Twenty years on, similar contrasts and contradictions continue to emerge.

For example, in 2014, Nigeria was being heralded as one of the world's fastest-growing economies, with the celebrated 'Nollywood' movie industry now second only to Bollywood (Bright, 2015). Yet Boko Haram militants operating in the north-east are undermining the state's authority with extremist violence linked to Islamic fundamentalism, threatening both regional and global security (Sulemana, 2014; Agbiboa, 2015; Sulemana and Azeez, 2015). This threat remains, despite well-meaning global social media campaigns, such as the #BringBackOurGirls Twitter feed (see BBC News, 2016).

Therefore, while it may be obvious to 'Africanists' (those who study Africa within academia) that 'Africa' is not just *one place*, and that each country has its own history, challenges and contradictions, this may not be obvious to the casual observer of post-independence Africa.

Post-independence and postcolonialism: clarifying history and definitions

The various terms colonialism and independence, and postcolonialism and post-independence, require some unpacking and clarification. Colonialism refers to that historical period in which European powers exercised direct rule over their new lands and peoples. A modern colonial state was created with a state apparatus and clearly defined boundaries. However, the authoritarian and exclusionary nature of the state led to struggles for independence. The time of independence may be referred to as the moment of the formal transfer of power from the colonial masters to the new independence leaders. However, while these independence leaders assumed political power, they also inherited the colonial state with its apparatus intact.

The post-independence period follows and often entails struggles over access to the institutions and resources of the colonial state. Often the post-independence period is characterised by the persistence of elements of the colonial state apparatus – the institutions and economic base – in a contemporary period of globalisation and neocolonialism. Hence, post-independence can be a period of persistent colonialism and the enduring colonial legacy. Therefore, the term 'post-independence' is used interchangeably with 'postcolonial', but as Mhlanga and Ndlovu-Gatsheni (2013) point out, despite being post-independent, 'postcolonial' may only occur after the vestiges of colonialism have disappeared. How did the transition from colonialism to formal independence occur?

Reasons for the transition to independence

The momentum for decolonisation and formal independence came from various post-Second World War pressures. First, the Second World War challenged the imperial discourse that viewed Africans as inferior. The Japanese defeat of the British at Singapore in 1941 challenged notions of colonial superiority and invincibility, pillars of colonial rule. Colonised peoples everywhere realised an alternative was possible. Consequently, there emerged post-war African struggles for national self-determination and independence. Second, in its call for democracy, liberalism and capitalism, the United States of America made clear its opposition to any moves towards the resumption of European colonialism. Finally, the administrative costs of colonialism were prohibitive, outweighing the spoils of colonial exploitation, and thus the colonisers had strong financial and moral incentives to withdraw (see Thomson, 2000: 32–35).

However, the transition to independence took different forms and paths, sometimes dependent on the colonial power and its mode of rule. There were peaceful handovers, armed struggles for independence, and more contemporary internationally mandated recognitions of independence and nationhood. For example, France and England had relinquished control of most of their colonies through peaceful transfers of power by 1960, while Portugal stubbornly clung to their colonies until 1975. More recently, in July 2011, South Sudan gained its independence after 30 years of civil war, with a relatively peaceful referendum that voted 'yes' to a separation from the state of Sudan (see Table 5.1).

Next, we turn briefly to the internal reasons for the transition to independence, emphasising the colonial state's loss of domestic legitimacy.

The colonial state: problems of legitimacy

The colonial project was essentially maintained through a heavy reliance on 'coercion'. This is in contrast to modern democratic states that tend to rely upon a combination of coercion and consent, with consent predominating. The colonial police and security forces were an essential part of the colonial state apparatus, and were often used to brutally 'quell expressions of dissatisfaction' and early independence/nationalist movements (see Chazan *et al.*, 1999: 43). The colonial state was by definition authoritarian and self-interested, and not particularly concerned with the political or economic inclusion of the majority (Kasfir, 1983, cited in Chazan *et al.*, 1999: 43). Consequently, the colonial state was not considered legitimate by the majority. However, when notions of colonial superiority and invincibility were smashed in the Second World War, Africans began to contemplate alternatives, particularly political independence (Cooper, 2003).

However, it is perhaps not surprising that when political independence was achieved, and the colonisers handed over the colonial state to independence leaders (the educated African elites) the promise of revolution a mantra of many nationalist movements – was thwarted. The illegitimate, coercive colonial state simply had new African masters (Fanon, 1968), and the

Table 5.1 List of African postcolonial states

Independent state name	*Colonial state name (if different)*	*Independent capital city*	*Colonial capital city (if different)*	*Former colonial administrator*	*Year independence achieved*	*How was independence achieved?*	*First head of state at independence*
Algeria		Algiers		France	1962	Armed liberation struggle	President Ahmed Ben Bella
Angola		Luanda		Portugal	1975	Armed liberation struggle	President Antonio Agostinho Neto
Benin	Dahomey	Porto-Novo		France	1960	Political transition	President Hubert Maga
Botswana	Bechuanaland	Gaborone		Britain	1966	Political transition	President Seretse Khama
Burkina Faso	Upper Volta	Ouagadougou		France	1960	Political transition	President Maurice Yameogo
Burundi	Urundi	Bujumbura	Usumbura (Ruanda-Urundi)	Germany, then Belgium from 1916	1962	Political transition	King Mwami Mwambutsa II and Prime Minister Andre Muhirwa
Cameroon	French Cameroons and British Southern Cameroon	Yaounde		Germany, then Britain and France from 1918	1960	Armed liberation struggle by the UPC, but eventually France relinquished power to more conservative party of Ahmadu Ahidjo	President Ahmadou Ahidjo
Cape Verde (West Africa)	Cape Verde Islands	Praia		Portugal	1975	Armed liberation struggle	President Aristides Pereira
Central African Republic	Ubangi Chari	Bangui		France	1960	Political transition	President David Dacko
Chad		Ndjamena		France	1960	Political transition	President Ngarta Tombalbaye

Comoros		Moroni		France	1975	Political transition	President Ahmed Abdallah
Côte d'Ivoire (Ivory Coast)		Yamoussoukro[1]		France	1960	Political transition	President Felix Houphouët-Boigny
Zaire (now Democratic Republic of Congo)	Belgian Congo, Congo-Kinshasa	Kinshasa	Leopoldville	Belgium	1960	Political transition (protesting led to rioting in 1958, 1959)	President Joseph Kasa-Vubu and Prime Minister Patrice Lumumba
Djibouti	French Somaliland, Afars and Issas	Djibouti		France	1977	Political transition	President Hassan Gouled Aptidon
Egypt		Cairo		Britain	1922	Violent liberation struggle (1919 Egyptian Revolution)	King Farouk
Equatorial Guinea	Fernando Po and Rio Muni	Malabo		Spain	1968	Political transition	President Francisco Macias Nguema
Eritrea		Asmara		Italy	1993 – federated to Ethiopia since 1952	Armed liberation struggle against Ethiopia	President Isaias Afewerki
Ethiopia		Addis Ababa		None	n/a	n/a	Emperors until 1974, then President Aman Mikael Andom
Gabon		Libreville		France	1960	Political transition	President Leon M'Ba
Gambia		Banjul	Bathurst	Britain	1965	Political transition	Prime Minister Dawda Jawara
Ghana	Gold Coast, British Togoland	Accra		Britain	1957	Political transition	President Kwame Nkrumah

(continued)

Table 5.1 Continued

Independent state name	*Colonial state name (if different)*	*Independent capital city*	*Colonial capital city (if different)*	*Former colonial administrator*	*Year independence achieved*	*How was independence achieved?*	*First head of state at independence*
Guinea	French Guinea	Conakry		France	1958	Political transition	President Ahmed Sekou Toure
Guinea-Bissau	Portuguese Island	Bissau	Medina do Boe	Portugal	1974	Armed liberation struggle	President Luiz De Almeida Cabral
Kenya		Nairobi (Machakos 1887, moved in 1899 to Nairobi)		Britain	1963	Armed liberation struggle	Prime Minister Jomo Kenyatta
Lesotho	Basutoland	Maseru		Britain	1966	Political transition	King Motlotlehi Moshoeshoe II
Liberia		Monrovia		n/a[2]	1847	n/a	President Joseph Roberts
Libya		Tripoli		Italy	1951	Political transition	King Idris
Madagascar	Malagasy Republic	Antananarivo		France	1960	Political transition	President Philibert Tsiranana
Malawi	Nyasaland	Lilongwe		Britain	1964	Political transition	Prime Minister Hastings Kamuzu Banda
Mali	Soudan	Bamako		France	1960	Political transition	President Modibo Keita
Mauritania		Nouakchott		France	1960	Political transition	President Mokhtar Ould Daddah
Mauritius		Port Louis		Britain	1968	Political transition	Prime Minister Seewoosagur Ramgoolam
Morocco		Rabat		Spain and France	1956	Political transition	King Mohammad V
Mozambique		Maputo	Laurenco Marques	Portugal	1975	Liberation struggle	President Samora Moises Machel
Namibia	South West Africa	Windhoek		German, then South African mandate from 1920	1990	Liberation struggle	President Sam Nujoma

Niger		Niamey		France	1960	Political transition	President Hamani Diori
Nigeria		Abuja	Lagos	Britain	1960	Political transition	Prime Minister Alhaji Abubakar Tafawa Balewa
Republic of the Congo	French Congo, Congo-Brazzaville	Brazzaville		France	1960	Political transition	President Fulbert Youlou
Rwanda	Ruanda	Kigali	Usumbura (Ruanda-Urundi)	Germany, then Belgium from 1916	1962	Communal violence due to disagreements on independence between Hutu and Tutsi	President Gregoire Kayibanda
Sahrawi Arab Democratic Republic (Western Sahara)		Laayoune		Spain/Morocco	TBA[3]	TBA	President Mohammed Abdelaziz
São Tomé and Príncipe		São Tomé		Portugal	1975	Political/peaceful	President Dr Manuel Pinto da Costa
Senegal		Dakar		France	1960	Political transition	President Leopold Sedar Senghor
Seychelles		Victoria		Britain	1976	Political transition	President James Mancham
Sierra Leone		Freetown		Britain	1961	Political transition	Prime Minister Sir Milton Margai
Somalia	Italian Somaliland and British Somaliland	Mogadishu/ Berbera[4]		Britain and Italy	1960	Political transition	President Aden Abdulla Osman and Prime Minister Abdirashid Ali Shirmarke
South Africa		Pretoria		Union of British colonies and Boer republics	1910, majority rule 1994	Liberation struggle	Prime Minister Dr Hendrik Verwoerd
South Sudan		Juba		Seceded from Sudan in 2011		Secession struggle	President Salva Kiir Mayardit

(continued)

Table 5.1 Continued

Independent state name	*Colonial state name (if different)*	*Independent capital city*	*Colonial capital city (if different)*	*Former colonial administrator*	*Year independence achieved*	*How was independence achieved?*	*First head of state at independence*
Sudan		Khartoum		Britain (Anglo-Egyptian condominium)	1956	Political transition	Prime Minister Ismail al-Azhari
Swaziland		Mbabane		Britain	1968	Political transition	King Sobhuza II
Tanzania	Union of Tanganyika and Zanzibar	Dar es Salaam	Zanzibar City[5]	Germany, then Britain from 1919	Tanganyika 1961, Zanzibar 1963, united in 1964	Political transition	Prime Minister Julius Nyerere
Togo	French Togoland	Lome		Germany, then Britain and France from 1919	1960 (British Togoland ceded to Ghana)		President Saylvanus Olympio
Tunisia		Tunis		France	1956	Political transition	Prime Minister Habib Bourghuiba
Uganda		Kampala		Britain	1962	Political transition	Prime Minister Apollo Milton Obote
Zambia	Northern Rhodesia	Lusaka		Britain	1964	Liberation struggle	President Kenneth Kaunda
Zimbabwe	Southern Rhodesia	Harare	Salisbury	Britain[6]	1980	Liberation struggle (against Rhodesia)	President Canaan Banana

Sources: Eyzaguirre (1989), Chazan *et al.* (1999), Thomson (2000), Cooper (2002), Edie (2003) and Reid (2012).

Notes:

1 Yamoussoukro was the birthplace of the first president and became the political capital; however, Abidjan is the economic capital and de facto political capital.

2 Liberia was not partitioned at the Berlin Conference, but established as a territory for freed slaves by the private company the American Colonization Society.

3 The Sahrawi Arab Democratic Republic (Western Sahara) remains a disputed territory. It was occupied by Morocco after the Spanish withdrawal in 1976.

4 Mogadishu was the existing colonial capital of Italian Somaliland; however, British Somaliland's capital was Berbera.

5 The capital of Zanzibar and Pemba yielded to Dar es Salaam after the merging of Tanganyika and Zanzibar.

6 Rhodesia declared a Unilateral Declaration of Independence in 1965 from Britain, which resulted in armed struggle against Rhodesia.

euphoria of independence with its promises of political freedom for Africans was in many cases short-lived. The colonial state was not completely dismantled, and this became an enduring obstacle for African social, political and economic development, which we refer to here as the *colonial hangover*. Indeed, the boundaries of the states themselves proved to be hurdles.

The colonial state: arbitrary boundaries

Africa's 'arbitrary boundaries' were drawn across the African continent by the European imperial powers. In 1885, then Chancellor of Germany Otto von Bismarck convened an international conference in Berlin to carve up the African continent among the competing European powers, England, France, Germany, Portugal, Italy and Spain, in order to avoid international conflict. These colonial boundaries were 'arbitrary', as Thomson (2000: 13) has noted, because they 'rarely matched existing pre-colonial political, social or economic divisions', but they were not 'random' boundaries because 'they reflected the short term strategic and economic interests of the imperial powers, [. . . just] not the interests of the Africans they housed'.

The sentiment of this exercise is revealed in the 1890 speech given by then British Prime Minister Robert Gascoyne-Cecil (also known as either the Third Marquis of Salisbury, or Lord Salisbury, and the capital of colonial Rhodesia was named after him). He noted the main feature of the colonial project – the creation of 'artificial' political boundaries – with some irony.

Given that his oft-quoted speech has been overly paraphrased, poorly cited and reinterpreted awkwardly over the last 125 years (more so since the advent of the Internet), his original words, as reported by *The Times* on 7 August 1890, are quoted in full below to relay the full context of the venue and occasion of their original utterance. Speaking to his ministers at a banquet at the Mansion House, London, the prime minister stated (emphasis added):

> Perhaps it is that very peacefulness of current affairs in Europe that has removed Europe almost entirely from the ken of those interested in foreign affairs. We think of Africa and of nothing else, not because I think Africa has become more interesting, but because Europe has become less interesting. But I am not surprised at the attention which has been given to the great and splendid discoveries of our explorers, and the explorers of other nations, and the development of industry and enterprise which has taken place in that vast and long neglected continent. Yet I believe that the deep interest which has been felt by political men in recent negotiations in Africa has not been entirely due to a hope that those vast unexplored regions would yield early and abundant fruit to the enterprise of the merchant, or the discoverer, or the colonist. I rather should be inclined to cite the deep interest that has been felt in Africa as another proof of the strong pacific feeling which is gradually gaining more and more undisputed influence over all the strongest and most ruling intellectuals in the world. Men have welcomed the agreements which we have made, or which we are making, with the principle nations of the world in regard to Africa, partly no doubt on account of the great field which is opened to English industry and enterprise, but much more, I believe, because they recognise that in those agreements we are removing the most probable and the most dangerous cause of possible quarrel between nations who ought always to be at peace. (Cheers.) *We have been engaged in what, perhaps, to a satirist may seem the somewhat unprofitable task of drawing lines upon maps where no human foot has ever trod. We have been giving away mountains and rivers and lakes to each other, but we have only been hindered by the small impediment that we never knew exactly where those mountains and rivers and lakes were.*

What is interesting about this speech is that Lord Salisbury clearly sees Africa as part of the future of his empire, and justifies colonisation and the carving up of the continent on the grounds that it would avoid wars between other European nations vying for its resources – hence the Berlin Conference.

This European colonisation was clearly racist in its mandate, as he made no mention of the peoples residing within Africa, clearly negating them as human, and seemed to think it somewhat amusing that the political boundaries were being marked out without knowing where they were. Furthermore, clearly Lord Salisbury did not have any care or understanding of the negative impact that colonialism would have in these newly carved out states.

What would this prime minister have made of contemporary Africa, plagued by civil war, ethnic genocide, religious conflict, recurrent military coups, massive political corruption, and ongoing poverty and underdevelopment?

Perhaps Salisbury and the other men at the Berlin Conference would have blamed the 'locals' – the indigenous political elite – for failing to take advantage of the colonial state to build an independent, modern, capitalist and democratic nation. Unsurprisingly, the 'locals' would disagree, maintaining that blame for Africa's woes sits squarely on the shoulders of men such as Salisbury – the colonisers. However, how useful is it, some 50 years after independence, to continue to blame the colonisers for the many post-independence or postcolonial problems African nations experience today?

On the eve of political independence in the early 1960s, it was the Organisation of African Unity that agreed that these 'arbitrary' boundaries were the best hope for African states to enter and complete the globalisation of states across the world, and thus be able to function and compete equally as nation states. The problem was that there were at least 14 states that were landlocked with no access to sea ports, and other states were left with little or no natural resources, or indeed a sense of nationhood and cohesive national identity, which are so important in order to avoid future conflicts (see Thomson, 2000: 13–15).

Colonial legacies

Thus, it has been that one of the defining features of postcolonial states in Africa is persistent conflict, military coups and civil wars, which have also contributed to the lack of human, economic, social and environmental development. It was because these arbitrary boundaries were not organically derived, but were created by imperial powers with no consideration of local conditions, that the independent and postcolonial African nations basically 'inherited weak links between state and society' (Young, 1988, cited in Thomson, 2000: 15). These weak links translated into fragile or weak states that created the conditions for conflict to thrive. One example, the Democratic Republic of Congo, formerly Zaire, has experienced continual conflict since independence, despite many ongoing interventions from the United States of America and the United Nations, and remains one of the world's most fragile states (Lefever, 1965; Marriage, 2011; Lemke, 2011; Fund for Peace, 2016). It is also noteworthy that of the top 10 fragile states as reported by the Fund for Peace in 2016, the top four are African states – Somalia, South Sudan, the Central African Republic and Sudan. Chad and the Democratic Republic of Congo are ranked seventh and eighth. These latter states are considered to be on 'very high alert' (Fund for Peace, 2016).

As Thomson (2000: 22) has summarised, there was simply 'no shared political culture between state and society', and this had a lot to do with the way in which political power was handed to an educated elite. Perhaps somewhat ironically, it was the same urban educated elites created by colonialism that led the nationalist movements that so effectively challenged colonialism

(Starkey, 2003). For example, Robert Mugabe was educated at a Jesuit mission school in Southern Rhodesia, and went on to complete degrees in Arts, Education and Science. In the 1960s, he formed the Zimbabwe African National Union, which was central to the political and later armed struggle against the white Rhodesian minority government that had declared a Unilateral Independence from Britain but had refused to allow political independence for the majority population (Russell, 1999). He went on to become the first prime minister in 1980, and later the president of independent Zimbabwe until 2017.

Yet when independence was granted, the colonial power structures were simply transferred to these African 'big men' (Russell, 1999), such as Mugabe, and in many cases continued on without challenge. However, because the power structures were not organically derived from any indigenous precolonial political institutions, the majority of the people remained as disenfranchised from the state as they did during the colonial period. Therefore, while the rural peasantry may have supported nationalism, decolonisation and independence, power clearly went into the hands of the new African elites – as Frantz Fanon put it, to 'black skin, white masks' (Fanon, 1968).

The colonial project was – simply put – about the extraction of raw materials from the continent to the imperial metropole, and 'it was never about the provision of public services for citizens' (Thomson, 2000: 15). As such, there was a 'deficit of legitimacy' (Thomson, 2000: 22), which often meant the state resorting to 'coercion' instead of the more democratic 'consent' that comes with the 'legitimacy' required for a state to house a nation. As such, the social contract between the government and the people was either limited or non-existent in many post-independence African nations, such as Zaire and Zimbabwe, and because there were no connections to civil society, there was no 'trust and shared political values . . . between the rulers and the ruled. State institutions never sought or gained the respect of the people' (Thomson, 2000: 15–16).

Soon after the Second World War, global development and political discourses required the colonisers to consider such provisions, the colonial project immediately became unprofitable, and the wave of decolonisation began. Decolonisation resulted in political independence for African states, but one-party states and dictatorships flourished to quell opposition and to maintain power and control of the post-independence states.

Therefore, from the above discussion of post-independence Africa and its colonial legacies and crisis of legitimacy, it is clear that the 'post' in postcolonial is not the simple prefix meaning 'after in time or order', but that it necessitates a more complex analysis of the legacies of colonisation and the structures and institutions that were handed down from their colonial designers and carried over to a new African political elite on the eve of independence for these African states.

Nationalism and persistent conflict

Therefore, as we continue here to understand the politics of postcolonial Africa, this chapter will now argue that 'nationalism' has been the dominant ideology that has influenced these postcolonial African political systems. However, has 'nationalism' enabled the postcolonial African state to function effectively? We know that in most African countries, nationalism was the mobilising force that united the people in the shared struggle against the colonisers and colonialism, and this fight included the 'desire to house the nation in its own sovereign state'. The problem for the new African leaders, however, was that they only inherited their states, not nations. For example, in the 1950s and 1960s, when African nationalist movements demanded self-determination for their states across Africa, the uniting aim was to expel the imperial and colonial rulers. However, it was a rather difficult task to bring together the different ethnic groups to focus on a collective nation state and engage in a nation-building project, particularly given the divisions of these ethnic groups encouraged by the colonisers, and the transfer of these arbitrary borders and state structures, including the acceptance of the former colonisers' language as the official language to be spoken at independence. Patrice Lumumba's efforts in former Zaire serve to illustrate this issue, in his attempts to unite the nation along nationalist lines. However, his plans were dramatically ended after his assassination while serving as the independent country's first democratically elected prime minister. Subsequently, this post-independence state remains scarred by the colonial legacy and postcolonial politics (see Young, 2003).

The late Samora Machel, the first president of independent Mozambique, famously stated, 'For the nation to live, the tribe must die'. His intention was not to advocate 'quasi-nationalism', a process that sets out to destroy the peoples that are in the way of national unity. For example, in postcolonial Zimbabwe in the mid-1980s, the ruling party, dominated by President Robert Mugabe's Zimbabwe African National Union (ZANU), set out to destroy the Ndebele ethnic group that mainly comprised the main opposition group, the Zimbabwe African People's Union. This kind of quasi-nationalism (as interpreted by a handful of African postcolonial leaders, such as Idi Amin in Uganda, Robert Mugabe in Zimbabwe and Juvenal Habyrimana in Rwanda) has been a catastrophe for the post-independence nation state. Machel's statement was a philosophical approach to answering the question: How do you create national unity out of a multicultural, multireligious, multi-ethnic or multiracial society? As Thomson has argued:

> 'Tribes' were the anti-thesis of a nation. They were portrayed as retrogressive, part of the past, and an obstacle to progress. In this respect, 'tribalism' was regarded as a sin against the post-colonial state and the freedoms that had been won. The nation was now the priority, not outdated 'tribal' associations, and state power would be used to promote this process of nation-building.
>
> *(Thomson, 2000: 37)*

During the liberation struggle, nationalism was the driving force, and after liberation nationalism became the dominant ideology. National unity was needed for economic development to occur along modernisation principles. However, without a common enemy – such as colonialism – to unite them, many post-independence nationalist parties began to develop rifts and internal conflicts. In order to maintain power, most post-independence African 'nations' relied heavily on one political party holding the reins, and as such, to maintain that control, they resorted to banning opposition groups and ethnic associations, co-opting or dismantling civil society groups, such as trade unions and youth and women's groups disregarding indigenous languages in favour of the official lingua franca (Mitchinson, 1984).

Therefore, post-independence African states had to develop their economies from a weak base of national unity. The economies that these states inherited from their colonisers were often based on the exportation of a few natural resources that continued to be dominated by trading relationships with their former colonisers. Transportation and communication networks were built for this trade during the colonial era, so it was no surprise that, for example, Côte d'Ivoire would trade more with its former coloniser France than its neighbour Ghana. Local and regional trading relationships were virtually non-existent, limiting the development of the post-independence African economies.

Furthermore, the former colonisers had also not developed the human resource capacities of their colonies. For example, in Guinea-Bissau, the Portuguese left the country with only 14 university graduates and a 97 per cent illiteracy rate. Thus, there were limited trading and economic opportunities handed over to the post-independence African state, and a limited number of people with the experience or education needed to manage a new nation state.

The crisis of legitimacy

It was difficult for these leaders who had 'cut their teeth' politically in an anti-colonial struggle to turn that experience into governing a nation state, for which they clearly had no experience or support. Thus, the *political institutions* that they inherited were weak, and it is certainly easier to then blame an inexperienced African leader for the troubles facing 'his' (it was never a 'her' in the immediate postcolonial period) state. However, can you expect a liberal democratic state to be born from an overly authoritarian bureaucratic, illegitimate, interventionist state? Since imperialism did not create legal, rational institutions, and, for example, the running of free and fair elections and democratic accountability, were the expectations of independence pitched too high?

What did the end of colonisation really mean for Africa? Did it mean *independence* for African nations? Liberation wars were fought, for example, in Kenya (e.g. Lonsdale, 2003; Anderson, 2005), Zimbabwe (e.g. Martin and Johnson, 1981) and Mozambique (eg. Funada-Classen, 2013), and after long and protracted struggles independence was eventually granted (see Table 5.1), and so now we think of African states as being 'post-independent' or 'postcolonial', in that they exist in the time period after the colonisers have left. Achille Mbembe defines this time period as the 'postcolony', where 'societies [have] recently emerg[ed] from the experience of colonisation and the violence which the colonial relationship . . . involve[d]' (Mbembe, 1992a: 3). Understanding this definition in relation to the *violence* of colonisation enables us to understand the forms of identities that shape modern-day Africa because the political violence and violations against humans shape and form those identities. For example, in the postcolonial period, ethnicity (replacing the older term of tribe) seems to be dominating identity in Africa. Yet there are multiple identities mobilised in Africa and, according to Mbembe, the 'postcolonial' African subjects must, for their own survival, constantly 'revise' their identities 'in order to achieve maximum instrumentality and efficacy' (Mbembe, 1992b: 5).

Postcolonialism and clientelism

In the context of Africa specifically, there are some 'pitfalls' in the literal use of postcolonial, in terms of a sequential timeline (McClintock, 1992; see also Rao, 2013). As Mbembe has argued:

> The younger generation of Africans have no direct or immediate experience of colonisation, whatever role it may have played as a foundational event in African history. Postcolonial, in other words, is applicable not to all of the postcolonial period, but only to that period after colonialism when among other things, a forgetting of its effects has begun to set in.
>
> *(Mbembe, 1992c: 137)*

We therefore need to problematise the 'postcolonial' because it is not just a linear progression of history from the precolonial to postcolonial. That is, the tendency to romanticise the precolonial, or indeed the colonial periods, as some kind of 'golden age' of the continent, obfuscates many of the harsh realities that already existed on the continent (e.g. the abuse of women as pawns or slaves in precolonial male-centred systems) (Allman *et al.*, 2002). However, some of the positives of colonisation, such as the liberation of many women from such traditional practices, for example through attachment to missions (Schmidt, 1992), should not be over-estimated because the violence of political repression and the curtailment of other freedoms dominated the colonial experience for most Africans (Bush, 2002).

Thus, it may be obvious that the *post* in postcolonial just means 'after', but it also means 'because of' and also 'inclusive of' the colonial. However, it also symbolises the 'anti-colonial'. That is, many postcolonial identities are based on an anticolonial ethic that encourages 'African solutions to African problems', and continues to resist the involvement of former colonisers or westerners in general (Mhlanga and Ndlovu-Gatsheni, 2013). Therefore, 'postcolonialism' refers to 'ideological orientations' rather than a 'historical stage', and as Vijay Mishra and Bob Hodge have argued, it is 'an always present tendency in any literature of subjugation marked by a systematic process of cultural domination through the imposition of imperial structures of power' (Mishra and Hodge, 1994: 284). Therefore, it is no surprise that postcolonial theory finds common ground with feminist theory (which critiques patriarchal oppression), or indeed postmodernism, which challenges the centralised structures of cultural and political power (Quayson, 2000).

Nonetheless, whether or not it is 'after' or 'because of' colonisation, post-independence Africa experienced a number of significant challenges that can be attributed to the colonial hangover, and the postcolonial international political economy. Apportioned blame must be laid at the feet of the leaders of these post-independence states (see Table 5.1) for the failures of these states, for example elite corruption and the disastrous implementation of structural adjustment programmes, introduced by the International Monetary Fund and the World Bank in Africa (Riddell, 1992). The African postcolonial state has thus either been considered weak or failed, using standard political and economic measures of international relations (Bates, 2008).

The postcolonial state and legitimacy

The postcolonial state's weakness is connected to its lack of legitimacy. Tunde Zack-Williams, referring to Mahmood Mamdani's term *decentralised despotism* – a form of 'postcolonial indirect rule' – argues that the 'nationalist coalition that brought independence was soon abandoned' (Zack-Williams, 2002: 2). That is, while the anti-colonial fighters were being demobilised, the new post-independence leaders (see Table 5.1), the traditional leaders that were needed to support them, and the petty bourgeoisie took control of power and access to the resources through a system of 'clientelism' and 'corruption' (Edie, 2003). While many of the new leaders blamed the former colonisers and their alien systems of governance for the lack of development or economic performance within their post-independence states, they still managed to enjoy the luxury goods and banking services that the inherited 'Western' system provided. The *Wabenzi* were thus created, referring to a term popularised in Kenya to describe the corrupt officials driving imported Mercedes-Benz cars. Carlene Edie (2003) describes this process of 'clientelism' emanating from the informal political institutions that shape state–society relations, and which enables civil servants and bureaucrats to gain access to power and resources for personal gain. Therefore, the state that was inherited at independence:

> was organized on the principles that were antithetical to the objectives of the nationalist movement. The new leaders continued to rule through the old colonial institutions, which emphasized personal authority. They were also unable to transform the state apparatus into one that was responsive to the socioeconomic needs of broad sectors of the population.
>
> *(Edie, 2003: 90)*

The result of all of this was economic marginalisation, poverty, poor health, political instability and ultimately the many civil wars that have plagued post-independence African states. The liberation fighters were often neglected in this relationship (e.g. Bhebe and Ranger, 1995).

The colonial hangover thus included powerful elites who only provided benefits to those often traditional leaders or chiefs who were given more localised powers, and thus who could exert some control over their own local subjects and who would secure and deliver votes for them, to maintain a facade of democratic legitimacy. Voting thus becomes an imaginary activity – as the structures inherited from the departing imperial powers are merely replaced by one-party states and dictatorships based on clientelism and patrimonialism (Thomson, 2004; Forje, 2009; Halupka and Star, 2012; Taylor, 2012).

Thus, for Chazan *et al.* (1999: 137–158), in order to understand the post-independence African state, we need to appreciate the different types of regimes that developed in Africa since the end of colonial rule, and in particular think about what this means for decision-making, policy application, societal reaction and consequent change. Ultimately, it comes down to the notion of 'good government', that is, how governments have responded to the needs of the people, linking sovereignty with responsible governance. Understanding post-independence Africa is determined by what emerges or develops after colonisation ends.

Therefore, we need to acknowledge that there is a debate on the issue of post-independence and postcolonialism. It is too simplistic to just refer to the period after colonisation has ended, as a period of independence. After all, how can this explain the multiple identities based on tribe and/or ethnicity that fuel quasi-nationalism, national disunity, or indeed the increase in civil wars and violence, for example Ethiopia in the mid-1980s, Sierra Leone and Liberia in the 1990s, Sudan in the 2000s, and even the genocidal events such as Rwanda in 1994 and Darfur in 2003.

For Chazan *et al.* (1999), Zack-Williams (2002), and Edie (2003), postcolonialism is the time after colonisation has ended whereby the newly independent regimes have made choices, often bad choices, that have led to the economic, political and social crises that are thus subsequently blamed upon colonisation, rather than any post-independence economic or political mismanagement.

Contemporary problems: colonial legacies or post-independence mismanagement?

The contemporary problems continuing to face African states are of concern to us all, and they can be seen most profoundly by the movement of people, through trade and other relations, but importantly through those forced to flee their homes due to civil wars and human insecurity, thus becoming refugees or internally displaced peoples. Somalia, Sudan, the Democratic Republic of Congo and Eritrea have remained among the top 10 source countries for refugees globally since 2010 (UNHCR, 2014). Images of overloaded boats heading north from the continent, hoping to reach the southern tips of Europe, have become commonplace. Sadly, the reports of many of these attempts being made in unseaworthy boats, commissioned by ruthless people smugglers, trading on human misery and resulting in hundreds of deaths at sea, are also commonplace (Loudon, 2015). Getting 'out of Africa' is now a survival strategy for thousands of individuals otherwise trapped in a cycle of civil war and poverty. This is not an issue caused by or affecting one state only, and cannot be solely blamed on former colonial rulers, or even the current leaders.

When the independent African states began to 'fall apart', from their colonial heritage, through the mismanagement and misrule of corrupt or simply bad leaders, constrained by the politics of the Cold War, restricted by their physical environments and economies and trading relationships (including many landlocked countries with no natural resources), suffering under the burdens of the foreign debt crisis, which then led to the disasters of the economic structural adjustment programmes imposed by the international financial institutions, it is curious that 'colonialism', the first on this long list of contributing factors, is usually solely blamed as the 'logical cause' of these African woes, and thus remains a convenient scapegoat for all of postcolonial Africa's problems – thus obfuscating other causes more directly linked to the postcolonial state. That is, when the postcolonial African state emerged, and, through postcolonial theories, challenges were made to Western discourses of Africa's past, present and future (Said, 1978), it was obvious that colonisation was the culprit.

Yet in the 2000s, we must consider that Africa is no longer beholden to the colonial state, and hasn't been for a long time, and that simply blaming the colonisers for all of the problems facing Africa today no longer has that much traction in international relations theory (Engel and Olsen, 2005; Murithi, 2014). The problems facing Africa are continent-wide issues and require solutions of similar magnitude. The international relations and connections within the continent are between *all* African states, and therefore the African Union needs to be part of that solution (Abass and Mystris, 2014).

Conclusion: Africa Rising – beyond colonialism and postcolonialism?

The establishment and formation of the African Union (AU) has been central to the 'Africa Rising' discourse. African states now have the ability, if not the political willingness, to 'intervene' in other states when required, and to demand a level of security across the continent within and between countries (Makinda *et al.*, 2015). For example, the fact that the African Union intervened in internal crises in Burundi, Sudan and Somalia clearly demonstrated its difference and departure from the former and more cumbersome Organisation of African Unity, which had a policy of non-intervention in the affairs of a sovereign state (Murithi, 2008).

There is still a need and role for the international community to respond to Africa's ongoing challenges. The Millennium Development Goals (MDGs) were once heralded as the key to Africa's future (e.g. see Easterly, 2007; Andrews *et al.*, 2015), and since 2015 the Sustainable Development Goals (SDGs) (Joshi *et al.*, 2015) have become the new mantra for the future. However, it is timely to ask: Has anything really changed for the better? To answer this question, as Severino and Ray (2011) have pointed out, we need to stop examining 'yesterday's Africa' and stop focusing on the usual 'African crises', and instead have a look at 'Africa today' and focus on the 'transformations' that have been witnessed across the continent. Among the growth in economies, urban populations and youth, there has been an increase in the number of middle-class people, albeit spread unevenly across the continent, and this will have massive (positive) implications for African development into the future (see Binns *et al.*, 2012).

The widespread adoption of modern information technologies (including phone banking, among other technologies) has enabled business and trade, and even social movements, to flourish (Okumu and Makinda, 2013). For example, the massive protests against authoritarian regimes witnessed across the North African states of Tunisia, Egypt and Libya in 2010 and 2011, collectively known as the 'Arab Spring', were also popularly labelled 'Twitter revolutions'. Mobile phone communications and social media enabled people to unite in a show of 'people power' (Barrons, 2012). Globalisation and the associated communications technologies have thus enabled Africans to impose themselves and make their mark upon the planet as a whole (Bratton, 2013; Brown and Harman, 2013; Edoho, 2013). The Marquis of Salisbury would surely have been intimidated by this ability of Africans to *unite*, when *divide and rule* had been such a successful colonial strategy!

Furthermore, and finally, the fact that African nations make up about one-quarter of the voting bloc in the United Nations, and near one-third of the Commonwealth members, means that their voices will be increasingly heard in international debates and within international relations. We must, however, concede that the optimism inspired by the 'Africa Rising' thesis is marred by the continuation of civil wars and conflicts in some countries, such as in South Sudan, the Democratic Republic of Congo, the Central African Republic and within north-eastern Nigeria. African states still dominate the top 20 listed countries in the Failed and Fragile States Index (Fund for Peace, 2015). The 2014 Ebola crisis across West Africa also tragically demonstrated that some African states have a long way to go to achieve adequate and sustainable development, and that there is no simple recovery from the violence of the past – colonial or postcolonial.

Note

1 Earlier versions of some sections of this chapter were originally published as Lyons *et al.* (2015), and have been included here with permission.

References

Abass, A. and Mystris, D. (2014) 'The African Union legal framework for protecting asylum seekers', in A. Abass and F. Ippolito (eds), *Regional Approaches to the Protection of Aslyum Seekers: An International Legal Perspective*, London and New York: Ashgate, pp. 19–44.

Agbiboa, D. (2015) 'Nigerian state responses to insurgency', in J.I. Lahai and T. Lyons (eds), *African Frontiers: Insurgency, Governance and Peacebuilding in Postcolonial States*, London: Ashgate/Routledge, pp. 81–103.

Allman, J., Geiger, S. and Musisi, N. (2002) *Women in African Colonial Histories*, Bloomington. IN: Indiana University Press.

Anderson, D. (2005) *Histories of the Hanged: The Dirty War in Kenya and the End of Empire*, New York: Norton.

Andrews, N., Khalema, N.E. and Assié-Lumumba, N.T. (eds) (2015) *Millennium Development Goals (MDGs) in Retrospect: Africa's Development Beyond 2015*, Cham: Springer.

Ayittey, G. (1998) *Africa in Chaos*, Basingstoke: Macmillan.

Barrons, G. (2012) 'Suleiman: Mubarak decided to step down #egypt #jan25 OH MY GOD: examining the use of social media in the 2011 Egyptian Revolution', *Contemporary Arab Affairs*, 5(1): 54–67, DOI: 10.1080/17550912.2012.645669.

Bates, R. (2008) *When Things Fell Apart: State Failure in Late-Century Africa*, New York: Cambridge University Press.

BBC News (2016) *Nigeria Chibok Girls: Boko Haram Video Shows Captives*, 14 August, available at: www.bbc.com/news/world-africa-37076644 (accessed 11 December 2017).

Bhebe, N. and Ranger, T. (1995) *Soldiers in Zimbabwe's Liberation War*, Portsmouth: Heinemann.

Binns, T., Dixon, A. and Nel, E. (2012) *Africa: Diversity and Development*, London: Routledge.

Bratton, M. (2013) 'Briefing: citizens and cell phones in Africa', *African Affairs*, 112(447), April: 304.

Bright, J. (2015) 'Meet "Nollywood": the second largest movie industry in the world', *Fortune*, available at: http://fortune.com/2015/06/24/nollywood-movie-industry/ (accessed 11 December 2017).

Brown, W. and Harman, S. (eds) (2013) *African Agency in International Politics*, London and New York: Routledge.

Bush, B. (2002) *Imperialism, Race and Resistance: Africa and Britain, 1919–1945*, Hoboken, NJ: Taylor & Francis.

Chazan, N., Lewis, P., Mortimer, R., Rothchild, D. and Stedman, S. (1999) *Politics and Society in Contemporary Africa*, 3rd edition, Boulder, CO: Lynne Reinner.

Conrad, J. (2013 [1902]) *Heart of Darkness*, London: Harper Press.

Cooper, F. (2002) *Africa since 1940: The Past and the Present*, Cambridge: Cambridge University Press.

Cooper, F. (2003) 'Conflict and connection: rethinking colonial African history', in J.D. Le Sueur (ed.), *The Decolonization Reader*, London and New York: Routledge, pp. 23–44.

Easterly, W. (2007) *How the Millenium Development Goals Are Unfair to Africa*, Brookings Global Economy and Development, Working Paper 14, November.

Edie, C. (2003) *Politics in Africa: A New Beginning?* Belmont: Wadsworth/Thomson Learning.

Edoho, F.M. (2013) 'Information and communications technologies in the age of globalization: challenges and opportunities for Africa', *African Journal of Economic and Management Studies*, 4(1): 9–33.

Engel, U. and Olsen, G.R. (2005) *Africa and the North: Between Globalization and Marginalization*, London: Routledge.

Eyzaguirre, P.E. (1989) 'The independence of São Tomé and Príncipe and agrarian reform', *The Journal of Modern African Studies*, 7(4): 671–678.

Fanon, F. (1968) *Black Skin, White Masks*, London: MacGibbon & Kee.

Forje, J. (2009) *State Building and Democracy in Africa: A Comparative and Developmental Approach*, New York: Nova Science Publishers.

Funada-Classen, S. (2013) *Origins of War in Mozambique: A History of Unity and Division*, Cape Town: African Minds.

Fund for Peace (2015) *Fragile States Index 2015*, available at: http://library.fundforpeace.org/library/fragilestatesindex-2015.pdf (accessed 11 December 2017).

Fund for Peace (2016) *Fragile States Index 2016*, available at: http://fsi.fundforpeace.org/rankings-2016 (accessed 11 December 2017).

Gascoyne-Cecil, Robert Arthur Talbot, the 3rd Marquis of Salisbury and British Prime Minister, (1890) as reported in 'The Mansion House Banquet to Her Majesty's Ministers', *The Times*, 7 August, p. 6.

Gordon, A. and Gordon, D. (eds) (2001) *Understanding Contemporary Africa*, 3rd edition, Boulder, CO: Lynne Reinner.

Halupka, M. and Star, C. (2012) 'Maintaining sovereignty in Africa: the role of external forces in warlord states', *Australasian Review of African Studies*, 33(2): 74–97.

Joshi, D., Hughes, B. and Sisk, T. (2015) 'Improving governance for the post-2015 Sustainable Development Goals: scenario forecasting the next 50 years', *World Development*, 70: 286–302, DOI: http://dx.doi.org/10.1016/j.worlddev.2015.01.013.

Lefever, E.W. (1965) *Crisis in the Congo: A United Nations Force in Action*, Washington, DC: Brookings Institution.

Lemke, D. (2011) 'Intra-national IR in Africa', *Review of International Studies*, 37(1): 49–70.

Lonsdale, J. (2003) 'Mau Maus of the mind: making Mau Mau and remaking Kenya', in J.D. Le Sueur (ed.), *The Decolonization Reader*, London and New York: Routledge, pp. 269–290.

Lorch, D. (1994) 'Heart of Rwanda's darkness: slaughter at a rural church (massacre of Tutsis, Nyarubuye, Rwanda)', *New York Times*, 3 June, 143, p. A1(N).

Loudon, B. (2015) 'Sea of chaos and despair for African refugees', *The Australian*, 21 April, available at: www.theaustralian.com.au/news/inquirer/sea-of-chaos-and-despair-for-african-refugees/story-e6frg6z6-1227312518392 (accessed 11 December 2017).

Lyons, T., Marlowe, J. and Thornton, A. (2015) 'Reflections on Africa and African studies: in memory of Cherry Gertzel', *Australasian Review of African Studies*, 36(2), December: 8–21.

Makinda, S.M., Okumu, F.W. and Mickler, D. (2015) *The African Union: Addressing the Challenges of Peace, Security, and Governance*, 2nd edition, London: Routledge.

Marriage, Z. (2011) 'Divisive "commonality": state and insecurity in the Democratic Republic of Congo', Third World Quarterly, 32(10), November: 1891–1910.

Martin, D. and Johnson, P. (1981) *The Struggle for Zimbabwe: The Chimurenga War*, London: Faber & Faber.

Mbembe, A. (1992a) 'Provisional notes on the postcolony', *Africa*, 62(1): 3–37.

Mbembe, A. (1992b) 'The banality of power and the aesthetics of vulgarity in the postcolony', *Public Culture*, 4(2), Spring: 1–30.

Mbembe, A. (1992c) 'Prosaics of servitude and authoritarian civilities', trans. J. Roitman, *Public Culture*, 5, Fall: 123–148.

McClintock, A. (1992) 'The angel of progress: pitfalls of the term "post-colonialism"', *Social Text*, 31/32: 84–98.

Mhlanga, B. and Ndlovu-Gatsheni, S.J. (2013) *Bondage of Boundaries and Identity Politics in Postcolonial Africa: The 'Northern Problem' and Ethno-Futures*, Pretoria: Africa Institute of South Africa.

Mishra, V. and Hodge, B. (1994) 'What is post(-)colonialism?', in P. Williams and L. Chrisman (eds), *Colonial Discourse and Postcolonial Theory: A Reader*, London and New York: Routledge, pp. 276–290.

Mitchinson, N. (1984) 'One party rule in Africa', *The Round Table*, 289: 38–44.

Murithi, T. (2008) 'The African Union's evolving role in peace operations: the African Union Mission in Burundi, The African Union Mission in Sudan and the African Union Mission in Somalia', *African Security Review*, 17(1): 69–82.

Murithi, T. (ed.) (2014) *Handbook of Africa's International Relations*, London: Routledge.

Ndlovu-Gatsheni, S.J. (2012) 'Coloniality of power in development studies and the impact of global imperial designs on Africa', *Australasian Review of African Studies*, 33(2): 48–73.

Ndlovu-Gatsheni, S.J. (2013) *Empire, Global Coloniality and African Subjectivity*, New York: Berghahn Books.

Ndlovu-Gatsheni, S.J. (2015) 'Decoloniality in Africa: a continuing search for a new world order', *Australasian Review of African Studies*, 36(2): 22–50.

Ofuho, C.H. (2003) 'The changing images of a continent: an eclectic survey of writings on Africa in the international system', *Global Society*, 17(2): 151–164.

Okumu, F.W. and Makinda, S. (2013) 'Engaging with contemporary Africa: key contexts for external actors', in D. Mickler and T. Lyons (eds), *New Engagement: Contemporary Australian Foreign Policy Toward Africa*, Melbourne: Melbourne University Press, pp. 17–33.

Quayson, A. (2000) *Postcolonialism: Theory, Practice, or Process?* Malden, MA: Polity Press.

Rao, R. (2013) 'Postcolonialism', in M. Freeden, L.T. Sargent and M. Stears (eds), *The Oxford Handbook of Political Ideologies*, Oxford: Oxford University Press, pp. 271–292.

Reid, R. (2012) *A History of Modern Africa 1800 to the Present*, 2nd edition, Oxford: Wiley-Blackwell.

Riddell, B. (1992) 'Things fall apart again: structural adjustment programmes in sub-Saharan Africa', *The Journal of Modern African Studies*, 30(1): 54–68.

Russell, A. (1999) *Big Men, Little People: The Leaders Who Defined Africa*, London: Macmillan.

Said, E. (1978) *Orientalism: Western Representations of the Orient*, London: Routledge and Kegan Paul.

Schmidt, E. (1992) *Peasants, Traders, and Wives: Shona Women in the History of Zimbabwe, 1870–1939*, London: James Currey.

Schraeder, P. (2004) *African Politics and Society: A Mosaic in Transformation*, 2nd edition, Belmont: Thomson.

Severino, J.-M. and Ray, O. (2011) *Le Temps de l'Afrique / Africa's Moment*, trans. D. Fernbach, Cambridge: Polity.

Starkey, H. (2003) 'Colonialism, character-building and the culture of nationalism in the Sudan 1898–1956', in J.D. Le Sueur (ed.), *The Decolonization Reader*, London: Routledge, pp. 218–235.

Sulemana, M. (2014) 'Centenary of failure? Boko Haram, jihad and the Nigerian reality', *Australasian Review of African Studies*, 35(2), December: 69–87.

Sulemana, M. and Azeez, G. (2015) 'Rethinking Islamism in western Africa', *Australasian Review of African Studies*, 36(1), June: 51–67.

Taylor, S. (2012) *Globalization and the Cultures of Business in Africa: From Patrimonialism to Profit*, Bloomington, IN: Indiana University Press.

Thomson, A. (2000) *An Introduction to African Politics*, London: Routledge.

Thomson, A. (2004) *An Introduction to African Politics*, 2nd edition, London: Routledge.

UNHCR (United Nations High Commissioner for Refugees) (2014) *War's Human Cost: Global Trends 2013*, available at: www.unhcr.org/5399a14f9.html (accessed 11 December 2017).

Wallace, B. and Gordon, N. (1994) 'The dawn of democracy: four days of voting in South Africa mark the end of apartheid and the beginning of a long road to equality', *Maclean's*, 107(19), 9 May: 22.

Young, C. (2003) 'Zaïre: the shattered illusion of the integral state', in J.D. Le Sueur (ed.), *The Decolonization Reader*, London and New York: Routledge, pp. 414–427.

Zack-Williams, T. (2002) 'Introduction: Africa at the millenium', in T. Zack-Williams, D. Frost and A. Thomson (eds), *Africa in Crisis: New Challenges and Possibilities*, London: Pluto Press, pp. 1–14.

Section 3
Social perspectives

6

Demographic characteristics, migration and employment-seeking behaviour

A case study of Ghana

Joseph K. Assan, Dinar D. Kharisma and Afia A. Adaboh

Introduction

By the year 2050, 70 per cent of the world's population will be urban. With 1.5 million people migrating into cities every week, managing and meeting their employment needs and expectations is one of the most important development challenges facing the world today (Wilson Center, 2017). Africa accounts for 1.2 billion people, representing 16 per cent of the world's population. Out of the total expected global population growth from 2015 to 2050, 50 per cent of the growth is expected to occur in Africa. The essential feature of Africa's population growth is that due to its large concentration of young adults who are yet to reach their reproductive prime, even if fertility rates fall, the sheer number of people of childbearing age will cause the population to increase. The Global Migration Group (2014) reveals that Africa's share of adolescents and youth aged 15–24 is about 205 million, and is projected to more than double by 2100, making it the region with the highest youth population in the world. Children under 15 years of age and persons between ages 15 and 24 make up 60 per cent of Africa's current population (UN DESA, 2015). The report explains that unemployment is higher among youth than other population groups. As this cohort ages and expands, greater pressure will be exerted on the labour markets of struggling economies.

Africa's population growth and accompanying changes in population dynamics adds a layer of complexity to the challenge of eradicating poverty and promoting livelihood and income security (UNFPA, 2013). These conditions are exacerbated by the absence of strong formal welfare programmes in most African countries (Assan, 2008, 2014).

Yet another implication of global demographic change is shifts in the dependency burden. The collective ability to cater for the increasing ageing population is linked to the ability of the youthful population to find 'productive and remunerative employment' (UNFPA, 2013: 17). Arguably, economic growth and social policy in many African countries lags behind changing population structures. According to Honwana (2012), many African youth are increasingly uncertain about the possibilities of obtaining employment, decent housing, having a stable family and being able to support themselves and their relatives after leaving school. Honwana argues that the characteristics of economically and socially independent adulthood for most young Africans appears to be a mirage.

According to the Global Migration Group (2014), West Africa hosts some of the most disadvantaged, marginalised and yet highly mobile youth in the world. The declining standard of living in some of the countries in the region is further inhibited by a lack of available education, healthcare and other social services. Generating industrious human capital represents a major challenge. As a consequence, internal migration has now gained a renewed focus as a labour-seeking mechanism of young adults and children in many African countries, including Ghana (Assan, 2008; Awumbila *et al.*, 2014). While internal migration is not a new phenomenon, a rapid acceleration in the shift from agriculture as the dominant income generation activity to non-agricultural economic activities within expansively liberalised economies is considered a factor in the observed mobility pattern. The process has been deepened by the World Bank and IMF-supported neo-liberal programmes and the continued adoption of extensive market liberalisation and fiscal policies (Gould, 2015).

This chapter will explore the impacts of the current demographic changes and shifts in population structure on internal migration and the labour-/employment-seeking behaviour of youth within the African context using Ghana as a case study. The objective of this study is to explore the welfare implications of the emerging demographic changes in Ghana on the employment-seeking behaviour of the country's youth, with particular attention paid to internal migration (rural–urban). Whether this pattern of mobility leads to better employment and welfare opportunities and a decline in extreme poverty, or conversely exacerbates socio-economic marginalisation and equalities, are issues this chapter seeks to explore.

The following discussion will examine rural–urban migration of youth in Ghana, and the association between internal migration, migrants' demographic characteristics, employment type and conditions, and their impacts on human well-being compared to their non-migrant counterpart. Awumbila *et al.* (2014) call for a more systematic understanding of the connections between the internal migration of young people to urban areas in Ghana and their prospects to exit poverty and improve their well-being.

The remainder of the chapter is structured into four parts. The second and third parts will discuss the literature on the concept of internal migration as a labour-seeking behaviour and the methodology employed to generate the study results. The fourth section discusses the findings from our analysis of the Ghana Living Standard Survey (2014) Round 6 data set. The final section will discuss some key recommendations for development policy and practice.

Literature review

Youth migration as an employment- and labour-seeking behaviour in Africa

The concept of internal migration (rural–urban) has a long provenance in antiquity. Migration is an important component of comprehensive livelihood strategies for rural African dwellers (Castles *et al.*, 2013). Peasant households often utilise outmigration as a means of diversifying income sources, improving well-being and building more sustainable livelihoods overall (Assan *et al.*, 2009). Often, outmigration represents the most viable means for such households to break intergenerational poverty traps and improve future outcomes (Boutin, 2016).

The shift by young Africans and their households to diversify income sources from primarily agricultural activities to non-farm work through internal migration is by no means arbitrary or irrational. According to the Harris and Todaro (1970) migration model, rural–urban migration, particularly the decision to migrate, is not only based on wage differentials between the two areas; it is also based on projected income 'differentials'. Ellis (1998) describes these shifts as 'rational and voluntary risk-management' decisions aimed at reducing 'co-variate risk' and ensuring consumption smoothing for households throughout the year. Assan (2008) highlights the importance of human agency in migration decisions.

Scholarship on the impacts of internal migration of young people can be organised into two main groups. The first group explores the efficacy of rural–urban migration as a household income diversification strategy, while the second assesses the broader impacts of the phenomenon on economic growth and national development. Within these groups, there are varied schools of thought on the actual impact of internal migration. Scholars who study internal migration as a livelihood strategy tend to assess its impacts on the household unit, while those interested in broader national socio-economic impacts focus on the effects of diversification and urbanisation on national development. This section of the chapter is organised around these themes.

Some scholars argue that internal migration is essential for ensuring adequate labour supply/distribution in industrialised urban areas and is also essential for the development of rural areas through the receipt of remittances from migrants (Cox-Edwards and Ureta, 2003; Pritchett, 2006; Yang, 2008; Awumbila *et al.*, 2014). In his seminal work, Kuznets (1964) argued that rural–urban migration of young people is a useful process through which underutilised or surplus labour in rural areas could be shifted from traditional agricultural production towards industry.

A third school of thought suggests that migration is not always a dual-point stream from rural to urban. In fact, there are two main types of rural–urban migration: agglomeration into cities and diversification into secondary towns with rural/non-farm economic activity source (Rondinelli, 1983; Christiaensen and Todo, 2014). Christiaensen and Todo (2014) argue that the former is more effective as a poverty reduction and well-being mitigation strategy.

Similar to scholarship on the impacts of rural–urban migration on economic growth, there is considerable diversity on the efficiency of rural–urban migration as a livelihood diversification process (Assan, 2008). Awumbila *et al.* (2014) argues that internal migration, especially to urban areas, can lead to positive outcomes on social capital accumulation, capacity development of relatives left behind and income diversification. Nevertheless, Mbatha and Roodt (2014) provide a counterargument on employment by revealing that recent South African migrants from rural to urban areas experience relatively lower rates of participation in formal labour markets compared to local residents in urban communities, and that these migrants are over-represented in the informal labour market and in the unemployment sector. This means that rural–urban migrants are less likely than locals to be found in formal employment and more likely to be found in informal employment and among the unemployed.

In the wake of declining agricultural activities and a lack of alternative income generation activities, rural populations in Africa are engaging in 'widespread occupational experimentation' more actively than ever (Bryceson, 2002: 725). Hence, migrants are now often composed of young adults in their productive and reproductive prime (Liviga and Mekacha, 1998). Internal migration of youth is caused by many of the same factors that encourage adult migration. However, it is more closely tied to a lack of access to land and unemployment, thus making migration the most viable option for creating a sustainable livelihood (Assan, 2008; Awumbila *et al.*, 2014). Migration has not only become a part of sustainable livelihood strategies for many Africans, as is the case in Ghana; it is woven into the sociocultural fabric of many rural communities. The changes in the age range of the 'typical' migrant are particularly evident in Ghana. In correlation with demographic changes in Ghana, there is a shift in the forms of migration previously witnessed. Youth between ages 5 and 17 are now actively migrating to urban centres in search of better livelihood opportunities (ILO, 2006).

Methodology

Data management

To illustrate the above trends, through a case study, this discussion utilises data from the Ghana Living Standards Survey (2014) (GLSS). The GLSS is a longitudinal survey of a nation-wide sample of 18,000 households (Ghana Statistical Service, 2014a). The first round of GLSS data was collected in 1987. The GLSS collects individual, household and community-level information on a range of socio-economic indicators. Multistage stratified sampling was used to ensure that the sample is both geographically and proportionally representative. The analysis relies heavily on cross-sectional data from Round 6 of the GLSS survey. These data were collected between October 2012 and October 2013 (Ghana Statistical Service, 2014b). Where appropriate, inferences will be drawn on long-term changes and trends using previous rounds of GLSS.

This chapter defines youth as individuals between the ages of 15 and 34, as employed by Ghana Statistical Service (2012, 2013). Birth region and current region of residence are used together as a proxy for migration. Hence, individuals who do not reside in their regions of birth are considered migrants. This analysis will focus on four out of the 10 administrative regions in Ghana that recorded the highest number of internal migrants: the Greater Accra, Western, Brong Ahafo and Ashanti regions. The study presented includes descriptive analysis of pertinent indicators, bivariate and multivariate analysis of demographic, economic and social characteristics, and regional data with respect to the youth population.

Multivariate analysis

The multivariate analysis enables the study to find the correlation between two specific factors while accounting for other factors and variation. This analysis is useful when several factors are believed to influence a single event altogether, as is the case in our study. In this study, multivariate analysis is mainly used to ascertain whether in-country migration status significantly influences youths' well-being in terms of access to education, health insurance and quality employment/livelihood. Many other factors, other than migration status, influence youths' well-being. These include gender, age, education and other demographic characteristics. The analysis mainly employs logistic regression, addressing the dependent variables that are mostly binary. The population weight is applied in the regressions to strengthen the statistical power and elevate the analysis to be nationally representative.

As indicated above, the multivariate analysis pools individual youth data from four regions with the biggest number of people born outside the regions: Western, Greater Accra, Ashanti and Brong Ahafo. According to the GLSS (2014), the youth population group consists of 8.7 million people, and they account for 33 per cent of the total Ghanaian population. About 5 million youths live in the four regions mentioned above, and 28 per cent of them were not born in the regions where they live now. To simplify the written analysis, the term 'migrant' will be used to refer to youth who were born in a different region than the current living place. Internal migration and rural–urban migration may also be used interchangeably.

The analysis uses several dependent variables reflecting youths' well-being, including education attainment, access to health insurance, as well as job quality indicators such as employment formality. Informal employment is usually defined as self-employed, domestic and family workers. Such employment often lacks employee protection measures such as a written contract, paid sick and maternity leave, paid holiday, pension and subsidised medical care. Furthermore, such employment may involve job-related risks such as injury, health hazards and risk of abuse at work. The independent variable of interest is migration status, which is proxied by region of birthplace if different than the current place of residence. Finally, a set of control variables are employed, consisting of education attainment and various demographic characteristics (gender, age, marital status, rank in household, parents' education attainment and disability status). The equation for this analysis is provided in the 'Appendix' section of the chapter.

Results and discussion

Internal migration

About 18 per cent of Ghanaians currently reside outside their region of birth. In line with changes in demographic structure and migratory trends, the highest percentages of internal outmigrants fall within the adult (24 per cent) and youth (21 per cent) population groups (Ghana Statistical Service, 2014a). Assan (2008) finds that young adults are often engaged in rural–urban migration while older male household heads often engage in rural–rural migration. Lower levels of education among older migrants and difficulties adjusting to urban life account for this variance in migration destination (Assan, 2008). Youth rural–urban migration has become a defining characteristic of internal migration in Ghana.

Another feature of internal migration of youth in Ghana is that it tends to trend southwards. The northern regions of Ghana were historically left largely undeveloped in order to maintain Ghana's residents as a ready source of cheap labour for the colonial enterprise (Abdulai, 2012). Despite attempts of post-independence governments to stimulate growth in the northern regions of Ghana, including the Ghana Poverty Reduction Strategy (GPRS I and II) and the Savannah Accelerated Development Authority (SADA) projects, the developed south versus the underdeveloped north dichotomy persists, and poverty reduction is mainly evidenced only in the south (World Bank, 2011; Abdulai and Hulme, 2015).

Figure 6.1 shows the percentage of the population in each region who are born in a different region. The Greater Accra region has the biggest proportion of residents not born in the region

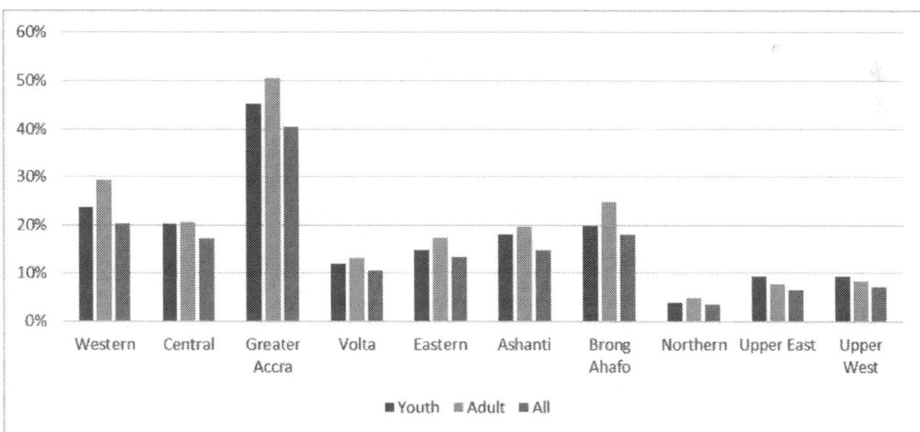

Figure 6.1 Migration status of residents of the 10 regions of Ghana

Source: GLSS Round 6 (Ghana Statistical Service, 2014b), authors' own calculation.

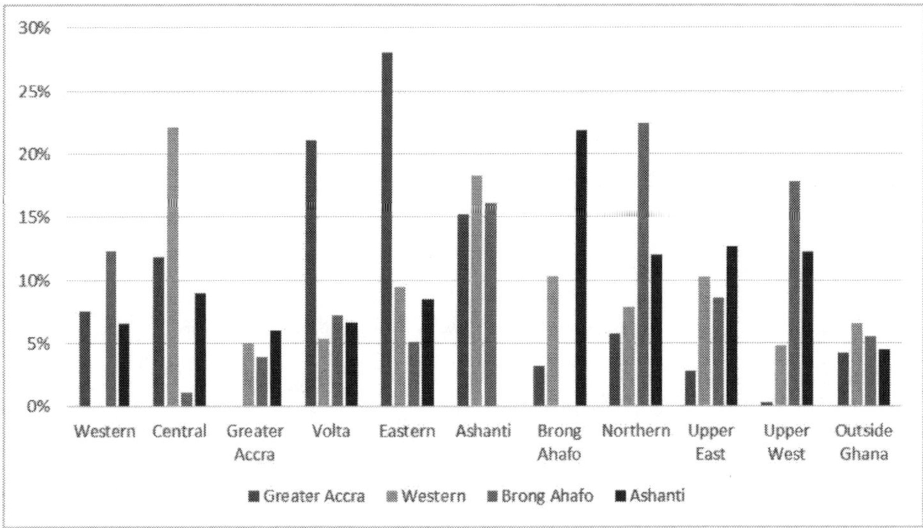

Figure 6.2 Birth origin of residents of Greater Accra, Western, Brong Ahafo, and Ashanti Regions who were born in other regions

Source: GLLS Round 6 (Ghana Statistical Service, 2014b), authors' own calculation.

(40 per cent) and the largest number of migrants in absolute terms (1.8 million). The Ashanti region has the second largest absolute number of residents not born in the region, with 778,000 migrants, while the Western and Brong Ahafo regions have the second and third largest proportions of residents not born in the region, at 20 per cent and 18 per cent, respectively (Ghana Statistical Service, 2014a).

Youth residents not born in the four regions of focus migrate internally from different parts of the country. Figure 6.2 shows the typical migration destination of individuals from each region. The pattern of migration shows that youth tend to move to the closest urbanised city from the region of birth. Hence, the largest proportion of migrants in the regions of focus originate from the regions with the closest geographical proximity, with a few exceptions. Youth residents not born in Greater Accra mostly came from the three neighbouring regions (Eastern, Volta and Central) and the Ashanti region. Similarly, youth residents not born in Western hail from the three adjacent regions (Central, Ashanti and Brong Ahafo) and the Upper East region. Youth residents not born in Brong Ahafo mostly came from the Northern, Upper West, Ashanti and Western regions. The only exception to the 'neighbouring region' norm is the Ashanti region, which has the largest proportion of non-local youth residents from the Upper East, Upper West and Northern regions, in addition to Brong Ahafo. However, as discussed above, youth migration tends to trend southwards, with youth moving into the closest urbanised cities. For the three northern regions, the closest urban metropolis in the south is Kumasi, the capital city of the Ashanti region.

Educational attainments of youth in Ghana

According to the analysis, about 20 per cent of Ghanaians who are 15 years and older are unschooled, with females being more likely to have never been in school (24.3 per cent) than men (14.6 per cent). According to GLSS Round 6 data, the average school attendance for individuals between age 6 and 25 is about 80 per cent. The Greater Accra region has the highest attendance rate of 92.0 per cent, while the three northern regions of the country have the lowest attendance rates at 50.4 per cent, 63.4 per cent and 63.6 per cent for the Northern, Upper East and Upper West regions, respectively. Generally, southern Ghana has higher attendance rates than northern Ghana due to the dichotomy of socio-economic inequality that continues to persist between the north and south of the country. Education status is a critical determinant of labour-seeking behaviour and the propensity to migration because it indicates the level of manpower available for economic activity and also shapes the employment prospects of migrant youth.

The study revealed that the typical youth migrant in Ghana tends to have fewer years of education and fewer employable skills compared to youth who live in the regions they are born in (see Table 6.1). An analysis of GLSS Round 6 data shows that youth migrants in all four regions of focus are more likely to have lower education status compared to non-migrant residents. In Western, Brong Ahafo and Ashanti, youth born in other regions are more likely to never attend school at all. They also tend to attain lower education grades compared to those born in the regions. Of the four regions of focus, Greater Accra has the lowest percentage of youth illiteracy.

Health

Ghana established a National Health Insurance Law in 2003 with the aim of making healthcare more accessible and affordable (Government of Ghana, 2003). In a study of household perceptions and enrolment decisions, Jehu-Appiah *et al.* (2012) found that despite the relatively low premiums (US$8 per year) and the absence of copays, unaffordability of premiums was cited by 72 per cent of uninsured as the reason for not enrolling. Nine per cent of households also cited being previously healthy as a reason for not enrolling. Figure 6.3 shows the differences in general health outcomes and insurance coverage for youth who are born in the Greater Accra, Western and Brong Ahafo regions compared to youth who were not born there but are currently resident.

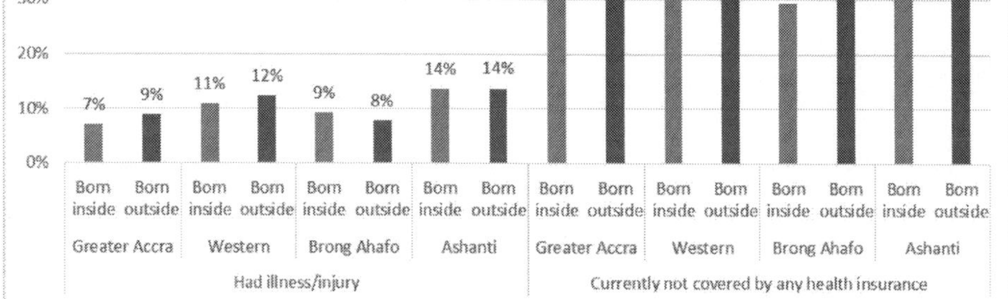

Figure 6.3 Differences in youth health and insurance status

Source: GLSS Round 6 (Ghana Statistical Service, 2014b), authors' own calculation.

Table 6.1 Education status of youth of four migrant-dense regions (absolute numbers and %)

	Western		*Greater Accra*		*Brong Ahafo*		*Ashanti*	
	Migrant youth	*Local born*	*Migrant youth*	*Local born*	*Migrant youth*	*Local born*	*Migrant youth*	*Local born*
Never attended school	28,600 15%	22,090 4%	31,473 4%	24,974 3%	47,699 29%	64,069 9%	60,608 20%	45,868 3%
Attended primary education but never finished	24,439 13%	80,381 13%	47,848 7%	52,844 6%	34,204 21%	106,821 16%	50,328 17%	154,096 11%
Finished primary school	29,097 15%	116,217 19%	84,394 12%	126,687 15%	28,447 17%	168,884 25%	41,599 14%	295,373 21%
Finished middle school	65,631 35%	253,966 41%	281,860 40%	289,996 34%	33,668 20%	228,223 34%	84,614 28%	549,894 39%
Finished secondary/high school	29,843 16%	106,938 17%	194,882 27%	251,888 30%	17,433 10%	90,385 13%	48,643 16%	289,712 21%
Finished post-secondary certification or pre-bachelor university diploma	6,555 3%	19,545 3%	30,120 4%	51,456 6%	3,041 2%	14,873 2%	11,315 4%	45,047 3%
Finished bachelor or postgraduate study	4,297 2%	15,950 3%	41,245 6%	55,043 6%	1,814 1%	4,515 1%	5,444 2%	24,583 2%
Total	188,463 100%	615,088 100%	711,823 100%	852,888 100%	166,305 100%	677,770 100%	302,551 100%	1,404,574 100%

Source: GLSS Round 6 (Ghana Statistical Service, 2014b), authors' own calculation.

In terms of health status, there is no striking difference between youth born in other or the same regions. But youth born in other regions in Greater Accra and Western are slightly more likely to experience illness or injury. In all regions, youth born in other regions were more likely to have no health insurance coverage at the time of survey.

Employment

Employment, or the lack thereof, in rural areas is cited as the main driver of youth urban migration (Assan, 2008). Youth who were not born in their current regions of residence have higher employment rates than their native-born counterparts in all four regions. Youth born outside the Greater Accra, Brong Ahafo and Ashanti regions but currently resident there have higher percentages of engagement in non-contract, non-unionised jobs.

Figure 6.4 shows employment status type and satisfaction of young adults in the four chosen regions. Unemployment rates among youth born in other regions are lower in all regions when compared to youth born in the same regions. Especially in Greater Accra and Brong Ahafo, youth coming from other regions are also slightly less likely to be engaged in informal sector work, such as self-employed, family work (agricultural and non-agricultural) and domestic work – formal employment includes paid employment, casual work and apprenticeship.

Nevertheless, migrant youth working in the formal sector are more likely to work as casual workers, arguably the least stable form of formal employment. In the four regions, only 7–11 per cent of local-born youths working in the formal sector are casual workers, while the figures are between 10 and 20 per cent for migrant youths (see Table 6.2).

Employment quality questions (written contract, trade union, paid holidays, paid sick leave, pension plan and medical care) were mostly answered only by employees who fall under the category of paid employee, casual worker and apprentice (Ghana Statistical Service, 2014a). Figure 6.5 shows the percentage of youth who are engaged in jobs without written contract and those in jobs with unionised workers. Among these employment categories, except in the Western region, youth employees born outside the regions are more likely to have a job with no written contract, especially in Great Accra and Brong Ahafo. This confirms that migration as a livelihood strategy for youth often entails them engaged in high-risk employments with no

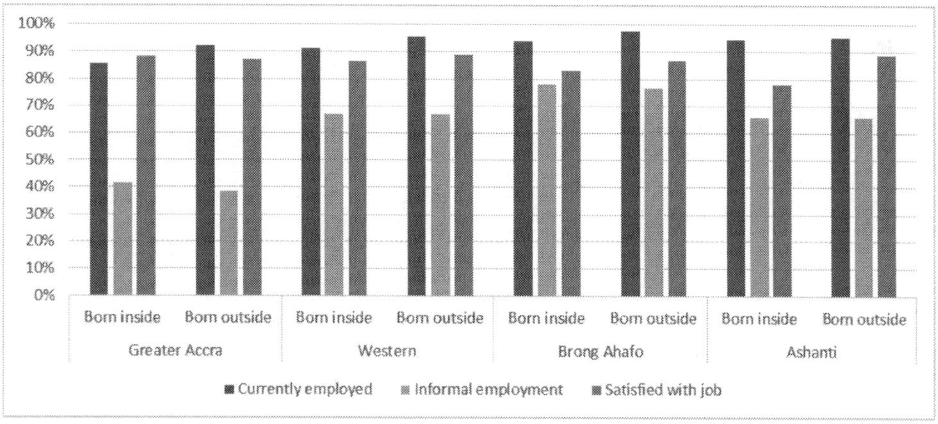

Figure 6.4 Youth employment status and job satisfaction

Source: GLSS Round 6 (Ghana Statistical Service, 2014b), authors' own calculation.

Table 6.2 Youth employment type and location of birth/residence (absolute numbers and %)

	Western		*Greater Accra*		*Brong Ahafo*		*Ashanti*	
	Born inside	*Born outside*	*Born inside*	*Born outside*	*Born inside*	*Born outside*	*Born inside*	*Born outside*
Formal employment								
Paid employee	96,809	37,313	202,310	231,862	65,896	15,461	227,424	59,999
	71%	73%	84%	82%	61%	50%	70%	81%
Casual workers	8,958	6,216	17,487	29,323	8,605	6,249	34,356	9,370
	7%	12%	7%	10%	8%	20%	11%	13%
Apprentice and others	31,336	7,374	21,072	23,127	33,194	9,258	61,866	5,099
	23%	14%	9%	8%	31%	30%	19%	7%
Total	137,103	50,902	240,868	284,312	107,695	30,968	323,646	74,468
	100%	100%	100%	100%	100%	100%	100%	100%
Informal employment								
Agricultural self-employed and family workers	160,286	58,585	21,800	16,963	295,725	80,772	319,496	71,903
	58%	57%	13%	10%	77%	79%	51%	50%
Non-agricultural self-employed and family workers	118,026	44,244	145,558	154,742	88,153	20,806	297,504	66,762
	42%	43%	85%	87%	23%	20%	48%	47%
Domestic workers	0	515	3,809	5,769	1,339	235	4,576	4,284
	0%	0%	2%	3%	0%	0%	1%	3%
Total	278,312	103,343	171,167	177,473	385,217	101,813	621,576	142,949
	100%	100%	100%	100%	100%	100%	100%	100%

Source: GLSS Round 6 (Ghana Statistical Service, 2014b), authors' own calculation.

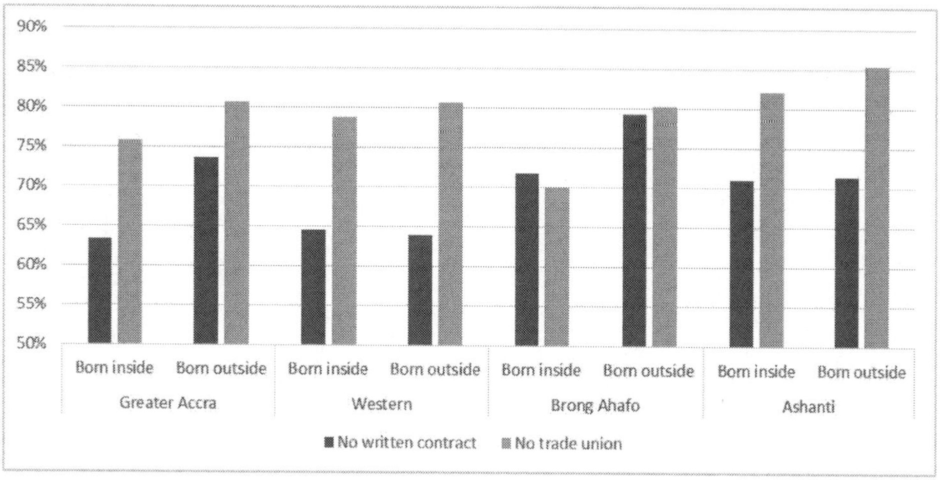

Figure 6.5 Contract and trade union membership

Source: GLSS Round 6 (Ghana Statistical Service, 2014b), authors' own calculation.

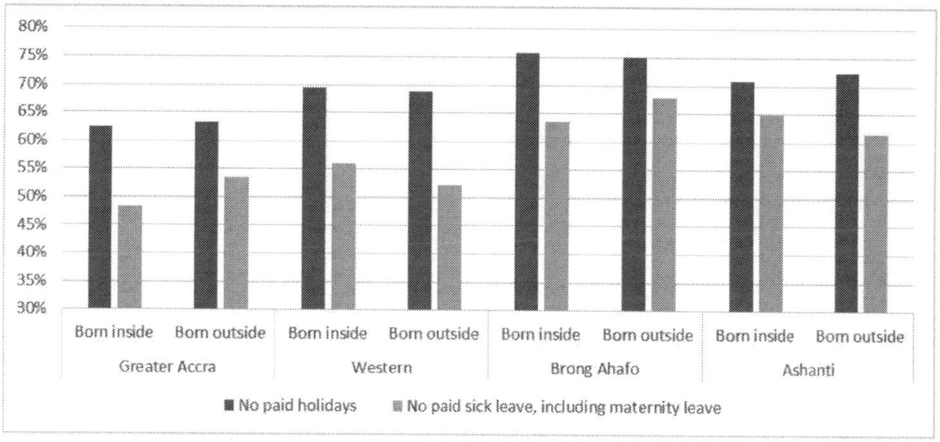

Figure 6.6 Paid holidays and sick leave

Source: GLSS Round 6 (Ghana Statistical Service, 2014b), authors' own calculation.

job and income security (Assan, 2008). In all regions, youth employees born outside the regions are more likely to be engaged in a job that does not have a trade union.

Figure 6.6 shows the percentage of youth born in and outside the regions of analysis who have unpaid holidays and sick leave, including maternity leave. The analysis shows no specific pattern on paid holidays and sick leave coverage. In Greater Accra, youth employees not born in the region are more likely to have no paid holidays and no paid sick leave. In the Western region, youth employees not born in the region are less likely to have no paid holidays and no paid sick leave when compared to youth employees born in the region. In Brong Ahafo, youth employees not born in the region are more likely to have no sick leave but slightly more likely to have paid holidays. In Ashanti, youth employees not born in the region are more likely to have no paid holidays but more likely to have paid sick leave.

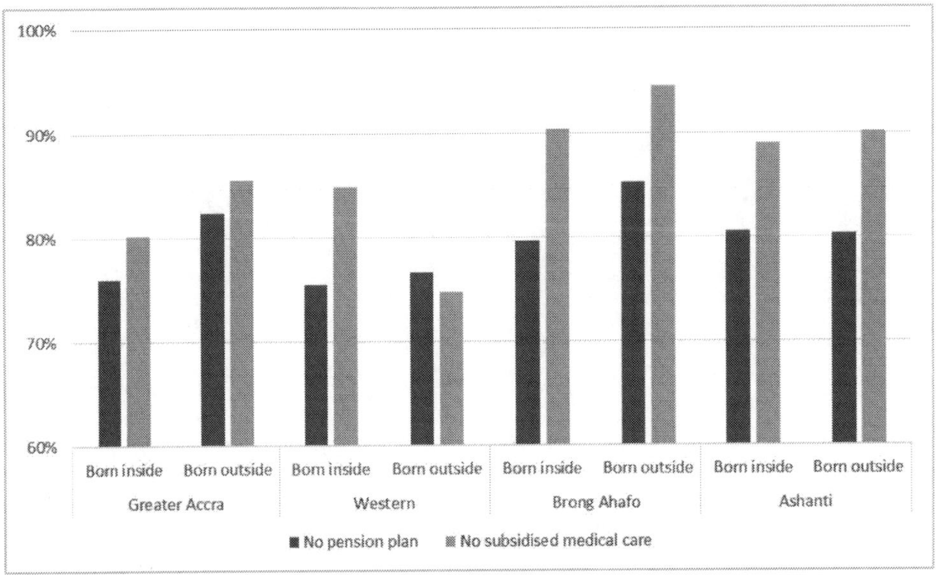

Figure 6.7 Pension plan and medical care

Source: GLSS Round 6 (Ghana Statistical Service, 2014b), authors' own calculation.

Figure 6.7 shows the percentage of youth in the Greater Accra, Western, Brong Ahafo and Ashanti regions who have jobs that provide pension plans and subsidised medical care. Young adult migrants are compared to counterparts residing in their region of birth. Except in Ashanti, youth employees not born in the regions are more likely to have a job with no pension/retirement fund. Youth employees not born in the regions are more likely to have a job with no medical care benefit, except in the Western region.

The findings of this analysis show that while migration is often the preferred strategy for youth hoping to secure a more sustainable livelihood, it often places young migrants in positions of disadvantage that can exacerbate the situations of poverty and marginalisation they aim to escape from. The final section of the chapter will discuss the implications of these findings.

Discussion of multivariate analysis

Our multivariate analysis suggests that gender and parents' educational status are strong predictors of youth education attainment. Holding everything else constant, female youths are 35 per cent less likely to finish high school and 45 per cent less likely to attain any university-level education. Having parents who finished high school increases the probability of finishing high school by up to 116 per cent, and of obtaining university-level education by up to 367 per cent. In terms of birthplace, migrants in general are less likely to have any university-level education, except those who were born in Greater Accra or Western. For instance, youth who were born in Greater Accra but currently live in other regions are 2.4 times more likely to have a university-level degree compared to the local-born. In contrast, youths who were born in the Northern, Upper East and Upper West regions have generally lower education status and are less likely to attain high school or university diplomas. For example, Upper East-born youths are 59 per cent less likely to have a high school diploma and 92 per cent less likely to have any university degree when compared to local-born youths (see Table 6.3).

Table 6.3 Determinants of education and health insurance status of youth in four migrant-dense regions

	I *Graduate high school*		*II* *Has some university-level degree*		*III* *Has health insurance*	
Female	0.65	***	0.65	***	2.03	***
Work in informal sector					1.38	***
Highest degree attained						
Elementary or middle school					1.63	***
High school					2.23	***
Some university					4.34	***
Father graduated high school or higher	2.16	***	2.16	***		
Mother graduated high school or higher	1.43	***	1.43	***		
Migration status						
Born in Western	1.49	***	1.46	***	1.00	***
Born in Central	1.24	***	0.66	***	0.36	***
Born in Greater Accra	0.75	***	2.45	***	0.74	***
Born in Volta	1.07	***	0.82	***	0.58	
Born in Eastern	1.50	***	0.95	***	0.75	***
Born in Ashanti	2.05	***	0.87	***	0.80	***
Born in Brong Ahafo	1.41	***	0.43	***	0.98	***
Born in Northern	0.28	***	0.41	***	1.58	***
Born in Upper East	0.41	***	0.08	***	0.71	***
Born in Upper West	0.37	***	0.27	***	1.12	***
Born outside Ghana	0.66	***	1.52	***	0.83	***
N	5,110		5,110		6,209	

The stars behind the numbers represent statistical significance: * $p < 0.05$, ** $p < 0.01$, *** $p < 0.001$.

The population is youth (15–34 years old) living in four within-country migrant-dense regions: Western, Greater Accra, Ashanti and Brong Ahafo. The N represents sample size, but population weight is applied in each regression.

The 'work in informal sector' category refers to self-employed, family workers and domestic workers.

The reference category for 'highest degree attained' is youth population with no education degree at all.

The reference category for 'migration status' is youth population currently residing in the same region as their birthplace (the local-born).

All equations (I–III) use logistic regression with constant (not shown). Values mentioned in the table are odd ratios. The cut-off point of odd ratio is 1. Odd ratios more than 1 represent positive correlation between dependent and independent variables, and values less than 1 represent negative correlation. For example, in II, 'born in Western' has an odd ratio of 1.46, meaning that migrant youth born in Western are 1.46 times more likely to have some university-level education compared to local-born youth. Also in II, 'born in Northern' has an odd ratio of 0.41, meaning that migrant youth born in Northern have less likelihood of only 0.41 times (fraction of) local born youth's likelihood to own a university-level education degree.

The presented coefficients for 'migration status' in all three regressions are the multiplication between two odd ratios: as a migrant in general (main effect) and as a migrant born in a certain region (interaction effect). The statistical significance displayed is based on the interaction effect's significance level.

Not all covariates are displayed in the table. All regressions also included individuals' rank in household (1 if household head or spouse), age in years, marriage status (1 if never married) and disability status (1 if disabled).

Source: GLSS Round 6 (Ghana Statistical Service, 2014b), authors' own calculation.

Health insurance

In addition, the regression analysis revealed that female youths and youths working in the informal sector are more likely to have health insurance coverage. This is probably because they are more likely to be part of a vulnerable population targeted by the government to receive a health insurance subsidy. Higher education status is also associated with higher likelihood of having health insurance, perhaps due to more well-informed decisions and better employment. Migrant youths are in general significantly less likely to have any health insurance coverage. Only migrant youths born in Northern and Upper West are more likely than the local-born to have health insurance. Migrant youths born in Central and Volta regions have the least probability of having insurance: they are 74 per cent and 42 per cent less likely, respectively, to be covered by any health insurance, when compared to local-born youths.

Employment

Female youths are less likely to be employed. However, when they do work, they are more likely to be in the informal sector and have slightly more hours per week. Highly educated youths are also less likely to have a job, but when they do have a job, they are significantly less likely to be in the informal sector and have significantly less weekly hours. University-educated youths, when compared to youths with no education at all, are 95 per cent less likely to work in the informal sector and only have to work about nine hours less per week. In general, migrant youths are more likely to have a job, less likely to be in the informal sector, and work more hours weekly, when compared to local-born youths. This might be due to the higher motivation of migrant youths living outside their home towns. Since informal employment is mostly in the form of self-employment, this finding might also indicate that starting a business is more feasible for local-born youths than for newcomers (see Table 6.4).

Job quality

In terms of job quality, youths working in the informal sector are less likely to have a monetary wage and written contract, but they tend to have greater job satisfaction. Youths with higher education are also significantly more likely to feel satisfied by their job, have a monetary wage and have a written contract. Local-born youths, when compared to migrant youths, are more likely to have a non-monetary paid job, except for youths who were born in the Upper West, Upper East and Northern regions. Female youths are also more associated with non-monetary paid employment. Non-monetary paid jobs are mostly within family business (76 per cent of non-monetary paid workers are family workers). This finding then reveals that females, rather than males, are more likely to be employed in family businesses. This also confirms that, in general, migrant youths live away from their family and are less likely to be engaged in family businesses.

However, youths born in the Upper West, Upper East and Northern regions might come to the current regions by 'employment invitation' from settled relatives who already own their businesses in these locations. This is quite consistent with the finding that working youths from Upper West and Upper East are more likely, compared to the local-born, to not have any written employment contract, similar to working youths born in the Ashanti, Central and Eastern regions.

Youths with higher education are more likely to have a better-quality job with paid holiday leave, paid sick/maternity leave, a pension plan and subsidised medical care. Youths with some university degree, for instance, have 9–65 times greater odds than their uneducated counterparts of having any of these job benefits. Most migrant youths are more likely to have paid holiday, maternal and sick leave, but they are significantly less likely to be covered by any pension plan when compared to local-born youths. Controlling for education and other factors, youths born

Table 6.4 Determinants of employment status of youth in four migrant-dense regions

	IV *Have a job*		V *Main job in informal sector*		VI *Number of work hours per week*	
Female	0.71	***	2.65	***	0.72	***
Work in informal sector					–9.93	***
Highest degree						
Elementary or middle school	0.85	***	0.72	***	–1.18	***
High school	0.40	***	0.27	***	–0.28	***
Some university	0.34	***	0.05	***	–9.06	***
Migration status						
Born in Western	2.40	***	0.89	***	1.71	***
Born in Central	0.85	***	0.70	***	2.70	***
Born in Greater Accra	1.21	***	1.25	***	–3.08	***
Born in Volta	0.87	***	0.48	***	4.77	***
Born in Eastern	0.91	***	0.48	***	1.22	***
Born in Ashanti	1.18	***	0.75	***	4.90	***
Born in Brong Ahafo	1.03	***	0.77	***	0.57	***
Born in Northern	1.39	***	0.74	***	2.62	***
Born in Upper East	2.67	***	0.60	***	–0.55	***
Born in Upper West	1.65	***	1.40	***	–1.02	***
Born outside Ghana	0.77	*	1.06	***	0.45	***
N	6,688		6,214		5,835	

The stars behind the numbers represent statistical significance: * $p < 0.05$, ** $p < 0.01$, *** $p < 0.001$.

The population for regression IV is youth (15–34 years old) living in four within-country migrant-dense regions: Western, Greater Accra, Ashanti and Brong Ahafo. For regression V and VI, the population is only the working youth in the regions. The N represent sample size, but population weight is applied in each regression.

The 'work in informal sector' category refers to self-employed, family workers and domestic workers.

The reference category for 'highest degree attained' is youth population with no education degree at all.

The reference category for 'migration status' is youth population currently residing in the same region as their birthplace (the local-born).

Equations IV and V use logistic regression with constant (not shown). Values mentioned in the table are odd ratios. The cut-off point of odd ratio is 1. Odd ratios more than 1 represent a positive correlation between dependent and independent variables, and values less than 1 represent a negative correlation. See the explanation in Table 6.3 for interpreting odd ratios.

Equation VI use linear regression with number of work hours per week as the dependent variable. The coefficients represent changes in weekly work hours due to changes in independent variables.

The presented coefficients for 'migration status' in regression IV are the multiplication between two odd ratios: as a migrant in general (main effect) and as a migrant born in a certain region (interaction effect). The presented coefficients for 'migration status' in regression VI are the addition between two coefficients: as a migrant in general (main effect) and as a migrant born in a certain region (interaction effect). The statistical significance displayed is based on the interaction effect's significance level.

Not all covariates are displayed in the table. All regressions also included individuals' rank in household (1 if household head or spouse), age in years, marriage status (1 if never married) and disability status (1 if disabled).

Source: GLSS Round 6 (Ghana Statistical Service, 2014b), authors' own calculation.

in Greater Accra are the ones most likely to get any employment benefit compared to other migrants. In the contrary, migrant youths born in Brong Ahafo are generally less likely to receive any of the employment benefits. Migrant youths born in Central are less likely to have paid holiday leave, paid sick/maternity leave and a pension plan, and youths originally from Upper West are significantly less likely to have either a pension plan or subsidised medical care. According to the Global Migration Group (2014: 8), youth in West Africa are more likely than adults to be in the informal, low-pay, low-productivity sector and work longer hours, under insecure contracts and for lower incomes (see Table 6.5).

Table 6.5 Determinants of employment amenities of youth in four migrant-dense regions

	VII Paid holiday leave		*VIII Paid sick and maternity leave*		*IX Pension plan*		*X Medical care subsidy*	
Female	1.00		1.06	***	1.26	***	1.24	***
Work in informal sector	0.66	***	2.22	***	0.14	***		
Highest degree								
Elementary or middle school	0.92	***	1.09	***	1.08	***	0.98	
High school	4.40	***	3.08	***	6.29	***	2.79	***
Some university	26.01	***	19.85	***	64.74	***	9.43	***
Migration status								
Born in Western	0.71	***	0.98		0.73	***	0.57	***
Born in Central	0.53	***	0.75	***	0.90	***	1.11	**
Born in Greater Accra	2.53	***	3.68	***	1.67	***	2.20	***
Born in Volta	3.26	***	2.12	***	0.61	***	1.24	
Born in Eastern	0.70	***	0.97		0.43	***	0.77	***
Born in Ashanti	1.15	***	0.90	**	0.67	***	1.51	***
Born in Brong Ahafo	0.51	***	0.70	***	0.36	***	0.46	***
Born in Northern	1.90	***	1.88	***	0.90	***	0.63	***
Born in Upper East	1.45	***	1.22	***	0.38	***	1.00	***
Born in Upper West	0.98		2.26	***	0.37	***	0.42	***
Born outside Ghana	1.28	***	0.67	***	0.88	***	1.04	***
N	2,236		2,236		2,236		2,217	

The stars behind the numbers represent statistical significance: * $p < 0.05$, ** $p < 0.01$, *** $p < 0.001$.

The population for all regressions is working youth (15–34 years old) living in four within-country migrant-dense regions: Western, Greater Accra, Ashanti and Brong Ahafo. However, for all regressions, responses came mostly from youth working in the formal sector (paid employee, apprentice and casual workers). The N represent sample size, but population weight is applied in each regression.

The 'work in informal sector' category refers to self-employed, family workers and domestic workers.

The reference category for 'highest degree attained' is youth population with no education degree at all.

The reference category for 'migration status' is youth population currently residing in the same region as their birthplace (the local-born).

All equations use logistic regression with constant (not shown). Values mentioned in the table are odd ratios. The cut-off point of odd ratio is 1. Odd ratios more than 1 represent positive correlation between dependent and independent variables, and values less than 1 represent negative correlation. See Table 6.3 for an explanation of interpreting odd ratios.

The presented coefficients for 'migration status' in all regressions are the multiplication between two odd ratios: as a migrant in general (main effect) and as a migrant born in a certain region (interaction effect). The statistical significance displayed is based on the interaction effect's significance level. For regression VIII, the main effect is not statistically significant, but some of the interaction effects are significant.

Not all covariates are displayed in the table. All regressions also included individuals' rank in household (1 if household head or spouse), age in years, marriage status (1 if never married) and disability status (1 if disabled).

Source: GLSS Round 6 (Ghana Statistical Service, 2014b), authors' own calculation.

Work-related risks

In terms of work-related risks, female and highly educated working youths are less likely to be exposed to risks of work-related abuse, health hazards or injury. However, youths working in the informal sector have more probability of having work-related injuries at 40 per cent and of facing work-related health hazards at 78 per cent, when compared to youths working in the formal sector. Migrant youths also face the same risks, with most of them being more likely to have health hazards or injuries at the workplace when compared to local-born colleagues. As an exception, youths born in the Ashanti or Northern regions are, in general, less likely to face health hazards or risk of injury at work, compared to local-born counterparts. This might be due to their comparatively better levels of education, which often allows them to obtain less manual/hazardous forms of employment.

Nevertheless, youths from three regions, Western, Upper West and Upper East, are those with the biggest risks of having an injury or health hazard at work. In comparison to the local-born, holding everything else constant, youths born in these regions have 2.3 times bigger odds of experiencing injury and 2.9 times larger likelihood of dealing with health hazards at work.

Conclusion and policy recommendations

Based on an in-depth analysis of GLSS 6 data, this study has assessed youth rural–urban migration as part of sustainable livelihood strategies for Ghanaian households. This analysis entailed an assessment of current migratory trends, as well as the educational, health and employment status of young migrants. The policy implications of these findings are discussed below.

The search for alternate income generation activity and a desire to create sustainable livelihoods are the principal motivation factors for internal migration of youth from rural areas to urban centres (Bryceson, 2002). This study shows that employment rates among migrant youth in urban destinations remains comparatively higher than those of non-migrants. Therefore, outmigration provides an avenue for rural unemployed youth to gain access to wider employment opportunities. Levels of employment satisfaction reported by youth in the GLSS sample also indicates youth migrants have higher levels of contentment with the jobs they find in their urban destinations.

On the other hand, the study also finds that young migrants are often engaged in less stable, non-contract, non-unionised jobs, making them vulnerable to employer abuses and exploitation. This finding is corroborated by case studies of migrants from Akuapem North and Dangme West and studies of female porters in Accra (Awumbila and Ardayfio-Schandorf, 2008; Assan, 2014). Labour laws are one policy response that can limit the arbitrariness of informal employment. Migrants appear to operate outside the scope of any labour regulations, making them easy targets for unscrupulous employers; hence, the promulgation of laws specifically aimed at protecting the rights of migrant workers will provide a safety net for them and also reduce vulnerability. Adequate enforcement of such laws will also be essential to ensure their effectiveness as a deterrent for labour abuses and exploitation.

Another important factor to note in assessing the effectiveness of youth migration as a sustainable livelihood strategy is educational attainment. This study finds that a large majority of young migrants are less educated and without job training or skills. Consequently, migrants are limited to less secure lower-wage jobs. Educational levels are directly associated with migrant abilities to find better employment opportunities. Also, Assan (2014) finds that migrants who can communicate in English have improved access to formal higher-wage jobs compared to those who are illiterate. Provision of job training skills for migrant youth is an essential policy solution to improve their employability skills. Adult literacy classes should also be made available to youth who are interested in pursuing formal education.

Health status of young migrants is another important factor to consider in assessing the effectiveness of migration as an income diversification strategy. While the difference in health status between young migrants and non-migrants is comparatively less, migrants in the Greater Accra and Western regions are more prone to illness and injury. This difference is attributable to the higher-risk employments that migrants are engaged in. Compared to migrants living in their region of birth, higher percentages of youth migrants do not have health insurance. In this regard, the scope and coverage of the Ghana Health Insurance Scheme needs to be reconsidered. Currently, children under 18, adults over 70 and pregnant women are exempted from making premium payments (Jehu-Appiah *et al.*, 2012). However, premium costs remain unaffordable for low-income Ghanaians (Asante and Aikins, 2008). This condition is likely to hold for young migrants as well due to the precarious financial situations they are often faced with. As a policy solution, subsidies can be made available to young migrants below a set income threshold to stimulate their enrolment in the Health Insurance Scheme and reduce their vulnerability to health shocks.

In conclusion, prevailing evidence suggests that internal migration and urbanisation can potentially lead to economic growth and overall national development (Cox-Edwards and Ureta, 2003; Pritchett, 2006; Yang, 2008). The importance of Ghana's youth population in this process cannot be overemphasised. With youth aged between 15 and 34 constituting close to 35 per cent of the population, their engagement in income generation activity can propel the country to new levels of socio-economic development and growth. As long as rural–urban migration remains ubiquitous to Ghanaian youth as a means of improving their livelihoods and well-being, development and social policy planning in the country must be expanded in tandem to address the potential challenges associated with this move. Innovative social protection programmes such as conditional cash transfer programmes, subsidies for low-wage workers, subsidised housing and supplemental nutrition programmes should be considered as a means of enabling young migrants to cater for themselves and also improve the well-being of their households.

Further reading

Agence Française de Développement and the World Bank (2014) *Youth Employment in Sub-Saharan Africa*, Washington, DC: International Bank for Reconstruction and Development, World Bank, available at: https://openknowledge.worldbank.org/bitstream/handle/10986/16608/9781464801075.pdf (accessed 6 June 2017). [This volume explores various policy approaches to expand opportunities for youth employment in sub-Saharan Africa.]

Boutin, D. (2016) *The Role of Internal Migration in Access to First Job: A Case Study of Uganda*, ILO Work4Youth Technical Brief No. 5, available at: www.ilo.org/wcmsp5/groups/public/—-ed_emp/documents/publication/wcms_531463.pdf (accessed 6 June 2017). [This report provides a study on school-to-work transition among youth in Uganda, with internal migration experience as the main indicator of interest.]

Hino, H. and Ranis, G. (eds) (2013) *Youth and Employment in Sub-Saharan Africa: Working but Poor*, London: Routledge. [This volume examines the nature of youth unemployment and underemployment in sub-Saharan Africa, with case studies on Ethiopia, Ghana, Kenya and South Africa.]

Honwana, A. (2012) *The Time of Youth: Work, Social Change, and Politics in Africa*, Boulder, CO: Kumarian Press. [This book provides a critical reading that explores African youths' everyday activities and coping strategies in the face of inadequate education, massive unemployment, poverty and HIV/AIDS.]

Mbatha, C. and Roodt, J. (2014) 'Recent internal migration and labour market outcomes: exploring the 2008 and 2010 National Income Dynamics Study (NIDS) panel data in South Africa, *South African Journal of Economic and Management Sciences*, 17(5): 653, available at: https://sajems.org/index.php/sajems/article/view/515 (accessed 6 June 2017). [This article offers insights into urban–rural migrants' labour market outcomes in South Africa.]

References

Abdulai, A.G. (2012) 'State elites and the politics of regional inequality in Ghana', unpublished PhD thesis, University of Manchester, UK.

Abdulai, A.G. and Hulme, D. (2015) 'The politics of regional inequality in Ghana: state elites, donors and PRSPs', *Development Policy Review*, 33(5): 529–553.

Asante, F. and Aikins, M. (2008) *Does the NHIS Cover the Poor?* Accra: DANIDA Health Sector Support Office.

Assan, J.K. (2008) 'Generational differences in internal migration: derelict economies, exploitative employment and livelihood discontent', *International Development Planning Review*, 30(4): 377–398.

Assan, J.K. (2014) 'Experiences of young migrants working in urban food markets in southern Ghana: implications for well-being and social policy', *Journal of Social Science for Policy Implications*, 2(3): 1–25.

Assan, J.K., Caminade, C. and Obeng, F. (2009) 'Environmental variability and vulnerable livelihoods: minimising risks and optimising opportunities for poverty alleviation', *Journal of International Development*, 21(3): 403–418.

Awumbila, M. and Ardayfio-Schandorf, E. (2008) 'Gendered poverty, migration and livelihood strategies of female porters in Accra, Ghana', *Norsk Geografisk Tidsskrift / Norwegian Journal of Geography*, 62(3): 171–179.

Awumbila, M., Owusu, G. and Teye, J. (2014) *Can Rural-Urban Migration into Slums Reduce Poverty? Evidence from Ghana,* Migrating out of Poverty Research Programme Consortium, Working Paper 13, available at: http://migratingoutofpoverty.dfid.gov.uk/files/file.php?name=wp-13—-awumbila-owusu-teye-2014-can-rural-urban-migration-into-slums-reduce-poverty-final.pdf&site=354 (accessed 6 June 2017).

Boutin, D. (2016) *The Role of Internal Migration in Access to First Job: A Case Study of Uganda*, ILO Work4Youth Technical Brief No. 5, available at: www.ilo.org/wcmsp5/groups/public/—-ed_emp/documents/publication/wcms_531463.pdf (accessed 6 June 2017).

Bryceson, D.F. (2002) 'The scramble in Africa: reorienting rural livelihoods', *World Development*, 30(5): 725–739.

Castles, S., de Haas, H. and Miller, M.J. (2013) *The Age of Migration: International Population Movements in the Modern World*, London: Palgrave Macmillan.

Christiaensen, L. and Todo, Y. (2014) 'Poverty reduction during the rural–urban transformation: the role of the missing middle', *World Development*, 63: 43–58.

Cox-Edwards, A. and Ureta, M. (2003) 'International migration, remittances, and schooling: evidence from El Salvador', *Journal of Development Economics*, 72(2): 429–461.

Ellis, F. (1998) 'Household strategies and rural livelihood diversification', *The Journal of Development Studies*, 35(1): 1–38.

Ghana Statistical Service (2012) *The 2010 Population and Housing Census: Summary Report of Final Results*, Accra: Ghana Statistical Service.

Ghana Statistical Service (2013) *The 2010 Population and Housing Census: National Analytical Report, GSS*, Accra: Ghana Statistical Service.

Ghana Statistical Service (2014a) *Ghana Living Standards Survey, Report on the Sixth Round*, Accra: Ghana Statistical Service.

Ghana Statistical Service (2014b) *Ghana Living Standards Survey Round 6*, data file and code book, Accra: Ghana Statistical Service.

Global Migration Group (2014) *Migration and Youth: Challenges and Opportunities*, United Nations Children's Fund, available at: www.globalmigrationgroup.org/migrationandyouth (accessed 6 June 2017).

Gould, W.T.S.G. (2015) *Population and Development*, 2nd edition, London: Routledge.

Government of Ghana (2003) *The National Health Insurance Act: Act 650*, Accra: Government of Ghana.

Harris, J.R. and Todaro, M.P. (1970) 'Migration, unemployment and development: a two-sector analysis', *The American Economic Review*, 60(1): 126–142.

Honwana, A. (2012) *The Time of Youth: Work, Social Change, and Politics in Africa*, Boulder, CO: Kumarian Press.

ILO (International Labor Organization) (2006) *Ghana: Multiple Indicator Cluster Survey, 2006*, Geneva: ILO.

Jehu-Appiah, C., Aryeetey, G., Agyepong, I., Spaan, E. and Baltussen, R. (2012) 'Household perceptions and their implications for enrolment in the National Health Insurance Scheme in Ghana', *Health Policy and Planning*, 27(3): 222–233.

Kuznets, S. (1964) 'Economic growth and the contribution of agriculture', *International Journal of Agrarian Affairs*, 3(2): 56–75.

Liviga, A.J. and Mekacha, R.D. (1998) *Youth Migration and Poverty Alleviation: A Case Study of Petty Traders (Wamachinga) in Dar es Salaam*, Research Report No. 98.5, Dar es Salaam: Research Programme on Poverty Alleviation.

Mbatha, C. and Roodt, J. (2014) 'Recent internal migration and labour market outcomes: exploring the 2008 and 2010 National Income Dynamics Study (NIDS) panel data in South Africa', *South African Journal of Economic and Management Sciences*, 17(5): 653, available at: https://sajems.org/index.php/sajems/article/view/515 (accessed 6 June 2017).

Pritchett, L. (2006) *Let Their People Come: Breaking the Gridlock on International Labor Mobility*, Washington, DC: Center for Global Development: Brookings Institution.

Rondinelli, D.A. (1983) *Secondary Cities in Developing Countries: Policies for Diffusing Urbanization*, London: Sage.

UN DESA (United Nations Department of Economic and Social Affairs) (2015) *World Population Prospects: The 2015 Revision, Key Findings and Advance Tables*, Working Paper ESA/P/WP.241.

UNFPA (United Nations Population Fund) (2013) *Population Dynamics in the Post-2015 Development Agenda: Report of the Global Thematic Consultation on Population Dynamics*, available at: www.worldwewant2015.org/file/313464/download/340868 (accessed 30 May 2017).

Wilson Center (2017) *Urban Perspectives: Climate Change, Migration, Planning and Financing – A New Generation of Ideas 2017*, Washington, DC: Woodrow Wilson International Center for Scholars.

World Bank (2011) *Tackling Poverty in Northern Ghana*, Report No. 53991-GH, Washington, DC: World Bank Africa Region.

Yang, D. (2008) 'International migration, remittances and household investment: evidence from Philippine migrants' exchange rate shocks', *The Economic Journal*, 118(528): 591–630.

Appendix

Equations for multivariate data analysis

This study employs two equations for the regressions. The first equation, a logit equation model, predicts the determinants of youths' ability to finish high school or obtain a university-level education degree:

$$logit\left(p\left(Ed\right)\right) = \log\left(\frac{p\left(Ed\right)}{1-p\left(Ed\right)}\right) = \beta_0 + \beta_1 F_i + \beta_2{}'PHS_i + \beta_3 M_i + \beta_4 M \star B_i + \beta_5{}'A_i \qquad (1)$$

$p(Ed)$: Probability of youth to finish high school or obtain some university-level education

F: Gender identifier (female/male)

'PHS: A vector describing if parents (mother and father) finished high school

M: Migration status identified by birthplace

*M***B*: Interaction term between migration status and specific birthplace region

'A: A vector of covariates (age, rank in household, marriage status, disability status)

The second equation, also a logit model, predicts the factors influencing youths' working condition, including the employment status, employment formality, working hours, availability of facilities and amenities as employees, as well as various employment-related risks. A linear equation using a similar form as Equation 2 is also used to predict the factor influencing youths' weekly working hours:

$$logit\left(p('Z)\right) = \log\left(\frac{p('Z)}{1-p('Z)}\right) = \beta_0 + \beta_1 F_i + \beta_2 'EA_i + \beta_3 M_i + \beta_4 M \star B_i + \beta_5 'C_i \quad (2)$$

$p('Z)$: Probability of youth to have/experience various employment-related conditions (be employed, work in informal sector, have various job amenities, face job-related risks)

$'EA$: A vector describing individuals' highest education degree attained (elementary/middle school, high school, some university)

$'C$: A vector of covariates (age, rank in household, marriage status, disability status, employment informality status)

Both equations include the individuals' migration status in the form of main effect (the effect of being someone not born in the current living area) as well as interaction effect between migration status and birthplace origin (the effect of being someone born in a specific region). Additionally, several covariates are included, including gender, education status (of parents and individuals), as well as other demographic characteristics.

7

Gender and development in Africa

A 50-year journey from women in development to gender justice

Hazel Barrett

Introduction

Until the 1970s, development policy, practice and evaluation in Africa had been gender-blind, with women's roles and contribution to development in particular being invisible (Benería and Sen, 1981). The dominant development paradigm of the time was one of modernisation. Based on Rostow's model of economic 'take-off', it focused on modernising the economy to achieve increases in economic growth, involving a movement from family production to specialised production of goods and services (Rostow, 1960; Boserup, 1990). The assumption was that the benefits of economic growth would trickle down and benefit all citizens (Rathgeber, 1990; Zwart, 1992; Mohindra and Nikiema, 2010). Thus, at best, modernisation would benefit all, and at worst it would be benign. However, in 1970, this myth was exploded by the publication of Boserup's influential book *Women's Role in Economic Development* (1970). This book was the first comprehensive overview of women's roles in the development process, and in particular investigated how development affected women's subordinate position in society (Benería and Sen, 1981). Boserup demonstrated that the benefits of modernisation and economic growth in Africa seemed to accrue more to men than women, and that women were often marginalised from or even forbidden access to productive innovations and commercial enterprises (Boserup and Liljencrantz, 1975; Boserup,1989; Moghadam, 1998; Mohindra and Nikiema, 2010). Boserup's book is acknowledged as the catalyst and foundation for the inclusion of feminist and gender perspectives in development thinking, policy and programming in Africa (Kilby, 2015). This chapter will trace the approaches to gender and development in Africa over the last 50 years, and explores the journey from a women in development (WID) focus in the 1970s through to the current gender justice approach of the Sustainable Development Goals.

Women in development (WID) and women and development (WAD): integrating women into development (1970–1985)

Prior to the 1970s, addressing the challenges that faced women in Africa was limited to a human rights agenda with a focus on basic needs (food, clothing, shelter, etc.) and maternal and child welfare (Podems, 2010), thus emphasising women's roles as mothers and homemakers. This approach did not challenge the social structures that placed women in a subordinate position, and thus has remained popular with donors and governments throughout the last 50 years.

However, Boserup's groundbreaking work challenged Western donors to integrate women into the development process as equal partners with men. In the USA, her powerful message led to the establishment in 1974 of a women in development (WID) office within USAID (Kilby, 2015). In 1975, the DAC convened an expert group on women in development. The UN designated 1975–1985 as International Women's Decade following the success of the first UN Women's Conference held in Mexico City in 1975. The conference led to a World Plan of Action around three broad objectives, one of which was the integration and full participation of women in development (Kilby, 2015). While the plan was criticised for stereotyping women as passive victims of development, it was the first internationally agreed set of guidelines for governments and donors to follow with a set of minimum targets to be met by 1980.

WID programmes were underpinned by the modernisation paradigm and inspired by liberal feminism (Mohindra and Nikiema, 2010; Davids *et al.*, 2013), and focused particularly on poverty and assumed that women were not efficient in what they were doing (Ostergaard, 1992; Moser, 1993). Interventions were initiated that aimed to increase women's agricultural production by providing them with access to resources and technology, as well as encouraging them to enter the cash economy by providing new opportunities. However, these initiatives increased the workload of women, adding to their already time- and energy-demanding triple workloads. They also led to gender discord and conflict within families and communities, as men perceived such interventions as challenging traditional male authority. Thus, by focusing on women, WID alienated men and failed to address the structural issues women faced. In fact, in some circumstances, these became worse, as men forbade their female relatives from participating in such interventions, or even trashed women's cash cropping plots. WID was adopted because it was a non-threatening approach that did not challenge existing social structures (Mohindra and Nikiema, 2010); at the same time, this reticence to tackle the underlying structures that disadvantaged women was seen as a major weakness of the approach.

From the late 1970s, in an attempt to address these criticisms, WID was gradually replaced by WAD. WAD, often characterised as underpinned by socialist-Marxist feminism, was an approach that, in addition to focusing on the economic position of women, acknowledged the political and social structures that were barriers to women's economic betterment, and called for them to be tackled (Bhavnani and Foran, 2008; Mohindra and Nikiema, 2010; Podems, 2010; Davids *et al.*, 2013). Whereas WID programmes had a microeconomic focus, WAD programmes assumed that if the macro-context improved, women would benefit (Podems, 2010).

WID and WAD conceptualise women's disadvantage coming from their exclusion from resources of production, gender-biased laws and a lack of recognition for the work they do (Moorosi, 2009). WID and WAD programmes tended to focus on women's productive roles rather than their reproductive roles, and thus attempted to incorporate women into existing economic initiatives or developed new programmes created exclusively for women (Zwart, 1992). WID/WAD projects were typically income-generating activities where women were taught a specific skill or craft and were often organised into marketing cooperatives (Rathgeber, 1990; Mohindra and Nikiema, 2010). Such initiatives have been criticised for placing an extra work burden on women while failing to challenge male-dominated power structures. WID/WAD was a non-confrontational approach that did not question the sources and nature of women's subordination, but advocated for more equal participation of women in modernisation (Rathgeber, 1990; Mohindra and Nikiema, 2010).

Gender and development (GAD): empowering women as agents for change (1985–2015)

The GAD perspective emerged in the 1980s as a response to the criticisms of WID and WAD, and was formally acknowledged at the Beijing Conference of 1995 (Kilby, 2015). GAD combines elements of WID and WAD, but is based on challenging existing systems of gender relations, which, together with the neo-liberal economic development paradigm that had replaced modernisation as a result of the imposition of structural adjustment programmes in most countries of Africa through the 1980s, limited the options available to poor men and women. In this approach, the relationship between gender relations and the development process is interactive (Moghadam, 1998), with the empowerment of women and marginalised men being an important element (Davids *et al.*, 2013). GAD uses gender relations as a unit of analysis, with men viewed as potential supporters of women, and promotes a gender-sensitive approach to development (Moorosi, 2009). Thus, as Rathgeber (1990: 494) neatly summarises, 'GAD is not concerned with women per se but with the social construction of gender and the assignment of specific roles, responsibilities, and expectations of women and men'.

In response to these calls, academics such as Moser (1989, 1993) and Molyneux and Steinberg (1995) have suggested that GAD should focus on two aspects of gender that are relevant to women: practical interests/needs and strategic interests/needs. Practical gender interests/needs are those defined by women as part of their given gender roles, and include both productive and reproductive roles. Strategic gender interests/needs arise from women's subordination, and include tackling the issue of women's subordination and inequality. GAD recognised that women experience subordination differently according to race, ethnicity, age, class and culture, and thus the GAD approach does not treat women as a homogenous category, as WID and WAD did. Thus, the GAD approach focuses on the intersectionality of gender, class and race, and the social construction of their defining characteristics (Podems, 2010). As Davids *et al.* (2013: 3) explain, 'The GAD approach moves away from women as a universal category and objects of change and introduces the concept of gender as complex power relations shaping peoples' lives, including development'.

GAD thus sees women as agents of change rather than as passive recipients of development, and encourages them to form political alliances and groups to question the underlying social, economic and political structures that lead to gender inequality (Rathgeber, 1990; Kilby, 2015).

During this period, the HIV/AIDS epidemic was a hindrance and challenge to development efforts in Africa, in particular highlighting the importance of gender relations in designing and implementing health policies. Proponents of GAD advocated that gender analysis should extend beyond productive roles to encompass reproductive roles, with a move to a human rights focus to give women rights over their body, health and reproductive health choices (Mohindra and Nikiema, 2010). Women's health had become a matter of social justice, not just an instrument in economic development. The International Conference on Population and Development in Cairo in 1994 confirmed that access to reproductive healthcare was a prerequisite for gender equity in health (Doyal, 2000). The Beijing Conference in 1995 stated 'that women's health is more than maternal health' and should encompass reproductive and sexual health (Mohindra and Nikiema, 2010: 558). The Beijing Plan called for action to address the underlying causes of women's poor health, including violence against women, work burdens and lack of decision-making power.

The HIV/AIDS epidemic in sub-Saharan Africa, with its associated high levels of morbidity and mortality that resulted in decreasing life expectancies of both men and women, brought to the fore the issue of women's disadvantage concerning property rights, access to land and inheritance patterns. The situation highlighted two fundamental problems regarding women's land rights in many parts of Africa. First, the absence of autonomous property rights for women, and second, where women's rights to inheritance were enshrined in law, the lack of enforcement (Joireman, 2008).

GAD is currently the discourse used by the World Bank and International Monetary Fund (Bhavnani and Foran, 2008: 321), which aims to 'not only integrate women into development, but [to] look for the potential in development initiatives to transform unequal social/gender relations and to empower women'.

It is difficult to find development projects that have been designed specifically from a GAD perspective (Rathgeber, 1990), as they would inevitably question traditional views of gender roles and responsibilities. As a result, many development practitioners have been wary of GAD, and where GAD has been adopted in practice, projects have often been depoliticised (McIlwaine and Datta, 2003). Many projects that address women's practical interests/needs are often heavily reliant on WID and WAD concepts. However, programmes that address women's strategic interests/needs tend to have a human rights focus, such as improving women's access to land and physical resources within customary and statutory law, reviews of legal frameworks that favour male ownership and inheritance, and enacting legislation concerning gender equality and equal opportunities.

One of the main achievements of GAD has been the introduction of a pro-gender development agenda articulated through the concept of 'gender mainstreaming', where issues of gender, and in particular women's marginalisation, have become central components of African government policy and development practice. It is perhaps no coincidence that this happened during the 1990s when the human development paradigm was beginning to challenge neo-liberal economics as the dominant philosophy of development following serious criticisms of the human cost on the poor and marginalised people of Africa after the imposition by the Breton Woods Institutions of global neo-liberal economics on the poor and marginalised people of Africa during the 1980s and 1990s (Cornia *et al.*, 1987).

The shift in development policy towards a human development framework was formally accepted by all 189 member countries of the UN in 2000 when they gathered at the UN to agree to fight poverty in its many human dimensions. That vision was translated into eight UN Millennium Development Goals (MGDs), with progress being assessed against 18 quantitative targets. These goals and targets were the overarching global development agenda from 2000 to 2015 (UN, 2015). While all eight goals have gender relevance, Goal 3, promote gender equality and empower women, is especially linked to the GAD approach. But it was also the one MDG with the weakest targets, namely to 'promote' gender equality and the empowerment of women. This was translated into parity in primary and secondary education and increasing the number of women sitting in national assemblies. As Mohindra and Nikiema (2010: 560) conclude:

> empowerment has disappeared, reinforcing MDG3's emphasis on education. There are no specific strategies relating to improving women's socioeconomic position or developing progressive gender-related public policies. While education is critical to improving women's status, focussing solely on education diverts attention from other dimensions of women's status, including control over resources and decision making powers . . .

According to the Economic Commission for Africa (UNECA, 2015), Africa has made 'considerable progress' towards achieving the MDGs, particularly with respect to moving forward to achieving universal primary education (Goal 2), combating HIV/AIDS, malaria and other diseases (Goal 6) and promoting gender equality and empowering women (Goal 3). But as Table 7.1 shows, Africa's performance in eradicating extreme poverty and hunger (Goal 1), reducing child

Table 7.1 Africa's MDG performance

MDG goals	*Status*	*Remarks*
Goal 1: Eradicate extreme poverty and hunger	*Off track*	• Poverty in Africa (excluding North Africa) declined from 56.5 per cent in 1990 to 48.4 per cent in 2010, and in North Africa from 5 per cent to 1 per cent. • Poverty is perpetuated by rising inequalities, unemployment, the youth bulge, unplanned urbanisation, lack of diversification, etc. • Hunger declined by 8 per cent in Africa (excluding North Africa) between 1990 and 2013. • Africa (excluding North Africa) is the most food-deficient of all global regions, with 25 per cent of its population facing hunger and malnutrition in 2011–2013.
Goal 2: Achieve universal primary education	*On track*	• In 2012, over 68 per cent of African countries had a net enrolment rate of at least 75 per cent in primary education. • Average primary completion rate stands at 67 per cent. • Youth literacy rate reached 69.61 per cent in 2012.
Goal 3: Promote gender equality and empower women	*On track*	• Gender parity in primary education increased from 0.82 to 0.96 in North Africa and 0.83 to 0.92 in the rest of Africa between 1990 and 2012. • Gender barriers evident in the low transition rates between education levels and into employment, and at higher education levels owing to pregnancy and other barriers. • Africa has made the most progress in increasing the number of seats held by women in national parliaments, with an average increase of 15 per cent between 2000 and 2014.
Goal 4: Reduce child mortality	*Off track*	• Under-five mortality rate fell by 55 per cent between 1990 and 2012, while infant mortality rates fell by 40 per cent. • Only Egypt, Liberia, Malawi and Tunisia have achieved both targets on reducing child mortality.
Goal 5: Improve maternal health	*Off track*	• By 2013, Africa had 289 maternal deaths per 100,000 live births, compared to the world average of 210 deaths per 100,000 live births.
Goal 6: Combat HIV/AIDS, malaria and other diseases	*On track*	• A downward trend in the incidence, prevalence and death rates associated with HIV/AIDS, malaria and tuberculosis, especially since 2000.
Goal 7: Ensure environmental sustainability	*Off track*	• Declining forest cover in Africa. • Increasing proportion of terrestrial and marine areas protected. • In 2012, only 64 per cent of the population of Africa (excluding North Africa) used an improved drinking water source. • The proportion of people with access to improved sanitation between 1990 and 2012 increased only moderately in Africa (excluding North Africa) from 24 per cent to 30 per cent, compared to 72 per cent to 91 per cent in North Africa.

Source: Adapted from UNECA (2015: 79–80).

mortality (Goal 4), improving maternal health (Goal 5) and ensuring environmental sustainability (Goal 7) has been disappointing. However, while the UNECA (2015) states that progress towards achieving gender equality and empowering women is 'on track', this is mainly due to improvements in gender parity in primary education and an increase in the seats held by women in national parliaments. However, the report does admit there are serious gender barriers preventing women progressing into post-primary education and wage employment.

Thus, for women in Africa, overall the progress towards gender parity in human development and the MDGs has been frustrating, with poor performances with respect to maternal and child health and mortality, tackling poverty, malnutrition and hunger, environmental issues such as access to safe drinking water and improved sanitation, as well as deforestation, and women's empowerment at a local level, including at family and community scales. So while there have been global and regional human development gains in this period (1990–2015), the then-Secretary-General of the UN, Ban Ki-Moon, recognised that inequalities persist and progress has been uneven, with women benefiting least from development: 'Progress tends to bypass women and those who are lowest on the economic ladder or are disadvantaged because of their age, disability or ethnicity' (UN, 2015: 3).

The UNECA report confirms the importance of the GAD approach and makes the following recommendations:

> Women's empowerment is central to the achievement of many other social development goals, notably in the areas of poverty, inequality, hunger, employment, education, and health. Hence countries should promote gender informed policy interventions . . . Countries should also take bold steps to remove structural and cultural impediments hindering women's economic and social opportunities through limited access to and control over productive resources.
>
> *(UNECA, 2015: 82)*

The example of women's legal rights to land and property

Women produce up to 80 per cent of the food crops in Africa (FAO, 2002) and their labour is crucial to both subsistence food and cash crop production, yet due to customary property ownership traditions that favour men, combined with women's subordinate position to men, they rarely have the right to own land and property in their own right (Joireman, 2008). Joireman (2008: 1242) summarises the issue:

> In sub-Saharan Africa, the view of women as embedded in the household without autonomous access to property has led to uncertainty and instability in their lives when their personal status changes due to the death of a husband, divorce, or polygamy. This uncertainty and lack of autonomous property rights are particularly problematic given the economic responsibilities of women in the rural and urban economies of Africa.

The UN Millennium Development Goals project highlighted the tenure insecurity of women in Africa as a key problem for women's development and called for special efforts to be made to 'ensure women's access to and control of land' (cited in Place, 2009: 1331). Economists agree that 'poorly defined property rights' are a barrier to economic growth and reducing poverty (de Soto, 2000; Joireman, 2008; Sjaastad and Cousins, 2008), but proponents of GAD argue that ensuring women's rights to ownership and inheritance of property is also a human rights issue. Both these perspectives and the pressures from donors and NGOs have ensured that land policy reform is once again on the agenda in Africa, with many African countries at various stages of reviewing or reforming their land policies and laws (Peters, 2004; Boone, 2007; Place, 2009). Between 1992 and 2006, 19 African countries implemented land law initiatives (Boone, 2007: 569). In Côte d'Ivoire, Kenya, Uganda, Tanzania, Senegal, Mozambique, Zimbabwe and South Africa, land policy and land law reform have been highly visible public issues (Boone, 2007) that have included debates on citizenship, community, social inequality and political authority.

The situation is complicated by the complex interplay of communal property rights that are enshrined in customary law and individual title to property protected by national legal frameworks (Joireman, 2008; Place, 2009), both of which disadvantage women. Place (2009: 1327) describes this land tenure situation as a 'bundle of land rights' held with 'rights', with tenure insecurity resulting from a lack of or change in 'rights'. It is changing 'rights' that appear to affect women disproportionately to men. Place (2009) found that the majority of men had unfettered rights to give land to family members in Burundi, Uganda and Zambia, but less than 5 per cent of women could do so. Even in matrilineal systems, women are rarely allowed to inherit land. The current land reform debates are based on the distinctions between statutory and customary law and formal and informal tenure systems, with the formalisation of land rights being seen as necessary for the modernisation of agriculture and economic development (Peters, 2009).

There are ongoing debates concerning what form the formalisation of property, and in particular land ownership, should take (Boone, 2007; Sjaastad and Cousins, 2008). The World Bank is an advocate of land titling and privatisation, individualisation and transferability of title being the goal (Boone, 2007). The alternative position is that the state should uphold historically grounded rights derived from communal membership (de Soto, 2000; Boone, 2007). Over 80 per cent of all arable land in sub-Saharan Africa is held under some form of customary tenure (Boone, 2007), but this is further complicated in that 'customary tenure' can include a mixture of communal rights and user rights. Boone (2007) therefore suggests there are three visions of land reform for Africa:

1 Reinforcing community rights (e.g. Côte d'Ivoire Land Law 1998, Ugandan Land Act 1998).
2 Promote private property rights. No African country has embraced this vision nationally, but law changes in Zambia (1995), Uganda (1998) and Malawi (2002) are seen as precursors. The Zambia 1995 Land Act, which attempted to strengthen the government's ability to issue freehold titles to land, is in dispute. Many countries allow private property rights in urban areas.
3 Institutionalising user rights. No African country has embraced this vision nationally, which is often seen as being in conflict with community rights. However, the Tanzania Village Land Act 1999 was underpinned by this vision.

Unfortunately, little research has been undertaken on the gender implications of these three visions of land rights, in particular their impact on women. The general consensus is that private property rights have strengths, including establishing clear boundaries and clearly identifying the rights of owners. The issue is how is communally managed property transitioned to a commodified legal status? And in a communal system where women have few if any property rights and no capital or collateral, how do they access private property? With respect to legally constituting community and user rights, the same challenge exists, as land and property are the basis of political power in African communities, and this is vested in the hands of elite men. Platteau (cited in Peters, 2009: 1319) states that customary land tenure systems 'embody considerable inequality, intra and inter-group conflict, illegal sales by traditional leaders, and appropriation for private use by representatives of the state'. Obeng-Odoom (2012) therefore argues that gender-aware democratic systems of land tenure should be the basis of land reform rather than the current focus on formalisation and legal title.

This example demonstrates how difficult it is to address gender strategic needs in Africa when traditional ingrained 'rights' are the focus of reform and require national government legislation and public approval. It is likely therefore that this issue will take some time to resolve and that women will continue to be discriminated against for many decades to come.

Transformative gender change: gender justice (post-2015)

The difficulties of confronting gender strategic needs and the disappointing outcomes for African women of the MDGs has sparked heated debates concerning the effectiveness and appropriateness of the GAD approach, and in particular gender mainstreaming. Verma (2014: 188) sums up much of this criticism:

> Experience shows that simply 'adding women and stirring' them into complex power-laden gender relations within development is unlikely to yield empower or meaningful results for women, or men . . . And yet this has been done for years in the name of 'gender mainstreaming'. One key reason for this is that both development and gender equality are transformative processes and, as such, inherently political and demanding. Rather than aiming for deep changes in objectives, outcomes, and gender power relations both in institutions and at the grassroots, gender equality has often been reduced to developing tools, logframes, mechanisms, bureaucratic targets and organizational exigencies.

There have been numerous critiques of gender mainstreaming (Davids *et al.*, 2013; Verma, 2014) mainly focusing on its lack of strategies for change and underperformance in achieving societal transformation. Davids *et al.* (2013: 7) suggest that the shortcomings of gender mainstreaming can be explained by 'gender being reduced to women and women to neo-liberal subjects' and 'the abuse of gender in other power disputes' that seem to 'veil specific anticipations of change that was hoped for'. In summary, Davids *et al.* (2013: 7) claim that 'It seems that, albeit not intentionally, gender mainstreaming resembles more the WID approach than intended'. This is consistent with Mohindra and Nikiema's (2010: 543) assessment that the MDG project 'seems to have relapsed into instrumentalism', with women being 'considered either passive recipients of welfare or objects of investment to achieve other ultimate goals' (Mohindra and Nikiema, 2010: 554). However, Davids *et al.* (2013: 9), while accepting the criticisms of gender mainstreaming, suggest that we should be patient and give the approach more time:

> Gender mainstreaming strategies therefore might partly subvert and partly comply with existing power relations but certainly not bring the revolutionary change as might have been anticipated by the inception of gender mainstreaming as a global strategy. Rather they could be seen as little movements, steps forward, backward and sideways in a very, very slow revolution and process of change.

Given the critiques of and practical difficulties implementing GAD, and the mainstreaming of gender in the development process, combined with the fact that gender transformation is so controversial and is meeting resistance from those with vested interests in the status quo (Verma, 2014), there have been calls to ramp up gender equality and empowerment within development in the post-MDG era. There is a growing call for a 'gender-analytical approach', which Verma (2014: 190) defines as 'the systematic examination of the multiple roles, relationships, social institutions, agency, and processes between and among women and men in different cultures and societies, focussing on imbalance of power, wealth, workloads, ownership, and access to resources'.

She argues that such an approach will lead to gender transformative change, which Sweetnam (2012: 393) claims will uphold:

> the right of women living in the global South to participate equally with men not only in development organisations, but, more widely, in all institutions which determine their lives and choices: including the household, community organisations, the market and the state.

Bearing in mind the variable outcomes, challenges and critiques of previous approaches, and building upon the positive elements of GAD and gender mainstreaming and incorporating a human rights approach, a gender transformative framework, which can be labelled as gender justice, can be seen to underpin the UN Sustainable Development Goals that came into effect in 2016.

In September 2015, 193 UN member states approved the Sustainable Development Goals (also known as the Global Goals). These 17 goals and 169 targets build on the progress of the MDGs, while including new areas such as peace and justice, climate change, economic inequality, innovation and sustainable consumption (UN Women, 2015; UNDP, n.d.). The SDGs came into effect in January 2016 and will run for 15 years. According to Global Goals (n.d.), if the goals are met, then it would mean an end to extreme poverty, inequality and climate change by 2030.

As with the MDGs, all SDGs are relevant to the poorest and most marginalised in Africa, the majority of whom are female. Through the SDG discussions, there appeared to be a general consensus on the need to promote gender equality, although agreement on promoting women's empowerment was less evident (Smee and Woodroffe, 2013). While gender mainstreaming is apparent in the SDGs, a standalone goal on gender equality was agreed. SDG5 aims to 'achieve gender equality and empower all women and girls'. This 'stand-alone' gender goal recognises that deep legal and legislative changes are needed to ensure women's rights around the world (UN Women, 2015), and includes targets to end all forms of discrimination against all women and girls everywhere, and to eliminate all forms of violence against all women and girls in the public and private spheres, including trafficking and sexual and other types of exploitation (UN Women, 2015). Smee and Woodroffe (2013: 5), in their report for the Gender and Development Network (GADN), argue that these targets should:

> provide voice and agency for marginalised women; address the root causes of gender inequality; recognise all the barriers that women face in accessing their rights, including those created by social norms and values; and address unequal power relations at all levels of society including within the household.

It is now widely recognised that gender inequality holds back progress on other development outcomes (Smee and Woodroff, 2013), such as reducing hunger and malnutrition, food security, reducing poverty, improving health, and raising education levels. However, women continue to face discrimination across economic, social and political spheres, and entrenched gender disparities remain a major driver of poverty (World Bank, 2012). The SDGs acknowledge that attaining gender equality and women's empowerment requires action on a broader range of issues than were included in the MDGs. For example, the SDGs include targets concerning violence against women and women's property ownership. By focusing on gender justice, the SDGs have transformative power. Time will tell if they have been successful!

The example of female genital mutilation (FGM) in Africa

It is estimated that approximately 150 million women and girls have been subjected to female genital mutilation (FGM) in 29 countries in Africa and the Middle East, with 3 million girls at risk of FGM each year (UNICEF, 2016). FGM is defined as 'all procedures involving partial or total removal for the female external genitalia or other injury to the female genital organs for non-medical reasons' (WHO, 2008: 4).

The type of FGM performed varies by nationality, but mainly by ethnicity. Current estimates (UNICEF, 2016) indicate that FGM types 1 (mainly clitoridectomy), 2 (excision) and 4 (symbolic forms) account for 90 per cent of cases, with type 3 (infibulation) accounting for 10 per cent of cases. Type 3 is the physically most severe form of FGM, and is mostly practised in Djibouti, Eritrea, Ethiopia, Somalia and Sudan. Within Africa, there are wide variations in FGM prevalence, with eight countries recording a prevalence rate of over 80 per cent in girls and women aged 15–49 (Egypt, Sudan, Somalia, Djibouti, Eritrea, Sierra Leone, Guinea and Mali). A further six countries recorded prevalence rates between 50 and 80 per cent (Ethiopia, Mauritania, Gambia, Liberia, Burkina Faso and Guinea-Bissau).

FGM is condemned by a number of international treaties and conventions such as the Universal Declaration of Human Rights, the UN Convention on the Elimination of All Forms of Discrimination Against Women and the Convention Against Torture and Other Cruel, Inhuman, or Degrading Treatment or Punishment. As FGM is performed mainly on children, it also violates the Convention on the Rights of the Child. FGM is therefore a human rights abuse. It violates a girl and woman's rights to health, security, physical integrity, the right to be free from torture and degrading or inhuman treatment, as well as to the right to life, when it results in death (Askew *et al.*, 2016). FGM is a serious violation of human rights and can have very serious physical and mental health outcomes that may last a lifetime (Finke, 2006; Reisel and Creighton, 2015). Mpinga *et al.* (2016), in a systematic review of research on the economic and social impacts of FGM, found that those who had been subjected to FGM type 3 had a shortened life expectancy, and the FGM-related obstetric complications in six African countries represented $3.7 million (0.1–1.0 per cent) of government spending on health for women aged 15–49. In Nigeria, the average management costs of medical complications of FGM per person was $120, and in Somalia the mean number of hospital days due to FGM complications was 16.5 days. FGM also contributes to high school dropout. There are also reports that in Egypt, fathers receive a higher bride price if FGM type 3 has been performed on their daughters.

Over the last few decades, global efforts to end FGM have intensified, and by 2009 16 of the 29 African states where FGM is prevalent had passed laws banning the practice (Mpinga *et al.*, 2016). In 2012, the UN General Assembly Resolution 'Intensifying Global Efforts for the Elimination of Female Genital Mutilations' called on all countries to enact legislation banning FGM. But FGM has proven to be a very persistent practice in Africa, and is often seen as an act of power within highly patriarchal societies that has become a strong social norm. FGM symbolises the control and subordination of women, and in regions where FGM is highly prevalent, women usually have little independence, few property rights or opportunities for education, and are often married at a young age (Finke, 2006). Although there is some evidence that FGM prevalence rates are declining and that attitudes of young people towards the practice are changing (UNICEF, 2016), not all countries have made progress, and the pace of decline has been uneven. Of concern is the fact that current progress is insufficient to keep up with increasing population growth. UNICEF are concerned that if trends continue, the number of girls and women subjected to FGM will 'rise significantly' over the next 15 years (UNICEF, 2016). Ending FGM has become a global gender justice issue. It is therefore no surprise that FGM features as a target in SDG5.

A long journey. . .

The road to achieving gender equality in Africa is a long one that began 50 years ago. There has been some progress, especially with respect to addressing the practical needs of women, in particular the enrolment of girls into primary education. But there have also been disappointments, such as the slow progress in reducing maternal and child mortality. However, tackling the strategic needs of women, including the economic, social and political structures that discriminate against women, has been difficult and slow. The shift in approaches from WID/WAD and GAD to gender justice illustrates how hard it is to tackle the structural inequalities affecting women. There is now a growing recognition that gender justice must be addressed if Africa is to develop not just in terms of human rights, but also economic success. The SDGs have mainstreamed gender and acknowledge the importance of gender justice to sustainable development. Let us hope that with this renewed international commitment, within the next 15 years inequality and injustice based on gender will be a thing of the past.

References

Askew, I., Chaiban, T., Kalasa, B. and Sen, P. (2016) 'A repeat call for complete abandonment of FGM', *Journal of Medical Ethics*, 42: 619–620.

Benería, L. and Sen, G. (1981) 'Accumulation, reproduction, and women's role in economic development: Boserup revisited', *Journal of Women in Culture and Society*, 7(2): 279–298.

Bhavnani, K.K. and Foran, J. (2008) 'Feminist futures: from dystopia to eutopia?', *Futures*, 40: 319–328.

Boone, C. (2007) 'Property and constitutional order: land tenure reform and the future of the African state', *African Affairs*, 106(425): 557–586.

Boserup, E. (1970) *Women's Role in Economic Development*, London: George Allen & Unwin.

Boserup, E. (1989) 'Population, the status of women and rural development', *Population and Development Review*, 15, Supplement: Rural Development and Population: Institutions and Policy: 45–60.

Boserup, E. (1990) 'Economic change and the roles of women', in I. Tinker (ed.), *Persistent Inequalities: Women and World Development*, New York: Oxford University Press, pp. 14–26, available at: www.popline.org/node/384076 (accessed 12 December 2017).

Boserup, E. and Liljencrantz, C. (1975) *Integration of Women in Development: Why, When, How*, New York: UNDP.

Cornia, G., Jolly, R. and Stewart, F. (eds) (1987) *Adjustment with a Human Face*, Oxford: Clarendon Press.

Davids, T., van Driel, F. and Parren, F. (2013) 'Feminist change revisited: gender mainstreaming as slow revolution', *Journal of International Development*, DOI: 10.1002/jid.2945.

de Soto, H. (2000) *The Mystery of Capital*, New York: Basic Books.

Doyal, L. (2000) 'Gender equity in health: debates and dilemmas', *Social Science and Medicine*, 51: 931–939.

FAO (Food and Agriculture Organization) (2002) *Gender and Access to Land (No. 4)*, Rome: FAO.

Finke, E. (2006) 'Genital mutilation as an expression of power structures: ending FGM through education, empowerment of women and removal of taboos', *African Journal of Reproductive Health*, 10(2): 13–17.

Global Goals (n.d.) *The Global Goals for Sustainable Development*, available at: http://globalgoals.org (accessed 12 December 2017).

Joireman, S.F. (2008) 'The mystery of capital formation in sub-Saharan Africa: women, property rights and customary law', *World Development*, 36(7): 1233–1246.

Kilby, P. (2015) *NGOs and Political Change*, Canberra: ANU Press.

McIlwaine, C. and Datta, K. (2003) 'From feminising to engendering development', *Gender, Place and Culture*, 10(4): 369–382.

Moghadam, V.M. (1998) 'Feminisms and development', *Gender and History*, 10(3): 590–597.

Mohindra, K.A. and Nikiema, B. (2010) 'Women's health in developing countries: beyond an investment?', *International Journal of Health Services*, 40(3): 543–567.

Molyneux, M. and Steinberg, D.L. (1995) 'Mies and Shiva's ecofeminism: a new testament?', *Feminist Review*, 49: 86–107.

Moorosi, P. (2009) 'Gender, skills development and poverty reduction', *Gender and Poverty Reduction*, 81: 110–117.

Moser, C. (1989) 'Gender planning in the Third World: meeting practical and strategic needs', *World Development*, 17(11): 1799–1825.

Moser, C. (1993) *Gender Planning and Development: Theory, Practice and Training*, New York: Routledge and Kegan Paul.

Mpinga, A.K., Macias, A., Hasselgard-Rowe, J., Kandala, N.B., Felicien, T.K., Verloo, H., *et al.* (2016) 'Female genital mutilation: a systematic review of research on its economic and social impacts across four decades', *Global Health Action*, 9, DOI: http://dx.doi.org/10.3402/gha.v9.31489.

Obeng-Odoom, F. (2012) 'Land reforms in Africa: theory, practice, and outcome', *Habitat International*, 36: 161–170.

Ostergaard, L. (1992) *Gender and Development: A Practical Guide*, London: Routledge.

Peters, P.E. (2004) 'Inequality and social conflict over land in Africa', *Journal of Agrarian Change*, 4(3): 269–314.

Peters, P.E. (2009) 'Challenges in land tenure and land reform in Africa: anthropological contributions', *World Development*, 37(8): 1317–1325.

Place, F. (2009) 'Land tenure and agricultural productivity in Africa: a comparative analysis of the economics literature and recent policy strategies and reforms', *World Development*, 37(8): 1326–1336.

Podems, D.R. (2010) 'Feminist evaluation and gender approaches: there's a difference?', *Journal of Multidisciplinary Evaluation*, 6(14): 1–17.

Rathgeber, E.M. (1990) 'WID, WAD, GAD: trends in research and practice', *Journal of Developing Areas*, 24: 489–502.

Reisel, D. and Creighton, S.M. (2015) 'Long-term health consequences of female genital mutilation (FGM)', *Maturitas*, 80: 48–51.

Rostow, W.W. (1960) *The Stages of Economic Growth: A Non-Communist Manifesto*, Cambridge: Cambridge University Press.

Sjaastad, E. and Cousins, B. (2008) 'Formalisation of land rights in the South: an overview', *Land Use Policy*, 26: 1–9.

Smee, S. and Woodroffe, J. (2013) *Achieving Gender Equality and Women's Empowerment in the Post-2015 Framework*, London: GADN.

Sweetnam, C. (2012) 'Introduction', *Gender and Development*, 20(3): 389–403.

UN (United Nations) (2015) *The Millennium Development Goals Report 2015*, New York: UN.

UNDP (United Nations Development Programme) (n.d.) *Sustainable Development Goals*, available at: www.undp.org/content/undp/en/home/sustainable-development-goals/resources.html (accessed 12 December 2017).

UNECA (United Nations Economic Commission for Africa) (2015) *MDG Report 2015: Lessons Learned in Implementing the MDGs – Assessing Progress in Africa Toward the MDGs*, Addis Ababa: UNECA.

UNICEF (United Nations Children's Fund) (2016) *Female Genital Mutilation/Cutting: A Global Concern*, New York: UNICEF.

UN Women (2015) *Women and the Sustainable Development Goals (SDGs)*, available at: www.unwomen.org/en/news/in-focus/women-and-the-sdgs (accessed 8 January 2018).

Verma, R. (2014) 'Business as unusual: the potential for gender transformative change in development and mountain contexts', *Mountain Research and Development*, 34(3): 188–196.

WHO (World Health Organization) (2008) *Eliminating Female Genital Mutilation: An Interagency Statement*, Geneva: WHO.

World Bank (2012) *World Development Report 2012: Gender Equality and Development*, New York: World Bank.

Zwart, G. (1992) 'From women in development to gender and development: more than a change in terminology?', *Agenda: Empowering Women for Gender Equity*, 14: 16–21.

8
Inequality in Africa

Philip Nel

Understanding inequality

Inequality refers to the skewed distribution of valued and scarce resources among people, either individually or organised as groups. Currently, we are especially concerned with the skewed distribution of wealth (assets) and income because these economic inequalities have to do with 'the fundamental disparity that permits one individual [or group] certain material choices, while denying another individual those very same choices' (Ray, 1998: 170). Worldwide, the number of people who live in absolute poverty has declined during the most recent era of intense globalisation, and a large new middle class has emerged in Asia. One result of this is that the level of inequality *between* states has moderated, measured as the distribution of per capita national income, weighted for population size (Milanovic, 2016). In stark contrast, *within-country* inequality has increased across high-, middle- and low-income states, with dire social, economic and political consequences (Nel, 2008a; Melamed and Samman, 2013; Piketty, 2014; Atkinson, 2015). Africa has not escaped these within-country trends, nor their negative consequences. This contribution reviews the available evidence concerning the level of within-country inequality in Africa, its dynamics, and its consequences. I follow the practice of focusing predominantly on income inequality data, given their availability, but also highlight wealth inequality where appropriate. First, some conceptual distinctions and methodological comments must be made.

Wealth resides in effective control over human and non-human capital assets. Income is the sum of earnings, interest, profits and transfers flowing into households. Consumption refers to the flow of resources from households, and is measured in terms of expenditures on consumables and durables, but also taxes paid (Davies, 2009; Folbre, 2009). All three of these are distributed unequally in all societies that rely to one or other extent on the market as the institution that determines the price of goods and services. In some societies, though, these inequalities are so pronounced that they undermine the market and the political and social fabric of society. Inequalities of wealth, income and consumption often overlap with absolute poverty, denying some people access to the means of existence despite the abundance of such means. Within-nation inequality can be considered *vertically*, focusing on the distribution between individuals and households. It can also be considered *horizontally*, looking at the distribution between groups of people, distinguished by gender, ethnicity, space and/or time. Inequality can be studied synchronically and diachronically. In the former, we study a distribution at one point in time, comparing it with other distributions at that time, for instance. Diachronically, we trace the evolution of distributions over a period of time, for instance over the lifetimes of individuals, or over the history of a state or region.

Much of the empirical work on inequality in Africa uses survey data on household consumption expenditures, regarded as a good indicator of current welfare (Ferreira and Ravallion, 2009). Survey data on consumption are scarce, though, and many analysts make do with data on market and net income, especially if they are interested in cross-national comparisons. Market income is the sum of all earned income (wages, profits, rents, dividends and private transfers) before direct taxes, while net income refers to household earnings after taxes are deducted and fiscal transfers added (Lustig, 2015). Net income is a close enough approximation of consumption. To overcome the problems of inconsistency that plague many survey-based data sets, this contribution complements the use of survey consumption data with estimations from the Standardized World Income Inequality Database (SWIID) version 5.1 (Solt, 2009, 2014), which is increasingly being favoured in cross-national studies (Acemoglu *et al.*, 2015). It uses data selected from the widest possible range of data sets, and relies on a large set of conversion factors for different income or consumption concepts to obtain comparable estimates based on market income and net income, and the difference between the two. We know that data on Africa can sometimes be suspect (Klasen and Blades, 2013), hence the importance of using the best possible data sources available.

How unequal is Africa?

The literature on inequality in Africa (well summarised in Sahn and Stifel, 2000; Okojie and Shimeles, 2006; Go *et al.*, 2007; van de Walle, 2009) emphasises two general findings. The first focuses on the surprisingly high level of inequality, especially in sub-Saharan Africa (SSA), while the second emphasises the developmental costs of inequality. We return to the second below, but first focus on how unequal Africa in fact is. Figure 8.1 shows the dynamics of net income

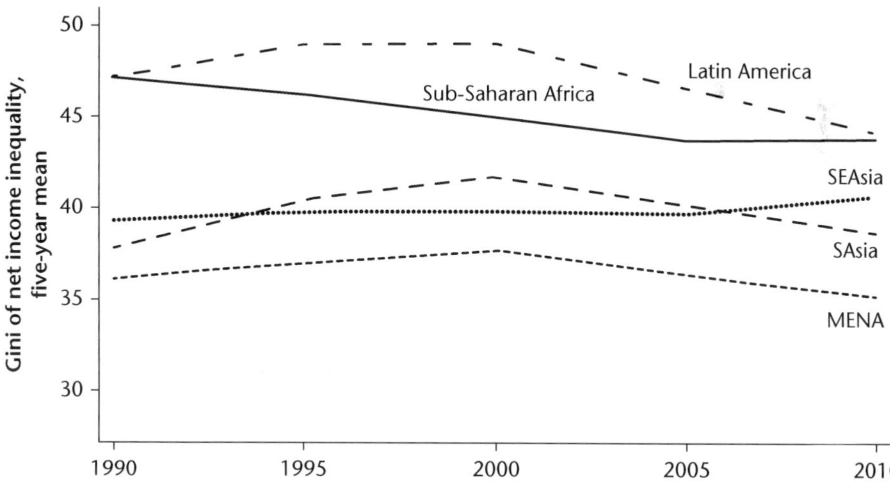

Figure 8.1 Dynamics of inequality in selected developing regions, 1990–2010

Source: Own calculations based on the Standardized World Income Inequality Database (Solt 2009, 2014).

inequality for a number of regions in the developing world. Africa has the dubious honour of being the second worst region over the period 1990–2010, which is the period for which we have good data. Note the marginal decline in recent years – something to which we will return later.

The means reported in Figure 8.1 hide a large variety. Table 8.1 provides a summary of the best data that we have on the distribution of wealth, household consumption expenditures, and net income within African states. All data are measured in the Gini index: zero equals full equality and a score of 100 means that all wealth and income are owned/received by one household. Below, we will see that these states do share some dimensions of inequality, and suffer in similar ways from the consequences of inequality. But African states and subregions also differ sufficiently to pose an explanatory challenge. Southern Africa is the most unequal subregion, followed by central Africa and West Africa. East Africa, the Sahel and North Africa have the lowest levels of inequality in Africa. Lower inequality in North Africa is largely due to the egalitarian ethos of Islam (Nel, 2008b) and the relative absence of manufacturing industries with their accompanying skills premium. Although low in comparison, the levels of inequality in North Africa are associated with the social tensions that boiled over in the region in 2011 and subsequently (AfDB, 2012). We will also see below (Figure 8.3) that most North African states experienced increasing inequality between 1990 and 2010.

Wealth in Africa is highly concentrated, with the top 10 per cent owning close to 78 per cent of all assets. This figure is just below that of Asia, but higher than in Latin America and in North America, where most of the extremely wealthy people of the world are concentrated (Credit Suisse, 2014). Despite marginal improvements in some African states in recent years – to which we will return – there is a persistent long-term pattern of higher-than-expected inequality in SSA in particular (Milanovic, 2003; Ravallion and Chen, 2012).

The data on wealth inequality in Table 8.1 are based on estimations and are much less reliable than the figures for consumption. Nevertheless, they do reflect trends that are reported in studies of land and asset inequalities (Booysen *et al.*, 2007; Frankema, 2010). Land inequality is higher in SSA than in comparable regions of the developing world, except Latin America and South Asia (Deininger and Squire, 1998; Frankema, 2005). Crucially, land inequality in Africa is associated with considerable land poverty (Shipton and Goheen, 1992). Again, there is considerable variation, with southern Africa the most unequal in terms of land inequality. Surprisingly, given its lower level of income inequality, North Africa also has a very unequal distribution of land ownership, which makes up most of its high wealth inequality levels, as reported in Table 8.1. Given the large supply of arable agricultural land in Africa, and the widespread practice of communal land use, general access should be less of a problem. Recent research shows that there is nevertheless a surprisingly high incidence of inequality-driven competition and conflict in Africa over land and natural resources, and the frequency and durability of such conflict fly in the face of the general assumption that customary African landholding institutions give poor smallholders security of access. More than 90 per cent of all rural land in SSA is undocumented and informally administered, and customary land use systems leave excessive room for discretion and favouritism on the part of land patrons. Most significantly, communal land use institutions largely exclude females from land use rights (Peters, 2004; FAO, 2011; Byamugisha, 2013).

Apart from consumption and wealth inequalities, Africa is known for disproportionate access to education and health and nutrition, access to and use of public services, labour market inequalities, and the ability to influence decision-making on public matters (Okojie and Shimeles, 2006). What is particularly striking about all these forms of inequality is that they have significant *ethnic, gender* and *spatial* dimensions, while racial dimensions also play a role in some parts of southern Africa. Often, these dimensions overlap. As far as ethnicity is concerned,

Table 8.1 Inequality of consumption and wealth in Africa: latest comparable data

State	*Comparable estimations*			*Surveys*		
	Gini net income 1990s	*Gini net income 2000s*	*Gini wealth*	*Gini consumption*	*Year of survey*	*Source of data*
Angola	48.2	54.2	75.5	58.1	2000	WIDER gross
Botswana	54.8	53.7	75.3	60.9	1994	WYD
Lesotho	60.7	53.3	73.7	52	2002	WYD
Malawi	n/a	42.3	67.2	38.8	2010	WYD
Mozambique	42	44.3	70.2	45.6	2008	WYD
Namibia	66.2	64.8	82.9	59.7	2010	WYD
South Africa	59.2	59.7	81.8	57.3	2000	WYD
Swaziland	56.8	49.8	73.5	51.4	2009	WYD
Zambia	54.2	52.3	71.5	57.4	2010	WIDER
Zimbabwe	52.1	51	81.3	42.3	2011	POVCAL
Mean SA	**54.9**	**52.5**	**75.3**	**52.4**		
Benin	n/a	36.6	66.5	34.1	2011	WYD
Burkina Faso	47.3	41.5	63.8	39.8	2009	WYD
Cameroon	43.2	41.1	67.7	39	2007	WYD
Cape Verde	52.8	53.5	71.3	73.5	2002	WYD
Côte d'Ivoire	40.5	45.3	72.1	44.2	2002	WYD
Ghana	35.9	39.9	66.8	42.8	2006	WYD
Liberia	n/a	47.8	65.9	38.1	2007	WYD
Niger	44.4	41.7	66.4	29.8	2011	WYD
Nigeria	47.3	44.6	80.3	46.8	2010	WYD
São Tomé and Príncipe	n/a	53.4	73.2	39	2010	WYD
Senegal	42.9	38.2	67.3	33.8	2011	WYD
Sierra Leone	55.7	42.3	66.2	32.1	2011	WYD
Togo	n/a	36.9	65.5	40.9	2011	WYD
Mean WA	**45.6**	**43.3**	**68.7**	**41.1**		
Burundi	37.1	35.7	n/a	33.1	2006	POVCAL
Central African Republic	52.9	49.3	71.3	56.2	2008	WYD
Congo, Dem. Rep.	n/a	44.7	68.6	44.4	2008	WYD
Congo, Rep.	n/a	42.9	71.1	38.4	2011	WYD
Gabon	n/a	42.4	74.7	41.1	2005	WYD
Gambia, The	44.4	47.6	70.0	46.9	2003	WYD
Guinea	39.9	41.3	66.3	39.6	2007	WYD
Guinea-Bissau	46.7	37.4	64.4	38.3	2005	WYD
Rwanda	39.4	48.6	72.2	50.2	2011	WYD
Mean CA	**43.4**	**43.3**	**69.8**	**43.1**		
Chad	n/a	39.9	66.5	42.1	2011	WYD
Mali	42.1	38.5	66.3	33	2010	WYD
Mauritania	38.6	37.9	67.1	40.5	2008	WYD
Sudan	n/a	35.6	64.2	34.4	2009	WYD
Mean Sahel	**40.4**	**38.0**	**66**	**37.5**		
Comoros	n/a	57.4	80.8	62.9	2005	WYD
Djibouti	37.2	39.5	67.4	39.7	2002	WYD

(continued)

Table 8.1 Continued

State	*Comparable estimations*			*Surveys*		
	Gini net income 1990s	*Gini net income 2000s*	*Gini wealth*	*Gini consumption*	*Year of survey*	*Source of data*
Ethiopia	36.1	30.7	62.4	33.3	2011	WYD
Kenya	49.1	43.9	70.7	29.9	2007	WYD
Madagascar	39.0	43.7	68.5	40.9	2010	WYD
Tanzania	37.0	36.6	64.5	37.6	2007	WYD
Uganda	38.4	41.7	68.8	42.9	2010	WYD
Mean EA	**39.5**	**41.9**	**69**	**41**		
Mean SSA	***46.1***	***44.6***	***70.2***	***44***		
Algeria	33.4	36.1	67.6	34.6	1995	WYD gross household
Egypt, Arab Rep.	33.6	33.4	80.7	31.8	2005	WYD
Morocco	38.6	40.5	79	40.7	2007	WYD
Tunisia	41.3	38	74	38.5	2010	WYD
Mean NA	**36.7**	**37.0**	**75.3**	**36.4**		
Mean Africa	***45.0***	***43.9***	***72.8***	***40.1***		

Sources:

SWIID = Standardized World Income Inequality Database (Solt 2009, 2014), available at: http://fsolt.org/swiid/ (accessed 13 December 2017).

WYD = World Income Distribution data set, maintained by Branco Milanovic, World Bank, available at: http://go.worldbank.org/IVEJIU0FJ0 (accessed 13 December 2017).

WIDER = UNU-WIDER, 'World Income Inequality Database (WIID3.0b)', September 2014, available at: www.wider.unu.edu/research/Database/en_GB/database/ (accessed 13 December 2017).

POVCAL = PovcalNet household survey data, World Bank, available at: http://iresearch.worldbank.org/PovcalNet/ (accessed 13 December 2017). Gini wealth is based on estimations reported in Credit Suisse's *Global Wealth Databook 2013*, available at: http://usagainstgreed.org/GlobalWealthDatabook2013.pdf (accessed 13 December 2017).

Notes:

No data for Mauritius, Libya, Seychelles and South Sudan.

Gini net income = measure of overall distribution of household per capita net income, which is income after taxes and transfers.

Gini consumption = measure of overall distribution of household net per capita consumption, except for Zimbabwe, which is gross household per capita consumption; Algeria and Burundi, which are household gross consumption; and Botswana, which is household gross income. All data based on national household surveys.

it is well known that Africa is extremely diverse and that ethnic favouritism plays a large distributional role, and that it is a significant factor explaining underdevelopment (Bates, 1983; Londregan *et al.*, 1995; Frank and Rainer, 2012).

The *gender* dimension of inequality is considerable. The mean level of human development (on the Human Development Index) among African women is 13 per cent lower compared to men, due to official and social discrimination, widespread violence against women, and also because of the perverse incentives that households have to favour investment in the health and education of the male rather than the female child (AfDB *et al.*, 2015). Studies commissioned by the African Academy of Sciences found that there is a gender imbalance, especially in rural areas, in access to and perseverance in education in countries as disparate as Botswana, Cameroon, the Democratic Republic of Congo, Kenya, Mali, Nigeria, South Africa, Tanzania and Uganda (Okojie and Shimeles, 2006).

There is also a significant *spatial* dimension to poverty and inequality in SSA (Sahn and Stifel, 2003; Bird *et al.*, 2010; Mveyange, 2015). Development is always a spatially unequal process, but this is particularly pronounced in SSA, where political, environmental, economic and social factors conspire to form spatial poverty traps that slow development down in areas that are far removed from harbours and main cities, and/or in regions that are not favoured by the political patrons in a particular state (AfDB *et al.*, 2015). Access to health services, clean water and sanitation is widely divergent between rural and urban areas in southern and East Africa, and spatial factors also determine a large portion of asset inequality. Spatial inequalities are important factors in perpetuating poverty, and the majority of extremely poor people are found in inland areas some distance from urban centres and/or harbours and poorly served by infrastructure. The nutritional status of children and the incidence of growth deficiencies differed markedly between urban and rural areas in a range of countries (Okojie and Shimeles, 2006). Spatial inequalities will continue to be important: the majority of the populations of SSA are rural, will remain so for at least two or more decades, and could grow by as much as two-thirds by mid-century (AfDB *et al.*, 2015).

The common impoverished conditions that characterise rural SSA, and the prevalence of small-scale subsistence farming there, have led many to assume that inequality within rural areas of SSA should be relatively low compared to the urban areas where wage differentials are larger. Recent findings show that inequality within rural areas is larger than would be expected. Otherwise diverse states such as Mozambique, Senegal and Mauritania display a significant degree of 'socio-economic differentiation' in rural areas: a process of increased inequality and fragmentation into, on the one hand, groups that own assets and groups that rely on incidental/insecure wage labour, and, on the other hand, fragmentation into groups that depend on agriculture and groups that have to rely on non-agricultural income. The latter groups are vulnerable, politically weak and constitute the bulk of the poor (Oya, 2010).

Urban-based inequality and poverty are also high in most of SSA, and more than half of all urban workers were surviving on less than US$1 a day in 2007, despite significant output growth in a number of SSA states in the preceding decade (De Vreyer and Roubaud, 2013). Inequality in the urban labour markets in SSA is determined by the low level of wage employment, the relatively small manufacturing sector, the large service sector and the dominance of the informal sector. As a percentage of those employed, less workers are in wage employment in SSA than anywhere else in the world today, and the informal sector employs two-thirds of all urban workers. The service sector in SSA is large, and is often dominated by a bloated civil service and the numerous own-account retailers and service providers who scrape out an existence in the informal economy. The low level of wage employment and the size of the informal sector imply that the urban workforce in general is weakly organised and fragmented. In states where there is a sizeable manufacturing and natural resource industry, such as South Africa, high wage inequality is maintained by a combination of skill shortages, unequal education opportunities and labour market rigidities that prevent the entry of new workers into the labour force (van der Berg, 2014; Seekings and Nattrass, 2015).

Explaining inequality

Why is SSA so unequal, given the fact that technology-backward subsistence agriculture and the informal sector are comparatively large (Milanovic, 2003; Anderson and McKay, 2004;

van der Walle, 2009)? Note that there is considerable variation in levels of inequality in the region, illustrated by Table 8.1. The high mean level is driven by a specific group of worse performers, and we will focus on them soon. In general, though, inequality in Africa reflects the concentration of privilege among urban-based, male-dominated racial and/or ethnic elites that have created or inherited social and political institutions that constitute 'limited access orders', in the felicitous phrase of North *et al.* (2009). Limited access orders are particular elite institutional responses to the threat of endemic violence, and consist of arrangements that allow those who potentially or actually have access to violent means of coercion to share economic rents and privileges among themselves, while excluding the largest parts of their populations from these privileges. The specifics of limited access orders differ from state to state, with some having more fragile arrangements than others, and with some having stronger democratic dimensions than others (Botswana, Cape Verde, Ghana and Mauritius, for instance). However, there is no African state that can lay claim to being a full 'open-access order' in which the rule of law and property rights are universally applicable and respected, there is effective civilian political control over the means of state violence (police and military), organisations are freely formed and have a life independently of tribal, ethnic or state political institutions, and in which improving standards of living in general are a significant purpose of state spending. Inequality, one can say, is an index of the restrictions on open access in African societies. Below, we will ask if the recent decline in inequality in some states in the region heralds a new institutional era.

To account for the variability noted in Table 8.1, we have to complement this general comment with a comparative historical-institutional approach, relying on distinctions made by Amin (1972), Engerman and Sokoloff (2000) and Acemoglu *et al.* (2001). According to this approach, it is the manner in which an economy is incorporated into the global division of labour that ultimately determines longer-term distributional patterns. The manner of incorporation is determined by the natural endowments of land, labour and natural resources, and the patterns of external settlement that these endowments encourage, and the way in which coercive power is exercised and institutionalised to affect the incorporation. This shapes the preservation and/or undermining of existing social, economic and political institutions, and crucially determines the creation of new political and economic institutions that are designed to structure and administer the extractive activities that characterise a particular form of incorporation.

The variability of inequality is most noticeable in SSA, and can be traced back to two forms of incorporation into the transnational colonial economies of the eighteenth and nineteenth centuries (Angeles, 2007; van der Walle, 2009; Frankema, 2010). Following Amin (1972) and Mkandawire (2010), we can distinguish between labour reserve economies and trade-extractive economies, a distinction that relates to two broad patterns of colonial settlement also found elsewhere (Acemoglu *et al.*, 2001; Lange, 2004; Frankema, 2005; Angeles, 2007; van der Walle, 2009; Frankema, 2010). In one pattern – labour reserves – large numbers of Europeans settled to make use of favourable climatic conditions, readily available agricultural land, enticing deposits of valuable natural resources and the low costs of suppressing a large supply of indigenous labour to facilitate capital accumulation. In the other – trade-extractive economies – colonists encountered unfavourable climatic conditions, settled only sparsely, and preferred to exploit favoured cash crops and other resources from a distance. Labour reserves have hierarchical educational institutions created to serve the relatively large number of settlers, and disenfranchisement and coercion of indigenous labour stimulated class formation along racial and urban–rural divides, and fed highly unequal patterns of landholding. These racial, class and spatial inequalities were reproduced in the state formation that followed independence as political and economic power remained concentrated among an elite that perpetuated exclusivist rent-seeking, but gave it a non-racial tinge. The tracks were thus set for high levels of income and wealth inequality that still characterise large parts of southern Africa, where most of the labour reserves are located. At the same time, the infrastructure, and the political and legal institutions that settler colonialists established to manage and protect their joint affairs, have imparted a legacy of bureaucratic competence, independence of the judiciary, and considerable transport and communication capacity.

In contrast, European settlers formed a small minority in the colonies of central, East and West Africa. In these regions, the natural endowments of the tropics and the absence of easily accessible mineral riches discouraged large-scale European settlement, and encouraged the cultivation of cash crops and exploitation of timber resources, among others. Cash crops provided an easily taxable resource, which the colonial authorities duly set out to do, using their coercive power to favour the production of crops that could secure the best international price, to raise tax revenue, and to compel labour from one area into a ready supply for the cultivation of crops such as tea, coffee, cacao and cotton in another. Colonial penetration was thin beyond intrusive revenue collection, however, with limited European land ownership, minimal colonial investment in infrastructure that favoured littoral and urban areas, a dependency on the local indigenous providers of services, and a large degree of reliance on favoured indigenous political leaders to assist in the administration and maintaining of law and order. Divide-and-rule tactics were more common in these hands-off extractive or predator peasant colonies compared to the settler colonies of southern Africa, with Nigeria, German East Africa (Rwanda) and the Belgian Congo being prime examples. Once decolonisation set in, the miniscule European minority universally lost political power, and the remnants of the extractive/predatory state became a bone of contention between rival ethnic groups that the former colonisers deliberately cultivated and exploited. The extractive colonial and postcolonial practices of favouring the urban and littoral regions and underinvesting in the infrastructure and public services in the hinterland helped to maintain a huge income and wealth gap between the urban areas and the rural hinterland, exemplified in the North–South divides of some West African states – Côte d'Ivoire, Ghana and Nigeria being prime examples. While spatial and horizontal (between-group) inequalities are high in trade-extractive economies, vertical (between-individual) income inequality on average is 11 to 13 Gini index points lower than in the labour reserve economies of the region. SSA is indeed one of the most unequal regions of the world, but not all inequalities of the region are similar, and income inequality is highly pronounced only in a subgroup.

Inequality's human costs

The human costs of inequality in Africa are high, and these costs are experienced across the continent. Once disregarded as an inevitable by-product of socio-economic development, inequality is today appreciated for what it really is: a *preventable* scourge that has deleterious effects on the growth and development prospects of a population as a whole (Nel, 2008a).

There is a growing body of evidence that inequality impacts negatively on development, first by undermining growth, second by limiting the potential of growth to be pro-poor, and third by stunting human development in general. Global cross-national studies have found that high levels of inequality undermine growth potential through a variety of mechanisms. The deleterious effect of inequality on growth is realised through a variety of channels. One is the associated failure of credit markets, which is particularly relevant in the African context. Access to credit is determined by the holding of assets that provide security, and asset inequality thus skews the provision of credit in favour of the asset-rich, discouraging the asset-poor from investing in education and physical capital expansion, and providing them with incentives to invest in a large number of children to assist in income generation and to provide old-age security. The human capital stock is thus suboptimally developed, and social institutions and practices that discriminate against the female child worsen the situation (Nel, 2003; Odedokun and Round, 2004).

A second channel through which inequality could affect growth is by fostering political instability, defined as the propensity for regime or government collapse. Political instability, often accompanied by violent civil conflict, destroys human lives and infrastructure, undermines productivity, affects domestic investment decisions, discourages foreign direct investment, undermines trade, and has rightly been called 'development in reverse'. An early study found little evidence of a direct link between vertical income inequality and political instability in Africa, although the former directly and significantly affects *perceptions* of instability (Nel, 2003). Since 2003, there has been a significant improvement in the range and quality of data especially on horizontal inequality in Africa. Making use of detailed subnationally indexed data on inequality and conflict, Fjelde and Østby (2014) found that regions of SSA with high levels of welfare and education inequality between individuals (vertical inequality) and between ethnic groups (horizontal inequality) are significantly more exposed to communal conflict than regions with lower inequality levels. These findings are echoed in the results of Peters (2004), which identify a significant level of communal conflict fuelled by differential and changing patterns of access to land in many parts of rural SSA. A study of the Niger Delta found that perceived inequality, both vertical and horizontal, may be as important in fuelling grievances and fostering pro-violence attitudes as actual inequality. It is also not the relatively deprived who are always the proponents of violence. In the case of inter-group (horizontal) inequality in the Niger Delta, it is the relatively privileged who are more likely to support violence to rectify the perceived inequality (Rushtad, 2015).

Third, inequality undermines the creation and sustaining of institutions that are conducive to economic growth, such as the rule of law and the control of corruption (Chong and Gradstein, 2004). Very little work on Africa has been done in this regard (Blundo and Olivier de Sardan, 2006), but a cross-national study of SSA by Gyimah-Brempong (2002) found that increased corruption is positively correlated with income inequality, and that it decreases economic growth directly by undermining productivity and indirectly by discouraging investment in physical capital. Slow growth and corruption, in turn, interact to increase income inequality (Uslaner, 2007).

The empirical testing of the growth-depressing effects of inequality in Africa is still in its infancy, and much remains to be done (Nel 2003; Odedokun and Round, 2004; Okojie and Shimeles, 2006; Thorbecke, 2013). The limited empirical results indicate that a growth-depressing effect of inequality can be detected in SSA in general, but that this does not apply everywhere and it is less significant in poorer states. This seems to also apply to poorer states elsewhere (Deininger and Squire, 1998).

The human costs of inequality are more significantly reflected in the negative effects of inequality on what is known as 'inclusive growth' (i.e. growth that is pro-poor). Pro-poor growth is associated with lasting employment opportunities and the provision of broader access to public services and education (Kakwani and Pernia, 2000). A number of studies have found that high initial inequality in SSA is a significant deflator of the likelihood that economic growth, if and when it occurs, will be pro-poor (Lopesa, 2005; Fosu, 2008, 2009, 2010; Thorbecke, 2013).

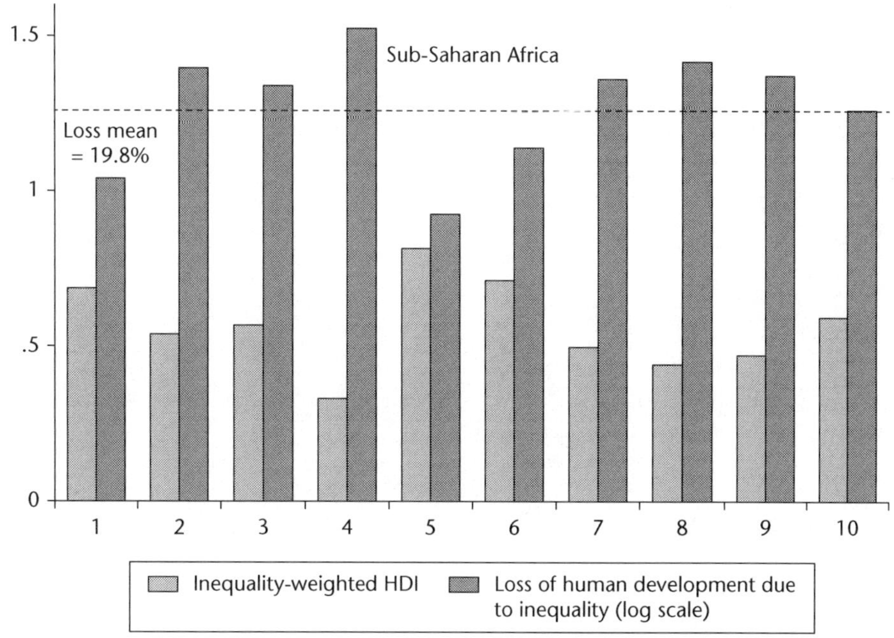

Figure 8.2 The human costs of inequality, 2013

Source: Own calculations, based on UNDP (2014).

Notes: 1 = Eastern Europe and ex-Soviet Union; 2 = Latin America; 3 = Middle East and North Africa; 4 = sub-Saharan Africa; 5 = Western Europe and North America; 6 = East Asia; 7 = South East Asia; 8 = South Asia; 9 = Pacific; 10 = Caribbean.

HDI = Human Development Index, as used by the UNDP.

However, nothing illustrates the human costs of inequality so dramatically as the findings by the UNDP in its 2014 *Human Development Report*. The people of SSA suffer the largest human development losses of all regions in the world due to uneven access to health services and quality education, both of which are closely correlated with wealth and income inequality. Figure 8.2 illustrates how precarious human development in SSA is due to the effect of inequality compared to other regions of the world.

Africa has undergone a significant wave of democratisation since the 1980s (Lynch and Crawford, 2011), and it is important to ask how inequality affects the initiation of democracy and its consolidation on the continent. Democratisation is incomplete in all but two states of Africa. Cape Verde and Mauritius are the only two states that score a full 10 out of 10 on Polity IV's 10-point democracy scale, with South Africa and Botswana at 9 out of 10, and Kenya and Nigeria at 8 out of 10 (Marshall, 2013). Most African states have hybrid regimes, mixing elections with authoritarian practices such as unlimited terms of offices for the head of state and state repression of opposition parties. This general absence of consolidated democracy can be partly attributed to the invidious effects of inequality and its accompanying concentration of political power. Broad comparative studies of democratisation have suggested that high levels of inequality may well trigger the democratic impulse, both *from below* as the poor and disenfranchised try to improve their position and *from above* when the costs of suppression outrun its benefits. However, it seems as if high inequality retards the translation of these reform impulses into durable democratic practices and the creation of open-access institutions (Sunde *et al.*, 2007).

Escaping from inequality?

The dangers of within-country inequality have finally been appreciated by African leaders. On 30 April 2014, 250 delegates from across Africa accepted the Accra Declaration on Inequalities in the Context of Structural Transformation. The declaration formed part of the common African position on the post-2015 development agenda, eventually reflected in the emphasis on the importance of reducing inequality as goal 10 of the United Nations Sustainable Development Goals to 2030. The Accra Declaration also supports 'Agenda 2063', the strategy of the African Union to achieve an equitable and humane future for the 2.5 billion Africans that there will be by mid-century.

Figure 8.1 suggests that many African states have made progress in recent years towards reducing income inequality, in contrast to what happened in South East Asia, and in most of the high-income members of the Organisation for Economic Co-operation and Development (OECD), where inequality continues to rise (Milanovic, 2016). Figure 8.3 summarises by state and type of economy what happened to net income inequality in Africa over two decades, with the solid line the prediction of inequality during the latter period based on earlier levels. Note the relative concentration of labour reserves (diamond markers) at the top end of the inequality spectrum and above the solid line. The almost 45 degrees of the slope of the solid line suggests that inequality in the 1990s was a good predictor of inequality in the 2000s. Nevertheless, in 17 of the 32 states for which we have data for both points in time, income inequality was lower than predicted (placed below the regression line). Nineteen had higher levels, including most of the labour reserves. In North Africa, only Tunisia experienced a decline in inequality.

Figure 8.3 Net income inequality in Africa, 1990s and 2000s

Source: Own calculations, based on UNDP (2014).

The downward trends await analysis, but it is likely that income windfalls for the middle and poorer classes explain most of what happened to inequality in SSA (Nel, 2016). Increased exports of natural resources accounted for the largest sectoral contribution to economic growth experienced by SSA in the period under review. Growth in the natural resource sector contributed twice as much as retail and wholesale commerce to growth, and considerably more than agriculture, telecommunications and manufacturing (Leke *et al.*, 2010). Natural resource rents in SSA are labour-intensive but do not demand the level of skilled labour associated with the manufacturing sector. Unskilled labour is an abundant/cheap factor of production, and its income is boosted by increased external demand, as standard trade theory suggests. One indication of this growth in external demand is the natural resource intensity of exports, which in the case of SSA increased by 70 per cent from the 1990s to the late 2000s (based on World Bank, 2015). Initial tests confirm that export growth in a labour-intensive, low-skilled environment boosted incomes of the lower-income groups sufficiently to lower income inequality in SSA (Nel, 2016). Given the volatility of international resource demands, such income windfalls are not a solid basis for inequality reduction over the longer term.

Increased remittance flows could also have acted as an income windfall. In the early 1990s, total remittance flows per capita to SSA averaged US$27. By 2010, this increased to just below US$50 (2005 constant US dollars), and overall some US$20 billion in remittances flowed to SSA in 2010 alone (Anyanwu, 2011; Mohapatra and Ratha, 2011). We have no reason to believe that remittances flow only to the poor, though, or more to them than to other income groups. The empirical evidence suggests that remittance flows actually marginally increase income inequality, i.e. the middle classes and the rich gain relatively more than the poor (Anyanwu, 2011).

A third source of increased relative income for the 'not rich' is the recent expansion of social protection programmes in the region. Social protection includes social insurance aimed at reducing risks due to economic shocks, and social assistance whose purpose is to augment consumption of vulnerable groups. Following examples of successful programmes in Latin America (Bastagli, 2010; Garcia and Moore, 2012; Barrientos, 2013), the focus in SSA in the 2000s shifted increasingly to broad-based, non-contributory social assistance. In Africa as a whole, the number of non-contributory insurance and assistance programmes tripled between 2000 and 2015 (Cirillo and Tebaldi, 2016), and in SSA 43 such programmes were launched in two-thirds of the states in the region between 1995 and 2015, only beaten by Latin America with 71 such programmes during the same era. The overwhelming majority are in the form of targeted or non-targeted *unconditional* cash transfers, primarily aimed at augmenting the consumption of vulnerable groups, including poor families, war veterans, widows, survivors of genocide, retirees, and children and their caregivers. Data collected by the World Bank suggest that these programmes significantly augment the net income of poorer households and account for declining net income inequality in those states below the solid line in Figure 8.3 (ASPIRE, 2016). Of all the factors mitigating inequality, unconditional cash transfers seem to hold the most promise, as long as they explicitly target the poorer segments of African societies.

Conclusions

Multidimensional inequality is deeply entrenched in most of Africa, and displays both vertical and horizontal dimensions that retard human development. The roots of inequality lie in the colonial past, and have been reinforced by the limited-access institutions that the colonisers had established and generations of African leaders since then have maintained. The evidence reviewed here gives us good reason to doubt that our extensive programmes of poverty reduction will ever be effective as long as inequality is not addressed as at least a co-determinant of poor human development. There are still too many gaps in our understanding of inequality in Africa to be able to do so systematically, though. In particular, we need to collect more information on the exact levels of the different dimensions of inequality in all of the countries of the region. More attention should also be paid to the diachronic dimensions of inequality, and especially whether and how inequality changes over the lifetime of individuals, and how intergenerational mobility is affected by inequality.

There are some indications that African decision-makers and their advisers are taking inequality more seriously than hitherto. One positive initiative that has flowed from this is the large number of conditional and unconditional cash transfers that some African states and international donors have started to finance as part of a programme of enhanced social protection. The full emergence of 'welfare states' that systematically reduce inequality through fiscal means depends on political will and the revenue generating capacity of African states (Addison and Ndikumana, 2001). Too much capital still exits Africa legally and illegally, but attempts are being made to address these drains on fiscal capacity (Kumar, 2014). It can also be expected that as effective democratic practices take firmer root on the continent, pressure for general and inclusive social redistribution will increase (Seekings, 2008).

References

Acemoglu, D., Johnson, S. and Robinson, J. (2001) 'The colonial origins of comparative development: an empirical investigation', *American Economic Review*, 91(5): 1369–1401.

Acemoglu, D., Naidu, S., Restrepo, P. and Robinson, J. (2015) 'Democracy, redistribution, and inequality', in A.B. Atkinson and F. Bourguignon (eds), *Handbook of Income Distribution*, Volume 2B, Amsterdam: Elsevier, pp. 1885–1966.

Addison, T. and Ndikumana, L. (2001) *Overcoming the Fiscal Crisis of the African State*, Discussion Paper 2001/12, UNU-WIDER, Helsinki, Finland.

AfDB (African Development Bank Group) (2012) 'Inequality and Arab Spring revolutions in North Africa and the Middle East', *Africa Economic Brief*, 3(7), July: 1–24.

AfDB (African Development Bank Group), OECD and UNDP (2015) *African Economic Outlook 2015: Regional Development and Spatial Inclusion*, available at: www.africaneconomicoutlook.org/sites/default/files/content-pdf/AEO2015_EN.pdf (accessed 13 December 2017).

Amin, S. (1972) 'Underdevelopment and dependence in black Africa: origins and contemporary forms', *The Journal of Modern African Studies*, 10(4): 503–524.

Anderson, E. and McKay, A. (2004) *Why Is Inequality So High, but Also So Variable in Sub-Saharan Africa?* London: Poverty and Public Policy Group, Overseas Development Institute.

Angeles, L. (2007) 'Income inequality and colonialism', *European Economic Review*, 51(5): 1155–1176.

Anyanwu, J. (2011) *International Remittances and Income Inequality in Africa*, Working Paper No. 135, Tunis: African Development Bank.

ASPIRE (2016) *ASPIRE: The Atlas of Social Protection Indicators of Resilience and Equity*, Washington, DC: World Bank, available at: http://datatopics.worldbank.org/aspire/ (accessed 13 December 2017).

Atkinson, A. (2015) *Inequality: What Can Be Done?* Cambridge, MA: Harvard University Press.

Barrientos, A. (2013) *Social Assistance in Developing Countries*, Cambridge: Cambridge University Press.

Bastagli, F. (2010) *Poverty, Inequality and Public Cash Transfers: Lessons from Latin America*, Background Paper for the European Report on Development (ERD) 2010 on Social Protection for Inclusive Development, Florence: European University Institute.

Bates, R. (1983) 'Modernization, ethnic competition, and the rationality of politics in contemporary Africa', in D. Rothchild and V. Olorunsola (eds), *The State Versus Ethnic Claims: African Policy Dilemmas*, Boulder, CO: Westview Press, pp. 152–171.

Bird, K., Higgins, K. and Harris, D. (2010) *Spatial Poverty Traps: An Overview*, Chronic Poverty Research Centre Working Paper No. 161, London: Overseas Development Institute.

Blundo, G. and Olivier de Sardan, J.-P. (2006) *Everyday Corruption and the State: Citizens and Public Officials in Africa*, London: Zed Books.

Booysen, F., Burger, R., Du Rand, G., van der Berg, S. and von Maltitz, M. (2007) *Trends in Poverty and Inequality in Seven African Countries*, PMMA Working Paper 2007/06.

Byamugisha, F. (2013) *Securing Africa's Land for Shared Prosperity*, Washington, DC: World Bank.
Chong, A. and Gradstein, M. (2004) *Institutions and Inequality*, Report of the Inter-American Development Bank, Washington, DC.
Cirillo, C. and Tebaldi, R. (2016) *Social Protection in Africa: Inventory of Non-Contributory Programmes*, Rio de Janeiro: International Policy Centre for Inclusive Growth, UNDP and Government of Brazil.
Credit Suisse (2014) *World Wealth Report 2014*, Zürich: Credit Suisse Research Institute.
Davies, J. (2009) 'Wealth and economic inequality', in B. Nolan, W. Salverda and T.M. Smeeding (eds), *The Oxford Handbook of Economic Inequality*, Oxford: Oxford University Press, pp. 127–149.
De Vreyer, P. and Roubaud, F. (2013) *Urban Labour Markets in Sub-Saharan Africa*, Washington, DC: World Bank.
Deininger, K. and Squire, L. (1998) 'New ways of looking at old issues: inequality and growth', *Journal of Development Economics*, 57: 259–287.
Engerman, S. and Sokoloff, K. (2000) 'History lessons: institutions, factor endowments, and paths of development in the New World', *Journal of Economic Perspectives*, 14(3): 217–232.
Ferreira, F. and Ravallion, M. (2009) 'Poverty and inequality: the global context', in B. Nolan, W. Salverda and T.M. Smeeding (eds), *The Oxford Handbook of Economic Inequality*, Oxford: Oxford University Press, pp. 599–636.
FAO (Food and Agriculture Organization) (2011) *The State of Food and Agriculture 2010–11: Women in Agriculture*, available at: www.fao.org/gender-landrights-database/resources/publications/publication-details/en/c/240130/ (accessed 13 December 2017).
Folbre, N. (2009) 'Inequality and time use in the household', in B. Nolan, W. Salverda and T.M. Smeeding (eds), *The Oxford Handbook of Economic Inequality*, Oxford: Oxford University Press, pp. 342–363.
Fosu, A. (2008) 'Inequality and the growth poverty nexus: specification empirics using African data', *Applied Economics Letters*, 15(7–9): 563–566.
Fosu, A. (2009) 'Inequality and the impact of growth on poverty: comparative evidence for sub-Saharan Africa', *Journal of Development Studies*, 45(5): 726–745.
Fosu, A. (2010) 'The effect of income distribution on the ability of growth to reduce poverty: evidence from rural and urban African economies', *American Journal of Economics and Sociology*, 69(3): 1034–1053.
Frank, R. and Rainer, I. (2012) 'Does the leader's ethnicity matter? Ethnic favoritism, education, and health in sub-Saharan Africa', *American Political Science Review*, 106(2): 294–325.
Frankema, E. (2005) *The Colonial Origins of Inequality: Exploring the Causes and Consequences of Land Distribution*, Discussion Paper, Ibero-America Institute of Economic Research, University of Goettingen, Germany.
Frankema, E. (2010) 'The colonial roots of land distribution: geography, factor endowments or institutions?', *Economic History Review*, 63(2): 418–451.
Garcia, M. and Moore, C.M.T. (2012) *The Cash Dividend: The Rise of Cash Transfer Programs in Sub-Saharan Africa*, Washington, DC: World Bank.
Go, D., Nikitin, D., Wang, X. and Zou, H. (2007) 'Poverty and inequality in sub-Saharan Africa: literature survey and empirical assessment', *Annals of Economics and Finance*, 8(2): 251–304.
Gyimah-Brempong, K. (2002) 'Corruption, economic growth, and inequality in Africa', *Economics of Governance*, 3: 183–209.
Fjelde, H. and Østby, G. (2014) 'Socioeconomic inequality and communal conflict: a disaggregated analysis of sub-Saharan Africa, 1990–2008', *International Interactions: Empirical and Theoretical Research in International Relations*, 40(5): 737–762.
Kakwani, N. and Pernia, E. (2000) 'What is pro-poor growth?', *Asian Development Review*, 18(1): 1–16.
Klasen, S. and Blades, D. (2013) 'Issues and challenges in measuring national income, wealth, poverty, and inequality in sub-Saharan Africa: an introduction', *Review of Income and Wealth*, 59, October: S1–S8.
Kumar, C. (2014) *Africa Rising? Inequalities and the Essential Role of Fair Taxation*, London: Christian Aid.
Lange, M.K. (2004) 'British colonial legacies and political development', *World Development*, 32(6): 905–922.
Leke, A., Lund, S., Roxburgh, C. and van Wamelen, A. (2010) *What's Driving Africa's Growth?*, available at: www.mckinsey.com/global-themes/middle-east-and-africa/whats-driving-africas-growth (accessed 8 January 2018).
Londregan, J., Bienen, H. and van de Walle, N. (1995) 'Ethnicity and leadership succession in Africa', *International Studies Quarterly*, 39(1): 1–25.
Lopesa, P. (2005) *The Disconcerting Pyramids of Poverty and Inequality in Sub-Saharan Africa*, IMF Working Paper No. 47, Washington, DC.

Lustig, N. (2015) *Inequality and Fiscal Redistribution in Middle Income Countries: Brazil, Chile, Colombia, Indonesia, Mexico, Peru and South Africa*, Working Paper No. 410, Washington, DC: Center for Global Development.

Lynch, G. and Crawford, G. (2011) 'Democratization in Africa: 1990–2010 – an assessment', *Democratization*, 18(2): 275–312.

Marshall, M.G., Gurr, T.R. and Jaggers, K. (2013) *Polity IV: Political Regime Characteristics and Transitions, 1800–2013*, available at: www.systemicpeace.org/polityproject.html (accessed 13 December 2017).

Melamed, C. and Samman, E. (2013) *Equity, Inequality, and Human Development in a Post-2015 Framework*, New York: UNDP.

Milanovic, B. (2003) *Is Inequality in Africa Really Different?* Washington, DC: World Bank.

Milanovic, B. (2016) *Global Inequality: A New Approach for the Age of Globalization*, Cambridge, MA: Harvard University Press.

Mkandawire, T. (2010) 'On tax efforts and colonial heritage in Africa', *The Journal of Development Studies*, 46(10): 1647–1669.

Mohapatra, S. and Ratha, D. (2011) 'Migrant remittances in Africa: an overview', in S. Mohapatra and D. Ratha (eds), *Remittance Markets in Africa*, Washington, DC: World Bank.

Mveyange, A. (2015) *Night Lights and Regional Income Inequality in Africa*, Working Paper No. 2015/085, Helsinki: UNU-WIDER.

Nel, P. (2003) 'Income inequality, economic growth, and political instability in sub-Saharan Africa', *The Journal of Modern African Studies*, 41(4): 611–639.

Nel, P. (2008a) *The Politics of Economic Inequality in Developing Countries*, Houndmills: Palgrave Macmillan.

Nel, P. (2008b) 'Dealing with inequality: a conversation with Islamic economics', in T. Shogimen and C.J. Nederman (eds), *Western Political Thought in Dialogue with Asia*, Lanham, MD: Lexington Books, pp. 135–159.

Nel, P. (2016) 'Redistribution with African characteristics', unpublished research paper available on request from author.

North, D., Wallis, J. and Weingast, B. (2009) *Violence and Social Orders: A Conceptual Framework for Interpreting Recorded Human History*, Cambridge: Cambridge University Press.

Odedokun, M. and Round, J. (2004) 'Determinants of income inequality and its effects on economic growth: evidence from African countries', *African Development Review*, 16(2): 269–284.

Okojie, C. and Shimeles, A. (2006) *Inequality in Sub-Saharan Africa: A Synthesis of Recent Research on the Levels, Trends, Effects, and Determinants of Inequality in Its Different Dimensions*, Report of the Inter-Regional Inequality Facility, London: Overseas Development Institute.

Oya, C. (2010) *Rural Inequality, Wage Employment and Labour Market Formation in Africa: Historical and Micro-Level Evidence*, Working Paper No. 97, Policy Integration Department, Geneva: International Labour Organization.

Peters, P. (2004) 'Inequality and social conflict over land in Africa', *Journal of Agrarian Change*, 14(3): 269–314.

Piketty, T. (2014) *Capital in the Twenty-First Century*, Cambridge, MA: Belknap Press.

Ravallion, M. and Chen, S. (2012) *Monitoring Inequality*, available at: http://blogs.worldbank.org/developmenttalk (accessed 13 December 2017).

Ray, D. (1998) *Development Economics*, Princeton, NJ: Princeton University Press.

Rushtad, S. (2015) 'Socioeconomic inequalities and attitudes toward violence: a test with new survey data in the Niger Delta', *International Interactions*, DOI: 10.1080/03050629.2015.1048856.

Sahn, D. and Stifel, D. (2000) 'Poverty comparisons over time and across countries in Africa', *World Development*, 28(12): 2123–2155.

Sahn, D. and Stifel, D. (2003) 'Urban–rural inequality in living standards in Africa', *Journal of African Economies*, 12(4): 564–597.

Seekings, J. (2008) 'Welfare regimes and redistribution in the South', in I. Shapiro, P.A. Swenson and D. Donno (eds), *Divide and Deal: The Politics of Redistribution in Democracies*, New York: New York University Press, pp. 19–42.

Seekings, J. and Nattrass, N. (2015) *Policy, Politics, and Poverty in South Africa*, Houndmills: Palgrave Macmillan.

Shipton, P. and Goheen, M. (1992) 'Introduction: understanding African land-holding – power, wealth, and meaning', *Africa: Journal of the International African Institute*, 62(3): 307–325.

Solt, F. (2009) 'Standardizing the World Income Inequality Database', *Social Science Quarterly*, 90(2): 231–242.

Solt, F. (2014) *The Standardized World Income Inequality Database*, available at: https://dataverse.harvard.edu/dataset.xhtml?persistentId=hdl:1902.1/11992 (accessed 13 December 2017).

Sunde, U., Cervelatti, M. and Fortunato, P. (2007) *Are All Democracies Equally Good? The Role of Interactions Between Political Environment and Inequality for Rule of Law*, Discussion Paper No. 2984, Bonn: Institute for the Study of Labor.

Thorbecke, E. (2013) 'The interrelationship linking growth, inequality and poverty in sub-Saharan Africa', *Journal of African Economies*, 22(Suppl. 1): 15–48.

UNDP (United Nations Development Programme) (2014) *Human Development Report 2014*, New York: UNDP.

Uslaner, E. (2007) *Corruption and the Inequality Trap in Africa*, AfroBarometer Working Paper No. 69, Cape Town: Institute for Democracy in South Africa.

van der Berg, S. (2014) 'Inequality, poverty and prospects for redistribution', *Development Southern Africa*, 31(2): 197–218.

van de Walle, N. (2009) 'The institutional origins of inequality in sub-Saharan Africa', *Annual Review of Political Science*, 12: 307–327.

World Bank (2015) *World Development Indicators 2015*, Washington, DC: World Bank.

9
Education

David Stephens

Introduction

This chapter explores the understandings and development of education in Africa. From a discussion of indigenous knowledge and schooling, the chapter moves on to examine seven key trends in African education since the independence of African nations from colonial rule in the 1960s: increasing access and enrolment in education, questions of quality in education, issues of language and literacy, the relationship between gender parity and equality in education, higher education, aid and education in Africa, and the role of education in conflict, reconciliation and reconstruction. The chapter concludes with a brief examination of the future of education on the continent and lessons that have been learnt.

Indigenous knowledge and education

It is worth beginning with a truism, namely that schooling and education are not necessarily the same thing. In fact, the history of education in Africa provides ample evidence of the divergences and great ideological divides between Western-style schooling, largely introduced onto the continent by Western missionaries, and what has become known as indigenous knowledge systems (IKS) (Odora Hoppers, 2002). It is worth reiterating from the start that whereas Western-style schooling was introduced into Africa relatively recently, well-established forms of traditional or indigenous education have existed in Africa for hundreds of years. The relatively recent attempt by the self-styled Islamic State to destroy the libraries of ancient documents in Timbuktu is evidence of one of Africa's centres of learning that well predates the arrival of the Europeans.

It is also a mistake to think of indigenous knowledge systems as solely existing in 'another country', but rather to evaluate the contribution traditional education might make in the solving of the continent's enduring developmental challenges.

Indigenous traditional knowledge has been defined as knowledge or knowing based upon a world view and a culture that is basically relational. The spirit of the African world view includes wholeness, community and harmony, which is deeply embedded in its cultural values. An individual becomes human only in the midst of others, and seeks both individual and collective harmony as the primary task in the process of becoming a true person (Sarpong, 1991, 2002).

African indigenous knowledge is essentially local rather than global, embedded in a particular cultural context, communal rather than individualistic, holistic rather than adhering to a subject/object dichotomy, and of course practical in its preoccupation with learning the skills and knowledge required for sustainable living. Its major aim is fostering a strong sense of inclusive, communal identity, particularly during the period of transition between childhood and adulthood (Agrawal, 1995). Such traditional forms of education are well described in two of Africa's greatest novels: Chinua Achebe's (1958) *Things Fall Apart* and the slightly lesser-known *The African Child* by Camara Laye (1959).

The failure of grand universalist theories of development in Africa have led to a resurgence of interest in the potential of IKS to contribute to development initiatives, particularly those that are local and concerned with sustainable development, for example in health, agriculture and animal husbandry (Eyong, 2007).

The introduction and establishment of Western-type schooling in Africa

In 1971, the esteemed African educationist E.A. Ayandele wrote this about the coming of Western education to Africa:

> The feverish energy with which practically all independent African countries have been sponsoring literary education in the last ten years, signally underlines their full recognition of the strategic and decisive role which this sine qua non dynamic of social change and progress should play in the programme of modernization.
>
> *(Ayandele, 1971: 21)*

As mentioned earlier, Western-type education in most African countries was established by Christian missionaries and subsequently supplemented by colonial governments. Formal schooling along Western lines was a major activity of all Christian missions since they considered schools a good medium through which to mould the attitudes of youth and a valuable source of future manpower for the spread of the gospel. Western-type schooling was not introduced and established in a uniform manner, however.

Because of geographic factors, such as the closeness of the coast or the existence of cool areas favourable to Europeans' living conditions, and because of different responses by chiefs to mission initiatives, some groups were influenced more strongly than others. Peoples who had previously existed more or less independently found themselves surrounded and defined by colonial boundaries. With the introduction of the cash economy, those who had received the most education came to dominate the rest. Examples of educational and economic imbalances that arose in this way can be found from across the continent. The Lagoon peoples of Ivory Coast, for example, were strongly influenced by colonial schooling because they inhabited a coastal region and because the French colonial administration was anxious to forestall the influence of the British in neighbouring Gold Coast, while educational developments among the Senoufolo in the north was much less. Similarly, in Southern Sudan, the Dinka people responded more favourably to the availability of Western-type schooling than the Nuer (Sommers, 2005). In both countries, the groups with the greater Western educational development came to dominate subsequent economic and political processes. Once these imbalances have been created, they tend to persist, and successive governments have found them very difficult to eliminate.

Colonial authorities were well aware of the potentially disruptive effect of the economic and social structures they and the missionaries were importing, and in both British and French colonies efforts were made to ensure the dominance of certain groups in both the old and new systems by making special provisions for the sons of chiefs (Clignet and Foster, 1964).

The idea was that by the time the sons took over from their fathers, they would have acquired the skills necessary for leadership in the new system as well as the old. In many cases, this policy of 'indirect rule' was successful, and it reinforced existing social structures while also ensuring traditional chiefs remained compliant. In Uganda, for example, both Protestant and Roman Catholic missionaries established schools for the ruling groups with the idea of converting lower classes to their religion through the chiefs. However, it also happened that traditional rulers were suspicious of European activities, and when placed under pressure some sent children of slaves instead of their own children (Stephens, 1982).

It must also be remembered that in these new forms of education, learning was largely conducted in the languages of the colonialists. In East and southern Africa, schools were also racially segregated, with educational institutions in Kenya and Tanganyika, for example, divided along racial grounds. The subsequent European, Asian and African schools could be ranked in order of quality, each possessing different curricula and each producing children destined for prescribed positions within the economy and society (Brock-Utne, 2001).

Education and development in Africa: seven key trends

As a general rule, upon independence from the colonial powers, all African countries inherited underdeveloped and racially segregated education systems characterised by inequalities and regarded as irrelevant to the needs of the country (Nieuwenhuis, 1997). An assumption that guided education policy formulation in postcolonial Africa was that education was one of the most important vehicles for bringing about development and change, that it was the way to ensure economic growth, to restructure the social order and to reduce the social ills of society in general. Countries were therefore committed to reforming or restructuring their education systems, adjusting the length of educational cycles, changing curriculum content, and linking education and training to perceived requirements for national socio-economic development (Thurlow, 1992).

Since the heady days of independence until now (2016), it is possible to identify seven major trends or challenges that have characterised the development of African education.

Increasing access and enrolment in education

One of the most important challenges facing newly independent African nations in the early 1960s was opening up education, particularly basic education, to a growing population eager to access some kind of schooling. Since the 1960s, this challenge – which has been called universal primary education, or UPE – has largely been met, but the road has not been easy. It is possible to identify three periods in enrolment growth between 1960 and 2005 (Fredricksen, 2016):

- *1960–1980*: The gross enrolment ratio (GER) in primary education in sub-Saharan Africa (SSA) grew from about 40 per cent in 1960 to 80 per cent in 1980, implying an enrolment growth of 230 per cent. This impressive growth meant that enrolment in 1980 exceeded by 25 per cent, suggested by the 1961 'Addis Ababa target' of attaining UPE by 1980. However, the UPE target was not reached because the population of primary school age almost doubled during this period (20 per cent foreseen).
- *1980–2000*: This was a period of stagnation, sometimes referred to as the 'lost decades', characterised by the imposition of structural adjustment and neo-liberalism. The gross

enrolment ratio declined to a low of 72 per cent in 1992, and only regained its 1980 level of 80 per cent in 2000. The 70 per cent enrolment growth achieved was just sufficient to keep pace with population growth.

- *1999–2005*: In recent years, in line with commitments to the Millennium Development Goals, strong enrolment growth has resumed. Primary enrolment grew by 36 per cent during this six-year period, and the GER increased from 80 per cent to 97 per cent.

Since 2005, we have seen this trend continue, apart from those sub-Saharan African countries recovering from recent periods of war or civil unrest. We will return to this issue when looking at our seventh trend, conflict, reconciliation and reconstruction.

The increase or stagnation in enrolment can be explained by the differing abilities of African countries to finance education (Fredricksen, 2016). Education budgets increased annually by only 2.3 per cent during the period 1980–1999 (i.e. slightly below population growth). This reflects the economic crisis that hit the continent from the mid-1970s, resulting in a decline in GDP per capita by 36 per cent during the 1970–1997 period. To substitute for stagnant public budgets, many countries introduced school fees, a cost parents could ill afford. The result was a major reduction in UPE. The number of countries with a GER above 100 per cent declined from 16 in 1980 (accounting for 50 per cent of sub-Saharan Africa's school-age population) to seven in 1995 (7 per cent of the population, most of which was in South Africa).

During the period 1999–2005, total education budgets in sub-Saharan Africa increased by 9.2 per cent annually, and school fees were abolished in many countries. The improved public financing is explained by a combination of factors:

- resumption of economic growth (4.8 per cent annual GDP growth during the 2000–2005 period as compared to 2.1 per cent during the period 1980–1999);
- increased share of GDP spent on education (from 3.7 per cent in 1999 to 5.0 per cent in 2005); and
- resumed growth in education aid (3.8 per cent annually between 1999 and 2005) as compared to a marked decline in the 1990s (Lewin, 2008).

A major conundrum, however, has been the challenge in raising enrolment through increased financial investment in school buildings, teachers' salaries, etc., while at the same time ensuring that questions of quality are attended to. It is to this second trend we now turn.

Questions of quality

In 1989, the author and a colleague at London University's influential Institute of Education wrote *Questions of Quality: Primary Education and Development* (Hawes and Stephens, 1989). This 'classic book' (Davies, 2013) reviewed the landscape of education and development worldwide, and argued that whereas great strides had been made in increasing the number of schools, huge challenges remained when it came to the quality of education most children received. Generally speaking, progress in terms of access to schooling has outpaced rises in the quality of education, the latter hampered too by a lack of standardised indicators to measure progress.

Once enrolled, the question is one of quality. As we wrote in 1989, our interpretation of quality is a relational one, namely:

> not only whether school content makes leavers immediately more employable but also whether it makes them, in the long term, more adaptable. Not only whether certain approaches in school make children perform better in conventional multiple choice tests but also whether such tests cause learners and their teachers to question what it is that these tests are measuring. Also not only whether the existence of trained teachers or textbooks makes a difference to achievement but also what kind of teachers and what kind of textbooks are desirable given the real needs of learners. And finally, not only whether school knowledge is efficiently transmitted but also whether such knowledge remains school knowledge or whether it transfers into the home and community once the pupils walk out of the school compound.
>
> *(Hawes and Stephens, 1989: 11)*

Our argument for raising the quality of primary education in Africa – and elsewhere – was grounded in a rights-based perspective. For us, quality schooling was the appropriate vehicle for the enactment of both negative and positive rights, from the right not to be abused or exploited at work to the positive rights around the celebration of creativity, the use of local languages in the classroom and pupil participation in democratic structures and debate, many of which could be modelled at school (Tikly and Barrett, 2011). Looking back, I think it is fair to say that our ideas were very much grounded in the spirit of the times; if writing now I, for one, would accord a greater role to not just educational *processes*, but cognitive and learning *outcomes*.

An alternative approach to quality drew upon human capital theory, arguing that improved levels of quality education led to increased earnings for the school leaver individually and to economic growth nationally (Pillay, 1994). Pillay's study looked at racially specific differences in South Africa, where he found that the quality of the learner's final examination, called the matriculation or matric, had a direct relationship to the level of earnings of that school leaver.

Since the late 1980s, a great deal has been written on strategies to improve the quality of schooling in Africa (Harber and Stephens, 2010; Davies, 2013). Though the human capital and rights-based approaches to quality remain popular, since the early years of this century we have seen the extension of a rights-based approach to that of capabilities and social justice approaches that derive from the work of Sen and Nussbaum (cited in Tikly and Dachi, 2009). Here, quality is understood in terms of capabilities, in particular those that promote the idea of agency freedom (i.e. that individuals can act to bring about changes they value). Quality schooling therefore becomes one of supporting the development of autonomy and the ability to make choices in later life, rather than simply providing individuals with the necessary resources to learn. The social justice approach to quality schooling argues that not all learners require access to the same kind of inputs. Gender equity is also very much part of a social justice approach (Unterhalter, 2007) in which cultural norms and values across the African continent need to provide the framework for a relevant education that will lead to political, economic and cultural change.

It is important to understand, however, that the quality of education is not improved in isolation. There are a number of factors that promote quality within the school, the language of instruction being an important one that we will turn to now. Other external factors concern levels of poverty and the commitment of national governments to prioritise teacher education and infrastructural support of schools and colleges of education. We will address these external factors later in this chapter.

Literacy and language education

An enduring educational trend since the early 1960s has been the debate concerning the role of language in education. Upon independence, most African countries continued to rely upon the language of their colonial master. The choice of language of instruction in Africa was – and still is – a political choice, a choice that may distribute power in a global context, as well as within an African country, between the elites and the masses. As Alexander (2000: 11) asserts:

> it is an indisputable fact that in the post-colonial situation, the linguistic hierarchy built into the colonial system led to knowledge of the conquerors' language becoming a vital component of the 'cultural capital' of the neo-colonial elite. It was and remains their knowledge of English and/or French (for example) that sets them apart from the vast majority of African compatriots and which keeps them and their offspring in the privileged middle and upper classes.

Since Alexander (2000) wrote this assertion, though the language of instruction issue has changed little, what we have seen is both the rise of technology, in particular the increase in affordability of the mobile phone, and the rise and reach of English as a global language. This presents staunch proponents of mother tongue education such as Brock-Utne – who publishes widely in her second language, English – with a dilemma. Many parents aspiring to achieve education as a force for social mobility for their offspring can readily see the advantages in acquiring such cultural and economic capital. However, education is not just about capital or social mobility, but also cultural identity and the promotion of world views shaped by indigenous languages. It is the 'glue' that binds past and present, art and representations of the world, and is a crucial part of a child's first three years of schooling (Fafunwa, 1990; Bambose, 1992; Brock-Utne, 2001).

It is important, however, not to treat the issues surrounding language in education in the African continent as a monolithic entity. The situation of Botswana and Tanzania is particularly interesting (Arthur, 2001) with the ex-colonial language English continuing to play a significant role in education as one of the official languages of instruction, but this role being challenged by a lingua franca with a wide national currency (i.e. Setswana in the case of Botswana, and Kiswahili in the case of Tanzania). These two countries provide compelling evidence that if properly resourced in terms of materials in first or other second languages, and if teachers are trained to use more communicative approaches to language learning (whether in English or the mother tongue), children can enjoy a bilingual education that is fit for purpose in a world that is both local and global.

Gender parity and equality

'Educate a man, and you educate an individual; educate a woman, and you educate a whole nation', so says the old adage. Girls' education is not only a fundamental human right, but is also an important driver for economic growth and human development (Shabaya and Konadu-Agyemang, 2004).

Despite progress in recent years, girls continue to suffer severe disadvantage and exclusion in education systems throughout their lives (UNICEF, 2017). An estimated 31 million girls of primary school age and 32 million girls of lower secondary school age were out of school in 2013. Sub-Saharan Africa has the lowest proportion of countries with gender parity: only two out of 35 countries. On current trends, it is projected that 69 per cent of countries will have achieved parity in primary education, and 48 per cent of countries will have achieved parity in lower secondary education by the 2015 deadline.

Girls' education is both an intrinsic right and a critical lever to reaching other development objectives. Providing girls with an education helps break the cycle of poverty: educated women are less likely to marry early and against their will, less likely to die in childbirth, more likely to have healthy babies, and more likely to send their children to school. When all children have access to a quality education rooted in human rights and gender equality, it creates a ripple effect of opportunity that influences generations to come.

Girls' education is essential to the achievement of quality learning relevant to the twenty-first century, including girls' transition to and performance in secondary school and beyond. Adolescent girls that attend school delay marriage and childbearing, are less vulnerable to disease, including HIV and AIDS, and acquire information and skills that lead to increased earning power. Evidence shows that the return to a year of secondary education for girls correlates to a 25 per cent increase in wages later in life (Morley, 2006).

While gender parity has improved, barriers and bottlenecks around gender disparities and discrimination remain in place, especially at the secondary school level and among the most marginalised children. There are various barriers to girls' education throughout the world, particularly in sub-Saharan Africa, ranging from supply-side constraints to negative social norms. Some include school fees, strong cultural norms favouring boys' education when a family has limited resources, inadequate sanitation facilities in schools such as lack of private and separate latrines, and negative classroom environments, where girls may face violence, exploitation or corporal punishment. Additionally, schools often lack sufficient numbers of female teachers.

Increasingly, adolescent girls also face economic and social demands that further disrupt their education, spanning from household obligations and child labour to child marriage, gender-based violence and female genital cutting/mutilation. Recent estimates show that one-third of girls in the developing world are married before age 18, and one-third of women in the developing world give birth before age 20 (UNICEF, 2015). If all girls had secondary education in sub-Saharan Africa and South and West Asia, child marriage would fall by 64 per cent, from almost 2.9 million to just over 1 million. Inadequate or discriminatory legislation and policies often inhibit girls' equal access to quality education. In countries such as Afghanistan and Pakistan, formal or written threats to close girls' schools or end classes for girls have fuelled gender-motivated attacks on schools.

When compounded by factors such as poverty, disability and locations, such barriers can become nearly insurmountable for young girls. Over the past few years, ministries of education on the African continent, universities in the North and South, and organisations such as the Forum for African Women Educationalists (FAWE) have made important strides in not only raising the profile of gender in education, but also in identifying ways forward (Stephens, 1998).

Research in 2004 in Ghana, Zimbabwe and Kenya on the relationships between unequal access and participation in schooling and the gender gap in education suggested a number of policy recommendations (Shabaya and Konadu-Agyemang, 2004). These included the need for African governments to rededicate their efforts towards giving female education the highest priority, creating girl-friendly school environments, helping resolve parental poverty issues that compel them to force girls into early marriages, legislating and enforcing laws that compel all children to stay in school for at least 12 years, and wiping out the long-existing spatial inequities that enable people in certain locations to have better access to education than others (Shabaya and Konadu-Agyemang, 2004).

Higher education

The second half of the last century will go down in the history of higher education as the period of its most spectacular expansion: an over sixfold increase in student enrolments worldwide, from 13 million in 1960 to 82 million in 1995. But it is also the period that has seen the gap widen between industrially developed nations, the so-called developing countries, and in particular the least developed countries, many of which are to be found in Africa, with regard to access and resources for higher learning and research (Materu, 2007). Many universities in sub-Saharan Africa struggle to fund high-quality research, with the result that a majority of teaching is at undergraduate level, and any research activity either taking the form of collaboration with well-funded external projects or as evaluation of development assistance initiatives.

It has also been a period of increased socio-economic stratification, of the collapse of the Soviet Union, globalisation and the rise of the 'market' as a driving force, and, some would argue, the consequent widening of the gap in educational opportunities within countries, including in some of the most developed and wealthiest nations (Daun, 2005). The argument goes that without adequate higher education and research institutions providing a critical mass of skilled and educated people, no African country can ensure genuine endogenous and sustainable development. In particular, without higher education, developing countries and the least developed countries cannot reduce the gap separating them from the industrially developed ones. Sharing knowledge, international cooperation and new technologies can offer new opportunities to reduce this gap.

As we survey the African continent, we can see an unprecedented demand for and a great diversification in higher education, as well as an increased awareness of its vital importance for sociocultural and economic development, and for building the future for which the younger generations will need to be equipped with new skills, knowledge and ideals. Higher education was not only viewed as being central to post-independence development, but was also as a symbol of nationhood. For many, the coming of the university in Africa was greeted with a sense of communal pride and identity (Ajayi *et al.*, 1996). Much hope was pinned on the university for a better and fuller life for Africa's children and future generations (Oketch, 2009). However, this hope was short-lived, as from the 1980s this sense of optimism began to diminish, and in the 1990s it simply collapsed (Oketch, 2009). A succession of crises ensued as higher education basically 'fell apart' (Neave, 2003). A number of factors have been associated with this crisis. Notable ones include increased expansion beyond the public's financial capacity, declining economic performance, weak institutional management, lack of serious donor interest in higher education, a refocus on universal basic or primary education, and political and military unrest (Ajayi *et al.*, 1996; Oketch, 2009).

A 'solution' to this 'problem' has been the emergence of a mix between public and private provision, with universities springing up all over the continent. There are those that argue that the emergence of the private provider in Africa and elsewhere will encourage greater diversity and enhanced accountability to the 'student customer', while those concerned about the erosion of the state in the financing and management of higher education worry that private institutions in Africa tend to offer programmes with high private but fewer social benefits. There is also growing evidence that private universities offer poor value for money (Wilkinson and Yussof, 2005; Oketch, 2009).

In many ways, the challenges facing higher education in Africa are similar to those in the economically richer North: the gradual domination of managerialism in the organisation of both teaching and research, the commercialisation of research, and the outsourcing of many services.

What is different is the emergence of a global higher education 'industry' in which universities in the North – strapped for cash – are turning to the South as a growing and ready market willing to send its children to Northern institutions if they can afford the costs, or if not to cheaper outreach campuses. Those who cannot afford this route to higher education must rely upon a mix of public and private provision that varies in both quality and cost.

And there are important policy issues to be addressed in terms of basic and higher education. Currently, most countries in sub-Saharan Africa spend at least 10 times more on a university student than on a primary school pupil, with on average 8 out of every 10 dollars spent on university education in Africa being subsidised by country governments. These figures raise a serious question of equity. In Lesotho, for example, spending on tertiary students exceeds 50 times expenditure per primary school student (UNESCO, 2011).

The challenge seems to be whether it is possible for African governments to shift more resources to lower levels of education, without threatening the growth of the tertiary sector (Mingat and Tan, 1996). Development assistance or aid has played a significant role in shaping the education landscape of Africa since the independence movements of the early 1960s. From the arrival of European and American volunteer teachers in the 1960s, through the emergence of the World Bank's structural adjustment programmes in the 1980s, to the 'arrival' of the Chinese more recently, aid, as it has become known, can be seen as both a blessing and a curse.

Before we examine the record on aid to education, it is worth getting a sense of the scale of assistance. Overall, sub-Saharan Africa spends approximately 5 per cent of its gross domestic product on education, which is second only to North America and Europe at 5.3 per cent, with external aid accounting for about 5.6 per cent of government education budgets (though in some countries, such as Guinea, Mali, Rwanda and Zambia, development assistance accounts for about 50 per cent of government education budgets). However, although this figure is relatively small in terms of funding for education, it is fair to say that aid has played a significant role in influencing the policy priorities of most governments in sub-Saharan Africa.

Analyses of the impact of aid on education in Africa (e.g. Reimers, 1994; Lauglo, 1996; Heyneman, 2003; Stephens, 2007) suggest four broad lessons:

1 In the early 1960s, at the insistence of the World Bank, newly independent countries over-invested in vocational and technical education, with the result that thousands of workshops and laboratories for the most part became useless 'white elephants' (Heyneman, 2003).
2 Those who determined aid policy – the 'bankers' – tended to look at education through quite different professional lenses from the 'educationists' who were charged with implementing those policies (Lauglo, 1996). Whereas the former – the bankers – tended to focus upon reducing the complexity of education to easily generalisable but limited aspects of inputs and outputs, the latter – the teachers on the ground – were much more concerned with raising the quality of educational processes.
3 Structural adjustment, the mantra of the 1980s, had a devastatingly negative effect upon education, particularly on both the direct financing of public education and indirectly on poorer households who were unable to pay 'user' fees. For Reimers (1994), structural adjustment can be blamed for lost opportunities not only in terms of what it did in marginalising education for all, but for what it failed to do in missing an opportunity in the 1980s to prioritise public educational development in Africa.
4 The arrival of the Chinese government as a leading player in development assistance to Africa raises a number of questions, the most important being the direct linkages between educational assistance and the economic benefits to the Chinese (Stephens, 2015). A second

> issue concerns the current 'stand-alone' bilateral policy of the Chinese in its relationships with target recipient countries. This policy ties the giving of aid directly to economic benefits to be accrued by the Chinese. However, the signing of the Declaration of Sharm El Sheikh in 2009 indicates that the Chinese government is willing, in principle at least, to gradually align its discourse and practices with the traditional donor groups. Here, 'trade for aid' is viewed as just one relationship, others being 'soft' loans and outright development assistance grants (Nordtveit, 2011).

Wedged between the recent economic crises and looming population growth, most African governments will need to make strategic decisions on how to budget for education. The population of sub-Saharan Africa's 5 to 14-year-olds is expected to grow by more than 34 per cent over the next 20 years, and the region will need to respond to the demands of 77 million new students (UNESCO, 2011). Aid will continue to shape education developments in Africa from unconditional grants, conditional loans to the work of non-governmental organisations and charities.

Conflict, reconciliation and reconstruction

In 2005, the influential journal *Compare* devoted a special issue to 'Education in the Twenty-First Century: Conflict, Reconciliation and Reconstruction'. The timing is important and appropriate. The author began his teaching career at the prestigious Wesley secondary school in Segbwema in Sierra Leone's Eastern Province, the site of the later terrible conflicts over blood diamonds in the 1980s. That school now lies in ruins, torched during the civil war and unlikely at the moment to be rebuilt. As the editor of the special edition notes, 'the number of countries experiencing war and armed conflict continues to increase' (Smith, 2005). The conflicts in Nigeria and neighbouring states as a result of Boko Haram, the post-conflict situations in North Africa following the Arab Spring of the past decade, and the enduring conflicts in Sudan, the Democratic Republic of Congo, Eritrea and the Al-Qaeda presence in the Maghreb areas of northern Africa illustrate the troubled nature of great swathes of the continent.

In brief, there are three challenges to Africa and her partners: first, an understanding of how conflict affects, and is influenced by, education systems; second, how education assistance can best be delivered during and after conflict; and third, what strategies are available for long-term conflict prevention and post-conflict reconciliation and reconstruction (Leach *et al.*, 2005). What is certain is that:

> protracted conflict, whatever its roots and justifications, has a devastating impact on education systems, with pupils and teachers killed or forced into fighting on one side or the other, infrastructure destroyed, and administration halted. As we know, civilians are increasingly the causalities of armed conflict; among them the vulnerable, children and women, are disproportionately affected.
>
> *(Leach* et al., *2005: 353)*

Strategies to deliver education during civil unrest include non-formal education, which can also be a cost-effective vehicle for conflict resolution and peace-building (Plonski, 2005). Clive Harber has long argued for a focus upon democratic education in African schools, arguing that many schools are 'toxic' in relation to corporal punishment and the modelling of authoritarian and undemocratic learning (Harber, 2008). Others have called for greater attention to be paid to global citizenship education, and more broadly for school knowledge in Africa and elsewhere to be seen as generative, participatory and empowering rather than as detached, authoritarian and hierarchical (Leach et al., 2005). What is clear is that the roots and causes of violence are to be found both within and without the African continent, as are the solutions to these enduring challenges.

The future of education in Africa

If the past is another country 'where we do things differently', perhaps the future is also an unknown landscape, but one we can be sure is very different from the one we encounter on the African continent and wherever Africans are educated? If the past has been about increasing access to schooling, raising the quality of education, improving the life chances of girls and women, and addressing issues of poverty and their impact upon education, the future appears – if the seeds are to be seen in the present – to be one in which the forces of education will be shaped by neo-liberalism, globalisation and the widening divide between rich and poor, the elites and the masses, and those enjoying better lives in the cities of Africa.

These are the lessons we can learn when we look back at the development of education in Africa. These include reappraising the contribution of indigenous forms of learning to modern education, continuing to prioritise the education of girls, ensuring that curricula and languages of instruction are relevant, employing modern technology in the classroom, and also ensuring that education is rightly used to reduce poverty and conflict (Adams and Page, 2003), and enhance social mobility.

References

Achebe, C. (1958) *Things Fall Apart*, London: Heinemann.

Adams, R. and Page, J. (2003) 'Poverty, inequality and growth in selected Middle East and North Africa countries, 1980–2000', *World Development*, 31(12): 2027–2048.

Agrawal, A. (1995) 'Dismantling the divide between indigenous and scientific knowledge', *Development and Change*, 26(3): 413–439.

Ajayi, J., Goma, L. and Johnson, G. (1996) 'The African experience with higher education', *Higher Education*, 35(4), June: 473–474.

Alexander, N. (2000) 'English unassailable but unattainable', *PRAESA Occasional Papers* 3, Cape Town: Project for the Study of Alternative Education in South Africa (PRAESA).

Arthur, J. (2001) 'Perspectives on educational languages policy and its implementation in African classrooms: a comparative study of Botswana and Tanzania', *Compare*, 31(3): 347–363.

Ayandele, E.A. (1971) 'The coming of Western education to Africa', *West African Journal of Education*, 15(1): 21–33.

Bambose, A. (1992) *Language Policy Options in Basic Education: Implications for Policy Formulation*, paper given at Language Policy, Literacy, and Culture Roundtable Discussion from the International Conference on Education, Geneva, 18 September.

Brock-Utne, B. (2001) 'Education for all – but in which language?', *Oxford Review of Education*, 27(1): 115–134.

Clignet, R. and Foster, P.J. (1964) 'French and British colonial education in Africa', *Comparative Education Review*, 8(2), October: 191–198.

Daun, H. (2005) 'Globalization of education and stigma: a Senegalese case study', in D.P. Baker and A.W. Wiseman (eds), *Global Trends in Educational Policy*, Bingley: Emerald Group Publishing, pp. 197–219.

Davies, L. (2013) 'International review of quality of education', unpublished manuscript, Centre for International Education and Research, University of Birmingham.

Eyong, C.T. (2007) 'Indigenous knowledge and sustainable development in Africa: case study on central Africa', in E.K. Boon and H. Hens (eds), *Indigenous Knowledge Systems and Sustainable Development: Relevance for Africa*, Delhi: Kamla-Raj, pp. 121–139.

Fafunwa, A.B. (1991) *A History of Education in Nigeria*, Ibadan: NPS.

Fredricksen, B. (2016) 'Funding education in sub-Saharan Africa: can the momentum be maintained during the current economic slowdown?', *Norrag News*, 27 April.

Harber, C. (2008) *Toxic Schooling: How Schools Became Worse*, Nottingham: Educational Heretics Press.
Harber, C. and Stephens, D. (2010) *From Shouters to Supporters: Quality Education Project – Final Evaluation Report*, Oslo: Save the Children.
Hawes, H. and Stephens, D. (1989) *Questions of Quality: Primary Education and Development*, Harlow: Longman.
Heyneman, S. (2003) 'The history and problems in the making of education policy at the World Bank 1960–2000', *International Journal of Educational Development*, 23: 315–337.
Lauglo, J. (1996) 'Banking on education and the uses of research: a critique of World Bank priorities and strategies for education', *International Journal of Educational Development*, 16(3): 221–233.
Laye, C. (1959) *The African Child*, London: Collins.
Leach, M., Scoones, I. and Wynne, B. (eds) (2005) *Science and Citizens: Globalization and the Challenge of Engagement*, London: Zed Books.
Lewin, K. (2008) *Strategies for Sustainable Education in Sub-Saharan Africa*, Washington, DC: World Bank.
Materu, P. (2007) *Higher Education Quality Assurance in Sub-Saharan Africa: Status, Challenges, Opportunities and Promising Practices*, Washington, DC: World Bank.
Mingat, A. and Tan, J.P. (1996) 'Subsidization of higher education versus expansion of primary enrolments: what can a shift of resources achieve in sub-Saharan Africa?', *International Journal of Educational Development*, 5(4): 259–268.
Morley, L. (2006) *Gender Equity in Commonwealth Higher Education: An Examination of Sustainable Interventions in Selected Commonwealth Universities*, London: DfID.
Neave, G. (2003) 'The Bologna Declaration: some of the historic dilemmas posed by the reconstruction of the community in Europe's systems of higher education', *Educational Policy*, 17(1): 141–164.
Nieuwenhuis, F.J. (1997) 'Can research into the development of education in post-colonial Africa shape education policies in South Africa?', *International Journal of Educational Development*, 17(2): 129–143.
Nordtveit, B.H. (2011) 'An emerging donor in education and development: a case study of China in Cameroon', *International Journal of Educational Development*, 31: 99–108.
Odora Hoppers, C.A. (2002) *Indigenous Knowledge and the Integration of Knowledge Systems: Towards a Philosophy of Articulation*, Cape Town: New Africa Books.
Oketch, M. (2009) 'Public–private mix in the provision of higher education in East Africa: stakeholders' perceptions', *Compare*, 39(1): 21–23.
Pillay, P. (1994) 'Quality of schooling, certification and earnings in South Africa', *International Journal of Educational Development*, 14(1): 13–23.
Plonski, S. (2005) 'Developing agency through peacebuilding in the midst of intractable conflict: the case of Israel and Palestine', *Compare*, 35(4): 393–411.
Reimers, F. (1994) 'Education and structural adjustment in Latin America and sub-Saharan Africa', *International Journal of Educational Development*, 14(2): 119–129.
Sarpong, P. (1991) *Ghana in Retrospect: Some Aspects of Ghanaian Culture*, Accra: African Books Collective.
Sarpong, P. (2002) *Peoples Differ: An Approach to Inculturation in Evangelization*, Accra: Sub-Saharan Publishers.
Shabaya, J. and Konadu-Agyemang, K. (2004) 'Unequal access, unequal participation: some spatial and socio-economic dimensions of the gender gap in education in Africa with special references to Ghana, Zimbabwe and Kenya', *Compare*, 34(4): 395–425.
Smith, A. (2005) 'Education in the twenty-first century: conflict, reconstruction and reconciliation', *Compare*, 35(4): 373–391.
Sommers, M. (2005) *Islands of Education: Schooling, Civil War and the Southern Sudan (1983–2004)*, Paris: IIEP UNESCO.
Stephens, B. (2015) 'China set for key role in post-2015 global health agenda', *The Guardian*, 10 June, p. 1
Stephens, D. (1982) 'A study of teacher education and attitudes over two generations in the Kano metropolitan area of northern Nigeria', unpublished PhD Thesis, University of Exeter, UK.
Stephens, D. (1998) 'Girls and basic education in Ghana: a cultural enquiry', *International Journal of Educational Development*, 20(1): 29–47.
Stephens, D. (2007) *Culture in Education and Development: Principles, Practice and Policy*, Oxford: Symposium Books.
Thurlow, M. (1992) 'Education policy in the African context: political posturing or potential for effective change?', *Mentor*, 74(3): 3–8.
Tikly, L. and Barrett, A. (2011) 'Social justice, capabilities and the quality of education in low income countries', *International Journal of Educational Development*, 31(1): 3–14.

Tikly, L. and Dachi, H.A. (2009) 'Social justice in African education in the age of globalization', in W. Ayers, T. Quinn and D. Stovall (eds), *Handbook of Social Justice in Education*, New York: Routledge, pp. 120–137.

UNESCO (United Nations Educational, Scientific and Cultural Organization) (2011) *Financing Education in Sub-Saharan Africa: Meeting the Challenges of Expansion, Equity and Quality*, Paris: UNESCO.

UNICEF (United Nations Children's Fund) (2015) *Gender Action Plan*, 16–19 June, Geneva: UNICEF.

UNICEF (United Nations Children's Fund) (2017) *Education in Emergencies*, available at: www.unicef.org/education/bege_70640.html (accessed 14 December 2017).

Unterhalter, E. (2007) *Gender, Schooling and Global Social Justice*, London: Routledge.

Wilkinson, R. and Yussof, I. (2005) 'Public and private provision of higher education in Malaysia: a comparative analysis', *Higher Education*, 50: 361–366.

10
Children

Nicola Ansell

Introduction

In demographic terms, Africa is the youngest continent on earth: children aged under 15 constitute 41 per cent of its population (PRB, 2015). However, until the late 1990s, there were relatively few academic studies of children or childhood in Africa. A few anthropologists and others had explored children's lives (examples include Katz, 1991; Katz and Monk, 1993; Schildkrout, 2002), but children were considered decidedly marginal to academic interest. In the 1990s, this began to change, driven in part by two distinct processes. First, people-centred approaches to development provided a rationale for attending to the lived experiences of Africa's populations, and in particular the lives of those deemed marginalised. Studies of women, minority ethnic groups, peasants and the urban poor were soon supplemented with studies of children. Second, the 1990s witnessed a surge of interest in childhood among social scientists. The so-called 'new social studies of childhood' (James *et al.*, 1998) focused in particular on children's social agency and on the social construction of childhood. African children and childhoods offered particularly enticing examples of both.

This chapter begins by outlining some of the research that was spawned from these trends: research that focused on the distinct understandings and practices of childhood in African societies and on groups of young people who were considered to exhibit unusual levels of agency such as street children and child heads of household. In light of emerging critiques of research that focuses on children in isolation and on exceptional categories of children, the latter part of the chapter refocuses on children in Africa through a relational lens. It examines how children's lives are strongly shaped by relationships with families and other adults in their communities, by national policies and cultural practices, and increasingly by international organisations and the operations of the global marketplace.

Childhoods in Africa: challenging Western ideals, recognising social agency

Early scholarship in the new social studies of childhood sought to deconstruct the Western ideal of childhood. Over several centuries, European thought had come to characterise children as typically innocent, passive, vulnerable and dependent; less than, but in preparation to become, adults; requiring intensive nurture and supervision; and inhabiting the private space of the nuclear family, or spaces set aside for children such as schools and playgrounds. While the ideal was particularly associated with middle-class childhoods and had shifted and intensified over time, the separation of childhood from adulthood could be traced to the emergence of industrialisation from the early modern period onwards (Aries, 1962).

Childhoods in Africa typically contrast markedly with this Western ideal. While most African children spend a considerable proportion of their time in school, there is often a greater spatial and functional blurring between children's and adults' lives. For instance, around 11.8 per cent of African children aged 5–14 are employed in economically productive work (ILO, 2013). Children are commonly expected to undertake chores for adults from a very early age, and to be able to survive without direct supervision for significant parts of the day. Rather than being held apart from adults in spaces of care and supervision, they are more likely part of the wider public sphere, working and playing in spaces that in some other societies might be reserved for adults.

Studies of African childhoods, then, have presented illustrations of the social constructedness of childhood. They illustrate that normative views of childhood not only change over time, but also differ from place to place. The meaning and importance attached to children varies culturally. Among the Dinka and Nuer of South Sudan, for instance, children are 'the reason for everything we do' (Jok, 2003: 3). For a man to die without surviving children is to die a 'complete death', with no permanent position within the male ancestral lineage. In much of Africa, children represent not only lineage continuity, but also the material survival of families and communities (Rwezaura, 1998). They are 'wealth in people', expected to provide labour and support the older generation and those who are sick or needy. There are cultural differences not just in the significance attached to children, but also in who children are. The use of chronological definitions of childhood is problematic in societies where adulthood is attained upon, for instance, initiation, marriage or childbearing.

Research focusing on African constructions of childhood is not merely of theoretical interest, but helps illuminate why the translation of policy ideas from Western settings or efforts at establishing universal standards have not always met with success. Boyden and Ennew (1997) describe how a 'global model of childhood' has circulated worldwide through the media and international institutions, and today informs many interventions aimed at supporting children in Africa. Particularly notable among these interventions are the Convention on the Rights of the Child (UN, 1989) and ILO Conventions 138 and 182 that aim to eliminate child labour. Various scholars (e.g. Liebel, 2007; Abebe and Bessell, 2011) have highlighted how the version of childhood on which such instruments are based (and which they seek to impose on African societies) is barely recognisable to many young Africans.

African children contest the 'global model' in part by living lives that demonstrate much greater social agency than is expected of the 'passive' children imagined in Western discourse. Working children clearly contribute to the economy, but there are other categories of children whose lives exhibit unusual levels of apparent agency. These include street children, living largely independently of adult control, and children heading households comprising only their sibling group. Such young people seemingly exercise a high degree of autonomy, taking decisions and responsibility for themselves and other children, and participate actively in social and economic life. These groups are also often targets of policy interventions, and thus further highlight the contradictions between the 'global model' and African discourses of childhood.

Research exploring the lives of African children over the past 20 years has focused primarily on the mismatch between global prescriptions and African childhoods and on African children's exercise of agency. Such a focus has contributed both to the development of childhood studies and to our understanding of the impacts of policy and practice on the lives of African children. In recent years, however, such research has begun to reveal the limitations of the dominant ideas within childhood studies. The distinction between the 'global model' of childhood expressed in international policy and the views prevalent in African societies is less binary than was once imagined. Moreover, the agency of African children is both constrained and arguably a product of, rather than a retort to, vulnerability (Mizen and Ofosu-Kusi, 2013).

Below, I consider three categories of African children whose lives appear to contravene Western childhood ideals and also reveal significant agency. These three groups have all received intense attention from researchers. Beyond demonstrating the value of research focused on social constructions of childhood and children's agency in illuminating the children's lives and the limitations of current policy approaches, I draw attention to newer, more critical perspectives.

Working children

Across Africa, 11.8 per cent of 5 to 14-year-olds are in employment (ILO, 2013). Such children undertake a wide variety of tasks, including household chores, farming, trading, shoeshining, brick-making, mining and handicrafts. Some children work for families, others for themselves or for employers; they may work in the home, on the street, in a factory or a mine; for pay, for other remuneration or for nothing; a few hours a week after school or 14 hours a day (Ansell, 2016a). Such activities run counter to the Western ideal that childhood is a time for learning and leisure, and not for work.

In Europe, children have been progressively marginalised from economic activity since the late nineteenth century. While there is an expectation that children should engage in some forms of work (chores for the benefit of their families; a few hours in exchange for 'pocket money' as they grow older), these are seen principally as ways of learning (to participate in the economy). The idea that either families or economic systems should be directly supported by the labour of children is viewed with a level of moral repugnance.

This moral sentiment opposing children's work now informs international discourse. Many African countries have committed to reducing, if not eliminating, children's participation in economic production in response to the efforts of international organisations, particularly the ILO. Rather than appealing to an inherent incompatibility between childhood and work, the ILO's argument today is that work is harmful to children and denies them the opportunity to attend school. The ILO considers 21.4 per cent of African children to be engaged in 'child labour' (ILO, 2013), which it defines as work that 'is mentally, physically, socially or morally dangerous and harmful to children; and interferes with their schooling' (ILO, 2014). Whether the work so categorised is actually harmful is, however, subject to debate (see Bourdillon *et al.*, 2009). Some forms of work are classed as labour because they are hazardous, but other work is deemed labour simply on account of the child's age and/or hours worked, a categorisation that clearly relates to an ideal that work at an early age is fundamentally wrong. Moreover, work within the context of the family is not classed as 'child labour' by the ILO, reflecting the Western idealisation of the family as the appropriate space of childhood. However, such work is not necessarily any less exploitative than that which takes place in factories (Abebe and Bessell, 2011).

The problematisation of children's work in international discourse is challenged by research that indicates some potential benefits for young people of engaging in work – even for quite young children. The income working children earn can enable them to buy food and clothing for themselves and their households, enabling them to support siblings, to buy luxuries for themselves, and even to pay school fees and buy uniforms (Abebe and Bessell, 2011). The majority of 7 to 14-year-olds engaged in child labour in sub-Saharan Africa also attend school (Gibbons *et al.*, 2005), and work enables some children to attend school where they would otherwise not be able to (Bourdillon, 2011). It also offers an alternative form of socialisation into adulthood: the skills and practical knowledge acquired through work may in fact be more important than learning through poor-quality and irrelevant schooling (Katz, 2004). Children also gain self-respect and a sense of purpose to their lives, particularly those who are less successful in school (Bourdillon *et al.*, 2009).

Problems often arise where children are excluded from opportunities to engage in work. Legislation based on international definitions of harmful work may fail to protect children from that which is harmful in their own contexts. Research in Ethiopia, for instance, found that while an export-oriented flower farm obeyed the national law and refused to employ children in what they viewed as desirable work, less scrupulous employers (without the pressure of overseas markets) escaped enforcement and employed children in forms of work the children considered harmful – those that were tiring, where they were treated unfairly or that interfered with school (Orkin, 2010). National laws inspired by international discourse also tend to deny children the opportunity to contribute to their own sustenance, again with harmful consequences. In South Africa, for instance, child labour legislation since 1994 has deepened chronic hunger and child poverty on farms where children previously worked seasonally or part-time (Levine, 2011).

Nonetheless, the view that child labour is harmful is gaining ground, particularly among middle-class Africans. There are fears that if global child labour standards are not imposed and enforced in Africa, African children will be exploited. Without the protection of child labour legislation, African elites fear that children's low-wage labour will support cheap exports for global consumers but deprive African children of the schooling needed to compete for more lucrative employment in the global economy.

Children's work is also a focus for researchers interested in children's agency. Arguments for allowing children to work often centre on working children's volition: they are understood to choose to work. However, the extent of children's choice of whether and where to work differs widely, and the concept of agency needs to be used cautiously. It is often difficult for children to refuse to work if their families wish them to, especially where the family depends upon their activities or income. Yet other than in situations of forced labour, it is rare for children to take no part in decisions regarding the work they will perform (Bourdillon, 2000). While their agency is structurally highly constrained, children are resourceful and make choices, even in the direst circumstances (Mizen and Ofosu-Kusi, 2013). In Berlan's (2013) research with cocoa producers in Ghana, children 'chose' to work because of malnutrition (they were swayed by the availability of food on farms or in school), lack of interest in or support for education, or poor experiences of schooling. Young Sierra Leonean 'bearing boys' responded to a water crisis in Freetown by building wooden carts ('bearings') that they used to ferry five-gallon plastic containers of water around the city. They were responsible for the design, regulating themselves, and creating new forms of entrepreneurship in response to local economic and social needs (Shepler, 2010). In Ethiopia, too, children have adapted to shifts in rural livelihoods by developing new entrepreneurial skills in domestic work, trade and migration (Abebe and Kjørholt, 2009). Through informal means of regulation, however, children get locked into types of working that become difficult to escape (Mizen and Ofosu-Kusi, 2012).

Street children

Children living on the streets in African cities (as in other cities of the Global South) became objects of concern in the 1990s, in part because they represent highly visible challenges to idealised Western notions of childhood. The juxtaposition of 'street' and 'child' generates anxiety because children are not deemed to belong in public space: by definition, Western thought defines street children as problematic. The apparent absence of adult supervision of such children also conflicts with the Western childhood ideal. Street children are generally assumed to have been abandoned and their families are represented as negligent. Street children themselves are portrayed as both victim and deviant. Discussions of their lives tend to focus on deficits – lack of shelter, food, education and healthcare (de Moura, 2002). They are represented as alien to mainstream society, defying moral values by indulging in drug use, sexual promiscuity, prostitution and crime. They are seen as feral and untamed, not fully responsible for their own behaviour (de Moura, 2002).

These representations are in many cases misguided. By no means are all street children abandoned; some maintain close relationships with their families (van Blerk, 2012). Many of those to whom the 'street child' label is attached actually sleep at home with their families most nights. Others maintain relationships with family members on or off the streets, as well as with other adults (Thomas de Benitez, 2011). While many street children do engage in criminal activity – often as a means of sustenance – research suggests that they develop their own moral codes (Mizen and Ofosu-Kusi, 2012) and may be better nourished and develop more advanced cognitive skills than their slum-dwelling peers.

Because association with the street is viewed as inherently harmful, policy responses to street children have often focused on removing children from the streets and placing them in shelters. This can cause problems, including breaking up their support networks (Ennew, 1994). Those organisations that break with normative expectations and view children as having a right to live on the streets, providing support through outreach teams, may be more effective, particularly if they offer an approach that is personalised, accepting, flexible and participatory (Thomas de Benitez, 2011).

While on the one hand street children are highlighted as a problem because their very existence confronts Western notions of childhood, on the other hand the fact that their lives in many ways fail to conform with normative expectations means that they are sometimes omitted from consideration in child policy. Ennew (2000) has criticised the Convention on the Rights of the Child (CRC) for its lack of relevance to street children's lives. Moreover, the move towards broad-based social protection, social welfare and education policies over the past decade has tended to bypass street children as a specific target group, with generic policies that often fail to apply to the lives of those living without parents or other adult guardians (Thomas de Benitez, 2011).

Most child-focused policies and interventions still regard children as passive beneficiaries of adult actions – for instance, understanding that the way to support children is to support their households. Yet street children are generally the key active agents in their own lives. In research, street children's agency has been a focus of attention. Research into the factors behind children turning to the streets has led many street children to recount how they decided to leave their families, to escape harm, or to improve their own or their families' socio-economic situations (Thomas de Benitez, 2011). Such children, once living on the streets, commonly exercise more choices than children in other environments (Connolly and Ennew, 1996), and survive by employing a range of coping strategies. Indeed, they are attracted to city centres partly by income-generating possibilities: selling newspapers, shining shoes, guarding cars, cleaning windscreens or guiding tourists. Research in Uganda has revealed how street children negotiate public and private, street and non-street locations (van Blerk, 2005), appropriating or creating niches in which to perform daily activities: work, leisure, washing, eating and sleeping (Young, 2004).

As with working children, however, the concept of agency might not be the most appropriate lens through which to understand street children's lives; while street children exercise a degree of autonomy, it is not altogether positive. Gigengack (2014) has argued that too much attention is given to celebrating street children's agency, which is frequently self-destructive. For Mizen and Ofosu-Kusi (2013), street children's agency is not an outcome of rational choice or the 'normative force of childhood', but rather a product of vulnerability. In their research, street children described leaving their homes and heading for the streets of Accra in terms of choice and self-determination, but they also framed the process in relation to their vulnerability, frailty and need.

Orphans and sibling-headed households

In the 1990s, adult mortality in southern Africa in particular rose sharply due to the AIDS pandemic. By 2001, there were 10 African countries where more than 15 per cent of children had lost a parent, around half of them due to AIDS (USAID *et al.*, 2002). Because the likelihood of orphanhood increases with age, among teenagers the proportion was much higher. Moreover, the nature of the pandemic meant that a high proportion of children lost not just one, but both parents.

Orphanhood is considered particularly problematic in Western society, where the nuclear family is deemed crucial to children's well-being. Unlike working or street children, orphans are arguably the 'quintessential vulnerable child' (Meintjes and Giese, 2006), clearly innocent of blame for their situation. A considerable amount of research activity has focused around the impact of AIDS on the well-being of African children. Sickness or death of a parent or other immediate family member has been understood to adversely affect children in a number of ways: declining access to resources, both material and emotional; psychosocial distress (Cluver *et al.*, 2012); an increased burden of domestic, agricultural or income-generating work (Urassa *et al.*, 1997; UNAIDS, 1999); an expectation that children would provide practical care for a sick individual, either at home or elsewhere (Robson, 2000; Robson and Ansell, 2000); withdrawal from school owing to impoverishment or work/care commitments (Bicego *et al.*, 2003); stigmatisation (Webb, 1997); break-up of the household; migration to an unfamiliar place (Ansell and van Blerk, 2004); and discrimination or abuse in the home of a grandparent or relative who is struggling to cope with several incoming children (Foster and Williamson, 2000).

There is some evidence in support of all of these impacts, but at an aggregate level there is little evidence of an association between AIDS orphanhood and individual disadvantage. A systematic review of literature on the impacts of orphanhood found most studies reported some negative effects across a wide range of physical, socio-economic and psychological outcomes, but often no differences were detected (Sherr *et al.*, 2008). Parikh *et al.* (2007) undertook a cohort study in South Africa that found no statistically significant differences in most education, health and labour outcomes between orphans and the non-orphans with whom they lived, although paternal orphans were more likely to be behind in school. Evidence from the Young Lives Ethiopian study reveals little difference in school enrolment, school attendance or body mass index between orphans and non-orphans, and where small differences exist, they sometimes favour orphans (Crivello and Chuta, 2012).

Understanding why orphans became such a focus of international attention and why the evidence appears to confound expectations relates to the mismatch between Western and African expectations of childhood (Ansell, 2016b). The attention paid to orphans is predominantly from media and organisations based in Western societies where the nuclear family is dominant and perceived as the 'natural' milieu for child-rearing. Where adult mortality rates are relatively high, it is not unusual for children to lose a parent before they enter adulthood. Moreover, in African societies, children have historically been raised by a more diffuse group of adults. In South Africa, fewer than half of non-orphans live with both parents (Meintjes and Giese, 2006). Similarly, research in Tanzania found that fewer than a quarter of foster children were orphans (Foster and Williamson, 2000). If orphans' living arrangements generally differ little from those of many other children, it is unsurprising that there are few sharp distinctions in other aspects of their lives.

Yet orphans appeal to donors because of their iconic status as child victims. Where orphan status attracts resources, there is a risk that the orphan identity becomes commodified (Meintjes and Giese, 2006). Children may claim to be orphans in order to access services, and families may adopt orphans for material gain. An extreme instance of the commodification of orphanhood is the establishment of tourism-focused orphanages (Richter and Norman, 2010).

While orphans are not necessarily deemed to exercise particular agency, a specific subgroup of children in this category has received attention for their active role in sustaining households. A number of African children take leadership roles in households, either in the absence of adult members or where any adult members are sick and unable to act as household heads. The number of such 'child-headed' or 'sibling-headed' households is relatively small (accounting for only 0.47 per cent of South African households according to Meintjes *et al.*, 2010), but often celebrated as demonstrating social agency.

Where children reside in their parental home without adults, this is indeed often an active choice rather than an indication of abandonment. Children do so in order to retain access to land and resources, retain a degree of autonomy or to keep siblings together (Evans, 2011). In South Africa, 92.1 per cent of 'sibling-headed' households actually have a living parent (Meintjes *et al.*, 2010). In some respects, then, sibling-headed households are part of the wider patchwork of non-nuclear family arrangements prevalent in African societies. However, they remain unconventional – not only from a Western perspective. African communities are seldom well adjusted to ensuring that such households' needs are known and represented. Kesby *et al.* (2006) describe these as 'other other childhoods', other to European expectations, but also to African norms.

Children in sibling-headed households do exercise agency. They distribute tasks among themselves and take on adult roles as necessary, exercising authority over younger siblings, as well as performing tasks such as cooking, cleaning, growing food and generating income (Bell and Payne, 2009; Evans, 2011). They are, however, seldom entirely unsupported by adult kin or neighbours, and are unlikely to be treated as fully 'adult' by their communities (Evans, 2011).

Towards a relational approach to African childhoods

Overall, research from the 1990s onwards, by focusing on particular categories of children, has highlighted the ways in which childhood is understood differently in Africa (and why international interventions based on Western childhood ideals sometimes have harmful outcomes). It has also demonstrated that African children exercise forms of social agency. However, there is growing recognition that Western and African childhood ideals are not necessarily binary opposites, and also that the nature and value of the agency that marginalised children exhibit is somewhat questionable. Moreover, the focus of research on constructions of childhood and autonomous agency, and on particular categories of children, whose lives appear relatively independent of adults, has led to a tendency to view children's lives in isolation. Emerging research is pointing to the value of recognising the relationships – familial and structural – in which African children's lives are embedded.

Personal relationships: age, gender and generation among peers and families

African children do not live isolated lives, but their situations and their agency are deeply intertwined with the lives of both peers and adults. As indicated above, children living in sibling-headed households have important and carefully structured relationships with siblings (or other children) within the household, shaped in part by sibling order, age and gender (Evans, 2010). Moreover, they generally maintain relationships with kin living in separate households, as well as with other community members. Such relationships involve dependencies that are not one-way. As with other orphaned children, there is often an expectation that young people will contribute, providing assistance to elderly or sick relatives, or undertaking household chores in exchange for food or school fees. Such expectations vary with a child's age and gender, and shift over time (Ansell and van Blerk, 2004). Understanding the lives of such young people requires research that extends beyond their own immediate situations and perspectives and explores the power-laden relationships that shape them.

Similarly, research indicates that street children's lives reflect their relationships with others (van Blerk, 2012). Peer relations among young people living on the streets have long been recognised as a key source of practical and emotional support. Many street children are part of 'street families' of largely unrelated young people of different ages. Street children's relationships with adults – both those living on the streets and those who they interact with on a daily basis – are also very significant. Moreover, there is an expanding body of research that recognises the significance of family relationships in the lives of street children – not simply in propelling them into street life, but on an ongoing basis (van Blerk, 2012). Many children on African streets communicate regularly with kin; some make frequent visits and may provide resources, or equally may be assisted by their families. It is important to recognise that although such relationships may be supportive, they are not always benign. Street life is often characterised by high levels of violence (Thomas de Benitez, 2011).

The lives of working children also depend heavily on relationships with others – both family and employers. Work needs to be seen in the context of generational relations. In societies with high ratios of children to healthy working-age adults, it is often necessary for children to begin to contribute to household sustenance at an earlier age than where numbers of children are relatively small (Bourdillon, 2006). Moreover, in wealthier societies, intergenerational transfers of resources – from workers to dependents – often take place at the national level via systems of taxation or private pensions. In Africa, there remains a higher level of interdependence within the family, so those who are able to contribute are expected to do so, if others are in need (Abebe and Bessell, 2011).

Wider structural relationships

African children's lives are shaped not only by personal relationships, but also by structural relations that operate nationally and globally. While there is little robust evidence that children are systematically economically disadvantaged as a result of orphanhood, many African children experience poverty, and that poverty (and its impacts on children) cannot be understood without attending to the economic relations that produce it. While there may be many personal situations that precipitate a move to the streets, street children are overwhelmingly from impoverished backgrounds. The abundance of street and working children in Africa reflects poverty and structural circumstances as much as it does specific familial situations. Indeed, the latter are often products of the former.

A number of scholars have pointed to ways in which child poverty in the Global South is functional to the global economy. The roles that children play as economic producers are perhaps the most obvious instance of this. It has been estimated, for instance, that tobacco producers British American Tobacco and Philip Morris save $1.2 billion in Malawi by using child labour because children are cheaper than adults: they consume less and most are 'subsidised' by adult household members (Otañez and Glantz, 2011). The availability of cheap child labour also means employers can keep adult wages low (and conversely, low adult wages mean households depend more on children's incomes) (Wahba, 2006). Moreover, because most children are supported by families, they can serve as a 'reserve army of labour', entering the workforce when labour demand increases, but easily dismissed when demand is low.

Evidence of the ways in which children's labour is intentionally incorporated into the global economy in the service of affluent nations and corporations may also be seen in the way that some forms of labour are more regulated than others. In 1919, the first ILO Convention on child labour related only to industrial work, allowing children to continue to contribute to agriculture and to the social reproduction of peasant households whose labour fuelled the global economy (Nieuwenhuys, 2007). Permitting child labour in the colonies arguably facilitated its abolition in Europe and North America (Nieuwenhuys, 2007). As African children have progressively been excluded from the formal sector labour market, they continue to contribute to social reproduction, serving, as in Nieuwenhuys' (2007) terms, a 'global womb'.

The existence of child poverty, then, encourages children to participate in economic production, subsidising export commodities and the production of wealth elsewhere. Increasingly, however, African children are also seen as a market for global corporations – from companies such as Pearson that own chains of low-cost schools (Ansell, 2015) to fast-food chains and mobile phone providers (Porter *et al.*, 2016). In order to understand the changing policy contexts of children's lives in Africa, and the development interventions shaping their contemporary lives, it is crucial to recognise the roles played by national and international political economy.

Conclusion

In summary, much research has been undertaken with African children over the past two decades. This research has built upon and contributed to broader research agendas in childhood studies, and has consequently focused on two key areas. First, research has demonstrated that childhood is constructed in African societies in diverse ways that tend to differ from the normative expectations of childhood in broadly Western societies. Second, research has revealed that African children – particularly those whose lives are most markedly different from Western ideals of childhood – are noteworthy exemplars of children's social agency. Research has focused on certain categories of children and revealed how the mismatch between 'global' and 'local' discourses has often resulted in interventions that have had unintended negative consequences for young people (see Twum-Danso Imoh and Ansell, 2014).

Understanding this mismatch has been helpful, but as research with African children has advanced it has become apparent that the binary categorisation of African and Western childhood ideals is an oversimplification, and that the agency of marginalised children is highly constrained and may not be worthy of celebration. Moreover, the focus on groups of marginalised children in isolation offers a distorted understanding of children's lives and the significance of their relationships with others – and also the lives of those who do not fall into such categories. There is a growing interest in the relationalities of African children's lives: the roles played by personal relationships and by wider structural relations.

References

Abebe, T. and Bessell, S. (2011) 'Dominant discourses, debates and silences on child labour in Africa and Asia', *Third World Quarterly*, 32(4): 765–786.

Abebe, T. and Kjørholt, A.T. (2009) 'Social actors and victims of exploitation: working children in the cash economy of Ethiopia's South', *Childhood*, 16(2): 175–194.

Ansell, N. (2015) 'Shaping global education: international agendas and governmental power', *International Development Planning Review*, 37(1): 7–16.

Ansell, N. (2016a) *Children, Youth and Development*, 2nd edition, London: Routledge.

Ansell, N. (2016b) '"Once upon a time . . ." Orphanhood, childhood studies and the depoliticisation of childhood poverty in southern Africa', *Childhood*, 23(2): 162–177.

Ansell, N. and van Blerk, L. (2004) 'Children's migration as a household/family strategy: coping with AIDS in Lesotho and Malawi', *Journal of Southern African Studies*, 30(3): 673–690.

Aries, P. (1962) *Centuries of Childhood*, New York: Vintage Press.

Bell, S. and Payne, R. (2009) 'Young people as agents in development processes: reconsidering perspectives for development geography', *Third World Quarterly*, 305: 1027–1044.

Berlan, A. (2013) 'Social sustainability in agriculture: an anthropological perspective on child labour in cocoa production in Ghana', *Journal of Development Studies*, 49(8): 1088–1100.

Bicego, G., Rustein, S. and Johnson, K. (2003) 'Dimensions of the emerging orphan crisis in sub-Saharan Africa', *Social Science and Medicine*, 56: 1235–1247.

Bourdillon, M. (2000) 'Introduction', in M. Bourdillon (ed.), *Earning a Life: Working Children in Zimbabwe*, Harare: Weaver Press, pp. 1–10.

Bourdillon, M. (2006) 'Children and work: a review of current literature and debates', *Development and Change*, 37(6): 1201–1226.

Bourdillon, M. (2011) 'A challenge for globalised thinking: how does children's work relate to their development?', *South African Review of Sociology*, 42(1): 97–115.

Bourdillon, M., Myers, W. and White, B. (2009) 'Reassessing working children and minimum-age standards', *International Journal of Sociology and Social Policy*, 29(3/4): 106–117.

Boyden, J. and Ennew, J. (1997) *Children in Focus: A Manual for Participatory Research with Children*, Stockholm: Radda Barnen.

Cluver, L., Orkin, M., Boyes, M. and Gardner, F. (2012) 'Persisting mental health problems among AIDS-orphaned children in South Africa', *Journal of Child Psychiatry and Psychology*, 53(4): 363–370.

Connolly, M. and Ennew, J. (1996) 'Introduction: children out of place', *Childhood*, 3: 131–145.

Crivello, G. and Chuta, N. (2012) 'Rethinking orphanhood and vulnerability in Ethiopia', *Development in Practice*, 22(4): 536–548.

de Moura, S.L. (2002) 'The social construction of street children: configuration and implications', *British Journal of Social Work*, 32: 353–367.

Ennew, J. (1994) 'Parentless friends: a cross-cultural examination of networks among street children and street youth', in F. Nestman and K. Hurrelman (eds), *Social Networks and Social Support in Childhood and Adolescence*, London: de Gruyter, pp. 409–425.

Ennew, J. (2000) 'Why the Convention is not about street children', in D. Fottrell (ed.), *Revisiting Children's Rights: 10 Years of the UN Convention on the Rights of the Child*, Dordrecht: Kluwer, pp. 169–182.

Evans, R. (2010) 'Children's caring roles and responsibilities within the family in Africa', *Geography Compass*, 4(10): 1477–1496.

Evans, R. (2011) '"We are managing our own lives . . ." Life transitions and care in sibling-headed households affected by AIDS in Tanzania and Uganda', *Area*, 43(4): 384–396.

Foster, G. and Williamson, J. (2000) 'A review of current literature of the impact of HIV/AIDS on children in sub-Saharan Africa', *AIDS*, 14: S275–S284.

Gibbons, E.D., Huebler, F. and Loaiza, E. (2005) 'Child labour, education and the principle of non-discrimination', in P. Alston and M. Robinson (eds), *Human Rights and Development: Towards Mutual Reinforcement*, New York: Oxford University Press, Chapter 10.

Gigengack, R. (2014) 'Beyond discourse and competence: science and subjugated knowledge in street children studies', *European Journal of Development Research*, 26(2): 264–282.

ILO (International Labour Organization) (2013) *Marking Progress Against Child Labour: Global Estimates and Trends 2000–2012*, Geneva: ILO.

ILO (International Labour Organization) (2014) *What Is Child Labour?*, available at: www.ilo.org/ipec/facts/lang—en/index.htm (accessed 7 May 2014).

James, A., Jenks, C. and Prout, A. (1998) *Theorizing Childhood*, Cambridge: Polity Press.

Jok, J.M. (2003) 'Sudan's prolonged civil war and the shifting value(s) of children', in J. Abbink (ed.), *Youth and the Politics of Generational Conflict in Africa*, Leiden: African Studies Centre, pp. 145–162.

Katz, C. (1991) 'Sow what you know: the struggle for social reproduction in rural Sudan', *Annals of the Association of American Geographers*, 81(3): 488–514.

Katz, C. (2004) *Growing up Global*, Minneapolis, MN: University of Minnesota Press.

Katz, C. and Monk, J. (1993) *Full Circles: Geographies of Women Over the Life Course*, London: Routledge.

Kesby, M., Gwanzura-Ottemoller, F. and Chizororo, M. (2006) 'Theorising other, "other childhoods": issues emerging from work on HIV in urban and rural Zimbabwe', *Children's Geographies*, 4(2): 185–202.

Levine, S. (2011) 'The race of nimble fingers: changing patterns of children's work in post-apartheid South Africa', *Childhood*, 18(2): 261–273.

Liebel, M. (2007) 'Opinion, dialogue, review: the new ILO report on child labour – a success story, or the ILO still at a loss? *Childhood*, 14(2): 279–284.

Meintjes, H. and Giese, S. (2006) 'Spinning the epidemic: the making of mythologies of orphanhood in the context of AIDS', *Childhood*, 13(3): 407–430.

Meintjes, H., Hall, K., Marera, D.-H. and Boulle, A. (2010) 'Orphans of the AIDS epidemic? The extent, nature and circumstances of child-headed households in South Africa', *AIDS Care: Psychological and Socio-Medical Aspects of AIDS/HIV*, 22(1): 40–49.

Mizen, P. and Ofosu-Kusi, Y. (2012) 'Engaging with a world outside of ourselves: vistas of flatness, children's work and the urban informal economy', *Sociological Research Online*, 17(2): 1–17.

Mizen, P. and Ofosu-Kusi, Y. (2013) 'A talent for living: exploring Ghana's "new" urban childhood', *Children and Society*, 27(1): 13–23.

Nieuwenhuys, O. (2007) 'Embedding the global womb: global child labour and the new policy agenda', *Children's Geographies*, 5(1–2): 149–163.

Orkin, K. (2010) 'The child's best interests? Legislation on children's work in Ethiopia', *Journal of International Development*, 22(8): 1102–1114.

Otañez, M. and Glantz, S.A. (2011) 'Social responsibility in tobacco production? Tobacco companies' use of green supply chains to obscure the real costs of tobacco farming', *Tobacco Control*, 20(6): 403–411.

Parikh, A., DeSilva, M.B., Cakwe, M., Quinlan, T., Simon, J.L., Skalicky, A., *et al.* (2007) 'Exploring the Cinderella myth: intrahousehold differences in child wellbeing between orphans and non-orphans in Amajuba District, South Africa', *AIDS*, 21(SUPPL 7) S95–S103.

Porter, G., Hampshire, K., Milner, J., Munthali, A., Robson, E., De Lannoy, A., *et al.* (2016) 'Mobile phones and education in sub-Saharan Africa: from youth practice to public policy', *Journal of International Development*, 28(1): 22–39.

PRB (Population Reference Bureau) (2015) *Population Reference Bureau World Population Data Sheet*, available at: www.prb.org/DataFinder/Topic/Rankings.aspx?ind=10&loc=246,247 (accessed 14 December 2017).

Richter, L. and Norman, A. (2010) 'AIDS orphan tourism: a threat to young children in residential care', *Vulnerable Children and Youth Studies*, 5(3): 217–229.

Robson, E. (2000) 'Invisible carers: young people in Zimbabwe's home-based healthcare', *Area*, 32: 59–70.

Robson, E. and Ansell, N. (2000) 'Young carers in southern Africa: exploring stories from Zimbabwean secondary school students', in S.L. Holloway and G. Valentine (eds), *Children's Geographies: Playing, Living, Learning*, London: Routledge, pp. 174–193.

Rwezaura, B. (1998) 'Competing "images" of childhood in the social and legal systems of contemporary sub-Saharan Africa', *International Journal of Law, Policy and the Family*, 12: 253–278.

Schildkrout, E. (2002) 'Age and gender in Hausa society: socio-economic roles of children in urban Kano', *Childhood*, 9(3): 342–368.

Shepler, S. (2010) 'Child labour and youth enterprise: post-war urban infrastructure and the "bearing boys" of Freetown', *Anthropology Today*, 26(6): 19–22.

Sherr, L., Varrall, R., Mueller, J., Richter, L., Wakhweya, A., Adato, M., *et al.* (2008) 'A systematic review on the meaning of the concept "AIDS orphan": confusion over definitions and implications for care', *AIDS Care*, 20(5): 527–536.

Thomas de Benitez, S. (2011) *State of the World's Street Children*, London: Consortium for Street Children.

Twum-Danso Imoh, A. and Ansell, N. (eds) (2014) *Children's Lives in an Era of Children's Rights: The Progress of the Convention on the Rights of the Child in Africa*, London: Routledge.

UN (United Nations) (1989) *Convention on the Rights of the Child, United Nations*, available at: www.ohchr.org/en/professionalinterest/pages/crc.aspx (accessed 20 December 2015).

UNAIDS (Joint United Nations Programme on HIV/AIDS) (1999) *A Review of Household and Community Responses to the HIV/AIDS Epidemic in the Rural Areas of Southern Africa*, Geneva: UNAIDS.

Urassa, M., Boerma, J.T., Ng'weshemi, J.Z.L., Isingo, R., Schapink, D. and Kumogola, Y. (1997) 'Orphanhood, child fostering and the AIDS epidemic in rural Tanzania', *Health Transition Review*, 7: 141–153.

USAID (United States Agency for International Development), UNICEF (United Nations Children's Fund) and UNAIDS (Joint United Nations Programme on HIV/AIDS) (2002) *Children on the Brink 2002: A Joint Report on Orphan Estimates and Program Strategies*, New York: USAID, UNICEF and UNAIDS.

van Blerk, L. (2005) 'Negotiating spatial identities: mobile perspectives on street life in Uganda', *Children's Geographies*, 3(1): 5–21.

van Blerk, L. (2012) 'Berg-en-See street boys: merging street and family relations in Cape Town, South Africa', *Children's Geographies*, 10(3): 321–336.

Wahba, J. (2006) 'The influence of market wages and parental history on child labour and schooling in Egypt', *Journal of Population Economics*, 19(4): 823–852.

Webb, D. (1997) *HIV and AIDS in Africa*, London: Pluto.

Young, L. (2004) 'The "place" of street children in Kampala, Uganda: marginalisation, resistance and acceptance in the urban environment', *Environment and Planning D: Society and Space*, 21: 607–627.

11
Diaspora

Ben Page and Claire Mercer

Introduction: what is diaspora? What is diaspora development?

In the year 722 BCE, Israel was destroyed by Assyria and the people fled to Judah, where they came to be known as Jews. When the history of this movement was written down between 640 and 610 BCE, it was decreed of the Jewish people that 'thou shalt be a diaspora in all kingdoms of the earth' (Deuteronomy, 28:25). From these very specifically Jewish origins, the term has spread to describe the general experience of large-scale geographical dispersion of human populations from a shared home place as a result of violent and traumatic events. So, the scattering of Greeks after the fall of Constantinople (1453), of Armenians after the First World War, or of Africans as a result of the transatlantic slave trade are all seen as archetypal diasporas.

However, the term has continued to evolve to cover any population that has dispersed across many countries for whatever reason, not just 'victim diasporas' scattered as a result of violent trauma. Robin Cohen (2009) also identifies labour diasporas (such as communities of Indian heritage found around the world and derived from the movement of contract labourers), imperial diasporas (such as populations of British heritage found around the world in the wake of colonial occupations), trade diasporas (such as Lebanese and Chinese traders found around the world pursuing their business interests) and deterritorialised diasporas (such as the Roma, who trace their origins back to India, but who have lost all contact with those geographical origins and have no 'home' state to which they could imagine returning).

Most recently, 'diaspora' has entered the lexicon of international development with key organisations such as the International Organization for Migration (IOM) using the term as a tool with which to frame a set of policies that developing countries, development organisations and donors should embrace in order to maximise the benefits of international migration (Dufoix, 2015). It is in this context that in 2000, the African Union declared in its constitution that the African diaspora was the 'sixth region' of the continent, and defined it as follows: 'people of African origin living outside the continent, irrespective of their citizenship and nationality and who are willing to contribute to the development of the continent and the building of the African Union' (African Union, 2000). In this extremely inclusive definition, the reason that someone left Africa (whether as a slave, a contract worker, a soldier, a trader, a refugee, a student or a professional) is irrelevant; so too is the date (they may be the descendent of a slave forced out of Africa many centuries ago or they may have left by plane yesterday), but a commitment to African development is seen as crucial. Individuals of African heritage are excluded from the diaspora by the AU if they are not committed to African development.

As readers of this book will be aware, definitions of 'development' are many and diverse. For the purpose of this discussion, however, we want to keep the idea of development as broad as possible. We take development to include not only the deliberate planned activities that are labelled 'development' (programmes that explicitly focus on social and economic betterment in the Global South), but also those activities (such as migrants sending money to family in Africa, or investing in businesses in Africa) that might not be explicitly called development or be strategically or formally organised, but which can lead to profound changes in Africa. In addition, it is crucial to remember that 'development' in this broad sense doesn't just occur in the places that the African diaspora leave behind, but also in the places that they move to. Research has shown, for example, how the African diaspora contributes to development in the places to which they move through entrepreneurial activities, paying taxes, and patronising local businesses (Mohan, 2002; Mazzucato, 2008; Smith and Mazzucato, 2009). In this chapter, we will briefly trace the history of the interest in diasporas in development studies and the making of the African diaspora over time, before considering the different ways in which the contemporary African diaspora engages with development on the African continent.

What is the history of the interest in diasporas in development studies?

In the 1960s, the Intergovernmental Committee for European Migration (which later became the IOM) was running schemes that placed highly skilled Europeans displaced by the Second World War or the Cold War in Latin America as part of a European development effort. In the following decade, this was extended to help members of the Latin American diaspora in Europe to return to their home countries, and in the 1980s the programme was extended to include those from Africa and Asia. The schemes were small-scale and focused purely on the return migration of the highly skilled to counteract the brain drain. In other words, they paid little attention to those members of diasporas whom we would now probably describe as 'transnational' because of their long-term investment in the place they have moved to. IOM's scheme continues up to the present under the label 'Migration for Development in Africa' (MIDA), which began in 2001. In the early 1990s, civil society organisations such as the London-based African Foundation for Development (AFFORD), started to lobby the British development sector to collaborate with the African diaspora. But although development policy has had an interest in migration that goes back at least 50 years (Iskander, 2010; de Haas, 2012), policy-makers did not start to think explicitly about how to formally enrol diaspora communities into international development until the early 2000s.

The emergence of formal diaspora engagement brought together multiple institutions. First, the European Council, meeting in Seville in 2002, started to develop the idea of co-development (i.e. policies that aim to benefit labour-exporting and labour-importing countries simultaneously), building on ideas that had emerged in France and Spain in the late 1990s (Ostergaard-Nielsen, 2011). In the USA, the Migration Policy Institute produced a highly influential report looking at the role of diasporas in poverty reduction in their countries of origin (Newland and Patrick, 2004). The UN's Global Commission on International Migration, with which the IOM and the World Bank were closely involved, ran from 2003 to 2005. It led to the formation of the High-Level Dialogue on International Migration and Development (HLD) at the UN in September 2006 and then to the establishment of the Global Forum on Migration and Development (GFMD) later the same year. This now meets annually to bring together politicians, policymakers, practitioners and analysts from governments, civil society and the private sector to identify opportunities, challenges and best practice for maximising development benefits from migration. The UN's flagship annual publication the *Human Development Report* focused on migration in 2009 (UNDP, 2009). The World Bank started to produce the biannual overview of global remittance data in 2006, as better data became available. There is now a dedicated Migration and Remittances team at the World Bank responsible for monitoring, forecasting and analysing migration and remittance flows. The World Bank has been at the forefront of developing policy measures to harness the potential of the African diaspora for African development, especially in the areas of remittances, investments and diaspora bonds.

The policies that have emerged from these 15 years of discussions and experiments can roughly be divided into two types: those associated with labour-importing countries (typically, though not exclusively, in the Global North) and those associated with labour-exporting countries. In labour-importing countries, development ministries have generally aimed to support resident diasporas from other countries by providing: matching funds for their projects; capacity-building workshops to improve their human-resources; inventories of diaspora organisations to improve communication; networking fora and research funds for identifying best practices; market information to reduce the cost of remitting; and platforms for effective lobbying of governments. Key examples of such schemes would include Comic Relief's Common Ground Initiative (largely funded by the UK's Department for International Development), which has to date provided GB£25million of funding to small and African diaspora organisations. These strategies have also shaped USAID's Diaspora Engagement Programme and the Diaspora Engagement Alliance originally associated with (then) US Secretary of State Hillary Clinton. It is, however, vital to emphasise that these policies represent a very small proportion of development spending, and often contradict far more politically influential policies that seek to restrict immigration to countries of the Global North.

Policies in labour-exporting countries are more varied and more significant in scale. Alan Gamlen's (2006) review of more than 70 migrant-sending countries identifies three different types of diaspora policy: *capacity building, extending rights* and *extracting obligations*. What he calls 'capacity building policies' attempt 'to produce a communal mentality amongst non-residents; a sense of common belonging to the home-state' (Gamlen, 2006: 7), for example by funding diaspora newspapers or websites, holding conventions for the diaspora at home, monitoring diaspora organisations overseas and ensuring that senior political figures meet with the diaspora when they travel overseas. These policies seek to actually *create* a national diaspora in order to then render them governable. The goal is to produce a means of communication through which a government can speak to its diaspora overseas. In contrast, the policies relating to 'extending rights' are about improving the political position of members of the diaspora in their home country, for example by allowing dual citizenship, voting rights at home, the possibility of standing for election and formal representation in parliament for the diaspora. The theory is that such improved political rights for the diaspora will generate goodwill and a positive relationship with those overseas, and thereby protect remittance and investment flows. Third, policies described as 'extracting obligations' seek to capture the economic potential of those in the diaspora. These include offering members of the diaspora preferential tax rates on investments at home, accelerated bureaucratic procedures in relation to land or other commercial acquisitions, and capturing remittances for public goods via matching fund schemes, such as the well-known Mexican scheme through which the Mexican state adds three dollars to every one remitted from a Mexican overseas for specified development projects (Aparacio and Meseguer, 2011). These policies can also leverage patriotism to extract mandatory payments and knowledge and skills transfers from members of the diaspora.

Overall, this is a field with a clear policy goal and a lively history of policy ideas reflecting changing attitudes to the development–migration relationship. It is useful to draw a distinction between the policies developed by 'donor' states and organisations who 'host' diasporas and the policies developed by African states. Ultimately, it is important, however, to keep these policies in perspective: they are often underpowered both in terms of financing and politics, and there is often a gulf between policy goals and delivery.

Where is the African diaspora?

The global dispersal of people beyond the African continent can be traced back as far as the fifth century BCE when traders, soldiers, domestics, slaves and others began to travel to Europe, the Middle East and Asia. Between 600 and 1900 CE, the trans-Saharan slave trade moved approximately 9 million Africans to the Muslim world, while at the height of the Indian Ocean slave trade in the nineteenth century, approximately 800,000 enslaved Africans from the interior were dispersed across the coastal plantations of the Middle East, Asia and East Africa (Akyeampong, 2000). However, the largest dispersal took place over the four centuries of the transatlantic slave trade, when approximately 200,000 Africans were moved to Europe and between 11 and 12 million to the Americas (Palmer, 1998). Brazil is now the country with the largest African diaspora, and there are significant African-descended communities throughout Latin America, Central America, the Caribbean and the USA.

Twentieth-century movements by seamen and students, among others, were small by comparison to the transatlantic slave trade. By the 1980s and 1990s, however, long-term economic decline on the African continent, precipitated by the debt crisis and the ensuing structural adjustment programmes, provided a new stimulus for emigration (Zeleza, 2005). This most recent and ongoing dispersal was dominated by movements to the Eastern bloc (before 1989), North America and Europe. However, following transformations in the world economy and increasingly restrictive immigration policies into the Global North, there are significant new movements of Africans both within and beyond Africa. These include those moving within the African continent, who have moved and established communities in African countries away from home as refugees, traders and workers (Van Dijk, 2003; Zeleza, 2005; Mercer *et al.*, 2008; Bakewell and Binaisa, 2016), those moving to Asia in response to new trade patterns and geopolitical developments (Bodomo and Ma, 2012; Bork-Hüffer *et al.*, 2016), and refugees fleeing countries such as the DRC, Libya, Somalia and Eritrea to seek entry to Europe. The relationship between 'refugee' communities and diasporas is complex; the two terms are not synonymous. Not all diasporas are comprised of refugees, though many refugees (or their descendants) might ultimately become part of the African diaspora.

The extent to which all of these people actually describe themselves as members of an 'African diaspora' is an open question. Even when the term is used, it might signify quite different histories and affiliations. In North America, for example, first-generation arrivals from Africa in the twenty-first century may well have quite a different understanding of the African diaspora from those who left Africa many generations previously as a result of the slave trade (Mohan, 2002; Waters *et al.*, 2014). Nevertheless, there have been important movements, such as pan-Africanism, that claim there is a shared universal 'African' experience among Africans living within and without Africa (Mazrui, 2005).

What are the developmental impacts of diasporas on the African subcontinent?

In the rest of this chapter, we consider how to analyse the actually existing development impact of diasporas in Africa. To do so, we follow Van Hear (2015) in distinguishing between three different sites of diaspora engagement (1) in the household/extended family sphere; (2) in the known community sphere; and (3) in the imagined community sphere. As often happens with typologies, it is important to emphasise that these spheres do not operate entirely independently of each other and there are often ways in which individual actors straddle these divides.

Household/extended family sphere

The household sphere refers to all developmental activities undertaken by the diaspora that are organised through kin networks. This includes both the everyday routines of diaspora life, such as the sending of remittances to family and friends at home, and also the involvement with more exceptional life course events such as weddings and funerals (Van Hear, 2015). Historically, these large-scale capital transfers tend to have been dismissed as non-developmental as the money is consumed in everyday expenditures (food, rent, clothes, taxes, bills, school fees). However, this flow of money can reasonably be considered as a source of development finance. For example, Mazzucato *et al.*'s (2006) study of a Ghanaian funeral shows the significant role played by the family's diaspora in the Netherlands in organising and funding the funeral events at home, thereby putting considerable resources into local businesses in Ghana (see also Page, 2007; Mbiba, 2010).

To this list of family-level diaspora development work through money transfers, we would explicitly draw attention to three other kin-based activities: first, housebuilding in Africa by members of the diaspora (Diko and Tipple, 1992; Ndjio, 2009; Sinatti, 2009; Smith and Mazzucato, 2009; Melly, 2010; Obeng-Odoom, 2010); second, investments in education (McGregor, 2008; Kufakurinani *et al.*, 2014); and third, investment in income-generating activities for family members (Mazzucato, 2008; Lindley, 2010). In doing so, our aim is to signal the wider developmental benefits of activities that are sometimes overlooked in discussions of diaspora contributions to development (Page and Mercer, 2012).

Officially recorded remittances to sub-Saharan Africa reached \$33 billion in 2016 (World Bank, 2017), although this is likely to be an underestimation of the scale of remittances sent via all channels (including unofficial ones, such as hand carries and *hawala* networks). The largest remittance recipient in sub-Saharan Africa is Nigeria, which received an estimated \$19 billion in 2016, accounting for 50 percent of remittances to the continent (World Bank, 2017). Official data on remittances reveal that there is a huge variation in remittance receipts between African countries. Nevertheless, according to Mohapatra and Ratha (2011), relative to other financial flows to Africa (including foreign direct investment and official development assistance), remittances are beneficial from a macroeconomic and a microeconomic perspective. They provide stable, countercyclical financing and access to foreign exchange, they smooth household consumption during economic downturns or other shocks (such as conflict or famine), and they are associated with reductions in poverty and higher levels of health and educational attainment.

More fine-grained research in southern Africa has demonstrated that the long-established circular migratory patterns from neighbouring countries to South Africa sustain poor households, but has not yet proved transformative in the developmental sense. Remittance money is not reinvested in entrepreneurial or other income-generating activities (Pendleton *et al.*, 2006; Crush and Frayne, 2007; Bolt, 2015). In Zimbabwe, McGregor (2014) shows that money invested by diasporans buying houses serves to inflate the local real estate market in Harare and provides a mechanism for middle-class reproduction. In contrast, Lindley's (2010) research in post-conflict Somaliland has shown how in the city of Hargeisa, remittances are invested into local small and medium-sized businesses, land and property, gifts and social networks, education, and investment in re-emigration – the sponsoring of family members. She argues that while those families receiving remittances tend to belong to the middle- and upper-income families, remittances nevertheless put money into the local economy that is recirculated through social networks. And yet although remittances account for an estimated 23 per cent of household income in Somaliland (Lindley, 2010), it has become much harder to send money home as international banks (in the US and UK, for example) have withdrawn retail banking and payment services from Somali money transfer companies because of the perceived risk that they may contravene new international regulations designed to prevent the financing of terrorism and money laundering. The efforts by donor governments to respond to pressure and find a 'secure' means of remitting money to Somalia is tacit recognition of their developmental significance.

The contrast between the southern Africa, Zimbabwe and Somaliland cases demonstrates the diversity of impacts that remittances have across Africa and alerts us to the reproduction of uneven geographical development that is embedded in patterns of remittance sending and receiving. Research in Cameroon and Tanzania, for example, shows that remittance recipients tend to be located in better-off households and in better-off parts of the country (Mercer *et al.*, 2008). Nevertheless, current development policy on migration is dominated by attempts to encourage more people in the diaspora to remit more money through more formal channels (Page and Mercer, 2012). Policymakers draw attention to the fact that the cost of sending remittances to sub-Saharan Africa is the highest in the world: the average sender paid 11.5 per cent transfer fees on a $200 remittance in 2014 (Watkins and Quattri, 2014; World Bank, 2015). Bringing down the costs of formal money transfers, as well as finding ways to encourage the diaspora to invest in government diaspora bonds (see below), currently tops the policy agenda (e.g. see Ratha *et al.*, 2011).

Overall, the family sphere is probably the most significant for development in terms of the volume of capital it entails and the number of people it enrols. Yet doubts persist about both the macroeconomic impacts of household remittances and the precise route by which this money turns into 'development' at the local level. Crucially, it is important to remember that remittances are both socially and spatially uneven.

Known community sphere

The known community sphere describes social contacts beyond kin, but which do involve personal contacts. The people in these networks know each other as individuals. Typically, this would be characterised as 'associational life' (Van Hear, 2015). In the African diaspora, the known community sphere includes professional associations, business associations, old school (alumni) networks, home town associations and specific religious congregations. Around the world, for example, there are many small-scale informal health professional associations that bring together doctors and nurses from the same African country to socialise and to raise money for specific health projects or campaigns back at home (Van Hear *et al.*, 2004; Chikanda and Dodson, 2013; Page and Tanyi, 2015). Similarly, many old school and university networks are enabled by social media, such as email discussion lists and Facebook, allowing diaspora members to share news and updates and to organise periodic social events and fundraisers for their alma mater (Hofman and Kramer, 2015).

Across the African diaspora, home town associations (which bring together those with ties to a shared home place) are the most common type of institution in the known community sphere. Home town associations have a long history on the African continent, as circular and rural–urban migration brought migrants from the same villages together when they were away from 'home' working in mines and on plantations from the turn of the twentieth century. With increased international migration over the last 30 years, these migrants' groups have been established across North America, Europe and beyond, in those places where there are sufficient members to come together for socialising, welfare support and (sometimes) for developmental efforts in support of the home place (Mercer *et al.*, 2008; Evans, 2010; Chauvet *et al.*, 2015).

Home town associations first emerged in Africa in the early twentieth century for the purposes of mutual support away from 'home', and only recently have they been recognised as development institutions that attempt to raise funds from their diaspora to support projects at home, such as school-building, health facilities or cultural heritage (e.g. for traditional rulers). It is impossible to estimate the scale of the resources that home town associations generate for their home places since they operate informally (see Chauvet *et al.*, 2015). Nevertheless, research has documented home town associations' contributions to, among other things, rural water supply projects, schools (Mercer *et al.*, 2009), mortuaries (Page, 2007), orphanages (Lampert, 2012) and medical equipment (Mohan, 2006). Home town associations provide an important linkage between the home town and the diaspora that can channel resources and knowledge between the two places, and given their voluntary nature (their members devote time and resources amid busy lives and multiple obligations), they can achieve small-scale successes.

The extent to which these associations can be effective agents of development has been much debated. Research has shown that home town associations can reproduce existing social inequalities of class and gender (Henry and Mohan, 2003; Lampert, 2012), as well as reproducing existing inequalities between places (Mercer *et al.*, 2008; Mazzucato and Kabki, 2009). They are also deeply interwoven with domestic political agendas at the local and national scale (Lampert, 2009; McGregor, 2009; Sall, 2010). Research has also questioned the assumption (which underpins the diaspora and development policy debate) that it is the diaspora who are located in North America and Europe that are most likely to transform their home communities in Africa. Rather, research has shown that it is more likely the diaspora groups located in towns and cities across Africa (the 'domestic diaspora') who are often more engaged in developmental activities at home (Mercer *et al.*, 2009; Lampert, 2014). This is an important point which reminds us that there are resourceful and engaged diaspora communities *within* Africa as well as outside of it (Bakewell and Binaisa, 2016).

The imagined community sphere

The imagined community sphere refers to those activities that engage members of the diaspora beyond immediate circles of acquaintance. In this sphere, the individuals who are engaged in development activities may not even meet each other, but can still participate in some kind of shared project or programme. Typically, this can be illustrated by reference to a nation. The members of the Sierra Leonean diaspora scattered around the world, for example, cannot possibly all know each other personally; however, they can imagine a solidarity based on a common relationship to the home nation (Anderson, 1991). This solidarity can then be mobilised (by a government or organisation) to achieve a particular developmental goal – here, we can consider diaspora taxes in Eritrea, diaspora bonds, and the challenges associated with mobilising these sentiments in a polarised nation such as Zimbabwe. However, in the diaspora development field, the imagined community is not limited just to questions of nationalism, but might also include business or religious identities – illustrated here with reference to the Mouride Islamic brotherhood in Senegal.

One of the best-known diaspora mobilisations in the interests of the imagined national community is that of Eritrea. The liberation war (with Ethiopia, 1962–1993, and the further war, 1998–2000) generated an Eritrean diaspora initially concerned to support the independence struggle they had left behind, and who contributed to it extensively. However, this practice of supporting the nation has since become institutionalised. Following independence, the Eritrean government introduced a 2 per cent income tax on all Eritreans working abroad. Initially a popular move, the coercive measures later undertaken by the Eritrean state to collect the tax have since attracted criticism (Tecle and Goldring, 2013; Campbell and Afework, 2015).

Nevertheless, the international development policy community has identified the diaspora as a significant potential source of development finance for Africa, in which it is assumed that diasporic identification with a home nation will encourage diasporans to invest in development projects for the good of the nation through vehicles such as diaspora bonds (Ketkar and Ratha, 2010). Diaspora bonds are debt instruments that are issued by a national government, typically with five- to ten-year maturities. They offer governments an alternative to raising money from international financial markets, and they offer the diaspora an opportunity to make a patriotic investment in their home country.

Diaspora bonds could be used to fund investments in public goods. The World Bank estimates that sub-Saharan African countries could potentially raise between US$5 and US$10 billion a year by issuing diaspora bonds. They suggest that this policy might be particularly well-suited to countries such as Ethiopia, Ghana, Kenya, Liberia, Nigeria, Senegal, Uganda and Zambia, which have large diasporas in high-income countries (Ratha *et al.*, 2011). Ethiopia was the first African country to issue diaspora bonds in 2008 and 2011, both of which attempted to raise financing for national electricity-generating projects. However, neither issuance matched expectations in terms of take-up in the diaspora, and the 2011 bond was more popular with domestic rather than diaspora investors (Kayode-Anglade and Spio-Garbrah, 2012).

The difficulties experienced by the Eritrean and Ethiopian governments' attempts to raise money from their diasporas underline the crucial importance of the politics of diaspora engagement, and specifically the diaspora's relationship to the political field in the home nation. JoAnn McGregor has shown this most clearly in her work on the Zimbabwean diaspora in the UK. She argues that the highly polarised nature of Zimbabwean politics between the ruling ZANU–PF party and the opposition Movement for Democratic Change has been reproduced in the diaspora, paralysing any broader attempt to engage in development at home for the national good (McGregor, 2009).

At the other end of the diaspora engagement spectrum is perhaps the best-known imagined community in the African diaspora, the Mouride brotherhood (Diouf and Rendall, 2000; Diedhiou, 2011), an Islamic Sufi order that considers the Senegalese city of Touba as its spiritual home place. Within this brotherhood are many *dahira* (self-help groups) that have increasingly come to act as powerful development agents in Touba. Foley and Babou (2011) recount the experience of the Matlaboul Fawzaini *dahira*, which drew on contributions from its 60,000 members in Africa, Europe and North America to build a large hospital in Touba. The *dahira* raised a staggering US$10 million over 10 years, built the hospital, and handed it over to local government. However, they also recount the project's limitations, which included lack of technical knowledge and expertise in hospital construction. Eighty per cent of the original building had to be demolished and rebuilt as it did not conform to hospital specifications (such as doorways wide enough for stretchers), and the finished building had insufficient space for hospital operations. As one local doctor put it, 'money and know-how are not the same' (Foley and Babou, 2011: 82).

Conclusion

The African diaspora is profoundly diverse. It has grown over many centuries for many reasons to produce different communities in different places. It is still changing. Debates about the development impacts of African diasporas began in the 1960s with the rise of interest in the 'brain drain' and policies designed to encourage return migration of highly skilled migrants. In the 1980s, the argument reversed and analysts began to draw attention to the positive development impacts of remittances. By the early 2000s, policymakers were thinking explicitly about how to formally enrol African diaspora communities into international development, a move that was often prompted by lobbying from diaspora groups themselves.

The metaphor of a pendulum (de Haas, 2012; Gamlen, 2014a, 2014b) swinging between positive and negative attitudes to international migration within the development community is a useful one, not only because it captures those changing ideas about what is progressive and what isn't, but also because it emphasises a sense of continuity in the debate. This is a field where the sense of the potential 'usefulness' of the African diaspora to development policymakers rubs up against the 'unruly' creativity and autonomy of the African diaspora themselves. The development work of African diasporas can usefully be assessed in terms of three different spheres: the household (e.g. remittances within families), the known community (e.g. diaspora associations) and the imagined community (e.g. diaspora bond issues). However, such a tidy typology doesn't really capture the sense that the policymakers and African diasporas often seem to talk past one another, and the *potential* development impact is greater than that which has been delivered. Policymakers' attempts to govern diaspora behaviour are often frustrated, leading to a constant sense that the development impact of African diasporas is always in the future.

Yet we would go still further here and insist that we look for ways to break out from the circular debate about how to harness the African diaspora for developmental purposes. As Parvati Raghuram (2009) has persuasively argued, there are bigger questions at stake here about the politics and the geography of development that are revealed by the way we imagine diaspora development debates. These revolve around questions of sacrifice and responsibility – the sacrifice that individual migrants must make in order to take on responsibility for the development of the countries they have left, and the blindness that renders migrants' contribution to development in the countries where they currently live invisible. There are other questions, too, about the inequalities that are reproduced by remittances, and the unknown consequences for migrants of enrolling them into formal remittance channels. The question 'What kind of development for what kind of diaspora?' is still often unaddressed.

References

African Union (2000) *The Diaspora Division*, available at: http://pages.au.int/cido/pages/diaspora-division (accessed 4 October 2016).

Akyeampong, E. (2000) 'Africans in the diaspora: the diaspora and Africa', *African Affairs*, 99(395): 183–216.

Anderson, B. (1991) *Imagined Communities: Reflections on the Origin and Spread of Nationalism*, London: Verso.

Aparacio, F.J. and Meseguer, C. (2011) 'Collective remittances and the state: the 3×1 program in Mexican municipalities', *World Development*, 40(1): 206–222.

Bakewell, O. and Binaisa, N. (2016) 'Tracing diasporic identifications in Africa's urban landscapes: evidence from Lusaka and Kampala', *Ethnic and Racial Studies*, 39(2): 280–300.

Bodomo, A. and Ma, E. (2012) 'We are what we eat: food in the process of community formation and identity shaping among African traders in Guangzhou and Yiwu', *African Diaspora*, 5(1): 3–26.

Bolt, M. (2015) *Zimbabwe's Migrants and South Africa's Border Farms: The Roots of Impermanence*, Cambridge: Cambridge University Press.

Bork-Hüffer, T., Rafflenbeul, B., Li, Z., Kraas, F. and Xue, D. (2016) 'Mobility and the transiency of social spaces: African merchant entrepreneurs in China', *Population, Space and Place*, 22(2): 199–211.

Campbell, J.R. and Afework, S. (2015) 'Ethiopian and Eritrean immigrants in Britain: refugee organizing, transnational connections and identity, 1950–2009', *African Diaspora*, 8: 98–119.

Chauvet, L., Gubert, F., Mercier, M. and Mesplé-Somps, S. (2015) 'Migrants' home town associations and local development in Mali', *Scandinavian Journal of Economics*, 117(2): 686–722.

Chikanda, A. and Dodson, B. (2013) 'Bandaid transnationalism: remittance practices of emigrant Zimbabwean doctors', *Migration and Development*, 2(1): 57–73.

Cohen, R. (2009) *Global Diasporas*, 2nd edition, London: Routledge.

Crush, J. and Frayne, B. (2007) 'The migration and development nexus in southern Africa', *Development Southern Africa*, 24(1): 1–23.

de Haas, H. (2012) 'The migration and development pendulum: a critical view on research and policy', *International Migration*, 50(3): 8–25.

Diedhiou, A. (2011) 'Remittances, transnational dahiras and governance in Senegal', *International Migration*, 53(1): 171–186.

Diko, J. and Tipple, G. (1992) 'Migrants build at home: long distance housing development by Ghanaians in London', *Cities*, 9: 288–294.

Diouf, M. and Rendall, R. (2000) 'The Senegalese Murid trade diaspora and the making of a vernacular cosmopolitanism', *Public Culture*, 12(3): 679–702.

Dufoix, S. (2015) 'The loss and the link: a short history of the long-term word "diaspora"', in N. Sigona, A. Gamlen, G. Liberatore and H. Neveu Kringelbach (eds), *Diasporas Reimagined: Spaces, Practices and Belonging*, Oxford: Oxford Diasporas Programme.

Evans, M. (2010) 'Primary patriotism, shifting identity: hometown associations in Manyu Division, south-west Cameroon', *Africa*, 80(3): 397–425.

Foley, E. and Babou, C.A. (2011) 'Diaspora, faith and science: building a Mouride hospital in Senegal', *African Affairs*, 110(438): 75–95.

Gamlen, A. (2006) *Diaspora Engagement Policies: What Are They, and What Kinds of States Use Them?* COMPAS Working Paper 06–32, University of Oxford.

Gamlen, A. (2014a) 'Diaspora institutions and diaspora governance', *International Migration Review*, 48(S1): S180–S217.

Gamlen, A. (2014b) 'The new migration-and-development pessimism', *Progress in Human Geography*, 38(4): 581–597.

Henry, L. and Mohan, G. (2003) 'Making homes: the Ghanaian diaspora, institutions and development', *Journal of International Development*, 15: 611–622.

Hofman, K. and Kramer, B. (2015) 'Human resources for research: building bridges through the diaspora', *Global Health Action*, 8, DOI: http://dx.doi.org/10.3402/gha.v8.29559.

Iskander, N. (2010) *Creative State: Forty Years of Migration and Development Policies in Morocco and Mexico*, New York: Cornell University Press.

Kayode-Anglade, S. and Spio-Garbrah, N. (2012) 'Diaspora bonds: some lessons for African countries', *Africa Economic Brief*, 3(13): 1–13.

Ketkar, S.L. and Ratha, D. (2010) 'Diaspora bonds: tapping the diaspora during difficult times', *Journal of International Commerce, Economics and Policy*, 1(2): 251–263.

Kufakurinani, U., Pasura, D. and McGregor, J. (2014) 'Transnational parenting and the emergence of "diaspora orphans" in Zimbabwe', *African Diaspora*, 7(1): 114–138.

Lampert, B. (2009) 'Diaspora and development: Nigerian organizations in London and the transnational politics of belonging', *Global Networks*, 9(2): 162–184.

Lampert, B. (2012) 'Diaspora and development? London-based Nigerian organisations and the transnational politics of socio-economic status and gender', *Development Policy Review*, 30(2): 149–167.

Lampert, B. (2014) 'Collective transnational power and its limits: London-based Nigerian organizations, development at "home" and the importance of local agency and the "internal diaspora"', *Journal of Ethnic and Migration Studies*, 40(5): 829–846.

Lindley, A. (2010) *The Early Morning Phone Call: Somali Refugees' Remittances*, Oxford: Berghahn Books.

Mazrui, A. (2005) 'Pan-Africanism and the intellectuals: rise, decline and revival', in T. Mkandawire (ed.), *African Intellectuals: Rethinking Politics, Language, Gender and Development*, London: Zed Books, pp. 56–77.

Mazzucato, V. (2008) 'The double engagement: transnationalism and integration – Ghanaian migrants' lives between Ghana and the Netherlands', *Journal of Ethnic and Migration Studies*, 34(2): 199–216.

Mazzucato, V. and Kabki, M. (2009) 'Small is beautiful: the micro-politics of transnational relationships between Ghanaian hometown associations and communities back home', *Global Networks*, 9(2): 227–251.

Mazzucato, V., Kabki, M. and Smith, L. (2006) 'Transnational migration and the economy of funerals: changing practices in Ghana', *Development and Change*, 37(5): 1047–1072.

Mbiba, B. (2010) 'Burial at home? Negotiating death in the diaspora and Harare', in J. McGregor and R. Primorac (eds), *Zimbabwe's New Diaspora Displacement and the Cultural Politics of Survival*, Oxford: Berghahn Books, pp. 144–163.

McGregor, J. (2008) 'Children and "African values": Zimbabwean professionals in Britain reconfiguring family life', *Environment and Planning A*, 40(3): 596–614.

McGregor, J. (2009) 'Associational links with home among Zimbabweans in the UK: reflections on long-distance nationalisms', *Global Networks*, 9(2): 185–208.

McGregor, J. (2014) 'Sentimentality or speculation? Diaspora investment, crisis economies and urban transformation', *Geoforum*, 56: 172–181.

Melly, C. (2010) 'Inside-out houses: urban belonging and imagined futures in Dakar, Senegal', *Comparative Studies in Society and History*, 52(1): 37–65.

Mercer, C., Page, B. and Evans, M. (2008) *Development and the African Diaspora: Place and the Politics of Home*, London: Zed Books.

Mercer, C., Page, B. and Evans, M. (2009) 'Unsettling connections: transnational networks, development and African home associations', *Global Networks*, 9(2): 141–161.

Mohan, G. (2002) 'Diaspora and development', in J. Robinson (ed.), *Development and Displacement*, Oxford: Oxford University Press, pp. 77–140.

Mohan, G. (2006) 'Embedded cosmopolitanism and the politics of obligation: the Ghanaian diaspora and development', *Environment and Planning A*, 38: 867–883.

Mohapatra, S. and Ratha, D. (2011) 'Migrant remittances in Africa: an overview', in S. Mohapatra and D. Ratha (eds), *Remittance Markets in Africa*, Washington, DC: World Bank, pp. 1–68, available at: http://siteresources.worldbank.org/EXTDECPROSPECTS/Resources/476882-1157133580628/RMA_FullReport.pdf (accessed 16 December 2015).

Ndjio, B. (2009) 'Migration, architecture, and the transformation of the landscape in the Bamileke Grassfields of West Cameroon', *African Diaspora*, 2: 73–100.

Newland, K. and Patrick, E. (2004) *Beyond Remittances: The Role of Diaspora in Poverty Reduction in Their Countries of Origin – A Scoping Study by the Migration Policy Institute for the Department of International Development*, Washington, DC: Migration Policy Institute.

Obeng-Odoom, F. (2010) 'Urban real estate in Ghana: a study of housing-related remittances from Australia', *Housing Studies*, 25(3): 357–373.

Ostergaard-Nielsen, E. (2011) 'Codevelopment and citizenship: the nexus between policies on local migrant incorporation and migrant transnational practices in Spain', *Ethnic and Racial Studies*, 34(1): 20–39.

Page, B. (2007) 'Slow going: the mortuary, modernity and the hometown association in Bali-Nyonga, Cameroon', *Africa*, 77(3): 419–441.

Page, B. and Mercer, C. (2012) 'Why do people do stuff? Reconceptualizing remittance behaviour in diaspora-development research and policy', *Progress in Development Studies*, 12(1): 1–18.

Page, B. and Tanyi, R. (2015) 'Engaging the African diaspora in the fight against malaria', in A. Christou and E. Mavroudi (eds), *Dismantling Diasporas: Rethinking the Geographies of Diasporic Identity, Connection and Development*, Surrey: Ashgate, pp. 189–201.

Palmer, C. (1998) 'Defining and studying the modern African diaspora', *Perspectives on History*, September, available at: www.historians.org/publications-and-directories/perspectives-on-history/september-1998/defining-and-studying-the-modern-african-diaspora (accessed 16 December 2015).

Pendleton, W., Crush, J., Campbell, E., Green, T., Simelane, H., Tevera, D., et al. (2006) *Migration, Remittances and Development in Southern Africa*, available at: www.queensu.ca/samp/sampresources/samppublications/policyseries/Acrobat44.pdf (accessed 16 December 2015).

Raghuram, P. (2009) 'Which migration, what development? Unsettling the edifice of migration and development', *Population, Space and Place*, 15(2): 103–117.

Ratha, D., Mohapatra, S., Ozden, C., Shaw, W. and Shimeles, A. (2011) *Leveraging Migration for Africa: Remittances, Skills and Investments*, Washington, DC: World Bank.

Sall, M. (2010) 'Straightforward critics or would-be candidates? International migrants and the management of local affairs and development: the case of the Senegal River Valley', *Environment and Urbanization*, 22(1): 81–89.

Sinatti, G. (2009) 'Home is where the heart abides: migration, return and housing in Dakar, Senegal', *Open House International*, 34(3): 49–56.

Smith, L. and Mazzucato, V. (2009) 'Constructing homes, building relationships: migrant investments in houses', *Tijdschrift voor Economische en Sociale Geografie*, 100(5): 662–673.

Tecle, S. and Goldring, L. (2013) 'From "remittance" to "tax": the shifting meanings and strategies of capture of the Eritrean transnational party-state', *African and Black Diaspora: An International Journal*, 6(2): 189–207.

UNDP (United Nations Development Programme) (2009) *Human Development Report 2009: Overcoming Barriers – Human Mobility and Development*, New York: UNDP.

Van Dijk, R. (2003) 'Localisation, Ghanaian Pentecostalism and the stranger's beauty in Botswana', *Africa*, 73(4): 560–583.

Van Hear, N. (2015) 'Diaspora engagement in war-torn societies', in R. Cohen, J. Story and N. Moon (eds), *The Impact of Diasporas*, Oxford: Oxford Diasporas Programme and The Impact of Diasporas on the Making of Britain: Evidence, Memories and Inventions, University of Leicester, pp. 18–21.

Van Hear, N., Pieke, F. and Vertovec, S. (2004) *The Contribution of UK-Based Diasporas to Development and Poverty Reduction*, Report by the ESRC Centre on Migration, Policy and Society (COMPAS), University of Oxford, for the Department for International Development.

Waters, M., Kasinitz, P. and Asad, A.L. (2014) 'Immigrants and African Americans', *Annual Review of Sociology*, 40: 369–390.

Watkins, K. and Quattri, M. (2014) *Lost in Intermediation: How Excessive Charges Undermine the Benefits of Remittances for Africa*, London: Overseas Development Institute.

World Bank (2017) *Migration and Development Brief 27*, 13 April, Migration and Remittances: Recent Developments and Outlook, April 2017, World Bank.

Zeleza, P.T. (2005) 'Rewriting the African diaspora: beyond the Black Atlantic', *African Affairs*, 104(414): 35–68.

Section 4
Health

12
Health challenges

Kenneth Lynch

Introduction

The 2014 World Health Organization (WHO) report on health in the African region described the widely held perception that there has been no progress on improving health in the region as a 'convenient untruth' (WHO, 2014: xx). Harmonization for Health in Africa (HHA), a consortium of international organisations working on health and well-being issues in the continent, demonstrated that Africa accounted for only 11 per cent of the global population but sub-Saharan Africa accounted for 49 per cent of maternal deaths, 50 per cent of child deaths under 5 years and 67 per cent of HIV/AIDS cases (HHA, 2011). Despite the wider economic and social challenges facing the continent that are discussed elsewhere in more detail in this book, in fact considerable achievements have been made. However, this said, there is considerable room for progress, both in terms of global health indicators and in terms of wider enhancements in public health outcomes for the people of Africa. This chapter sets out to briefly discuss the major developments in health and identify the health challenges facing the continent. This comprises a brief discussion of the background to health issues in Africa, examining some of the latest data, in order to discern some patterns or trajectories. The chapter then explores the relationships between health and development, before considering some of the latest trends in the area of health.

Background

The relationship between health and development in Africa, and indeed globally, has been considered over a long period. After Dr Ronald Ross' discovery in 1898 of the life cycle of the mosquito and of its link with malaria, he levelled criticism at the British West Africa administration's woeful neglect of sanitation in the cities of the colonies (Nye and Gibson, 1997). Ross visited Freetown in Sierra Leone in 1899 and in 1902, when he substantially advanced the understanding of the malarial virus. As a result of this, Ross made three main recommendations that were to have an impact on future urban planning:

> These are (1) scrupulous drainage of the soil; (2) pending this, the persistent treatment of Anopheles' breeding-pools by culicicides; (3) the segregation of Europeans. We also recommended the protection of public buildings, such as barracks, gaols, hospitals, and rest-houses by wire gauze screens; the isolation of the sick; and the habitual employment of mosquito-nets and *punkahs* by individuals.
>
> *(Ross, 1902: 92)*

Culicicides refers to insecticides specifically used to target the *culicinae* mosquitoes, the family of mosquitoes known by then to pass on malaria. The *punkah* refers to a fan that is hand-operated by an unskilled labourer – also known at that time as a 'coolie' – hired to move the fan by pulling on a string to give the fan motion, and by doing so cool the colonial inhabitants of the buildings and keep mosquitoes from flying. Ross recognised the link between the environment and the living conditions of the people in Freetown and their vulnerability to disease.

Of course, the history of African health and medicine does not begin with the colonialists. David Livingstone, for example, recorded in his journals numerous observations of his encounters with healthy Africans and with African healers (Rijpma, 2015). More recent discourse around health and development focuses around two bodies of theory. The first is that research among health economists has found a relationship between life expectancy and economic growth. As Hans Rosling (n.d.) puts it in one of his Gapminder videos, there are no high-income countries with a life expectancy below 74 years and no low-income countries where it is above 64. This said, the middle-income countries, which account for a large proportion of the global population, do not follow a pattern. Acemoglu and Johnson (2007) argue that the relationship between life expectancy and economic growth suggests there is no evidence of a relationship between these two variables at this level of economic development.

Apart from the fact that African countries were excluded from Acemoglu and Johnson's (2007) analysis, Hussein (2012) argues that the development of a general model linking life expectancy and economy requires plausible causal explanations to complement any statistical patterns. He questions the underlying assumptions, in particular assumptions about the timing of countries entering an epidemiological transition (Omran, 1971). The epidemiological transition is a theory of unilinear transition from widespread infectious diseases to extended life expectancy and a rise in non-infectious diseases. Acemoglu and Johnson (2007) suggest that this began happening first in what Omran (1971) termed the 'classical model transition', the countries who entered the transition early, shortly after the end of the Second World War. Hussein (2012) cautions that the evidence suggests that this transition began in a small and inconsistent way from 1900 to the 1940s. Although available in the immediate post-war period and more widely used in advanced economies, as well as what Omran calls the accelerated epidemiological transition model, it was not until the 1970s and even the 1980s, argues Hussein, that widespread use of vaccines for polio, measles, mumps and rubella had substantial and immediate effects in the poorer countries of the world, including those in Africa. And this is when it begins to impact on these societies, or those following Omran's contemporary (or 'delayed') epidemiological transition model. More recent interpretations of Omran's theory suggest that the unidirectional model simplifies complex relationships between health, disease, technological, environmental and sociocultural changes, and that these relationships may result in reverse and overlap. This characterisation of the relationship between health and development as complex has significant implications for our understanding of health and development in Africa.

The most recent global and regional collection of health and development indicators suggests that most countries in Africa have made considerable progress on key areas of health concern. For example, the indicators summarised in Table 12.1 show significant progress across the WHO

Table 12.1 Health indicators for the WHO Africa region

Indicator	*2000*	*2015*	*MDG target*
Under-five mortality rate (per 1,000 live births)	154	81	59
Infant mortality (per 1,000 live births)	94	55	n/a
Maternal mortality rate (per 1,000 live births)	840	542	205
Female life expectancy at birth (years)	51	60*	n/a
Male life expectancy at birth (years)	49	57*	n/a
Measles-containing vaccine coverage	54	73*	90
Population of underweight children (%)	23.6	17.7	17

Source: Adapted from WHO (2016, 2017).

Notes: Morocco, Tunisia, Egypt, Sudan, Djibouti and Somalia are excluded as they are part of the WHO Eastern Mediterranean region.

* 2013 data provided as latest available.

Africa region. However, what Table 12.1 also shows is that the progress has fallen short of the targets linked to the Millennium Development Goals (MDGs) set in 2000 with a deadline for 2015. When examining the record of the countries' performance against the MDGs, it is evident that many countries failed to meet several targets, but some countries failed on more targets than others. The more affluent countries such as Algeria, Botswana and Mauritius met more targets, while less affluent countries such as Chad, the DRC and Togo only met two health-related targets each. There were, however, one or two surprises. Rwanda met seven of the 13 targets, while Benin and Cameroon met only four. Cape Verde met six targets, while Equatorial Guinea, Kenya and Nigeria met only one. Only two countries (Cape Verde and Rwanda) achieved the target of reducing maternal mortality, while only six countries failed to meet the target for reducing the HIV/AIDS incidence, Algeria, Cape Verde, Equatorial Guinea, Mali, South Sudan and Uganda (WHO, 2016).

What is the context for this mixed performance? The following discussion explores some of the evidence on health across the African continent – the trends, as well as some links with poverty and development.

There are a wide range of international and national agencies working in the African health sector that are responding to the health challenges facing the continent. The agencies include government, non-governmental organisations (NGOs), international non-governmental organisations (INGOs) and international agencies. This makes the institutional landscape highly complex. One of the things that distinguishes Africa from other WHO regions is a significant proportion of the investment in health is from external agencies. This can act in the continent's favour, such as in the coordinated response to the HIV/AIDS epidemic, which provides a partial explanation of the widespread achievement across the continent on this MDG target. In the response to the HIV/AIDS epidemic, the initial approach, which was developed in Western countries where the risk groups were very different, was unsuccessful and had to be revised (De Cock *et al.*, 2002). Iliffe (2006), in a major review of the history of the HIV/AIDS epidemic, argues that the African continent has suffered the most from the HIV/AIDS epidemic. Africa had the epidemic first, and when its significance was realised, its emergence was largely found to be the result of social and sexual networks. The African response has evolved and changed along with developments in medical governance, relationships between health donors and recipients, and the emerging understanding of the means of transmission.

The result has been a substantial and sustained effort in coordinating such health initiatives focused on a particular disease. There are a number of international and regional agreements that underpin these initiatives. For example, Table 12.2 lists a selection of key regional initiatives

Table 12.2 African regional health initiatives

Year	*Initiatives*
2001	Abuja Declaration on the allocation of 15 per cent of public expenditure to health
2005	Maputo Declaration on tuberculosis as an emergency
2006	Year of Acceleration of HIV Prevention in Africa
2008	Ouagadougou Declaration on Primary Health Care and Health Systems
2010	Universal Access to HIV and AIDS Prevention, Treatment, Care and Support in Africa

Table 12.3 Major health challenges and opportunities facing the African region

Challenge	*Opportunity*
Chronic under-resourcing of health provision	HHA (2011) demonstrated that investing an additional US$21–36 per person in sub-Saharan Africa could save states US$100 billion by 2015.
Complex inequalities, leaving some populations more vulnerable to health problems	Urban–rural inequalities not as large as inequalities affecting children between educated and uneducated mothers. Wealth inequalities have significant impact on access to health.
Africa bears a disproportionate burden of several global health crises	Strategic investments in selected evidence-based interventions and targeted health systems strengthening in some areas (Heaton *et al.*, 2016).
Changes taking place, leading to improvements in some areas and new challenges (epidemics, non-communicable disease, climate change)	The adoption of integrated approaches providing an interdisciplinary response that can yield good results to complex health challenges (Smith *et al.*, 2015; Lebov *et al.*, 2017).

that were intended to reinforce the focus on the health-related MDGs. This has resulted in two major trends. The first is that coordination of health sector institutions has been focused around an increasing number of time-bound and coordinated sets of intentions. This provides a framework for all national health systems, as well as local and international agencies operating in the African health sector. The evidence above suggests that this is having measurable impacts, ensuring progress on specific targeted health issues. However, Makuta and O'Hare (2015) argue that the relationship between public health and health outcomes is weak. They go on to argue that there is evidence that this weak impact of public health spending may be partly explained by the quality of health governance. For example, their analysis can detect variations in governance that show where good governance can lead to improved outcomes for health indicators such as under-five mortality.

A second major trend illustrated in the regional health initiatives mentioned above relates to the way in which the influence of these initiatives poses barriers to African countries being able to take responsibility for integrated healthcare provision and establish governance over their own public health service. The initiatives outlined in Table 12.2 suggest that Africa is rising to meet specific major regional challenges, but some of these can pose a barrier to achieving widespread health and well-being for all Africans. For example, Table 12.3 illustrates a number of major health-related challenges facing the region that require more sustained and integrated approaches than a campaign to tackle a single disease might be able to achieve. Some of these challenges are exacerbated by the widespread involvement and influence of so many organisations. As the continent with the most number of low-income countries and a high proportion of the world's fragile states, this evidence demonstrates that there is still considerable work that needs to be done at the national level. For example, the WHO estimates that approximately one-third of Africa's disease burden can be attributed to environmental hazards, particularly relating to lack of access to safe water, indoor air pollution, and lack of access to sanitation and hygiene. These can combine with what are described as modern environmental health hazards, such as water pollution, air pollution and emerging infectious diseases. These traditional and modern disease burdens place a kind of double health burden on poor but developing societies (Nweke and Sanders, 2009).

There is some evidence therefore of beneficial impacts on the economy of improved health in a population. For example, the WHO (2014) estimates that for every 10 per cent increase in life expectancy at birth, there is a corresponding rise in economic growth of 0.4 per cent per year. If correct, this suggests that the beneficial economic impact of improved morbidity would cover the costs of investment in bringing about an improved situation. In fact, the HHA (2011) estimated that improved health investment of between US$28 and 30 billion per year over the period 2011 to 2015 could save the lives of 3.1 million people, prevent stunting in approximately 4 million children, increase the number of health workers by almost 1 million, and bring about economic benefits of almost US$100 billion in 2015 alone. In other words, provide a fourfold economic return on investment and save lives, employ health workers and improve the health prospects of millions of children (WHO, 2014). It can be argued that an African child has as much right to health and well-being as a child growing up in England, the US or New Zealand. Improved health and well-being is therefore beneficial to the slippery concept of human rights, as well as to economic development.

There has been a considerable decline in child, maternal and adult mortality rates, and substantial decreases in the burdens of several diseases.

Two major population trends are likely to have a significant impact on the continent's ability to respond to the health challenges it faces. The first is that the declining levels of population growth and parallel declines in female fertility rates mean that health outcomes for African infants, children and mothers have been improving. The second is that as the population works through a transition to a majority urban-based population, a larger proportion of the population will be closer to primary healthcare, pharmacy and hospital facilities. This will result broadly in quicker and more accessible medical assistance and greater access to specialist medical treatments, though clearly proximity does not necessarily mean easy access. It will also provide the potential for public health and planning administrations to provide healthier living conditions for a larger share of the population.

The evidence shows that the continent is still some way behind the rest of the world, with health indicators lagging. For example, life expectancy at birth is around 15 years behind the global situation. While enormous improvements in child mortality have been achieved, Africa still has the highest levels of any region – with almost double the rates of the nearest other WHO regions (WHO, 2016). Similar patterns exist on many of the health indicators, but as discussed above, some of the positive achievements are the result of approaches tackling particular diseases or health issues rather than achieving system-wide healthcare provision. Small-scale successes do also achieve impact that can be challenging to scale up to national-level health systems.

Infectious diseases

In Africa, communicable diseases account for two-thirds of the disease burden. There have been significant successes in efforts to tackle this burden, resulting in halving malarial deaths through the application of simple and affordable interventions such as increased availability and use of insecticide-treated bed nets, diagnosis-based treatment, community-based vector control and innovative combination therapies (Nweke and Sanders, 2009). Similarly, there are a number of other elements of the infectious disease burden that can be reduced with relatively inexpensive solutions, focused on access to clean water and sanitation facilities. However, more complex infectious diseases, particularly newly emergent diseases such as HIV/AIDS and Ebola (see Box 12.1), pose more complex challenges to health systems. In all these cases, integrated and universal health provision and prevention is cheaper and more effective, but ultimately requires the institutional support from national health services.

Box 12.1 Zoonotic disease in African urban areas: the case of Nairobi

Zoonotic diseases are diseases that affect animals. The risk to human health arises from the possibility of them jumping from animals to humans. A number of recent epidemics that have been categorised as Public Health Emergency of International Concern (PHEIC) have been diseases that fall within this category, including Ebola, avian influenza and swine flu. One example of a response to this emergence of the risk to human health of zoonotic diseases has been the rise of the One Health (OH) approach. It has been adopted because of the challenges experienced while addressing the enormous burden of endemic zoonotic diseases, including in the areas of laboratory diagnosis, surveillance and response to outbreaks.

For example, in Kenya, the OH approach has gained support and been successfully applied in response to disease outbreaks of human African trypanosomiasis (April 2012), rabies (March 2012) and anthrax (October 2012). Kenyan prevention and control strategies for rabies and Rift Valley fever are also at advanced stages of development.

Another related approach illustrating the complexity of zoonotic disease transmission is in the UrbanZoo research project, which adopts an interdisciplinary research approach that addresses the links between social-environmental and spatial conditions and the microbial diversity to which people are exposed in rapidly growing urban and peri-urban areas, particularly where large proportions of the population are living in slum conditions where livestock, dogs and rodents are present. The project also addresses planning, economic development and the institutional context in which zoonotic diseases develop in Nairobi and how this is shaped by social and spatial fragmentation. This involves the collection of samples of food, water and stools in order to carry out a battery of biochemical, biological and genetic tests to understand the disease transmission routes and barriers. For example, strands of the research focus on goat's milk value chains, human nutrition in poor urban residents, molecular phylogenetics of bacteria, food chain risk assessment, and delivery of integrated surveillance and disease control activities. The project also mapped the movement of livestock in order to gain a better understanding of exposure of children to animals in two of Nairobi's slums. Over several years, the project aims to understand the link between keeping livestock (farm animals: cows, chicken, pigs, etc.), consuming livestock products (milk, meat, eggs, etc.) and children getting diarrhoea in informal settlements.

Nairobi has been transformed into a city with a growing human population. This population growth and the pace of change raises many questions about the vulnerability of this population to zoonotic disease risk. In essence, the livestock food system context (the 'milieu') is a major piece of the puzzle of understanding disease emergence and planning improved systems of managing livestock food systems. Livestock health and human health are therefore intricately linked.

Source: Hassell *et al.* (2016) and Alarcon *et al.* (2017).

Healthcare provision

Herrick (2017) points out that while Africa bears 24 per cent of the world's disease burden, it has only 3 per cent of the world's health workers. This places the region at a significant disadvantage in terms of the human resources available to tackle its health burden. Heaton *et al.* (2016) argue that the presence of a skilled health attendant at birth is associated with better health of the child. However, they also found that higher levels of wealth increased the likelihood of a medical attendant at birth, suggesting that wealth benefits health. So, this results in the irony that increased spending on training birth attendants may improve the health prospects of the wealthy before it benefits the poor. Heaton *et al.* (2016) argue therefore that significant reductions in health disparities will not necessarily occur unless this becomes an explicit goal of health policy.

Ruckert and Labonté (2014) argue that one response to the health challenges facing the lowest-income countries, and related to the change in aid architectures, has been a rise in global health partnerships (GHPs). A key strand of this response has been the rise of public–private partnerships (PPPs) as a tool for global health. Ruckert and Labonté (2014) argue that on the one hand, the rise of GHPs and PPPs have resulted in additional resources that would otherwise be unavailable to the health systems. On the other hand, such initiatives have served to undermine the alignment of aid, international health interventions and local health policies, leading to a complex governance of health responses. Herrick (2017) argues that a further effect of this is to focus initiatives around campaigns and specific diseases at a global level, rather than fostering the development of integrated national-level health policy and governance. Such an approach assumes that the only response to such diseases relies on clinical, market-based solutions, and a focus on diseases that are a global-level concern, and makes response to national- and local-level health issues more challenging if they are not covered by GHPs. This does not build national healthcare systems that are integrated, and it can interfere with this, for example by distracting from system-wide initiatives, or by recruiting health professionals out of national health systems. A stark example of this is illustrated in the 2013 response to a TB epidemic in Sierra Leone. This epidemic was brought under control and declared as finished through the combined efforts of international TB partnerships and the national health service. The following year, the country's health system was found wanting in the face of the Ebola epidemic of 2014–2015.

Patel (Chapter 15, this volume) discusses international partnerships focused around access to medicines. This is particularly successful in the establishment of vaccination programmes. However, some GHPs have failed to coordinate with national health systems or even donor countries. Ruckert and Labonté (2014) report on a study in which almost half a Tanzanian medical officer's time was spent writing reports for 11 different funders.

Rising issues

The latest research in health-related issues in Africa suggests a small number of key developments in health risks, research and treatment. This section will now briefly touch on each of these as areas that merit future attention.

Vulnerability to epidemics

Iliffe (2006) maintains that the history of HIV/AIDS is part of Africa's history of human colonisation on the continent. The zoonotic dimension is also important because in the case of these diseases, an integrated approach to health will optimise the interventions, in particular where a high proportion of the population lives in close proximity to animals that could form the zoonotic disease pool or vector, which leaves them vulnerable to infectious diseases. In Africa, a high proportion of the population is directly dependent on agriculture, which can include husbanding animals among their agricultural activities. In urban areas, a high proportion of the population – particularly those living in slums and squatter settlements – live close to disease vectors, such as livestock, dogs, fowls, insects and rodents (see Box 12.1). The major health challenges summarised in Table 12.4 illustrate the importance of combining human and environmental factors and the need for integrated public health systems.

One Health approach

Calvin Schwabe, known as the 'father of modern veterinary epidemiology', traced the origins of Western medicine to the 'one medicine' of ancient Egyptian and Dinka priests. He argued for a return to those roots, identifying in a detailed case study of southern Sudanese specialists treating animals and people (Schwabe and Kuojok, 1981). More recently, Lebov *et al.* (2017) outlined the One Health approach as a complex and multidisciplinary approach that acknowledges the integrated nature of human, animal and environmental health. In 2015, the WHO designated 15 diseases that were at risk of severe outbreak, two-thirds of which involve an animal pool or transmission vector, indicating the importance of the need for more research on this type of health risk (WHO, 2015; see also Box 12.1). Leach *et al.* (2017) present analysis of a range of disease cases studies from across Africa that link ecosystems, livelihoods and well-being. They argue that this evidence suggests that these links represent synergies, tensions and trade-offs that need to be better understood in order to inform One Health approaches to reduce disease burdens and maximise access to well-being.

Social and economic circumstances

The social and economic development in Africa is leading to an increasing proportion of the population dwelling in urban areas. It is well documented that the population of the continent will generally shift from a physically demanding agricultural livelihood, consuming subsistence-produced agricultural commodities, to predominantly urban scenarios, where a significant proportion of the population rely on access to low-cost processed foods, sometimes pre-prepared, and spend long periods of the day physically inactive as a result of long commutes, or pursuing jobs that do not require physical labour. The increased consumption of tobacco and alcohol, as well as sodium and sugar in processed foods, contributes to this trend. This change in lifestyle, combined with improved access to water and sanitation, means that there is an emerging burden of non-communicable diseases (NCDs), including strokes, hypertension and diabetes. In fact, some of the latest research on these describe an NCD epidemic. This can also result in a double burden where residents of low-income slum areas – normally defined by their lack of services, including sanitation, water and electricity – suffer from communicable diseases associated with higher levels of poverty and higher vulnerability and exposure to environmental health risks, combined with the NCD risks associated with their urban lifestyle (Ramin, 2009; Joubert, 2015).

Introduction of continental centre for disease control based on the model of the US CDC

The examples of the health challenges facing the continent, some of which are illustrated in Table 12.4, show the scale of the future challenges. There is therefore a need to share practice, research and resources across the continent. The inception in January 2017 of the Pan-African

Table 12.4 One Health approach to researching diseases of high outbreak risk in Africa

Health threat	*Environmental factors*	*Animal factors*	*Human factors*	*African*
Crimean-Congo haemorrhagic fever	Tick habitat – bushes and tall grasses	Ticks Livestock infections	Use of insect repellant Contact with livestock	Endemic throughout Africa – co-located with the *Hyalomma* tick
Ebola virus and Marburg virus	Climatic conditions associated with outbreaks	Fruit bats are reservoirs Seroprevalence in dogs as sentinels Impacts on animal populations	Cultural practices in caring for the sick Treatment of the deceased	Guinea, Sierra Leone and Liberia (2014–2015), DR Congo
Lassa fever	Household conditions Poor sanitation	Rodent urine or faeces	Grain storage practices Hygiene Disposal of contaminated materials in hospitals	Endemic in Benin, Ghana, Guinea, Liberia, Mali, Sierra Leone
Middle East respiratory syndrome (MERS) coronavirus	Persistence of MERS on environmental surfaces and air	Dromedary camels	Early diagnosis Medical countermeasures (such as isolation) to prevent transmission Personal protective equipment for healthcare staff	Egypt, Tunisia
Nipah virus	Haze deforestation and drought forced bats to migrate to areas where pig farming was common	Pigs affected and may be hosts Bats are reservoirs	Pig slaughtering methods Consumption of raw date palm sap (liquor) Prevention of nosocomial infection	Ghana
Rift Valley fever	Heavy rainfall affected by ocean temperatures Land use degradation (crop irrigation providing mosquito breeding sites)	Mosquitoes Livestock	Contact with animal fluids (e.g. birth or slaughter) Irrigation practices	Niger (2016), Mauritania (2012), South Africa (2010)

Source: Adapted from WHO (2015) and Lebov *et al.* (2017).

Africa Centres for Disease Control and Prevention (Africa CDC) in Addis Ababa, Ethiopia is the latest response to this need. The centre is modelled on and supported by the US Center for Disease Control. Its main functions are:

1 health surveillance and improving capacity for surveillance, disease prediction, and decision-making and action on public health;
2 clinical and public health laboratory networks across Africa;
3 supporting preparedness and response planning for public health emergencies; and
4 strengthening the science underpinning public health practice.

This will build on the successful experience of peer support during the epidemics, for example the support provided by the African Union to Sierra Leone, Guinea and Liberia during the 2014–2015 Ebola epidemic. Some 862 African volunteers from 20 countries were deployed under the Africa Union Support to the Ebola Outbreak in West Africa (ASEOWA) programme (AAEST, 2016). Their work comprised the tracing of disease contacts, training over 6,500 medical personnel, the management of emergency treatment units, achieving a 52 per cent survival rate, with no medical personnel affected with the disease.

Conclusion

This chapter has identified a number of strategic challenges that need to be overcome to improve the health outcomes for Africans. These include increasing national investment levels and increasing the number of health professionals per head of the population. Some of this will be made easier as a greater percentage of Africans live in the city. As African economies have grown over the last decade, more efforts have gone into raising tax revenue so there is money to invest in improved health infrastructure.

This chapter discusses contradictory evidence of both considerable progress in health outcomes for Africans and evidence that simply by the accident of being born an African, health outcomes are still considerably worse than in any other region. Much of the progress has been on health challenges focused around single issues such as child mortality, HIV/AIDS or recent epidemics. These have benefited from global and regional partnerships mobilising human and financial resources for targeted outcomes. There is no doubt this has saved lives and improved health outcomes. However, as African societies modernise and grow their economies over the next 20 years, they will also experience health transitions, changes in social and demographic structures, and climate change that will require development of robust public health services that focus as much on prevention and well-being as on responding to health crises.

References

AAEST (Africa Against Ebola Solidarity Trust) (2016) *Africa Against Ebola Solidarity Trust Contribution to African Union Support to Ebola Outbreak in West Africa*, Midrand: AAEST.

Acemoglu, D. and Johnson, S. (2007) 'Disease and development: the effect of life expectancy on economic growth', *Journal of Political Economy*, 115(6), December: 925–985.

Alarcon, P., Fèvre, E., Murungi, M., Muinde, P., Akoko, J., Dominguez-Salas, P., et al. (2017) 'Mapping of beef, sheep and goat food systems in Nairobi: a framework for policy making and the identification of structural vulnerabilities and deficiencies', *Agricultural Systems*, 152: 1–17.

De Cock, K.M., Mbori-Ngacha, D. and Marum, E. (2002) 'Shadow on the continent: public health and HIV/AIDS in Africa in the 21st century', *The Lancet*, 360, 6 July: 67–72.

Hassell, J.M., Begon, M., Ward, M.J. and Fèvre, E.M. (2016) 'Urbanization and disease emergence: dynamics at the wildlife–livestock–human interface', *Trends in Ecology & Evolution*, 32(1): 55–67.

Heaton, T.B., Crookston, B., Pierce, H. and Amoateng, A.Y. (2016) 'Social inequality and children's health in Africa: a cross sectional study', *International Journal for Equity in Health*, 15: 92–107.

Herrick, C. (2017) 'The strategic geographies of global health partnerships', *Health & Place*, 45: 152–159.

HHA (Harmonization for Health in Africa) (2011) *The Case for Strengthening Systems for Better Health Outcomes*, Geneva: HHA, in association with WHO, World Bank, UNFPA, USAID, UNICEF, UNAIDS, JICA, African Development Bank, available at: www.aho.afro.who.int/sites/default/files/publications/5266/Atlas-2016-en.pdf (accessed 23 May 2017).

Hussein, M.J. (2012) 'Alternative estimates of the effect of the increase of life expectancy on economic growth', *Economics Bulletin*, 32(4): 3025–3035.

Iliffe, J. (2006) *The African AIDS Epidemic: A History*, Athens, OH: Ohio University Press.

Joubert, J. (2015) 'Chronic non-communicable diseases: where to now in Africa?', *Journal of the Neurological Sciences*, 348(1): 3–5.

Leach, M., Bett, B., Said, M., Bukachi, S., Sang, R., Anderson, N., et al. (2017) 'Local disease–ecosystem–livelihood dynamics: reflections from comparative case studies in Africa', *Philosophical Transactions of the Royal Society B*, 372: 1–18.

Lebov, J., Grieger, K., Womack, D., Zaccaro, D., Whitehead, N., Kowalcyk, B., et al. (2017) 'A framework for One Health research', *One Health*, 3: 44–50.

Makuta, I. and O'Hare, B. (2015) 'Quality of governance, public spending on health and health status in sub-Saharan Africa: a panel data regression analysis', *BMC (BioMed Central) Public Health*, 15: 932–943.

Nweke, O.C. and Sanders, W.H. (2009) 'Modern environmental health hazards: a public health issue of increasing significance in Africa', *Environmental Health Perspectives*, 117(6): 863–870.

Nye, E. and Gibson, M. (1997) *Ronald Ross: Malariologist and Polymath – A Biography*, Basingstoke: Macmillan.

Omran, A.R. (1971) 'The epidemiologic transition: a theory of the epidemiology of population change', *The Milbank Quarterly*, 49(4): 509–538.

Ramin, B. (2009) 'Slums, climate change and human health in sub-Saharan Africa', *Bulletin of the World Health Organisation*, 87(12): 886–887.

Rijpma, S. (2015) *David Livingstone and the Myth of African Poverty and Disease: A Close Examination of His Writing on the Pre-Colonial Era*, Leiden: Brill and Afrika-Studiecentrum.

Rosling, H. (n.d.) *How Does Income Relate to Life Expectancy?*, available at: www.gapminder.org/answers/how-does-income-relate-to-life-expectancy/ (accessed 18 May 2017).

Ross, R. (1902) *Researches on Malaria*, Nobel Lecture, 12 December, available at: www.nobelprize.org/nobelprizes/medicine/laureates/1902/ross-lecture.pdf (accessed 9 May 2016).

Ruckert, A. and Labonté, R. (2014) 'Public–private partnerships (ppps) in global health: the good, the bad and the ugly', *Third World Quarterly*, 35(9): 1598–1614.

Schwabe, C. and Kuojok, I.M. (1981) 'Practices and beliefs of the traditional Dinka healer in relation to provision of modern medical and veterinary services for the southern Sudan', *Human Organization*, 40(3): 231–238.

Smith, J., Taylor, E.M. and Kingsley, P. (2015) 'One World-One Health and neglected zoonotic disease: elimination, emergence and emergency in Uganda', *Social Science & Medicine*, 129: 12–19.

WHO (World Health Organization) (2014) *African Regional Health Report: The Health of the People – What Works*, available at: www.afro.who.int/en/rdo/annual-and-biennial-reports/african-regional-health-report-2014.html (accessed 14 July 2015).

WHO (World Health Organization) (2015) *WHO Publishes List of Top Emerging Diseases Likely to Cause Major Epidemics*, available at: www.who.int/medicines/ebola-treatment/WHO-list-of-top-emerging-diseases/en/ (accessed 15 July 2015).

WHO (World Health Organization) (2016) *African Atlas of Health Statistics 2016*, Geneva: WHO.

WHO (World Health Organization) (2017) *Global and Regional Trends by UN Regions, 1990–2025: Joint Child Malnutrtition Estimates*, Geneva: WHO, with UNICEF and the World Bank, available at: http://apps.who.int/gho/data/node.main.ngest?lang=en (accessed 3 January 2018).

13

The African HIV/AIDS epidemic

Is the end in sight?

Hazel Barrett

Introduction

This chapter will explore the epidemiology of the African HIV/AIDS epidemic over the last 35 years, and discusses the measures and policies implemented to treat those infected with HIV and prevent new infections during that period. The watershed moment for tackling the HIV/AIDS epidemic in Africa was the acceptance of the Millennium Development Goals by the global community, which recognised the disease as a development issue and accelerated the international response, which became dominated by biomedical technologies. The result has been some very dramatic reductions in mortality due to HIV/AIDS in the region and sustained decreases in new infections since 2000. This has led many to suggest that the end of the epidemic is in sight.

Morbidity and mortality in Africa

Africa has the worst mortality and morbidity indicators of any global region. Table 13.1 shows that when using life expectancy at birth, maternal mortality and under-five mortality rates, Africa performs poorly, with life expectancy 11 years below the global average and maternal mortality and under-five mortality being approximately twice the global average. It is widely accepted that these poor statistics can be partly explained by the HIV/AIDS epidemic, which has devastated the region for 35 years. As the WHO states, 'Progress on life expectancy stalled in Africa because of the rising HIV epidemic' (WHO, 2016a: 7). In fact, between 1990 and 2008, adult mortality rates increased 'largely as a result of the havoc wreaked by HIV/AIDS, which killed large numbers of young adults' (WHO, 2014a: 7). However, since 2005, life expectancy in the region has risen by 10 years (WHO, 2016a), twice the global average, as the African HIV/AIDS epidemic begins to show signs of decline and the roll-out of antiretroviral therapy (ART) prolongs the life expectancy of those infected with HIV (WHO, 2014a).

The mortality statistics suggest that the African region has made little if any progress concerning health and well-being since the 1970s. Africa is the only region yet to start to progress through the epidemiological transition (Omran, 1971), which describes changing patterns of

Table 13.1 Mortality statistics for 2015 by WHO region

WHO region	*Life expectancy at birth (years)*	*Maternal mortality per 100,000 live births*	*Under-five mortality per 1,000 live births*
Africa	60	542	81
The Americas	77	52	15
South East Asia	69	164	43
Europe	77	16	11
East Mediterranean	69	166	52
Western Pacific	77	41	14
Global	71	216	43

Source: Compiled from WHO (2016a).

Table 13.2 The top five causes of death and disability-adjusted life years (DALYs) for the WHO African region, 2000 and 2012

Causes of death (%)	*2000*	*2012*	*Proportion of DALYS* (%)*	*2000*	*2012*
1 HIV/AIDS	13.9	11.7	1 HIV/AIDS	11.6	10.1
2 Lower respiratory infections	11.10	11.2	2 Lower respiratory infections	10.1	9.9
3 Diarrhoeal disease	8.6	6.5	3 Injuries	7.5	9.3
4 Malaria	8.2	6.0	4 Malaria	10.1	7.6
5 Stroke	3.4	4.7	6 Diarrhoeal disease	9.2	7.0

Source: Compiled from WHO (2016a).

Note: * Number of years of life lost plus years of life lived with a disability.

population distributions in relation to changing patterns of mortality, fertility, life expectancy and leading causes of death (McKeown, 2009). It is a transition in which communicable diseases are replaced by non-communicable diseases as principal causes of ill health and mortality (Caldwell, 2001). According to the WHO (2014b, 2016b), in 2012, communicable diseases accounted for 71 per cent of deaths in Africa, with non-communicable diseases responsible for 19 per cent of deaths and the remainder accounted for by injuries.

Communicable diseases are responsible for two-thirds of the total disease burden of the region, measured by disability-adjusted life years (DALYs), with non-communicable diseases responsible for 25 per cent of DALYs and the remainder accounted for by injuries. Table 13.2 shows the leading causes of death and DALYs in Africa in 2000 and 2012. This shows that although the HIV/AIDS epidemic appears to be declining, as shown by the reduction in deaths and DALYS caused by HIV/AIDS between 2000 and 2012, it is still the leading cause of death and morbidity in the region, and is predicted to remain so until at least 2030 (WHO, 2008).

Epidemiology of the African HIV/AIDS epidemic: 1980–2000

It is believed that HIV/AIDS first became epidemic during the 1970s (before HIV was identified in 1983) in western equatorial Africa, where it had probably existed since the 1950s (Iliffe, 2006; Le Vasseur *et al.*, 2014). The later testing of blood samples taken in 1980–1981 from antenatal clinic attenders in Kinshasa showed an HIV prevalence of 3 per cent among that demographic group. As Iliffe states, 'The world's first HIV epidemic among a heterosexual population had begun before the existence of the virus was even suspected' (Iliffe, 2006: 12). The epidemic rapidly spread eastwards to East Africa, including Burundi, Rwanda, Uganda, Kenya and Tanzania. The epidemic also intensified and developed into a generalised epidemic, putting all members of the population at risk. During this period, the Great Lakes region became the nucleus of the African HIV/AIDS epidemic, with adult prevalence rates (15–49 years) in some countries exceeding 10 per cent. The spread of the disease was blamed on the 'vigorous commercial economies' around Lake Victoria and mobility along the trans-African highway (Iliffe, 2006).

During the 1990s, the HIV/AIDS epidemic diffused west and southwards from the Great Lakes region, principally following major transport routes, in particular air and sea transport hubs and along the transcontinental road network (Iliffe, 2006). The 'shift south' of the epidemic resulted in southern Africa quickly eclipsing East Africa as the centre of the African HIV/AIDS epidemic, with adult prevalence rates surpassing 20 per cent in many countries. By the end of the 1990s, South Africa had the highest number of infected people globally, with estimates that over 5 million people had been infected by 2003 (Iliffe, 2006: 43). North Africa remained largely isolated from the epidemic, with West Africa's concentrated epidemic dominated by the less virulent HIV-2 strain of the disease. HIV-2 is three times more difficult to transmit through sexual intercourse and 10 times more difficult to pass from mother to child than HIV-1, with mortality from HIV-2 thought to be only a third of HIV-1, and progression from HIV to AIDS taking as much as 25 years, compared to the 2–15 years for HIV-1 (Iliffe, 2006; WHO, 2015c). HIV-1 met HIV-2 in Abidjan in the mid-1980s, and HIV-1 quickly established itself in the region, with Côte d'Ivoire becoming the epicentre of the HIV/AIDS epidemic in West Africa (Barnett and Whiteside, 2006; Iliffe, 2006). The prevalence rates in the region, while high among sex workers (often above 50 per cent), remained relatively low in the general population at less than 5 per cent.

During this period, the feminisation of the epidemic became apparent, with more adolescent girls and women infected and dying of the disease than men. Across Africa, women made up 57 per cent of adults living with HIV, with on average 13 women living with HIV for every 10 infected men (Barnett and Whiteside, 2006). Young women aged 15–24 were particularly vulnerable, with studies suggesting that on average, there were 36 young women living with HIV for every 10 young men (Barnett and Whiteside, 2006). Over 75 per cent of all HIV-infected women worldwide lived in Africa. Biologically, girls and women are more susceptible to HIV infection, which, when combined with their economic, social and cultural vulnerability, puts them at greater risk of HIV infection than men. As a result, mother-to-child transmission became a serious issue. As HIV can be passed on by infected women to their unborn and newly born children, the numbers of babies and children born infected with HIV peaked in this period at about 0.5 million a year (UNAIDS, 2016a).

It is perhaps not surprising that the HIV/AIDS epidemic spread so quickly in Africa during these 20 years, as this was a period when the region was suffering severe financial crisis with the imposition of harsh structural adjustment policies by the international financial community, political turmoil and conflict (Barnett and Whiteside, 2006; Prados de la Escosure, 2013) that negatively affected poverty levels as well as the social determinants of health, such as education, health systems and public health. These, combined with the fact that HIV/AIDS is a long-wave epidemic (Barnett and Whiteside, 2006), with infected people remaining asymptomatic for many years (for HIV-1, on average 2–15 years depending on the individual; WHO, 2015a) and the denial of the need to take action by policymakers, led to a devastating situation. A continent that was home to only 10 per cent of the world's population bore the burden of 70 per cent of all cases of HIV/AIDS.

In terms of treatment, most countries promoted living positively with HIV/AIDS, often based on the principle of home-based care, which meant that those needing treatment were treated at home by family members rather than in hospital. Antiretroviral therapy (ART) was very expensive and only available to the very wealthy (Barnett and Whiteside, 2006). Instead, patients were treated for opportunistic infections to make life more comfortable for them as the disease progressed. At the same time, campaigns to reduce stigma and discrimination were initiated to try to give those infected and their families some dignity. With no effective treatment, HIV infection was a death sentence, and as a result mortality rates peaked at the turn of the twenty-first century.

With no cure, vaccine or affordable treatments, the only prevention options available in Africa during this period were based on the public health principles of knowledge, attitude, practice and behaviour (KAPB) to change people's sexual behaviour. This meant a focus on awareness-raising and public education about the disease and its transmission (Kabwama and Berg-Beckhoff, 2015), often conducted in public spaces using posters, and in schools, clinics and the workplace (Barrett, 2007a). These aimed to change sexual behaviour, in particular encouraging the use of condoms, reducing the number of sexual partners, promoting abstinence and the delay in sexual debut. The most well known of these campaigns was the ABC programme developed in Uganda (Barrett, 2007b) and then rolled out to most parts of the continent. This campaign stressed: A = abstinence; if you cannot abstain, then B = be faithful; if you cannot be faithful, then C = use a condom; failing to adhere to ABC leads to D = death. In addition to public health campaigns to promote behaviour change, health authorities began to screen blood, use clean disposable syringes, set up voluntary testing and counselling (VTC) services and scale up the treatment of sexually transmitted infections (STIs).

It is generally recognised that during this period, while knowledge about HIV and AIDS increased among the general public as a result of education approaches being employed in HIV/AIDS prevention programmes, this knowledge largely failed to translate into actual risk-reduction behaviour change (Barnett and Whiteside, 2006; Adelekan, 2015; Krishnaratne *et al.*, 2016), as evidenced by the very high rates of annual new infections in the region, which reached 2.3 million in 2000 (UNAIDS, 2015a, 2015b). It is clear that knowledge alone does not always translate into behavioural change (Adelekan, 2015) due to low perceptions of individual risk (Kabwama and Berg-Beckhoff, 2015) and/or feelings of powerlessness, as in the case of married African women who are expected to be submissive and dependent on their husbands. They indeed might be faithful, but their husbands might not be, but these women lack the authority within the relationship to insist on the use of condoms. Oster (2012: 47), in her review, concludes that 'behavior change is no more likely in areas with a lot of knowledge than areas without'.

The outcomes of KAPB interventions have been disappointing, rarely reducing HIV infections or prevalence (Oster, 2012; Krishnaratne *et al.*, 2016). Even if there is a behaviour response, this is often slight and short-lived (Oster, 2012). Explanations for this limited behaviour change tend to focus on Africa-specific cultural barriers to changing behaviour, such as fatalism, culture, lack of female empowerment in issues of sexual bargaining, such as insisting on the use of condoms (Oster, 2012), and high levels of intergenerational sexual mixing and sexual violence against women. However, Oster (2012) has suggested that in regions with low non-HIV adult life expectancy and where individuals expect to die early from non-HIV causes (e.g. maternal mortality and malaria), people are less responsive to HIV prevention messages. Her conclusion is that focusing on decreasing the disease burden in Africa in general, and improving health and well-being, will have positive 'spillovers' for HIV prevention.

The international response

Despite the high levels of infection and death from HIV/AIDS in Africa during the 1980s and 1990s, the response from national and international authorities are generally considered as inadequate (Barnett and Whiteside, 2006; Iliffe, 2006). Most African governments were slow to acknowledge the epidemic and were struggling to address economic and political challenges. Many countries tried to distance themselves from the epidemic, fearing that being associated with the epidemic would damage the image of the country and thus repel foreign investment and tourists. The WHO, the obvious body to take the lead in tackling the African epidemic, was suffering weak leadership and was focused on the HIV/AIDS epidemics in the developed world.

Recognising the failure to tackle the HIV/AIDS pandemic, the international community set up UNAIDS in 1996 to coordinate the global fight against this disease, which was acknowledged to be having a severe impact on the economic and human development of those countries in Africa that were hardest hit. In 2000, the international community agreed the UN Millennium Development Goals (MDGs), which included a number of goals and targets related to health, including lowering the incidence of HIV/AIDS by 2015 (MDG6A). Thirty-seven of the 47 countries that comprise the WHO Africa region have progressed or achieved the target of reducing infections (African Health Observatory, 2015), but the epidemic is far from over. The UN Sustainable Development Goals (SDGs), which have replaced the MDGs, reaffirm the need to tackle the HIV/AIDS epidemics, and have set the target (SDG3.3) of ending the AIDS epidemics by 2030 (African Health Observatory, 2015; UNAIDS, 2016b). UNAIDS state, 'The extraordinary accomplishments of the last 15 years have inspired global confidence that this target can be achieved' (UNAIDS, 2016b: 1). However, in 2015, the Africa region still recorded 1.4 million new HIV infections, representing 67 per cent of the 2.1 million global infections that year. While new HIV infections declined in sub-Saharan Africa, they rose in the Middle East and North Africa by 4 per cent between 2010 and 2015 (UNAIDS, 2016c).

Epidemiology of the African HIV/AIDS epidemic: 2000–2016

Since the turn of the millennium, and with the added momentum to fight the epidemic associated with the MDG to tackle HIV/AIDS, there have been promising signs that the fight against HIV/AIDS in sub-Saharan Africa is beginning to have positive results. Between 2000 and 2014, new HIV infections in Africa fell from 2.3 million to 1.4 million a year, an impressive drop of 41 per cent (UNAIDS, 2015a). Over the same period, AIDS-related deaths declined from 1.2 million to 0.79 million a year, a decrease of 34 per cent (UNAIDS, 2015a). This has led UNAIDS to claim, 'The world has halted and reversed the spread of HIV. The epidemic has been forced into decline' (UNAIDS, 2015a: 3). However, while HIV infections have declined in Africa, they still account for 70 per cent of all new global infections, and the decline in deaths related to AIDS is largely due to more people accessing ART, which enables them to live longer, healthier lives (UNAIDS, 2015a). In the words of UNAIDS, 'there is still an unacceptably high number of new HIV infections and AIDS-related deaths occurring each year' (UNAIDS, 2015a: 3).

Of concern is the fact that between 2010 and 2015, these gains do not appear to be accelerating. UNAIDS (2015b) state that new infections in sub-Saharan Africa fell by 13 per cent in 2000–2004, by 12 per cent in 2005–2009, and between 2010 and 2015 new infections declined by 13 per cent (UNAIDS, 2016b). Adolescent girls and women still have an elevated risk and vulnerability to HIV infection, with 63 per cent of new infections in the region affecting this demographic group. In 2014, of the 25.8 million people living with HIV/AIDS in sub-Saharan

Africa, 13.8 million were women (53.5 per cent), 9.7 million were men (37.5 per cent) and 2.3 million were children (9 per cent). In 2014, African women living with HIV who gave birth accounted for more than 90 per cent of the total number globally (UNAIDS, 2015b), explaining the large number of children infected with HIV in the African region. In 2014, 190,000 children were newly infected with HIV in sub-Saharan Africa, reflecting a decline of 47 per cent since 2009 (UNAIDS, 2015b).

So, what are the factors that account for this decline in new HIV infections and AIDS-related deaths over the last 15 years? The answer is the availability of affordable antiretroviral therapy (ART). ART suppresses HIV and stops its progression to AIDS, it also prevents onward transmission of the virus. In 2015, ART coverage in Africa was 54 per cent, up from almost nil in 2000. A person living in Africa diagnosed with HIV in 2000 knew it was a death sentence and could expect only to live for a few years. This was because ART was either not available or too expensive. However, today, the life expectancy of a person living with HIV and who is receiving treatment is the same as that of a person who is not infected with HIV (UNAIDS, 2015a: 40). ART is not a cure for HIV, but it has transformed a death sentence into a chronic manageable infection (Mayer *et al.*, 2016).

Many countries in Africa have made substantial progress in providing access to ART, although the MDG target of universal coverage (MDG6B) has not been achieved in any country (African Health Observatory, 2015). ART coverage increased from 24 per cent in 2010 to 54 per cent in 2015, reaching 12 million HIV-positive people, 10.3 million in East and southern Africa (a coverage of over 60 per cent) and 1.8 million in West and central Africa, representing 28 per cent coverage (UNAIDS, 2016b). South Africa has more people on ART than any other country in the world, reaching 3.4 million people. Within Africa, Kenya has the second largest ART programme (after South Africa), with nearly 0.9 million people receiving treatment. Between 2010 and 2015, 12 countries in Africa increased ART coverage by more than 25 per cent (UNAIDS, 2016b). Levels of ART coverage are shown in Table 13.3, which reveals that 10 sub-Saharan African countries have 50 per cent plus ART coverage. The five countries with the highest ART coverage in 2014 were Rwanda (68 per cent), Botswana (62 per cent), Swaziland (59 per cent), Zambia (57 per cent) and Algeria (57 per cent) (WHO, 2016b). It is also interesting to note that those countries with less than 25 per cent coverage tend to be located in West and central Africa, a region with lower HIV prevalence rates than in East and southern Africa. However, West and central Africa is densely populated and includes Nigeria, Africa's most populous country. Hence, while prevalence rates might be low, actual numbers could be high.

ART has revolutionised the treatment of HIV (Le Vasseur *et al.*, 2014; MacCarthy *et al.*, 2015). ART has clinical benefits for all individuals infected with HIV, including those with asymptomatic infection and high CD4 counts (Medland *et al.*, 2015). The introduction of triple therapy ART in 1996 led to a downturn in mortality from HIV infection worldwide (Kanters *et al.*, 2014; Koff, 2016). There is clear evidence that ART reduces mortality and increases life expectancy for those infected with HIV.

In the early 2000s, southern African countries began to offer ART through public sector programmes, often with support from international donors. A study undertaken in rural KwaZulu-Natal, South Africa, a region with a 29 per cent adult HIV prevalence rate, indicates that between 2003 (a year before ART became available in the public sector health system) and 2011, life expectancy increased from 49 to 60.5 years, a gain of over 11 years (Bor *et al.*, 2013). Both men and women experienced gains in life expectancy, with men gaining nine years while women experienced an increase of 13.3 years. These results are consistent with other studies, such as that undertaken by Reniers *et al.* (2014) in six eastern and southern African countries. They found that there had been a reduction of mortality in HIV-infected people following the expansion of ART of between 58 and 84 per cent. They also found that women had benefited more from the expansion of ART, either through higher uptake of ART or better treatment outcomes, or both, with mortality declines being larger for women than men. This may be explained by women's greater engagement with HIV services and beginning ART at an early stage of their infection.

ART can also reduce transmission by supressing the viral load in blood, semen and the genital tract (Le Vasseur *et al.*, 2014; Medland *et al.*, 2015), and thus is being used to underpin prevention (Medland *et al.*, 2015). Studies demonstrate that ART in HIV-discordant couples showed unequivocally that treatment of the HIV-infected partner with ART substantially lowers the probability of HIV transmission to the uninfected partner by 96 per cent (Tanser *et al.*, 2013). In a study in KwaZulu-Natal, South Africa, where the adult prevalence rate was 24 per cent, Tanser *et al.* (2013) found that after controlling for differences in age and sex distribution, the risk of an HIV-uninfected person acquiring HIV was substantially lower in areas of high ART coverage. In areas where ART coverage was 30–40 per cent, an uninfected person was on average 38 per cent less likely to acquire HIV infection than an HIV-uninfected person living in an area with less than 10 per cent ART coverage. In summary, they found that a 1 per cent increase in ART coverage is associated with a 1.4 per cent decline in the risk of acquisition of new HIV infection. Predictive mathematical models suggest that under certain conditions, high coverage of ART could lead to a substantial decrease in the rate of new HIV infections (Tanser *et al.*, 2013).

One of the most successful prevention stories is the control of mother-to-child transmission, that is the transmission of HIV from an HIV-infected mother to her child during

Table 13.3 Antiretroviral therapy (ART) coverage among adults living with HIV aged 15 years and over in sub-Saharan Africa, 2014

ART coverage less than 25%	*ART coverage 25–49%*	*ART coverage of over 50%*
Cameroon	Angola	Botswana
Central African Republic	Benin	Eritrea
Congo	Burkina Faso	Ethiopia
Gambia	Burundi	Kenya
Guinea-Bissau	Cape Verde	Malawi
Liberia	Chad	Rwanda
Madagascar	Côte d'Ivoire	Swaziland
Nigeria	Equatorial Guinea	Uganda
Sierra Leone	Gabon	Zambia
Democratic Republic of Congo	Ghana	Zimbabwe
Mauritania	Guinea	
	Lesotho	
	Mali	
	Mauritius	
	Mozambique	
	Namibia	
	Niger	
	São Tomé and Príncipe	
	Senegal	
	South Africa	
	Togo	
	United Republic of Tanzania	

Source: UNAIDS (2015b: 67).

pregnancy, labour, delivery or breastfeeding. Without intervention, mother-to-child transmission can be between 15 and 45 per cent (WHO, 2015c), but can be fully prevented by ART. Mother-to-child prevention programmes using ART have reduced the number of new global infections in children by 58 per cent between 2000 and 2014 (Sued *et al.*, 2016). In 2015, Cuba, and in 2016 Thailand, Armenia and Belarus, have received validation from WHO that mother-to-child transmission of HIV has been eliminated (The Lancet HIV, 2016). Within Africa, substantial progress is being made. In 2014, 77 per cent of pregnant women living in the 21 countries with the highest prevalence rates (UNAIDS priority countries) received ART (UNAIDS, 2015a). In these same countries, the number of children acquiring HIV fell by 48 per cent from 2009 to 2014 (UNAIDS, 2015a). While the Global Plan target to reduce the number of newly infected children by 90 per cent by 2015 has not been achieved, the target is in sight.

However, the emergence of drug-resistant HIV could compromise the effectiveness of ART as a treatment and preventative technology and threaten the progress of the fight against HIV/AIDS. With the roll-out of ART, drug resistance is slowly increasing. In East Africa, for example, drug resistance to first-line ART (a once daily triple therapy recommended by the WHO, 2013, 2014a) has been reported at 10 per cent (WHO, 2015b). Cameroon, Rwanda, Zambia, Uganda and Tanzania have all reported increases in HIV drug resistance (Le Vasseur *et al.*, 2014). Testing for viral load is an indicator of treatment failure from drug resistance (Le Vasseur *et al.*, 2014). Drug resistance most commonly occurs as a result of non-compliance or non-adherence in taking ART, which can result in HIV mutation. The WHO have a recommended second-line ART for patients with drug resistance, but it is more toxic, therefore having more side effects for the patient, and is much more expensive, thus making a strategy for ensuring ART adherence essential.

The cascade approach for fighting HIV/AIDS

The HIV treatment cascade

The cascade approach is now the internationally agreed framework for planning and monitoring HIV treatment programmes (WHO, 2014a, 2015b). 'The term "cascade" emphasizes that a sequence of services is needed to achieve desired impacts' (WHO, 2015c: 26). This approach is a linear strategy that comprises five elements: testing and diagnosis; getting an infected person enrolled in care for counselling and treatment advice; initiating ART; sustaining ART; and viral load suppression (having a low level of HIV in the blood) (USAID, 2015: 7). It is designed to ensure those who are diagnosed with HIV remain engaged with the treatment continuum from the initial stage of getting tested through to achieving the suppression of the virus. It is not a cure, but allows those infected to live longer and reduces the chance of passing HIV to others.

The HIV treatment cascade (sometimes called the HIV care cascade) has the advantage that it highlights where there are service gaps or where HIV-infected people drop out of the system, and thus informs where programmes need enhancing and can improve outcomes (Isbell *et al.*, 2016). As USAID states:

> It powerfully identifies 'leaks' in the system, so that implementers at site, district, provincial, or national levels can target limited resources toward effective interventions that improve the health of HIV-positive individuals, lower the amount of virus in vulnerable communities, and prevent new infections in the long term.
>
> *(USAID, 2015: 3)*

The main leakage points tend to be getting people to be tested, and then once diagnosed getting them enrolled on a programme of treatment (USAID, 2015). Data presented by USAID (2015) for people aged over 15 years in sub-Saharan Africa suggest that in 2014, only 48 per cent of the estimated 22.5 million people living with HIV in the region went for testing. Of these, 81 per cent (11 million) were put onto ART, with 83 per cent (9 million) of these continuing to take ART. Of these, 80 per cent (6 million) had viral load suppression. The poor retention of people on the cascade in the region is explained by 'substantial stigma and discrimination and adverse policy and legal environments' (USAID, 2015: 24).

The HIV treatment cascade underpins the UNAIDS 90-90-90 target and will provide the information needed to assess progress towards the targets (Medland *et al.*, 2015; WHO, 2015b). The target set in 2014 is that 90 per cent of all people infected with HIV will know their HIV status, 90 per cent of those diagnosed with HIV infection will receive ART, and 90 per cent of all people receiving ART will have durable viral load suppression by 2020 (USAID, 2015; WHO, 2015b). The target consists of three sequentially linked goals that, if achieved, would lead to 73 per cent of people living with HIV being effectively treated and their virus suppressed (Godfrey-Faussett, 2016).

However, the HIV treatment cascade has received criticism. Researchers such as MacCarthy *et al.* (2015) express concerns about the simplicity of the HIV treatment cascade and ask that the cascade is adapted to capture the complexity of multiple testing of CD4 counts and viral loads, the fact that many HIV infected people will enter the cascade late when their disease is well progressed, and many infected people may move in and out of care over the life course. Others query the comparability of the measurement of viral suppression (MacCarthy *et al.*, 2015; Medland *et al.*, 2015), which limits national comparisons.

The HIV prevention cascade

While prevention is stressed in the HIV treatment cascade, the development of an HIV prevention strategy is less advanced. Despite improvements in HIV prevention science since 2000, we have yet to see substantial declines in new infections (Isbell *et al.*, 2016; The Lancet HIV, 2016). Many argue that while good progress is being made with HIV treatment and care, prevention is receiving less attention (The Lancet HIV, 2016). In November 2015, the WHO issued a press release in which they expressed concern 'about a slow-down – or even reversal – in the decrease of new infections in some countries and among some of the most affected population groups' (WHO, 2015c). The HIV prevention cascade is emerging as a new approach to guide the design and monitoring of HIV prevention programmes (Hargreaves *et al.*, 2016).

The adoption of an HIV prevention cascade approach has been proposed by USAID (2015), which is similar to the HIV treatment cascade (Hargreaves *et al.*, 2016). This includes identifying key populations, giving prevention advice, making biotechnologies available, ensuring adherence, and regular HIV testing (USAID, 2015). Those who test positive for HIV will then transfer to the HIV treatment cascade. The HIV prevention cascade:

> encompasses primary prevention and highlights the need for constant re-engagement to promote combination prevention approaches that include rights-based, peer-led behaviour counselling, condom use, regular HIV testing, sexually transmitted infection (STI) services, harm reduction interventions and pre-exposure prophylaxis (PrEP) as appropriate.
>
> *(USAID, 2015: 8)*

In a systematic review, Krishnaratne *et al.* (2016) highlight the importance of combination HIV prevention as proposed in the HIV prevention cascade. According to one mathematical model (cited in Sued *et al.*, 2016), a combined prevention programme that includes HIV testing, early ART, male circumcision and the use of PrEP can avoid 62 per cent of new HIV infections, while programmes that rely on only HIV testing and early ART would reduce infections by just 34 per cent. There is strong evidence for the efficacy of voluntary medical male circumcision and PrEP (Krishnaratne *et al.*, 2016).

Randomised clinical trials in Kenya, Uganda and South Africa have shown that medical male circumcision reduces the risk of heterosexually acquired HIV infection in men by approximately 60 per cent (WHO, 2015a). In 2007, the WHO and UNAIDS recommended male circumcision as an HIV preventative method in settings where there is high HIV prevalence and low male circumcision rates. An increasing number of African countries are adopting large-scale male circumcision programmes supported by the WHO and international agencies. By 2014, 9.1million men had voluntary medical male circumcision performed in eastern and southern Africa. In 2014, 3.24 million men were circumcised in 14 African countries, a 22 per cent increase from the previous year, and a 750 per cent increase since 2010 (WHO, 2015b). Kenya and Tanzania have over 80 per cent coverage, with Mozambique and Uganda reaching between 50 and 79 per cent coverage (WHO, 2015b). The WHO target is an 80 per cent coverage across the region. In addition to contributing to the target of reducing new HIV infections, male circumcision is an opportunity to link men with relevant healthcare services such as HIV testing.

Researchers are agreed that pre-exposure prophylaxis, or PrEP, is highly effective when administered to healthy people at risk of contracting HIV, although few countries worldwide have implemented PrEP (Koff, 2016; Mayer *et al.*, 2016; Sued *et al.*, 2016). Studies have shown that when PrEP is used by heterosexual discordant couples, it was associated with a 67 per cent and 75 per cent reduction in male and female HIV transmission, respectively (Sued *et al.*, 2016). In a randomised clinical study involving men who have sex with men, PrEP taken before and after sexual encounters showed an 86 per cent efficacy in reducing HIV acquisition (Sued *et al.*, 2016). However, when used with women in studies in sub-Saharan Africa, the use of PrEP was ineffective. It was found that the participants were not taking the drugs as instructed. Fear of adverse effects, stigma and falsely being labelled as HIV-positive were factors that affected adherence (Sued *et al.*, 2016). This reminds us that biomedical prevention methods must take account of behavioural and structural factors.

There is, however, the question of the funding of the HIV prevention cascade and the feasibility of implementing the full range of prevention technologies in low-income countries where resources are limited. A systematic review of prevention studies (Smith *et al.*, 2016) suggests the optimum mix of preventative methods for South Africa would be the scaling up of male circumcision and early ART initiation with outreach testing, as these are low-cost and highly effective. PrEP was found to have the potential to be cost-effective if delivered to populations at highest risk of HIV infection such as sex workers and intravenous drug users.

The overdependence of the HIV prevention cascade on biomedical interventions has been raised by researchers such as Hargreaves *et al.* (2016), who are concerned that there is a lack of appreciation that behavioural and structural interventions are essential for biomedical technologies (and non-medical methods) to have a population-wide impact on HIV infections. They believe this can lead to excessive optimism about the efficacy of biomedical interventions. They argue that if the HIV prevention cascade is to be successful, the social determinants of HIV behaviour must be integrated into the strategy and an enabling policy environment created. They argue that the HIV prevention cascade is different from the HIV treatment cascade as the treatment cascade focuses on the steps needed to reduce the viral load of the individual, whereas the prevention cascade focuses on the population-level effect of reducing infection. Nevertheless, Hargreaves *et al.* (2016) argue that 'Together, treatment and prevention efforts guided by widespread adoption of a cascade approach have the potential to rapidly accelerate the decline of HIV prevalence using currently available methods' (Hargreaves *et al.*, 2016: e321).

Is the end of the African HIV/AIDS epidemic in sight?

Over the last 15 years, there have been renewed efforts to fight the HIV/AIDS epidemic in Africa, spurred on by the international obligation, and resources committed, as a result of the signing of the Millennium Development Goals, and in particular the target to halt and reverse the spread of HIV/AIDS. While progress has been impressive, helped by improvements in medical research and biomedical technologies, with new infections dropping by 41 per cent during this period and a decrease of 34 per cent in AIDS-related deaths, the goal of ending the HIV/AIDS epidemic is yet to be achieved. Encouraged by these remarkable gains, the international community has pledged in the Sustainable Development Goals to end the HIV/AIDS epidemic by 2030.

Since 2000, ART has become the mainstay of both treatment and prevention programmes in Africa, and is central to the cascade approach to the HIV/AIDS epidemic that has been introduced by UNAIDS. This roll-out of ART and other biomedical technologies, such as male circumcision, across the region is yielding 'profound health benefits' (UNAIDS, 2015b). These benefits include increased life expectancy and reduced mother-to-child HIV transmission. However, the reliance on biomedical technologies is not without some risk, including the development of drug-resistant HIV that could stall the efforts to eradicate HIV in Africa by 2030.

UNAIDS and the WHO have labelled the period 2000–2015 in the fight against HIV/AIDS as 'Africa Rising': 'In the past 15 years, Africa has built its capacity on a variety of fronts to manage the AIDS crisis' (UNAIDS, 2015b: 41). However, the WHO warned in a press release issued in November 2015 that 'HIV remains a major health challenge – drawing sharp attention to health system weaknesses and gaps in universal health coverage. Addressing these issues will be critical to meeting the new global targets for AIDS' (WHO, 2015c). The end of the HIV/AIDS epidemic might be in sight, but we must not become complacent. The fight against HIV/AIDS in Africa must continue until the disease has been eradicated.

References

Adelekan, M. (2015) 'A critical review of the effectiveness of educational interventions applied in HIV/AIDS prevention', *Patient Education and Counseling*, DOI: http://dx.doi.org/10.1016/j.pec.2015.12.004.

African Health Observatory (2015) *Health MDGs and Post-2015 Sustainable Development Goals (SDGs)*, Brazzaville: WHO.

Barnett, T. and Whiteside, A. (2006) *AIDS in the Twenty-First Century*, 2nd edition, Basingstoke: Palgrave Macmillan.

Barrett, H.R. (2007a) 'Too little, too late: responses to the HIV/AIDS epidemics in sub-Saharan Africa', *Geography*, 92(2): 87–96.

Barrett, H.R. (2007b) 'As easy as ABC . . .? Uganda's fight against HIV/AIDS', *Geography*, 92(2): 154–157.

Bor, J., Herbst, A.J., Newell, M.L. and Barnighausen, T. (2013) 'Increases in adult life expectancy in rural South Africa: valuing the scale-up of HIV treatment', *Science*, DOI: 10.1126/science.1230413.

Caldwell, J.C. (2001) 'Population health in transition', *Bulletin of the World Health Organization*, 79(2): 159–170.

Godfrey-Faussett, P. (2016) 'The HIV prevention cascade: more smoke than thunder?', *The Lancet HIV*, 3: e286–e288, available at: www.thelancet.com/hiv (accessed 18 December 2017).

Hargreaves, J.R., Delaney-Moretlwe, S., Hallett, T.B., Johnson, S., Kapiga, S., Bhattacharjee, P., *et al.* (2016) 'The HIV prevention cascade: integrating theories of epidemiology, behavioural, and social science into programme design and monitoring', *The Lancet HIV*, 3: e318–e322, available at: www.thelancet.com/hiv (accessed 18 December 2017).

Iliffe, J. (2006) *The African AIDS Epidemic: A History*, Oxford: James Currey.

Isbell, M.T., Kilonzo, N., Mugurungi, O. and Bekker, L.G. (2016) 'We neglect primary HIV prevention at our peril', *The Lancet*, 3: e284–e285, available at: www.thelancet.com/hiv (accessed 18 December 2017).

Kabwama, S.N. and Berg-Beckhoff, G. (2015) 'The association between HIV/AIDS-related knowledge and perception of risk for infection: a systematic review', *Perspectives in Public Health*, 135(6): 299–308.

Kanters, S., Mills, E.J., Thorlund, K., Bucher, H.C. and Ioannidis, J.P.A. (2014) 'Antiviral therapy for initial human immunodeficiency virus/AIDS treatment: critical appraisal of the evidence from over 100 randomised trials and 400 systematic reviews and meta-analyses', *Clinical Microbiological Infections*, 20: 114–122.

Koff, W.C. (2016) 'A shot at AIDS', *Current Opinion in Biotechnology*, 42: 147–151.

Krishnaratne, S., Hensen, B., Cordes, J., Enstone, J. and Hargreaves, J.R. (2016) 'Interventions to strengthen the HIV prevention cascade: a systematic review or reviews', *The Lancet HIV*, 3: e307–e317, available at: www.thelancet.com/hiv (accessed 18 December 2017).

Le Vasseur, M.T, Goldstein, N.D. and Welles, S.L. (2014) 'A public health perspective on HIV/AIDS in Africa: victories and unmet challenges', *Pathophysiology*, 21: 237–256.

MacCarthy, S., Hoffman, M., Ferguson, L., Nunn, A., Irvin, R., Bangsberg, D., *et al.* (2015) 'The HIV care cascade: models, measures and moving forward', *Journal of the International AIDS Society*, 18: 193–195.

Mayer, H.H., Shisana, O. and Beyer, C. (2016) 'AIDS 2016: from aspiration to implementation', *The Lancet*, 387: 2484–2485.

McKeown, R.E. (2009) 'The epidemiologic transition: changing patterns of mortality and population dynamics', *American Journal of Lifestyle Medicine*, 3(1): S19–S26, DOI: 10.1177/1559827609335350.

Medland, N.A., McMahon, J.H., Chow, E.P.F., Elliot, J.H., Hoy, J.F. and Fairley, C.K. (2015) 'The HIV care cascade: a systematic review of data sources, methodology and comparability', *Journal of the International AIDS Society*, 18(1): 1–8.

Omran, A.R. (1971) 'The epidemiologic transition: a theory of the epidemiology of population change', *The Milbank Memorial Fund Quarterly*, 49(4): 509–538.

Oster, E. (2012) 'HIV and sexual behaviour: why not Africa?', *Journal of Health Economics*, 31: 35–49.

Prados de la Escosure, L. (2013) 'Human development in Africa: a long-run perspective', *Explorations in Economic History*, 50: 179–204.

Reniers, G., Slaymaker, E. and Nakiyingi-Miiro, J. (2014) 'Mortality trends in the era of antiretroviral therapy: evidence from the network for analysing longitudinal population based HIV/AIDS data on Africa (ALPHA)', *AIDS*, 28 (Supplement 4): S533–S542.

Smith, J.A., Anderson, S.J., Harris, K.L., McGillen, J.B., Garnett, G.P. and Hallett, T.B. (2016) 'Maximising HIV prevention by balancing the opportunities of today with the promises of tomorrow: a modelling study', *The Lancet HIV*, 3: e289–e296, available at: www.thelancet.com/hiv (accessed 18 December 2017).

Sued, O., Figueroa, M.I. and Cahn, P. (2016) 'Clinical challenges in HIV/AIDS: hints for advancing prevention and patient management strategies', *Advanced Drug Delivery Reviews*, 103: 5–19.

Tanser, F., Barnighausen, T., Grapsa, E., Zaidi, J. and Newell, M.L. (2013) 'High coverage of ART associated with decline in risk of HIV acquisition in rural KwaZulu-Natal, South Africa', *Science*, 339(6122): 966–971.

The Lancet HIV (2016) 'Editorial: the prevention issue', *The Lancet HIV*, e283, available at: www.thelancet.com/pdfs/journals/lanhiv/PIIS2352-3018(16)30061-3.pdf (accessed 8 January 2018).

UNAIDS (Joint United Nations Programme on HIV/AIDS) (2015a) *AIDS by the Numbers: 2015*, Geneva: UNAIDS.

UNAIDS (Joint United Nations Programme on HIV/AIDS) (2015b) *The Global AIDS Epidemic Report 2015*, Geneva: UNAIDS.

UNAIDS (Joint United Nations Programme on HIV/AIDS) (2016a) *Children and HIV: Fact Sheet*, Geneva: UNAIDS.

UNAIDS (Joint United Nations Programme on HIV/AIDS) (2016b) *Global AIDS Update 2016*, Geneva: UNAIDS.

UNAIDS (Joint United Nations Programme on HIV/AIDS) (2016c) *Fact Sheet 2016: Global Statistics – 2015*, Geneva: UNAIDS.

USAID (United States Agency for International Development) (2015) *HIV Cascade Framework for Key Populations*, Washington, DC: USAID, www.fhi360.org/sites/default/files/media/documents/linkages-hiv-cascade-framework-oct15.pdf (accessed 8 January 2018).

WHO (World Health Organization) (2008) *World Health Statistics 2008*, Paris: WHO.

WHO (World Health Organization) (2013) *Consolidated Guidelines on the Use of Antiviral Drugs for Treating and Preventing HIV Infection*, Geneva: WHO.

WHO (World Health Organization) (2014a) *The Health of the People: What Works – The African Regional Health Report 2014*, Luxembourg: WHO.

WHO (World Health Organization) (2014b) *Consolidated Guidelines on HIV Prevention, Diagnosis, Treatment and Care for Key Populations*, Geneva: WHO.

WHO (World Health Organization) (2015a) *HIV/AIDS Fact Sheet Number 360*, November, available at: www.who.int/mediacentre/factsheets/fs360/en (accessed 18 December 2017).

WHO (World Health Organization) (2015b) *Consolidated Strategic Information Guidelines for HIV in the Health Sector*, Geneva: WHO.

WHO (World Health Organization) (2015c) *Accelerate Expansion of Antiretroviral Therapy to All People Living with HIV*, 30 November, available at: www.who.int/mediacentre/news/statements/2015/antiviral-therapy-hiv/en/ (accessed 18 December 2017).

WHO (World Health Organization) (2016a) *Atlas of African Health Statistics, 2016*, Brazzaville: African Health Observatory.

WHO (World Health Organization) (2016b) *World Health Statistics 2016: Monitoring Health for the SDGs*, Paris: WHO.

14
Food insecurity and famine

Stephen Devereux

Introduction

What is food security? One of the most enduring definitions is also one of the most succinct: 'Secure access by all people at all times to enough food for a healthy, active life' (World Bank, 1986). Despite its brevity, this definition encapsulates several concepts. 'Secure access' recognises various sources of food, not just production; 'all people' extends to multiple levels, from global to national to household and individual; 'all times' recognises that food insecurity can be permanent (chronic), or temporary (acute, transitory or seasonal); 'enough food' does not mean only macronutrients (carbohydrate, protein, fat), but also micronutrients – 'hidden hunger' refers to a diet that is deficient in vitamins and minerals that are essential for good health.

This definition also reflected a significant shift away from earlier supply-side framings that focused on food supplies at aggregate – national or global – levels. The 1974 World Food Summit had defined food security as 'availability at all times of adequate world food supplies of basic foodstuffs . . . to sustain a steady expansion of food consumption . . . and to offset fluctuations in production and prices' (UN, 1975).

Instead of asking 'Is there enough food?', contemporary definitions focus on who gets food and who does not. This paradigm shift was influenced by Amartya Sen's 'entitlement approach' (Sen, 1981), which identified several sources of food, including production, but also market purchases and transfers such as food aid. Sen analysed several famines in Africa and Asia to reveal that people can starve even when there is more than enough food in the country – the key issue is who is 'entitled' to access that food.

The 1996 World Food Summit adopted a slightly longer definition that introduced a focus on nutrition and food safety. 'Food security exists when all people, at all times, have physical and economic access to sufficient, safe and nutritious food that meets their dietary needs and food preferences for an active and healthy life' (World Food Summit, 1996). It is now understood that 'food security' and 'nutrition security' are not necessarily synonymous, and the emerging concept of 'food and nutrition security' recognises that good nutrition requires not only sufficient food, but also adequate sanitation facilities and healthcare.

This chapter starts by reviewing trends in food insecurity, globally and in Africa. It then examines the causes of persistent food insecurity in Africa, using the 'four pillars' of food availability, access, stability and utilisation. Next, proposed solutions to food insecurity are presented, ranging from increased food production to social protection and nutrition interventions, to improved governance. Finally, famines are examined as a particularly virulent form of food insecurity to which Africa remains uniquely vulnerable.

Food insecurity

Trends in food insecurity

Africa is the most food-insecure continent. Based on the Global Hunger Index, which combines three equally weighted national-level indicators (the proportion of the population that is undernourished, the prevalence of underweight under-five children, and the under-five mortality rate), five of only seven countries in the world with 'alarming' levels of hunger in 2016 were in Africa – the Central African Republic, Chad, Madagascar, Sierra Leone and Zambia. On the world hunger map, most of sub-Saharan Africa is coloured brown, signifying 'serious' or 'alarming' levels of hunger. Two African countries with no data in 2016 – Burundi and Eritrea – were coloured red in 2012, meaning 'extremely alarming' levels of hunger. Apart from South Asia, most of the rest of the world was coloured green in 2016, signifying 'moderate' or 'low' hunger (IFPRI *et al.*, 2016).

In 1996, the World Food Summit (WFS) set a target to halve the number of undernourished people, from 1 billion to 500 million, between 1990 and 2015. In 2000, the first Millennium Development Goal (MDG1c) set a target to halve the proportion of hungry people, from 24 per cent to 12 per cent, between 1990 and 2015. Because of population growth, the WFS target was more ambitious, being based on absolute numbers rather than a proportion of the population. Figure 14.1a shows that the MDG1c target was close to being achieved at the global level, having fallen to 13 per cent by 2015, but the WFS target was very far from being achieved, with an estimated 780 million people still undernourished. At the national level, 72 of 129 countries monitored reached the MDG target by 2015, but only 29 achieved the WFS goal (FAO *et al.*, 2015: 4).

In Africa, the trends were much less positive. Neither the WFS target nor the MDG1 target was achieved for Africa by 2015. The proportion of undernourished people fell, but only from 28 per cent to 20 per cent, still far above the MDG1c target of 14 per cent. In terms of absolute numbers, sub-Saharan Africa is the only world region where the number of hungry people is rising rather than falling – up from 182 million in 1990 to 232 million in 2015 (see Figure 14.1b).

Causes of food insecurity

A simple conceptual framework identifies four dimensions of food insecurity (Pieters *et al.*, 2013):

- *Availability (how much food is there?).* Food supply can be measured at national level by the *food balance sheet*, which adds up total food production after the annual harvest, plus carry-over stocks, plus commercial imports, plus concessional food aid, and subtracts exports and post-harvest losses in handling and storage. When converted into kilocalories and divided by the national population, this gives an estimate of per capita 'dietary energy supply' (DES) for the country at a point in time.

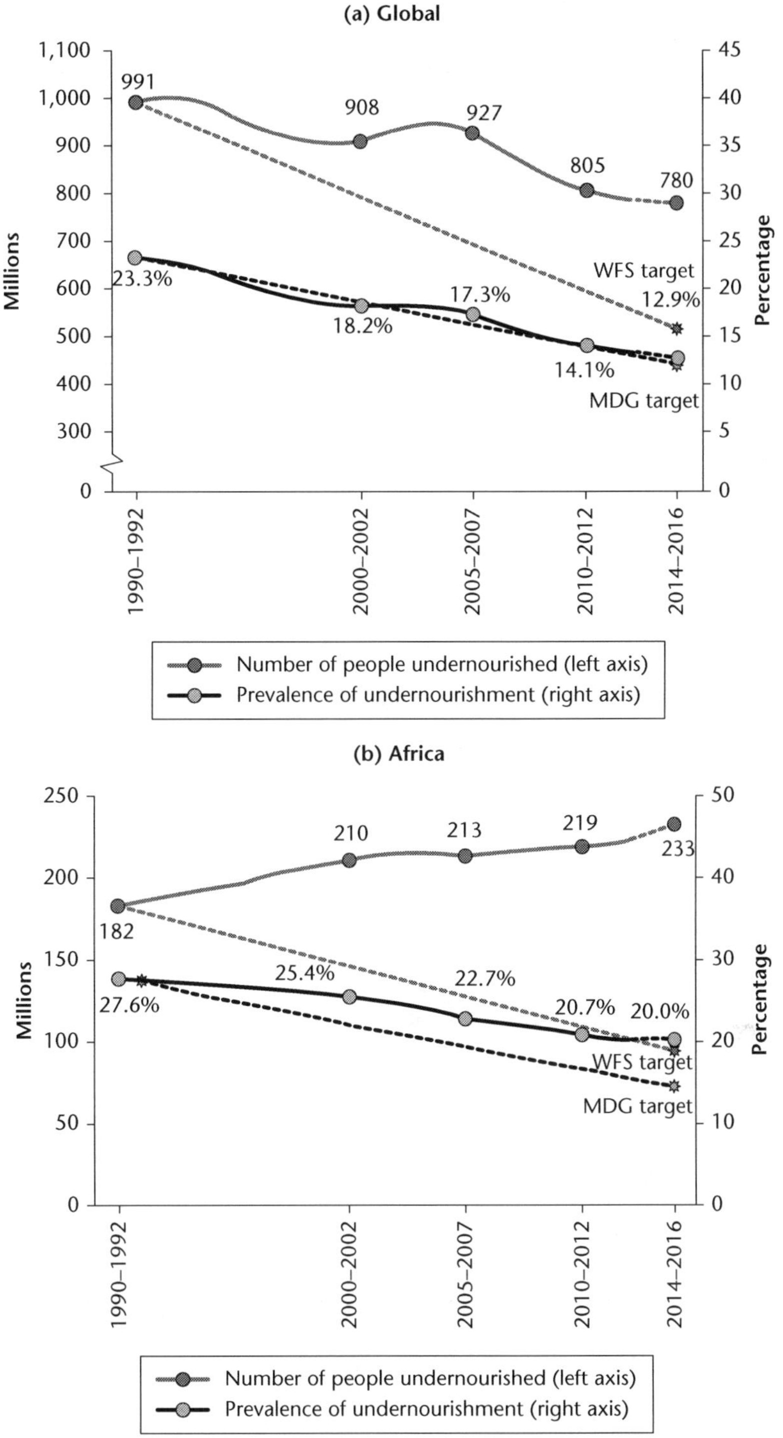

Figure 14.1 Trends in hunger, 1990–2015

Source: (a) FAO *et al.* (2015: 9); (b) FAO *et al.* (2015: 11).

- *Access to food (who has access to available food?).* Access is both *physical*, which is dictated by market supplies and infrastructure (transport, etc.), and *economic*, which is determined by incomes and food prices. Fragmented markets and lack of effective demand – low incomes or high food prices – restrict economic access.
- *Stability of food (is access to food stable and secure over time?).* Stability focuses on the temporal dimension of food security. Food supplies or access to food can be disrupted by natural disasters, seasonality (the 'hungry season'), food price volatility, conflict, and other shocks.
- *Utilisation of food (do individuals derive full nutritional value from the food they consume?).* Utilisation focuses on dietary quality, food preparation, hygiene and sanitation behaviour, health status, and other factors that determine whether nutrients will be properly absorbed and metabolised in the body.

These four dimensions represent four pathways to food insecurity. Each of these is examined below.

Availability

The primary source of food in rural communities is from agriculture, and many low-income African countries aim for self-sufficiency in food production. However, crop yields have remained persistently low in sub-Saharan Africa, which has undermined the capacity of agriculture to feed the continent, leaving Africans – especially urban consumers – dependent on imports. Between 1960 and 2006, staple crop yields more than doubled in the United States, from under 3 MT/ha to over 6 MT/ha, and almost trebled in Asia, from around 1 MT/ha to 3 MT/ha. Africa started with the lowest staple crop yields, at under 1 MT/ha in the 1960s, and rose to only slightly above 1 MT/ha by the early 2000s (see Figure 14.2). Although some of the gap between low-yield and high-yield farming systems is explained by agroecological constraints, such as poor soils and water scarcity, much of it represents lost potential food production that could be narrowed with well-designed investments and support to farmers.

Low agricultural productivity is a driver of chronic food insecurity. Globally, yields of staple cereals are projected to continue to increase, to about 4.5 MT/ha by 2030, but yield growth rates will be lowest in Africa, which will average only 1.42–1.73 MT/ha by 2030. Per capita cereals production is estimated to increase globally, from 357 kg/year in 2010 to 375 kg/year by 2030, but over the same period it will fall in Africa, from 134 kg/year to 122 kg/year, just one-third of the global average (Zhang *et al.*, n.d.: 4–6).

There are several reasons why crop yields are persistently low in Africa. One is the predominance of 'low input, low output' smallholder agriculture – family-based farming with high labour intensity but low capital intensity. Another factor is limited access to yield-enhancing inputs such as chemical fertilisers and improved seed varieties – unlike parts of Asia, Africa never experienced a 'green revolution'. Related to this are the liberalisation policies that were introduced across the developing world in the 1980s, and imposed on many African governments as conditions for IMF and World Bank loans, which led to underinvestment in agricultural research and extension services. Finally, women control food production and feeding in much of Africa, but are often denied access to land and farm inputs, and are neglected in agricultural policy, and by extension services that tend to target male farmers (Hunt and Lipton, 2011).

Food supply is the most basic determinant of nutrition outcomes. There is a strong statistical correlation between food availability and undernutrition. In countries where average dietary energy supply (DES) is higher than the amount of food required by the population, the prevalence of undernourishment tends to be low (see the case of Ghana in Figure 14.3). Conversely, where food availability is lower than national requirements – or even slightly above 100 per cent, because of unequal distribution of food (see Uganda in Figure 14.3) – indicators of hunger and malnutrition are high (FAO *et al.*, 2013: 25).

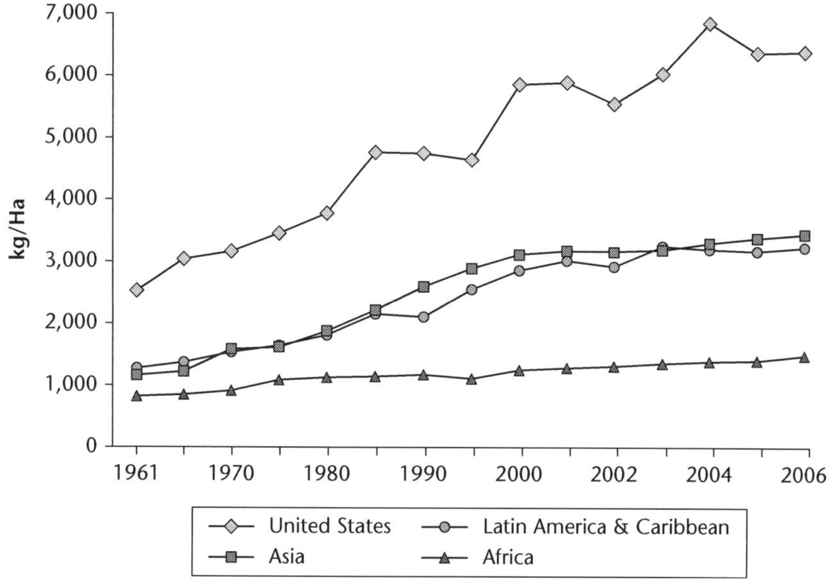

Figure 14.2 Staple crop yields by region, 1960–2006

Source: FAOSTAT (n.d.)

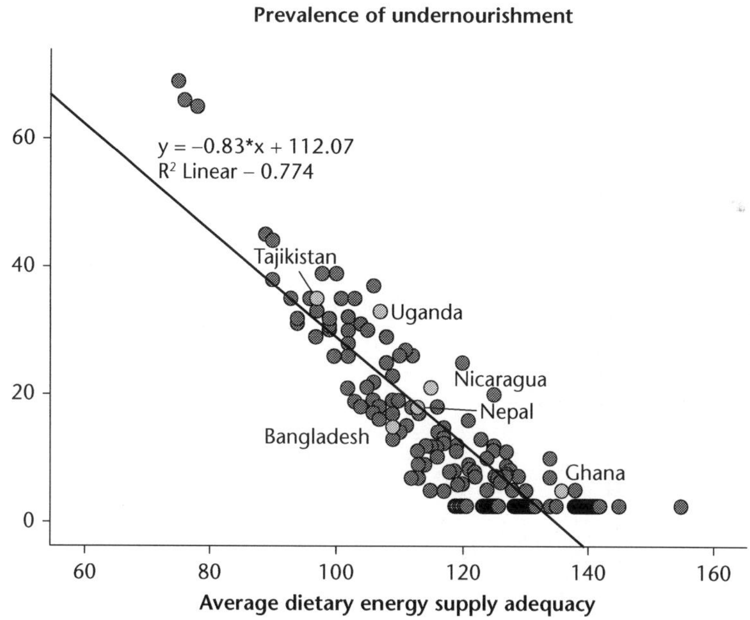

Figure 14.3 Food availability and undernutrition

Source: FAO *et al.* (2013: 25).

This might suggest that growing more food is the obvious solution, but it is important to recognise that food security does not require production self-sufficiency, at either the household or the national level. Several high-income countries do not produce enough food to feed their population. Instead, they import food, which they finance by exporting high-value goods and services. Japan's food self-sufficiency ratio in 2013 was only 39 per cent, for instance. National food insecurity in many African countries derives from a combination of low agricultural production, on the one hand, and low export earnings to finance food imports, on the other.

Access

Food security requires economic access to food. People who don't grow their own food, and smallholder farmers who don't grow enough food to feed their families, depend mainly on markets for subsistence needs. Economic access to food is determined by income, so poverty and hunger are closely correlated. At the national level, the most food-insecure countries are also among the world's poorest.

Global poverty has fallen faster than global hunger since the 1990s – the poverty MDG was achieved by 2010, when poverty had already halved, to 22 per cent from 47 per cent in 1990 (see Figure 14.4). Rising incomes and lower poverty are necessary but not sufficient to eradicate hunger. Food insecurity and income poverty are related, but they are not synonymous.

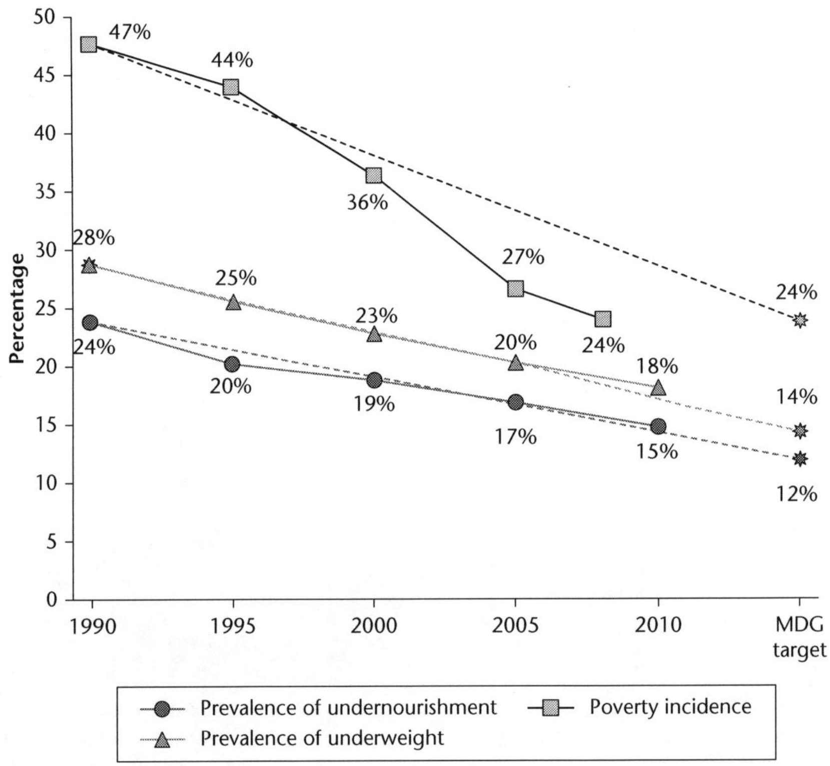

Figure 14.4 Global poverty and hunger, 1990–2015

Source: FAO *et al.* (2013: 20).

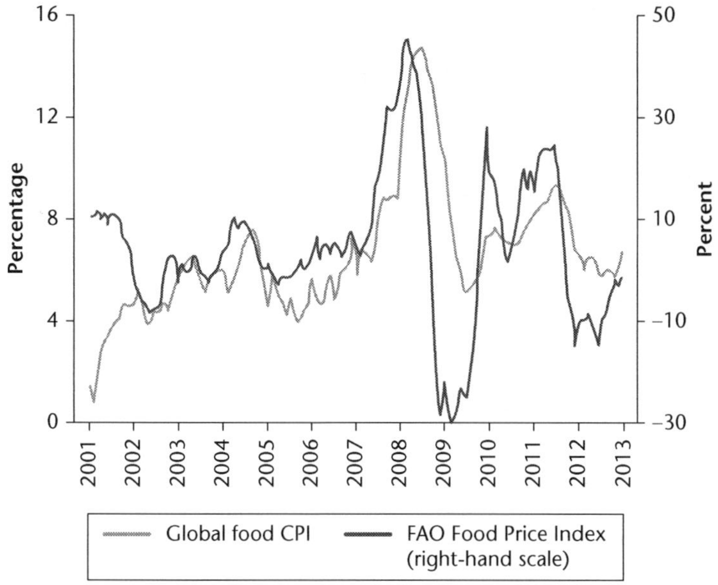

Figure 14.5 Global food prices, 2001–2013

Source: FAO *et al.* (2013: 14).

For people who buy their food, prices are a major determinant of food security. Food price inflation and food price volatility undermine access to food. The 'triple F' (food, fuel and financial) crises of 2007–2008 had the effect of causing the number of poor or hungry people globally to increase by an estimated 60 to 160 million people (Headey, 2011). The spike in food prices (see Figure 14.5) was triggered by a convergence of several adverse factors, including slow food production growth compounded by two successive harvest failures in major cereal exporters (including Australia and Ukraine), export bans by major rice producers (especially India and Vietnam), rising oil prices (which, inter alia, increased transport costs), and rising production of biofuels (which diverted agricultural investment towards ethanol production instead of food for human consumption) (von Braun, 2009).

Stability

Throughout rural Africa, where livelihoods derive mainly from agriculture and there is one rainy season each year (a unimodal rainfall system), farming families face an annual 'hungry season' or *soudure*. The agricultural cycle starts with the planting season, when farmers plough and sow their seeds, then watch the skies and pray for rain. If the rains are adequate and well distributed, they will reap a good harvest and fill their granaries with grain, which will feed their families for the next several months. When the granary is emptied, they will turn to the market for food, but by now market supplies are scarce while demand is increasing, so prices start to rise. By the time the next rainy season starts, many of these families are rationing consumption and adopting other 'coping strategies', but their children are going to bed hungry, and malnutrition rates will peak just before the next harvest comes in. In northern Ghana and southern Malawi, for instance, child malnutrition rates track millet and maize prices (respectively) through the year, rising together pre-harvest and falling together post-harvest (Devereux, 2010).

Regular or cyclical instability, like seasonality, can be predicted and therefore (in theory) managed. Various interventions have been implemented to smooth food consumption across the year. At the production end, efforts to stimulate production to close the annual 'hunger gap' are dominated by farm input subsidy programmes, which have had some success in countries such as Malawi and Zambia. At the consumption end, several countries run public works programmes providing temporary employment on community infrastructure projects for up to six months each year. Ethiopia's 'Productive Safety Net Programme' (PSNP) is the largest of these in Africa. By timing the work opportunities for the off-season, these food-for-work or cash-for-work projects deliver counter-seasonal access to food (Devereux, 2010).

Irregular instability or shocks – such as weather-induced harvest failure or food price spikes – cannot be predicted, and are therefore more difficult to manage than predictable cycles of food (in)security. In extreme cases, disrupted access to food can be so protracted and severe that it results in famine (as discussed below).

Utilisation

Utilisation of food relates to the biological drivers of malnutrition. Food intake is not the only determinant of nutrition status. The quality of the diet also matters. A powerful indicator of food insecurity and malnutrition is low dietary diversity. People whose consumption is dominated by two or three food groups (typically cereals, pulses and tubers) are more likely to be malnourished than those with a more diverse food intake (including meat, fish, leafy vegetables and fresh fruit) (Hoddinott and Yohannes, 2002). This relationship can be clearly seen in Figure 14.6, which reveals that globally, children whose diets derive mainly from starchy foods (cereals, roots and tubers) are more likely to have stunted growth than children who consume a wider variety of food items.

Recently, food security and nutrition have become more closely linked, partly as a response to evidence that poverty has fallen, and food security indicators have improved in many low-income countries, but with little or no improvement in nutrition outcomes. This is because there are several pathways to malnutrition, and inadequate food intake is only one – also important are caring and feeding practices and the disease environment, which is determined more by access to improved sanitation and clean water than by access to food.

Malnutrition takes three forms, which together form the 'triple burden of malnutrition'. Undernutrition or hunger follows from inadequate food intake, and can be either short term (transitory or acute) or long term (chronic). Being overweight or obese results from excessive food intake or an unbalanced diet (e.g. cheap fatty foods). 'Hidden hunger' is a consequence of micronutrient deficiencies – iron, vitamin A, iodine, zinc – and leads to nutrition-related diseases.

Although hunger is usually seen as a symptom of poverty, and obesity as a product of affluence, being overweight or obese can also be associated with being poor. Globally, 37 per cent of men and 38 per cent of women are now classified as overweight or obese,[1] and 62 per cent of obese individuals live in developing countries (Ng *et al.*, 2014: 777). Obesity prevalence rates are rising rapidly across the world, mainly due to a shift from active to sedentary lifestyles and the 'nutrition transition' – from organic foods (high in cereal and fibre) to highly processed foods (high in sugars, fat and chemical additives).

Obesity is reaching epidemic proportions in much of Africa. In southern Africa and North Africa, the prevalence of being overweight or obese in adults exceeds the global average, and is

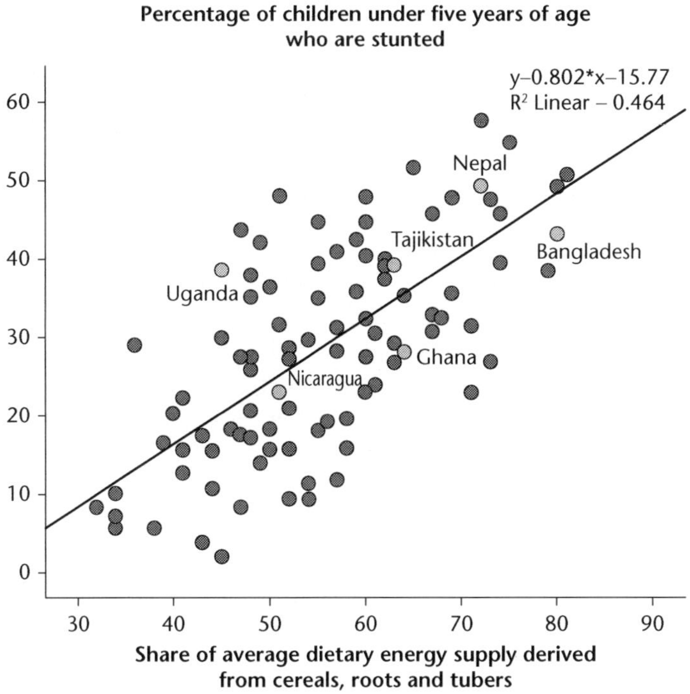

Figure 14.6 Undiversified diets cause children to be malnourished

Source: FAO *et al.* (2013: 26).

generally higher among women than men. In South Africa and Swaziland, for instance, 69 per cent of women are overweight or obese, but the prevalence among men is 39 per cent and 34 per cent, respectively, compared to a world average of 36.9 per cent for men and 38 per cent for women. In Egypt, 79 per cent of women and 71 per cent of men are overweight or obese. In Libya, 30 per cent of men are obese, and the rate for women is one of the world's highest at 57 per cent (Ng *et al.*, 2014).

Solutions to food insecurity

Food insecurity was originally conceptualised as a food availability deficit or a failure of food supply. It followed that the solution to food insecurity was to boost food production through 'extensification' (expanding the land under cultivation) or 'intensification' (raising crop yields per unit of arable land). The logic of the 'input–output' model was simple: increase the supply of agricultural inputs to increase the volume of agricultural outputs. This logic drove the success of the green revolution in Asia, but Africa never experienced a green revolution of its own, due mainly to weak postcolonial states (Frankema, 2014), so growth in output and yields has consistently lagged behind the rest of the world.

As the concept of food security evolved from a supply-side concept to one that incorporates the demand side, other approaches emerged. Here, we consider four proposed solutions derived from different problem analyses and policy perspectives: production, social protection, nutrition and governance.

The new productivist agenda

The most direct way to increase food availability is to grow more food. Contemporary versions of this approach could be called 'the new productivist agenda'.

In 2003, the New Partnership for Africa's Development (NEPAD) launched the Comprehensive Africa Agriculture Development Programme (CAADP), a policy framework that aimed to achieve food security for Africa through agricultural transformation – strengthening agricultural research, farmers' access to markets, rural infrastructure and sustainable agriculture. In 2014, the African Union announced the 'Year of Agriculture and Food Security', and adopted the Malabo Declaration, which resolved that member states should increase public funding of agriculture to 10 per cent of government spending per annum, with objectives that include doubling agricultural productivity, halving post-harvest losses, and reducing stunting in children to 10 per cent, all by 2025 (African Union, 2014).

Public–private partnerships will be strengthened to include smallholder farmers. Women account for almost half of all smallholder farmers in sub-Saharan agriculture, and the Malabo Declaration also commits to involving women in state-led agricultural initiatives and to empower women farmers. African leaders will be held accountable by the African Union for achieving these goals, through mutual accountability mechanisms – regular monitoring and a systematic peer review process.

The new social protection agenda

Social protection is still a relatively new policy agenda that boosts household incomes with social transfers, provides safety nets and social insurance against risks and shocks, and promotes livelihoods through input support to farmers or holistic packages such as 'graduation model' programmes. A report titled 'Social Protection for Food Security', commissioned by the Committee on World Food Security (HLPE, 2012), argued that food insecurity is a multidimensional problem requiring an integrated suite of interventions, and it advocated for a 'food security floor', with three platforms (see Figure 14.7):

1 *supply-side* support to food production (agricultural livelihoods development);
2 *demand-side* boost to access to food (a 'social protection safety net'); and
3 *humanitarian interventions* against food security shocks (emergency assistance).

There is persuasive evidence that social protection programmes, especially food or cash transfers that target the poor, have positive impacts on a range of food security indicators, such as food consumption, meals per day and dietary diversity. Because poor people spend a high proportion of any incremental income on food, cash transfers boost their effective demand for food and finance additional food purchases, leading to improved household food security in impact evaluations across Africa, including in Kenya, Lesotho, Malawi and Zambia (UNICEF, 2015). Impacts on individual nutrition status have been less positive, because the pathways to nutrition are more complex and include health and sanitation, as well as feeding practices, which are not directly influenced by income transfers.

The world's poor remain acutely vulnerable to future potential food price shocks, as occurred in 2007–2008. This has led some to propose stronger interventionist measures to stabilise food prices. Galtier (2013: 72) notes that 'Managing agricultural price instability has been a major concern for policymakers since time immemorial: public grain stocks were already in use in ancient Egypt'. When the neo-liberal 'Washington Consensus' achieved policy dominance in the 1980s,

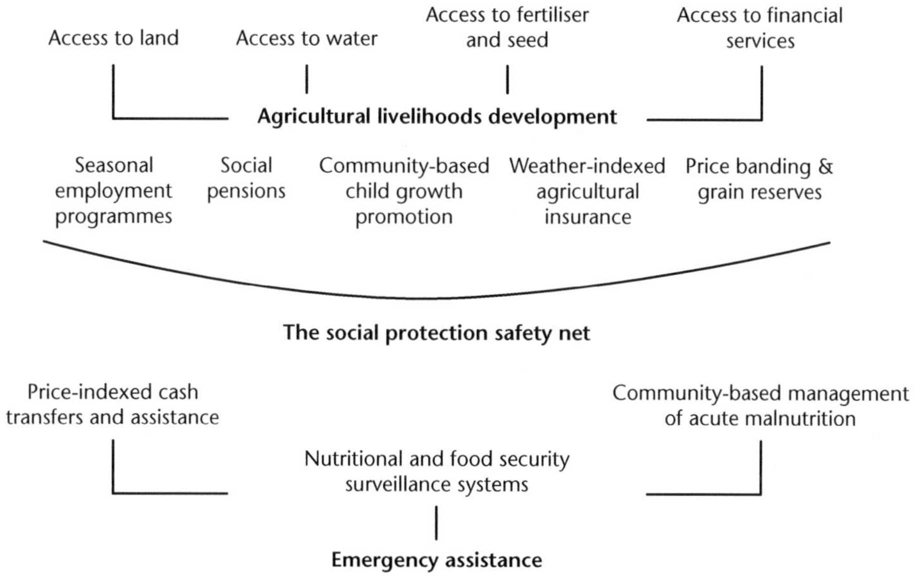

Figure 14.7 The 'food security floor'

Source: HLPE (2012: 59).

anti-interventionist sentiment backed up with financial leverage exercised by the Bretton Woods Institutions compelled the state to withdraw from agricultural production and marketing across the developing world. Many aid-dependent African governments reluctantly phased out food price subsidies for consumers and producers, commercialised their agricultural marketing parastatals, and scaled down or closed their national grain reserves. While it is true that these policies and institutions were often expensive and poorly managed and distorted local markets, food security was compromised during the transition period, and governments were left with few policy levers to protect their citizens against food security shocks such as the 'triple F' crises.

Galtier (2013) argues that the 'doctrine' of non-interventionism by the state went too far and was based on erroneous assumptions about the capacity of markets to either reduce or buffer price instability, especially in contexts of weak or missing markets (e.g. for insurance), as is still the case in much of rural Africa. Given that governments and the international community have a mandate to protect food security and uphold the right to food for all, Galtier (2013) concludes that there is a strong case for reintroducing policies at national and international levels to stabilise food prices, through the use of instruments such as price subsidies or buffer stock management, to regulate market supplies and defend a 'price band' between a minimum 'floor' price for producers and a maximum 'ceiling' price for consumers.

The new nutrition agenda

The 'new nutrition agenda' asserts that malnutrition cannot be tackled only by reducing poverty. Rising incomes are not sufficient to eradicate malnutrition, because nutrition status is driven by several factors, of which food intake is only one. The 'Scaling Up Nutrition' (SUN) movement argues that nutrition-specific interventions are needed, including the following (see SUN, n.d.):

1 *Feeding practices and behaviours*: Promoting exclusive breastfeeding up to 6 months of age, and breastfeeding plus complementary foods up to 2 years of age.
2 *Fortification of foods*: Incorporating essential nutrients into food and water supplies.
3 *Micronutrient supplementation*: Direct provision of extra nutrients.
4 *Treatment of acute malnutrition*: Ensuring that people with moderate or severe malnutrition receive effective treatment.
5 *Clean water and sanitation*: Universal access in order to reduce infection and disease.

It is now well known that a substantial proportion of malnutrition globally is related to open defecation and the transmission of faecal infections such as diarrhoea. 'Diarrhoeal episodes adversely affect the nutritional status of children: nutrients are evacuated; less is eaten during illness; . . . absorption of nutrients is reduced and energy [is] diverted to producing antibodies to fight off pathogens' (Chambers and von Medeazza, 2014: 14). It follows that improving water quality and sanitation practices, for example through 'community-led total sanitation' (CLTS), will be sufficient to reduce diarrhoeal diseases and improve children's nutrition status – without any increase being needed in incomes or food intake.

'Behaviour change communication' (BCC) and 'communication for development' (C4D) are innovative approaches to sharing ideas and knowledge that aim to empower individuals and communities to take positive actions to improve their lives. BCC and C4D are being used to communicate information about breastfeeding, dietary diversity, good sanitation practices such as handwashing, and so on, sometimes on their own and sometimes as part of social protection programmes – conditional cash transfers (CCTs), 'graduation model' programmes and 'cash+' programmes. One experimental study from Bangladesh found that 'cash + nutrition BCC' improved children's nutrition status more significantly than other interventions such as cash only, food only or 'food + nutrition BCC' (Ahmed *et al.*, 2016). These insights are starting to be applied in pilot projects in Africa, for example in Burundi and Rwanda (Sabates and Devereux, 2015).

The new governance agenda

Eradicating hunger and food insecurity is a political challenge as much as a technical or financial one. One way to build political commitment is to make access to food rights-based. The right to food has existed in international law since the Universal Declaration of Human Rights in 1948. In 1999, the UN Committee on Economic, Social and Cultural Rights clarified what this right means in practice: 'The right to adequate food is realized when every man, woman and child, alone or in community with others, has physical and economic access at all times to adequate food or means for its procurement' (UN CESCR, 1999). Although *Voluntary Guidelines on the Progressive Realization of the Right to Adequate Food in the Context of National Food Security* (FAO, 2005) exist, this right has never been enforced.

In general, Africa does not have a strong rights-based culture. Human rights are sometimes seen as luxuries that only higher-income economies can afford, or as impositions from the West. But the right to food does exist in some countries, for example in South Africa's Constitution: 'Everyone has the right to have access to sufficient food' (section 27). In attempting to achieve the progressive realisation of this right, the National Food and Nutrition Security Plan 2017–2022 sets out ambitious targets, including to halve by 2022 the percentage of households and individuals vulnerable to hunger, and to reduce the proportion of under-five children who are stunted from 26 per cent in 2013 to less than 15 per cent by 2022 (Republic of South Africa, 2016: 11).

The Hunger and Nutrition Commitment Index (HANCI) measures political commitment to tackling hunger and undernutrition in 45 developing countries, 30 of which are in Africa. HANCI uses 22 indicators that include legal frameworks (e.g. constitutional rights to food, social security and women's access to agricultural land), policies and programmes (e.g. access to improved drinking water, sanitation and vitamin A coverage) and public expenditures (e.g. percentages of government budget spent on agriculture and health). The 2014 HANCI report found that three of the eight countries that demonstrated 'high political commitment' were in Africa – Malawi, Madagascar and Burkina Faso. At the bottom end of the ranking, 11 of the 16 countries that demonstrated 'very low political commitment' were also African – Zambia, Liberia, DR Congo, Togo, Mauritania, Burundi, Lesotho, Nigeria, Angola, Sudan and Guinea-Bissau (te Lintelo and Lakshman, 2015). Nigeria's lowly position (40 out of 45 countries), and the fact that South Africa recorded only 'moderate commitment' despite its extensive social grants, reveals that the wealthiest countries in Africa do not necessarily display the highest commitment to ending hunger.

Famine

A famine is a form of acute food insecurity, triggered by a drastic disruption to food availability and access. Sub-Saharan Africa, especially the Horn and the Sahel, has historically been famine-prone. There were more than 20 major famines in Africa during the twentieth century that together killed more than 4 million people (Devereux, 2001). In the early twenty-first century, Africa has already suffered four mass mortality food crises, in Ethiopia (2000), Malawi (2002), Niger (2005) and Somalia (2011).

Causes of famine

All four twenty-first-century famines to date were triggered by droughts that reduced crop harvests and livestock herds in rural communities. However, 'drought causes famine' is no longer an adequate explanation – if it ever was. Following Amartya Sen's entitlement approach (Sen, 1981), food can be legally acquired from four sources: production (grow food), trade (buy food), labour (work for food) or transfers (receive food). Contemporary famines are caused by a series of failures that usually start with a poor harvest, but markets and institutional responses (by governments and humanitarian agencies) also need to fail for a crop failure to evolve into a famine.

Production failure?

The famine in Ethiopia in 2000 occurred in a pastoralist area (the Somali region) and mainly affected livestock, not crops – it was a livelihood shock for livestock keepers and traders, not a food crop failure for farmers. The famine in Malawi in 2002 followed a 32 per cent fall in the harvest, triggered by drought, but this followed two bumper years – the national maize harvest in 2002 was actually 6 per cent higher than the 10-year average. Niger's famine in 2005 was triggered by drought plus locusts that reduced the grain harvest by a moderate 11 per cent below the five-year average, but part of the deficit was offset by commercial food imports (Devereux, 2009). The famine in Somalia was preceded by a very severe drought – the lowest recorded rainfall for 50 years – which reduced the maize and sorghum harvests to less than half the five-year average. Even in this case, however, the famine could have been prevented if markets and institutional responses had not also failed (see Figure 14.8).

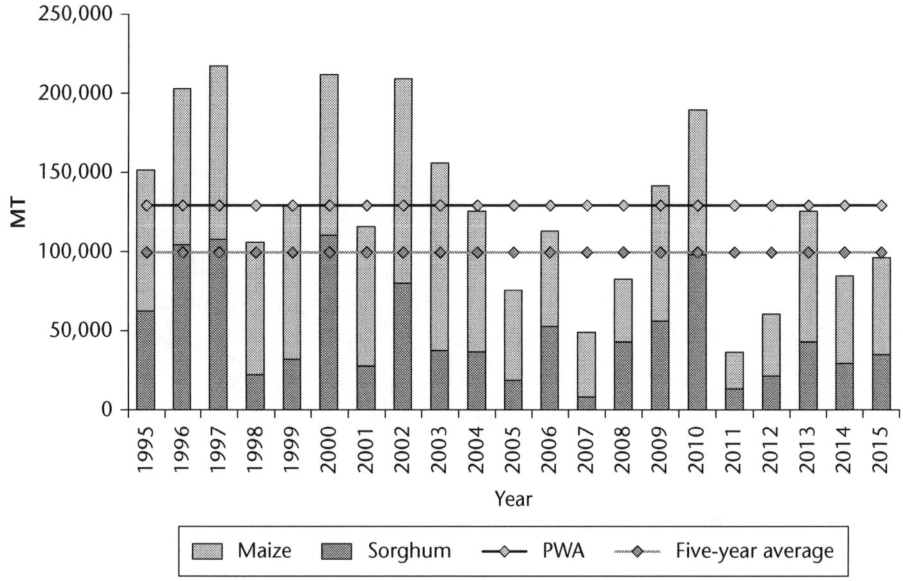

Figure 14.8 Trends in *Gu* cereal production in southern Somalia, 1995–2015

Source: FSNAU (2015: 10).

Market failure

The four most recent African famines occurred in rural areas with weak markets. Reduced food supplies coupled with a slow response by traders resulted in food price spikes that left poor families unable to purchase the food they needed. In Malawi, for instance, a poor harvest in 2001 led to depleted granaries and rapid price rises, as self-provisioning farmers turned to local markets that were unable to meet the surge in demand. Retail maize prices increased by 500 per cent or more in less than a year before the next harvest boosted supplies and drove prices down (see Figure 14.9).

This pattern is fairly typical, and similar price spikes have been observed in most famines. The failure of traders to move food into drought-affected areas compromised the ability of local people to access food at affordable prices, and precipitated migration out of famine zones, as hungry people went in search of cheaper food elsewhere.

Response failure

In all four cases, national governments lacked either the capacity or the political will to intervene. In Ethiopia, for instance, a famine early warning system was set up after the 1984 famine to predict and respond to food crises in the northern highlands, but the famine of 2000 occurred in the southern lowlands, in the politically marginalised Somali region. Neighbouring Somalia has not had a functioning government since the 1990s. When famine was declared in July 2011, the transitional federal government was engaged in conflict with the Islamist insurgent group Al-Shabaab, which effectively controlled the drought-affected regions of south-central Somalia, and restricted humanitarian access (Maxwell and Fitzpatrick, 2012). In Niger in 2005, the government responded to rising food prices by selling grain at subsidised prices, but this intervention was ineffectual as it covered only a fraction (about 20 per cent) of the national food deficit and was not enough to stop retail prices from doubling (ODI, 2005).

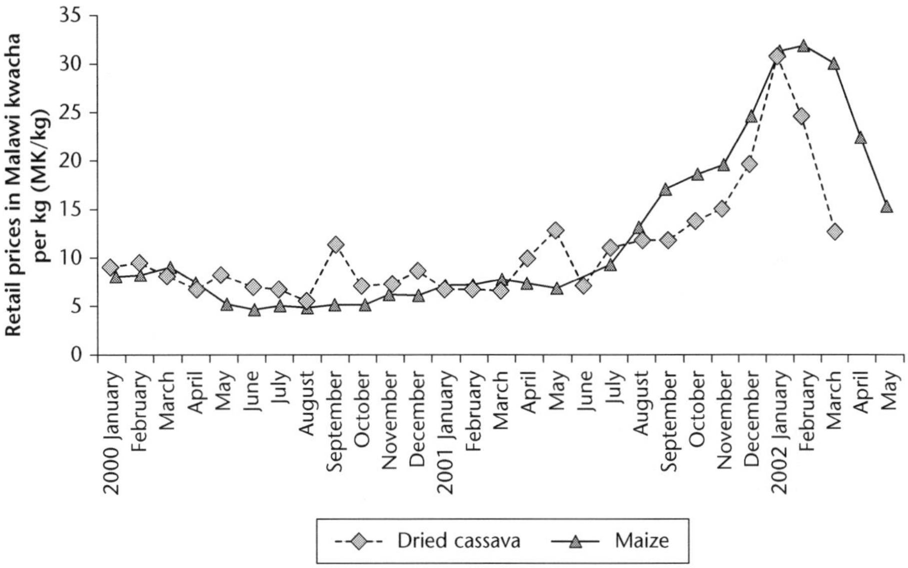

Figure 14.9 Average maize and cassava prices in Malawi, 2000–2002

Source: Devereux and Tiba (2007: 156).

Because of weak government capacity, responsibility for famine prevention has become the mandate of the international community. Technical agencies such as FAO and USAID monitor food security to predict crises, and operational agencies such as the World Food Programme channel food aid from donor countries to people assessed as being at risk of starvation. However, in all four recent African famines, there were problems at critical moments in the relationship between national governments and international partners or relief agencies that compromised the humanitarian response.

In Ethiopia, the government was involved in a border war with Eritrea from 1998 to 2000. During this time, donor agencies stopped replenishing the Emergency Food Security Reserve because they suspected the government was drawing down the reserve to feed its soldiers. When the food crisis struck the Somali region in 2000, the donors were slow to deliver food aid for the same reason – they feared this food would be diverted to the army. Similar distrust contributed to a delayed drought relief response in Malawi in 2002. After the IMF advised the government to sell its maize stocks in the strategic grain reserve (SGR) to pay off a debt, officials profiteered from the sale, leaving the reserve empty in late 2001, and donors refused to deliver food aid until the government explained what had happened to 175,000 MT of SGR maize.

In Somalia, security concerns and Al-Shabaab's hostility to external actors left humanitarian agencies facing severe information and access challenges before and during the 2011 famine. Most agencies operating in Somalia were based in Nairobi, and found it difficult to mobilise food and non-food aid and to deliver emergency relief to famine-affected populations in south-central Somalia (Hobbs *et al.*, 2012). Al-Shabaab had already banned CARE from working in areas under its control in 2009, followed by the World Food Programme in 2010. Compounding this, US counterterrorism legislation such as the Patriot Act prohibited giving US aid to any designated terrorist organisation, which 'essentially criminalised almost all flows of remittances and aid into areas controlled by Al-Shabaab' (Menkhaus, 2012: 32). The possibility that agency officials could be legally liable if Al-Shabaab seized relief supplies resulted in the suspension of humanitarian aid through USAID and US-based NGOs, and fatally delayed the international response.

Solutions to famine

In the past, famine was seen as a 'Malthusian' crisis – too many people and not enough food – and the solution was seen as growing more food, or even population control. Nowadays, a famine is understood more as an economic and sociopolitical crisis, rather than either a natural disaster or a crisis of underproduction or overpopulation. As with other forms of food insecurity, contemporary famines are about inadequate access to food, rather than an absolute lack of food. Famines tend to disappear as economies develop – markets integrate, transport and communications infrastructure reaches into remote rural areas, livelihoods diversify away from dependence on rain-fed agriculture, and poverty falls. Conversely, poverty, food insecurity and vulnerability to famine are closely connected, and all remain endemic in parts of sub-Saharan Africa.

Most contemporary famines could have been avoided through timely public intervention. It follows that the solutions to preventing future food supply shocks evolving into famines are mainly political. Democratic institutions such as multiparty elections and a free press are powerful antidotes to famine, because a government that allows a famine to occur can be held to account by the electorate and the media. According to Sen (1999: 178), 'there has never been a famine in a functioning multiparty democracy'. Sen cites Botswana and Zimbabwe in the 1980s as positive case studies in Africa, where timely intervention by democratic governments prevented potential famines following severe droughts.

Equally important is to strengthen accountability in relationships between national governments and the international community, especially development partners and humanitarian agencies mandated with protecting food security and preventing famines, in countries with limited capacity to do so without external assistance. Howe and Devereux (2004) argued that specific famine prevention roles and responsibilities must be clearly designated among relevant actors in each country (national government, international donors, early warning agencies, NGOs and others), and that the principle of proportionate accountability should be applied when a food crisis or famine occurs.

Edkins (2007) went further, arguing that contemporary famines are preventable and should therefore be considered not as natural disasters, but as crimes against humanity. In the absence of an effective 'anti-famine contract' (de Waal, 1997) between governments and people, the international community should criminalise 'mass starvations' as an incentive to prevent famines and a deterrent to deliberately causing them. In this context, it is striking that no individuals, agencies or governments have been held accountable for the four twenty-first-century famines discussed above.

Conclusion

In recent years, Africa has made some progress in tackling the related challenges of hunger, chronic food insecurity and vulnerability to famine that have compromised well-being and ended millions of African lives prematurely for centuries. As we have seen, the proportion of the population that is undernourished is falling, but remains at around 20 per cent, or one in five, while the absolute number of Africans who are classified as hungry continues to rise, and is currently estimated at approximately 232 million.

Sub-Saharan Africa is the last remaining world region still considered to be famine-prone, having suffered four mass starvations within the first 12 years of the twenty-first century. On the other hand, the magnitude and severity of contemporary famines are much less than in the past, and it seems unlikely that a catastrophic event on the scale of the Ethiopia famine of 1984 will recur in the future. Food crises are easier to predict and prevent than ever before, and response capacities of governments and international agencies are constantly improving.

The key factor explaining the persistence of hunger and famine in Africa is probably not a lack of financial resources and technical capacities, but how available resources and capacities are allocated and managed. This is a political and governance issue. As the 'HANCI' analysis reveals, some of the poorest countries in Africa are demonstrating the highest political commitment to ending hunger and undernutrition, while some of the richest countries are displaying the lowest commitment.

Two shifts in thinking are crucial. The first is the recognition that food security is as much about access to food as availability of food. It might seem paradoxical that the most food-insecure people in Africa are smallholder farmers who produce food, but farming is a highly risky business. Especially in low-potential water-stressed areas, where climate change threatens to undermine future harvests even further, the pathway to food security might not lie in intensification of agriculture, but in diversification of livelihoods out of agriculture.

The second insight is that good nutrition outcomes, especially for children, are determined not only by food consumption, but by other factors such as access to clean drinking water, hygienic sanitation practices, exclusive breastfeeding and the nutrition status of mothers during the 'first 1,000 days' of a child's life. Food security interventions have come a long way from subsidising fertiliser and seeds for farmers. Ensuring access to inputs remains an important part of the solution, but tackling food insecurity in Africa requires a comprehensive multisectoral approach.

Acknowledgement

The writing of this chapter was supported by the National Research Foundation of South Africa (Grant Number: 98411) and the Newton Fund, administered by the British Council.

Note

1 Clinically, an overweight adult has a body mass index (BMI) ≥ 25 to <30 kg/m^2, while obesity in adults is defined as having a BMI ≥ 30 kg/m^2.

References

African Union (2014) *Malabo Declaration on Accelerated Agricultural Growth and Transformation for Shared Prosperity and Improved Livelihoods*, Addis Ababa: African Union Commission.

Ahmed, A., Hoddinott, J., Roy, S., Sraboni, E., Quabili, W. and Margolies, A. (2016) *Which Kinds of Social Safety Net Transfers Work Best for the Ultra-Poor in Bangladesh? Operation and Impacts of the Transfer Modality Research Initiative*, Dhaka: IFPRI and WFP.

Chambers, R. and von Medeazza, G. (2014) *Reframing Undernutrition: Faecally-Transmitted Infections and the 5 As*, IDS Working Paper 450, Brighton: Institute of Development Studies.

Devereux, S. (2001) 'Famine in Africa', in S. Devereux and S. Maxwell (eds), *Food Security in Sub-Saharan Africa*, London: ITDG Publishing, pp. 117–148.

Devereux, S. (2009) 'Why does famine persist in Africa?', *Food Security*, 1(1): 25–35.

Devereux, S. (2010) 'Seasonal food crises and social protection in Africa', in B. Harriss-White and J. Heyer (eds), *Comparative Political Economy of Development: Africa and South Asia*, London: Routledge, pp. 111–135.

Devereux, S. and Tiba, Z. (2007) 'Malawi's first famine: 2001–2', in S. Devereux (ed.), *The New Famines*, London: Routledge, pp. 143–177.

de Waal, A. (1997) *Famine Crimes: Politics and the Disaster Relief Industry in Africa*, Oxford: James Currey.

Edkins, J. (2007) 'The criminalization of mass starvations: from natural disaster to crime against humanity', in S. Devereux (ed.), *The New Famines*, London: Routledge, pp. 50–65.

FAO (Food and Agriculture Organization) (2005) *Voluntary Guidelines on the Progressive Realization of the Right to Adequate Food in the Context of National Food Security*, Rome: FAO.

FAO (Food and Agriculture Organization), IFAD (International Fund for Agricultural Development) and WFP (World Food Programme) (2013) *The State of Food Insecurity in the World 2013: The Multiple Dimensions of Food Security*, Rome: FAO.

FAO (Food and Agriculture Organization), IFAD (International Fund for Agricultural Development) and WFP (World Food Programme) (2015) *The State of Food Insecurity in the World 2015: Meeting the 2015 International Hunger Targets – Taking Stock of Uneven Progress*, Rome: FAO.

FAOSTAT (Food and Agriculture Organization Corporate Statistical Database) (n.d.) *Food Crop Production*, available at: www.fao.org/faostat/en/#search/crop%20yields (accessed 27 September 2017).

Frankema, E. (2014) 'Africa and the green revolution: a global historical perspective', *Wageningen Journal of Life Sciences*, 70–71(6): 17–24.

FSNAU (Food Security and Nutrition Analysis Unit) (2015) *Somalia Food Security and Nutrition Analysis: Post Gu 2015*, Technical Series Report No. VII.60, Nairobi: Food Security and Nutrition Analysis Unit.

Galtier, F. (2013) 'Managing food price instability: critical assessment of the dominant doctrine', *Global Food Security*, 2(2): 72–81.

Headey, D. (2011) *Was the Global Food Crisis Really a Crisis? Simulations Versus Self-Reporting*, IFPRI Research Brief, 17, Washington, DC: IFPRI.

HLPE (High Level Panel of Experts on Food Security and Nutrition) (2012) *Social Protection for Food Security: A Report by the High Level Panel of Experts on Food Security and Nutrition of the Committee on World Food Security*, Rome: CFS.

Hobbs, C., Gordon, M. and Bogart, B. (2012) 'When business is not as usual: decision-making and the humanitarian response to the famine in south central Somalia', *Global Food Security*, 1(1): 50–56.

Hoddinott, J. and Yohannes, Y. (2002) *Dietary Diversity as a Food Security Indicator*, FCND Discussion Paper 136, Washington, DC: IFPRI.

Howe, P. and Devereux, S. (2004) 'Famine intensity and magnitude scales: a proposal for an instrumental definition of famine', *Disasters*, 28(4): 353–372.

Hunt, D. and Lipton, M. (2011) *Green Revolutions for Sub-Saharan Africa*, Chatham House Briefing Paper, London: Royal Institute of International Affairs.

IFPRI (International Food Policy Research Institute) (2016) *2016 Global Hunger Index*, Washington, DC: IFPRI.

Maxwell, D. and Fitzpatrick, M. (2012) 'The 2011 Somalia famine: context, causes, and complications', *Global Food Security*, 1: 5–12.

Menkhaus, K. (2012) 'No access: critical bottlenecks in the 2011 Somali famine', *Global Food Security*, 1(1): 29–35.

Ng, M., Fleming, T., Robinson, M., Thomson, B., Graetz, N., Margono, C., *et al.* (2014) 'Global, regional, and national prevalence of overweight and obesity in children and adults during 1980–2013: a systematic analysis for the Global Burden of Disease Study 2013', *The Lancet*, 384: 766–781.

ODI (Overseas Development Institute) (2005) *Humanitarian Issues in Niger*, HPG Briefing Note, London: ODI.

Pieters, H., Guariso, A. and Vandeplas, A. (2013) *Conceptual Framework for the Analysis of the Determinants of Food and Nutrition Security*, FOODSECURE Working Paper No. 13, Wageningen: FOODSECURE.

Republic of South Africa (2016) *National Food and Nutrition Security Plan for South Africa 2017–2022*, Pretoria: Government of the Republic of South Africa.

Sabates, R. and Devereux, S. (2015) 'Evidence on graduation in practice: Concern Worldwide's graduation programme in Rwanda', *IDS Bulletin*, 46(2): 64–73.

Sen, A. (1981) *Poverty and Famines*, Oxford: Clarendon Press.

Sen, A. (1999) *Development as Freedom*, Oxford: Oxford University Press.

SUN (Scaling Up Nutrition) (n.d.) *The Vision and Principles of SUN*, available at: http://scalingupnutrition.org/about-sun/the-vision-and-principles-of-sun/ (accessed 8 January 2018).

te Lintelo, D. and Lakshman, R. (2015) *The Hunger and Nutrition Commitment Index (HANCI 2014): Measuring the Political Commitment to Reduce Hunger and Undernutrition in Developing Countries*, Brighton: Institute of Development Studies.

UN (United Nations) (1975) *Report of the World Food Conference*, New York: United Nations.
UN CESCR (United Nations Committee on Economic, Social and Cultural Rights) (1999) *Substantive Issues Arising in the Implementation of the International Covenant on Economic, Social and Cultural Rights, General Comment 12, The Right to Adequate Food (Art. 11), E/C.12/1999/5*, Geneva: United Nations.
UNICEF (United Nations Children's Fund) (2015) *Social Cash Transfers and Children's Outcomes: A Review of Evidence from Africa*, Nairobi: UNICEF-ESARO.
von Braun, J. (2009) 'Addressing the food crisis: governance, market functioning, and investment in public goods', *Food Policy*, 1: 9–15.
World Bank (1986) *Poverty and Hunger*, Washington, DC: World Bank.
World Food Summit (1996) *Rome Declaration on World Food Security*, Rome: FAO.
Zhang, W., Liu, G. and Bai, C. (n.d.) *A Forecast Analysis on Global Production of Staple Crops*, Guangzhou: Sun Yat-sen (Zhongshan) University.

15

Access to essential medicines in Africa

Aarti Patel

Introduction

Pretty Buthelezi, 27 years old and working as a domestic worker in Durban, South Africa, has just found out she is pregnant. Last month, she was sent home by the local clinic without receiving her Depo-Provera injection because the clinic did not have any stock. Pretty is also on medicines for tuberculosis that she gets from the local community health centre. This is not a planned pregnancy. Pretty thought she was safe as she was using the injection, she has not been seeing her boyfriend for too long and her job is also quite new. What went wrong?

Pretty is being treated for tuberculosis (TB). These TB medicines affect the efficacy of her contraceptive by reducing the duration of its protective cover. The injection is administered once every three months, but in Pretty's case the injection should be given every two months. However, Pretty was not given her Depo injection during her recent clinic visit as this was unavailable. She now faces an unplanned pregnancy and a very uncertain future for herself and her unborn child.

Pretty is fictional, but there are many Prettys across Africa who find themselves in challenging situations because health systems, and specifically access to medicines, have failed them. This chapter attempts to address the issue of access to essential medicines and diagnostics in the African context informed by examples from mainly southern and eastern African countries. The chapter is developed around two frameworks: the medicines management cycle and the WHO access framework. In keeping the focus on Pretty, readers are reminded to consider these issues from the perspective of the person/people affected – not the product (medicine), the prescriber and dispenser (professionals), the supplier (manufacturers, wholesalers, retailers), or the system (policymakers, researchers, technical agencies). Keeping the focus on Pretty requires different approaches to understanding the problems and finding people-centred solutions.

Background

The concept of the 'essential (drug) medicine' arose in the 1970s, linked to the global movement for primary healthcare. Medicines were recognised as critical to improve health, well-being and dignity. Since then, all major declarations and global and regional initiatives related to well-being, poverty reduction and sustainable development have included access to essential medicines in one form or another (see Box 15.1).

Box 15.1 Examples of declarations and initiatives embracing essential medicines

- Alma Ata Declaration: health for all by 2000 (WHO, 1978).
- Health sector reform, including the Bamako Initiative for Africa (McPake *et al.*, 1993).
- United Nations' Millennium Development Goals (UN, 2000).
- Southern African Development Community's Pharmaceutical Business Plan (SADC, 2007).
- Pharmaceutical Manufacturing Plan for Africa (African Union, 2012).
- 2030 Agenda for Sustainable Development (UN, 2015): 'Achieve universal health coverage, including financial risk protection, access to quality essential health-care services and access to safe, effective, quality and affordable essential medicines and vaccines for all'.

The essential medicine concept embraces those medicines needed to treat the priority health needs of a given population. The first major conference on essential medicines policies was held in Nairobi in 1985, resulting in the WHO Revised Drugs Strategy adopted by the World Health Assembly in 1986. Since then, many international, regional and national policies have been developed and implemented relating to the selection, procurement, supply, pricing, prescribing, dispensing and use of medicines. In November 2016, just over 30 years from the 1985 Nairobi Conference, the report of the Lancet Commission on Essential Medicines Policies was produced (The Lancet, 2016). This timely publication allows us to perform a stock take of what has been achieved during the past 30 years, as well as how we need to proceed to address the challenges of equitable access, as 'Without essential medicines, no health system can ensure that the population it serves progressively realises its right to health. Yet essential medicines policies have received insufficient attention' (The Lancet, 2016). Equitable access to essential medicines still eludes many people, as demonstrated by Pretty and others like her from the developing world, where it is estimated that between 1.3 and 2.1 billion people remain without access (WHO, 2011a).

Definition: essential medicines

Essential medicines are those medicines that meet the priority health needs of the population. They are selected on the basis of the role with respect to disease prevalence and public health relevance, evidence of clinical efficacy and safety, and comparative costs and cost-effectiveness. These medicines must be available, affordable, their quality assured, and they must be used appropriately (Quick, 2003). Quality, safety, availability and affordability are key features of an essential medicine. There are many medicines on the market – what makes one essential is through its ability to treat prevalent diseases that are of public health relevance, its efficacy and safety profile, and affordability to the people who will be procuring these, as well as those using them.

How does Pretty benefit from an essential medicine?

Medicines are not ordinary commodities – they are entities produced by international and innovative pharmaceutical manufacturers as well as local manufacturers, using ingredients produced in different parts of the world, procured by internal agencies, donors, governments and private organisations and recommended for use by clinicians. Pretty, who swallows the pills for her TB or gets the injection for her contraception, did not decide which products to use. It was the doctor at the

TB hospital and the nurse at the family planning clinic that decided. The prescriber, dispenser and Pretty all relied on the Medicines Regulatory Authority of South Africa to ensure the medicines they use are safe and of approved quality. The decision to actually use the medicines is informed by socio-economic factors that affect Pretty. Both her TB medicines and her contraceptive are on the list of essential medicines for South Africa. Pretty benefits by not having to pay for these medicines as essential medicines are provided free of charge in South Africa. What went wrong for Pretty? The non-availability of the contraceptive injection meant reduced contraceptive cover and resulted in her pregnancy. The supply, management and use of medicines is complex and interlinked. These issues are explored in the sections below as part of the medicines management cycle.

Medicines management cycle

The medicines management framework was proposed by Management Sciences for Health (MSH) and the World Health Organization (WHO), which was discussed in the seminal book *Managing Drug Supply* (MSH, 2012), based on work that continues to be done in developing countries. This framework, presented below in Figure 15.1, demonstrates the complexities of ensuring the availability and affordability of quality-assured essential medicines that are used correctly by showing the linkages between the different aspects of medicines from their selection through to their use.

Four basic functions contribute to the cycle: selection, procurement, distribution and use of medicines. *Selection* is concerned with reviewing the prevalent health problems within a particular setting, identifying treatments of choice using principles of evidence-based medicine, choosing drugs and dosage forms, and identifying which drugs will be available at each level of the healthcare system, i.e. primary (mainly clinics and community health centres managed by nurses), secondary (mainly hospitals) and tertiary (specialist services) (MSH, 2012).

Procurement is about quantifying drug requirements for the healthcare system and is informed by the decisions around selection. Procurement processes include identifying procurement methods, managing tenders, identifying suppliers, managing contracts and assuring drug quality.

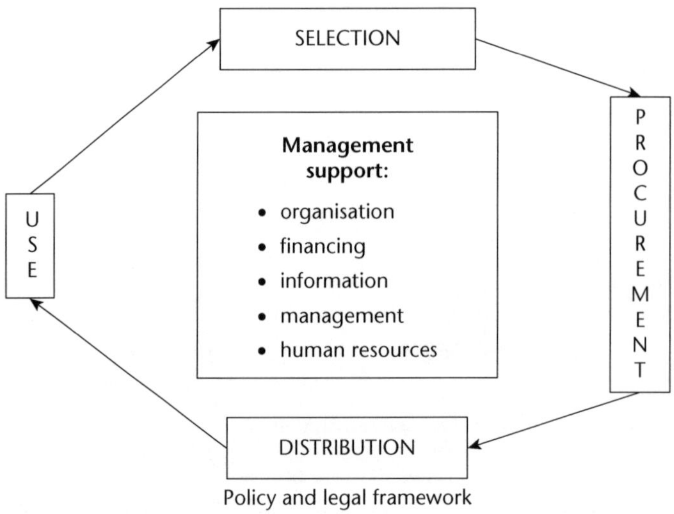

Figure 15.1 Medicines management cycle

Source: MSH (2012).

Distribution includes all processes that get the medicine from the manufacturer to the patient. The processes in distribution include stock control, stock management and delivery to healthcare facilities. *Use* is about diagnosing, prescribing, dispensing and appropriate use by the consumer.

This cycle is underpinned by the relevant legal and policy framework of the particular country. To ensure proper functioning of the cycle, at the centre of the cycle is the core of management support systems, including organisation, financing, information and human resources.

In Pretty's case, the medication management cycle failed between procurement, distribution and use. This failure could be as a result of poor financing, poor procurement and distribution.

Even though work continues to ensure the sustainable supply of essential medicines in African countries, we have not seen research addressing the medicines management cycle holistically. There are opportunities to apply this framework to specific countries and regional settings following a few people and their experiences.

National medicines policies (NMPs)

The WHO supports countries to formulate NMPs that are intended to serve as the 'guide to action' for the different stakeholders involved in selecting, procuring, distributing and using medicines (WHO, 2001). Historically, NMPs were developed but implementation was slow. To overcome this, countries are supported to simultaneously develop an implementation plan alongside the NMP. This plan addresses the time frames, resources and responsible actors needed for implementation of the policy. Supportive legislation and regulations are needed to ensure that policy goals, purpose and objectives are met. Across Africa, the majority of countries have formulated NMPs. However, implementation remains poor. The WHO uses indicators to track the progress of countries regarding their pharmaceutical profiles. From these profiles, the present status of 43 African countries regarding availability of NMPs and implementation plans is presented in Table 15.1. Data for 11 countries were not available.

Five out of 43 countries do not have NMPs and eight of the remaining countries do not have implementation plans, with information missing for Sudan and Tunisia. Some NMPs are still in their first version, including South Africa, considered to be one of the most developed countries in southern Africa. The information from Table 15.1 provides some insights into why people such as Pretty cannot receive the right medicine when needed: policies are not available, or if available they are often outdated, or are without a costed implementation plan to guide action in the country.

The main components of a national medicines policy generally consist of the following:

- *Selection*: Evidence-based, focusing on morbidity patterns, EMLs.
- *Supply*: Local production, procurement mechanisms, distribution and storage, disposal of unwanted or expired medicines.
- *Rational use*: STGs, medicines information, rational medicine use for training, education, promotion.
- *Affordability*: Taxes or tariffs on essential medicines, pricing, use of generics, TRIPS mechanisms.
- *Financing*: User charges, health insurance, donor assistance.
- *HR development*: Education, training, continuing education.
- *Monitoring and evaluation*: Baseline surveys, indicators for monitoring, periodic monitoring, independent external evaluation.
- *Research*.
- *Technical cooperation among countries*.

- *Legislative and regulatory framework*: DRAs, good governance for medicines, legislation and regulation, medicines registration and licensing, quality assurance (inspection and enforcement), regulation of prescription and distribution.

Table 15.1 NMPs and implementation plans for African countries

Country	*NMP*	*Year*	*Implementation plan*
Angola	N	N	N
Burkina Faso	Y	2011 (update)	Y
Burundi	N	Draft	N
Cameroon	Y	2008	n/a
Comores	Y	2007 (update)	Y (2004)
DRC	Y	2007 (update)	N
Ghana	Y	2004 (2016 update)	2005
Guinea-Bissau	Y	2009	Y (2009)
Guinea	Y	2007	1992
Kenya	Y	2010	n/a
Lesotho	Y	2005	2005
Madagascar	Y	2010	2005
Mauritius	N	N	N
Nigeria	Y	2005	N
Seychelles	N	N	N
South Africa	Y	1996	2006
Swaziland	Y	2008	N
Zimbabwe	Y	1995	2005
Botswana	Y	2002	2011
Central African Republic	Y	2010	N
Chad	Y	2009	2009
Côte d'Ivoire	Y	2008	N
Eritrea	Y	2010	2007
Ethiopia	Y	2007	2007
Gabon	Y	2009	1999
Gambia	Y	2007	2009
Liberia	Y	2011	2010
Malawi	Y	2009	2009
Mali	Y	1998	2009
Mauritania	Y	2011	N
Mozambique	Y	1997	2007
Namibia	Y	1998	2000
Niger	Y	1995	2005
São Tomé and Príncipe	N	N	N
Senegal	Y	2006	2008
Sierra Leone	Y	2004	2004
Uganda	Y	2002	2010
Tanzania	Y	1991	1992
Zambia	Y	2010	2010
Egypt	Y	2005	N
Sudan	Y	2005	N
Morocco	Y	2008	2008
Tunisia	Y	2007	N

Source: WHO (2011b).

Table 15.2 Missing policies in NMPs

NMP aspects	Countries with NMPs but missing *policies* in the following areas
Selection	Tunisia
Financing	Comores
Pricing	Comores, Ghana, Ethiopia, Sierra Leone, Zambia
Procurement	
Distribution	
Regulation	
Pharmacovigilance	Guinea, Botswana, Ethiopia, Sierra Leone
Use	
HR	Guinea-Bissau, Sierra Leone
Research	Comores, Guinea-Bissau, Egypt
M&E	Botswana, Mozambique, Namibia
Traditional medicines	Guinea-Bissau, Mozambique, Egypt, Morocco, Tunisia

Source: WHO (2011b).

The scope of NMPs generally covers pharmaceuticals for human and veterinary use,[1] as well as herbal products. Some countries, for example Kenya, Tanzania and Ghana, have extended scopes to also include medical devices and technologies, food products, tobacco products, cosmetics and emerging health technologies.

Table 15.2 summarises the gaps in NMPs for those African countries with NMPs.

The data from Tables 15.1 and 15.2 show that African countries have NMPs with legislation and regulations to support implementation, but challenges lie in the capacities of these countries to ensure policies are updated regularly to remain relevant. Updating NMPs also require updates in medicines legislation and regulations. Often, legislation and regulations are outdated and not enabling; in some instances, the legislation is the bottleneck for implementation. Box 15.2 provides a few examples of where NMPs, legislation and regulations need to be updated.

Box 15.2 National examples

South Africa

NMP introduced in 1996. By 2016, the policy is yet to be reviewed and updated. South Africa's NMP has three main objectives: health, economic and national development. South Africa also has the most advanced pharmaceutical market in Africa, which is estimated to be worth US$3.34 billion (Business Monitor International, 2007). Most of the major multinational pharmaceutical companies are in South Africa, either with their own manufacturing plants or through licensing agreements with local producers. South African firms are the leading generic manufacturers in the country. Around 47 per cent of the country's pharmaceutical needs are met by domestic production (Business Monitor International, 2007). Based on the pharmaceutical manufacturing presence in South Africa, it is clear that the economic and national development objectives of the NDP are being met. However, the health objectives – in relation to Pretty – were not met.

Uganda

Outdated medicines legislation from the 1970s places the National Drugs Authority as the overall body responsible for both the policy and the regulator. For most countries, NMPs remain within the policy arm of ministries of health, and not with the regulatory authorities. Regulators regulate and ensure safety of the population. Regulators are not expected to draft policies.

Kenya

Kenya's Pharmacy and Poisons Act of 1957 (Chapter 244) regulates both the products and the practice, while the NMP calls for the separation of these functions. Kenya's current legislation contributes to confusion regarding the autonomy of the regulatory body as it places the Chief Pharmacist of the Ministry of Health as the Registrar and the Director of Medical Services and as Chair of the Board. Therefore, in Kenya, the provisions in the 2012 policy to elevate the status of pharmaceutical services from under the medical directorate cannot be achieved within the existing legislation.

Most African countries have included local production as a key policy objective in their NMPs. To implement this recommendation, countries require that a certain percentage of state tenders for essential medicines be awarded to local manufacturers. Through this requirement, national economic development is supported.

As local production across Africa remains in its infancy, countries rely largely on pharmaceutical imports. These imports are mainly from China and India. Where pharmaceutical manufacturing does occur, this involves the production of non-complex, high-volume, essential products such as basic analgesics, simple antibiotics and vitamins. Kenya, South Africa, Ghana and Egypt are some of the African countries with large pharmaceutical manufacturing capacity.

Pharmaceutical manufacturers are either local or multinationals. Few multinationals have local manufacturing plants; rather, there are local agents who distribute their products and maintain marketing offices. Multinational firms generally have brand name products in the market, while local manufacturers provide lower-priced generics.

Thus, while Africa is promoting the local production of medicines, this may come at a cost to the health sector when local manufacturers are not able to meet quality standards (e.g. good manufacturing practice), potentially resulting in the manufacture and supply of substandard medicines.

As we shall see from the WHO access framework, African countries have to ensure that their health systems are functioning effectively through sustainable financing and adequate capacity for the appropriate selection and procurement of medicines. Adding trade imperatives to the medicines policy landscape starts to blur boundaries and drive the focus away from where it matters most – the people. Research is needed to determine whether, and to what extent, local production is improving access to essential medicines.

The medicines management cycle helps with ensuring that medicines are available, affordable and used appropriately. The focus is on the medicine. The WHO access framework goes beyond this cycle to address the wider issue of improving access for people such as Pretty. This framework is shown in Figure 15.2, with its components explained in Table 15.3.

Table 15.3 briefly describes each access component and summarises the WHO-recommended strategies to ensure equitable access to essential medicines.

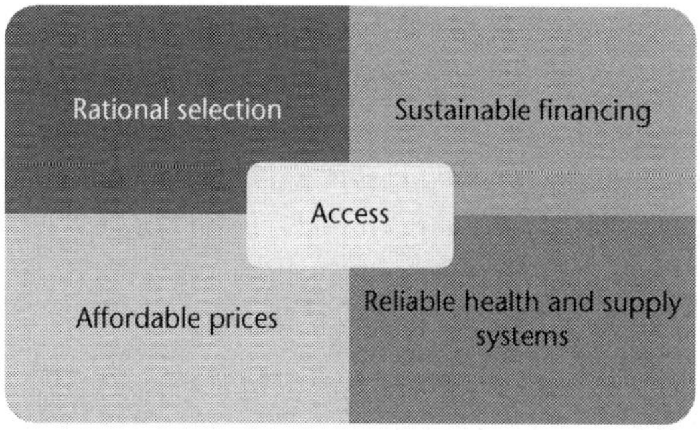

Figure 15.2 WHO Access Framework

Source: Bigdeli (2013).

Table 15.3 WHO-recommended strategies to increase access to essential medicines

Access component	*Recommended strategies*
Rational selection means choosing appropriate medicines for the health situation based on safety and cost-effectiveness.	• Development of treatment guidelines based on the best available evidence regarding efficacy, safety, quality and cost-effectives. • Development of essential medicines list based on these treatment guidelines. • Use of the EML for procurement, reimbursement, training, donations and supervision.
Sustainable financing through adequate funding levels and equitable payment options ensures that the poor are not facing higher costs for their medicines.	• Increasing public funding for health, including for essential medicines. • Reducing out-of-pocket spending, especially by the poor. • Expanding health insurance through national, local and employer schemes. • Targeting external funding. • Exploring other financing mechanisms such as debt relief and solidarity funds.
Affordable prices ensure that governments, providers and consumers can pay for their medicines.	• Using available and impartial price information. • Allowing price competition in the local market. • Promoting bulk procurement. • Implementing generics policies. • Negotiating equitable pricing for newer essential medicines for priority diseases. • Undertaking price negotiations for newly registered essential medicines. • Eliminating duties, tariffs and taxes on essential medicines. • Reducing mark-ups through more efficient distribution and dispensing systems. • Encouraging local production of essential medicines of assured quality when appropriate and feasible. • Including WTO/TRIPS-compatible safeguards into national legislation and applying these.

(continued)

Table 15.3 Continued

Access component	*Recommended strategies*
Reliable health and supply systems have two components: procurement and distribution. Reliable procurement and efficient distribution ensure that medicines are available at all levels of the healthcare system.	• Integrating medicines in health sector development. • Creating efficient public–private–NGO mix approaches in supply delivery. • Assuring quality of medicines through regulatory control. • Exploring various purchasing schemes and procurement cooperatives. • Including traditional medicines in healthcare provision.

Source: WHO (2017).

Even though countries are sovereign with their own governments, policies and practices, Africa also has a vision for regional integration and development. This is achieved through regional economic communities (RECs). The RECs are the Southern African Development Community (SADC), the East African Community (EAC) and the Economic Community of West African States (ECOWAS). North Africa is in the early stages of developing its regional integration strategy. Partner states that make up the regional economic communities chose which region to belong to (e.g. Tanzania belongs to both SADC and EAC). The RECs have regional headquarters and a secretariat that oversees regional integration activities, with funding coming from partner states and donors. Botswana hosts SADC in Gaborone, Tanzania hosts EAC in Arusha, and Nigeria hosts ECOWAS in Abuja.

The general approach to working together in RECs in Africa is to first recognise and respect the sovereignty of the individual partner states. This is followed by identifying areas of cooperation for mutual benefit. Secretariat staff assist countries to harmonise policies, practices and regulations, and promote information sharing and adoption of best practices for strengthening systems.

The sections below discuss the WHO access framework from the perspective of SADC.

Rational selection

Rational selection ensures that the right medicines are chosen for the setting, taking into account best evidence, epidemiology and available resources. In 2010, prior to the implementation of the Southern African Regional Programme on Access to Essential Medicines and Diagnostics (SARPAM), a baseline survey was undertaken of the 14 Southern African Development Community (SADC) member states.[2] Thirteen countries had standard treatment guidelines (STGs) and essential medicines lists (EMLs). Seven of these countries restrict public sector procurement of medicines to their EMLs. At the national level, SADC member states are following global recommendations for the selection and rational use of medicines by having NMPS, STGs and EMLs. While it is unclear whether common treatment guidelines are used across the member states, a 2010 review did find gaps in the quality and variable alignment with evidence for available guidelines for five priority diseases, including adult HIV (Kredo *et al.*, 2012). Member states in 2010 were not keen to have harmonised regional STGs and EML. Instead, countries preferred to have regional standards in place for the development of national STGs. However, countries did agree to having harmonised STGs in relation to specific diseases such as HIV and AIDS for regional initiatives such as the Global Fund Cross-Border Project.

The region has identified HIV/AIDS, malaria and TB as priority communicable diseases, and cardiovascular diseases, diabetes, asthma and some of the cancers as priority non-communicable diseases. While the ADB-funded Communicable Diseases Project has developed minimum standards for HIV/AIDS, malaria and TB, these do not focus on actual treatment. Countries are presently being supported to either adapt or adopt these minimum standards into national processes.

Sustainable financing

Data on pharmaceutical budgets and expenditure are generally difficult to obtain for African countries. Many countries are still reliant on donors for the supply of some of their essential medicines, especially the ARVs. For example, of the total US$3.1 billion disbursed by the Global Fund for SADC countries, 47 per cent has been spent on the procurement of essential medicines and supplies (SARPAM, 2012). Donors are also decreasing their support for the procurement of ARVs. The percentage of total PEPFAR funds that goes towards the supply of adult and paediatric ARVs decreased from 35 to 28 per cent in 2010 (Besada, 2012). SADC countries such as Namibia, Mozambique and Tanzania lost between 10 and 30 per cent of their total treatment budgets (Besada, 2012). Once donor funding starts to dry up, countries are finding that they cannot meet the demand for essential medicines. This results in poor availability at health facilities and increased out-of-pocket expenditure when medicines have to be purchased from the private sector; ultimately, it is the poor and vulnerable that suffer the consequences. At the same time, African leaders are being recognised for their efforts to mobilise resources for ARVs. Angola, Botswana and South Africa are financing more than 80 per cent of ARV treatments from domestic spending (UNAIDS, 2013).

All SADC countries have provisions allowing for free medicines for HIV and TB patients. Lesotho, Tanzania and Zimbabwe have implemented national health insurance, with Tanzania and Zimbabwe also charging dispensing fees at outpatient public health facilities.

Without proper data, it is unclear what countries are spending on essential medicines, nor are we clear on the total contribution from donors. What is clear is that both donors and member states are not meeting the Abuja target of increased funding for health and essential medicines.

Affordable prices

During the 2010 pharmaceutical market analysis (PMA) undertaken as part of the inception phase of the Southern African Regional Programme on Access to Medicines (SARPAM, 2010), attempts were made to assess the efficiency of national public sector procurement agencies by comparing the prices paid for a list of tracer medicines to international reference prices. Some key findings and conclusions were made:

- Agencies cannot easily provide price and other purchase information. Most of them are not used to sharing or analysing this information.
- Undertaking regional or international price comparisons are difficult across SADC as member states use different Incoterms.
- After adjusting for 'Free on Board' prices, the PMA found significant differences between SADC countries for public procurement prices for tracer items. For the majority of products, the ratio between the highest price among study countries and the lowest price is greater than 5 (i.e. one country is paying five times as much as another for the same products).

In 2012, SADC approved and adopted the regional strategy for the Pooled Procurement of Essential Medicines and Health Commodities. This strategy provides a phased approach to implementation, starting with information and work sharing and moving on to group contracting.

While the majority of SADC countries use generic medicines in the public sector, few have policies that explicitly promote their use. There is political commitment at the regional level for generics, with the pharmaceutical business plan promoting their use as well as supporting the local production of generic medicines for HIV/AIDS, malaria and TB. SADC is overly reliant on imported medicines for both patented and generics. For example, around 85 per cent of generic ARVs are imported from India, with the remaining 15 per cent produced by SADC countries, including South Africa, Zambia, Zimbabwe and Tanzania. At the same time, SADC member states are also pushing for the local production of essential medicines. For instance, all SADC member states have provisions in national legislation, giving priority to domestic manufacturers for public procurement. To date, the feasibility study on local production is still to be completed.

All SADC countries except for the Seychelles are signatories to the World Trade Organization (WTO) and must meet their obligations to the TRIPS Agreement, including introducing patent protection for pharmaceuticals. Patented medicines have been shown to be more expensive and are unaffordable for developing countries, especially when needed to treat diseases such as HIV/AIDS. In 2001, the DOHA Declaration of the TRIPS Agreement and public health was adopted by the WTO to address the public health concerns and the impact of access to affordable essential medicines. Flexibilities were identified where public health considerations can be used to govern the extent to which patents are enforced. SADC countries have been encouraged to include TRIPS flexibilities into national legislation to ensure affordability of essential medicines. The majority of SADC countries have done this, but evidence of actual use of these flexibilities remains scarce.

Affordability of medicines is generally directed at the patient within the community. SADC's regional initiatives tend to focus on national structures such as central medical stores. Little is known about what happens closer to the patient at facility level.

Reliable health systems

Well-functioning and reliable health systems integrate essential medicines into the overall health sector development. The situation across Africa is variable, with many countries suffering from unreliable supply systems. The impact of unreliable health systems was clearly evident in West Africa through the crisis created by the Ebola virus. Sierra Leone lost most of its capability and capacity to deliver any form of health service due to the disaster.

The median number of pharmaceutical personnel, including pharmacists, pharmaceutical assistants, pharmaceutical technicians and related occupations, in member states is 0.7 per 10,000 inhabitants (range 0.3 to 7) (SARPAM, 2012). While no global norm exists for this ratio, the median for SADC shows the serious shortages of qualified personnel in SADC countries for the pharmaceutical sector.

Four of 14 SADC countries do not have medicines regulatory authorities. Regionally, SADC medicines regulators have established the regulators forum, where information could be shared and discussions held regarding the registration and overall regulation of medicines. The regulators were supported by the NEPAD agency to draft a proposal for harmonisation of medicines registration as part of the African Medicines Registration Harmonisation Initiative. The draft proposal still needs to be discussed and approved by the SADC ministers of health.

Countries have varying levels of private–public partnerships regarding the supply of pharmaceuticals. In a few countries, the non-governmental sector also procures and supplies medicines. However, no regional platform currently exists for including both the private and not-for-profit sectors.

Conclusions: summary and way forward

The pharmaceutical sector of African countries encompasses research, products, trade, personnel and services – all linked in a complex and dynamic matrix of health, economic and political issues. There are many stakeholders involved, each with their own agenda, which do not always align with the overall public health goal of improving access to essential medicines and health commodities. Pretty often gets lost in these wider issues and agendas.

NMPs have been accepted by the majority of African countries as the policy instrument to guide the national pharmaceutical sector. In terms of regional harmonisation, existing NMPs of partner states allow for technical and regional cooperation.

There is strong support for the regional harmonisation agenda by partner states and partners. Two key gaps present the bottlenecks. First, none of the RECs have regional pharmaceutical policies. These policies are needed to guide the harmonisation process, as well as providing platforms for collaboration and cooperation in the pharmaceutical sector. Second, national and regional regulatory and enforcement mechanisms are missing, resulting in countries participating in regional initiatives but not always domesticating the regional decisions into national policy and regulatory instruments.

Some of the issues that contribute to Pretty not receiving her medicines at the time she needs them are due to the following:

- Countries having inappropriate institutional structures for policy direction and governance that fail to recognise and address the complexities of the manufacture, supply and use of pharmaceuticals. The Office of the Chief Pharmacist, generally tasked with the NMP, is not supported to fulfil this mandate.
- Outdated pharmaceutical laws that fail to provide the regulatory authorities with the legal instruments to address and adapt to new trends.
- Limited resources or political will to enforce pharmaceutical legislation.
- Policy development and implementation not evolving with other national and regional developments, resulting in underperformance and stagnation.
- Conceptualisation of pharmaceutical services limited to procurement and supply, and to some extent prescribing and dispensing.
- Lack of sustainable strategies for implementation, weak management and programing of pharmaceutical services.

African countries have a long way to go towards ensuring equity of access (for all our Prettys), strengthening institutions (to manage another Ebola-like crisis), and promoting accountability (to manage corruption that steals from our Prettys).

Notes

1 Normally, vaccines, blood products and other biologicals are considered to be within the framework of medicinal products for human use. This is also true for herbal products for human use.

2 SADC countries: Angola, Botswana, DRC, Lesotho, Malawi, Mauritius, Mozambique, Namibia, Seychelles, South Africa, Swaziland, Tanzania, Zambia and Zimbabwe. Madagascar, under expulsion in 2010, is now back in SADC.

References

African Union (2012) *Pharmaceutical Manufacturing Plan for Africa*, Addis Ababa: African Union.

Besada, D. (2012) 'PEPFAR problems: patients to pay the price', *NSP Review*, 2, available at: www.nspreview.org/download/NSP%20Review%202%20web.pdf (accessed 4 April 2017).

Bigdeli, M. *et al.* (2013) 'Access to medicines from a health systems perspective', *Health Policy and Planning*, 28: 692–704.

Business Monitor International (2007) *South Africa Pharmaceuticals and Healthcare Report Q3*, London: BMI.

Kredo, T., Gerritsen, A., van Heerden, J., Conway, S. and Siegfried, N. (2012) 'Clinical practice guidelines within the Southern African Development Community: a descriptive study of the quality of guidelines development and concordance with best evidence for five priority diseases', *Health Research Policy and Systems*, 10(1), DOI: 10.1186/1478-4505-10-1.

MSH (Managing Sciences for Health) (2012) *MDS-3: Managing Access to Medicines and Health Technologies*, Arlington, VA: MSH.

McPake, B., Hanson, K. and Mills, A. (1993) 'Community financing of health care in Africa: an evaluation of the Bamako Initiative', *Social Science and Medicine*, 36(11): 1383–1395.

Quick, J. (2003). 'Editorial: essential medicines twenty-five years on – closing the access gap', *Health Policy and Planning*, 18(1): 1–3.

SADC (Southern African Development Community) (2007) *SADC Pharmaceutical Business Plan (2007–2013)*, available at: www.sadc.int/themes/health/pharmaceuticals/ (accessed 28 December 2016).

SARPAM (Southern African Regional Programme on Access to Medicines) (2010) 'Pharmaceutical market analysis, volumes I and II', unpublished report, Johannesburg: SARPAM.

SARPAM (Southern African Regional Programme on Access to Medicines) (2012) 'SADC pooled procurement of essential medicines and medical supplies: situational analysis and feasibility study', unpublished report, Johannesburg: SARPAM.

The Lancet (2016) *Essential Medicines for Universal Health Coverage: The Lancet Commission on Essential Medicines Policies*, London: The Lancet.

UN (United Nations) (2000) *United Nations Millennium Development Goals*, New York: United Nations.

UN (United Nations) (2015) *Sustainable Development Goals*, New York: United Nations.

UNAIDS (Joint United Nations Programme on HIV/AIDS) (2013) *UNAIDS Report on the Global AIDS Epidemic*, available at: www.unaids.org/en/resources/documents/2013/20130923_UNAIDS_Global_Report_2013 (accessed 8 January 2018).

WHO (World Health Organization) (1978) *Declaration of Alma Ata*, available at: www.who.int/publications/almaata_declaration_en.pdf (accessed 5 December 2016).

WHO (World Health Organization) (2001) *How to Develop and Implement a National Drug Policy*, Geneva: WHO.

WHO (World Health Organization) (2011a) *The World Medicines Situation*, Geneva: WHO.

WHO (World Health Organization) (2011b) *Country Pharmaceutical Profiles*, available at: www.who.int/medicines/areas/coordination/coordination_assessment/en/index1.html (accessed 1 February 2017).

WHO (World Health Organization) (World Health Organization) (2017) *Essential Medicines and Health Products Information Portal*, available at: http://apps.who.int/medicinedocs/en/ (accessed 8 January 2018).

Section 5
Environment

16

Water resources and development

Clive Agnew

Introduction: scarcity and surplus

Africa has extensive water resources that experience much lower levels of exploitation than the rest of the world, but the people of Africa continue to suffer from water scarcity. The UN announced that as we approached the end of the UN Decade for Water 2005–2015, there remained many challenges worldwide, with 700 million people in 43 countries suffering from water scarcity, forecast to worsen such that by 2030, half the world's population will be living under high water stress, including up to 250 million in Africa. Sub-Saharan Africa has the largest number of water-stressed countries of any region (UN DESA, 2014). Yet the FAO (2014) estimates that only 5.5 per cent of renewable freshwater resources are actually withdrawn in Africa set against a global average of 8.8 per cent and 21.5 per cent for Asia. The levels of exploitation are uneven, and withdrawals are increasing through the demands made by a growing urban population and intensification of irrigation. UNEP (2008) estimates world increases in water withdrawals of 1.38 times between 1995 and 2025, but for Africa this is expected to be in excess of 1.5.

Africa's average annual precipitation of 677 mm ranges from only 96 mm in northern Africa to 1,425 mm in central Africa, set against a world average of 814 mm. We have known for two decades that the annual renewable water resources of around 4,000 km^3 is derived from just 20 per cent of this precipitation (FAO, 1995: 22). Aquastat data sheets calculate that this yields 4,527 $m^3c^{-1}y^{-1}$, ranging from 20,000 in central Africa to 325 in northern Africa (FAO, 2014). Given that Howard and Bartram (2003) suggest an average household water consumption of at least 40 $m^3c^{-1}y^{-1}$ would be sufficient to meet all health and hygiene requirements, while Twort *et al.* (2000) note in major cities a consumption of 200 $m^3c^{-1}y^{-}1$, we must ask why many households in Africa are facing water shortages?

Figure 16.1 shows that the water resources for Africa are much less than other continents, and many regions of Africa face the problems of water supply through aridity, drought and uncertain surface flows. Northern Africa and the Sudano-Sahelian regions are comparable to the levels of water scarcity found in the Middle East, while central Africa has abundant resources, equivalent to the USA. Furthermore, when water resources are compared with demand through renewable fresh water per person, Figure 16.2 demonstrates Africa's relative water paucity. There is a marked contrast between withdrawal rates in excess of 100 per cent of supply in the arid North

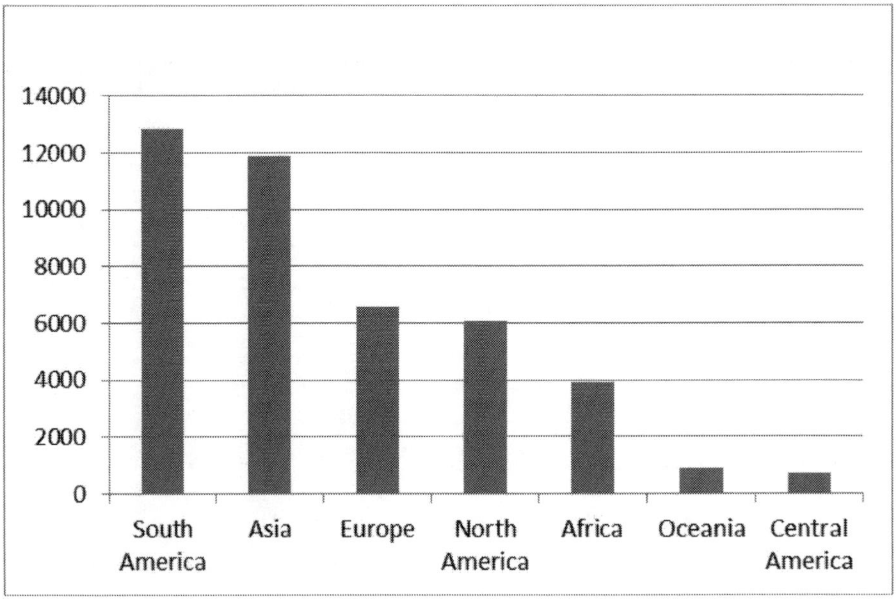

Figure 16.1 Renewable freshwater resources (km^3y^{-1})

Source: FAO (2014).

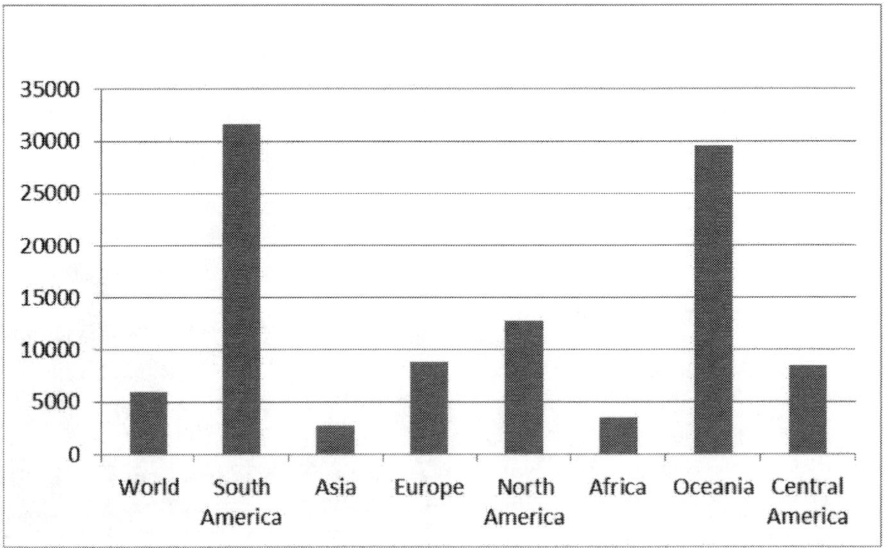

Figure 16.2 Freshwater per capita ($m^3c^{-1}y^{-1}$)

Source: FAO (2014).

Table 16.1 Africa's major river basins

	Area	*Density*	*Large cities*	*Water supply*	*Dams in basin*	*Dams in basin*
Name	*km²*	*Person per km²*	*>100,000*	$m^3cap^{-1}yr^{-1}$	*>15 m*	*>150 m*
Congo	3,730,881	15	18	22,752	11	0
Nile	3,254,853	46	25	2,207	11	0
Lake Chad	2,497,738	11	6	7,922	0	0
Niger	2,261,741	31	12	4,076	30	0
Zambezi	1,332,412	18	6	10,000	12	1
Orange	941,351	10	4	1,050	37	1
Okavango	721,258	2	1	N/A	1	0
Jubba	497,626	12	1	1,076	0	0
Limopopo	421,123	32	4	716	25	0
Senegal	419,575	10	1	5,775	2	0
Volta	407,093	43	3	2,054	3	0
Shaballe	336,604	29	1	1,700	0	0

Source: FAO (2007).

compared to much lower rates in central, eastern and southern regions. A continent of water resource contrasts is caused by uneven distributions of surface and subsurface supplies.

Africa hosts some of the world's major river basins. The FAO (2007) provides details on 19 river basins, of which even the smallest are comparable to those regarded as major in Europe. Details of the 12 largest listed in Table 16.1 reveals the variation in size, relative population and exploitation. At a time when the construction of mega dams (see below) has dominated discussions of water resources development, it is clear that many of Africa's rivers have comparatively small amounts of dams. For example, the Mississippi River basin in the USA covers 3 million km^2, so is comparable with Africa's largest basins, but in this single basin there are 63 small dams and four mega dams. This is rapidly changing. There are then opportunities for the development of surface waters, but the challenges of securing investment and effective management are compounded by the international nature of Africa's water resources. Goulden *et al.* (2010) report that each of the Congo, Niger, Nile, Zambezi and Lake Chad river basins are shared by at least eight countries, and 30 are shared by at least two countries.

Over 90 per cent of the world's fresh water is subsurface (UNEP, 2008), but it has proved difficult to get reliable regional assessments of this resource and their current recharge rates, which has hampered assessments of climate change impacts, for example. A recent initiative to address this shortcoming (MacDonald *et al.*, 2012) to map the hydrogeology of Africa notes the total volume of groundwater lies between 0.36 and 1.75 million km^3, which they believe is 100 times more than annual renewable freshwater resources and 20 times greater than the water stored in Africa's lakes (see Figure 16.3). As with surface resources, these are unevenly distributed, but there are extensive aquifers in Libya, Algeria, Egypt and Sudan, while local boreholes and springs can be highly signifcant in supporting local communities. They note that many of the large seditmentary aquifers in northern Africa are not currently being recharged, and the water currently being extracted may be linked to a wetter period some 5,000 years ago.

A threshold of less than 1,000 $m^3c^{-1}y^{-1}$ is used to denote water scarcity. When applied to Africa, this includes the arid areas of North Africa and Sudano-Sahel, but the list of water-scarce countries also includes Kenya, Burkina Faso and even South Africa (FAO, 2014). There is a danger with such aggregations, and if we look to the countries where water supply per person is in excess of the water stress threshold of 1,700 $m^3c^{-1}y^{-1}$, this not only includes the expected

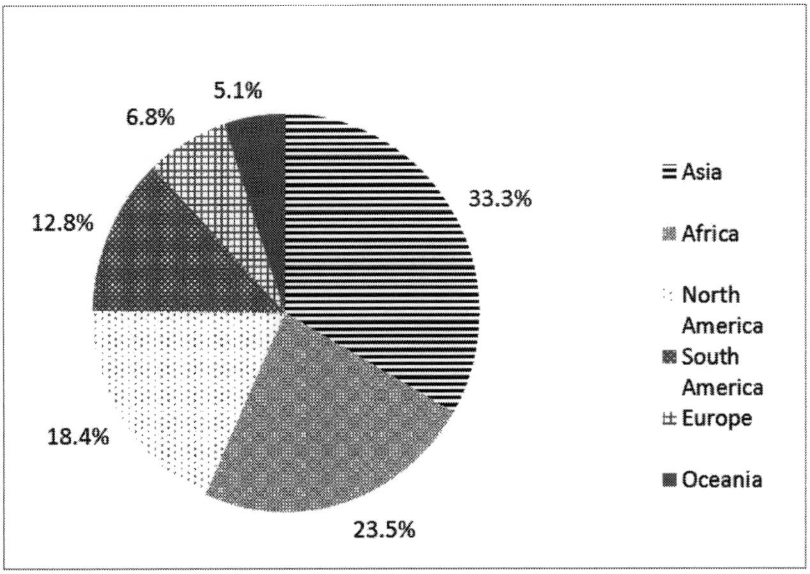

Figure 16.3 Distribution of groundwater resources

Source: UNEP (2008).

Congo and Gabon, but also those in semi-arid regions such as Chad, Mali, Mauretania and Senegal. Gassert *et al.* (2014) use 12 indicators, including river basin seasonal flows and sector water withdrawals, to generate maps of water risk that illustrate the huge variations across the continent and within countries with the highest risk in northern, eastern and south-western regions of Africa balanced by surplus central and western regions.

It is worth noting that following Allan's (2011) work on virtual water, the concept of a 'water footprint' has been developed to include the total volume of fresh water used to produce the goods and services consumed by the individual or community or produced by the business (Water Footprint Network, n.d.). Hence, Africa is a huge exporter of water to other, wetter parts of the world through the agricultural goods it trades globally. In Europe, over two-thirds of the national water footprint is generated by imported goods, while in Africa it is much lower (e.g. 29 per cent of Egypt's footprint is external, and for South Africa 22 per cent, Kenya 17 per cent, and Congo 3 per cent).

Even though the total utilisation of renewable waters at a continental level is low, there are examples of insufficient supply to meet demand, which is predicted to become worse. Of the UN's (2013) estimate of the 800 million who live on the African continent, more than 300 million live in a water-scarce environment. This is not simply a question of climate (i.e. aridity), but also the management of resources in the face of competing and growing demand. For example, 5 per cent of total water withdrawals are for industry, 13 per cent is extracted for municipal centres and a staggering 82 per cent is required by agriculture (FAO, 2014). This publication noted the area equipped for irrigation grew by 70 per cent between 1970 and 2012, with the highest rate of increase in southern Africa and the Indian Ocean islands. There have been many attempts to exploit these resources during the twentieth century, but if successful these have often been a short-lived success, as described decades ago by Therkildsen (1988) and Adams (1992). The challenge is not just to understand the variation in water resources, but also to find strategies that facilitate sustainable use and governance that engage with the population.

The African water crisis

Changing focus from volumes of water to how it supports health, well-being and economic development, it is interesting to note the recognition of the key role water plays in sustainable development has grown since only one paragraph was devoted to the topic in the 'Brundtland' report (UN, 1987). Today's perspective is summarised by Ban Ki-moon (WWDR, 2015):

> Water issues have risen in prominence in recent years, reflecting growing understanding of water's centrality as well as the world's success in achieving the Millennium Development Goal target of halving the proportion of people without sustainable access to safe drinking water.

Despite this optimistic statement by the then-UN Secretary-General, there continue to be numerous appeals to deal with a growing global water crisis. The WWC (2010) explain:

> While the world's population tripled in the 20th century, the use of renewable water resources has grown six-fold. Within the next fifty years, the world population will increase by another 40 to 50%. This population growth – coupled with industrialization and urbanization – will result in an increasing demand for water and will have serious consequences on the environment.

We have already noted the important role for water in Africa due to the expansion of irrigation, but rivers can provide hydroelectric power (e.g. new dams on the Nile in Uganda and Ethiopia), while potable water is key to human health and well-being. The adoption of water filtration and chlorination led to a 50 per cent increase in USA life expectancy during the twentieth century (Cutler and Miller, 2005). Newsom (1994) noted improvements in safe water supply and sanitation could reduce water-related infectious diseases by 80 to 90 per cent, although Gleick (1993) cautioned that water developments can also result in an immediate increase in infectious diseases. The need to take into account safe sanitation alongside safe drinking water can be exemplified by the Kanyike Project (Agnew and Woodhouse, 2011) in rural Uganda (see Table 16.2). Contamination of pond water is to be expected, but the high levels of contamination of the spring sources can be attributed to latrines in close proximity, whereby human waste is leaching into the water source.

Table 16.2 Water quality for pond collection point and spring source Kanyike, Uganda (faecal coliform per 100 ml)

Spring	*Pond*	*Spring source*
Kakoma	7,000	27
Kanyike	2,000	139
Sowola	228	2
Nakasozi	119	9
Nakamebe	19	15

Source: Agnew and Woodhouse (2011).

A combined approach to safe water and sanitation was evident in the UN International Drinking Water Supply and Sanitation Decade (1981–1990), and more recently UNICEF (2016) have strengthened awareness of the essential links between water, sanitation and hygiene through their WASH programme.

Recognising the need to address the water crisis at the start of the twenty-first century, the UN identified a number of Millennium Development Goals (MDGs), which included Target 7C: to halve, by 2015, the proportion of the population without sustainable access to safe drinking water and basic sanitation. The UN report notes the world has met this target for drinking water, and between 1990 and 2015 2.6 billion people gained access to improved drinking water sources, 1.9 billion gained access to piped drinking water on premise, and over half of the global population (58 per cent) now enjoys this higher level of service (UN, 2015). There was also progress on safe sanitation, with 2.1 billion more people having access to improved sanitation, yet 2.4 billion are still using unimproved sanitation facilities, including 946 million people who are still practising open defecation. Many of the countries not meeting MGD targets are to be found in Africa (UNICEF and WHO, 2015).

Figure 16.4 shows the regional variations in the numbers of people in Africa who lack access to safe drinking water and sanitation. Only four (out of 48) African countries have piped drinking water for more than 70 per cent of their population (Algeria, Egypt, Mauritius and South Africa), although this proportion rises to two-thirds when improved water sources are included. Note that this refers to delivery and not necessarily water quality, so the supply of potable water is probably overestimated. Still over half of African countries have a third of their population reliant on surface or unimproved water supplies, those in the worst position being the Democratic Republic of Congo, Ethiopia, Madagascar, Mozambique and Somalia. There is no simple correlation between safe water and sanitation with aridity, and the starkest contrasts in water access are between urban and rural inhabitants.

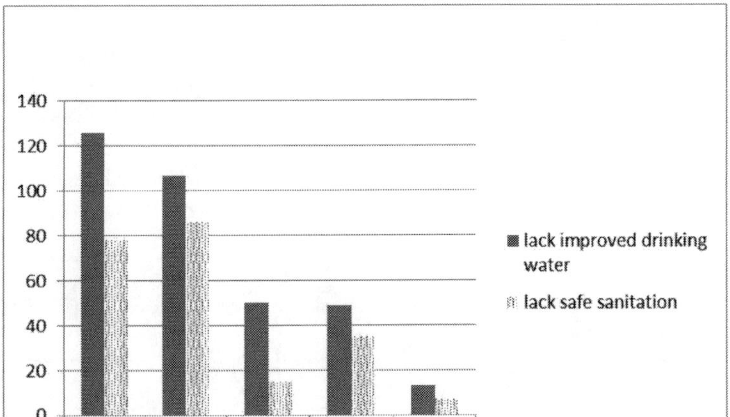

Figure 16.4 Regional populations (millions) lacking safe water or sanitation

Source: UNICEF and WHO (2012).

The WHO/UNICEF Joint Monitoring Programme for Water Supply and Sanitation (AMCOW, 2012) estimates that since 1990, 322 million Africans gained access to an improved drinking water source and 189 million gained access to an improved sanitation facility. Yet Figure 16.5 shows many still lacking access, in particular in rural areas, with 65 million more people in Africa lacking access to an improved drinking water source in 2010 than in 1990. The population, without an improved sanitation facility, increased by 197 million since 1990. While the provision is worse in rural than urban areas, the urban provision is still inadequate as 25 per cent of urban dwellers rely on shared/public sanitation, while 20 per cent practise open defecation. Open defecation is found for a third of the population in a third of countries. Those with the lowest provision are Niger, Chad, Burkina Faso, Benin and Mauritania.

The challenges to provide safe water and sanitation are daunting, but there has been progress, albeit uneven, with at least 10 countries making some improvement in sanitation. UNICEF and WHO (2015: 13) further note:

> In sub-Saharan Africa, access to sanitation is highly correlated with wealth and residence in urban areas . . . The poorest 60% of the population in sub-Saharan Africa are largely denied the comforts and health benefits of a piped drinking water supply on premises.

The key point is access to water and sanitation infrastructure rather than supply or climate and technology (i.e. the prominence of social and economic inequalities when assessing the provision of safe water and sanitation).

It is evident that progress in Africa towards safe sanitation has been slower than safe drinking water, and this, coupled with increasing water scarcity, has led to the belief that Africa is facing a water crisis. The idea of a global water crisis is not new, and two decades ago Postel (1992) argued for greater emphasis to be placed on management and conservation of existing resources, a point reiterated more recently in the third UN World Water Development Report (WWDR, 2009):

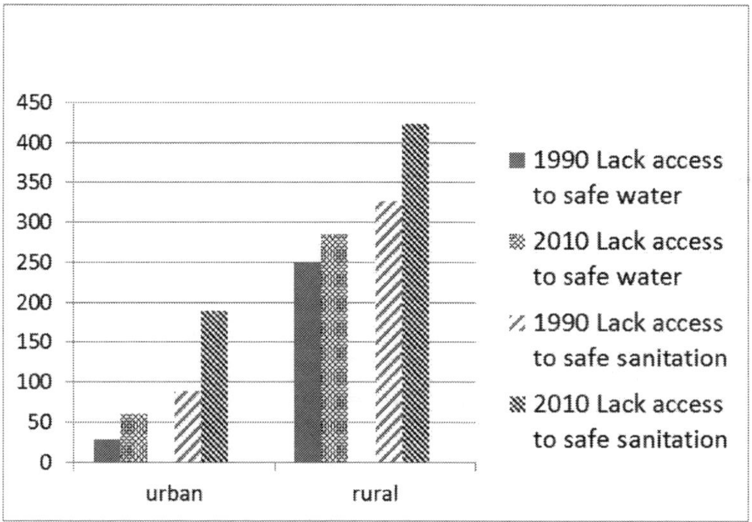

Figure 16.5 Urban and rural populations (millions) lacking safe water or sanitation across Africa

Source: UNICEF and WHO (2012).

> At the beginning of the twenty-first century, the Earth, with its diverse and abundant life forms, including over six billion humans, is facing a serious water crisis. All the signs suggest that it is getting worse and will continue to do so, unless corrective action is taken. This crisis is one of water governance, essentially caused by the ways in which we mismanage water.

In the next section, we will examine the strategies that have evolved to develop water resources and use the growth in dams and irrigation to exemplify some of the challenges facing Africa.

Water and development

Water is bulky, supply is often seasonal, and it has questionable economic value until it becomes scarce. Harnessing water resources requires a degree of collaboration and investment; hence, development of water is interlinked with economic and social development. For OECD countries, Swyngedouw (2006) identified a series of transitions from local to state-driven water developments:

- *Up to 1850s*: Small private (commercial) water supply companies.
- *Up to 1920s*: Municipal bodies investing in water and sanitation measures due to health and environmental concerns.
- *Up to 1970s*: Large-scale, national control of water investment and regulation with goals of economic development and social welfare improvement.
- *Post-1970s*: Period of privatisation and decline of state-led economic development.

Africa continues to witness centralised state-driven initiatives, but the phase of privatisation has been forecast rather than observed. The UN (1987) recognised 20 years ago that environmental, economic and social issues are interlinked. In addition to integrated water resources management, the UN Millennium Development Goals identified five additional key themes underpinning current approaches to water development (Moriarty and Butterworth, 2003), encompassing demand management, cost recovery, decentralisation, community management and equity. This list emphasises the move away from regarding water resources development as a series of solely technological challenges, as critically reported by Postel (1992) as 'engineering's promise'. Across the world, there is a growing emphasis upon demand management, for example in the face of climatic change, but in Africa the move from supply enhancement to demand management has yet to take place, which is partly explained by poorly developed infrastructure and two-thirds of people living in rural areas.

UNECA (2006) identified three phases in responding to water scarcity in Africa, an initial emphasis upon storage and extraction where engineering dominates, then a growing attention for strategies that seek to conserve where efficiency is stressed, and finally greater emphasis upon the value of water for society, often with an urban focus. All of these can be seen in some form across Africa, but it is argued (Agnew and Woodhouse, 2011) that the Dublin 1992 Water Conference presented the key principles upon which water management is based, subsequently reinforced at the Rio Earth Summit in the same year:

1. The ecological principle recognises the need for holistic (integrated) management of the water resources, often driven by a state or central controlled approach where science and hydrological information is key for decision-making.
2. The instrument principle recognises water as having economic value, supports a neo-liberal market-driven approach whose resurgence in the 1980s coincided with the concept of

'sustainable economic development', and desires to switch away from state-sponsored projects with more private investment.

3 The institutional principle calls for decision-making on water resources to be decentralised to the smallest scale feasible. Following the criterion of subsidiarity, importance is attributed to participatory decision-making processes, including the role of women.

Across the twentieth century, Africa's adherence to these principles can be found in different forms, which also reflects UNECA's (2006) approach, with an initial emphasis upon using science and engineering to harness water resources and growing state and 'top-down' organisation as scale increases from local communities to the river basin. Inevitably, the investment for large-scale developments increases the need for benefits assessment, alternative finance models and efficiencies. It is also possible to recognise a newly emerging alternative to the state-driven approach and for local communities to be engaged with stakeholders. This can be exemplified through examination of Africa's attempts to increase water supply for irrigation.

Water supply enhancement: dams and irrigation

Africa has a long history of water resource utilisation for agriculture, from traditional swamp rice cultivation, run-off harvesting, flood recession farming on the Nile, and even in northern Africa subsurface qanats for irrigation. Bernstein et al. (2000) argue that these traditional practices can yield higher returns than many more modern interventions. It is widely asserted that today, Africa lags behind the rest of the world in terms of harnessing surface water resources via dam construction and irrigation expansion. The WCOD (2000) reports that around 3,800 km^3 of fresh water is withdrawn annually from the world's lakes, rivers and aquifers, which is twice the volume extracted 50 years ago. There were 5,000 large dams in 1949, but by 2000 this had increased to 45,000, but mainly in Asia (notably China), the Americas and Europe, with Africa well behind. There are some notable exceptions. Agnew and Woodhouse (2011) describe the expansion of dams along the Nile following barrages in 1861, to be followed 40 years later by the Aswan Dam, raised in 1912, and the Sennar Dam in the 1920s, leading to the vast Gezira irrigation scheme, which at 880,000 ha is one of the largest in the world (World Bank, 2000). The majority of dams have been built for irrigation, but this is also much less developed in Africa than for other continents. The FAO (2007) estimates that 70 per cent of Africa's irrigation is in just five countries, mostly in northern Africa. In sub-Saharan Africa (with the exception of South Africa), irrigation is weakly developed, with just 3.5 per cent of the cultivated area being irrigated, compared to over 40 per cent in South Asia. But this is changing. A twenty-first-century 'African land grab' has been reported, which Laishley (2014) describes as a second scramble for Africa, whereby foreign investors are buying up land for agriculture, but The Economist (2016) suggests in some parts it is the African wealthy urbanites who are turning to farming investments. New dams are being constructed across the continent, notably on the Nile, but these are still hugely reliant upon external funding, from China in particular.

Since the nineteenth-century and early twentieth-century developments on the Nile, there have been a number of major projects across Africa during the latter half of the twentieth century, mostly for irrigation or urban demand, although HEP from these dams produces 80 per cent of the total power generation in 19 African countries (FAO, 2007). African dams are normally of a modest height, often below the 150 m threshold used to denote 'mega dams', but they impound some of the largest lakes in the world. The Kariba Dam (Zimbabwe) is reported to impound a lake of 5,580 km^2 and a volume of 188 billion m^3, making it the largest in the world, alongside the Akosombo Dam on the Volta in Ghana, which created a lake of 148 billion m^3, and the Aswan Dam, 162 billion m^3. These dams also provide HEP, 1,470 MW for Kariba and 2,100 MW for Aswan (Water Technology, 2016). The FAO (2007) estimates that there are now around 1,300 large to medium-sized African dams, but 40 per cent are in South Africa, and just 53 of these large dams account for 90 per cent of Africa's total volume, revealing an uneven distribution.

Despite a century of large-scale dam building across the world, Africa is seen to have untapped potential. It only uses 5.5 per cent of renewable water resources, and for sub-Saharan Africa this is much less. The Economist (2010) noted a number of new dam building projects with significant investments (e.g. Merowe Dam in Sudan costing an estimated US$1.8 billion and GIBE III Dam in Ethiopia at US$2 billion). International Rivers (2010) lists a further 20 African countries with dam building projects supported by funding from China. This accelerated development comes at a time when some are questioning the costs and benefits of large dams (Altinbilek, 2002; Biswas, 2004). Reviewing progress, the WCOD (2000: XXVIII) noted the substantial contributions dams can make to human development, but balanced this with a critical assessment: 'In too many cases an unacceptable and often unnecessary price has been paid to secure those benefits, especially in social and environmental terms, by people displaced, by communities downstream, by taxpayers and by the natural environment'. The adverse impacts have led to predictions at the end of the twentieth century that the construction of large dams was at an end. Agnew and Woodhouse (2011) summarised the environmental costs, which include downstream impacts on flow regime, upstream increases in sedimentation resulting in reduced dam longevity, health impacts locally due to open water bodies, evaporation and seepage losses, and even possible increased seismic activity. There are also social costs caused by the disruption through resettlement of peoples displaced, and often the lack of benefits to local population cause resentment and lack of participation, leading to further alienation.

Yet a few years ago, the *New Scientist* reported that 'Mega-dams were back on the agenda' (Pearce, 2006), with the World Bank identified as now supporting this strategy. Despite the concerns for adverse impacts of large dams, the need for hydropower in a continent of low fossil fuel resources, the demand from a growing urban population and the need to combat desiccation due to climate changes mean that dam construction is being progressed alongside irrigation expansion.

It has already been established that much of Africa's water is withdrawn for agriculture and dams are built mainly for irrigation, but in 2014 only 4.9 per cent (14.74 million ha) of the world's irrigated land was found in Africa (ICID, 2016). The FAO (2014) estimates that the total potential for irrigation in Africa is 38 million ha, but only 7.7 million ha is currently being used, so there is potential for growth. The distribution is uneven, with most irrigated land in Egypt (3.7 million ha), followed by South Africa (1.6) Morocco (1.5) Algeria (0.5) and Tunisia (0.5). The lowest proportion of land that is irrigated is found in sub-Saharan Africa, with only 3 per cent being used, compared to 21 per cent globally.

The spatial growth of the irrigation area in Africa at the end of the twentieth century was modest (see Figure 16.6) compared to some other continents, but the rate of increase, 41 per cent between 1980 and 2003 (FAO, 2016), was comparable to the world average, albeit from a low base with an annual rate of growth in 1961–2000 of 2 per cent for sub-Saharan Africa (World Bank, 2009), compared to 1.8 per cent for the world, 1.4 per cent in China and 2.6 per cent in India.

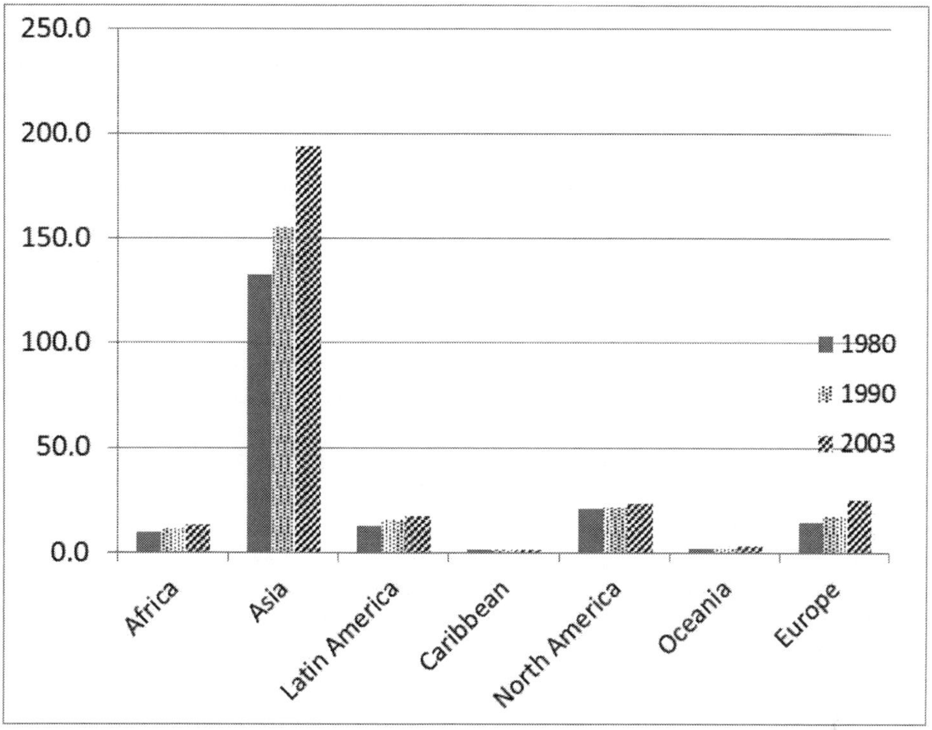

Figure 16.6 Irrigation area (million ha), 2006

Source: FAO (2016).

The potential that irrigation offers is demonstrated by irrigated cereal yields being 60 per cent higher than rain-fed crops (Rosegrant *et al.*, 2010). The same authors argued that future food production will depend to a large extent on irrigation, with a substantial increase in cereal demand by 2025, particularly in developing countries.

The FAO (2016) notes that only 85 per cent of the world's 324 million ha equipped for irrigation is actually being used. Irrigation is notorious for using water in a profligate manner, with low water use efficiencies for flood and sprinkler applications. Across the world, 56 per cent of water withdrawn is actually used, while in sub-Saharan Africa this falls to only 28 per cent. The World Bank (2000) reports that the Gezira scheme used 35 per cent of the Nile's flow, but with an efficiency of only 50 per cent. The reality is that irrigation is 'thirsty', e.g. 1 kg of cereals requires around 1,600 kg of water, while the water footprint for pulses is much higher, and meat products can be 5 to 10 times more (Mekonnen and Hoekstra, 2011). It is worth noting that water withdrawn for domestic and industrial use is largely recycled, but this is not the case for agriculture, where consumption is much closer to extraction.

Examination of the twentieth-century development of dams and irrigation across Africa reveals an emphasis upon 'top-down' enterprises driven by revenue generation with secondary goals of sustainable development. This exemplifies adherence to the ecological and instrument principles for water development, with limited attention being paid to impacts on local communities and the environment (Brautigam *et al.*, 2015). Ansar *et al.* (2014) also conclude the costs for hydropower dam construction are underestimated and the benefits exaggerated. Africa is then in danger of heading along the same path that led to cessation of funding for major projects in other parts of the world at the end of the twentieth century. The WWDR (2015: Chapter 7) notes:

> By 2050, agriculture will need to produce 60% more food globally, and 100% more in developing countries . . . Current growth rates of agricultural demands on the world's freshwater resources are unsustainable. Inefficient use of water for crop production depletes aquifers, reduces river flows, degrades wildlife habitats.

There are opportunities to use water more efficiently, and the same publication exemplifies how deficit irrigation practices can lead to substantial water savings of 25 per cent, while Kenny et al. (2009: 44) note that US water withdrawals for irrigation peaked in the 1980s and since then have remained fairly stable. The demand for more dams and more irrigation in Africa does not, however, appear to be diminishing.

A changing approach to water in Africa?

At a time when the threats posed by climate change are most significant for those living in Africa, the continent is witnessing an unprecedented increase in population, most notably in urban areas with a similar increase in the demand for food. As a consequence, huge resources are being invested across Africa in dam building, expansion of irrigation and provision of HEP, which should herald economic growth. Yet each year, more households are facing water scarcity and more of Africa's population lacks access to safe sanitation.

In the developing world, Africa has experienced the highest urban growth rate of nearly 4 per cent during the last two decades, but its 335 million urban population represents just one-third of the total (Henderson et al., 2014). The AfDB (2012) reports that the proportion of those who live in urban areas is forecast to increase to 60 per cent by 2050, but adds that 'urbanization in Africa has failed to bring about inclusive growth which, in turn, has resulted in proliferation of slums, urban poverty and rising inequality'. Increased poverty and growing inequalities as manifest through access to water has been evident in other parts of the world. Writing about Mumbai, Gandy (2008: 122) notes, 'array of technological, scientific and architectural innovations . . . enable wealthy households to insulate themselves from the environmental conditions of the poor'.

This need for greater inclusivity and recognition of growing inequalities has led to a rise in interest in the adoption of the institution principle for water resources development. Poff *et al.* (2003) report that an alternative model of collaboration between scientists, managers and other stakeholders is emerging around the world. Lane *et al.* (2011) argue for 'doing flood risk science differently'. They discount the top-down approach epitomised by the public education and public debate models of stakeholder engagement, instead promoting the co-production of knowledge. This is being applied to examine the water ecosystems for Dar es Salaam (Roy, 2016), but it is unlikely that this move to greater community engagement will simply replace the 'engineer's promise'. Where stakeholder engagement has been attempted, Brown (2013) refers to a move away from top-down 'technocratic management', with a shift to greater stakeholder participation and decentralisation, and yet concludes for South Africa:

> There has been little change in the redistribution of water rights in the Inkomati that is the result of participatory processes. Although it is fair to say that participation has aided discourse between blacks and whites, its potential to transform inequalities in society is nothing but a pipedream . . . This encourages stability rather than change.
>
> *(Brown, 2013: 278)*

There is only limited evidence of the use of the institutional principle in water supply enhancement. This is not unique. Ortolano and Cushing (2002) examined the lessons of the Grand Coulee Dam in Washington state 70 years after its construction, and pointed to failed attempts to deal with the impacts on local communities. The recent history of water resources development in Africa is dominated by engineering solutions and state-driven development for national priorities. This is yet to be replaced by wider consideration of environmental and social costs, efficiency gains and sustainable funding.

References

Adams, W.M. (1992) *Wasting the Rain: Rivers, People and Planning in Africa*, London: Earthscan.

AfDB (African Development Bank) (2012) *Urbanization in Africa*, available at: www.afdb.org/en/blogs/afdb-championing-inclusive-growth-across-africa/post/urbanization-in-africa-10143/ (accessed 19 December 2017).

Agnew, C.T. and Woodhouse, P. (2011) *Water Resources Development*, London: Routledge.

Allan, T. (2011) *Virtual Water: Tackling the Threat to Our Planet's Most Precious Resource*, London: I.B. Tauris.

Altinbilek, D. (2002) 'The role of dams in development', *International Journal of Water Resources Development*, 18(1): 9–24.

AMCOW (African Ministers' Council on Water) (2012) *A Snapshot of Drinking Water and Sanitation in Africa: 2012 Update*, WHO/UNICEF Joint Monitoring Programme for Water Supply and Sanitation, available at: www.wssinfo.org/fileadmin/user_upload/resources/Africa-AMCOW-Snapshot-2012-English-Final.pdf (accessed 19 December 2017).

Ansar, A., Flyvbjerg, B., Budzier, A. and Lunn, D. (2014) 'Should we build more large dams? The actual costs of hydropower megaproject development', *Energy Policy*, DOI: http://dx.doi.org/10.1016/j.enpol.2013.10.069i.

Bernstein, H., Woodhouse, P. and Hulme, D. (2000) *African Enclosures? The Social Dynamics of Wetlands in Drylands*, Oxford: James Currey.

Biswas, A.K. (2004) 'Dams: cornucopia or disaster?', *International Journal of Water Resources Development*, 20(1): 3–14.

Brautigam, D. Hwang, J. and Wang, L. (2015) *Chinese-Financed Hydropower Projects in Sub-Saharan Africa*, Policy Brief, SAIS China Africa Research Initiative, available at: www.internationalrivers.org/campaigns/chinese-dams-in-africa (accessed 19 December 2017).

Brown, J. (2013) 'Can participation change the geography of water? Lessons from South Africa', *Annals of the Association of American Geographers*, 103(2): 37–41.

Cutler, D. and Miller, G. (2005) 'The role of public health improvements in health advances: the twentieth-century United States', *Demography*, 42(1): 1–22.

FAO (Food and Agriculture Organization) (1995) *Water Resources of African Countries: A Review*, Rome: FAO.

FAO (Food and Agriculture Organization) (2007) *Dams in Africa*, Rome: FAO, available at: www.fao.org/3/a-bc815e.pdf (accessed 18 December 2017).

FAO (Food and Agriculture Organization) (2014/2016) *Aquastat Global Water Recourses Database*, Rome: FAO, available at: www.fao.org/nr/water/aquastat/main/index.stm (accessed 18 December 2017).

Gandy, M. (2008) 'Landscapes of disaster: water, modernity, and urban fragmentation in Mumbai', *Environment and Planning A*, 40: 108–130.

Gassert, F., Luck, M., Landis, M., Reig, R. and Shiao, T. (2014). *Aqueduct Global Maps 2.1: Constructing Decision-Relevant Global Water Risk Indicators*, Working Paper, Washington, DC: World Resources Institute, available at: www.wri.org/publication/aqueduct-globalmaps-21-indicators (accessed 18 December 2017).

Gleick, P.H. (1993) *Water in Crisis*, Oxford: Oxford University Press.

Goulden, M., Conway, D. and Persechino, A. (2010) 'Adaptation to climate change in international river basins in Africa: a review', *Hydrological Sciences Journal*, 54(5): 805–828, DOI: 10.1623/hysj.54.5.805.

Henderson, J.V., Storeygard, A. and Deichmann, U. (2014) *50 Years of Urbanization in Africa: Examining the Role of Climate Change*, World Bank Policy Research Working Paper 6925.

Howard, G. and Bartram, J. (2003) *Domestic Water Quantity, Service Level and Health*, Geneva: WHO.

ICID (International Commission on Irrigation and Drainage) (2016) *The International Commission on Irrigation and Drainage Data Base*, available at: www.icid.org/icid_data.html (accessed 19 December 2017).

International Rivers (2010) *Africa Dams*. Berkeley, CA: International Rivers, available at: www.internationalrivers.org/sites/default/files/attached-files/afrdamsbriefingjune2010.pdf (accessed 19 December 2017).

Kenny, J.F., Barber, N.L., Hutson, S.S., Linsey, K.S., Lovelace, J.K. and Maupin, M.A. (2009) *Estimated Use of Water in the United States in 2005*, U.S. Geological Survey Circular 1344, available at: http://pubs.usgs.gov/circ/1344/pdf/c1344.pdf (accessed 19 December 2017).

Laishley, R. (2014) 'Is Africa's land up for grabs?', *Africa Renewal: Special Edition on Agriculture*, available at: www.un.org/africarenewal/magazine/special-edition-agriculture-2014 (accessed 19 December 2017).

Lane, S.N., Odoni, N., Landstrom, C., Whatmore, S.J., Ward, N. and Bradley, S. (2011) 'Doing flood risk science differently: an experiment in radical scientific method', *Transactions of the Institute of British Geographers*, 36(1): 15–36.

MacDonald, A.M., Bonson, H.C., Dochartaigh, O. and Taylor, R.G. (2012) 'Quantitative maps of groundwater resources in Africa', *Environmental Research Letters*, 7(2), DOI: 10.1088/1748-9326/7/2/024009.

Mekonnen, M.M. and Hoekstra, A.Y. (2011) 'The green, blue and grey water footprint of crops and derived crop products', *Hydrology and Earth System Sciences*, 15: 1577–1600.

Moriarty, P. and Butterworth, J. (2003) *The Productive Use of Domestic Water Supplies*, The Hague: IRC International Water and Sanitation Centre.

Newsom, M. (1994) *Hydrology and the River Environment*, Oxford: Clarendon.

Ortolano, L. and Cushing, K.K. (2002) 'Grand Coulee Dam 70 years later: what can we learn?', *International Journal of Water Resources Development*, 18(3): 373–390.

Pearce, F. (2006) 'Mega-dams back on the agenda', *New Scientist*, 191(2569): 10.

Poff, N.L., Allan, J.D., Palmer, M.A., Hart, D.D., Richter, B.D., Arthington, A.H., *et al.* (2003) 'River flows and water wars: emerging science for environmental decision making', *Frontiers in Ecology and the Environment*, 1: 298–306.

Postel, S. (1992) *The Last Oasis: Facing Water Scarcity*, London: Earthscan.

Rosegrant, M.W., Cline, S.A. and Valmonte-Santos, R.A. (2010) 'Global water and food security: megatrends and emerging issues', in C. Ringler, A.K. Biswas and S. Cline (eds), *Global Change: Impacts on Water and Food Security*, Berlin: Springer, pp. 17–47.

Roy, M. (2016) *Institutions for Urban Poor's Access to Ecosystem Services: A Comparison of Green and Water Structures in Bangladesh and Tanzania*, available at: www.espa.ac.uk/projects/ne-l001616-1 (accessed 19 December 2017).

Swyngedouw, E. (2006) *Power, Water and Money: Exploring the Nexus – Human Development Report 2006*, Occasional Paper, New York: UNDP.

The Economist (2010) *Dams in Africa: Tap That Water*, 6 May, available at: www.economist.com/node/16068950 (accessed 19 December 2017).

The Economist (2016) *City Slickers on the Farm: Africa's Real Land Grab*, 23 July, available at: www.economist.com/printedition/2016-07-23 (accessed 19 December 2017).

Therkildsen, O. (1988) *Watering white elephants? Lessons from Donor Funded Planning and Implementation of Rural Water supplies in Tanzania*, Centre for Development Research Publications No. 7, Uppsala: Scandinavian Institute of African Studies.

Twort, A.C. Ratnayaka, D.D. and Brandt, M.J. (2002) *Water Supply*, 5th edition, Oxford: Butterworth-Heinemann.

UN (United Nations) (1987) *Report of the World Commission on Environment and Development: Our Common Future*, New York: United Nations.

UN (United Nations) (2013) *UN International Year of Water Cooperation: Water Facts and Figures*, available at: www.unwater.org/water-cooperation-2013/water-cooperation/facts-and-figures/en/ (accessed 18 December 2017).

UN (United Nations) (2015) *The Millennium Development Goals Report*, New York: United Nations.

UN DESA (United Nations Department of Economic and Social Affairs) (2014) *International Decade for Action Water for Life 2005–2015*, available at: www.un.org/waterforlifedecade/scarcity.shtml (accessed 18 December 2017).

UNECA (United Nations Economic Commission for Africa) (2006) *Water in Africa: Management Options to Enhance Survival and Growth*, Addis Ababa: UNECA.

UNEP (United Nations Environment Programme) (2008) *An Overview of the State of the World's Fresh and Marine Waters: The Coming Water Scarcity in Africa*, 2nd edition, Nairobi: UNEP, available at: www.unep.org/dewa/vitalwater/article83.html (accessed 18 December 2017).

UNICEF (United Nations Children's Fund) (2016) *Water Sanitation and Hygiene*, available at: www.unicef.org/wash/ (accessed 18 December 2017).

UNICEF (United Nations Children's Fund) and WHO (World Health Organization) (2012) *A Snapshot of Drinking Water and Sanitation in Africa: 2012 Update for AMCOW*, WHO/UNICEF Joint Monitoring Programme for Water Supply and Sanitation, available at: www.wssinfo.org/fileadmin/user_upload/resources/Africa-AMCOW-Snapshot-2012-English-Final.pdf (accessed 18 December 2017).

UNICEF (United Nations Children's Fund) and WHO (World Health Organization) (2015) *25 Years: Progress on Sanitation and Drinking Water Update and MDG Assessment*, Geneva: WHO, available at: www.wssinfo.org/fileadmin/user_upload/resources/JMP-Update-report-2015_English.pdf (accessed 18 December 2017).

Water Footprint Network (n.d.) *Water Footprint Network*, available at: http://waterfootprint.org/en (accessed 3 January 2018).

Water Technology (2016) *Top 10 Biggest Dams*, available at: www.water-technology.net/features/feature-ten-largest-dams-in-the-world-reservoirs/ (accessed 19 December 2017).

WCOD (World Commission on Dams) (2000) *Dams and Development: A New Framework*, London: Earthscan.

World Bank (2000) *Options for the Sustainable Development of the Gezira Scheme*, Report No. 20398-SU, prepared jointly by Government of Sudan and Macroeconomics 2 Country Department 6 – Eritrea, Ethiopia, Somalia, Sudan, Washington, DC: World Bank.

World Bank (2009) *Water Data and Statistics*, available at: www.worldbank.org/en/topic/water (accessed 19 December 2017).

WWC (World Water Council) (2010) *World Water Council*, available at: www.worldwatercouncil.org/index.php?id=25 (accessed 18 December 2017).

WWDR (World Water Development Report) (2009) *Water in a Changing World*, available at: www.unesco.org/new/en/natural-sciences/environment/water/wwap/wwdr/wwdr3-2009/ (accessed 19 December 2017).

WWDR (World Water Development Report) (2015) *Water for a Sustainable World*, available at: www.unesco.org/new/en/natural-sciences/environment/water/wwap/wwdr/2015-water-for-a-sustainable-world/ (accessed 18 December 2017).

17
Wetlands

Alan Dixon

Introduction

This chapter examines Africa's wetland environments and the often-contested nature of people–wetland relationships that exist around the continent. Through their provision of various ecosystem services, wetlands have played a key role in the development of people's livelihoods since time immemorial, and yet in recent years the global discourse on wetlands has been dominated by two contrasting perspectives. The first is that wetlands and their critically important biodiversity continue to be threatened by exploitative human actions, and hence require urgent conservation. The second argues that wetlands are unproductive wastelands that require drainage and development in order to transform them into more productive environments. After outlining the evidence for wetlands' importance around the continent, this chapter discusses the ways in which these different views have shaped wetland management and policy in Africa, and how there is an emerging consensus of the need to recognise wetlands as multifunctional socioecological systems, in which ecosystem services and livelihood benefits can be balanced and sustained.

Wetlands in Africa: nature and distribution

Wetlands are diverse environments, both spatially and temporally, but also in terms of their physical size, ecology, hydrology and geomorphology. A myriad of literature has debated the definition, characterisation and functions of wetlands in different environments (Dugan, 1990; Roggeri, 1995; MA, 2005; Maltby and Barker, 2009), but the most widely accepted definition adopted in conservation and planning continues to be that established by the Ramsar Convention in 1971. This embraces the diversity of wetlands by grouping together a variety of landscape units whose ecosystems share the fundamental characteristic of being strongly influenced by water:

> areas of marsh, fen, peatland or water, whether natural or artificial, permanent or temporary, with water that is static or flowing, fresh, brackish or salt, including areas of marine water the depth of which at low tide does not exceed six metres.
>
> *(Davis, 1994: 3)*

While the Ramsar Bureau developed a classification of 42 distinct wetland types, it later simplified these into five broad categories:

- marine (coastal wetlands, including coastal lagoons, rocky shores and coral reefs);
- estuarine (including deltas, tidal marshes and mangrove swamps);
- lacustrine (wetlands associated with lakes);
- riverine (wetlands along rivers and streams); and
- palustrine (meaning 'marshy' – marshes, swamps and bogs).

(Ramsar, 2009)

Each of these wetland 'types' is found throughout Africa, and estimates suggest that they constitute somewhere between 1 and 16 per cent of the total land area, depending upon whether the larger lakes are included (Hughes, 1996; Shuyt, 2005) (see Figure 17.1). This is likely to be an underestimate, however, simply because data on wetlands are far from comprehensive

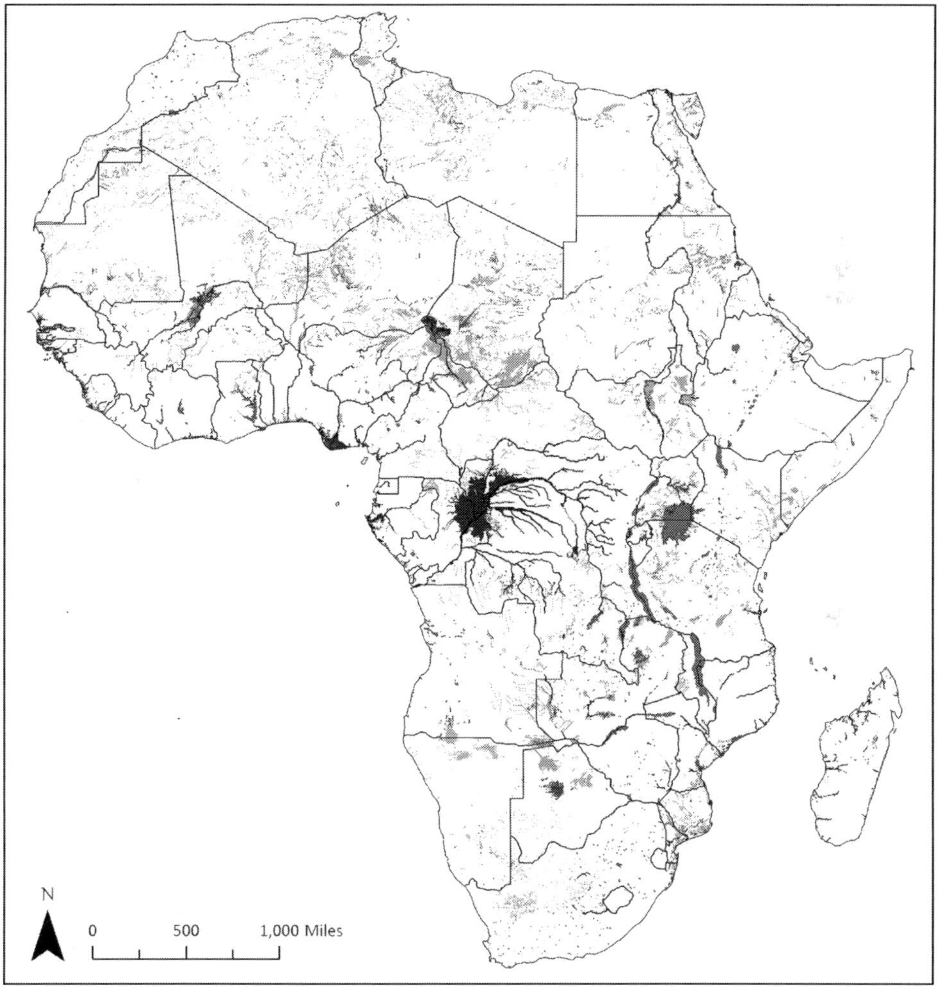

Figure 17.1 The distribution of wetlands (indicated by shaded areas) in Africa

Source: Adapted from MA (2005).

due to inconsistencies in the wetland terminology used around the continent, and the logistical constraints of carrying out continental-scale surveys (Denny, 2001). The occurrence of wetlands reflects the variation in climate and geomorphology, and Denny (1993) identifies two broad physiographic units: 'low Africa', situated to the north and west and characterised by sedimentary basins, upland plains below 600 m above sea level and abundant floodplain wetlands, and in contrast, 'high Africa' to the south and east, characterised by extensive mountainous areas with deep valleys, highland plateaus and abundant lakes and swamps. Seasonally inundated floodplains and swamps account for almost half of the continent's total wetland area (Drijver and Marchand, 1985), and research has suggested that these areas are critical resources for Africa's human population, especially in areas where rainfall is seasonal and wetlands represent a reliable source of water throughout the year (Hollis, 1990).

Wetlands, livelihoods and development

Wetlands have played a critical role in the livelihoods of people in Africa for millennia, not least because they have been sources of food and water for people living in often dry and semi-arid

Table 17.1 Ecosystem services provided by, or derived from, wetlands

Services	*Comments and examples*
Provisioning	
Food	Production of fish, wild game, fruits and grains
Fresh water	Storage and retention of water for domestic, industrial and agricultural use
Fibre and fuel	Production of logs, fuelwood, peat and fodder
Biochemical	Extraction of medicines and other materials from biota
Genetic material	Genes for resistance to plant pathogens, ornamental species, etc.
Regulating	
Climate regulation	Source and sink for greenhouse gases; influence local and regional temperature, precipitation and other climate processes
Water regulation (hydrological flows)	Groundwater recharge/discharge
Water purification and waste treatment	Retention, recovery and removal of excess nutrients and other pollutants
Erosion regulation	Retention of soils and sediments
Natural hazard regulation	Flood control and storm protection
Pollination	Habitat for pollinators
Cultural	
Spiritual and inspirational	Source of inspiration; many religions attach spiritual and religious values to aspects of wetland ecosystems
Recreational	Opportunities for recreational activities
Aesthetic	Many people find beauty or aesthetic value in aspects of wetland ecosystems
Educational	Opportunities for formal and informal education and training
Supporting	
Soil formation	Sediment retention and accumulation of organic matter
Nutrient cycling	Storage, recycling, processing and acquisition of nutrients

Source: Adapted from MA (2005).

environments. Indeed, much has been made in the literature of the last 30 years of the capacity of wetlands to provide a diverse range of functions and services that have supported people, ecological systems and the physical environment (Maltby, 1986; Dugan, 1990; Barbier, 1993; Davis, 1994; Roggeri, 1995; Barbier *et al.*, 1997; MA, 2005). As greater understanding of the role wetlands play in ecological and hydrological cycles has emerged, however, there has been a concurrent shift in attitudes away from wetlands being viewed as unproductive wastelands to one of seeing wetlands as multifunctional resources. This has most recently been evident in the Millennium Ecosystem Assessment's classification of 'wetland ecosystem services that contribute to human well-being and poverty alleviation' (MA, 2005: 1). Table 17.1 shows the range of services associated with wetlands, although there continues to be some debate concerning which services manifest in specific wetland types, and the evidence base on which assumptions are made (Bullock and Acreman, 2003).

It is in their capacity to provide 'provisioning services' that arguably makes the contribution of wetlands to people's livelihoods so important in Africa. In particular, wetlands have been, and continue to be, critical sources of food, fresh water, fibre and fuel for poorer people in many of the most marginal areas (Trapnell and Clothier, 1937; Hollis, 1990; Scoones, 1991; Woodhouse *et al.*, 2000; Wood *et al.*, 2013). In northern Nigeria, for example, seasonally flooded low-lying areas known as *fadamas* support the livelihoods of hundreds of thousands of smallholder farmers (Tarhule and Woo, 1997; Dan-Azumi, 2010; Tanko, 2013). Throughout central and southern Africa, seasonally waterlogged riverine and palustrine depressions known as *dambos* are common landforms that retain soil moisture into the dry season, and hence represent valuable agricultural resources (Turner, 1986). In Malawi in particular, *dambos* account for around 12 per cent of the total cultivable land area (Mloza-Banda, 2005), and have long supported water provision, fishing, grazing, seasonal agriculture and wild plant collection (Roberts, 1988; Noble, 1996; Wood and Thawe, 2013) (see Figure 17.2). Similar benefits to people have been documented across eastern and southern Africa (Crafter *et al.*, 1992; Kamukala and Crafter, 1993; Abebe and Geheb, 2003; FAO, 2004).

The provision of moisture during periods of the year usually associated with drought or even famine (the so-called 'hungry season') has rendered wetlands critically important in the agricultural and broader livelihood strategies of many. As Adams (1993: 210) puts it, 'within semi-arid Africa, wetlands have a strategic importance out of all proportion to their size . . . and there are many examples of communities, with one foot in the wetland and another in the dry'.

In parts of western Ethiopia, for example, wetlands are cultivated with maize and vegetables early in the dry season, enabling crops to be harvested at a time when the upland crops are only in their early growing phase (Dixon, 2003). Critically, such strategies also represent a wider process of ongoing livelihood diversification and socioecological adaptation that farmers around the continent are engaged in. Research has drawn attention to the ways in which people's use of wetlands has led to an enhancement in human and social capital, and reduced vulnerability, as well as the more tangible improvements to food security, nutrition and the household economy (Dixon, 2005, 2008).

Indeed, some assessments of the economic contribution of wetlands confirm the continued importance of these areas. In Zambia, the total use value of wetlands (with fish production and floodplain recession agriculture accounting for the main share) was estimated to be the equivalent of approximately 5 per cent of Zambia's GDP in 1990 (Seyam *et al.*, 2001). Across the whole Zambezi River basin, the economic value of wetlands in terms of crops alone was estimated to be close to US$50 million a year, with fisheries contributing a further US$80 million (UNEP, 2006) and floodplain grasslands support of livestock production contributing another over US$70 million annually.

Figure 17.2 Vegetable cultivation in a Malawian Dambo

Source: Author.

Research on wetland contributions to livelihoods at the local community level has similarly highlighted their significance. McCartney and van Koppen (2004) report that wetland cultivation contributes 80 per cent of the total annual cash income of the poorest households in the Kilombero Valley in Tanzania. In Ga-Mampa wetland in South Africa, the average annual value of cultivation per household was estimated at US$93 (Adekola *et al.*, 2008); in Nakivubo urban wetland in Kamapala, Uganda, it was US$300 per household (Emerton, 2005); in the Barotse floodplain in Zambia, US$109 per household (Turpie *et al.*, 1999); in the Lower Shire, Malawi, US$363 per household (Turpie *et al.*, 1999); and in the Chipala Ibenga wetland, Zambia, US$19–107 per household (Masiyandima *et al.*, 2004).

Beyond these direct and more tangible contributions to people's livelihoods, many wetlands across Africa contribute regulating and supporting services that facilitate the provisioning services discussed above, but also make indirect contributions to people's livelihoods at a range of different scales. For example, the links between wetlands' capacity to regulate floods and recharge groundwater, thereby supporting people's livelihoods, have been well documented in the Hadejia-Jama'are River basin in Nigeria (Hollis *et al.*, 1993; Acharya and Barbier, 2000; Shuyt, 2005). Similarly, in Uganda, a country where wetlands comprise 13 per cent of the total land area (Mafabi, 2005), government conservation policies recognise the critical contribution of the country's wetlands to water regulation and purification (Emerton *et al.*, 1998; Namaalwa *et al.*, 2013). In economic terms, these contributions are highly significant; Brander *et al.* (2013) estimate that the total value of wetland regulating services across Africa is around US$9 billion per year, and US$4–5 billion in West Africa alone due to its agricultural dependency on the groundwater recharge and flood control services associated with wetlands (Zwarts *et al.*, 2006).

Culturally and aesthetically, wetlands similarly play a significant role in the lives of many. Spiritual values may range from simply imparting a sense of place and shared history among local populations (Bouahim *et al.*, 2015), to wetlands being inextricably linked with the evolution of people's culture, as evidenced by the Lozi's annual Kuomboka ceremony on the Barotse floodplain in Zambia (Flint, 2006). In the case of the latter, there is added significance in that Kuomboka has also become a tourist attraction, and hence has wider economic significance. This is evident elsewhere throughout Africa; for example, in the Great Lakes region, the Botswana's Okavango delta, South Africa's St Lucia Wetland Park and Kenya's Lake Nakuru, wetland conservation and ecotourism are seen as mutually reinforcing goals. However, while the revenue generated from wetland tourism can be significant, concerns have been raised about who actually benefits from this, and whether local people have an equitable share in both wetland management decision-making and where revenue is spent (Ramsar and UNWTO, 2012).

Wetlands under threat?

Even a cursory analysis of the global literature on wetlands during the past 40 years reveals that the narrative of 'wetlands under threat' has dominated the discourse and driven subsequent wetland policy (Maltby, 1986; Dugan, 1990; Farrier and Tucker, 2000; Barker, 2009). There is of course a rationale for this; while acknowledging that rates of loss in Africa are especially difficult to measure, Davidson (2014) suggests that between 54 and 57 per cent of the world's wetlands have been lost since 1900. There is little doubt that due to a lack of appreciation for the economic significance of wetlands and their ecosystem services, many policymakers in Africa have, in the past at least, driven forward unsustainable and destructive developments that include agricultural drainage, irrigation schemes, dam construction and conversion for residential or industrial use (Maltby, 1986; Adams, 1992; Hughes, 1996; Mafabi, 2005; Shuyt, 2005; Wood *et al.*, 2013).

Although many of these pressures have been mitigated and reduced in recent years, not least as a result of the successful awareness-raising and legislative actions of large INGOs such as the Ramsar Convention Bureau, the drivers behind these pressures remain. This is a particular issue for the smaller and more ubiquitous wetlands that have evaded the attention of policymakers seeking to designate and conserve high-profile wetlands as being of 'international importance' (Wood *et al.*, 2013). The desire for sustained economic growth and poverty reduction from the national to household level, coupled with high rates of population growth across the continent, continue to place pressure on such wetland areas once considered by many to be marginal environments. Even where wetland use has a long history of sustainable utilisation, whether for subsistence agriculture, fishing or foraging, the demand for wetland livelihood benefits is likely to increase. Indeed, Dixon and Wood (2003: 119) describe wetlands as a 'new agricultural frontier' in Africa due to rural population growth and subsequent land shortages, the degradation of uplands and a reduction in quality and quantity of rain-fed harvests, and also the increased needs for earning cash. The development and affordability of technologies such as treadle irrigation pumps have also exacerbated smallholder wetland use, as has been seen throughout southern and West Africa (Kay and Brabben, 2000; Tanko, 2013; Wood and Thawe, 2013). Furthermore, as the localised impacts of climate change begin to manifest and the search for water intensifies in many areas, wetlands will become a critical arena in the process of adapting to climate change and mitigating its impact – a process that has become increasingly evident in many areas (Boko *et al.*, 2007; Osbahr *et al.*, 2008).

There is little doubt that wetlands will become increasingly important resources in Africa in years to come, providing both a safety net for the poor and landless, and a livelihood diversification option for wealthier farmers. A critical challenge, as ever, is how to ensure that wetlands are used and managed in a manner that does not degrade the natural resource base from which the various livelihood benefits and ecosystem services are derived and sustained.

Changing approaches to wetland management

Wetland utilisation and management throughout Africa is characteristically diverse, and reflects a myriad of dynamic socioecological contexts. It is useful to situate these within the framework presented in Figure 17.3, where wetland management can be located on one axis ranging from traditional, locally adapted systems, to externally driven government or NGO/INGO management interventions. A second axis reflects the nature of these approaches, and in particular where they can be placed within the classic conservation versus development dichotomy. While empirical studies from around the continent illustrate that a range of utilisation and management scenarios continue to exist (not least because of the cross-sectoral position of wetlands within government planning) (Woodhouse *et al.*, 2000; Wood *et al.*, 2013), it is clear that people's relationship with wetlands has been shaped by a changing global (mainly 'northern') wetlands discourse, which has until fairly recently been dominated by a conservation agenda that has prioritised the protection of wetlands, their species and biodiversity (Maltby and Barker, 2009). In theory, there are some marked differences between 'conservationist' and 'preservationist' approaches (Brockington *et al.*, 2008; Adams, 2009) in that the former seeks to maintain the ecological character of an area through regulating human interaction with it, while the latter emphasises the exclusion of humans. In practice, however, and in the context of wetland management in Africa, these differences have often become blurred.

The modern-day wetland conservation movement had its origins in 'northern' concerns about endangered bird species during the 1940s. Emerging from discussions among various NGOs, including IUCN and IWRB, the Ramsar Convention on Wetlands of International Importance Especially as Waterfowl Habitat was established in 1971, and set out a process of accession for countries wishing to designate and list wetlands of international importance for conservation purposes (Matthews, 1993; Davis, 1994). The original text of the Ramsar

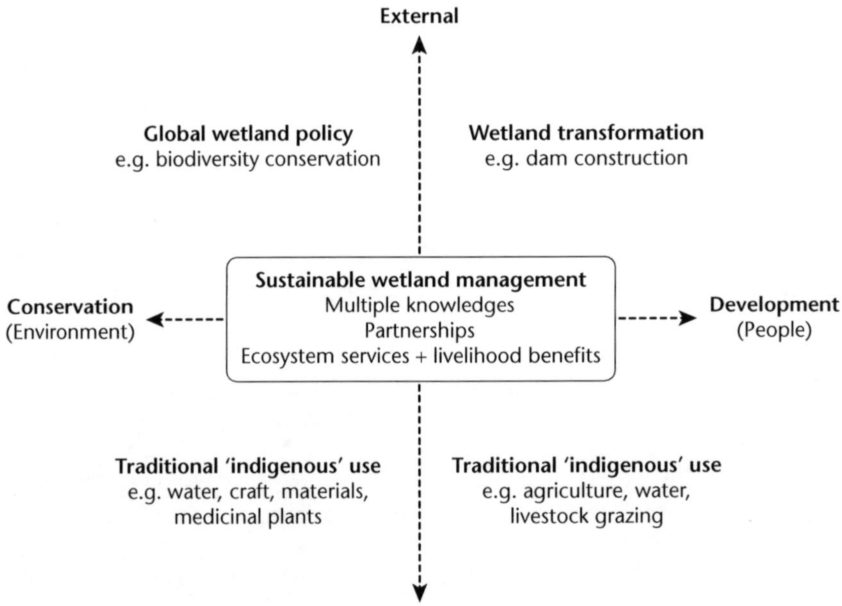

Figure 17.3 A framework of wetland management scenarios in Africa

Source: Author.

Convention makes it clear that wetlands and their various ecological functions should be protected from human encroachment for the benefit of waterfowl, and that protection should be achieved via the 'conservation of the wetlands included in the list' (Ramsar, 1971: Article 3). Although it suggests that countries should also promote 'as far as possible the wise use of wetlands in their territory', thus hinting at some acceptable form of human intervention, it would be a further 16 years before the concept of 'wise use' was formally defined and recommended for adoption by contracting parties:

> The wise use of wetlands is their sustainable utilization for the benefit of humankind in a way compatible with the maintenance of the natural properties of the ecosystem.
>
> Sustainable utilization is defined as human use of a wetland so that it may yield the greatest continuous benefit to present generations while maintaining its potential to meet the needs and aspirations of future generations.
>
> Natural properties of the ecosystem are defined as those physical, biological or chemical components, such as soil, water, plants, animals and nutrients, and the interactions between them.
>
> *(Ramsar, 1987: Recommendation 3.3)*

While in many respects this was a significant step forward in acknowledging the importance of people in wetland utilisation (occurring during a period of unprecedented global interest in sustainable development), the wise use principle remained problematic for several reasons. First, in terms of attempting to embrace sustainable development and bring together conservation and the needs of people, it remained heavily skewed in favour of environmental concerns, not least because of the perceived incompatibility between human use and the maintenance of 'natural properties'. As Farrier and Tucker (2000) point out, 'naturalness' is a state that is difficult to define with regards wetlands, especially since very few wetlands will have evaded the influence of humans in one way or another. A second related issue is the nebulous nature of natural properties themselves (what Ramsar reinterpreted as 'ecological character' in 1999, and later linked to the MA's 'ecological services' in 2005). Although one of the great successes of the global wetland conservation movement has been the promotion and acceptance of the idea of wetland values and functions rooted in their natural properties, the empirical scientific evidence remains sketchy and often contradictory (Bullock and Acreman, 2003; McCartney *et al.*, 2010).

Finally, the adoption of the wise use principle served to legitimise some forms of human interaction with wetlands but not others, and hence raised issues of equity, particularly regarding access to these resources – an issue with particular ramifications for poor people throughout Africa. Consequently, the forms of acceptable 'wise use' that were being promoted widely in wetland conservation initiatives during the late 1980s and early 1990s tended to consist of non-transformational uses of wetlands such as fishing and reed collection, and argued against any intrusion of agriculture into wetlands (Marchand and Udo, 1989; Dugan, 1990). These approaches, often operationalised and branded as Integrated Conservation and Development Plans (ICDPs), actively sought the participation of local people in wetland management strategies, and while they were successful in conserving the natural status of many wetlands, there is little evidence to suggest that these initiatives have had a major impact on poverty reduction or addressed other critical needs such as food security (Kellert *et al.*, 2000; Fisher *et al.*, 2005; Sellamuttu *et al.*, 2008).

Since the late 1990s, however, there has been a marked reorientation towards human development in wetland conservation, which has been reflected in the themes of the Ramsar COP meetings. Of particular note is the inclusion of 'sustaining livelihoods' in the title of COP9, the first to be held in Africa, and significant for several reasons. First, it integrated the MA's ecological services concept into a new definition of wise use: 'Wise use of wetlands is the maintenance of their ecological character, achieved through the implementation of ecosystem approaches, within the context of sustainable development' (Ramsar, 2005: 6). Second, the meeting was notable for its attention to river basin planning, a hitherto neglected area of the convention, which critics suggested had resulted in wetlands erroneously being viewed in isolation from upstream and downstream influences. Third, and perhaps most significantly, the convention recognised the importance of the Millennium Development Goals, and consequently set out specific actions for contributing to poverty reduction. In 2008, the focus changed to 'poverty eradication', with a further recognition of the importance of wetlands for human health, the inclusion of traditional knowledge into wetland management strategies, and the potential role that financial incentives could play in wetland management.

This gradual shift in Ramsar thinking has been mirrored in other wetland conservation organisations. For example, in 2005, Wetlands International established its Wetlands and Poverty Reduction Project (WPRP), and working in four target areas within Africa the project sought to demonstrate how sustainable wetland management, based on 'wise use', could be integrated into poverty reduction strategies (Wetlands International, 2005). Like Ramsar, the WPRP highlighted its strategic alignment with the Millennium Development Goals, which have become increasingly influential within conservation organisations, largely as a result of pressure from the international donor community. Indeed, the Millennium Declaration and the universal acceptance of the eight goals and 21 development targets have significantly altered the landscape of international aid, to the extent that it has raised some difficult questions for the conservation movement over whether it is morally justifiable to conserve wetlands in ways that deny people an opportunity to develop their livelihoods (Fisher *et al.*, 2005).

In response, organisations such as Ramsar and Wetlands International have taken steps to reposition (and rebrand) their wetland conservation activities to embrace development goals and livelihood outcomes, but there are also concerns over tokenism and whether the issue of livelihoods and poverty reduction are merely being 'tagged on' to conservation programmes as a convenient way of accessing funding. Critically, what even the most enlightened wetland conservation initiatives still arguably lack is a truly participatory approach to exploring wetland-based livelihoods, i.e. one that examines the development needs and aspirations of local people, and then seeks to identify how these can be achieved alongside the sustainable use of ecosystem services. Wood *et al.* (2013) suggest that such a reluctance to engage emanates from a realisation that local people will ultimately demand to use their wetlands for agriculture – long cited as the greatest threat to wetlands and their conservation or sustainable use throughout Africa – because food security concerns are higher up their development needs than an appreciation of biodiversity.

Conservation organisations such as Ramsar have, laudably, made significant progress in developing and adapting their strategies on wetland use in response to changes in the wider environment development policymaking arena, and arguably one of their greatest successes has been instilling an almost universal precautionary ethic among wetland users and planners, not least throughout Africa (see Box 17.1). For Africa, though, the challenge of how to mainstream agriculture-based livelihoods into wetland management strategies that can sustain social, economic and environmental outcomes remains pressing. Externally driven wetland management that places restrictions on how and what poor people can use wetlands for can perhaps only result in what could be termed 'sustainable miserable livelihoods' at best, and must be replaced by those that actually facilitate real livelihood development.

Box 17.1 Uganda: a pioneer of wetlands policymaking

Uganda has been a pioneer of wetland management policymaking within Africa. After a series of agricultural intensification policies during the 1970s and 1980s led to increasing evidence of environmental degradation, it created a National Wetlands Programme in 1989 with assistance from IUCN in order to identify how Ramsar's 'wise use' principles could be integrated into wetland policies. This kick-started a period of public consultation, awareness-raising, and the collection of data to establish a national inventory of wetlands and their uses around the country. This National Wetlands Policy was subsequently launched in 1995, with an aim to 'promote the conservation of Uganda's wetlands in order to sustain their ecological and socio-economic functions for the present and future wellbeing of the people' (MNR, 1995: 3). Indeed, a key impact of the NWP was that for the first time in Africa, the conservation and management of wetlands were enshrined in a country's constitution and 'held in trust for the common good of all citizens'.

Uganda's Wetland Inspection Division, the body responsible for implementing wise use management strategies in the field, clearly prioritised the maintenance of wetland hydrological and ecological integrity in their early days; people were permitted to exploit the provisioning services of wetlands, but only if the quantity and quality of the wetland resources were maintained or enhanced, or if the wetland was deemed 'dispensable' in the 'Kampala Matrix' decision-making tool. In effect, this largely precluded any agricultural transformation, and for many years the extractive use of wetlands was limited to water, fish, craft material and cattle fodder.

The Kampala Matrix: prioritising wetland management in Uganda

		Status		
		Threatened	*Not threatened*	*Destroyed*
Importance	Vital	Restore	Monitor strictly	Restore
	Valuable	Ensure wise use	Monitor	Restore?
	Dispensable	Encourage wise use	Monitor?	Forget for the time being

Source: Mafabi (2005).

This precautionary approach has gradually evolved over time, both as a result of an increase in the development needs of the population, but also arguably as a result of wider shifts in wetland thinking, which has focused more on the role of local communities in adaptive co-management. Recent initiatives such as the Community-Based Wetlands Biodiversity conservation project (COBWEB) have sought to devolve ownership and monitoring of wetlands to local people, so that management can be more adaptive to the range of unique and dynamic wetland socioecological contexts.

New directions: people-centred wetland management

In exploring new directions for wetland management in Africa, Wood *et al.* (2013) recently examined the ways in which ideas from development provide alternative conceptualisations of people–wetland interactions, placing people and livelihoods at the forefront of strategies that seek to balance environment and development outcomes. This, it is argued, starts with the need for policymakers to view wetlands as holistic and adaptive socioecological systems, in which wetland ecosystem services underpin livelihood benefits and where livelihood strategies and wetland interactions shape the availability of ecosystem services. Rather than being excluded from wetland use on account of their destructive tendencies, as traditional conservation approaches advocate, the socioecological approach argues that people should be seen as key actors capable of managing and sustaining wetlands and their benefits, not least because it is in their best interests to do so. The evidence here lies in the vast body of literature that argues that in many cases, poor people can accumulate detailed knowledge of environmental systems (Brokensha et al., 1980; Chambers, 1983; Richards, 1985; Warren *et al.*, 1995; Ellen *et al.*, 2000; Berkes, 2012), adapt their resource use to environmental and socio-economic pressures (Batterbury and Forsyth, 1999; Berkes *et al.*, 2000; Reij and Waters-Bayer, 2001; Berkes *et al.*, 2004), establish their own institutional arrangements for managing wetlands (van Koppen *et al.*, 2007; Pollard and Cousins, 2008; Dixon *et al.*, 2013), and manage their environment in a way that sustains social, economic and environmental benefits. Moreover, and to re-emphasise a previous point, there is certainly evidence that this has been the case throughout Africa with respect to wetlands and the development of locally adapted strategies for wetland agriculture. Adams (1993: 211), for example, asserts that 'Africa has a strong and diverse tradition of irrigation', and highlights extensive evidence of traditional agricultural practices in the many floodplain wetlands of West Africa where farmers have a detailed understanding of ecological and hydrological variability, and have subsequently adapted their farming to this. In the Lower Tana River, Kenya, Terrer *et al.* (2004) suggest that the Pokomo and the Wardei people have similarly developed traditional, adaptive wetland agricultural practices that are rooted in their environmental knowledge. In the western highlands of Ethiopia, local strategies for wetland agriculture have developed within an appreciation of the wider range of provisioning and regulating services that small, valley bottom wetlands supply to the community; maize and vegetables are cultivated downstream or adjacent to areas of *cyperus* sedge, reserved because farmers need the sedges for thatching and craft materials, but also because they understand that these areas of 'natural' vegetation help regulate the supply of water downstream (Dixon, 2003). This latter example illustrates the importance here of socioecological trade-offs; local people value a range of different ecosystem services, and rarely will they tolerate the complete loss of one. Moreover, there is an implicit understanding that these services are mutually reinforcing, and hence many traditional wetland management interactions reflect this need to balance the benefits.

This is, of course, not to say that every local community living near a wetland will automatically engage in its sustainable use. As discussed earlier, wetland use (and certainly wetland agriculture) in many places represents a new form of resource use for which local people may have little experience and knowledge. What we can draw from the literature, however, is that there is great potential for a more nuanced, progressive and inclusive approach to wetland management in Africa that empowers wetland users themselves and reflects their development needs. With reference to Figure 17.3, this is the middle ground – balancing conservation with development needs, and drawing on both local and external needs and knowledge. Throughout Africa, new approaches to wetland management that address these different dimensions are beginning to emerge (see Box 17.2).

Box 17.2 A functional landscape approach to wetland management: the Striking a Balance project in Zambia and Malawi

In the last two decades, periodic droughts in Malawi and Zambia have led to the intensification of dry season dambo cultivation by many households in order to supplement reduced upland harvests (Chisinga and Kayuni, 2011; Wood and Thawe, 2013). The Striking a Balance (SAB) project, funded by Wetlands International in 2005, sought to identify the ways in which local people use and manage dambos, identify potential threats to their sustainability, but critically identify ways in which their ecosystem services could be sustained in the face of increasing pressure to transform these areas.

The SAB approach draws upon a holistic, socioecological systems view of the dynamic relationship between people and the environment, in both space and time, and adopts a functional landscape approach (FLA) that recognises how different landscape units and livelihood activities are linked, and how specific interventions and management strategies can support and sustain interrelated ecosystem services and livelihoods (Wood and Thawe, 2013; Dixon and Carrie, 2015). Following a period of participatory research and consultation with dambo stakeholders in six project areas, during which the socioecological significance of dambos were assessed, communities were invited to identify what management practices, both in the dambo and catchment, could be feasibly implemented in order to maintain multiple ecosystem services that were economically, socially and environmentally beneficial. Typical practical interventions included afforestation, agroforestry, contour terracing, composting beds, and wetland buffer zones being implemented in the catchment. Within the dambo, cultivation of vegetables was prohibited around the central channel to prevent gullying, natural vegetation was retained in some demarcated areas, fish ponds developed, and over-abstraction prohibited. Indeed, a key part of the SAB included the development of locally specific by-laws governing the extent of these interventions and dambo use generally, which were overseen by a locally elected village natural resource management committee.

Post-project evaluations suggest that people's livelihoods benefited significantly from this more strategic approach to utilising dambos; many were able to achieve food security throughout the year, and some sold their surplus at local markets, enabling them to purchase bicycles, livestock and mobile phones. At the same time, there was little evidence that the increased use of dambos facilitated by the SAB project had led to any serious detrimental environmental impacts, with Kotze (2013: 138) remarking: 'Overall, it appears that the multiple demands on these areas can be met and that it is possible to have well-managed agriculture in the wetlands, which supports livelihoods while also sustaining multiple ecosystem services'.

Conclusions

This chapter has presented an overview of the emerging importance of wetland environments in Africa's development landscape. It has argued that rather than constituting benign places where rare ecosystems exist and humans avoid contact, wetlands are, and have increasingly become, contested socioecological spaces at the interface of people–environment interactions. While this continues to be a source of tension between conservation and development policymakers, there has been some acknowledgement in recent years of the need to develop more holistic ways of thinking about people's interactions with wetlands. Of particular significance has been the emergence of new approaches that recognise how wetland ecosystem services can be maintained, or even enhanced, through local management strategies that incorporate a more nuanced understanding of the mutually supportive relationship between ecosystem integrity and human well-being and development. It is at this grassroots level throughout Africa where many local people appreciate all too well that the development of their livelihood rests on careful management of their wetlands, and a wealth of institutional knowledge and experience of wetlands exists.

A key priority for the future, therefore, must be to incorporate the many local-level experiences and community structures that engage in adaptive co-management into what have traditionally been top-down and centralised wetland policy and decision-making arenas at the national and regional levels. Indeed, the challenges here are not dissimilar to those of sustainable development more broadly, which has, for many years, emphasised the need for local representation, multiple knowledges and partnerships, cross-sectoral collaboration, and equity. With the repositioning of development as the Sustainable Development Goals, however, there is undoubtedly a significant opportunity in the years ahead to make further progress in developing and implementing a more inclusive, pluralistic and systemic approach to sustaining wetlands and their benefits for all.

References

Abebe, Y. and Geheb, K. (2003) *Wetlands of Ethiopia*, Nairobi: IUCN.

Acharya, G. and Barbier, E.B. (2000) 'Valuing groundwater recharge through agricultural production in the Hadejia-Nguru wetlands in northern Nigeria', *Agricultural Economics*, 22: 247–259.

Adams, W.M. (1992) *Wasting the Rain: Rivers, People and Planning in Africa*, London: Earthscan.

Adams, W.M. (1993) 'Indigenous use of wetlands and sustainable development in West Africa', *The Geographical Journal*, 159: 209–218.

Adams, W.M. (2009) *Green Development: Environment and Sustainability in a Developing World*, London: Routledge.

Adekola, O., Morardet, S., de Groot, R. and Grelot, F. (2008) *The Economic and Livelihood Value of Provisioning Services of the Ga-Mampa Wetland, South Africa*, Montpellier: World Water Congress.

Barbier, E.B. (1993) 'Sustainable use of wetlands: valuing tropical wetland benefits – economic methodologies and applications', *The Geographical Journal*, 159: 22–32.

Barbier, E.B., Acreman, M. and Knowler, D. (1997) *Economic Valuation of Wetlands: A Guide for Policy Makers and Planners*, Gland: Ramsar Convention Bureau.

Barker, T. (2009) 'Introduction: sustainable wetlands in a global context', in E. Maltby and T. Barker (eds), *The Wetlands Handbook*, Chichester: Wiley-Blackwell, pp. 821–828.

Batterbury, S. and Forsyth, T. (1999) 'Fighting back: human adaptations in marginal environments', *Environment*, 41: 6–9.

Berkes, F. (2012) *Sacred Ecology*, London: Routledge.

Berkes, F., Colding, J. and Folke, C. (2000) 'Rediscovery of traditional ecological knowledge as adaptive management', *Ecological Applications*, 10: 1251–1262.

Berkes, F., Colding, J. and Folke, C. (eds) (2004) *Navigating Social-Ecological Systems: Building Resilience for Complexity and Change*, Cambridge: Cambridge University Press.

Boko, M., Niang, I., Nyong, A., Vogel, C., Githeko, A. and Medany, M. (2007) 'Africa', in M.L. Parry, O.F. Canziani, J.P. Palutikof, P.J. Linden and C.E. van der Hanson (eds), *Climate Change 2007: Impacts, Adaptation, Vulnerability – Contribution of Working Group II to the Fourth Assessment Report of the Intergovernmental Panel on Climate Change*, Cambridge: Cambridge University Press, pp. 333–367.

Bouahim, S., Rhazi, L., Ernoul, L., Mathevet, R., Amami, R., Er-Riyahi, S., et al. (2015) 'Combining vulnerability analysis and perceptions of ecosystem services in sensitive landscapes: a case from western Moroccan temporary wetlands', *Journal for Nature Conservation*, 27: 1–9.

Brander, L., Brouwer, R. and Wagtendonk, A. (2013) 'Economic valuation of regulating services provided by wetlands in agricultural landscapes: a meta-analysis', *Ecological Engineering*, 56: 89–96.

Brockington, D., Duffy, R. and Igoe, J. (2008) *Nature Unbound: Conservation, Capitalism and the Future of Protected Areas*, London: Earthscan.

Brokensha, D., Warren, D. and Werner, O. (1980) *Indigenous Knowledge Systems and Development*, Lanham, MD: University Press of America.

Bullock, A. and Acreman, M. (2003) 'The role of wetlands in the hydrologic cycle', *Hydrology and Earth System Sciences*, 7: 358–389.

Chambers, R. (1983) *Rural Development: Putting the Last First*, London: Longman.

Chisinga, B. and Kayuni, H. (2011) *Hedging Food Security Through Winter Cultivation: The Agronomy of Dimba Cultivation in Malawi*, paper presented at 'From Scarcity to Access and Rights: Changing Perspectives on Food Insecurity' workshop, Center for Development and the Environment, University of Oslo.

Crafter, S.A, Njuguna, S.G. and Howard, G.W. (eds) (1992) *Wetlands of Kenya, Proceedings of the KWWG Seminar on Wetlands of Kenya, Nairobi, Kenya, 3–5 July 1991*, Gland: IUCN.

Dan-Azumi, J. (2010) 'Agricultural sustainability of fadama farming systems in northern Nigeria: the case of Karshi and Baddeggi', *International Journal of Agricultural Sustainability*, 8: 319–330.

Davidson, N. (2014) 'How much wetland has the world lost? Long-term and recent trends in global wetland area', *Marine and Freshwater Research*, 65: 934–941.

Davis, T.J. (1994) *The Ramsar Convention Manual: A Guide to the Convention on Wetlands of International Importance Especially as Waterfowl Habitat*, Gland: Ramsar Convention Bureau.

Denny, P. (1993) 'Wetlands of Africa: introduction', in D. Whigham, D. Dykjova and S. Hejny (eds), *Wetlands of the World: Inventory, Ecology and Management. Volume 1: Handbook of Vegetation Science*, Dordrecht: Kluwer Academic Publishers, pp. 1–31.

Denny, P. (2001) 'Research, capacity-building and empowerment for sustainable management of African wetland ecosystems', *Hydrobiologia*, 458: 21–31.

Dixon, A.B. (2003) *Indigenous Management of Wetlands: Experiences in Ethiopia*, Aldershot: Ashgate.

Dixon, A.B. (2005) 'Wetland sustainability and the evolution of indigenous knowledge in Ethiopia', *The Geographical Journal*, 171: 306–323.

Dixon, A.B. (2008) 'The resilience of local wetland management institutions in Ethiopia', *Singapore Journal of Tropical Geography*, 29: 341–357.

Dixon, A.B. and Carrie, R. (2015) 'Creating local institutional arrangements for sustainable wetland socio-ecological systems: lessons from the "Striking a Balance" project in Malawi', *International Journal of Sustainable Development and World Ecology*, 23: 40–52.

Dixon, A.B. and Wood, A.P. (2003) 'Wetland cultivation and hydrological management in eastern Africa: matching community and hydrological needs through sustainable wetland use', *Natural Resources Forum*, 27: 117–129.

Dixon, A.B., Hailu, A. and Semu, T. (2013) 'Local institutions, social capital and sustainable wetland management: experiences from western Ethiopia', in A.P. Wood, A.B. Dixon and M. McCartney (eds), *Wetland Management and Sustainable Livelihoods in Africa*, London: Earthscan, pp. 85–111.

Drijver, C.A. and Marchand, M. (1985) *Taming the Floods: Environmental Aspects of Floodplain Development in Africa*, Leiden: Centre for Environmental Studies, State University of Leiden.

Dugan, P.J. (1990) *Wetland Conservation: A Review of Current Issues and Action*, Gland: IUCN.

Ellen, R., Parkes, P. and Bicker, A. (eds) (2000) *Indigenous Environment Knowledge and Its Transformations: Critical Anthropological Perspectives*, London: Harwood Academic Publishers.

Emerton, L. (2005) *Values and Rewards: Counting and Capturing Ecosystem Water Services for Sustainable Development*, IUCN Water, Nature and Economics Technical Paper No. 1, Gland: IUCN.

Emerton, L., Lyango, L., Luwum, P. and Malinga, A. (1998) *The Present Economic Value of Nakivubo Urban Wetland, Uganda*, Nairobi: IUCN.

FAO (Food and Agriculture Organization) (2004) *Wetland Contributions to Livelihoods in Zambia*, Netherlands Partnership Programme: Sustainable Development and Management of Wetlands, Rome: FAO.

Farrier, D. and Tucker, L. (2000) 'Wise use of wetlands under the Ramsar Convention: a challenge for meaningful implementation of international law', *Journal of Environmental Law*, 12: 21–42.

Fisher, R.J., Magginnis, S. and Jackson, W.J. (2005) *Poverty and Conservation: Landscapes, People and Power*, Gland: IUCN.

Flint, L. (2006) 'Contradictions and challenges in representing the past: the Kuomboka Festival of western Zambia', *Journal of Southern African Studies*, 32: 701–717.

Hollis, G.E. (1990) 'Environmental impacts of development on wetlands in arid and semi-arid lands', *Hydrological Sciences Journal*, 35: 411–428.

Hollis, G.E., Adams, W.M. and Aminu-Kano, M. (1993) *The Hadejia-Nguru Wetlands: Environment, Economy and Sustainable Development of a Sahelian Floodplain*, Gland: IUCN.

Hughes, F.M.R. (1996) 'Wetlands', in W.M. Adams, S.A. Goudie and A.R. Orme (eds), *The Physical Geography of Africa*, Oxford: Oxford University Press, pp. 267–286.

Kamukala, G.L and Crafter, S.A (eds) (1993) *Wetlands of Tanzania, Proceedings of a Seminar on the Wetlands of Tanzania. Morogoro, Tanzania, 27–29 November 1991*, Gland: IUCN.

Kay, M. and Brabben, T. (2000) *Treadle Pumps for Irrigation in Africa*, International Programme for Technology and Research in Irrigation and Drainage, Rome: FAO.

Kellert, S.R., Mehta, J.N., Ebbin, S.A. and Lichtenfeld, L.L. (2000) 'Community natural resource management: promise, rhetoric, and reality', *Society and Natural Resources*, 13: 705–715.

Kotze, D. (2013) 'Assessing the ecological sustainability of wetland cultivation: experiences from Zambia and Malawi', in A.P. Wood, A.B. Dixon and M. McCartney (eds) *Wetland Management and Sustainable Livelihoods in Africa*, London: Earthscan, pp. 138–159.

MA (Millennium Ecosystem Assessment) (2005) *Ecosystems and Human Well-Being: Wetlands and Water Synthesis*, Washington, DC: World Resources Institute.

Mafabi, P. (2005) *From Conversion to Conservation: Fifteen Years of Managing Wetlands for People and the Environment in Uganda*, Gland: IUCN.

Maltby, E. (1986) *Waterlogged Wealth: Why Waste the World's Wet Places?* London: Earthscan.

Maltby, E. and Barker, T. (2009) *The Wetlands Handbook*, Chichester: Wiley-Blackwell.

Marchand, M. and Udo, H.A. (eds) (1989) *The People's Role in Wetland Management*, Proceedings of the International Conference in Leiden, the Netherlands, 5–8 June, Leiden University, Centre for Environmental Studies.

Masiyandima, M., McCartney, M.P. and van Koppen, B. (2004) *Wetland Contributions to Livelihoods in Zambia*, Netherlands Partnership Programme: Sustainable Development and Management of Wetlands, Rome: FAO.

Matthews, G.V.T. (1993) *The Ramsar Convention on Wetlands: Its History and Development*, Gland: Ramsar Convention Bureau.

McCartney, M.P. and van Koppen, B. (2004) *Wetland Contributions to Livelihoods in United Republic of Tanzania*, Netherlands Partnership Programme: Sustainable Development and Management of Wetlands, Rome: FAO.

McCartney, M., Rebelo, L., Sellamuttu, S.S. and de Silva, S. (2010) *Wetlands, Agriculture and Poverty Reduction*, IWMI Research Report 137, Colombo: IWMI.

Mloza-Banda, H.R. (2005) 'Integrating new trends in farming systems approaches in Malawi', *African Crop Science Conference Proceedings*, 7: 961–966.

MNR (Ministry of Natural Resources) (1995) *National Policy for the Conservation and Management of Wetland Resources*, Kampala: Ministry of Natural Resources.

Namaalwa, S., van Dam, A.A., Funk, A., Ajie, G.S. and Kaggwa, R.C. (2013) 'A characterization of the drivers, pressures, ecosystem functions and services of Namatala wetland, Uganda', *Environmental Science and Policy*, 34: 44–57.

Noble, R. (1996) 'Wetland management in Malawi as a focal point for ecologically sound agriculture', *ILEA Newsletter*, 12: 9.

Osbahr, H., Twyman, C., Thomas, D.S.G. and Adger, W.N. (2008) 'Effective livelihood adaptation to climate change disturbance: scale dimensions of practice in Mozambique', *Geoforum*, 39: 1799–2132.

Pollard, S. and Cousins, R. (2008) *Community-Based Governance of Freshwater Resources in Southern Africa*, Research Report TT 328/08, Pretoria: Water Research Commission.

Ramsar (1971) *Convention on Wetlands of International Importance Especially as Waterfowl Habitat*, Final Text adopted by the International Conference on the Wetlands and Waterfowl at Ramsar, Iran, 2 February, available at: www.ramsar.org/sites/default/files/documents/library/original_1971_convention_e.pdf (accessed 13 October 2015).

Ramsar (1987) *Convention on Wetlands (Ramsar, Iran, 1971) 3rd Meeting of the Conference of the Contracting Parties Regina, Canada 27 May–5 June 1987. Recommendation 3.3: Wise Use of Wetland*, available at: www.ramsar.org/sites/default/files/documents/library/key_rec_3.03e.pdf (accessed 13 October 2015).

Ramsar (2005) *9th Meeting of the Conference of the Parties to the Convention on Wetlands (Ramsar, Iran, 1971) Kampala, Uganda, 8–15 November 2005. Resolution IX.1 Annex A: A Conceptual Framework for the Wise Use of Wetlands and the Maintenance of Their Ecological Character*, available at: www.ramsar.org/sites/default/files/documents/pdf/res/key_res_ix_01_annexa_e.pdf (accessed 13 October 2015).

Ramsar (2009) *What Are Wetlands?*, available at: http://archive.ramsar.org/cda/es/ramsar-about-faqs-what-are-wetlands/main/ramsar/1-36-37%5E7713_4000_2_ (accessed 12 October 2015).

Ramsar and UNWTO (United Nations World Tourism Organization) (2012) *Destination Wetlands: Supporting Sustainable Tourism*, Gland and Madrid: Secretariat of the Ramsar Convention on Wetlands Switzerland and World Tourism Organization.

Reij, C. and Waters-Bayer, A. (eds) (2001) *Farmer Innovation in Africa: A Source of Inspiration for Agricultural Development*, London: Earthscan.

Richards, P. (1985) *Indigenous Agricultural Revolution: Ecology and Food Production in West Africa*, London: Hutchinson.

Roberts, N. (1988) 'Dambos in development: management of a fragile ecological resource', *Journal of Biogeography*, 15: 141–148.

Roggeri, H. (1995) *Tropical Freshwater Wetlands: A Guide to Current Knowledge and Sustainable Management*, Dordrecht: Kluwer Academic Publishers.

Scoones, I. (1991) 'Wetlands in drylands: key resources for agricultural and pastoral development in Africa', *Ambio*, 20: 366–371.

Sellamuttu, S.S., de Silva, S., Nguyen Khoa, S. and Samarakoon, J. (2008) *Good Practices and Lessons Learned in Integrating Ecosystem Conservation and Poverty Reduction Objectives in Wetlands*, Colombo and Wageningen: IWMI and Wetlands International.

Seyam, I.M., Hoekstra, A.Y. and Ngabirano, H.H.G. (2001) *The Value of Freshwater Wetlands in the Zambezi Basin*, Delft: UNESCO-IHE, Institute for Water Education.

Shuyt, K.D. (2005) 'Economic consequences of wetland degradation for local populations in Africa', *Ecological Economics*, 53: 177–190.

Tanko, A.I. (2013) 'Agriculture, livelihoods and fadama restoration in northern Nigeria', in A.P. Wood, A.B. Dixon and M. McCartney (eds), *Wetland Management and Sustainable Livelihoods in Africa*, London: Earthscan, pp. 205–228.

Tarhule, A. and Woo, M. (1997) 'Characteristics and use of shallow wells in a stream fadama: a case study in northern Nigeria', *Applied Geography*, 17: 29–42.

Terrer, T., Ndiritu, G.G. and Gichuki, N.N. (2004) 'Socio-economic values and traditional strategies of managing wetland resources in Lower Tana River, Kenya', *Hydrobiologia*, 527: 3–15.

Trapnell, C.G. and Clothier, J.N. (1937) *The Soils, Vegetation and Agriculture Systems of North-Western Rhodesia*, Lusaka: Government of Northern Rhodesia.

Turner, B. (1986) 'The importance of dambos in African agriculture', *Land Use Policy*, 3: 343–347.

Turpie, J.K., Smith, B., Emerton, L. and Barnes, J. (1999) *Economic Value of the Zambezi Basin Wetlands*, Harare: IUCN ROSA.

UNEP (United Nations Environment Programme) (2006) *Africa Environment Outlook 2: Our Environment, Our Wealth*, Nairobi: UNEP.

van Koppen, B., Giordano, M. and Butterworth, J. (2007) *Community-Based Water Law and Water Resource Management Reform in Developing Countries*, Wallingford: CABI.

Warren, D.M., Slikkerveer, L.J. and Brokensha, D. (1995) *The Cultural Dimension of Development: Indigenous Knowledge Systems*, London: ITDG Publishing.

Wetlands International (2005) *The Wetlands and Poverty Reduction Project: Linking Wetland Conservation and Poverty Alleviation*, Wageningen: Wetlands International.

Wood, A.P. and Thawe, P. (2013) 'Catchments and wetlands: a functional landscape approach to sustainable use of seasonal wetlands in central Malawi', in A.P. Wood, A.B. Dixon and M. McCartney (eds), *Wetland Management and Sustainable Livelihoods in Africa*, London: Earthscan, pp. 63–84.

Wood, A.P., Dixon, A.B. and McCartney, M. (eds) (2013) *Wetland Management and Sustainable Livelihoods in Africa*, London: Earthscan.

Woodhouse, P., Bernstein, H. and Hulme, D. (2000) *African Enclosures? The Social Dynamics of Wetlands in Drylands*, Oxford: James Currey.

Zwarts, L., Beukering, P.V., Kone, B., Wymenga, E. and Taylor, D. (2006) 'The economic and ecological effects of water management choices in the Upper Niger River: development of decision support methods', *Water Resources Development*, 22: 135–156.

18

Conservation and people

Fiona Nunan

Introduction

Conservation in Africa is most commonly associated with wildlife and national parks. Images of untamed landscapes and the 'big five' must-see mammals support calls for land and wildlife to be protected. It has long been recognised, however, that landscapes are not always as 'untamed' and uninhabited as at times portrayed. People live in the same areas as wildlife, seeking to graze their livestock on the same pastures and farm their crops in areas where wild animals roam (Adams and Hulme, 2001; Brockington, 2002; Fabricius, 2004). There are, in addition, many other natural resources people draw on and interact with, such as forests, fisheries and grazing land, with their own conservation and livelihood dilemmas (Nelson, 2010). This chapter focuses on the experience of conservation in relation to wildlife and protected areas, and draws, in places, on broader literature, bringing insights from other areas of natural resource governance and livelihoods.

The chapter is informed by two key interlinked questions that have dominated conservation thinking and debate in Africa in relation to people:

1 To what extent should land be 'protected' or 'fenced off' to prevent use by people and provide land for wildlife?
2 Can the involvement of local people in the management of protected areas improve both conservation outcomes and livelihoods, thereby creating 'win–win' situations?

These themes are explored in this chapter through sections on the evolution of conservation approaches, community-based natural resource management, poverty and poaching, and contemporary thinking and approaches.

'Protected area' (PA) is the generic term used for areas that have some degree of protection to support conservation. The International Union for the Conservation of Nature (IUCN) defines a PA as 'a clearly defined geographical space, recognised, dedicated and managed, through legal or other effective means, to achieve the long-term conservation of nature with associated ecosystem services and cultural values' (Dudley, 2013: 8). The IUCN has led on the development of common categories and definitions of PAs, though terms and definitions may vary between countries. The use of common categories enables comparison across countries and regions and sets standards for each category of protection. There are currently six categories, with increasing human influence and presence through the categories, ranging from Category 1 for strict nature reserves and wilderness areas to Category 6 for protected areas with sustainable use (Dudley, 2013).

The evolution of conservation approaches

The conservation of wildlife through designation of PAs is usually traced back to the creation of the Yellowstone National Park in the USA in 1872. However, tracing protected areas back to this park has been seen by some as problematic in two ways: it neglects the history of conservation through designated areas prior to that, and promotes a particular model of conservation associated with untouched landscapes and keeping people out (Brockington *et al.*, 2008). It was this 'model' and approach that dominated conservation approaches in Africa in the twentieth century up until around the 1980s.

The model is characterised as taking a 'fences and fines' approach – where fences are erected to keep people and their livestock out, and sanctions are imposed on people who encroach on, or allow their animals to graze in, fenced-off parks; this has also been referred to as 'fortress conservation' (Adams and Hulme, 2001). This type of approach often led to the displacement of people from their land and to prevention of their use of certain land for grazing and moving livestock.

From the 1980s, there was a shift in thinking towards the inclusion of local people in the management of conservation areas, in what became known as community-based natural resource management (CBNRM) (Adams and Hulme, 2001). While community-based approaches have been taken up to some degree, elements of fortress conservation remain, and even where CBNRM approaches have been adopted, there is still significant involvement of government ministries and agencies, at national and local levels (Benjaminsen and Svarstad, 2010). In addition to CBNRM, there may be collaborative arrangements, sometimes referred to as 'co-management' or 'joint management', with governments at national and local levels working in conjunction with local people and other resource users to manage natural resources (Barrow and Murphree, 2001).

CBNRM has been defined as 'the management of resources such as land, forests, wildlife and water by collective, local institutions for local benefit' (Roe and Nelson, 2009: 5). CBNRM is premised on the assumption that people have a greater sense of ownership and responsibility by being involved in decision-making, and this would therefore lead to better protection of wildlife and less poaching. It is also premised on the belief that involving local people in conservation will lead to livelihood benefits through employment in guarding and tourism and empowerment, through participation in decision-making. CBNRM, it is argued, should lead to 'win–win' situations of better protection and greater sustainability at the same time as improving livelihoods and reducing poverty. CBNRM initiatives are particularly associated with wildlife management (national parks and protected areas), with benefits accruing from park entrance fees, trophy hunting and related tourism services. Reviews of CBNRM initiatives note achievements in income generation, empowerment of disadvantaged groups and in the productivity of resources, as well as challenges in implementation (e.g. see Roe *et al.*, 2009).

The widespread adoption of CBNRM has often been supported or fuelled by growth in tourism. Tourism may be encouraged to support conservation, through park entrance fees, safaris, accommodation and trophy hunting. The development of tourism to support conservation has, however, been viewed as part of the wider 'neo-liberalising nature' endeavour, with nature being seen to be 'commodified', as an economic resource alone, to be capitalised upon. Duffy and Moore (2010) see the growth in global tourism as one of the drivers of neo-liberalism, referring to elephant-backed tourism in Botswana and Thailand as an example of how nature has been 'neo-liberalised' – reduced to economic value and a purveyor of economic growth driven by market forces rather than the state.

This neo-liberal approach, with encouragement of the market and private sector, is also seen in the increasing designation of privately held land as protected areas. Private reserves can tap into the tourism and trophy hunting markets and employ guards, at the same time as providing greater land and corridors for wildlife and reducing demand on government budgets for conservation. However, such initiatives raise concerns about regulating activities, access to land and resources for local people, and leaving conservation to the vagaries of market forces.

Community-based natural resource management

CBNRM and joint management approaches are informed by common property theory, largely associated with Elinor Ostrom, developed partly in response to Garrett Hardin's *Tragedy of the Commons* (1968). Hardin argued that a resource cannot be sustainably maintained over time where there is common ownership because people will want to maximise their own benefit, even at the expense of others. Eventually, the resource will become degraded and of little productive use. He put forward a metaphor of herders grazing common land, with each herder facing the incentive of grazing as many of their own livestock as possible to gain the most out of the land. Eventually, this disregard for the common good will lead to the land being overgrazed and degraded. This tragedy scenario has informed many natural resource management approaches, with adoption of state ownership and privatisation of land and other resources resulting. This kind of thinking supports the 'fences and fines' or 'fortress conservation' approach introduced above (Nelson, 2010).

Ostrom and others, such as Fikret Berkes (1989), developed a research agenda during the 1980s focused on resources managed as commons, analysing examples where resource users have cooperated over time and developed complex institutions to govern resource use. Such examples demonstrated that common property regimes could lead to sustainable resource use and, it was argued, that Hardin's metaphor had more in common with open-access regimes, where there is no cooperation or rules to govern access and use. Ostrom (1990) put forward eight 'design principles' that are associated with successful commons governance, leading to sustainable use. These design principles informed the design of CBNRM policies and initiatives, including identifying the area to be governed, who should be involved, what rules there should be, and how conflict over resource use should be resolved (Fabricius, 2004; Jones, 2010). The design principles are set out and explained in Table 18.1.

These design principles have been critiqued by some as being too prescriptive (e.g. see Cleaver, 2012), and Ostrom herself later emphasised the need for context-specific solutions and approaches, rather than taking a panacea perspective and seeing community-based management, or certain common property regimes, as the solution for all natural resource dilemmas (Ostrom and Cox, 2010). This implies that there should be many forms of CBNRM or joint management approaches and that CBNRM will not be suitable for all situations.

Table 18.1 Ostrom's design principles for effective common property regimes

Design principle	*Explanation*
1 Clearly defined boundaries	There should be agreement about who – individuals, groups of people, households – has the right to use the natural resource and what the boundaries are for the property held in common.
2 Rules governing use or provision of the resource must be appropriate to local conditions	Rules are needed about where the resource can be extracted, what can be taken, by whom, when and how much, using what technology and approaches. These should be appropriate to the local situation.
3 Collective choice arrangements	People affected by the rules should be able to participate, or have a say in, what those rules are. This is possible through arrangements for collective decision-making, often through a user group or committee.
4 Monitoring of rules and use: by users or accountable to the users	There should be a system in place to monitor that the rules are being abided by. This may involve the resource users themselves or others who are accountable to the users.
5 Graduated sanctions	Sanctions, such as fines, exclusion or even being taken to court are applied in a progressive manner, depending on the seriousness of the offence and how many times it has happened. The sanction may be applied by the users or appropriate officials, who should be accountable to the users.
6 Conflict resolution mechanisms	Mechanisms should be in place to resolve conflicts between users or between users and officials.
7 Recognition of legitimacy	There should be formal recognition for the users to create their own collective choice arrangements and rules.
8 Nested enterprises	Where common property resources are part of larger systems, there may be multiple levels of structures and rules, with one nested in the next – reporting and resource sharing may take place.

Source: Adapted from Ostrom (1990: 90).

One of the first examples of CBNRM was the CAMPFIRE programme in Zimbabwe. CAMPFIRE – the Communal Areas Management Programme for Indigenous Resources – was set up by the then Department of National Parks and Wildlife Management in the mid-1980s, with authority devolved to the rural district councils (RDCs) and revenue shared with communities through ward-level committees. Most of the revenue accruing through CAMPFIRE comes from trophy hunting and ecotourism (Taylor, 2009). The programme remains active despite challenges over the years, but has been critiqued for the lack of devolution of rights to the community level. As a consequence of rights going to the RDCs, communities were dependent on the RDCs sending on revenue rather than receiving it directly until reforms in 2005 allowed for direct payments to communities (Taylor, 2009; Matema and Andersson, 2015). In some areas, participatory decision-making structures have not been sustained over time, with capture of key positions and committees by elite members of communities, limiting the resulting benefits to the wider community further (Balint and Mashinya, 2006).

A further example of CBNRM is the community conservancy programme in Namibia, the key features of which are set out in Box 18.1. It can be seen that the approach in Namibia differs from Zimbabwe in that revenue goes directly to community-based conservancies rather than through government, meaning that the communities receive more income and are in greater control of decision-making. The Namibian conservancy approach is widely seen as being successful as it delivers on the much sought-after 'win–win' objectives of CBNRM. It generates revenue for communities and delivers on conservation outcomes through increasing the diversity and number of animal species (Binot *et al.*, 2009).

Despite some examples of success, CBNRM is largely seen as having failed to deliver on conservation and livelihood outcomes, with factors contributing to failure including the top-down nature of initiation, limited devolution of power and lack of economic return (Murphree, 2009; Measham and Lumbasi, 2013). Measham and Lumbasi (2013) suggest, however, that CBNRM is more likely to succeed if the initiative is driven by the community themselves rather than by government, NGOs or donors. Meta-analysis of 55 cases by Andrade and Rhodes (2012: 5) demonstrated that, indeed, the inclusion of local people in the management of PAs is 'likely to be a key determinant of the level of compliance with PA conservation strategies'.

CBNRM approaches have faced many challenges. These can be summarised as: definitions of a 'community'; elite capture; and incomplete decentralisation of power, responsibilities and resources. The term 'community' within CBNRM has been criticised for portraying communities as homogenous groups of people with shared interests and concerns (Agrawal and Gibson, 1999; Blaikie, 2006; Brockington *et al.*, 2008). As Agrawal and Gibson (1999: 633) observe:

> this vision fails to attend to differences within communities, and ignores how these differences affect resource management outcomes, local politics, and strategic interactions within communities, as well as the possibility of layered alliances that can span multiple levels of politics.

Elite capture refers to situations where more powerful actors within communities take control of the governance system. They may do this to capture the benefits from natural resource management, prevent change that is not in their interest and promote change that will benefit them. Elite capture has been identified as a common problem facing CBNRM and collaborative management, with such capture often reflecting social and power relations that existed prior to the introduction of the new governance approach. Muyengwa *et al.* (2014) identify five ways that elites may benefit: establishing entitlements to benefits (e.g. cash transfers to elites); making demands; influencing allocation decisions; stating preferences (for committee members, use of funds and projects to be initiated); and misappropriation of funds. They cite examples as including traditional leaders claiming allowances, regular payments and meat in Zimbabwe's CAMPFIRE, and the prevention of record-keeping and accountability of committees in Botswana. Evidence on how elite capture may be countered suggests that it is likely to take time for others to be sufficiently organised and experienced to gain power, though it is also noted that 'elite capture' does not necessarily imply that local people are dissatisfied with the arrangements (Lund and Saito-Jensen, 2013). In addition, the presence of an external agency in decentralised natural resource governance arrangements can reduce the risk of elite capture (Persha and Andersson, 2014).

Box 18.1 Community conservancies of Namibia

Community conservancies were created in the 1990s, enabled by the Nature Conservation Amendment Act of 1996. A conservancy should be registered with the Ministry of Environment and Tourism, and have:

- a representative committee;
- a legal constitution which provides for the sustainable management and utilisation of game in the conservancy;
- a method for the equitable distribution of income from the sustainable use of wildlife and from tourism;
- a defined membership; and
- defined boundaries agreed by neighbouring communities.

(Jones, 2010: 22)

By March 2013, 79 conservancies were registered (Jones *et al.*, 2015). The conservancies receive revenue directly (e.g. from tourism activities and trophy hunting). Decision-making over how revenue is used and how the conservancy is managed is undertaken by an elected committee, which has a fixed term of office. Much of the revenue is used to pay staff salaries – with game guards being the main type of employee. Committee members do not receive salaries, but may receive an allowance. Instances of misappropriation of funds have occurred, though action against committee members rarely happens (Bollig and Schwieger, 2014).

Humavindu and Stage (2015) report on an evaluation of the financial and economic sustainability of the first 59 conservancies set up in Namibia, using data from the Namibian Association of CBNRM Support Organisations (NACSO). They selected Namibia as an example of CBNRM due to the long duration and scale of the programme. The analysis confirmed that almost all of the revenue was related to tourism, the social net benefit is consistently higher than the financial benefits, and the financial viability of many conservancies was very questionable. Costs are often higher than revenue, and without government and/or donor support, the authors question whether the conservancy approach is sustainable into the long term in Namibia.

Finally, many examples of CBNRM projects have demonstrated little evidence of power and responsibility being devolved to communities. 'Community involvement' has, at times, been seen as amounting to little more than consultation (Shackleton *et al.*, 2010). This reflects the broader experience of the decentralisation of natural resource governance to lower levels of government and to resource users, where limited power and resources have meant that structures at this level are not able to deliver on conservation outcomes (Ribot, 2002).

Poverty and poaching

There are two key narratives on poverty in relation to conservation. These are: (i) illegal hunting and encroaching onto protected land for livestock grazing is caused by poor people lacking alternative options and having short-term perspectives; and (ii) sustainable use of natural resources can contribute towards poverty reduction, as assumed in the thinking behind CBNRM (Adams et al., 2004). However, delivering on 'win–win' – positive conservation and poverty reduction outcomes – is very difficult in practice, with poor people often losing out as a result of establishing a protected area. This may be through displacement, loss of grazing land or loss of crops from animal damage. From local people's perspectives, one of the most significant challenges resulting from protected areas is destruction of their livelihoods through crop destruction and livestock predation by wildlife. Barua et al. (2013) argue that hidden costs associated with people–wildlife conflicts, such as impacts on psychosocial well-being and food security, should be taken into consideration, as well as the more obvious costs on guarding crops and loss of food crops.

In terms of understanding illegal hunting, two types are usually identified. Commercial poaching refers to illegal extraction of wildlife for financial gain, whereas subsistence poaching involves hunters taking game for home consumption. These categories can be blurred in practice, with hunting containing elements of both and interactions between the two categories. Commercial poaching has also been associated with armies and rebel groups (Duffy, 2000), as well as demand for wildlife products in East Asia driving the practice. Blaming 'the poor' for poaching, however, is unhelpful, as well as not always accurate. Duffy *et al.* (2016) argue that a much broader understanding of poverty, and the causes of poverty, is needed to inform anti-poaching approaches. The historical, economic and social context of poverty and poaching should be investigated to identify factors such as the perceived legitimacy of regulations and agencies, claims over resources, and wider networks involved in wildlife trade.

Contemporary thinking and approaches

Two areas of contemporary thinking and practice prevalent in conservation in Africa are the increasing re-centralisation of conservation and the adoption of payments for ecosystem services schemes, to provide additional revenue to support conservation through compensating local resource users.

While community-based management approaches have been prevalent in conservation discourse and policy since the 1980s, it has been observed that there has been a reversal of this, with states taking back control and revenue (Benjaminsen and Svarstad, 2010; Benjaminsen *et al.*, 2013; Nelson, 2010). In addition to concern that states are re-centralising conservation management, some conservationists have advocated for separation of conservation objectives from poverty reduction objectives, arguing that CBNRM has failed to deliver on conservation objectives (Adams and Hutton, 2007; Torquebiau and Taylor, 2010).

One of the challenges for CBNRM has been to deliver on adequate economic returns to local people. A relatively new mechanism may provide a solution to this. Payments for ecosystem services (PES) schemes have been developed to provide incentives or compensation to the direct users of natural resources to conserve or undertake different practices to deliver on environmental outcomes such as more sustainable land management, biodiversity conservation or carbon storage. PES schemes do this through generating payments from the beneficiaries of the ecosystem services (e.g. downstream water users or tourists viewing wildlife or enjoying a landscape). There are few examples of conservation of wildlife through PES schemes, though some exist, such as the the Simanjiro PES initiative in Tanzania, where a consortium of tourism operators negotiated with a village to make an annual payment in exchange for no agricultural cultivation or permanent settlement in a key wildlife dispersal area outside Tarangire National Park (Nelson *et al.*, 2010).

Other commentators have drawn on experience from CBNRM to inform PES and associated 'payments for wildlife services' schemes, such as learning from the CAMPFIRE programme in Zimbabwe that schemes need to be flexible and adaptive (Frost and Bond, 2008). In a similar vein, Naidoo *et al.* (2011) demonstrate how Namibia's CBNRM programme, described in Box 18.1, can be considered as a PES scheme by analysing the programme through the five components expected of PES schemes. They identify sellers, buyers and what it is that is being sold (photographic safaris, trophy hunting and products), and assess the degree to which the transactions can be considered 'voluntary' and the extent of 'conditionality' of the programme (i.e. whether the transaction will lead to an improvement in ecosystem conditions). From their analysis, they suggest that PES has much to learn from CBNRM, and recognising some examples of CBNRM as PES schemes would encourage such lesson learning.

Key points

Conservation is a challenging area for development in Africa, with many contentious debates concerning concepts, drivers, objectives and approaches. Key points from this introduction to conservation include:

- Conservation was associated with keeping local people and their livestock out of designated areas during the colonial period through 'fences and fines' approaches and 'fortress conservation'. Elements of this thinking can still be seen in some conservation approaches today.
- From the 1980s, there was widespread rhetoric and movement towards community-based natural resource management (CBNRM), reflecting broader moves towards decentralisation within many governments in Africa and adoption of participatory approaches in development.
- CBNRM initiatives seek to deliver on conservation and livelihood improvement outcomes, referred to as 'win–win', but have rarely done so in a substantial way. This lack of delivery has been attributed to limited devolution of power and creation of space for participation, elite capture and illegal hunting.
- The relationship between poverty and poaching is complex, with many drivers and actors involved. There is no simple one-way relationship. Understanding of the historical, socio-economic and political context of people and conservation is essential.
- While CBNRM approaches are still widely adopted in conservation in Africa, there are increasing examples of re-centralisation of conservation by the state, more use of private land for conservation, and interest in payments for ecosystem services schemes to support conservation and livelihoods.

The approach taken to conservation and the level and nature of participation of local people differs over time and space; there is no panacea approach that should be adopted. There are, however, many lessons and insights from a wealth of experience that should inform a more inclusive, adaptive and effective response to the need for conservation and improved livelihoods.

References

Adams, W. and Hulme, D. (2001) 'Conservation & community: changing narratives, policies & practices in African conservation', in D. Hulme and M. Murphree (eds), *African Wildlife & Livelihoods: The Promise and Performance of Community Conservation*, Oxford: James Currey, pp. 9–23.

Adams, W.M. and Hutton, J. (2007) 'People, parks and poverty: political ecology and biodiversity conservation', *Conservation and Society*, 5(2): 147–183.

Adams, W.M., Aveling, R., Brockington, D., Dickson, B., Elliott, J., Hutton, J., *et al.* (2004) 'Biodiversity conservation and the eradication of poverty', *Science*, 306: 1146–1149.

Agrawal, A. and Gibson, C.C. (1999) 'Enchantment and disenchantment: the role of community in natural resource conservation', *World Development*, 2(4): 629–649.

Andrade, G.S.M. and Rhodes, J.R. (2012) 'Protected areas and local communities: an inevitable partnership toward successful conservation strategies?', *Ecology and Society*, 17(4): 591–606.

Balint, P.J. and Mashinya, J. (2006) 'The decline of a model community-based conservation project: governance, capacity, and devolution in Mahenye, Zimbabwe', *Geoforum*, 37: 805–815.

Barrow, E. and Murphree, M. (2001) 'Community conservation: from concept to practice', in D. Hulme and M. Murphree (eds) *African Wildlife & Livelihoods: The Promise and Performance of Community Conservation*, Oxford: James Currey, pp. 24–37.

Barua, M., Bhagwat, S.A. and Jadhav, S. (2013) 'The hidden dimensions of human–wildlife conflict: health impacts, opportunity and transaction costs', *Biological Conservation*, 157: 309–316.

Benjaminsen, T.A. and Svarstad, H. (2010) 'The death of an elephant: conservation discourses versus practices in Africa', *Forum for Development Studies*, 37(3): 385–408.

Benjaminsen, T.A., Goldman, M.J., Minwary, M.Y. and Maganga, F.P. (2013) 'Wildlife management in Tanzania: state control, rent seeking and community resistance', *Development and Change*, 44(5): 1087–1109.

Berkes, F. (1989) *Common Property Resources*, London: Belhaven Press.

Binot, A., Blomley, T., Coad, L., Nelson, F., Roe, D. and Sandbrook, C. (2009) 'Community involvement in natural resources management in Africa: regional overviews', in D. Roe, F. Nelson and C. Sandbrook (eds), *Community Management of Natural Resources in Africa: Impacts, Experiences and Future Directions*, London: IIED, pp. 113–154.

Blaikie, P. (2006) 'Is small really beautiful? Community-based natural resource management in Malawi and Botswana', *World Development*, 34(11): 1942–1957.

Bollig, M. and Schwieger, D.A.M. (2014) 'Fragmentation, cooperation and power: institutional dynamics in natural resource governance in north-western Namibia', *Human Ecology*, 42: 167–181.

Brockington, D. (2002) *Fortress Conservation: The Preservation of the Mkomazi Game Reserve, Tanzania*, Oxford: James Currey.

Brockington, D., Duffy, R. and Igoe, J. (2008) *Nature Unbound: Conservation, Capitalism and the Future of Protected Areas*, London: Earthscan.

Cleaver, F. (2012) *Development Through Bricolage: Rethinking Institutions for Natural Resource Management*, London: Earthscan.

Dudley, N. (ed.) (2013) *Guidelines for Applying Protected Area Management Categories*, Gland, Switzerland: IUCN. Including S. Stolton, P. Shadie and N. Dudley (2013) *IUCN WCPA Best Practice Guidance on Recognising Protected Areas and Assigning Management Categories and Governance Types*, Best Practice Protected Area Guidelines Series No. 21, Gland, Switzerland: IUCN.

Duffy, R. (2000) *Killing for Conservation: Wildlife Policy in Zimbabwe*, Oxford: James Currey.

Duffy, R. and Moore, L. (2010) 'Neoliberalising nature? Elephant-back tourism in Thailand and Botswana', *Antipode*, 42(3): 742–766.

Duffy, R., St John, F.A.V., Büscher, B. and Brockington, D. (2016) 'Towards a new understanding of the links between poverty and illegal wildlife hunting', *Conservation Biology*, 30(1): 14–22.

Fabricius, C. (2004) 'The fundamentals of community-based natural resource management', in C. Fabricius, E. Koch, H. Magome and S. Turner (eds), *Rights, Resources & Rural Development: Community-Based Natural Resource Management in Southern Africa*, London: Earthscan, pp. 3–43.

Frost, P.G.H. and Bond, I. (2008) 'The CAMPFIRE programme in Zimbabwe: payments for wildlife services', *Ecological Economics*, 65: 776–787.

Hardin, G. (1968) 'The tragedy of the commons', *Science*, 162(3859): 1243–1248.

Humavindu, M.N. and Stage, J. (2015) 'Community-based wildlife management failing to link conservation and financial viability', *Animal Conservation*, 18: 4–13.

Jones, B. (2010) 'Ostrom and Namibian conservancies', *Current Conservation*, 4(3): 21–23.

Jones, B.T.B., Diggle, R.W. and Thouless, C. (2015) 'From exploitation to ownership: wildlife-based tourism and communal area conservancies in Namibia', in R. van der Duim, M. Lamers and J. van Wijk (eds), *Institutional Arrangements for Conservation, Development and Tourism in Eastern and Southern Africa: A Dynamic Perspective*, Dordrecht: Springer, pp. 17–37.

Lund, J.F. and Saito-Jensen, M. (2013) 'Revisiting the issue of elite capture of participatory initiatives', *World Development*, 46: 104–112.

Matema, S. and Andersson, J.A. (2015) 'Why are lions killing us? Human–wildlife conflict and social discontent in Mbire District, northern Zimbabwe', *The Journal of Modern African Studies*, 53: 93–120.

Measham, T.G. and Lumbasi, J.A. (2013) 'Success factors for community-based natural resource management (CBNRM): lessons from Kenya and Australia', *Environmental Management*, 52: 649–659.

Murphree, M.W. (2009) 'The strategic pillars of communal natural resource management: benefit, empowerment and conservation', *Biodiversity Conservation*, 18: 2551–2562.

Muyengwa, S., Child, B. and Lubilo, R. (2014) 'Elite capture: a comparative case study of meso-level governance in four southern Africa countries', in G. Barnes and B. Child (eds), *Adaptive Cross-Scalar Governance of Natural Resources*, London: Routledge, pp. 179–202.

Naidoo, R., Weaver, C., De Longcamp, M. and Du Plessis, P. (2011) 'Namibia's community-based natural resource management programme: an unrecognized payments for ecosystem services scheme', *Environmental Conservation*, 38(4): 445–453.

Nelson, F. (2010) 'Introduction: the politics of natural resource governance in Africa', in F. Nelson (ed.), *Community Rights, Conservation and Contested Land: The Politics of Natural Resource Governance in Africa*, London: Earthscan, pp. 3–31.

Nelson, F., Foley, C., Foley, L.S., Leposo, A., Loure, E., Peterson, D., *et al.* (2010) 'Payments for ecosystem services as a framework for community-based conservation in northern Tanzania', *Conservation Biology*, 24(1): 78–85.

Ostrom, E. (1990) *Governing the Commons: The Evolution of Institutions for Collective Action*, Cambridge: Cambridge University Press.

Ostrom, E. and Cox, M. (2010) 'Moving beyond panaceas: a multi-tiered diagnostic approach for social-ecological analysis', *Environmental Conservation*, 37(4): 451–463.

Persha, L. and Andersson, K. (2014) 'Elite capture risk and mitigation in decentralized forest governance regimes', *Global Environmental Change*, 24: 265–276.

Ribot, J.C. (2002) *Democratic Decentralization of Natural Resources: Institutionalizing Popular Participation*, Washington, DC: World Resources Institute.

Roe, D. and Nelson, F. (2009) 'The origins and evolution of community-based natural resource management in Africa', in D. Roe, F. Nelson and C. Sandbrook (eds), *Community Management of Natural Resources in Africa: Impacts, Experiences and Future Directions*, London: IIED, pp. 2–12.

Roe, D., Nelson, F. and Sandbrook, C. (eds) (2009) *Community Management of Natural Resources in Africa: Impacts, Experiences and Future Directions*, London: IIED.

Shackleton, C.M., Willis, T.J., Brown, K. and Polunin, N.V.C. (2010) 'Reflecting on the next generation of models for community-based natural resource management', *Environmental Conservation*, 37(1): 1–4.

Taylor, R. (2009) 'Community based natural resource management in Zimbabwe: the experience of CAMPFIRE', *Biodiversity Conservation*, 18: 2563–2583.

Torquebiau, E. and Taylor, R.D. (2009) 'Natural resource management by rural citizens in developing countries: innovations still required', *Biodiversity Conservation*, 18: 2537–2550.

19

Ecosystem services in Africa

Rebecca Kariuki, Rob Marchant and Simon Willcock

Introduction

Ecosystem services (ES) are the benefits that societies derive from the ecosystems comprising of the flow of materials, energy and information from the natural environment to societies (Costanza *et al.*, 1997; Wangai *et al.*, 2016; Vaz *et al.*, 2017). The definitions of ES, however, vary among authors who apply varying emphasis on the ecological or economic importance of ES, though all definitions acknowledge a link between ecosystem processes and human utility and well-being (Braat and de Groot, 2012; Müller and Burkhard, 2012).

The concept of ES was introduced in the late 1970s (Gómez-Baggethun *et al.*, 2010; Braat and de Groot, 2012) as a bridging concept between natural and social science views of environmental pollution, resource scarcity and the subsequent views of managing economic development in a sustainable way (Braat and de Groot, 2012). ES emerged as a means of showing the dependency of societies on the biosphere (Gómez-Baggethun *et al.*, 2010; Daw *et al.*, 2011; Braat and de Groot, 2012) and was promoted after the publication of the Millennium Ecosystem Assessment (MA, 2005) (Gómez-Baggethun *et al.*, 2010; Daw *et al.*, 2011; Egoh *et al.*, 2012b; Shackleton *et al.*, 2016; Wangai *et al.*, 2016). The MEA was developed as a tool for informing the public and decision-makers (Dooley, 2005) about the sustainable management of ecosystems for human well-being (Yang *et al.*, 2013). The MEA viewed ecosystems through the services they provided to humans, the impact of humans on the provision of these services, and the consequences of ecosystem change for human well-being, subsequently developing a framework for understanding these interactions (Carpenter *et al.*, 2009; Yang *et al.*, 2013). The MEA highlighted the importance of ES to the well-being of poor societies and the relevance of the ES concept in understanding issues on environmental conservation, socio-economic development and poverty alleviation (Daw *et al.*, 2011; Whelan *et al.*, 2015). Research on ES has been growing globally since the publication of the MEA, with the first studies on ES in Africa done in South Africa (2005), Kenya (2006) and Tanzania (2007); collectively, these three countries account for 61.5 per cent of all ES publications on Africa (Wangai *et al.*, 2016).

Although the ES approach has been viewed as a tool for promoting the understanding of the biophysical environment (Ericksen *et al.*, 2012); it has been criticised as being concerned only with the positive effects of ecosystems while ignoring negative and harmful ecosystem effects

(Vaz *et al.*, 2017). These effects are sometimes referred to as ecosystem disservices (EDS), and include ecosystem functions, processes and attributes that have perceived or actual negative impacts on human well-being (Shackleton *et al.*, 2016; Vaz *et al.*, 2017). The EDS concept stemmed from research on ES provided by large cities, as well as the EDS associated with them, such as allergens from plants and trees blocking sunlight (Vaz *et al.*, 2017). Disservices can be caused by natural events such as wildfires, earthquakes and floods, as well as by anthropogenic actions such as release of toxic substances (Lyytimäki and Sipilä, 2009). However, the distinction between natural and anthropogenic causes of EDS is unclear (Lyytimäki and Sipilä, 2009), particularly because many EDS are aggravated by anthropogenically induced disturbances in natural environments (Shapiro and Báldi, 2014; Villa *et al.*, 2014) and tend to be most acute when ecosystems are poorly managed (Hohenthal *et al.*, 2015), such as in flash floods following forest clearance or in intensive fires in exotic plantations. The use of the EDS concept is, however, minimal and its analytical value can be challenged (Ango *et al.*, 2014).

Africa has a highly varied and variable climatic regime, with resulting ecosystems types and human population settlement patterns leading to varying levels of ES dependence across the continent. Most communities in Africa live in rural areas where employment opportunities are limited and natural resources underpin livelihoods. Though the financial rewards from these household-scale rural activities may be relatively small, it may be sufficient to alleviate poverty, especially where social support structures are missing (Egoh *et al.*, 2012a). Consequently, due to their dependence on natural resources, and their associated vulnerability to natural disasters, many rural African communities are reliant on ES (Daw *et al.*, 2011; Boafo *et al.*, 2016) and their well-being is determined by the ES they can access. For example, communities in humid areas of West and central Africa rely on farming and harvesting non-timber forest products for their livelihoods, while those in semi-arid areas of northern, eastern and southern Africa rely on tourism and grazing (Egoh *et al.*, 2012b). Conflict over resource management, interacting governance issues, climate and land use change can result in access to ES becoming challenging to communities that depend on them (Enfors and Gordon, 2008; Shackleton *et al.*, 2008). This is particularly so because areas providing ES, and the people benefiting from these, are unevenly distributed (Wangai *et al.*, 2016).

Ultimately, the varying dependence, perceptions and uses of ES among different communities promotes or hinders the supply of ES. When the supply of ES is higher than their demand, their value goes down, and when the demand is higher than supply, environmental degradation can occur (Wangai *et al.*, 2016). The application of market forces to ecosystems has been continually debated. If threshold levels of ES are considered essential to human well-being, then marginal analysis and monetary valuation of ES is inappropriate (Farley, 2012). However, the ES concept allows for the consideration of natural capital stocks and flows of ES, alongside traditional measures of economic capital (Turner and Daily, 2008). As such, the cost of traditional economic activities via the degradation of ecosystems may become more apparent, leading to more sustainable decision-making.

Types of ecosystem services

Four categories of ES characterise most frameworks: provisioning, regulating, cultural, and supporting services. Provisioning ES are material benefits that can be harvested from ecosystems to be directly used by humans. In communities where most households do not have access to other assets besides what the ecosystem provides, provisioning ES are important livelihood assets and food sources (Boafo *et al.*, 2016), especially during natural hazards such as floods and droughts. Provisioning ES include food, traditional medicine, bushmeat, raw materials, water, fuelwood and timber, among others, and they may be used for domestic purposes or sold and used as a source of income. For example, miombo woodlands in central, eastern and southern Africa cover an estimated area of 2.7 million km^2 (Kalaba *et al.*, 2013) and directly support livelihoods of 39 million people through the provision of food, fibre, fuelwood and charcoal to a further 15 million people who live in urban areas (Sileshi *et al.*, 2007). Miombo woodlands protect households from poverty by providing alternate food sources, fuelwood and medicine from the woodland products (Jew *et al.*, 2016), as well as food sources, largely fruits, during famines (Sileshi *et al.*, 2007). Water is another provisioning ES and is regarded as an essential natural resource (Mulatu *et al.*, 2014), but one that is highly unevenly distributed. For example, the Mara River basin extends from the headwaters on the Mau Escarpment in south-western Kenya through north-western Tanzania into Lake Victoria. As the Mara river flows through the Maasai Mara and Serengeti protected areas, the water supports wildlife and major economic activities such as pastoralism, agriculture, mining and tourism (Dessu *et al.*, 2014). Other water sources such as lakes also contribute significantly to national economies and support several livelihoods. For example, over half a million people in 11 countries in Africa are employed in fisheries, with fishing alone in Lake Chilwa in Malawi valued at US$18 million per year (Kafumbata *et al.*, 2014).

Regulating ES contributes to an environment that humans and other organisms can live in by protecting them from extreme environmental events and facilitating ecosystem functions that allow them to live in that environment (Braat and de Groot, 2012). Climate regulation is an important regulating ES and is often performed via carbon storage and sequestration across vegetation types, e.g. miombo woodlands (Jew *et al.*, 2016), as well as the fynbos and thickets in South Africa (Egoh *et al.*, 2009). The carbon sequestered and stored in vegetation can have a significant impact on atmospheric concentrations of carbon dioxide. Agricultural expansion in eastern Tanzania resulted in an emission of 0.9 Pg C over the twentieth century, a rate of 0.3 Mg C ha^{-1} yr^{-1} (Willcock *et al.*, 2016b). Ecosystems also play an important role in regulating the water flow across watersheds. In Kenya, the catchment area of the Tana River covers over one-sixth of the country, transports between 2.7 and 10.2 billion cubic metres of water annually, and controls flooding (Leauthaud *et al.*, 2013). Wetlands in sub-Saharan Africa occupy 4.7 per cent of the area, with about 65 per cent of the wetland area covered by the four largest basins – the Chad, Congo, Niger River and Nile River basin (Rebelo *et al.*, 2010) – and are also associated with EDS, such as providing habitats for invertebrate vectors of parasites such as malaria, which is transmitted by female anopheles mosquitoes, which causes 1.5–2.7 million deaths annually (Malan *et al.*, 2009). The disease burden caused by malaria is huge; for example, half of all consultations that occur in Malawi are classified as malaria (Mathanga *et al.*, 2012). Other incidences of impacts of regulating ES on human health include flooding events in Mozambique in 2000 that led to a rise in infectious diseases (Egoh *et al.*, 2012b) and the 1998 El Niño floods that led to an increase in cholera and typhoid, as well as damage to food crops, in Africa (Wangai *et al.*, 2016).

Cultural ES are non-material benefits that humans get from ecosystems (Sileshi *et al.*, 2007), and include recreational opportunities, spiritual importance and inspiration for art and music (Wenny *et al.*, 2011), as well as natural sites used as protected areas, natural heritage sites and nature used for education purposes (Egoh *et al.*, 2012b). Africa holds about 40 per cent of global biodiversity, and some may be of cultural importance (Egoh *et al.*, 2012b). For example, some African communities consider certain tree species sacred. In Kenya, *Ficus natalensis* and *Ficus thonningii* (Mũgumo) are sacred to the Gĩkũyũ tribe, symbolising life, power and fertility, with the most important sacrifices held under it being circumcision, transfer of traditional governments between generations, prayers for rains and cure from diseases (Karangi, 2008). In Tanzania, communities in Arusha place grass or flowers under *Ficus sycomorus* as an offering to God and ask for blessings, while in the Bamileke region of Cameroon, fig trees are used as sites of family worship (Wilson and Wilson, 2013). Some culturally important sites in Africa are considered sacred and are protected from human interference by restricting their access to religious leaders. For example, in Kenya, Mount Kenya is regarded as a holy mountain by the Gĩkũyũ and Meru, and in Malawi, Nyika National Park has several sacred sites used for rainmaking ceremonies. In Nigeria, the sacred Groves of Onshogbo have sacred spaces, worship points, shrines and two palaces, and are considered a World Heritage Site. In Tanzania, Misali Island has a shape that points towards Mecca, and is considered sacred by Muslims (Dudley *et al.*, 2009). One of the key ES stems from the scenic and biodiverse landscapes in Africa, tourism contributes significantly to national economies. In 2011, tourism contributed 2.7 per cent of gross domestic product (GDP) for sub-Saharan countries, directly and indirectly employed 12.8 million people, and directly contributed to 2.6 per cent of GDP in Ghana, 5.4 per cent in Kenya, 5 per cent in Tanzania and 2.9 per cent in South Africa (Christie *et al.*, 2013).

Supporting ES provide processes, such as nutrient cycling, biomass production and seed dispersal. These supporting ES are necessary for the production of other ES and for ecological communities and agricultural ecosystems (Sileshi *et al.*, 2007; Wenny *et al.*, 2011). They are considered as a fundamental basis for other categories of ES, and as such are often termed 'ecosystem processes' and not regarded as final services in their own right to avoid 'double counting' when final ES are evaluated (Fisher *et al.*, 2008).

While regulating ES can function without human intervention, provisioning and cultural ES only reach and benefit humans with some form of energy investment such as labour (Braat and de Groot, 2012). Provisioning ES require harvesting and gathering, while cultural ES require sensory and intellectual abilities to absorb and process information conveyed by ecosystems. It is possible for communities to derive multiple benefits from several ES at the same time (termed non-rival), such as the cultural value of scenic landscapes and the provision of pasture for livestock. ES benefits can also be mutually exclusive (termed rival), such as using forests for firewood/timber but depleting carbon stores.

While the rationale of the four categories of ES have been criticised (Daw *et al.*, 2011), and the process of studying ES is challenged by issues such as ES identification, quantification, interpretation and uncertainties (Wangai *et al.*, 2016), ES are clearly important to people across Africa. Furthermore, with the current 2.55 per cent annual rate of population growth in Africa (UN, 2015), the demand for ES will increase, and trade-offs and synergies between ES are likely to occur with varying spatial and temporal scales and intensities.

Modelling and mapping ecosystem services

ES models and maps aim to understand the stocks, demands and flows of ES across space and time (Burkhard *et al.*, 2013). Multiple ES have been mapped at numerous scales, including global (Naidoo *et al.*, 2008), continental (Schulp *et al.*, 2012), national (Bateman *et al.*, 2010) and subnational (Nelson *et al.*, 2009). Historically, the majority of ES maps focus on the biophysical potential of the landscape to provide a service, while more recent studies have shifted to include the demand and use of ES (Burkhard *et al.*, 2012). A number of reviews have been produced summarising the availability of ES data (Egoh *et al.*, 2012a; Martínez-Harms and Balvanera, 2012; Crossman *et al.*, 2013; Laurans *et al.*, 2013; Ruckelshaus *et al.*, 2013; McKenzie *et al.*, 2014; Wong *et al.*, 2015), and the main findings can be generalised as follows:

- ES stocks have received more attention than ES flows.
- Provisioning and regulating ES are more commonly mapped than cultural ES. Specifically, services related to climate regulation (carbon storage and sequestration), food production, water provision and recreation are most commonly studied.
- Secondary data (e.g. remotely sensed land cover maps) are most commonly used to create ES maps. However, these proxy methods are known to have the highest potential for error (Eigenbrod *et al.*, 2010a, 2010b).
- The most common scale of mapping ES is subnational.

Broadly, ES models can be divided into a three-'tier' system, analogous to the reporting system used by the Intergovernmental Panel on Climate Change. Tier 1 is the simplest method, using global values via look-up tables, e.g. benefit transfer (Costanza *et al.*, 1997). The intermediate Tier 2 level improves on Tier 1 by using country-specific data and locally validated regression models. Tier 3 is the most rigorous approach, using a time series of local data to support process-based models. The models are generally thought to become progressively more robust from Tier 1 to 3, however Tier 3 methods are more expensive and some nations may lack the capacity to adopt and apply such methods (Willcock *et al.*, 2014). Practically, it may be worthwhile to merge data and modelling activities from multiple tiers, using Tier 3 methodologies for highly variable and/or substantive ES flows, while Tier 1 and 2 methodologies may be sufficient for ES of lower importance. Thus, the types of ES maps and models available may be highly constrained by data deficiency and a lack of understanding surrounding the unique local situation. However, there are several widely available modelling frameworks designed to support the production of ecosystem maps and models to ultimately aid decision-making. Some of the leading ES modelling and mapping frameworks are described in Box 19.1.

Ecosystem services and decision-making

Decisions and policy regarding land and water management can be improved through the provision of quantitative ES models and maps derived from robust, spatially explicit data (Fisher *et al.*, 2009; McKenzie *et al.*, 2011; Bastian *et al.*, 2012). This could be especially relevant in developing countries, where the rural poor are often highly dependent on ES for their livelihoods, especially as a safety net during crises (Enfors and Gordon, 2008; Shackleton *et al.*, 2008). ES have the potential for reducing poverty as improved flows of ES lead to improved well-being of societies (Daw *et al.*, 2011), although this notion may be vulnerable to threat multipliers such as climate change, population growth and poor governance, and may be too simple for global application (Raudsepp-Hearne *et al.*, 2010).

Globally, ES science has been relatively successful at gaining traction within society, and as such has shown broad uptake into decision-making processes by a range of stakeholders. For example, some businesses (e.g. Unilever), development agencies (e.g. the World Bank), governments (e.g. China) and non-governmental organisations (NGOs) (e.g. Conservation International) have incorporated ES into their practices (Ruckelshaus *et al.*, 2013; Wong *et al.*, 2015). This relatively rapid uptake into policies and management may indicate that ES science has a high potential to alter decision-making practices, leading to more ecologically sound decisions that value economic, social and natural capital.

One reason that ES science has been successfully incorporated into decision-making is derived from their human-centric focus and their ability to be economically valued. As described above, ES are the benefits that societies derive from the ecosystems comprising of the flow of materials, energy and information from the natural environment to societies, and as such these services can be valued by calculating the relative contribution of ecosystems to the benefit received by societies. Valuation, either implicitly or explicitly, is a necessary step to aid decision-making as it allows the assessment of trade-offs in order to evaluate the path to a specified goal, e.g. the Sustainable Development Goals (UN, n.d.). However, there are ongoing challenges to overcome when valuing ES. For example, some of the methods used to derive value are based on individuals' perceptions of the benefits they derive, but an individual's perceptions may be limited and biased, neglecting benefits to whole communities or to future generations via increased sustainability and/or resilience. Thus, ES valuation studies combine stated and revealed preference methods to better capture the true value of each ecosystem (Costanza *et al.*, 2014).

As with ES maps and models, ES valuation methods can be performed at multiple times and scales, and show large ranges of complexity. For example, benefit transfer assumes a constant unit value per hectare of ecosystem type and multiplies that value by the area of each type to arrive at aggregate totals (Costanza *et al.*, 1997). As such, benefit transfer is analogous to a Tier 1 method and can be improved by adjusting to local conditions (Tier 2). As with other Tier 1 methods, benefit transfer fails to capture many of the complexities involved in the flows of ES to beneficiaries, and more spatially explicit and dynamic approaches (Tier 3) are often preferable (e.g. Bateman *et al.*, 2013). The tier required for ES valuation is partially dependent on the intended use for the valuation itself, with potential uses ranging from relatively simple raising of awareness to detailed analysis of various policy choices and scenarios (see Table 19.1) (Costanza *et al.*, 2014). Payments for ecosystem services (PES) schemes aim to maintain ES through actors who are willing to transact and pay for particular ES (Daw *et al.*, 2011), and often draw particular attention as a use for ES valuation due to these direct benefits on sustainability, conservation and development (Gómez-Baggethun *et al.*, 2010; Mulatu *et al.*, 2014). However, multiple methods are likely needed to achieve the goal of sustainable human well-being and sustainable provision of ES.

While the valuation of ES has helped integrate ES science into some decision-making practices, there are surprisingly few documented examples demonstrating how ES concepts have changed decision-making outcomes (Laurans *et al.*, 2013; Ruckelshaus *et al.*, 2013). This 'implementation gap' between the potential impact of ES research and its utilisation in practice has

Table 19.1 Range of uses for ecosystem service valuation

Use of valuation	*Appropriate values*	*Precision needed*
Raising awareness and interest	Total values, macro-aggregates	Low
National income and well-being accounts	Total values by sector and macro-aggregates	Medium
Specific policy analyses	Changes by policy	Medium to high
Urban and regional land use planning	Changes by land use scenario	Low to medium
Payments for ecosystem services	Changes by actions due payment	Medium to high
Full-cost accounting	Total values by business, product or activity, and changes by business, product or activity	Medium to high
Common asset trusts	Totals to assess capital and changes to assess income and loss	Medium

Source: Adapted from Costanza *et al.* (2014).

recently been investigated within sub-Saharan Africa. Willcock *et al.* (2016a) surveyed decision-makers from across sub-Saharan Africa to estimate levels of use of ES data in decision-making, as well as current and future requirements of this information, the first and only time this has been performed for the African continent, although other more localised studies do exist (Sitas *et al.*, 2013, 2014). Over 90 per cent of those surveyed reported incorporating ES information into their decision-making; however, the majority (over 70 per cent) require additional data, indicating a need for demand-driven research across the continent (Willcock *et al.*, 2016a). Reflecting data availability, respondents focus on provisioning and regulating services at national and sub-national scales (Willcock *et al.*, 2016a). The authors followed up with technical experts who were unanimous in highlighting the importance of scenarios to the decision-making process, emphasising the need for Tier 3 models and maps that are able to reflect the dynamic nature of ES flows (Willcock *et al.*, 2016a). This study provides evidence that stakeholders within sub-Saharan Africa are actively engaging with ES research and using the information to support policy development. However, more reliable information across more ES is required to better support sustainable development efforts in Africa. Capacity building should be widely adopted among ES researchers, minimising ambiguity at the science–policy interface while maximising the ability of research to support sustainable policies and increase human well-being.

Conclusion

The concepts of ES flows and stocks are increasingly useful ways to highlight, measure and value the interaction and interdependence between humans and ecosystems. The ES approach is complementary with other approaches to sustainable development, but provides conceptual and empirical tools that the others lack, and it communicates with different audiences for different purposes. ES science may be particularly important in Africa due to the high levels of poverty prevalent on the continent and, particularly in rural regions, the dependence on ES for well-being and sustaining livelihoods. Encouragingly, decision-makers across Africa are engaging with the ES concept, and its ability to evaluate trade-offs through policy scenarios will be key if the Sustainable Development Goals are to be achieved.

Box 19.1 An introduction to some of the leading ecosystem service modelling and mapping frameworks

Integrated Valuation of Environmental Services and Tradeoffs (InVEST) (www.naturalcapitalproject.org/InVEST.html) is a set of models designed to map and value trade-offs between multiple ecosystem services. InVEST adopts a tiered approach to the modelling of ES using a GIS interface: Tier 1 models use globally available data to form first estimates of ES values; Tier 2 models, require additional local data; and Tier 3 models are increasingly complex and data-intense. Currently, InVEST is able to model a larger range of ES than the other leading modelling platforms, with several more in development. Additionally, beta versions of InVEST are in existence and are reviewed through an open-source community, so they may be more reliable than other modelling systems. However, Tiers 2 and 3 (analogous to many of the ARIES models) are in early development. In summary, InVEST is well placed to capture changes in ES value and can represent alternative scenarios for identifying trade-offs. However, it has little functionality to explore feedbacks between ES and beneficiaries, and so future estimates may be highly uncertain.

Co$ting Nature (www.policysupport.org/costingnature) is a web-based tool to quantify ES, identify beneficiaries and assess the impacts of proposed human interventions. The models are analogous to Tier 1 InVEST models, but are available online without the need for GIS skills. However, the software is only able to model a limited range of ES in comparison to the other leading frameworks. Co$ting Nature is well suited for scenario assessment of impacts on ES delivery; however, the uncertainty associated with the simplistic models is not well understood, nor is it communicated to the end user.

Artificial Intelligence for Ecosystem Services (ARIES) (http://aries.integratedmodelling.org/) is an open-source technology that has the ability to select and run models to quantify and map ecosystem service provision, including biophysical generation, flow and extraction by sinks and beneficiaries. ARIES can be applied in a broad range of physical, social and economic contexts, providing access to a library of ecosystem service models and spatial data sets at multiple scales, ranging from global to local. Decision-making rules using artificial intelligence enable addition or removal of components as appropriate, allowing accurate results to be accessible to users without extensive skills or training. ARIES also uses Bayesian network modelling and Monte Carlo simulation to supply uncertainty estimates for the model outputs. At present, the ARIES models are mostly limited to a few case study locations. Thus, expert model development is required to run ARIES in most regions. However, global models that run without user data may be available by 2020. These will provide initial estimates of ES values in data-deficient regions. As with InVEST Tiers 2 and 3, ARIES plans integration of existing ecological process models in future releases, so allowing complex relationships to be better described. As such, ARIES is well suited for scenario assessment for different future climate and land use/cover conditions.

The Economics of Ecosystems and Biodiversity (TEEB) (www.teebweb.org/) model of ES was created under the United Nations Environment Programme (UNEP) with an aim of assessing the importance of ecological systems to societies. It places natural science in one category, and human, social and economic science in another category. This approach has ES flowing from natural science components to social science components with indications of relative importance of the links between various ecosystem components (Braat and de Groot, 2012). TEEB uses benefit transfer, and as such are Tier 1 ecosystem service models.

Further ecosystem service mapping tools of note are the Social Values for Ecosystem Services (SolVES) (https://solves.cr.usgs.gov/) and the Global Unified Metamodel of the BiOsphere (GUMBO). SolVES is a GIS tool to assess, map and quantify the perceived social values for ecosystems, such as aesthetics, biodiversity and recreation. GUMBO uses simulation modelling to model global dynamics and interactions of natural capital with built, social and human capital.

References

Ango, T.G., Börjeson, T., Senbeta, F. and Hylander, K. (2014) 'Balancing ecosystem services and disservices: smallholder farmers' use and management of forest and trees in an agricultural landscape in southwestern Ethiopia', *Ecology and Society*, 19(1): 30.

Bastian, O., Grunewald, K. and Syrbe, R.-U. (2012) 'Space and time aspects of ecosystem services, using the example of the EU Water Framework Directive', *International Journal of Biodiversity Science, Ecosystem Services & Management*, 8(1–2): 5–16, available at: http://dx.doi.org/10.1080/21513732.2011.631941 (accessed 17 November 2014).

Bateman, I.J., Mace, G.M., Fezzi, C., Atkinson, G. and Turner, K. (2010) 'Economic analysis for ecosystem service assessments', *Environmental and Resource Economics*, 48(2): 177–218.

Bateman, I.J., Harwood, A.R., Andrews, B., Binner, A., Day, B.H., Dugdale, S., *et al.* (2013) 'Bringing ecosystem services into economic decision-making: land use in the United Kingdom', *Science*, 341(6141): 45–50.

Boafo, Y.A., Saito, O., Jasaw, G.S., Otsuki, K. and Takeuchi, K. (2016) 'Provisioning ecosystem services-sharing as a coping and adaptation strategy among rural communities in Ghana's semi-arid ecosystem', *Ecosystem Services*, 19: 92–102.

Braat, L.C. and de Groot, R. (2012) 'The ecosystem services agenda: bridging the worlds of natural science and economics, conservation and development, and public and private policy', *Ecosystem Services*, 1(1): 4–15.

Burkhard, B., de Groot, R.S., Costanza, R., Seppelt, R., Jorgensen, S.E. and Potschin, M. (2012) 'Solutions for sustaining natural capital and ecosystem services', *Ecological Indicators*, 21: 1–6.

Burkhard, B., Crossman, N., Nedkov, S., Petz, K. and Alkemade, R. (2013) 'Mapping and modelling ecosystem services for science, policy and practice', *Ecosystem Services*, 4: 1–3.

Carpenter, S.R., Mooney, H.A., Agard, J., Capistrano, D., Defries, R.S., Díaz, S., *et al.* (2009) 'Science for managing ecosystem services: beyond the Millennium Ecosystem Assessment', *Proceedings of the National Academy of Sciences of the United States of America*, 106(5): 1305–1312.

Christie, I., Fernandes, E., Messerli, H. and Twining-Ward, L. (2013) *Tourism in Africa: Harnessing Tourism for Growth and Improved Livelihoods*, Washington, DC: World Bank.

Costanza, R., d'Arge, R., de Groot, R., Farberk, S., Grasso, M., Hannon, B., *et al.* (1997) 'The value of the world's ecosystem services and natural capital', *Nature*, 387(6630): 253–260, DOI: http://dx.doi.org/10.1038/387253a0.

Costanza, R., de Groot, R., Sutton, P., van der Ploeg, S., Anderson, S.J., Kubiszewski, I., *et al.* (2014) 'Changes in the global value of ecosystem services', *Global Environmental Change*, 26: 152–158.

Crossman, N.D., Burkhard, B., Nedkov, S., Willemen, L.L.J.M., Petz, K., Palomo, I., *et al.* (2013) 'A blueprint for mapping and modelling ecosystem services', *Ecosystem Services*, 4: 4–14.

Daw, T., Brown, K., Rosendo, S. and Pomeroy, R. (2011) 'Applying the ecosystem services concept to poverty alleviation: the need to disaggregate human well-being', *Environmental Conservation*, 38(4): 370–379, DOI: http://dx.doi.org/10.1017/S0376892911000506.

Dessu, S.B., Melesse, A.M., Bhat, M.G. and McClain, M.E. (2014) 'Assessment of water resources availability and demand in the Mara River basin', *Catena*, 115: 104–114.

Dooley, E.E. (2005) 'EHPnet: Millennium Ecosystem Assessment', *Environmental Health Perspectives*, 113(9): A591, available at: www.ncbi.nlm.nih.gov/pmc/articles/PMC1280419/ (accessed 4 January 2018).

Dudley, N., Higgins-Zogib, L. and Mansourian, S. (2009) 'The links between protected areas, faiths, and sacred natural sites', *Conservation Biology*, 23(3): 568–577.

Egoh, B., Reyers, B., Rouget, M., Bode, M. and Richardson, D.M. (2009) 'Spatial congruence between biodiversity and ecosystem services in South Africa', *Biological Conservation*, 142(3): 553–562.

Egoh, B.N., O'Farrell, P.J., Charef, A., Josephine Gurney, L., Koellner, T., Nibam Abi, H., *et al.* (2012a) 'An African account of ecosystem service provision: use, threats and policy options for sustainable livelihoods', *Ecosystem Services*, 2(3): 71–81.

Egoh, B., Drakou, E.G., Dunbar, M.B., Maes, J. and Willemen, L. (2012b) *Indicators for Mapping Ecosystem Services: A Review*, Ispra: European Commission, Joint Research Centre, JRC Scientific and Policy Reports, EUR 25456 EN.

Eigenbrod, F., Armsworth, P.R., Anderson, B.J., Heinemeyer, A., Gillings, S., Roy, D.B., *et al.* (2010a) 'Error propagation associated with benefits transfer-based mapping of ecosystem services', *Biological Conservation*, 143(11): 2487–2493.

Eigenbrod, F., Armsworth, P.R., Anderson, B.J., Heinemeyer, A., Gillings, S., Roy, D.B., *et al.* (2010b) 'The impact of proxy-based methods on mapping the distribution of ecosystem services', *Journal of Applied Ecology*, 47(2): 377–385.

Enfors, E.I. and Gordon, L.J. (2008) 'Dealing with drought: the challenge of using water system technologies to break dryland poverty traps', *Global Environmental Change*, 18(4): 607–616, available at: www.sciencedirect.com/science/article/pii/S0959378008000575 (accessed 17 November 2014).

Ericksen, P., de Leeuw, J., Said, M., Silvestri, S. and Zaibet, L. (2012) 'Mapping ecosystem services in the Ewaso Ng'iro catchment', *International Journal of Biodiversity Science, Ecosystem Services & Management*, 8(1–2): 122–134.

Farley, J. (2012) 'Ecosystem services: the economics debate', *Ecosystem Services*, 1(1): 40–49, available at: www.sciencedirect.com/science/article/pii/S2212041612000071 (accessed 11 May 2017).

Fisher, B., Turner, K., Zylstra, M., Brouwer, R., de Groot, R.S., Farber, S., *et al.* (2008) 'Ecosystem services and economic theory: integration for policy-relevant research', *Ecological Applications*, 18(8): 2050–2067.

Fisher, B., Turner, R.K. and Morling, P. (2009) 'Defining and classifying ecosystem services for decision making', *Ecological Economics*, 68(3): 643–653, available at: www.sciencedirect.com/science/article/pii/S0921800908004424 (accessed 9 July 2014).

Gómez-Baggethun, E., de Groot, R.S., Lomas, P. and Sotelo Montes, C. (2010) 'The history of ecosystem services in economic theory and practice: from early notions to markets and payment schemes', *Ecological Economics*, 69(6): 1209–1218.

Hohenthal, J., Owidi, E., Minoia, P. and Pellikka, P. (2015) 'Local assessment of changes in water-related ecosystem services and their management: DPASER conceptual model and its application in Taita Hills, Kenya', *International Journal of Biodiversity Science, Ecosystem Services & Management*, 11(3): 225–238.

Jew, E.K.K., Dougill, A.J., Salluh, S.M., O'Connell, J. and Benton, T.G. (2016) 'Miombo woodland under threat: consequences for tree diversity and carbon storage', *Forest Ecology and Management*, 361: 144–153.

Kafumbata, D., Jamu, D. and Chiotha, S. (2014) 'Riparian ecosystem resilience and livelihood strategies under test: lessons from Lake Chilwa in Malawi and other lakes in Africa', *Philosophical Transactions of the Royal Society of London. Series B: Biological Sciences*, 369(1639): 20130052, DOI: http://dx.doi.org/10.1098/rstb.2013.0052.

Kalaba, F.K., Quinn, C.H. and Dougill, A.J. (2013) 'The role of forest provisioning ecosystem services in coping with household stresses and shocks in Miombo woodlands, Zambia', *Ecosystem Services*, 5: 143–148.

Karangi, M.M. (2008) 'Revisiting the roots of Gĩkũyũ culture through the sacred Mũgumo tree', *Journal of African Cultural Studies*, 20(1): 117–132.

Laurans, Y., Rankovic, A., Billé, R., Pirard, R. and Mermet, L. (2013) 'Use of ecosystem services economic valuation for decision making: questioning a literature blindspot', *Journal of Environmental Management*, 119(3): 208–219.

Leauthaud, C., Duvail, S., Hamerlynck, O., Paul, J.-L., Cochet, H., Nyunja, J., *et al.* (2013) 'Floods and livelihoods: the impact of changing water resources on wetland agro-ecological production systems in the Tana River Delta, Kenya', *Global Environmental Change*, 23(1): 252–263.

Lyytimäki, J. and Sipilä, M. (2009) 'Hopping on one leg: the challenge of ecosystem disservices for urban green management', *Urban Forestry and Urban Greening*, 8(4): 309–315.

Malan, H.L., Appleton, C.C., Day, J.A. and Dini, J. (2009) 'Wetlands and invertebrate disease hosts: are we asking for trouble?', *Water SA*, 35(5): 753–768.

Martínez-Harms, M.J. and Balvanera, P. (2012) 'Methods for mapping ecosystem service supply: a review', *International Journal of Biodiversity Science, Ecosystem Services & Management*, 8(1–2): 17–25, available at: http://dx.doi.org/10.1080/21513732.2012.663792 (accessed 13 July 2014).

Mathanga, D.P., Walker, E.D., Wilson, M.L., Ali, D., Taylor, T.E. and Laufer, M.K. (2012) 'Malaria control in Malawi: current status and directions for the future', *Acta Tropica*, 121(3): 212–217.

McKenzie, E., Irwin, F., Ranganathan, J., Hanson, C., Kousky, C., Bennett, K., *et al.* (2011) 'Incorporating ecosystem services in decisions', in P. Kareiva, H. Tallis, T.H. Ricketts, G.C. Daily and S. Poloansky (eds), *Natural Capital: Theory and Practice of Mapping Ecosystem Services*, Oxford: Oxford University Press, pp. 339–355.

McKenzie, E., Posner, S., Tillmann, P., Bernhardt, J.R., Howard, K. and Rosenthal, A. (2014) 'Understanding the use of ecosystem service knowledge in decision making: lessons from international experiences of spatial planning', *Environment and Planning C: Government and Policy*, 32(2): 320–340, DOI: 10.1068/c12292j.

MEA (Millennium Ecosystem Assessment) (2005) *Millennium Ecosystem Assessment*, Washington, DC: World Resources Institute, UNEP, World Bank and UNDP, available at: http://millenniumassessment.org/en/ (accessed 8 January 2018).

Mulatu, D.W., van der Veen, A. and van Oel, P.R. (2014) 'Farm households' preferences for collective and individual actions to improve water-related ecosystem services: the Lake Naivasha basin, Kenya', *Ecosystem Services*, 7: 22–33.

Müller, F. and Burkhard, B. (2012) 'The indicator side of ecosystem services', *Ecosystem Services*, 1(1): 26–30.

Naidoo, R., Balmford, A., Costanza, R., Fisher, B., Green, R.E., Lehner, B., *et al.* (2008) 'Global mapping of ecosystem services and conservation priorities', *Proceedings of the National Academy of Sciences*, 105(28): 9495–9500, DOI: 10.1073/pnas.0707823105.

Nelson, E., Mendoza, G., Regetz, J., Polasky, S., Tallis, H., Cameron, D.R., *et al.* (2009) 'Modeling multiple ecosystem services, biodiversity conservation, commodity production, and tradeoffs at landscape scales', *Frontiers in Ecology and the Environment*, 7(1): 4–11.

Raudsepp-Hearne, C., Peterson, G.D. and Bennett, E.M. (2010) 'Ecosystem service bundles for analyzing tradeoffs in diverse landscapes', *Proceedings of the National Academy of Sciences of the United States of America*, 107(11): 5242–5247.

Rebelo, L.M., McCartney, M.P. and Finlayson, C.M. (2010) 'Wetlands of sub-Saharan Africa: distribution and contribution of agriculture to livelihoods', *Wetlands Ecology and Management*, 18(5): 557–572.

Ruckelshaus, M., McKenzie, E., Tallis, H., Guerry, A., Daily, G., Kareiva, P., *et al.* (2013) 'Notes from the field: lessons learned from using ecosystem service approaches to inform real-world decisions', *Ecological Economics*, 115(14): 11–21, DOI: 10.1016/j.ecolecon.2013.07.009.

Schulp, C.J.E., Alkemade, R., Klein Goldewijk, K. and Petz, K. (2012) 'Mapping ecosystem functions and services in Eastern Europe using global-scale data sets', *International Journal of Biodiversity Science, Ecosystem Services & Management*, 8(1–2): 156–168.

Shackleton, C., Shackleton, S., Gambiza, J., Nel, E., Rowntree, K. and Urquhart, P. (2008) *Links Between Ecosystem Services and Poverty Alleviation: Situation Analysis for Arid and Semi-Arid Lands in Southern Africa*, available at: www.unpei.org/sites/default/files/PDF/preliminaryassessments/Links-Ecosystem-Services-and-Poverty-Alleviation.pdf (accessed 19 December 2017).

Shackleton, C.M., Ruwanza, S., Sinasson Sanni, G.K., Bennett, S., De Lacy, P., Modipa, R., *et al.* (2016) 'Unpacking Pandora's box: understanding and categorising ecosystem disservices for environmental management and human wellbeing', *Ecosystems*, 19(4): 587–600.

Shapiro, J. and Báldi, A. (2014) 'Accurate accounting: how to balance ecosystem services and disservices', *Ecosystem Services*, 7: 201–202.

Sileshi, G., Akinnifesi, F.K., Ajayi, O.C., Chakeredza, S., Kaonga, M. and Matakala, P.W. (2007) 'Contributions of agroforestry to ecosystem services in the miombo eco-region of eastern and southern Africa', *African Journal of Environmental Science and Technology*, 1(4): 68–80.

Sitas, N., Prozesky, H.E., Esler, K.J. and Reyers, B. (2013) 'Opportunities and challenges for mainstreaming ecosystem services in development planning: perspectives from a landscape level', *Landscape Ecology*, 29(8): 1315–1331, DOI: 10.1007/s10980-013-9952-3.

Sitas, N., Prozesky, H.E., Esler, K.J. and Reyers, B. (2014) 'Exploring the gap between ecosystem service research and management in development planning', *Sustainability*, 6(6): 3802–3824, available at: www.mdpi.com/2071-1050/6/6/3802/htm (accessed 1 December 2014).

Turner, R.K. and Daily, G.C. (2008) 'The ecosystem services framework and natural capital conservation', *Environmental and Resource Economics*, 39(1): 25–35, available at: http://link.springer.com/10.1007/s10640-007-9176-6 (accessed 11 May 2017).

UN (United Nations) (2015) *World Population Prospects: The 2015 Revision, Key Findings and Advance Tables*, New York: UN DESA.

UN (United Nations) (n.d.) *Sustainable Development Knowledge Platform*, New York: UNDESA, available at: https://sustainabledevelopment.un.org/ (accessed 4 January 2018).

Vaz, A.S., Kueffer, C., Kull, C.A., Richardson, D.M., Vicente, J.R., Kühn, I., *et al.* (2017) 'Integrating ecosystem services and disservices: insights from plant invasions', *Ecosystem Services*, 23(Suppl 1): 94–107, DOI: 10.1016/j.ecoser.2016.11.017.

Villa, F., Bagstad, K.J., Voigt, B., Johnson, G.W., Athanasiadis, I.N. and Balbi, S. (2014) 'The misconception of ecosystem disservices: how a catchy term may yield the wrong messages for science and society', *Ecosystem Services*, 10: 52–53, DOI: 10.1016/j.ecoser.2014.09.003.

Wangai, P.W., Burkhard, B. and Muller, F. (2016) 'A review of studies on ecosystem services in Africa', *International Journal of Sustainable Built Environment*, 5(2): 225–245.

Wenny, D.G., DeVault, T.L., Johnson, M.D., Kelly, D., Sekercioglu, C.H., Tomback, D.F., *et al.* (2011) 'Perspectives in ornithology: the need to quantify ecosystem services provided by birds', *Auk*, 128(1): 1–14, DOI: 10.1525/auk.2011.10248.

Whelan, C.J., Şekercioğlu, Ç.H. and Wenny, D.G. (2015) 'Why birds matter: from economic ornithology to ecosystem services', *Journal of Ornithology*, 156(S1): 227–238.

Willcock, S., Phillips, O.L., Platts, P.J., Balmford, A., Burgess, N.D., Lovett, J.C., *et al.* (2014) 'Quantifying and understanding carbon storage and sequestration within the Eastern Arc Mountains of Tanzania, a tropical biodiversity hotspot', *Carbon Balance and Management*, 9(2), DOI: 10.1186/1750-0680-9-2.

Willcock, S., Hooftman, D., Sitas, N., O'Farrell, P., Hudson, M.D., Reyers, B., *et al.* (2016a) 'Do ecosystem service maps and models meet stakeholders' needs? A preliminary survey across sub-Saharan Africa', *Ecosystem Services*, 18: 110–117, DOI: 10.1016/j.ecoser.2016.02.038.

Willcock, S., Phillips, O.L., Platts, P.J., Swetnam, R.D., Balmford, A., Burgess, N.D., *et al.* (2016b) 'Land cover change and carbon emissions over 100 years in an African biodiversity hotspot', *Global Change Biology*, 22(8): 2787–2800, DOI: 10.1111/gcb.13218.

Wilson, D. and Wilson, A. (2013) 'Figs as a global spiritual and material resource for humans', *Human Ecology*, 41(3): 459–464.

Wong, C.P., Jiang, B., Kinzig, A.P., Lee, K.N., Ouyang, Z. and Knops, J. (2015) 'Linking ecosystem characteristics to final ecosystem services for public policy', *Ecology Letters*, 18(1): 108–118, DOI: 10.1111/ele.12389.

Yang, W., Dietz, T., Liu, W., Luo, J., Liu, J. and Cebrian, J. (2013) 'Going beyond the Millennium Ecosystem Assessment: an index system of human dependence on ecosystem services', *PLoS ONE*, 8(5): 64581, DOI: 10.1371/journal.pone.0064581.

Section 6
Development issues

20

African development

'Hopeless Africa', 'Africa Rising', or somewhere in between?

Tony Binns and Etienne Nel

Introduction

Kayima is a small and remote rural town of 2,000 people situated in the far north-east of Sierra Leone in West Africa, some three hours of difficult driving from the nearest metalled road. One of the authors spent a year in Kayima in 1974 and then revisited 40 years later in January 2014. Virtually nothing had changed – the houses, the streets and the people seemed just the same. Except for some burnt-out houses on the main street – a legacy of the civil war – a new community bank and a cell phone mast, everything else seemed to be frozen in time. Houses built of mud bricks, no electricity, and water drawn from a few outside standpipes were just as they were in 1974. In the case of Kayima at least, it seemed that not much development had happened in four decades (Binns and Maconachie, 2005; Maconachie and Binns, 2007; Bateman *et al.*, 2017).

But is the rest of Africa like Kayima, we must ask? Certainly, there are many Kayimas across the continent, but there are also skyscrapers, bourgeoning cities and communities that have undoubtedly been uplifted. One certainty is that it is not safe to generalise when talking about Africa, either today, in the twenty-first century, or in the past. The 54 countries reveal tremendous diversity in their cultures, economies and levels of development.

This chapter reflects on key dimensions of Africa's development and associated challenges. Consideration is first given to the global context before questioning whether there is 'growth without development', and the role that resource exploitation plays within this context. After examining what the Millennium Development Goals tell us about the continent, the chapter concludes with a call to pay closer attention to the very real needs of people on the ground in Africa.

Readers of *The Economist* may be critical of the generalisations made about Africa in the last 20 years. At the start of the new millennium in 2000, an *Economist* article titled 'Hopeless Africa' started with a depressing statement about Sierra Leone's capital city, Freetown, in the midst of the decade-long civil war, with the writer concluding that 'Sierra Leone manifests all the continent's worst characteristics' (The Economist, 2000: 6). In 2011, a second *Economist* article was much more upbeat about the continent. Titled 'Africa Rising', the article reported on bustling local markets, a commodities boom, improvements in healthcare, peaceful and democratic governance, and China's involvement in upgrading infrastructure and stimulating the manufacturing sector (The Economist, 2011: 3).

Pascal Lamy, Director-General of the World Trade Organization, in a speech at the University of Nairobi, Kenya, in May 2013, asserted that:

> Africa, both as a continent and as the sum of individual sovereign states, is poised to lead the new patterns of growth for the foreseeable future . . . Africa has changed from the land of pessimism to the land of opportunity.
>
> *(Lamy, 2013: 1)*

Indeed, since the start of the new millennium, some African countries have experienced the highest growth rates in per capita gross domestic product (GDP) in the world. Between 2000 and 2005, the United Nations Development Programme reported growth rates as high as 21.3 per cent in Equatorial Guinea, whereas Sierra Leone's growth rate was a more modest, yet by world standards, very respectable rate of 7.9 per cent, one of the highest in Africa (UNDP, 2007). While Equatorial Guinea's soaring growth rate was due to oil exploration, Sierra Leone was engaged in a period of national reconstruction from a very low base following the devastating civil war (1991–2002). However, we need to be acutely aware of the reality that national aggregate growth often reflects greater polarisation rather than equality within a society (OECD, 2014). Further, by 2015, the GDP per capita growth rate in Sierra Leone had turned negative (−22.3 per cent), as the country was forced into a lockdown mode trying to cope with the deadly Ebola epidemic (2014–2016).

However, the quality of the data on which such assessments are made in relation to African countries needs to be questioned. Many African countries cannot afford the cost of regular household and other statistical surveys, and often have limited qualified staff to both administer and analyse them. Where data do exist, they can often be unreliable. A 2016 World Bank report commented, 'The statistics suggest that Africa's people are faring better and that poverty has come down. But scrutiny of these statistics has raised doubts about the quality of the underlying data and the exact magnitude of Africa's progress' (Beegle *et al.*, 2016: 25).

In terms of inequality, it seems that Africa has some of the most unequal countries in the world, notably in southern Africa, but household surveys do not seem to record cases of extreme wealth. The 2014 *Forbes* list reported 19 African billionaires with an average net wealth of US$5.2 billion per billionaire (Beegle *et al.*, 2016: 135). The growth of an African middle class has been a feature of the last two decades, with the African Development Bank estimating that in 2010, the middle class represented 34.3 per cent of the population, or nearly 326 million people (AfDB, 2011). According to the African Development Bank, 'the middle class is usually defined as individuals with annual income exceeding US$3,900 in purchasing power parity (PPP) terms, or with a daily per capita expenditure between US$2 and US$10' (AfDB, 2011: 2). But these surprisingly low figures must be called into question as to whether individuals can really be regarded as 'middle class' on such low incomes (Melber, 2016), and whether international bodies are possibly misrepresenting the reality on the ground for political ends.

Growing inequality, as noted elsewhere in the world (OECD, 2014), also manifests itself spatially, as we see in a country such as South Africa, with significant regional variations in poverty levels that are often deeply entrenched and a legacy of apartheid-era policies (Rogerson and Nel, 2015). For example, in Limpopo Province, in the far north of the country, at the 2011 census it was recorded that 41.5 per cent of households were living below the food poverty line, whereas in Western Cape the figure was 23.2 per cent. Average household income levels in Western Cape, at 143,460 rand (US$20,494) were more than twice as high as Limpopo at 56,844 rand (US$8,120) (Statistics South Africa, 2015).

The changing global context

Since the heady days of the immediate post-independence era in the 1960s, and the associated beliefs in the capacity of modernisation and Keynesism to fundamentally reshape and 'advance' national economies, the last few decades have delivered a rather lacklustre set of outcomes for much of Africa (Hettne, 1995). While there have been clear economic success stories, such as Botswana, Namibia, Mauritius and Seychelles, and not insignificant social development advances have taken place in countries such as Morocco and South Africa, for much of the continent and its people, social and economic advances and 'development' have been difficult to attain, as the discussion below will illustrate. While many countries have experienced economic growth, this often translates into rising inequality within society (Bond, 2006; Potter *et al.*, 2017).

From the perspective of applied development, the belief in the role that large, externally funded projects could play in driving development, which held sway in the 1960s, gradually gave way to a more 'basic needs' focus in the 1970s and 1980s, while hopes of Africa's development levels matching those of the West evaporated after the 1980s debt crisis and the debilitating effects of structural adjustment imposed by the 'Washington Consensus'. Far from improving conditions, as neo-liberal orthodoxy proclaimed, the evidence across Africa was rather one of retarded development, lost employment and social service cuts, causing many commentators to refer to the 1980s as the 'lost decade' in Africa (Escobar, 1995; Binns *et al.*, 2012; Thornton, 2014).

Championed from the early 1980s by Margaret Thatcher (UK) and Ronald Reagan (USA), the rise of neo-liberalism as a hegemonic discourse, and associated processes of globalisation and the increasing reach of powerful multinational corporations, have fundamentally shaped Africa's position in the world's economy and determined its future prospects. Broad-based development has given way to market facilitation and the rather naïve notion that 'a rising tide lifts all boats' and everyone will benefit from neo-liberal policies, or, alternatively, the benefits of growth will 'trickle down' to the remotest regions and their communities. Experience has shown, however, that far more realistic is the recognition that neo-liberalism thrives in the context of 'uneven geographical development', with spatial, economic and social inequalities being fundamental to its existence and reproduction (Harvey, 2006, 2014). Neo-liberalism has favoured pockets of development in the continent and in selected countries – often based on the exploitation of resources. While aggregate national growth figures improve, it seems that inequality generally rises, with few deriving tangible benefit from the narrow focus of investment on selected enclaves, as the well-documented 'resource curse' thesis argues (Ross, 2001; Maconachie and Binns, 2007), and as the Kayima example vividly illustrates. Increasingly, Africa's development prospects are shaped by the whims and selective investment of international corporates, while nations have to contend with what are often weak and corrupt governments. The impact of the global financial crisis in 2008 aggravated conditions in many countries as export markets contracted, a situation that was exacerbated by the rise of economic nationalism in the Global North. Added to this, continuing rapid population growth, growing aid fatigue in the West and significant cuts in aid budgets have together created a situation of some concern and a questioning of the 'Africa Rising' concept (Bond, 2006; see also Hickel, 2014, 2016).

Growth without development?

But what does a statistic such as growth in per capita GDP actually tell us about the quality of life in Africa's remotest communities? The answer is very little. It is a national-level statistic, based on aggregated and often unreliable data, which generally masks social and spatial inequalities and tells us nothing about the nature of poverty, or access to basic needs such as healthcare, food and clean water, particularly at the local level. But is economic growth leading to poverty alleviation and a reduction in equality? As Dulani *et al.* (2013: 4) assert:

> Recent evidence suggesting that African countries are registering the fastest economic growth rates globally has led many to predict that the continent is on the verge of a development renaissance. While there is compelling evidence in support of Africa's strong economic performance, questions about whether this growth is contributing to the reduction of poverty remain highly contested.

Time and time again, it has been shown that rapid economic growth, as indicated by variables such as gross domestic product, does not automatically translate into improving livelihoods among the bulk of the population, and particularly those living in remote rural areas. This was demonstrated over 50 years ago in a book by Robert Clower *et al.* titled *Growth Without Development: An Economic Survey of Liberia*, in which the authors concluded that despite considerable multinational involvement in the production of rubber and iron ore, the country's social and political institutions prevented the bulk of the population from benefiting from the overseas investment and export of raw materials (Clower *et al.*, 1966).

The same conclusion might be reached about present-day Equatorial Guinea, which, thanks to the discovery of oil in the Gulf of Guinea in the 1990s, has the highest GDP per capita of any African country (US$18,308) (AfDB, 2016), yet under-five mortality is at 94.1 per thousand live births, and 26 per cent of under-five children are underweight. Furthermore, a household survey conducted in 2006 concluded that 76.8 per cent of the population was 'poor' (AfDB, 2008). Clearly, Equatorial Guinea is also a case of 'growth without development'. Following a coup in 1979, President Teodoro Obiang Nguema Mbasogo and his family have ruled with almost total control. Supposedly democratic elections have been criticised for being seriously flawed, and governance is characterised by a lack of transparency, widespread corruption and oppression (Binns *et al.*, 2012).

Unequally distributed oil wealth is a feature of other countries in Africa, and indeed elsewhere. In the decade following the end of the civil war in 2002, Angola's economy grew at an average annual rate of 7 per cent. A 2013 report from the Africa Progress Panel drew attention to the situation in Angola, where:

> The oil wealth that has delivered fortunes for the few has left most Angolans — including Luanda's slum dwellers — in grinding poverty. After a decade of rapid growth, half of the country – 10 million people — still lives on less than US$1.25 a day.
>
> *(Africa Progress Panel, 2013: 20)*

The report refers to the privileged position of the country's elite, who 'have benefitted not just from the opportunity to enrich themselves . . . [but] . . . they have also worked assiduously to ensure that the country's oil revenue is geared towards their interests' (Africa Progress Panel, 2013: 20).

A fair price for Africa's resources?

Inequality is a feature of many resource-rich African countries. In Angola, the poorest 10 per cent of the population account for 0.6 per cent of the national income, and the figure is less than 2 per cent in Nigeria, South Africa and Zambia (Africa Progress Panel, 2013). So, why is the wealth from exploiting natural resources not benefiting local people?

In the case of resource-rich African countries, it is vital that governments negotiate appropriate price levels for the exploitation and sale of their resources, and that the revenue that is received should be managed efficiently and channelled into meaningful development initiatives that will benefit the whole population. This needs to be done to avoid the challenges associated with nations that over-depend on a limited number of resources, skewing income and equality levels, and frequently stifling the growth of other sectors, a situation that theorists have variously referred to as 'resource curse' or 'Dutch disease' (Ross, 2001; Smith, 2015). The granting of concessions and the taxation of foreign companies are complex, but limited revenue collection is undoubtedly the case in many African countries. It has been suggested that the Democratic Republic of Congo, for example, is actually losing revenue due to tax evasion, corrupt practices and poor management of concessions (Africa Progress Panel, 2013).

Does this perhaps explain why a country such as Zambia, with its wealth of mineral resources, is only ranked 139 out of 188 countries according to the Human Development Index (UNDP, 2016), with life expectancy in 2014 estimated at only 60 years? Zambia is the seventh largest world producer of copper and is ranked second for cobalt, yet in 2013 UNICEF reported that:

> Over the past three decades, incomes in Zambia have fallen steadily. Poverty is widespread – 64% of the total population lives below the poverty line, rising to 80% in rural areas, meaning they do not have adequate income to meet their basic food requirements.
>
> *(UNICEF, 2013)*

A detailed analysis of mineral pricing and revenue indicates that when copper prices were high, Zambia had 'one of the lowest royalty rates in the mining sector in Africa under an agreement negotiated with two mining companies in the late 1990s' (Africa Progress Panel, 2013: 63). The country also has one of the world's least transparent budgets, and it has been estimated that between 1970 and 2010, Zambia lost US$17.3 billion in 'illicit capital flight' (Simpasa *et al.*, 2013). Furthermore, compared with other low-income copper producers, Zambia has achieved low levels of revenue generation under both private and state ownership. It has been suggested that with more effective revenue generation from mineral production in Zambia, 'it would have been enough to finance activities needed to ensure achievement of all the health targets under the Millennium Development Goals' (Simpasa *et al.*, 2013: 17). A UNDP report in 2013 advocated the need for better governance, transparency and accountability over Zambia's mineral revenues if they are going to make a significant contribution to both long-term economic and social development (Simpasa *et al.*, 2013: 18).

The idea of an Extractive Industries Transparency Initiative (EITI) was launched in September 2002 by UK Prime Minister Tony Blair, and at a conference held in London in June 2003, a Statement of 12 EITI Principles was adopted, designed to increase the transparency of payments and revenues in the extractive sector (EITI, 2016). The EITI Principles recognise that good management of natural resources should lead to sustainable economic growth, sustainable development and poverty reduction, and that transparency by governments and companies is needed, with better public financial management and accountability. The EITI Principles have evolved into a global standard and have led to reform in many countries. By 2013, 11 African countries were EITI-compliant, while a further seven countries were registered as EITI candidate countries. By 2013, some 70 of the largest global oil, gas and mining companies had agreed to support and participate in the initiative at the country level (Africa Progress Panel, 2013). But by 2016, the list of compliant countries still did not include Angola or Equatorial Guinea, although Zambia has been a member since 2009 (EITI, 2016).

Fluctuating commodity prices can have a significant effect on government revenue and economic growth. In recent years, commodity prices started to fall in 2013, and fell sharply in the second half of 2014, with the decline continuing into 2015 and 2016. By mid-February 2016, the price of crude oil had fallen to less than US$28 per barrel. With a worldwide oversupply of oil, the price fell by over 70 per cent between the middle of 2014 and early 2016. By late 2016, the oil price had risen to US$55 per barrel. The prices of some other commodities also declined, such as iron ore and cotton, and to a lesser extent coffee. Copper prices almost halved between February 2011 and February 2016, then some recovery had occurred by February 2017, although prices were still well below early 2011 levels. The price of gold recovered during 2016 and the cocoa price remained high in 2015, but the latter declined significantly from the middle of 2016.

Scoring the Millennium Development Goals (MDGs)

The Millennium Development Goals (MDGs) were introduced by the United Nations in 2000 to focus attention on development in poor countries, to provide motivation to achieve eight goals and 18 targets, and to measure performance in different quality of life indicators. The target date for achieving the MDGs was 2015, and in September of that year the UN announced the launch of the Sustainable Development Goals from January 2016. As Vandemoortele (2007: 1) points out, the choice of MDGs was based on two factors: first, on a series of internationally agreed indicators that could be measured; and second, whether 'reasonably good data were available to document global trends'.

It is perhaps useful to review the situation in 2015 and the levels of achievement attained by the different African countries at the end of the MDG period. While most commentaries have praised Africa for making progress during the 15-year period, they have often also indicated that sub-Saharan Africa (SSA) is lagging behind other world regions and still has some way to go in meeting certain MDG targets. A brief review of the record of achievement for each of the eight goals should indicate where progress still needs to be made. The results are summarised in Table 20.1.

MDG1 focused on eradicating extreme hunger and poverty, but in sub-Saharan Africa poverty levels only fell from 56.5 per cent in 1990 to 48.4 per cent in 2010, well off the MDG target of 28.25 per cent. Fluctuating commodity process and disease outbreaks such as Ebola in Guinea, Sierra Leone and Liberia from 2014 to 2016 undoubtedly affected the rate of poverty reduction (UNECA and UNDP, 2015). Despite falling poverty levels, however, there were an estimated 330 million Africans living in poverty in 2012, compared with *c.*280 million in 1990, due to high rates of population growth (Beegle *et al.*, 2016).

MDG2 was concerned with achieving universal primary education, and with a generally good level of resourcing for education. By 2012, over two-thirds of African countries had a net enrolment rate of at least 77 per cent in primary education. This led to an increase in youth literacy to 69.61 per cent in 2012. However, failure to complete primary education and a shortage of qualified teachers remain widespread problems.

MDG3 advocated promoting gender equality and empowering women, and some useful progress has been made in this area. During the 15-year period, increasing numbers of girls were attending primary school and literacy rates among women and girls improved. However, gender barriers are still evident, and reflected in girls' non-completion of primary schooling and poor transition rates between primary, secondary and tertiary education sectors and into wage employment. Many African countries have made impressive progress in gaining representation for women in national parliaments. By 2014, 20 per cent of seats in national parliaments were held by women, with the East Africa region showing the greatest representation (27 per cent) and North Africa being the lowest (16 per cent). Rwanda, for example, had over 63 per cent of women representatives in its national parliament (UNECA and UNDP, 2015).

MDG4 focused on reducing child mortality, and in African countries there has been some progress, but there is still some way to go. Between 1990 and 2012, under-five mortality rates in Africa as a whole fell by over 55 per cent, from 146 deaths per 1,000 live births in 1990 to 65 deaths in 2012. But this was short of the MDG target of a two-thirds reduction. In 2015, some African countries had the world's highest rates of child (under-five) mortality, with 156.9 deaths per 1,000 live births in Angola, 138.7 in Chad, 130.1 in the Central African Republic and 120.4 in Sierra Leone (UNDP, 2017).

MDG5 was concerned with improving maternal health, particularly before, during and after childbirth. In 2015, Africa had some of the highest maternal mortality ratios in the world, although data on maternal health in some African countries are often non-existent or unreliable. In the period between 1990 and 2013, only four countries met the target of reducing the maternal mortality ratio by three-quarters: Cape Verde, Equatorial Guinea, Eritrea and Rwanda. Countries that have suffered, or are still suffering, from political instability and conflict have experienced persistently high maternal mortality ratios, such as Somalia, the Central African Republic and Sierra Leone. Early childbearing and the availability of skilled health personnel, particularly to be present at births, are key factors in affecting levels of maternal mortality.

MDG6 aimed to combat HIV/AIDS, malaria and other diseases, and there is evidence of some significant progress being achieved during the MDG period. All countries except Angola and Uganda are now experiencing falling HIV infection rates. Between 2005 and 2013, HIV prevalence in sub-Saharan African countries fell from 5.6 per cent to 4.7 per cent, and there was a 40 per cent decline in AIDS-related deaths. Southern Africa is still the world's worst affected region for HIV/AIDS, but between 2005 and 2013 South Africa achieved a 48 per cent reduction in AIDS-related deaths, and Zimbabwe and Botswana 57 per cent and 58 per cent reductions, respectively. Progress has also been made in combating malaria. Malaria cases in SSA have fallen by 34 per cent and the death rate by 54 per cent since 2000. Spraying in homes and use of insecticide-treated bed nets have undoubtedly made an impact (UNECA and UNDP, 2015). Progress in reducing the incidence of tuberculosis has been less successful, with countries such as Cameroon, Liberia, Lesotho and South Africa actually experiencing an increase on the 1990 rates.

MDG7 related to environmental sustainability, and involves reducing biodiversity loss, improving access to safe drinking water and sanitation, and improving the lives of slum dwellers. The reduction of the forest area, both globally and in Africa, is of concern, with a decline in the forested area in SSA from 31.2 per cent of the land cover in 1990 to 28.1 per cent in 2010. The provision of a supply of clean drinking water has been less successful in SSA than in other parts of the world, with only 16 per cent of the population having access to piped drinking water in 2015. Provision of improved sanitation in SSA has also been slow and variable, with an increase from 24 per cent to 30 per cent between 1990 and 2012, some way from the MDG target of 66 per cent by 2015. With rapid urban growth outstripping general population growth rates in Africa, UN-Habitat (2013) has shown that in 2012, the proportion of people living in slum conditions in urban areas was highest in SSA.

MDG8 focused on developing a global partnership for development, and involved improving trading and financial systems, addressing the special needs of poor countries, issues concerning landlocked and small island states, dealing with debt problems, and spreading the benefits of new technologies. In Africa, there is still a wide gap between the eight MDG targets and levels of achievement. The continent's share of global merchandise exports fell from 3.55 per cent in 2012 to 3.3 per cent in 2013, while overseas development assistance (ODA) from DAC countries (Development Assistance Committee of the OECD) fell in real terms from 2013 to 2014, with many OECD countries allocating less than the target of 0.7 per cent of gross national income advocated by the UN. A report from UNECA and UNDP has argued that the Paris Declaration on Aid Effectiveness signed in 2005:

> has not worked, since it involved a commitment to provide more aid to LDCs (Least Developed Countries). The stated objective of ODA is to improve the livelihoods substances of the most vulnerable of the world. This would be particularly important from an African perspective since 34 of the 54 countries in the continent are classified as LDCs.
>
> *(UNECA and UNDP, 2015: 54)*

In terms of reducing the debt burden of African countries, the Heavily Indebted Poor Countries (HIPC) Initiative has made some impact, such that by September 2014, some 29 African countries were receiving debt relief. With regard to spreading new technology, the growth of mobile phone subscriptions in Africa has been staggering in the period of the MDGs, with many phone users having multiple subscriptions. By 2013, there were an estimated 823.7 million cell phone subscriptions in Africa (UNECA and UNDP, 2015). As with cell phone subscriptions, so the growth of Internet users in Africa has been much faster than in the rest of the world, with an average annual growth of 21.7 per cent, compared with a global average of 10.2 per cent. But it should be recognised that many parts of Africa, particularly in rural areas, still lack a reliable source of electric power, thus making the use of computers and the Internet difficult.

Having undertaken a brief review of the extent to which African countries have achieved the targets set by the MDGs, it should be acknowledged that some writers are critical of the MDGs and their relevance for African countries. For example, Vandemoortele (2007: 1) asserts that the MDGs are:

> essentially an extrapolation of global trends of the 1970s and 1980s and projected forward till 2015. Thus, assessing whether progress is 'on track' for meeting the 2015 targets can only be done at the global level . . . It is erroneous, for instance, to lament that sub-Saharan Africa will not meet the MDGs. These targets were not set specifically for that region.

He suggests that it is possibly more important to look at the performance trajectories of individual countries, and that perhaps development assistance and debt relief should be provided according to an individual country's performance. He suggests that it is demoralising to be categorised as a 'poor' country that is failing to meet the MDG targets (Vandemoortele, 2007: 5).

William Easterly, a former Senior Adviser at the World Bank, is also concerned about the possible demoralising effects of MDG achievements and rankings, particularly when African countries fail to meet one or more of the targets. He argues that 'The negative picture matters because it is demoralising to African leaders and activists, and because it might have real consequences for things like private foreign investment to reinforce the stereotype that "Africa always fails"' (Easterly, 2009: 33). Easterly concludes that:

> it seems undesirable to exaggerate the 'Africa as failure' image, which in turn exaggerates the role of 'The West as Savior' for Africa (as the MDG campaign has often played out in practice). It is demoralizing to have goals for Africa that can only be attained with progress that is nearly without historical precedent from other regions or in Africa itself. Africa has enough problems without international organizations and campaigners downplaying African success when it happens.
>
> *(Easterly, 2009: 34)*

In January 2016, the UN replaced the MDGs with the Sustainable Development Goals (SDGs). The SDGs seem much more complex than the MDGs, with the number of goals more than doubling to 17, while the targets have increased from 18 to no fewer than 169. It is difficult to see how governments in poor countries will be able to appreciate, let alone implement, so many goals and targets when funding is tight, expertise is limited and national priorities have already been identified. Some of the SDG goals seem remarkably opaque and will undoubtedly need further clarification. For example, Goal 9 urges national governments to 'Build resilient infrastructure, promote inclusive and sustainable industrialization and foster innovation', while Goal 12 proposes 'Ensure sustainable consumption and production patterns'. These statements will need to be dissected and carefully examined in the context of local needs, resources and pressures. Measuring the extent to which countries achieve the new goals is also likely to be much more difficult than with the MDGs. Specific indicators will need to be identified to evaluate such variables as 'resilient infrastructure' and 'sustainable consumption patterns', and with poor capacity to generate accurate data, many African governments are likely to struggle to do this.

However, some commentators, such as Freistein and Mahlert (2016), have welcomed the SDGs for potentially providing useful evidence on inequality. They argue that 'The SDGs can be seen as important milestones because they function as discursive nodes that tie together different and as yet only loosely related ideas about inequality, thus becoming a reference point for future policies' (Freistein and Mahlert, 2016: 2144). By contrast, other authors question whether setting defined numeric targets is not setting oneself up for failure, and they also question whether many countries will realistically be able to pursue so many interventions simultaneously (Pogge and Sengupta, 2014, 2015).

Listening to the people

While evaluating economic indicators and progress in achieving a range of development goals might give some indication of whether the quality of life in African countries is improving, there is probably no better measure of development progress than getting to the grassroots and asking ordinary people about their perceptions of changing fortunes at the household, community and national levels. The 2017 *World Happiness Report* (Helliwell *et al.*, 2017) ranked 155 countries according to levels of happiness generated by a number of variables such as per capita GDP, social support, healthy life expectancy at birth, freedom to make life choices, generosity, and perceptions of corruption. While Norway was ranked first for the period 2014–2016, African countries were well down the list, with 16 of the 20 lowest-ranking countries being in Africa, and the Central African Republic ranked 155, at the bottom of the table. Clearly, many Africans are not content with their current lifestyles and future livelihood prospects.

It seems that the benefits of independence after a long period of colonial rule have in many cases not reached the bulk of Africa's people, and there is much concern about standards of

Table 20.1 Summary of Africa's performance in meeting the Millennium Development Goals (MDGs), 2015

Goals	*Achievement status*	*Comments*
Goal 1: Eradicate extreme poverty and hunger	Further progress needed	**SSA:** Poverty declined from 56.5 per cent in 1990 to 48.4 per cent in 2010 **NA:** Poverty declined from 5 per cent to 1 per cent **SSA:** Hunger declined by 8 per cent between 1990 and 2013 **SSA:** World's most food-deficient region; 25 per cent of population faced hunger and malnutrition, 1990–2013
Goal 2: Achieve universal primary education	Progress made	2012: 68 per cent of African countries had net enrolment of at least 75 per cent in primary education Average primary completion: 67 per cent Youth literacy: 69.6 per cent in 2012 with better access to primary education
Goal 3: Promote gender equality and empower women	Progress made	Gender Parity Index (GPI) increase, 1990–2012 **North Africa:** 0.82 to 0.96 **SSA:** 0.83 to 0.92 Good progress in creating women MPs
Goal 4: Reduce child mortality	Further progress needed	1990–2012: Under-five mortality rate fell by 55 per cent Infant mortality rate fell by 40 per cent, but still high
Goal 5: Improve maternal health	Further progress needed	2013: Africa, 289 maternal deaths per 100,000 live births World average: 210 deaths
Goal 6: Combat HIV/AIDS, malaria and other diseases	Progress made	Since 2000, reduction in prevalence and deaths from HIV/AIDS, malaria and TB
Goal 7: Ensure environmental sustainability	Further progress needed	Declining forest cover **SSA:** Only 64 per cent of population used improved drinking water source **SSA:** 1990–2012, access to improved sanitation increase from 24 per cent to 30 per cent **NA:** 72 per cent to 91 per cent

Source: Adapted from UNECA and UNDP (2015).

Note: NA = North Africa; SSA = sub-Saharan Africa.

governance, even though most African countries now have (in theory at least) democratic regimes. As Helliwell *et al.* (2017: 110) comment:

> people in many African countries are still waiting for changes to improve their lives and to make them happy. In short, African people's expectations that they and their countries would flourish under self-rule and democracy appear not to have been met.

Despite this widespread feeling, there is an amazing degree of resilience among people in Africa's rural and urban communities and a high level of optimism, particularly among young people, many of whom struggle to get paid employment.

Surveys undertaken by Afrobarometer aim to place a 'finger on the pulse' of local feelings and aspirations. Afrobarometer is a 'pan-African, non-partisan research network that conducts public attitude surveys on democracy, governance, economic conditions, and related issues in more than 35 countries in Africa' (Afrobarometer, 2017). A survey released in October 2013 revealed widespread dissatisfaction with economic conditions, and reinforced earlier comments about 'growth without development', that 'trickle-down' of wealth and benefits to the grassroots has not occurred. As the Afrobarometer report comments, 'despite high reported growth rates, lived poverty at the grassroots remains little changed' (Dulani *et al.*, 2013: 1). It seems that improvements in social services (health, education) and infrastructure (electricity, water, paved roads, sewage systems and health clinics) can have a significant impact on levels of lived poverty. The report concludes that 'One in five Africans still experiences frequent ("many times" or "always") deprivation with respect to their most basic needs for food (17%), clean water (21%), and medicines and medical care (20%)' (Dulani *et al.*, 2013: 2). While there were some improvements in poverty reduction over the decade to 2013 in Cape Verde, Ghana, Malawi and Zambia, the survey revealed increases in poverty in Botswana, Mali, Senegal, South Africa and Tanzania. In conclusion, it seems the Afrobarometer survey findings:

> suggest that either economic growth is not trickling down to average citizens and translating into poverty reduction (and in fact, is instead leading to growing income inequality), or that there is a reason to question whether reported growth rates are actually being realized.
>
> *(Dulani* et al., *2013: 16)*

An uncertain future?

The euphoria associated with revelations that 'Africa is Rising' appears to have been somewhat short-lived. Superficially, there is evidence of economic growth rates that are impressive by world standards. But dig deeper, and at the grassroots level there is widespread poverty and increasing inequality in many, if not most, African countries, particularly in SSA. Neo-liberal top-down policies seem not to have benefited the bulk of Africa's population. As Andreou (2014: 1) suggests, 'Trickle-down economics is the greatest broken promise of our lifetime'. While some progress has been achieved in the last two decades, as an evaluation of MDG achievements shows, there is still a long way to go if African countries are going to catch up with high-income countries in Western Europe and North America.

So, what are the prospects for remote African communities such as Kayima in north-east Sierra Leone, where there was little change in the 40-year period up to 2014? Sadly, on the basis of previous experience, it seems likely that such communities will be largely the same after the next 40 years. A poor country such as Sierra Leone has so many urgent priorities, not least in catching up on long-awaited improvements after the civil war and Ebola lockdown. In Freetown, the capital city, new roads are being built and the main hospital has been refurbished since the war. Electricity is marginally more reliable than it used to be, but water shortages are still common, particularly in the dry season, and sewerage and sanitation systems are struggling to keep pace with urban growth. Out in the provinces, the two main trunk roads have been rehabilitated and there are now solar-powered street lights in some of the main towns. But remote Kayima, which is a long and tedious drive from the main road, is essentially just as it was in the 1970s (Bateman *et al.*, 2017). While Sierra Leone has achieved some success in meeting the Millennium Development Goals, the challenging targets for the new Sustainable Development Goals may be well out of reach for such poor countries. And while national-level statistics (though often unreliable) might indicate reductions in poverty and child mortality, at the local level the situation is often very different. Such places as Kayima desperately need better connectivity, improved healthcare, education, water supply and sanitation, but it is likely to be a long time before such improvements are implemented. In the meantime, livelihoods will probably continue much as they have for decades, with traditional farming systems hopefully ensuring food security and household survival from one year to the next. It is a sobering fact that rural Africa has many such places as Kayima, where economic growth and any benefits of 'Africa Rising' have seemingly bypassed these communities, and there have been only meagre improvements in the quality of life from one generation to the next.

It seems there is need for a new agenda and a better understanding of the targets!?

References

AfDB (African Development Bank) (2008) *Selected Statistics on African Countries, 2008*, Abidjan: AfDB.

AfDB (African Development Bank) (2011) *The Middle of the Pyramid: Dynamics of the Middle Class in Africa*, Market Brief, 20 April, available at: www.afdb.org/fileadmin/uploads/afdb/Documents/Publications/The%20Middle%20of%20the%20Pyramid_The%20Middle%20of%20the%20Pyramid.pdf (accessed 26 March 2017).

AfDB (African Development Bank) (2016) *African Statistical Yearbook, 2016*, Abidjan: AfDB.

Africa Progress Panel (2013) *Africa Progress Report 2013: Equity in Extractives*, available at: www.africapro gresspanel.org/wp-content/uploads/2013/08/2013_APR_Equity_in_Extractives_25062013_ENG_LR.pdf (accessed 19 March 2017).

Afrobarometer (2017) *Afrobarometer*, available at: www.afrobarometer.org/ (accessed 12 March 2017).

Andreou, A. (2014) *Trickle-Down Economics Is the Greatest Broken Promise of Our Lifetime*, 20 January, available at: www.theguardian.com/commentisfree/2014/jan/20/trickle-down-economics-broken-prom ise-richest-85 (accessed 2 April 2017).

Bateman, J., Binns, T. and Nel, E. (2017) 'Flirting with food security: resilience in the face of conflict, climate change and communicable diseases in rural Sierra Leone', in E. Thomas-Hope (ed.), *Climate Change and Food Security: Africa and the Caribbean*, London: Routledge/Earthscan, pp. 128–142.

Beegle, K., Christiansen, L., Dabalen, A. and Gaddis, I. (2016) *Poverty in a Rising Africa*, Washington, DC: World Bank Group, available at: www.worldbank.org/en/region/afr/publication/poverty-rising-africa-poverty-report (accessed 26 March 2017).

Binns, T. and Maconachie, R.A. (2005) '"Going home" in post-conflict Sierra Leone: diamonds, agriculture and re-building rural livelihoods in the Eastern Province', *Geography*, 90(1): 67–78.

Binns, T., Dixon, A. and Nel, E. (2012) *Africa: Diversity and Development*, London: Routledge.

Bond, P. (2006) *Looting Africa*, London: Zed Books.

Clower, R.W., Dalton, G., Harwitz, M. and Walters, A.A. (1966) *Growth Without Development: An Economic Survey of Liberia*, Evanston, IL: Northwestern University Press.

Dulani, B., Mattes, R. and Logan, C. (2013) *After a Decade of Growth in Africa, Little Change in Poverty at the Grassroots*, Policy Brief No. 1, Afrobarometer, available at: http://afrobarometer.org/sites/default/files/publications/Briefing%20paper/ab_r5_policypaperno1.pdf (accessed 12 March 2017).

Easterly, W. (2009) 'How the Millennium Development Goals are unfair to Africa', *World Development*, 37(1): 26–35.

EITI (Extractive Industries Transparency Initiative) (2016) *The EITI Principles*, available at: https://eiti.org/document/eiti-principles (accessed 19 March 2017).

Escobar, A. (1995) *Encountering Development*, Princeton, NJ: Princeton University Press.

Freistein, K., and Mahlert, B. (2016) 'The potential for tacking inequality in the Sustainable Development Goals', *Third World Quarterly*, 37(12): 2139–2155.

Harvey, D. (2006) *Spaces of Global Capitalism*, New York: Verso.

Harvey, D. (2014) *Seventeen Contradictions and the End of Capitalism*, New York: Oxford University Press.

Helliwell, J., Layard, R. and Sachs, J. (eds) (2017) *World Happiness Report 2017*, New York: Sustainable Development Solutions Network, available at: http://worldhappiness.report/ (accessed 25 March 2017).

Hettne, B. (1995) *Development Theory and the Three Worlds*, New York: Wiley.

Hickel, J. (2014) *The Death of International Development*, Al Jazeera, 20 November.

Hickel, J. (2016) 'The true extent of global poverty and hunger', *Third World Quarterly*, 37(5): 749–767.

Lamy, P. (2013) *Africa is the Growth Continent for the 21st Century*, speech at the University of Nairobi, 22 May, available at: www.wto.org/english/news_e/sppl_e/sppl283_e.htm (accessed 26 March 2017).

Maconachie, R.A. and Binns, T. (2007) 'Beyond the resource curse? Diamond mining, development and post-conflict reconstruction in Sierra Leone', *Resources Policy*, 32: 104–115.

Melber, H. (2016) *Africa's Rising Middle Class: Time to Sort out Fact from Fiction*, available at: https://theconversation.com/africas-rising-middle-class-time-to-sort-out-fact-from-fiction-59797 (accessed 30 March 2017).

OECD (Organisation for Economic Co-operation and Development) (2014) *Focus on Inequality and Growth*, Paris: OECD.

Pogge, T. and Sengupta, M. (2014) 'Rethinking the post-2015 development agenda: eight ways to end poverty now', *Global Justice: Theory Practice Rhetoric*, 7: 3–11.

Pogge, T. and Sengupta, M. (2015) 'The Sustainable Development Goals (SDGs) as drafted: nice idea, poor execution, *Washington International Law Journal*, 24(3): 571–587.

Potter, D., Binns, J.A., Elliott, J. and Nel, E. (2017) *Geographies of Development*, 4th edition, London: Routledge.

Rogerson, C. and Nel, E. (2015) 'Redressing inequality in South Africa: the spatial targeting of distressed areas', *Local Economy*, 3(1–2): 28–41.

Ross, M. (2001) 'Does oil hinder democracy?', *World Politics*, 53(3): 325–361.

Simpasa, A., Hailu, D., Levine, S. and Tibana, R.J. (2013) *Capturing Mineral Revenues in Zambia: Past Trends and Future Prospects*, New York: UNDP, available at: www.un.org/en/land-natural-resources-conflict/pdfs/capturing-mineral-revenues-zambia.pdf (accessed 19 March 2017).

Smith, B. (2015) 'The resource curse exorcised: evidence from a panel of countries', *Journal of Development Economics*, 116: 57–73.

Statistics South Africa (2015) *Census 2011: Income Dynamics and Poverty Status of Households in South Africa*, Pretoria: Statistics South Africa.

The Economist (2000) *Hopeless Africa*, 11 May, available at: www.economist.com/node/333429 (accessed 11 March 2017).

The Economist (2011) *Africa Rising*, 3 December, available at: www.economist.com/node/21541015 (accessed 11 March 2017).

Thornton, P. (2014) *The Great Economists*, Harlow: Pearson.

UNDP (United Nations Development Programme) (2007) *Human Development Report, 2007/2008*, Basingstoke: Palgrave Macmillan.

UNDP (United Nations Development Programme) (2016) *Human Development Report, 2015*, New York: UNDP.

UNDP (United Nations Development Programme) (2017) *Human Development Report, 2016*, New York: UNDP.

UNECA (UN Economic Commission for Africa) and UNDP (United Nations Development Programme) (2015) *MDG Report: Assessing Progress in Africa Toward the Millennium Development Goals*, Addis Ababa: Economic Commission for Africa, available at: www.uneca.org/publications/mdg-report-2015-assessing-progress-africa-toward-millennium-development-goals (accessed 21 March 2017).

UN-Habitat (United Nations Human Settlements Programme) (2013) *State of the World's Cities, 2012/2013*, Nairobi: UN-Habitat.

UNICEF (United Nations Children's Fund) (2013) *Poverty*, available at: www.unicef.org/zambia/children_5157.html (accessed 12 March 2017).

Vandemoortele, J. (2007) *The MDGs: 'M' for Misunderstood?* United Nations World Institute for Development Economics Research, available at: www.wider.unu.edu/publication/mdgs-%E2%80%98m%E2%80%99-misunderstood (accessed 20 March 2017).

21

Regional integration[1]

Olivier Walther

Introduction

More than 50 years have passed since the Organisation of African Unity (OAU) was created in Addis Ababa. Over this half century, several regional bodies have been created, with much pomp and circumstance, to foster integration between newly independent African states. However, despite the many charters signed, the numerous meetings called, and the countless strategic frameworks drawn up on paper, progress made towards effective regional integration has been slow on the continent.

Inter-state agreements that should ensure the free movement of goods still have little effect on business operators, who continue to rely on informal negotiations with state authorities in order to cross national boundaries, contributing to make Africa one of the most expensive places in the world to do business (OECD, 2009a; AfDB, 2015). On a continent divided into 54 states, regional integration also suffers from a partial elimination of tariff barriers, lack of coherence between development frameworks, and divergent macroeconomic policies – three factors that tend to increase the negative impact of the fragmentation of markets (World Bank, 2012). The mobility of people within and between regional blocs is another area where progress made on regional integration has been disappointing, often for lack of political will. In spite of protocols signed at the highest level, customs hassles are still a fact of life for migrants and travellers who make their way across the continent.

The lack of concrete results in past regional policies does not appear to reduce African leaders' enthusiasm for regional summits and treaties. Nor does it seem to dampen the enthusiasm of the policymakers and scientific community. Regional integration remains a highly popular research topic among economists, political scientists, geographers and other social scientists, and has always been the subject of ardent debates between the policy world and academia. As each discipline has approached regional integration from a slightly different angle, many of the debates have been fuelled by the relative absence of a strong consensus on what 'regional' and 'integration' really mean, contributing to obscure its geographical *scale* and functional or institutional *nature*.

The objective of this chapter is to discuss the social, economic and political factors that explain the gap between regional integration as an institutional project (*regionalism*) and regional integration as an everyday reality (*regionalisation*). After a short discussion on the scales and nature of integration, the second part of the chapter reviews the literature and discusses the historical roots and contemporary activities of the institutions created to promote regional integration on the continent. The third part discusses the policy implications of the limited achievements of regional integration institutions, while the conclusion stresses the need for an African model of integration.

Scales and nature of regional integration

For many political scientists and economists, the 'regional' level refers to supranational organisations at the continental or subcontinental scale. The African continent is often considered as a macro-region divided into several subregions, such as southern, central or West Africa (UN, 2015). Subcontinental regions constitute the privileged scale of regional blocs such as the Southern African Development Community (SADC), the Economic Community of Central African States (ECCAS) or the Economic Community of West African States (ECOWAS).

The term 'regional' is also used in regional science and geography to refer to administrative subdivisions between states and local authorities, or to functional areas in which social and economic interactions are particularly developed, possibly across borders (see Table 21.1). In West Africa, for example, 19 such potential cross-border functional regions have recently been identified based on the existence of border markets (OECD, 2014). To complicate matters further, many scholars now consider regions as social and historical constructs characterised by a territorial and symbolic shape, a number of institutions, and an established identity anchored in social practices (Paasi, 2002). The Dogon Country in Mali is a good example of a region that encompasses a (rather) homogeneous linguistic and ethnic group with a long history of resistance against enslaving states. Such regions approximate what political scientists have defined as micro-regions, i.e. entities that 'exist between the "national" and the "local" level' and can be subnational or cross-border (Söderbaum and Taylor, 2008: 13).

Another ambiguity of the concept of regional integration is that its *nature* is interpreted in very different ways, depending on whether the focus is on institutional or functional actors. Regionalism is a process of voluntary agreements between state and other public or private bodies that aim to address the negative consequences of national segmentation and promote political and economic integration between countries. In Africa, its overall objective, envisioned in the Treaty Establishing the African Economic Community, known as the Abuja Treaty, is to establish a continental common market by 2025 (OAU, 1991). African states are all, to varying degrees, institutionally integrated in various regional bodies such as the African Union (AU),

Table 21.1 Scale and nature of integration

Integration			
Scale		*Nature*	
The political science and economist view	The regional science and geography view	Regionalism (institutional, formal, or policy-driven integration)	Regionalisation (functional, informal, or market-driven integration)
global > macro-regional > national > micro-regional > local	global > national > regional > local	cooperative relationships between public and other bodies	social or economic relationships between private actors, firms, institutions

the Common Market for Eastern and Southern Africa (COMESA) or the Arab Maghreb Union (UMA). This institutional process is also known as *institutional, formal* regionalism (Söderbaum, 2011) or *policy-driven* integration (Perkmann, 2007) because it mainly involves the state and the conduct of political and policy affairs.

Regionalisation refers to the social and economic relationships that develop between individuals and firms across time and space, and which lead to the emergence of functional regions. In Africa, some micro-regions have achieved an exceptional level of functional integration due to the intense interactions between petty traders and long-distance merchants. The rapid growth of border markets (Dobler, 2016) illustrates how regional integration can also come from the activities of a skilled community of traders that successfully exploit border differentials. In the current literature, regionalisation is sometimes referred to as *functional, informal* or *market-driven* integration to distinguish it from the formal process of building supranational organisations.

Key themes

Moving beyond disciplinary lines

Disciplinary lines between regional integration scholars have proved nearly as resistant to change as African international boundaries. Until the late 1980s, there was very little dialogue between the policy approaches focusing on formal inter-state frameworks at the macro level and African studies literature focusing on functional interactions at the micro level. Supported by international financial institutions, the policy approach long stressed the institutional principles and theoretical benefits of regional integration (World Bank, 1989). State-centric and influenced by the European experience, this strand of literature could not have been further from the daily realities of African traders, migrants and entrepreneurs that were documented by the African studies approach (MacGaffey, 1987; Igué, 1989), which gave priority to microanalysis that showed how regional integration took place from below.

It was only in the late 1990s that the gap between the normative and empirical literature documenting regional integration in Africa started to be filled in. This period experienced a revival of regionalism due to the end of many one-party governments and an increased liberalisation of markets, both of which favoured intergovernmental arrangements (Hanson, 2015). Building on a conceptual framework inspired from international political economy and a number of detailed case studies taken from various regions of Africa, the new regionalism literature (Hettne and Söderbaum, 1998; Bøås, 2001; Hettne and Söderbaum, 2008; Söderbaum and Taylor, 2008; Shaw *et al.*, 2011) made an important contribution towards a more comprehensive understanding of regional integration that combines regionalism and regionalisation. This literature emphasised that regional integration resulted from the interdependencies between various geographical scales, which could no longer be considered as autonomous levels. The development of markets and transport corridors in certain micro-regions, for example, reflects the unequal state of liberalisation of the continent at the macro level, which in turn provides incentives to traders to exploit disparities between trade policies (Nixdorf, 2013).

Another important contribution to the study of region-building in Africa has been made by the border studies literature, whose interdisciplinary approach to borderlands and borderlanders has documented how regions were built from below (Asiwaju and Adeniyi, 1989; Nugent and Asiwaju, 1996; Feyissa and Hoehne, 2010; Hüsken and Klute, 2010; Korf and Raeymaekers, 2013; Miles, 2014). Border scholars have highlighted the vitality of social ties, political patronage and trade networks that exploit colonial partitioning and contribute to regionalisation in the so-called national peripheries, often in contradiction with the institutional initiatives developed by states and regional organisations (Meagher, 2003; Nugent, 2008; Titeca and de Herdt, 2010; Nugent, 2012; Walther, 2015; Zeller, 2015). This literature has demonstrated that the analytical distinction between regionalism and regionalisation has become increasingly blurred in Africa. Contemporary African bureaucracies are mixed with patronage and clientelism, and tend to support regional initiatives that benefit state elites and their allies (Gibb, 2009). Such networks of governance cut across public and private spheres and lead elites to pursue their own interests rather than the public good or a pan-Africanist vision (Raeymakers, 2014; Reyntjens, 2014). As shown by Bach (1999, 2016), a general understanding of the failures of regional integration can only be achieved if one considers *regionalism* and *regionalisation* simultaneously, because the process of building new regions in Africa results from relationships between the state and society that are constantly recombined.

Historical roots and current regional institutions

The idea of reducing regional disparities is hardly new in Africa. Since the end of the nineteenth century, European powers campaigned actively to create federations that could reduce the cost of colonial management or, as in East Africa, to tighten the grip of white settlers over their immediate neighbourhood. Established in 1895, French West Africa (AOF) was a federation of eight colonies extending from Senegal to Niger. Its counterpart was French Equatorial Africa (AEF), a federation of five colonial territories founded in 1910 and stretching from the Sahara to the Congo River. The two federations were dismantled in 1958, but the CFA franc common currency survived. Further south, the British established the Southern African Customs Union (SACU) between South Africa and today's Botswana, Lesotho and Swaziland. The SACU eventually led to the oldest working customs union worldwide (Erasmus, 2014). The British also set up the short-lived Central African Federation (1953–1963) between the colony of Southern Rhodesia and the protectorates of Northern Rhodesia and Nyasaland (Butler, 2000).

This colonial heritage was interpreted both as a legacy and a constraint by the political leaders of the 1960s (Makinda *et al.*, 2016). On the one hand, as myriad new countries became independent, the need for regional institutions that could remedy the negative effects of territorial partition had never been so pressing. Several new states were landlocked and dependent on road and port infrastructures located in a foreign country thousands of kilometres away. Other states had access to the sea, but were either too small or too disconnected from their neighbours by decades of colonial policies to develop a viable national economy. On the other hand, many of the new African leaders were also 'reluctant, if not totally unwilling, to support policies likely to restrain state sovereignty and, consecutively, their power' (Bach, 1999: 43). The creation of the OAU in 1963 reflects this tension: while leaders of the Casablanca Group led by Kwame Nkrumah of Ghana, Sékou Touré of Guinea and Gamal Abdel-Nasser of Egypt were in favour of a pan-African political union, other leaders from the Monrovia Group, mostly from the French colonies, Ethiopia and Nigeria, supported sovereignty, non-interference between states, and the principle of inviolability of borders as they existed upon independence that forms one of the foundations of the AU today (Khadiagala, 2008; Murithi, 2012).

As regards economic integration, while pan-Africanists called for continental unity, the Monrovia Group called for the creation of regional institutions as building blocks for African unity. Eight such regional blocs have been created between 1967 and 1998, and in the years following the Abuja Treaty in 1991 have been officially recognised as regional economic communities (RECs) by the African Union (see Table 21.2). RECs have achieved various degrees of integration: *free trade areas* such as COMESA were established to further reduce trade barriers; *customs unions*, such as the East African Community (EAC) and SACU (Drummond *et al.*, 2015), add another layer of integration by introducing a common external tariff on the external borders of the regional bloc; *common markets* further extend free trade areas by allowing services, capital and labour to circulate freely across countries; and *customs and monetary unions* guarantee a common external tariff and a single currency. In Africa, customs and monetary unions are subsets of larger regional bodies. Eight ECOWAS members formed the West African Economic and Monetary Union (UEMOA) following the devaluation of the West African CFA franc (XOF) in 1994 (Seck, 2013; Bossuyt, 2015). The Economic and Monetary Community of Central Africa (CEMAC) is also a grouping of six ECCAS countries using the Central African CFA franc (XAF) (Nono, 2014). In South Africa, SACU comprises five SADC countries, four of which base their currencies on the South African rand within a Common Monetary Area (CMA) (Lorenz and Cornelissen, 2012).

Other subregional organisations have not been formally recognised by the AU but play an important role in addressing environmental issues (World Bank, 1994). Some of them focus exclusively on water resources development within the perimeter of the main continental river basins. In West Africa, several countries have played a leading role in the implementation of such organisations since the mid-1960s, with the establishment of the Niger Basin Authority, the Senegal River Basin Development Authority, the Mano River Union and the Gambia River Basin Organisation (OECD, 2009c). Similar organisations were created in southern Africa with the Lesotho Highlands Development Authority, the Zambezi River Authority and the Komati Basin Water Authority (Söderbaum and Granit, 2014). Other organisations aim at covering a broader range of activities, including agriculture, infrastructure, fisheries and natural resources, such as the Lake Chad Basin Development Authority, created to improve land and water management, preserve ecosystems and promote peace-building and security, or the Kagera Basin Organisation, which was supposed to reduce poverty and improve socio-economic development in the Great Lakes region. Environmental issues have also motivated the creation of the Integrated Development Authority of the Liptako-Gourma region, which

Table 21.2 Regional economic communities

Name	*Date of creation*	*Headquarters*
East African Community (EAC)	1967, dissolved 1977, re-established 2000	Arusha, Tanzania
Economic Community of West African States (ECOWAS)	1975, revised 1993	Abuja, Nigeria
Southern African Development Community (SADC)	1980 as SADCC, 1992 as SADC	Gaborone, Botswana
Economic Community of Central African States (ECCAS)	1983	Libreville, Gabon
Intergovernmental Authority for Development (IGAD)	1986	Djibouti City, Djibouti
Arab Maghreb Union (UMA)	1989	Rabat, Morocco
Common Market for Eastern and Southern Africa (COMESA)	1993	Lusaka, Zambia
Community of Sahel-Saharan States (CEN-SAD)	1998	Tripoli, Libya

Source: Organisations' treaties.

brings together three landlocked Sahelian countries with similar problems of land degradation and access to water supply, and of the Permanent Interstate Committee for Drought Control in the Sahel (CILSS), created in the aftermath of the first great droughts of the 1970s to promote food security, develop renewable energies and fight against desertification between 13 Sahelian countries (OECD, 2016).

Main current activities

The overall objective of promoting regional integration in Africa is translated into several policies that aim at improving both hard and soft infrastructure (OECD, 2009a; AfDB, 2012a; World Bank, 2013). Policies promoting hard infrastructure aim to extend and rehabilitate the physical network of roads, railways, ports, airports, satellites, mobile phones and fibre optic cables (Deen-Swarray *et al.*, 2014). Also known as *network-enhancing policies*, these policies improve both the internal connectivity of economic actors at the local level and their external connectivity with the rest of the world. The Trans-African Highway Network envisioned by the United Nations Economic Commission for Africa and the African Development Bank's Aid for Trade Trust Fund, for example, plans to develop the roads connecting the major urban centres of the continent, and hence reduce transport costs. In many parts of Africa, transport corridors are presented as successful examples of public–private partnerships (Teravaninthorn and Raballand, 2009). The best-documented example is probably the Maputo Corridor (Taylor, 2011; Bowland and Otto, 2012; Sequeira *et al.*, 2014; Hanson, 2015), which connects the Gauteng, Limpopo and Mpumalanga provinces in South Africa with the capital city of Mozambique, but other examples can be found in other parts of southern and West Africa (AfDB *et al.*, 2015).

Several regional groupings also aim to establish one-stop border posts (OSBPs) to accelerate border clearance (AfDB, 2012b). No less than 51 OSBPs are currently planned, in construction or already operational in Africa, 39 of which are included within the framework of the Programme for Infrastructure Development in Africa (PIDA) in charge of the strategic vision for the development of regional and continental infrastructure. Most of these are located along the Gulf of Guinea and on the borders of landlocked countries (AU *et al.*, 2011). In West Africa, OSBPs are mushrooming between Benin, Burkina Faso, Niger, Nigeria and Togo under the supervision of ECOWAS and UEMOA and with the support of the European Development Fund. Between Niger and Benin, the recent OSBP of Malanville should facilitate the daily operations of petty traders and long-distance merchants, who used to stop twice on the bridge spanning the Niger River. There has also been a rapid development of OSBPs in eastern Africa since the end of the 2000s, facilitated by the adoption of the EAC One-Stop Border Posts Bill in May 2010. As well as making cross-border trade easier, border posts are supposed to reduce corruption between state authorities, companies and private actors, and help to formalise informal trade.

Policies targeting soft infrastructure aim to strengthen the institutions that underpin regional development. Also known as *growth-enhancing policies*, these policies aim to improve the qualifications of the local workforce, supporting existing entrepreneurs and creating the institutional conditions for the development of new economic activities (Walther, 2015). A crucial aspect of such policies is the improvement of the body of rules and regulations governing regional trade, which are currently seen as an obstacle to regional integration. ECOWAS's Regulatory Informal Trade Programme, for example, aims to incorporate current business practices that circumvent state regulations by simplifying administrative, tax and customs procedures, and encouraging traders to register their businesses. UEMOA, USAID and ECOWAS have also established a programme to eliminate corruption and illegal payments at border posts and along trade routes (USAID, 2014). Another aspect related to growth-enhancing policies is to promote informal trade activities, now regarded as 'the most efficient, organized and deep-rooted system of trade in the region' (AfDB, 2012a: 9). Long considered as speculators, African traders are increasingly seen by international financial institutions, regional organisations and aid agencies as crucial actors for alleviating poverty due to their ability to allocate supplies during food crises and provide employment opportunities for poor households (CILSS, 2014).

Thus far, most regional policies have promoted certain sectors (industry, livestock) or certain categories of population (the poor) without necessarily taking into account the spatial dimension of regional development. As a consequence, regional disparities have remained largely unaddressed, despite their crucial role in hindering inclusive growth (AfDB *et al.*, 2015). Development is, however, a rather unequal process, and the idea is gaining momentum within the academic and policy communities that more attention should be paid to the local circumstances in which regional integration occurs. Place-based policies, many argue, could enhance the economic potential and competitive advantage of regions (Barca *et al.*, 2012). Unlike spatially blind policies that apply without explicit consideration to space, place-based policies are based on the idea that local actors and institutions can be mobilised to foster regional development (OECD, 2009b). While most regional policies tend to rely on sectoral, top-down and subsidy-based interventions, place-based policies promote spatial inclusion by encouraging infrastructure and public goods adapted to the specific challenges of each region. Densely populated cross-border regions, such as between Niger and northern Nigeria, for example, will have different needs than regions with low population densities, while coastal and industrialised regions, such as Lagos, will require policies that are of little use in regions dominated by agriculture. Place-based policies could also support cooperation between infra-national authorities that share similar interests or constraints. Until now, African regions and municipalities are still marginally involved in cross-border cooperation due to the relative ineffectiveness of decentralisation policies that have rarely deconcentrated power and resources to newly created public authorities.

Implications

Disappointing outcomes

Generally speaking, regional integration in Africa has achieved the highest objectives when institutions have targeted very specific areas, have been heavily supported by external donors and member states, and have brought together countries that share a similar currency. Relying on sectoral cooperation between (mainly) francophone countries, CILSS, for example, has addressed some of the most urgent needs of the Sahelian area (food security, natural resources) with a regional focus (CILSS, 2014). The success of SACU and the Franc Zone also illustrate that a high level of regional integration can be achieved between countries where formal arrangements existed well before decolonisation.

Elsewhere in Africa, the progress made by regional organisations has been rather limited since the 1960s (AU, 2009, 2013; UNECA *et al.*, 2010, 2012, 2013). This is particularly evident when one compares the objectives of the Abuja Treaty with its current achievements. Coming into force in 1994, the Abuja Treaty envisioned the gradual establishment of the African Economic Community in six stages until 2034 at the latest. The first stage (1994–1999), which establishes economic communities in regions where these do not exist, was completed with a few exceptions. The second stage (1999–2007), which should have removed intra-regional barriers within RECs and harmonised customs duties in relation to third states, has recorded progress, but is not completed. The third stage (2007–2017), in which each regional economic community would establish a free trade area and a customs union, is far from complete. It is unknown whether the next stages, which should ultimately lead to a pan-African economic and monetary union, central bank and currency, will ever be attained.

Regional organisations have also marginally contributed to increased intra-regional trade, which represents approximately 12 per cent of total trade flows in sub-Saharan Africa (AfDB *et al.*, 2015). This share is well below the levels of other regions in the world, where intra-regional trade can represent up to 66 per cent of total trade, as in Europe. In some parts of the continent, as in southern Africa, intra-regional trade is actually declining, and only represented 15 per cent of total trade in 2008 against 22 per cent in 2002 (Mbekeani, 2013). Regional policies frequently refer to these figures, which are based on official data, to encourage additional efforts to strengthen regional integration. However, much of African trade is not recorded by official figures, and adding informal cross-border trade to existing official figures could lead to a significant re-evaluation of the share of intra-regional trade in sub-Saharan Africa. According to recent estimates, unrecorded cross-border trade could represent from 30 to 40 per cent of total trade in the SADC to more than 75 per cent in such countries as Benin or Uganda (UNCTAD, 2013). At the continental level, the real proportion of intra-African trade is probably similar to what is observed in Latin America or the Caribbean (20 per cent).

Shared responsibilities

One of the most frequent criticisms regarding region-building in Africa refers to the gap between how regional policies are designed by supranational bodies and the concrete application of these policies on the ground. Most regional bodies have so far been unable to enforce the application of the agreements signed between states and to translate these into forces for structural change for African regions and countries. In the words of Bach (2016), there is a mismatch between regionalism as it should be and regionalisation as it is experienced on a daily basis. It would be misleading, however, to attribute the disappointing outcomes and slow process of regional integration in Africa to the supranational organisations alone. Nation states and private actors also share part of the responsibility for the lack of practical implementation of integration policies on the continent.

First, many countries have few incentives to effectively engage in deeper institutional integration with their neighbours. In a patrimonial system that nurtures interpersonal relations across society and the state, regional initiatives often go against the very interests of the state and its clients in the private sector, for which borders are a resource rather than a constraint (Walther, 2012, 2015). As a result, regional agreements remain on paper, not only because the state is unable to apply them in the field, but because they threaten informal arrangements. Bach's comment, formulated more than 15 years ago, that 'trans-state integration is stimulated by market distortion, not trade liberalization' (Bach, 1999: 13), is still true today. In Nigeria, for example, the persistence of informal trade and the reluctance of the government to fully liberalise the market and reform customs have less to do with the protection of national industries than with the profits generated by illegal re-export trade (Raballand and Mjekiqi, 2010; Golub, 2012). In Benin, the existence of national border differentials has long been a guarantee of state and private revenues because most of what is imported from the world markets is ultimately destined to be sold informally to neighbouring countries (Igué and Soulé, 1992).

Second, regional integration also has a disappointing record due to the large number of organisations that exist with similar or competing purposes. Nearly all African states belong to multiple regional groupings, which leads to high coordination costs, competition between policies and confusion among international donors (Dirar, 2010; Tavares and Tang, 2011). Countries such as Burkina Faso, Côte d'Ivoire, Guinea, Liberia, Mali, Niger and Sierra Leone belong to five different groupings. In East and southern Africa, overlapping affiliations between COMESA, SADC, SACU and EAC are also particularly numerous (see Figure 21.1). The proliferation of regional institutions with overlapping responsibilities is seen by many African leaders as an opportunity for consolidating state power (Herbst, 2007). African regional organisations provide fertile ground for paradiplomacy at the global level, and are actively used by political elites to promote national interests within international financial institutions, the European Union and Western donors. Limiting the number of regional organisations would also mean limiting the opportunities for African leaders to reach potentially important sources of revenue and political support.

Third, regional integration has been greatly affected by political crises and conflicts. While the number of global conflicts has declined since the end of the Cold War, Africa has experienced an increase in political violence since the 1990s, as many one-party leaders suddenly lost their external support. As a consequence, several regional organisations have ceased their

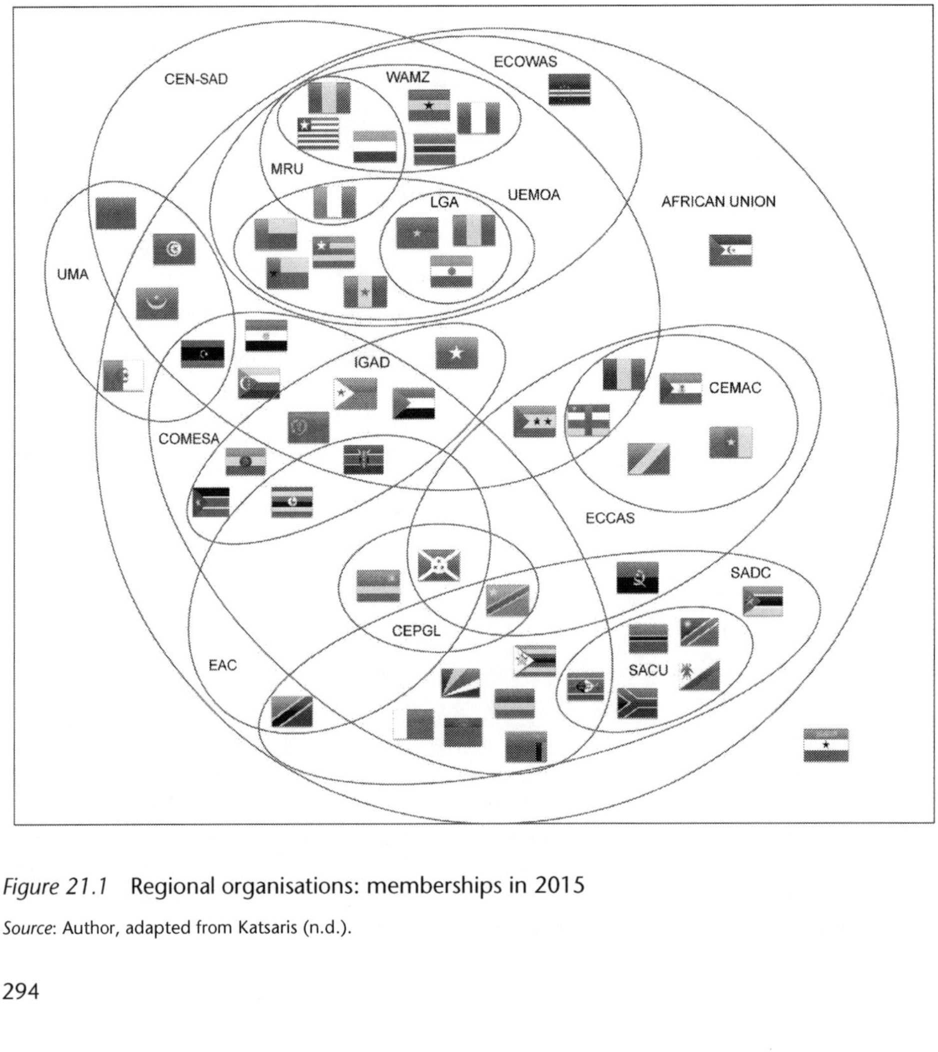

Figure 21.1 Regional organisations: memberships in 2015

Source: Author, adapted from Katsaris (n.d.).

activities more or less permanently. The Economic Community of Great Lake Countries (CEPGL), for example, was suspended in 1996 due to the First Congo War, and then reactivated in 2010. At a more local level, the Mano River Union (MRU) experienced a long period of inactivity resulting from the civil wars in Liberia and Sierra Leone before its reactivation in 2004. The Liberian Civil War prompted the creation of the ECOWAS Monitoring Group (ECOMOG), perhaps the most credible attempt to establish a regional security force thus far, later deployed in Sierra Leone and Guinea-Bissau (Adejumobi, 2016). More recently, the Intergovernmental Authority on Development's (IGAD) progress towards regional integration in eastern Africa has been slowed down by conflicts in Somalia and Sudan. In the northern part of the continent, regional integration in the Arab Maghreb Union (UMA) is slowed by the enduring hostility between Morocco and Algeria related to the status of Western Sahara. The Community of Sahel-Saharan States (CEN-SAD) created by the late Colonel Muammar Gaddafi, whose generous contributions financed by petrodollars had led the regional group to swell to 28 countries, was also put to a standstill by the Libyan crisis.

Conclusion

Nowhere in the world can a strict correspondence between the institutional and the functional be found, but the mismatch between *regionalism* and *regionalisation* is particularly pronounced in Africa. This is due to a combination of factors that include clientelist strategies of accumulation that favour informal arrangements over formal ones, overlapping responsibilities among regional bodies, and recurrent political unrest and conflicts.

Thus far, regional economic policies are dominated by modes of thought and actions in which the role of the state is to ensure that the rules of free trade are guaranteed. Regional policies argue that trade should be facilitated and regulation should be adopted to encourage that business is conducted without friction, because Africa is fragmented into so many countries and does not have the luxury to ignore the benefits of regional integration. With regard to the political sphere, most regional policies promote an agenda based on the principle of subsidiarity that holds that decisions should be taken as closely as possible to the citizen, on a transfer of sovereignty from the state to supranational bodies, and on multilevel governance, three key elements modelled on the European experience.

The economic and political agenda of regional organisations collides, however, with the system put in place within African states characterised by various business cultures and clientelism. While African regional organisations and their external donors struggle to formalise the informal, politicians invest in private companies, and businessmen financially support their political campaigns, or enter into politics themselves, making the division between state and society increasingly blurred. The fact that clientelism has proved resilient to virtually all free-market policies designed since the 1990s also means that African elites and their clients produce distinct regimes of integration that can hardly be called liberal or Western, and which do not necessarily lead to greater transparency, competition and social equality.

In countries such as Kenya, Nigeria, Senegal or South Africa, the liberalisation of markets tends to enrich large-scale merchants who have adopted global ways of entrepreneurship and accumulation while maintaining strong connections to the political elite of the country, rather than the small-scale traders that often live from the existence of border differentials. Transport corridors are another example of how public–private partnerships formulated by African states and international organisations can impact on the daily life of small traders across the continent, by making unrecorded trade paradoxically more difficult. As long as African regional policies follow an imported model, in which the specificities of the continent and the potential of

African business to promote regional development are largely ignored, the gap between the normative view of regional integration and its concrete manifestation on the ground is likely to remain.

Note

1 The author thanks Daniel Bach for his comments. Parts of the chapter build on a recent report written by the author for the OECD (2016).

References

Adejumobi, S. (2016) 'Region-building in West Africa', in D.H. Levine and D. Nagar (eds), *Region-Building in Africa: Political and Economic Challenges*, London: Palgrave Macmillan, pp. 213–230.

AfDB (African Development Bank) (2012a) 'Informal cross border trade in Africa: implications and policy recommendations', *African Development Bank Economic Brief*, 3(10): 1–13.

AfDB (African Development Bank) (2012b) *Border Posts, Checkpoints, and Intra-African Trade: Challenges and Solutions*, Tunis: AfDB Chief Economist Complex.

AfDB (African Development Bank) (2015) *Regional Integration Policy and Strategy (RIPoS) 2014–2023*, Abidjan: AfDB.

AfDB (African Development Bank), OECD (Organisation for Economic Co-Operation and Development) and UNDP (United Nations Development Programme) (2015) *African Economic Outlook: Regional Development and Spatial Inclusion*, Paris: OECD.

Asiwaju, A.I. and Adeniyi, P.O. (1989) *Borderlands in Africa: A Multidisciplinary and Comparative Focus on Nigeria and West Africa*, Lagos: University of Lagos Press.

AU (African Union) (2009) *Report of the Commission on the Implementation of the African Union Border Programme*, Addis Ababa: Executive Council of the AU.

AU (African Union) (2013) *Status of Integration in Africa (SIA IV)*, Addis Ababa: AU.

AU (African Union), NEPAD (New Partnership for Africa's Development) and AfDB (African Development Bank) (2011) *Study on Programme for Infrastructure Development in Africa (PIDA): Africa's Infrastructure Outlook 2040*, Addis Ababa: AU.

Bach, D. (ed.) (1999) *Regionalisation in Africa: Integration and Disintegration*, Oxford: James Currey.

Bach, D. (2016) *Regionalism in Africa: Genealogies, Institutions and Trans-State Networks*, London: Routledge.

Barca, F., McCann, P. and Rodríguez-Pose, A. (2012) 'The case for regional development intervention: place-based versus place-neutral approaches', *Journal of Regional Science*, 52(1): 134–152.

Bøås, M. (2001) *Regions and Regionalisation: A Heretic's View*, Discussion Paper 11, Uppsala: Nordic Africa Institute.

Bossuyt, J. (2015) *The Economic Community of West African States (ECOWAS) Report*, Brussels: EDCPM.

Bowland, C. and Otto, L. (2012) *Implementing Development Corridors: Lessons from the Maputo Corridor*, Policy Briefing 54, Johannesburg: South African Institute of International Affairs.

Butler, L.J. (2000) 'Britain, the United States and the demise of the Central African Federation, 1959–1963', *Journal of Imperial and Commonwealth History*, 28(3): 131–151.

CILSS (Permanent Interstate Committee for Drought Control in the Sahel) (2014) *Cross-Border Trade of Agricultural Products and Livestock in West Africa*, Ouagadougou: CILSS and USAID.

Deen-Swarray, M., Adekunle, B. and Odularu, G. (2014) 'Policy recipe for fostering regional integration through infrastructure development and coordination in West Africa', in D. Seck (ed.), *Regional Economic Integration in West Africa*, Heidelberg: Springer, pp. 29–56.

Dirar, L. (2010) 'Common market for eastern and southern African countries: multiplicity of membership issue and choices', *African Journal of Comparative and International Law*, 18(2): 217–232.

Dobler, G. (2016) 'The green, the grey and the blue: a typology of cross-border actors in Africa', *Journal of Modern African Studies*, 54(1): 145–169.

Drummond, P., Wajid, S.K. and Williams, O. (2015) *The East African Community: Quest for Regional Integration*, Washington, DC: IMF.

Erasmus, G. (2014) *SACU After Ten Years: Taking Stock and Looking Ahead*, Research Brief 14, Stellenbosch: Trade Law Centre (Tralac).

Feyissa, D. and Hoehne, M.V. (2010) *Borders and Borderlands as Resources in the Horn of Africa*, London: James Currey.

Gibb, R. (2009) 'Regional integration and Africa's development trajectory: meta-theories, expectations and reality', *Third World Quarterly*, 30(4): 701–721.

Golub, S. (2012) 'Government policies, smuggling, and the informal sector', in N. Benjamin and A.A. Mbaye (eds), *The Informal Sector in Francophone Africa: Firm Size, Productivity, and Institutions*, Washington, DC: World Bank.

Hanson, K.T. (ed.) (2015) *Contemporary Regional Development in Africa*, Farnham: Ashgate.

Herbst, J. (2007) 'Crafting regional cooperation in Africa', in A. Acharya and A. Johnston (eds), *Crafting Cooperation: Regional International Institutions in Comparative Perspectives*, Oxford: Oxford University Press, pp. 129–144.

Hettne, B. and Söderbaum, F. (1998) 'The new regionalism approach', *Politeia*, 17(3): 6–21.

Hettne, B. and Söderbaum, F. (2008) 'The future of regionalism', in A. Cooper, C. Hughes and P. De Lombaerde (eds), *Regionalization and Global Governance: The Taming of Globalization*, London: Routledge, pp. 61–95.

Hüsken, T. and Klute, G. (2010) 'Emerging forms of power in two African borderlands: A theoretical and empirical research outline', *Journal of Borderlands Studies*, 25(2): 107–121.

Igué, J. (1989) 'Le développement des périphéries nationales en Afrique', in B. Antheaume, C. Blanc-Pamard, J.-L. Chaleard, A. Dubresson, V. Lassailly-Jacob, J.-Y. Marchal, *et al.* (eds), *Tropiques. Lieux et Liens*, Paris: ORSTOM, pp. 594–605.

Igué, J. and Soulé, B. (1992) *L'Etat-Entrepôt au Bénin. Commerce Informel ou Solution à la Crise?* Paris: Karthala.

Katsaris, A. (n.d.) *Supranational African Bodies*, available at: https://commons.wikimedia.org/wiki/File:Supranational_African_Bodies-fr.svg (accessed 20 December 2017).

Khadiagala, G.M. (2008) *Governing Regionalism in Africa: Themes and Debates*, Policy Brief 51, Johannesburg: Centre for Policy Studies.

Korf, B. and Raeymaekers, T. (eds) (2013) *Violence on the Margins: States, Conflict, and Borderlands*, New York: Palgrave Macmillan.

Lorenz, U. and Cornelissen, S. (2012) 'Regional organisation, regional arena: the SADC in southern Africa', in T. Shaw, A. Grant and S. Cornelissen (eds), *The Ashgate Research Companion to Regionalisms*, Farnham: Ashgate, pp. 241–254.

MacGaffey, J. (1987) *Entrepreneurs and Parasites: The Struggle for Indigenous Capitalism in Zaire*, Cambridge: Cambridge University Press.

Makinda, S.M., Wafula Okumu, F. and Mickler, D. (2016) *The African Union: Addressing the Challenges of Peace, Security, and Governance*, London: Routledge.

Mbekeani, K.K. (2013) *Intra-Regional Trade in Southern Africa: Structure, Performance and Challenges*, Tunis: African Development Bank Regional Integration Policy Paper 2.

Meagher, K. (2003) 'A back door to globalisation? Structural adjustment, globalisation & transborder trade in West Africa', *Review of African Political Economy*, 30(95): 57–75.

Miles, W. (2014) *Scars of Partition: Postcolonial Legacies in French and British Borderlands*, Lincoln, NE: University of Nebraska Press.

Murithi, T. (2012) 'The African Union at ten: an appraisal', *African Affairs*, 111(445): 662–669.

Nixdorf, L. (2013) 'Regional integration and informal cross-border trade in the East African community', in U. Lorenz-Carl and M. Rempe (eds), *Mapping Agency: Comparing Regionalisms in Africa*, Farnham: Ashgate, pp. 133–147.

Nono, G.M. (2014) *Integration Efforts in Central Africa: The Case of CEMAC*, Working Paper 16, Bruges: UNU-CRIS.

Nugent, P. (2008) 'Border anomalies: the role of local actors in shaping spaces along the Gambia-Senegal and Ghana-Togo borders', in A. Bellagamaba and G. Klute (eds), *Besides the State: Emergent Powers in Contemporary Africa*, Cologne: Rudiger Koppe, pp. 101–115.

Nugent, P. (2012) 'Border towns in comparative perspective', in T. Wilson and H. Donnan (eds), *A Companion to Border Studies*, Hoboken. NJ: Wiley-Blackwell, pp. 557–572.

Nugent, P. and Asiwaju, A.I. (eds) (1996) *African Boundaries: Barriers, Conduits and Opportunities*, London: Pinter.

OAU (Organisation of African Unity) (1991) *Treaty Establishing the African Economic Community*, Abuja: OAU.

OECD (Organisation for Economic Co-Operation and Development) (2009a) *Informal Cross-Border Trade and Trade Facilitation Reform in Sub-Saharan Africa*, Trade Policy Working Paper 86, Paris: OECD.

OECD (Organisation for Economic Co-Operation and Development) (2009b) *How Regions Grow*, Paris: OECD.
OECD (Organisation for Economic Co-Operation and Development) (2009c) *Atlas on Regional Integration in West Africa*, Paris: OECD.
OECD (Organisation for Economic Co-Operation and Development) (2014) *An Atlas of the Sahara-Sahel: Geography, Economics and Security*, Paris: OECD.
OECD (Organisation for Economic Co-Operation and Development) (2016) *Cross-Border Cooperation in West Africa: Mapping Policy Networks*, Paris: OECD.
Paasi, A. (2002) 'Bounded spaces in the mobile world: deconstructing regional identity', *Tijdschrift voor Economische en Sociale Geografie*, 93(2): 137–148.
Perkmann, M. (2007) 'Policy entrepreneurship and multilevel governance: a comparative study of European cross-border regions', *Environment and Planning C: Government and Policy*, 25(6): 861–879.
Raballand, G. and Mjekiqi, E. (2010) 'Nigeria's trade policy facilitates unofficial trade but not manufacturing', in V. Treichel (ed.), *Putting Nigeria to Work: A Strategy for Employment and Growth*, Washington, DC: World Bank, pp. 203–252.
Raeymakers, T. (2014) *Violent Capitalism and Hybrid Identity in the Eastern Congo: Power to the Margins*, Cambridge: Cambridge University Press.
Reyntjens, F. (2014) 'Regulation, taxation and violence: the state, quasi-state governance and cross-border dynamics in the Great Lakes region', *Review of African Political Economy*, 41(142): 530–544.
Seck, D. (2013) 'Fifteen years of WAEMU: results and strategies for the future', in. E. Ayuk and S. Kaboré (eds), *Wealth Through Integration, Insight and Innovation in International Development*, Ottawa: IDRC & Springer, pp. 19–42.
Sequeira, S., Hartmann, O. and Kunaka, C. (2014) *Reviving Trade Routes: Evidence from the Maputo Corridor*, Discussion Paper 14, Washington, DC: Africa Transport Policy Program.
Shaw, T., Grant, A. and Cornelissen, S. (eds) (2011) *The Ashgate Research Companion to Regionalisms*, Farnham: Ashgate.
Söderbaum, F. (2011) 'Formal and informal regionalism', in T. Shaw, A. Grant and S. Cornelissen (eds), *The Ashgate Research Companion to Regionalisms*, Farnham: Ashgate, pp. 51–67.
Söderbaum, F. and Granit, J. (2014) *The Political Economy of Regionalism: The Relevance for International Waters and the Global Environment Facility*, Washington, DC: Global Environment Facility.
Söderbaum, F. and Taylor, I. (eds) (2008) *Afro-Regions: The Dynamics of Cross-Border Micro-Regionalism in Africa*, Uppsala: Nordiska Africainstitutet.
Tavares, R. and Tang, V. (2011) 'Regional economic integration in Africa: impediments to progress?', *South African Journal of International Affairs*, 18(2): 217–233.
Taylor, I.C. (2011) 'Spatial development initiatives: two case studies from southern Africa', in T. Shaw, A. Grant and S. Cornelissen (eds), *The Ashgate Research Companion to Regionalisms*, Farnham: Ashgate, pp. 325–338.
Teravaninthorn, S. and Raballand, G. (2009) *Transport Prices and Costs in Africa: A Review of the International Corridors*, Washington, DC: World Bank.
Titeca, K. and de Herdt, T. (2010) 'Regulation, cross-border trade and practical norms in West Nile, north-western Uganda', *Africa*, 80(4): 573–594.
UN (United Nations) (2015) *Composition of Macro Geographical (Continental) Regions, Geographical Sub-Regions, and Selected Economic and Other Groupings*, available at: http://unstats.un.org/unsd/methods/m49/m49regin.htm (accessed 20 December 2017).
UNCTAD (United Nations Conference on Trade and Development) (2013) *Economic Development in Africa: Intra-African Trade – Unlocking Private Sector Dynamism*, Geneva: UNCTAD.
UNECA (United Nations Economic Commission for Africa), AU (African Union) and AfDB (African Development Bank) (2010) *Assessing Regional Integration in Africa IV: Enhancing Intra-African Trade*, Addis Ababa: UNECA.
UNECA (United Nations Economic Commission for Africa), AU (African Union) and AfDB (African Development Bank) *Assessing Regional Integration in Africa V*, Addis Ababa: UNECA.
UNECA (United Nations Economic Commission for Africa), AU (African Union) and AfDB (African Development Bank) (2013) *Assessing Regional Integration in Africa VI: Harmonizing Policies to Transform the Trading Environment*, Addis Ababa: UNECA.
USAID (United States Agency for International Development) (2014) *Borderless Alliance. West Africa Trade Hub*, available at: www.borderlesswa.com/ (accessed 20 December 2017).

Walther, O. (2012) 'Traders, agricultural entrepreneurs and the development of cross-border regions in West Africa', *Entrepreneurship and Regional Development*, 24(3–4): 123–141.

Walther, O. (2015) 'Business, brokers and borders: the structure of West African trade networks', *Journal of Development Studies*, 51(5): 603–620.

World Bank (1989) *Sub-Saharan Africa: From Crisis to Sustainable Growth*, Washington, DC: World Bank.

World Bank (1994) *International River Basin Organizations in Sub-Saharan Africa*, Technical Paper 250, Washington, DC: World Bank.

World Bank (2012) *De-Fragmenting Africa: Deepening Regional Trade Integration in Goods and Services*, Washington, DC: World Bank.

World Bank (2013) *Estimating Trade Flows, Describing Trade Relationships, and Identifying Barriers to Cross-Border Trade Between Cameroon and Nigeria*, Report 78283, Washington, DC: World Bank.

Zeller, W. (2015) *What Makes Border Real in the Namibia-Zambia and Uganda-South Sudan Borderlands*. Helsinki: Unigrafia.

22

The Millennium Development Goals

John Briggs

Introduction

The Millennium Development Goals (MDGs) were approved by the General Assembly of the United Nations on 18 September 2000 and over 190 countries signed up to them. The MDGs to some extent were a product of their time in that they picked up the spirit of the new millennium, and in a sense were a new start. Here was a commitment by the nations of the world to reduce poverty and improve the lives of the global majority population. There was a great sense of optimism surrounding the MDGs, a new start and 'a break with business as usual', as the then Secretary-General of the United Nations, Kofi Annan, put it (cited in Rigg, 2014).

It was agreed that the MDGs should comprise eight goals to be measured against 21 targets (see Table 22.1), although the targets become less easily measurable and more vague as one moves further down the table. MDGs 1 to 5 have clear measurable targets, while the targets for MDG6 are a little less precise, and those for MDGs 7 and 8 are difficult to measure with any certainty. The MDGs deliberately do not have absolute numerical or statistical targets to them, but are rather expressed by 'halving' and 'reducing by two-thirds', and so on, which allows for progress to be measured at various scales from the local to the national and to the regional, or even local, if there is sufficient accuracy and granularity to available data at these latter scales. Although the MDGs were agreed in 2000, the baseline data were set at the 1990 levels and the deadline year for achievement was set at 2015, giving a 25-year period to demonstrate progress or otherwise.

What made the MDGs different from what had gone before was that a clear agenda for poverty reduction had been established by the international community. Until that point, poverty reduction interventions had been somewhat piecemeal in approach, and this was seen as a substantial weakness that would be addressed by the establishment of a clear and internationally agreed agenda. Moreover, the target-driven approach lent credibility in that the international community would have a yardstick against which to measure progress, as well as having a sharper focus to the deployment of official development assistance. By rallying around the MDGs, there was a greater political buy-in, perhaps because the MDGs were short and clear in their expression, although in practice this political buy-in was uneven internationally. Nonetheless, a number of donor agencies aligned their investment priorities to meeting the MDGs, a good

Table 22.1 Millennium Development Goals and targets

Goals	*Targets*
Eradicate extreme hunger and poverty	1a Halve, between 1990 and 2015, the proportion of people whose income is less than $1.25 a day[1]
	1b Achieve full and productive employment and decent work for all, including women and young people
	1c Halve, between 1990 and 2015, the proportion of people who suffer from hunger
Achieve universal primary education	2a Ensure that, by 2015, children everywhere, boys and girls alike, will be able to complete a full course of primary schooling
Promote gender equality and empower women	3a Eliminate gender disparity in primary and secondary education preferably by 2005, and in all levels of education no later than 2015
Reduce child mortality	4a Reduce by two-thirds, between 1990 and 2015, the mortality rate of children under five
Improve maternal health	5a Reduce by three-quarters, between 1990 and 2015, the maternal mortality ratio
	5b Achieve, by 2015, universal access to reproductive health
Combat HIV/AIDS, malaria and other diseases	6a Halt and begin to reverse, by 2015, the spread of HIV/AIDS
	6b Achieve universal access to treatment for HIV/AIDS for all those who need it
	6c Halt and begin to reverse, by 2015, the incidence of malaria and other major diseases
Ensure environmental sustainability	7a Integrate the principles of sustainable development into country policies and programmes and reverse the loss of environmental resources
	7b Reduce biodiversity loss, achieving by 2010, a significant reduction in the rate of loss
	7c Halve, by 2015, the proportion of the population without sustainable access to safe drinking water and basic sanitation
	7d Achieve, by 2020, a significant improvement in the lives of at least 100 million slum dwellers
Develop a global partnership for development	8a Develop further an open, rules-based, predictable, non-discriminatory trading and financial system
	8b Address the special needs of least developed countries
	8c Address the special needs of landlocked developing countries and small island developing states
	8d Deal comprehensively with developing countries' debt
	8e In cooperation with pharmaceutical companies, provide access to affordable, essential drugs in developing countries
	8f In cooperation with the private sector, make available benefits of new technologies, especially ICTs

Source: UN (2015).

Note: (1) The international poverty line was initially set at $1 per day when the MDGs were established, but this figure was raised to $1.25 in 2008.

example of this being the Department for International Development (DFID) in the United Kingdom. There was a strong feeling that the definition of eight clear goals and 21 targets would aid the monitoring of policy, practice and accountability, providing a clear focus to poverty reduction in the Global South, but there were alternative views. Hayman (2007), for example, suggests that the relatively limited list of eight MDGs allows donor agencies to focus on these at the expense of other pressing poverty reduction issues that are not captured by the MDGs, but which recipient countries might define as being more important in their own local context.

From the start, the MDGs were focused predominantly on the Global South, and only one of them, MDG8, explicitly acknowledged the role of other players outside the Global South, but even then only in rather vague terms. Although MDG8 set the goal of developing a global partnership for development, the six related targets, unlike those for MDGs 1 to 6 in particular, were extremely vague and largely unmeasurable. For example, what does dealing comprehensively with developing countries' debt (Target 8d) actually mean? Writing off 50 per cent, rescheduling, or even cancelling completely? Similarly, what does providing access to affordable, essential drugs in developing countries (Target 8e) mean? What may be affordable to one person may not be to another, and who defines the essential drugs anyway? Measurable targets appeared to be acceptable for countries of the Global South, but rather less so when countries from the North are included. The difficulty with MDG8, therefore, is that it is less of a measurable end in itself and rather more of a means to an end.

There was a further concern from the outset related to the quality of data that would be available to measure progress towards the agreed targets in any meaningful way. Many countries in the Global South had, and still do have, significant capacity issues in terms of collecting data. In addition, financial resources are frequently inadequate to fund data collection programmes, especially at the household level, and, in any case, there were other competing and greater priorities for financial investments in those countries. At the local level, and especially in the rural areas of Africa, many people are born and die without ever being formally registered, and so data accuracy on such basic information can be compromised. Given that the MDGs are formulated at the global scale, there is the challenge of different data collection methods between countries, which makes comparability difficult and the compilation of global tables of data particularly challenging.

Despite these early dissonant notes, there was nonetheless a sense of optimism, at least in the corridors of power at the United Nations and in some national governments, and their associated donor agencies, that real measurable change was on the way, and that the lives of the world majority would improve materially for the first time.

An evaluation of success

In 2015, the United Nations published its last MDGs report with the aim of 'document[ing] the 15-year effort to achieve the aspirational goals set out in the Millennium Declaration' (UN, 2015: 9). The use of the word 'aspirational' is interesting in this context as there has been mixed success in achieving the targets within each of the eight goals. Although the data, achievements and outcomes are well documented in the report, the discussion can seem quite diffuse, and Tables 22.2 and 22.3 are an attempt to summarise progress and outcomes relating to each of the targets in tabular form. Analysis is not helped, as in some instances the data refer to the entire world, but in others only to countries in the developing world. Note 1 in both Tables 22.2 and 22.3 signals where the data refer only to the developing world. MDG8 has been left out of the tables altogether because of difficulties in establishing measurable targets, something that the UN report acknowledges. In other instances, surrogate variables have been added at a later date by the UN to some of the targets, and hence did not figure in the initial MDG targets. For example, Target 7a (integrate the principles of sustainable development into country policies

Table 22.2 Global MDG targets and achievements

Targets	*1990 base*	*2015 target*	*2015 actual*	*Outcome*
1a Halve, between 1990 and 2015, the proportion of people whose income is less than $1.25 a day	36%	18%	12%	**Target achieved**
1b Achieve full and productive employment and decent work for all, including women and young people[1]	64% of population in employment	100% of population in employment	61% of population in employment	Target not achieved
1c Halve, between 1990 and 2015, the proportion of people who suffer from hunger[1]	23%	12%	13%	Target not achieved, but close
2a Ensure that, by 2015, children everywhere, boys and girls alike, will be able to complete a full course of primary schooling[1]	80%	100%	91%	Target not achieved
3a Eliminate gender disparity in primary and secondary education preferably by 2005, and in all levels of education no later than 2015[1]	Primary 87% Secondary 78% Tertiary 69%	97% to 103% for all three levels	Primary 99% Secondary 97% Tertiary 101%	**Target achieved** for all three levels
4a Reduce by two-thirds, between 1990 and 2015, the mortality rate of children under five	90 per 1,000 live births	30 per 1,000 live births	43 per 1,000 live births	Target not achieved
5a Reduce by three-quarters, between 1990 and 2015, the maternal mortality ratio	380 per 100,000 live births	95 per 100,000 live births	210 per 100,000 live births	Target not achieved
5b Achieve, by 2015, universal access to reproductive health[1]	35% of women with access	100% of women with access	52% of women with access	Target not achieved
6a Halt and begin to reverse, by 2015, the spread of HIV/AIDS	3,340,000 new infections		1,940,000 new infections	**Target achieved**
6b Achieve universal access to treatment for HIV/AIDS for all those who need it	*c.*1 million with access to antiretrovirals	15 million with access to antiretrovirals	13.6 million with access to antiretrovirals	Target not achieved, but close
6c Halt and begin to reverse, by 2015, the incidence of malaria and other major diseases	Malaria incidence 147 per 1,000 population		Malaria incidence 92 per 1,000 population	**Target achieved**

(continued)

Table 22.2 Continued

Targets	*1990 base*	*2015 target*	*2015 actual*	*Outcome*
7a Integrate the principles of sustainable development into country policies and programmes and reverse the loss of environmental resources[2]	CO_2 emissions 21.6 billion tonnes		CO_2 emissions 33.0 billion tonnes	No target set in the MDGs for this variable
7b Reduce biodiversity loss, achieving by 2010, a significant reduction in the rate of loss[2]	% protected areas 8.7%		% protected areas 8.7%	No target set in the MDGs for this variable
7c Halve, by 2015, the proportion of the population without sustainable access to safe drinking water and basic sanitation	76% with access to clean water and 54% access to sanitation	88% with access to clean water and 78% access to sanitation	91% with access to clean water and 68% access to sanitation	Clean water **target achieved;** sanitation target not achieved
7d Achieve by 2020 a significant improvement in the lives of at least 100 million slum dwellers[2]	39% population in slums		30% population in slums	Reduction in slum population, but not a target

Source: UN (2015).

Notes: (1) Refers only to countries in developing regions, not the whole world. (2) Surrogate measures being used.

and programmes and reverse the loss of environmental resources), as it is expressed, is difficult to measure in any meaningful way, and a surrogate measure of carbon dioxide emissions was eventually used. Similarly, Target 7b (reduce biodiversity loss, achieving by 2010, a significant reduction in the rate of loss) employs the percentage of land that has protected area status as a surrogate measure for the reduction of biodiversity loss. These are marked by reference to note 2 in Tables 22.2 and 22.3.

The data in Table 22.2 show some impressive outcomes. The headline MDG of halving the percentage of people in the world living on $1.25 per day (Target 1a) has been easily achieved and has been duly celebrated. From 36 per cent of the world's population living in extreme poverty in 1990 to a figure of around 12 per cent in 2015 is a spectacular achievement, and much better than the target of 18 per cent, although there is still work to be done for the remaining 12 per cent. A key driver of this achievement at the global level has been the performance of China in lifting large numbers of its population out of poverty on the back of the economic transformation of that country over the last 25 years. From a figure of 61 per cent of China's population living on $1.25 per day or less in 1990, the highest proportion of any of the world's regions (sub-Saharan Africa was at 57 per cent in 1990), China reduced this figure to only 4 per cent by 2015. Given that China accounts for almost 20 per cent of the world's population, such an achievement in reducing extreme poverty was bound to have a significant impact on the global extreme poverty figure. Every world region met its MDG target except for sub-Saharan Africa, which reduced the percentage of people living on $1.25 a day or less from 57 per cent in 1990 to only 41 per cent in 2015, significantly missing the MDG1 target of 29 per cent (see Table 22.3).

There were other notable successes at the global scale. Reducing gender disparities at all three education levels of primary, secondary and tertiary (Target 3a) has been achieved, even to the extent that tertiary education, which had the lowest female participation of the three levels in 1990, had by 2015 the highest, with slightly more women than men participating. The target of halting and beginning to reverse the spread of HIV/AIDS (Target 6a) has been achieved, with a reduction from 3.3 million new infections in 1990 to 1.9 million in 2015, as has the equivalent target on halting and reversing malaria infections (Target 6c). The final success has been with clean water provision. With another two targets being close but not quite making it (Targets 1c and 6b), there are grounds for some degree of satisfaction on the part of the UN. However, a number of targets were missed (see Table 22.2), and some quite spectacularly so; for example, the targets for reducing maternal mortality (Target 5a) and universal access to reproductive health (Target 5b) were missed by large margins (see Table 22.2), and there is clearly much work to be done there.

Table 22.3 summarises the data for sub-Saharan Africa only, and it makes for depressing reading. Whereas five MDG targets were met at the global scale, and two were near misses (see Table 22.2), the story for sub-Saharan Africa is rather different, with only two MDG targets being met in the region. These were halting and reversing the spread of HIV/AIDS (Target 6a) and the percentage of the population now having access to clean water (Target 7c) (see Table 22.3). In addition, although the target was not achieved in relation to reducing infant mortality (Target 4a), good progress was made nonetheless, in that the infant mortality rate was halved between 1990 and 2015 for sub-Saharan Africa. Unfortunately, the UN report does not disaggregate Target 6c (halting and reversing the incidence of malaria) to the major regional levels, but it seems inconceivable that the gains made in halting and reversing the incidence of malaria at the global scale from 147 per 1,000 population in 1990, down to 92 per 1,000 in 2015 could not have been achieved without significant progress in sub-Saharan Africa.

Table 22.3 African MDG targets and achievements

Targets	*1990 base*	*2015 target*	*2015 actual*	*Outcome*
1a Halve, between 1990 and 2015, the proportion of people whose income is less than $1.25 a day	57%	29%	41%	Target not achieved
1b Achieve full and productive employment and decent work for all, including women and young people[1]	63% of population in employment	100% of population in employment	65% of population in employment	Target not achieved
1c Halve, between 1990 and 2015, the proportion of people who suffer from hunger[1]	33%	17%	23%	Target not achieved
2a Ensure that, by 2015, children everywhere, boys and girls alike, will be able to complete a full course of primary schooling[1]	52%	100%	80%	Target not achieved
3a Eliminate gender disparity in primary and secondary education preferably by 2005, and in all levels of education no later than 2015[1]	Primary 84% Secondary 77% Tertiary 52%	97% to 103% for all three levels	Primary 95% Secondary 88% Tertiary 62%	Targets not achieved
4a Reduce by two-thirds, between 1990 and 2015, the mortality rate of children under five	179 per 1,000 live births	60 per 1,000 live births	86 per 1,000 live births	Target not achieved, but good progress
5a Reduce by three-quarters, between 1990 and 2015, the maternal mortality ratio	990 per 100,000 live births	248 per 100,000 live births	510 per 100,000 live births	Target not achieved

5b Achieve, by 2015, universal access to reproductive health[1]	47% of women with access	100% of women with access	49% of women with access	Target not achieved
6a Halt and begin to reverse, by 2015, the spread of HIV/AIDS	2,710,000 new infections		1,464,000 new infections	**Target achieved**
6b Achieve universal access to treatment for HIV/AIDS for all those who need it	No separate data for Africa in report			
6c Halt and begin to reverse, by 2015, the incidence of malaria and other major diseases	No separate data for Africa in report			
7a Integrate the principles of sustainable development into country policies and programmes and reverse the loss of environmental resources[2]	CO_2 emissions 0.7 billion tonnes		CO_2 emissions 1.2 billion tonnes	No target set in the MDGs for this variable
7b Reduce biodiversity loss, achieving by 2010, a significant reduction in the rate of loss[2]	% protected areas 10.6%		% protected areas 15.3%	No target set in the MDGs for this variable
7c Halve, by 2015, the proportion of the population without sustainable access to safe drinking water and basic sanitation	48% with access to clean water and 24% access to sanitation	75% with access to clean water and 65% access to sanitation	68% with access to clean water and 30% access to sanitation	Clean water **target achieved**; sanitation target not achieved
7d Achieve by 2020 a significant improvement in the lives of at least 100 million slum dwellers[2]	65% population in slums		55% population in slums	Reduction in slum population, but not a target

Source: UN (2015).

Notes: (1) Refers only to countries in developing regions, not the whole world. (2) Surrogate measures being used.

Compared to the other major world regions, it is sub-Saharan Africa that has made the least progress in addressing and meeting the MDG targets. However, it should be noted that progress has nonetheless been made, and in most areas performance has improved. Easterly (2009), however, has explained the relatively poor performance of sub-Saharan Africa in the context of the MDGs being 'poorly and arbitrarily designed to measure progress against poverty and deprivation' (Easterly, 2009: 26), and he goes on to suggest that the MDGs 'were actually set up in a way that made it more unlikely that Africa will attain them than other regions' (Easterly, 2009: 32). He illustrates this by arguing that it was always going to be difficult for sub-Saharan Africa to meet the various targets because of the challenging baselines from which the continent started, and the low levels of resources available to tackle the problems. In Easterly's view, for example, it was always going to be difficult, or even impossible, to meet the two-thirds reduction in child mortality because of the very starting figure of child mortality of 179 per thousand in 1990.

Critique of the MDGs

A key criticism of the MDGs has been that the focus has been too much on the welfare agenda to the exclusion of economic and other drivers of poverty reduction. In some ways, this may seem unfair because by promoting universal primary education, reducing child mortality, reducing the disease burden, and so on, lives will be improved. Vandermoortele (2009) has thus described the MDGs as a kind of welfare colonialism. This may seem a harsh criticism of what are very worthy goals, but there is a concern that unless the economic structures that underpin welfare are also taken forward, then such measures to reduce poverty are not going to be sustainable. It is economic growth and the transformation of economies that will make a difference to poverty reduction and the improvement of welfare and living standards. It can be argued that the MDGs, as they were constructed, are unlikely to deliver such outcomes, and unless major challenges, such as infrastructure developments, strategies to promote economic growth, mitigation and adaptation to climate change scenarios, and the provision of meaningful human rights are centrally addressed, then the MDGs are unlikely to make significant inroads into poverty reduction on a sustainable basis.

In many ways, this is related to the criticism that the MDGs are a prime example of top-down goal-setting, led by the major donors. Although involving governments from the Global South, the agenda, in terms of the goals and targets, was driven from the North. There is the view that the MDGs, as formulated, distort the poverty reduction agenda by allowing donors to focus just on the MDGs at the expense of what might be seen as greater priorities at the national or local levels (Hayman, 2007). Those countries, for example, who took the view that sustainable infrastructural development in support of promoting economic growth to provide for a more sustainable economic future was needed, found themselves at the back of the queue for official development assistance from donors, because this was not one of the eight MDGs, and therefore not a priority for many donor agencies and governments in the North. Barnes and Brown (2011) go even further by suggesting that the MDGs themselves are actually somewhat unambitious when set against the background of the sheer volume of basic needs for the majority of the population of the Global South. The counterargument to this, of course, is that there has to be a start somewhere, and if the MDGs had been overly ambitious at the outset, then this would have run the risk of failure, which subsequently would have achieved nothing at all in terms of reducing poverty.

There are the further charges that the MDGs comprise a largely technical fix to a set, however limited, of very real poverty challenges, and civil and political challenges and human rights remain untouched, ignored in deference to the supremacy of the technical solution (Ziai, 2011).

In some ways, this is understandable, in that technical solutions are frequently seen to be apolitical by negotiators. In the debates to establish the MDGs, there were some strong political stances taken; for example, there was considerable political resistance towards some of the proposals relating to women's rights in childbearing, and in the end these proposals were dropped. Discussions around environmental protection, climate change and biodiversity destruction were seen to be too 'political'. Hence, anything that seemed at all political was unlikely to make it to the final list of MDGs, although some targets that related to women's empowerment did make it. The result of this meant that the successful meeting of many of the MDGs by 2015 would depend to a great extent on the transfer of science and technology from the North to the Global South, even an uncritical transfer, but one that did little to address pressing issues such as governance, human rights, and social and economic transformation.

The MDG goals and targets themselves attracted criticism. Reflecting the dominance of the North in the establishment of the MDGs, Saith (2007) rather drily suggests that the MDGs are 'goals set for the poor, [with] goalposts set by the rich'. The goalposts are of course the 21 targets, and there have been charges that many of these are unmeasurable and/or unattainable in the time span allowed. Not even some of the rich nations of the North were able to achieve universal primary education in only 25 years, or make the radical improvements to maternal health in that same time span that the MDGs aspire to do. Many of the targets were about achieving numbers, such as that for universal primary education, and a good number of countries, as we saw above, have met the target. But drilling down into the data and assessing performance can reveal a rather more nuanced picture of reality. Although in many countries all, or nearly all, children are in primary school, many of these schools lack resources and sufficiently well-qualified teachers to deliver a meaningful education, such that it is not uncommon to come across educationalists and others in Africa who talk about 'functional illiteracy' because of this focus on numbers rather than achievement. But for those who do receive, and achieve, an effective primary education, and are therefore qualified to move onto secondary education, there is the frustration that, because of the focus on primary education at the expense of secondary (and incidentally tertiary) education, driven by the MDGs, there is a shortage of places at secondary level, as well as the continuing challenge of not enough well-qualified teaching staff being available. Clearly, a consequence of the focus of the MDGs on primary education, to the exclusion of secondary and tertiary levels, and not seeing education as a continuum, has distorted the system and has resulted in further challenges for governments going forward in the post-MDGs era. Indeed, if countries in the Global South are to make sustainable progress in reducing poverty, then they have to produce substantially more than just primary school leavers.

It is interesting that progress with MDG8, the only one out of the eight MDGs aimed specifically at high-income countries, has been disappointing. Since 2008, global official development assistance has fallen, perhaps as a result of the Global Financial Crisis (Fehling *et al.*, 2013), although there have been exceptions, such as the United Kingdom, which in 2014 reached the target of 0.7 per cent of gross national income being used for development assistance. Nonetheless, there is the view that an analysis of the targets of MDG8 are vague and largely difficult to measure in any meaningful way, and perhaps seem to be addressing the symptoms of poverty rather than its roots, with not enough emphasis on sustainable development.

The agreement not to set targets as absolute figures, but as relative ones, was welcomed at the outset. No country or region was therefore burdened with having to achieve an absolute target, thus recognising that geography matters and that places differ. Consequently, many targets have aspirations of 'halving' or 'reducing by two-thirds', set against a baseline of where that country or region was in 1990. Success, or otherwise, would be measured in this relative way in 2015. However, what became clear in the post-2000 period was that by using this approach, for very understandable reasons, the poorest countries will nonetheless remain the poorest, assuming that all countries meet their targets. The poorest countries have fewer resources to tackle the targets and so fall further behind in relative terms, as is demonstrated in the case of Africa from the discussion above. There have been some serious questions asked of the MDGs in that they may actually do least for the poorest and most vulnerable of the world's population. Easterly (2009) suggests that it is harder for poor countries with low baselines to show progress, and that one size does not fit all countries, and so the MDGs do not pay enough attention to the local context. Interestingly, over the last few years, perhaps as a response to this, there has developed the mantra of 'leave no one behind'. If that is indeed the case, do the MDGs actually militate against the poorest countries? So, should there be more locally led strategies? There are clear tensions here, because when the MDGs were established, they were set up essentially to measure global progress in poverty reduction, not progress by individual countries. But, of course, it is at this national level where the data are collected to inform progress towards achieving the MDGs at a global scale, and hence it is a small step for the gaze of observers to focus on the achievement of the MDGs at this national scale.

The discussion so far has been largely concerned with the operation and outcomes of the MDGs. However, there is a strong critique of the basic concept of the MDGs per se. The 1980s saw the emergence of the so-called Washington Consensus, comprising the World Bank, the International Monetary Fund and the United States Government, and the launch of neo-liberal structural adjustment programmes for a large number of countries in the Global South, which had profound and largely negative effects on the economies of those countries and on the livelihoods of their populations. Amin (2006) sees the establishment of the MDGs as an extension of this process, promoting a further deepening of neo-liberalism at a global scale. He suggests that the triad of the USA, Europe and Japan, supported by the World Bank, the IMF and the OECD, developed the MDGs essentially for their own neo-liberal ends of further privatisation and further commercial opportunities for capitalism, underpinned by maximum deregulation and uncontrolled movements of capital at a global scale (Amin, 2006). It is also argued that the MDGs are about 'catching up', implying a globalised world constructed very much in the interests of neo-liberalism. Hence, as Saith (2007) puts it, the formula of neo-liberal globalisation plus the MDGs equals development. Taken to its conclusion, for countries to reduce poverty among their populations and to raise living standards, the implication is that this is the (only) way forward.

The post-2015 agenda

As 2015 approached, it had become clear that the MDGs were not going to be fully met, and there was discussion about extending the deadline beyond 2015. However, the international community, through the United Nations, preferred to think along the lines of recasting the global development goals, and in September 2015 at the UN Sustainable Development Summit, the Sustainable Development Goals (SDGs) were adopted. Whereas the MDGs were seen by some to be modest and limited in scope, as discussed above, this charge cannot be levelled at the SDGs, which comprise 17 goals and 169 targets (see Table 22.4). Ironically, there is now the opposite criticism that the SDGs and their targets are too numerous to make them worthwhile and measurable in any meaningful way. Like the MDGs, the SDGs are also time-limited, in this case to 2030.

In truth, some of the SDGs are little more than extensions to the MDGs. Inspection of SDGs 1 to 6, as well as SDG17, suggests a strong resemblance to some of the previous MDGs, relating to the elimination of poverty and hunger, improving educational opportunities and outcomes, a commitment to gender equality, improving the availability of clean water and sanitation, and even the commitment to a global partnership. But there are new goals. The MDGs were

Table 22.4 Post-2015 Sustainable Development Goals (SDGs)

	Sustainable Development Goal
1	Poverty: end poverty in all its forms everywhere
2	Hunger and food security: end hunger, achieve food security and improved nutrition and promote sustainable agriculture
3	Good health and well-being: ensure healthy lives and promote well-being for all at all ages
4	Education: ensure inclusive and equitable quality education and promote lifelong learning opportunities for all
5	Gender equality and women's empowerment: achieve gender equality and empower all women and girls
6	Water and sanitation: ensure availability and sustainable management of water and sanitation for all
7	Energy: ensure access to affordable, reliable, sustainable and clean energy for all
8	Economic growth: promote sustained, inclusive and sustainable economic growth, full and productive employment and decent work for all
9	Infrastructure, industrialisation: build resilient infrastructure, promote inclusive and sustainable industrialisation and foster innovation
10	Inequality: reduce inequality within and among countries
11	Cities: make cities and human settlements inclusive, safe, resilient and sustainable
12	Sustainable consumption and production: ensure sustainable consumption and production patterns
13	Climate change: take urgent action to combat climate change and its impacts
14	Oceans: conserve and sustainably use the oceans, seas and marine resources for sustainable development
15	Biodiversity, forests, deforestation: protect, restore and promote sustainable use of terrestrial ecosystems, sustainably manage forests, combat desertification, and halt and reverse land degradation and halt biodiversity loss
16	Peace and justice: promote peaceful and inclusive societies for sustainable development, provide access to justice for all and build effective, accountable and inclusive institutions at all levels
17	Partnerships: strengthen the means of implementation and revitalise the global partnership for sustainable development

Source: UN (n.d.).

criticised for being overly social in character, with economic development and growth being ignored. This is no longer the case, with three SDGs (8, 9 and 12) being overtly economic. Human rights are also part of the SDG agenda (SDG16), and environmental concerns, including the challenge of climate change, are foregrounded.

A significant difference between the MDGs and SDGs is that whereas the former were essentially aimed at the Global South to achieve, the SDGs are very much aimed at the global community as a whole, and therefore include the richer countries of the North in the agenda. Indeed, there has been an initiative to rename the Sustainable Development Goals as the UN Global Goals to emphasise their global nature and to mark them out as something to be achieved by the entire global community, and not just the Global South. However, at the time of writing, this change has not been universally accepted. There is also a commitment to 'leave no one behind', and the use of language in the 17 SDGs is interesting in this context. The word 'inclusive' appears six times as if to emphasise the point about leaving no one behind (incidentally, the word 'sustainable' appears 11 times).

Inevitably, the SDGs have already drawn criticism, many of them the same ones as for the MDGs, for example data quality, which has been an issue all along for measuring progress towards meeting the MDGs. This challenge remains, but is even more critical, given the enormous expansion of the targets to be measured from 21 to 169, and some of the subtleties of the data collection process are going to prove very challenging, and not just for countries of the Global South. The concern about local context also remains, and in the same way that the MDGs took little notice of the local, so the SDGs do the same.

Conclusion

Although the MDGs were not all met in terms of meeting the targets set in 2000, it cannot be denied that progress was made between 2000 and 2015 in reducing the burden of poverty for much of the world's population, and even where targets were missed, there was still progress made, at least to some extent. However, although there is a relatively good news story at the global level, this is not the case for Africa, where performance against the MDG targets was disappointing, with only two targets being achieved. This does, though, mask the fact that progress was made in most countries in the continent, but whether these gains can be sustained, and improved upon, in the post-MDGs era remains to be seen.

The MDGs did, however, provide a focus for the global poverty reduction effort, which resulted in greater political buy-in, and also engaged, to an extent, the general public in understanding the challenges of world poverty. However, the charges of distorting the development agenda and ignoring local development priorities because they did not fit in with the MDGs, as well as the MDGs being an extension of the neo-liberal agenda, are hard to refute. The SDGs provide a continuing way forward to 2030, but meeting them is going to be challenging, perhaps even more so than for the MDGs, given the scope and reach of the SDGs. Whether the SDGs will be any more successful between 2015 and 2030 than the MDGs were between 2000 and 2015, and whether the political commitment to support them is maintained, remain to be seen of course, but what is clear is that they are not going away.

References

Amin, S. (2006) 'The Millennium Development Goals: a critique from the South', *Monthly Review*, 57(10): 1–15.

Barnes, A. and Brown, G. (2011) 'The idea of partnership within the Millennium Development Goals: context, instrumentality and the normative demands of partnership', *Third World Quarterly*, 32: 165–180.

Easterly, W. (2009) 'How the Millennium Development Goals are unfair to Africa', *World Development*, 37(1): 26–35.

Fehling, M., Nelson, B. and Venkatapuram, S. (2013) 'Limitations of the Millennium Development Goals: a literature review', *Global Public Health*, 8(10): 1109–1122.

Hayman, R. (2007) 'Are the MDGs enough? Donor perspectives and recipient visions of education and poverty reduction in Rwanda', *International Journal of Educational Development*, 27: 371–382.

Rigg, J. (2014) 'The Millennium Development Goals', in V. Desai and R. Potter (eds), *The Companion to Development Studies*, London: Routledge, pp. 67–73.

Saith, A. (2007) 'Goals set for the poor, goalposts set by the rich', *International Institute of Asian Studies Newsletter*, 45: 12–13, available at: https://iias.asia/the-newsletter/newsletter-45-autumn-2007 (accessed 4 January 2018).

UN (United Nations) (2015) *The Millennium Development Goals Report 2015*, New York: UN, available at: www.un.org/millenniumgoals/poverty (accessed 11 September 2015).

UN (United Nations) (n.d.) *Sustainable Development Knowledge Platform*, New York: UN DESA, available at: https://sustainabledevelopment.un.org/ (accessed 4 January 2018).

Vandermoortele, J. (2009) 'The NDG conundrum; missing the targets without missing the point', *Development Policy Review*, 27(4): 355–371.

Ziai, A. (2011) 'The Millennium Development Goals: back to the future?', *Third World Quarterly*, 32: 27–43.

23

Africa's natural resource corner and BRICS

Ian Taylor

Introduction

Since around 2000, there has been greater engagement between Africa and the Global North, reflected in various initiatives largely focused on the issue of poverty with a strong emphasis on 'good governance' (Cargill, 2011). With this, conditionalities have been applied, often in a fairly static and dogmatic fashion – a continuation of a long-standing pattern. This has often been bitterly resented by African elites, even though Africa's own NEPAD project placed standard liberal definitions of governance at the centre of its project (see Taylor, 2005).

At the same time, emerging new poles of accumulation have arisen that have, at times, been postulated to represent a new trading geography. These emerging centres have also been cast as new sites of alternative economic and political models. The most emblematic grouping of such economies is captured in the term BRICS (Brazil, Russia, India, China and South Africa). The original BRIC acronym (*sans* South Africa) was originally coined in 2001 by Jim O'Neill, chief economist for Goldman Sachs. At this time, the global investment banking and securities firm advanced the argument that these emerging economies were likely to surpass the traditional economic powerhouses of the global economy by 2040. A normative argument that such countries would reorder the global order was also advanced.

What is interesting with regard to the BRICS is that they provide options beyond the traditional North–South axes. An emphasis on developing infrastructure has been notable in this new set of relations, in themselves issues that have been neglected by traditional actors. Politically, this has also introduced new competitive dynamics into Africa's international relations: Africa has 'never been in such a strong bargaining position' than at the present, with numerous 'suitors' (Cargill, 2011: viii). The growing diversity of partners potentially offers a 'tremendous opportunity', 'as each country brings with it an array of capital goods, developmental experience, products and technology as well as new opportunities to trade goods, knowledge and models' (World Economic Forum *et al.*, 2011: 105). Trade with – and investment from – emerging economies such as the BRICS potentially reduces the North's political leverage and economic dominance in Africa (Southall, 2009: 31).

These developments may be interpreted in alternative ways. It may be put forward that these new actors now emerging are merely exploitive and self-interested, overall just as damaging to Africa as the extant and well-established set of relations with the traditional powers. Alternatively, these new relationships may be seen as somehow reflecting South–South values (whatever that may mean) and contributing to Africa's developmental goals. This appears to be what many African elites believe. Yet it seems obvious that Africa is the weaker partner in these new relationships. Specifically, regarding the BRICS, actors from those states are in Africa not because of some notional love of Africa or Africans, but for reasons based on capitalist logics. Interest in gaining access to natural resources in Africa is often central (Naidu *et al.*, 2009: 3). As Kimenyi and Lewis (2011: 20) put it, the attention of emerging economies towards Africa 'is not based on an altruistic goal to improve the economic well-being of Africans', but rather, *just like most other external actors*, actors from emerging economies are 'trying to maximise their own strategic economic and political interests by engaging with African countries'. Their relationships with SSA do not exhibit any notable 'exceptionalism', displaying patterns that are 'broadly similar to those of SSA "traditional" partners and mostly reinforce existing commodity-based export structures' (Sindzingre, 2013: 45). This contrasts with the diplomatic claims made by the emerging economies that their engagement with Africa is qualitatively different – and *better* – than that of the North, with relentless incantations about 'South–South' ties, 'solidarity', 'mutual benefits', 'win–win relations' and 'partnerships'.

Given the 'growing expectations among the citizens' in countries targeted by the emerging economies 'of the immediate upswings in their livelihoods and improvements in quality of life' (Aryeetey and Asmah, 2011: 22), the solidarity rhetoric may backfire. It is obvious that the quality of a country's governance institutions are crucial determinants in development and growth, and though there have been *some* improvements in governance in Africa of late, the incidence of corruption and general pathologies of maldevelopment remains high. Possibly compounding this situation is the 'non-interference' practised by some of the new partners. What this means in practical terms is that until and unless the elites in Africa themselves promote pro-development policies, no such standards will be adopted. In such a milieu, the perpetual question will then be: How might Africa engage with and exploit the increased engagement by new partners in order to benefit ordinary people and promote development?

Institutions, Africa and the BRICS

Recently, a revitalised diplomacy has been initiated towards Africa. With regard to the emerging economies specifically, various summits, institutions and agreements have been established that have witnessed a proliferation of fora: the Korea-Africa Forum, the Turkey-Africa Partnership, the Africa-Singapore Business Forum, the Malaysia-Africa Business Forum, the Taiwan-Africa Summit, the Brazil-Africa Forum, and so on. All of these have (consciously or not) replicated the Chinese example set by the Forum on China-Africa Cooperation (FOCAC), established in 2000.

The background to the FOCAC can be traced to the visit by Chinese Premier Jiang Zemin to Africa in 1996, when he publicly unveiled a new Chinese approach to Africa. According to a Chinese report, 'The guiding principle that China follows in developing relations with African countries in the new situation is: "to treat each other as equals, develop sincere friendship, strengthen solidarity and cooperation, and seek common development"' (PRC, n.d.). During a keynote speech to the Organisation of African Unity (OAU), entitled 'Toward a New Historical Milestone of Sino-African Friendship', Jiang advanced a five-point proposal for a new relationship between China and Africa:

1 fostering a sincere friendship between China and Africa and both sides becoming each other's reliable 'all-weather friends';
2 treating each other as equals and respecting each other's sovereignty and not interfering in each other's internal affairs;
3 seeking common development on the basis of mutual benefit;
4 enhancing consultation and cooperation in international affairs; and
5 looking into the future to create a better world.

Jiang's proposal was warmly received by the OAU, and may be seen as laying the foundation for current Sino-African relations. The FOCAC has subsequently been the official vehicle to realise these ambitions.

In October 2000, a Forum on China-Africa Cooperation Ministerial Conference in Beijing was held that culminated in the formation of the FOCAC. Previously, in October 1999, President Jiang Zemin had written to all heads of African states, as well as the Secretary-General of the OAU, to propose the convening of a Sino-Africa forum. When this was greeted with a favourable reception, the Chinese established a preparatory committee comprising of 18 ministries, with the Ministry of Foreign Affairs and Ministry of Foreign Trade and Economic Cooperation (MOFTEC) assigned the roles of anchormen. Interestingly, Chinese sources claim that it was African leaders who initiated and asked for a summit. He Wenping (2007: 147) asserts that:

> At the end of the 1990s, some African countries proposed that as the US, Britain, France, Japan and Europe had established mechanisms for contact with Africa, it was necessary for China and Africa to establish a similar mechanism to fit in with the need to strengthen relations. After earnest study, China decided to echo the suggestions of African countries, and proposed to hold the Forum in 2000.

Whether or not it was Beijing or African states that called for and initiated the summit, the FOCAC has quickly proved to be a major feature in Africa's international relations.

The meeting in October 2000 was attended by 80 ministers charged with foreign affairs and international trade and economic development, from 45 African states. Representatives of international and regional organisations also attended, as did delegates from two African countries that did not then have diplomatic ties with China (Liberia and Malawi). Discussions were organised into four separate sessions: trade, economic reform (with China's programme being showcased as a possible model), poverty eradication and sustainable development, and cooperation in education, science technology and healthcare.

At the meeting, Jiang Zemin gave the keynote speech, starting off with the implicit claim that China was the leader of the developing world, with the oft-reported refrain that 'China is the largest developing country in the world and Africa is the continent with the largest number of developing countries' (PRC, 2000). This Third Worldism was then made explicit by Jiang's claim that the meeting was a tangible example of South–South linkages that would produce an equitable and just new international political and economic order. Notably, Jiang cast Sino-African relations within an international context that was moving towards multipolarity and where the international situation was on the whole supposedly improving. This was seen as providing new opportunities for trade and cooperation. These favourable conditions, however, were potentially threatened as hegemonism from the West and power politics were still existent. Conflict and instability in the developing world was squarely blamed on the 'many irrational and inequitable factors in the current international political and economic order [which] are detrimental not only to world peace and development, but also to the stability and development of the vast number of developing countries' (PRC, 2000).

Jiang then went on to outline four key ways that China and Africa could, working together, help establish a new global order:

1. *'Strengthen solidarity and promote South-South cooperation'*. South–South cooperation was seen as the main way that developing countries could 'give full play to their advantages in natural and human resources, tap to the full their respective productive and technological potential, take advantage of the others' strengths to make up for their own weaknesses, and achieve common improvement'.
2. *'Enhance dialogue and improve North-South relations'*. According to Jiang, 'Developed countries should take full account and care of the interests of the less privileged developing countries and increase financial investment and technology transfer to them to help build up their capacity for development'. Intrinsic to this point was the assertion by Jiang that 'A smaller development gap and better political and economic relations between the North and the South is an important foundation for a just and equitable new international political and economic order'.
3. *'Take part in international affairs on the basis of equality and in an enterprising spirit'*. According to Jiang, China and Africa needed to increase consultation and cooperation on both 'the bilateral and multilateral fronts', and vigorously participate in international affairs and the formulation of international rules. Central to this was the promotion of reform of the international economic system, as through this, 'a fair international environment will be created and the legitimate rights and interests of developing countries will be effectively safeguarded'.
4. *'Look forward into the future and establish a new long-term stable partnership of equality and mutual benefit'*. Jiang stated that increased exchanges, 'especially direct contacts between top leaders of both China and African countries', would be pursued as central to this goal.

The FOCAC now meets every three years (alternately in Africa and China) and is a formalisation of China's engagement with Africa. Its model has been copied by others.

India, for example, instigated the India-Africa Forum Summit in 2008, which 'marked the culmination of India's renewed focus on Africa' (Kragelund, 2011: 596). Fourteen African countries attended the summit, which gave rise to two declaratory documents: the India–Africa Framework for Cooperation Forum and the Delhi Declaration. Both documents stressed South–South cooperation, capacity-building and mutual interests. A plan of action was launched, a clear replication of the FOCAC's own institutional framework (see Taylor, 2011). Subsequent to the summit, New Delhi committed a US$5.4 billion credit line over the next five years (rising from US$2.15 billion in the past five years), grants worth US$500 million and a unilateral opening of the Indian domestic economy to exports from all least developing countries (LDCs). Similarly, in April 2012, the first 'Brazil-Africa Forum 2012' met in Johannesburg.

This growing interest in Africa has arguably also been reflected by a changing attitude towards a greater inclusion of African voices in international financial institutions. Though ultimately unsuccessful, the very fact that Ngozi Okonjo-Iweala led a credible campaign to become president of the World Bank speaks volumes. Setting aside her decidedly orthodox neo-liberal position and status as a World Bank insider, along with former Colombian finance minister Jose Antonio Ocampo, Okonjo-Iweala helped create the bank's first ever competitive race for the presidency. That an African was one of the candidates – *and was taken seriously* – is noteworthy. Previously, the World Bank Group's annual meeting in 2008 agreed on reforms that created an additional Chair at the World Bank Board for Africa. The continent has become 'increasingly assertive in international forums and aware of its influence' as a region making up nearly 25 per cent of the world's countries, and thus is a potentially influential bloc (Cargill, 2011: 43).

For its part, the IMF has been discussing reforming its voting structure in order to better reflect the contemporary world, rather than the world as it was when the IMF was founded. As part of this, it was proposed that emerging economies would be granted increasing voting weight. Of interest, African countries 'reacted furiously' to such proposals, and argued that this would give undue priority to emerging countries, 'while delaying action to give the world's poorest countries greater influence over the body that often dictates their economic policies' (Elliott, 2006). Rather than endorse the proposal in the spirit of South–South solidarity, African elites argued that such plans would leave them in an even weaker and more dependent position than ever. Although they ultimately stalled (the US Congress refused to ratify the quota increase), such discussions do reflect a changing global reality: the plan is to make China the third largest voting member and revise the IMF's board to reduce Europe's dominance, 'part of a broader plan by the IMF to recognise within the organisation the growing economic clout of emerging economies' (Wroughton, 2012).

However, the reforms, supposed to be introduced in late 2012, were, at the time of writing, held up by interminable wrangling over the formula used to decide voting weight. What such developments indicate, however, is that emerging economies' elites are more and more pressing for some reform of global relations, albeit in problem-solving terms. Africa's support in such questions is actively sought, although there remains no common African position, and the AU continues to have no serious strategy for managing the continent's burgeoning relationships.

Towards the African century?

As noted, Africa's generalised robust economic performance (in terms of growth) has coincided with increasing engagement with emerging economies. The diversification of Africa's international relations has been increasingly influenced by the shift in relative capabilities to emerging economies. The financial crisis of 2008 had potentially significant repercussions for the international system in that a shift in material capabilities from traditional to emerging powers appeared evident, not least in the absolute need by the guardians of the global liberal order to incorporate new partners from the South to legitimise the overall system. The debate over IMF quota shares and a reliance on emerging actors to provide capital injections in order to stabilise the global economy reflects this (Kose and Prasad, 2010: 7). The very decision to expand the G8 to the G20 as the key international institution to discuss future economic global governance was a further manifestation of these processes.

Until the turn of the century, it would be fair to say that many African economies were dependent on the Northern-based IFIs for establishing key ideas and approaches to their development models *and* for access to capital and policy advice. This has now changed somewhat. The emerging economies' rise in material capabilities and their incorporation into the key global governance architecture has given rise to the notion that Africa's international relations are in a process of change, perhaps away from the North and towards the South, with attendant debates over the possibility of alternative models of development. Certainly, the potential ability to access different methodologies and new ideas concerning developmental thinking could possibly lessen Africa's dependence on the IFIs and their conditionalities (Cargill, 2011: vii). While conditionalities can be seen as reflecting neocolonial impulses – and the policy advice has been rigidly doctrinaire in its application of neo-liberal prescriptions – it is uncertain that shifting to *no* conditions is better, given the governance modalities of many African states. Equally, the environmental and social models that the emerging economies base their rise on (intensified labour and environmental exploitation and a free reign to capital) is hardly a superior alternative.

As Africa is routinely ranked the most corrupt region of the world, a hands-off approach by the BRICS over matters related to governance is not helpful. Furthermore, a set of new relationships based on the intensification of natural resource extraction will be equally problematic. One of the key lessons for Africa from the financial crisis was that those countries that were more diversified generally tended to be more resilient than those that were highly dependent on a few primary commodities (Mutenyo, 2011: 29). Reinscribing African dependence on commodities hardly offers any novel framework to emerging relationships with Africa and undermines the BRICS claims to be somehow 'different'. Even if the emphasis placed by some of the BRICS on addressing structural bottlenecks in Africa has been beneficial for the continent, new roads and railways in the absence of serious reforms will hardly make a sustainable and long-term contribution.

This returns us to the question as to whether emerging economies' increasing engagement with Africa is exploitive or benign. This question can only be answered in a contextual manner, dependent upon which actors from which emerging economy, and which sector of which country in Africa, are being discussed. But it is important to remember that actors such as the BRICS have increased engagement with Africa as a means to achieve their own economic and political goals, and that overall Africa remains the weaker partner. The weakness is usually ascribed to the continent's dependent relationship in the international system and Africa's historic insertion into the global capitalist economy. However, dependence is 'a historical process, a matrix of action', that permits the prospect of alteration stemming from changes in the dynamics, processes and organisation of the international system and the fundamental tendencies within Africa's political economy (Bayart, 2000: 234). Current emergent trends, such as robust economic growth and an increasing diversification of the continent's international relations, may play important roles in this regard, yet massive challenges remain. Africa's world market share in processing industries is extraordinarily low: SSA exports just 0.9 and 0.3 per cent of world light and heavy manufacturing exports, respectively (World Economic Forum *et al.*, 2011: 15). The bulk of the growth in African exports in the last decade or more has been heavily underpinned by mining-related commodities, deeply problematic in terms of development. After all, the export growth that the Asian economies used to leapfrog development was based on an increasing list of manufactures. Africa is nowhere near that position.

Yet it is true that actors from both the Global North and South are now actively pursuing closer engagement with Africa. This provides the elites of the continent with opportunities to extract leverage in return for access. This may or may not be a good thing, depending on the conjectural circumstances in each state formation and the nature of the external partners. It cannot be taken for granted that actors such as the BRICS are interested in furthering Africa's developmental priorities. Though Africa has possibly never been in a stronger bargaining position than at present, the key question remains: How can African leaders take advantage for the benefit of the ordinary citizen? Currently, this does not seem to be happening. A recent Afrobarometer survey revealed that despite a decade of strong GDP growth and the incessant narrative of an 'Africa Rising', there is 'a wide gap in perceptions between ordinary Africans and the global economic community', where 'a majority (53%) rate the current condition of their national economy as "fairly" or "very bad"', and only 'one in three Africans (31%) think the condition of their national economies has improved in the past year, compared to 38% who say things have gotten worse' (AfricaFocus Bulletin, 2013). Notably, when it came to their own elites:

> Africans give their governments failing marks for economic management (56% say they are doing 'fairly' or 'very badly'), improving the living standards of the poor (69% fairly/very badly), creating jobs (71% fairly/very badly), and narrowing income gaps (76% fairly/very badly).

As Hofmeyr (2013: 1) notes, 'popular opinion is thus increasingly out of sync with the "Africa Rising" narrative that has been gaining traction among government officials and international investors'.

'A hopeful continent'?

As mentioned, the trope around Africa has shifted from one extreme to another. Now, it is 'A hopeful continent' (The Economist, 2013). The mood swing about Africa is due, directly or indirectly, to the increasing global demand for the continent's resources: notably for oil, but also for gas, minerals, and other energy sources. This was driven, above all, by the sudden appearance of China as a world economic actor, whose dramatic burst of late industrialisation fuelled a global upswing. This has been missed by the 'Africa Rising' mantra, which prefers to construct endogenous factors as drivers. Yet, as Bond (2011: 31) notes:

> Ongoing resource extraction by Western firms was joined, and in some cases overtaken, by China [and others] . . . Still, Africa's subordinate position did not change, and aside from greater amounts of overseas development aid flowing into fewer than 15 'fragile states', the North-South flows were not to Africans' advantage. One would not know this from reading reports by the elite multilateral institutions in 2011, which celebrated the continent's national economies as among the world's leading cases of post-meltdown economic recovery.

The flip-flop regarding the continent has, to a certain extent, refuted the familiar media images of fly-blown children that so dominates much discussion of Africa. This *is* a good thing. Yet equally, the narrative has swung almost entirely in the opposite direction, with little critical reflection. Growth in GDP and opportunities for investors are the new intonations in a crude binary construction of Africa that has shifted almost overnight from basket case to bonanza.

The 'Africa Rising' discourse neglects a most fundamental context:

> only for nine of the forty-three [sub-Saharan] countries were growth rates during 1980-2008 high enough to double per capita income in less than thirty years, and in sixteen countries it was less than one hundred years. Performance would have been considerably worse had it not been for the brief years of relatively rapid growth in the mid-2000s.
>
> *(Weeks, 2010: 3)*

Africa needs to grow at least 7 per cent a year for the next 20 or 30 years if any serious tackling of continental poverty is to be realised. However, growth induced by commodity prices increases, new discoveries of natural resources or increase in sources of foreign capital 'is simply not sustainable' (Amoako, 2011: 24).

What GDP growth has occurred is overwhelmingly characterised by the deployment and inflow of capital-intensive investment for the extraction and exportation of natural resources. There is a conspicuous lack of value added on the African side:

> The principal focus of this activity is in oil, which not only offers limited opportunities for local employment, but also deliberately and actively seeks to avoid the hiring of African labour for fear of encountering resistance and the costs of appeasing affected local communities.
>
> *(Southall, 2008: 148)*

Problematically, while the hope of the development literature has been that higher rates of inflow of capital investment will have downstream effects on employment, there is little evidence that this will take place on a substantial scale. The main reason is that the growth rests heavily on the engagements of foreign governments and corporations with African elites. In *most* neopatrimonial administrations, of which Africa has many, sustainable and broad-based development is unlikely to occur (cf. Kelsall, 2013).

In late 2012, the Deputy Executive Secretary of the Economic Commission for Africa noted that the relatively good economic growth performance over the past decade had been driven mostly by non-renewable natural resources and high commodity prices. Alongside this, he noted, deindustrialisation had been a key feature, with the share of manufacturing in Africa's GDP falling from 15 per cent in 1990 to 10 per cent in 2008, going hand-in-hand with an increase in unemployment (UNCTAD, 2012). McMillan and Rodrik (2011) in fact show that since 1990, Africa has experienced a relative shift in the composition of employment towards sectors that create too few high-productivity jobs. Manufacturing growth has been near the bottom in 12 growth sectors – only public administration lagged behind.

This, of course, is not to write off the recent growth as devoid of any value at all. At the minimum, improved fiscal space is being generated. Retail sectors are growing, with revenue increasing by around 4 per cent per year, and there is growing investment in infrastructure (McKinsey Global Institute, 2010). Given that there is a correlation 'between infrastructure and export diversification, and the current low levels and distorted composition of exports from SSA are partly due to poor trade infrastructure', it can be stated that the improvement in infrastructure 'has per se a positive impact on SSA growth and trade capacity' (Sindzingre, 2013: 44). Africa's debts have fallen, partly thanks to the Heavily Indebted Poor Countries Initiative (HIPC) and the Multilateral Debt Relief Initiative (MDRI), and partly due to improved management – although note that 'in spite of the HIPC initiative, only half of SSA countries have witnessed a temporary reduction of their annual debt service' (Petithomme, 2013: 119). In social sectors, indicators vary, but increases in the years of schooling are reported across the continent, albeit unevenly. Health outcomes, particularly life expectancy at birth, have also generally improved, in some countries substantially. These are all obviously to be welcomed.

Table 23.1 Correlation between GDP growth for SSA and the Commodity Price Index (CPI)

	1990	*1991*	*1992*	*1993*	*1994*	*1995*	*1996*	*1997*	*1998*	*1999*
GDP growth	1.3	0.9	–0.9	0.4	1.8	3.7	4.9	3.7	2.3	2.4
CPI			52.6	52.5	50.06	58.6	58.7	64.6	51.0	43.3

	2000	*2001*	*2002*	*2003*	*2004*	*2005*	*2006*	*2007*	*2008*	*2009*
GDP growth	3.7	3.7	3.5	4.4	6.4	5.8	6.1	6.9	5.0	1.9
CPI	59.4	61.9	50.1	66.3	69.1	86.7	113.1	112.9	162.4	102.4

	2010	*2011*	*2012*
GDP growth	4.9	4.5	4.2
CPI	146.1	182.1	188.4

Source: World Bank, IMF data.

However, it is a contention of this chapter that there is a desperate need to convert natural resources and high commodity prices into structural change:

> defined as an increase in the share of industry or services in the economy, or as the diversification and sophistication of exports . . . or as the shift of workers from sectors with low labour productivity to those with high labour productivity.
>
> *(Sindzingre, 2013: 26)*

This is not happening. Instead, with the arrival of emerging economies in Africa, alongside traditional trade associates, the historical process of underdevelopment is in danger of being further entrenched. There has been a huge rise in commodity prices, and this has contributed in a big way to 'Africa Rising', if taken as an increase in GDP per capita. But the benefit to African economies in terms of providing a sustained platform for development is far more muted.

Indeed, the drivers of Africa's 'recovery' 'during the second half of the 2000s appear to have been a commodity price boom, debt relief and a decline in domestic conflicts' (Weeks, 2010: 10). World Bank figures with regard to the annual percentage growth rate of GDP at market prices, based on constant local currency (for all income levels, rounded up), compared to the movement of the Commodity Price Index (CPI) reveals this intimate link (see Table 23.1).

The years when SSA's growth figures surpassed 1996 levels (2004–2008) can be demonstrably linked to the period when emerging economies began to hugely demand commodities, as reflected in the CPI. In the energy realm, concern over predicted declines in petroleum reserves, apprehensions over the so-called 'peak oil' scenario, instability in the Middle East, and oil price speculation placed further upward pressure on prices, peaking in 2008. This reality is qualitatively different from the picture of 'Africa Rising', where 'spectacularly right' policies have driven growth. Official reports from international organisations have at times bolstered this latter interpretation, postulating Africa's 'economic resurgence' as being hinged on the ability of the continent to recover from the global crisis relatively quicker than other areas of the world (World Economic Forum et al., 2011: v). While true in and of itself, Africa's growth record over the last 10 years or so has occurred within the context of *overall* global growth. In this regard, Africa's growth has only been around 1 per cent higher than the world average: credible, but not fantastic (African Development Bank, 2012).

Despite the celebration of improved governance across the continent and the attempts to link this to Africa's recent growth spurt, there is little evidence that overall the quality of Africa's democracies are improving or that governance is dramatically on the up and up across the continent. The composite Mo Ibrahim Index of African Governance had a continental average of 47 per cent in 2000 – by 2013 it had increased to 51.6 per cent – hardly seismic, and in fact less than half (43 per cent) of people living in Africa live in a country that has shown overall governance improvement since 2010 (Mo Ibrahim Foundation, 2013: 24). This makes nonsense of strident claims that:

> What took the UK centuries can now be a matter of decades, even years . . . Today Africa has the greatest room to boom on the back of two centuries of global progress . . . In other words, Africa is ideally poised to leapfrog centuries of industrial development . . . It has an added advantage in that it does not have to carry baggage from the past.
>
> *(African Business, 2013: 19)*

Thus (yet another) commodity-driven boom in Africa, this time in part propelled by emerging economies, wipes the historical slate clean, makes dependent relationships and unequal terms of trade vanish instantaneously, and positions the continent to reach OECD status virtually overnight.

Of course, not all emerging economies' involvement in Africa revolves around commodities; that would be a crude caricature. But commodities certainly dominate BRICS–Africa trade (the balance on the tables are the non-commodity exports) (see Table 23.2).

Such a situation *further* reinforces and helps underpin the overall structure of Africa's insertion into the global economy. The BRICS certainly did not create this milieu, but their current trade profile with the continent promotes the reification of existing and ongoing developments.

Yet it is true that actors from both the Global North and South are now actively pursuing closer engagement with Africa. This provides the elites of the continent with opportunities to extract leverage in return for access. This may or may not be a good thing, depending on the

Table 23.2 Key product composition of BRICs imports from Africa (% share, 2010)

	Brazil	*Russia*	*India*	*China*	*South Africa*
Mineral fuels, oil, etc.	85		71	65	76
Ores, slag, ash		3	2	14	
Precious stones, metals	1		13	4	6
Copper				6	3
Fertilisers	5		1		
Edible fruit and nuts		29	2		
Cocoa		16			
Tobacco		9			
Inorganic chemicals	1	8	4	1	

Source: Fundira (2012: 11, 18, 24, 31, 38).

conjectural circumstances in each state formation and the nature of the external partners. It cannot be taken for granted that actors from the emerging economies are genuinely interested in furthering Africa's developmental priorities.

The natural resource corner

Resource-based commodities form the bulk of African exports. While this has certainly led to an increase in income for some African countries (or their elites), by redirecting resources from non-raw material sectors and contributing to real exchange rate appreciation, a price boom runs the risk of further locking commodity exporters into what Leamer called the 'raw material corner', with little scope for industrial progress or skills advancement. Leamer's corner came from his illustrating both relative factor endowments and relative factor intensities with three factors and any number of goods (Leamer, 1987). Given Africa's factor endowments being concentrated in commodities and the export profile and sector concentration being the same, the raw material corner has been the continent's broad fate:

> During colonisation and the period immediately after, the structure of external trade of African countries was mainly determined by the needs of the colonial masters. African countries mainly exported natural resources such as timber and minerals and imported manufactured goods. About six decades later, this structure of trade has not been significantly altered. Invariably, African countries have continually and consistently not managed to diversify trade into manufactured products.
>
> *(Afari-Gyan, 2010: 63)*

The result has been what Shivji (2009: 59) terms 'structural disarticulation', where Africa exhibits a 'disarticulation between the structure of production and the structure of consumption. What is produced is not consumed and what is consumed is not produced'.

While low commodity prices foster palpable problems, even high commodity prices can generate major difficulties and fashion a trap of sorts, obliging countries and producers to decide between immediate profits (highly attractive for unstable and predatory regimes) and future long-term sustainability. For instance, both Algeria and Nigeria:

> have fallen prey to over-optimistic spending habits during commodity booms, using current and expected profits to finance social and/or politically motivated projects. Such programmes can quickly become unsustainable when commodity prices drop, but are typically very tricky for politicians to cut, and so tend to get funded out of borrowed money, adding to a country's debt burden.
>
> *(Brown* et al.*, 2008)*

There is no doubt that the exponential growth of the emerging economies has helped stimulate the global commodity booms of the past decade (Wang, 2007). This is important given that labour-intensive agricultural and manufactured goods do not feature significantly in the exports of African countries to these economies. This dependence is a two-edged sword. Countries with the highest economic integration with the BRICS generally managed to sustain growth during the global downturn, compared to a contraction observed in countries with the least ties. Interestingly, the risk analysis company Maplecroft released in 2011 its *Emerging Powers*

Integration Index Series, assessing the economic integration of 180 countries with each of the BRICS (Maplecroft, 2011). According to Maplecroft, the countries most integrated with the BRICS are resource-rich developing economies, which provide the raw materials to fuel economic growth back in the BRICS domestic economies. Of these, many are located in Africa. Zimbabwe was ranked joint 1st, Liberia 5th, Guinea-Bissau 6th, Zambia 7th, DR Congo 10th, Mozambique 12th, Mauritania 15th, Congo 18th and Sudan 20th. While the data showed which countries stood to gain most from the economic rise of the BRICS, it also revealed just how exposed some countries were to economic contagion should this growth weaken. As Alyson Warhurst, CEO of Maplecroft noted, 'Should growth in the BRICs economies falter or lead to internal unrest and repression, we could see contagion spread to those countries that are most highly integrated with the emerging powers' (Verisk Maplecroft, n.d.). Clearly:

> [T]he positive effect of the world business cycle suggests that the economic performance of African countries is sensitive to world markets. Specifically, this result provides strong support for the hypothesis of the dependence of African countries' economic growth on the economic growth of industrialised nations. This implies that a relatively high degree of integration of African countries with the world economy carries some benefits in as far as the industrialised countries continue to grow. However, should industrialised countries suffer economic setbacks, this could have adverse impacts on the African economies.
>
> *(Bangwayo-Skeete, 2012: 312)*

Exports from Africa to both traditional and non-traditional trading partners exhibit a very clear and continuous pattern in terms of commodity structure, which is consistent with Africa's Ricardian advantage in commodity production, with extractive commodities dominating. In short, such processes are simply the diversification of dependency, with Africa being further trapped into low value-added production structures. This is hardly congruent with the idea of 'Africa Rising'.

References

Afari-Gyan, N. (2010) 'Transforming Africa's structure and composition of trade after the global economic crisis', *Global Trade Alert*, 5, May: 63–73.

AfricaFocus Bulletin (2013) *Africa: Whose 'Africa Rising'?* 18 October, available at: www.africafocus.org/docs13/econ1310.php (accessed 5 January 2018).

African Business (2013) *Why Africa Will Rule the 21st Century*, 7 January, available at: http://africanbusinessmagazine.com/uncategorised/why-africa-will-rule-the-21st-century/ (accessed 5 January 2018).

African Development Bank (2012) *African Economic Outlook 2012*, Paris: OECD.

Amoako, K. (2011) 'Transforming Africa: start now, we can't wait', *African Business*, July: 3–4.

Aryeetey, E. and Asmah, E. (2011) 'Africa's new oil economies: managing expectations', in Brookings Institute (ed.), *Foresight Africa: The Continent's Greatest Challenges and Opportunities for 2011*, New York: Brookings Institute.

Bangwayo-Skeete, P. (2012) 'Do common global economic factors matter for Africa's economic growth?', *Journal of International Development*, 24: 304–315.

Bayart, J.-F. (2000) 'Africa in the world: a history of extraversion', *African Affairs*, 99: 217–267.

Bond, P. (2011) 'Africa's "recovery": economic growth, governance and social protest', *Africa Insight*, 41(3): 30–45.

Brown, O., Crawford, A. and Gibson, J. (2008) *Boom or Bust: How Commodity Price Volatility Impedes Poverty Reduction, and What to Do About It*, Regina: International Institute for Sustainable Development.

Cargill, T. (2011) *Our Common Strategic Interests: Africa's Role in the Post-G8 World*, London: Royal Institute of International Affairs.

Elliott, L. (2006) 'Africa calls on Brown to block IMF reforms', *The Guardian*, 31 August, available at: www.theguardian.com/business/2006/aug/31/politics.internationalaidanddevelopment (accessed 5 January 2018).

Fundira, T. (2012) *Brazil Africa Relations: 2011 Highlights*, Stellenbosch: TRALAC.

He Wenping (2007) 'China's perspective on contemporary China–Africa relations', in C. Alden, R. Soares de Oliveira and D. Large (eds), *China Returns to Africa: A Rising Power and a Continent Embrace*, Cambridge: Cambridge University Press, pp. 143–166.

Hofmeyr, J. (2013) '"Africa Rising"? Popular dissatisfaction with economic management despite a decade of growth', *Afrobarometer Policy Brief*, 2, October.

Kelsall, T. (2013) *Business, Politics, and the State in Africa: Challenging the Orthodoxies on Growth and Transformation*, London: Zed Books.

Kimenyi, M. and Lewis, Z. (2011) 'The BRICs and the new scramble for Africa', in Brookings Institute (ed.), *Foresight Africa: The Continent's Greatest Challenges and Opportunities for 2011*, New York: Brookings Institute.

Kose, M.A. and Prasad, E.S. (2010) 'Emerging markets come of age', *Finance and Development*, December: 7–10.

Kragelund, P. (2011) 'Back to BASICs? The rejuvenation of non-traditional donors' development cooperation with Africa', *Development and Change*, 42(2): 585–607.

Leamer, E. (1987) 'Paths of development in the three-factor, *n*-good general equilibrium model', *Journal of Political Economy*, 95: 961–999.

Maplecroft (2011) *Emerging Powers Integration Report*, Bath: Maplecroft.

McKinsey Global Institute (2010) *Lions on the Move: The Progress and Potential of the African Economies*, Seoul, London and Washington, DC: McKinsey & Co, available at: www.mckinsey.com/global-themes/middle-east-and-africa/lions-on-the-move (accessed 5 January 2018).

McMillan, M. and Rodrik, D. (2011) *Globalization, Structural Change and Productivity Growth*, NBER Working Paper No. 17143, June.

Mo Ibrahim Foundation (2013) *2013 Ibrahim Index of African Governance: Data Report*, London: Mo Ibrahim Foundation.

Mutenyo, J. (2011) 'Driving Africa's growth through expanding exports', in Brookings Institute (ed.), *Foresight Africa: The Continent's Greatest Challenges and Opportunities for 2011*, New York: Brookings Institute.

Naidu, S., Corkin, L. and Herman, H. (2009) 'Introduction', *Politikon*, 36(1): 1–8.

Petithomme, M. (2013) 'Much ado about nothing? The limited effects of structural adjustment programmes and the Highly Indebted Poor Countries Initiative on the reduction of external debts in sub-Saharan Africa: an empirical analysis', *African Journal of Political Science and International Relations*, 7(2): 107–120.

PRC (People's Republic of China) (2000) *China and Africa Usher in the New Century Together: Speech by President Jiang Zemin of the People's Republic of China at the Opening Ceremony of the Forum on China–Africa Cooperation*, Beijing: Ministry of Foreign Affairs of the People's Republic of China, available at: www.fmprc.gov.cn/zflt/eng/zyzl/zyjh/t157712.htm (accessed 5 January 2018).

PRC (People's Republic of China) (n.d.) *President Jiang Zemin's Visit to Six African Countries*, Beijing: Ministry of Foreign Affairs of the People's Republic of China, available at: www.fmprc.gov.cn/mfa_eng/ziliao_665539/3602_665543/3604_665547/t18035.shtml (accessed 5 January 2018).

Shivji, I. (2009) *Accumulation in an African Periphery: A Theoretical Framework*, Dar es Salaam: Mkuki na Nyota.

Sindzingre, A. (2013) 'The ambivalent impact of commodities: structural change or status quo in sub-Saharan Africa?', *South African Journal of International Affairs*, 20(1): 23–55.

Southall, R. (2008) 'The "new scramble" and labour in Africa', *Labour, Capital and Society*, 41(2): 128–155.

Southall, R. (2009) 'Scrambling for Africa? Continuities and discontinuities with formal imperialism', in R. Southall and H. Melber (eds), *A New Scramble for Africa? Imperialism, Investment and Development*, Scottsville: University of KwaZulu-Natal Press.

Taylor, I. (2005) *NEPAD: Towards Africa's Development or Another False Start?* Boulder, CO: Lynne Rienner.

Taylor, I. (2011) *The Forum on China-Africa Cooperation (FOCAC)*, London: Routledge.

The Economist (2013) *A Hopeful Continent? Special Report: Emerging Africa*, 2 March, available at: www.economist.com/sites/default/files/20130203_emerging_africa.pdf (accessed 5 January 2018).

UNCTAD (United Nations Conference on Trade and Development) (2012) *The Economic Development in Africa Report 2012: Structural Transformation and Sustainable Development in Africa*, New York: UNCTAD.

Verisk Maplecroft (n.d.) *Risk Calculators and Dashboards: Latest Product News*, available at: https://maplecroft.com/about/news/emerging_powers.html (accessed 5 January 2018).

Wang, J.-Y. (2007) *What Drives China's Growing Role in Africa?* IMF Working Paper, WP/07/211, Washington, DC: IMF, available at: www.imf.org/external/pubs/ft/wp/2007/wp07211.pdf (accessed 5 January 2018).

Weeks, J. (2010) 'A study for trade and development report 2010: employment, productivity and growth in Africa south of the Sahara', unpublished paper, Centre for Development Policy and Research, School of Oriental and African Studies, University of London.

World Economic Forum, World Bank and African Development Bank (2011) *The Africa Competitiveness Report 2011*, Geneva: World Economic Forum.

Wroughton, L. (2012) 'IMF vote reform bogged down by delays, deadlock', *Reuters*, 8 October, available at: www.reuters.com/article/us-imf-governance/imf-vote-reform-bogged-down-by-delays-deadlock-idUSBRE8970B120121008 (accessed 5 January 2018).

24
Aid and debt

John Serieux

Introduction

To say that Africa's experiences have been turbulent on the economic front since the mid-1970s would be to state the obvious. The region's experiences with aid and debt also do not belie that description. Though Africa remains the highest aid-receiving region, the flow of aid to the continent has been neither consistent nor singular in purpose. The evolution of aid receipts has involved a sharp rise during the period of the first oil crisis (1973–1975), a peak in the 1980s, continuous decline during the 1990s, and a recovery since 2000. The primary intent of aid giving also appears to have shifted over time: from the exigencies of Cold War politics to the financing requirements for the Millennium Development Goals and the war on terror. Through all this, though aid has not been irrelevant to economic outcomes, it has not been central (Museru *et al.*, 2014; Nwaogu and Ryan, 2015).

The fact that the region is home to the overwhelming majority of formerly heavily indebted poor countries (HIPCs) also means that a majority of the countries in the region had to endure crushing debt burdens through the 1980s and 1990s as the extent of their debt-related challenges remained largely unacknowledged (Serieux, 2001a). After reaching a peak in 2003, the region's debt levels were brought down quite sharply through the implementation of the HIPC Initative and the Multilateral Debt Relief Initative (MDRI). Nevertheless, exceptions to that general (debt reduction) experience still remain (with Zimbabwe as one example). This new, post-debt relief era has spawned a greater presence of African governments on international bond markets. While this brings new options for resource mobilisation, it also exposes these countries to a new and more unforgiving global market that many may not be ready to deal with (te Velde, 2014).

The levels and evolution aid flows to Africa

It is not an exaggeration to assert that in the postcolonial era (roughly from 1970 onwards),[1] aid has been the principal means through which external financial flows reached African countries. From 1970 to 2004, flows of aid (official development assistance)[2] to Africa exceeded all other financial flows, and often by a significant margin (see Figure 24.1). In fact, from the mid-1980s to the mid-1990s, the annual value of aid flows was typically greater than the annual value of all other financial flows combined. Since 2005, however, foreign direct investment (FDI) or personal remittances have exceeded aid flows in absolute value, even though aid flows to the region have been generally increasing (see Figure 24.1).

By most absolute or relative measures, Africa also receives more aid than any other region. In terms of absolute value, Africa has annually received more aid than any other region in the post-colonial period.[3] When aid is measured relative to national income, African countries have also received, on average, more aid than countries in any other region (see Table 24.1). However, when aid receipts are measured relative to population (as aid per capita), Africa has typically been second to the Middle East in terms of annual aid receipts (see Table 24.2). Nonetheless, even in per capita terms, Africa has always received more technical assistance than any other region, and until 1999 more food and humanitarian aid.[4]

The high levels of aid receipts for the region have not, however, meant uniformly high aid receipts at the country level. When aid receipts are measured relative to the size of the recipient economies, sub-Saharan African countries have generally received more aid than North African countries, but when aid is measured relative to the size of country populations, North Africa typically received more aid up to the mid-1990s (and thereafter, sub-Saharan Africa received more aid). Relative to both income and population, poorer (low-income) African countries have typically received more aid than richer (middle-income) countries, and smaller African countries have typically received more aid than larger ones. However, neither of these relationships (level of income and population size) can always predict relative aid receipts in any given year. For example, Senegal, with a population of nearly 15 million, has often received more aid (relative to income) than Togo, with half its population and a significantly lower per capita income.[5]

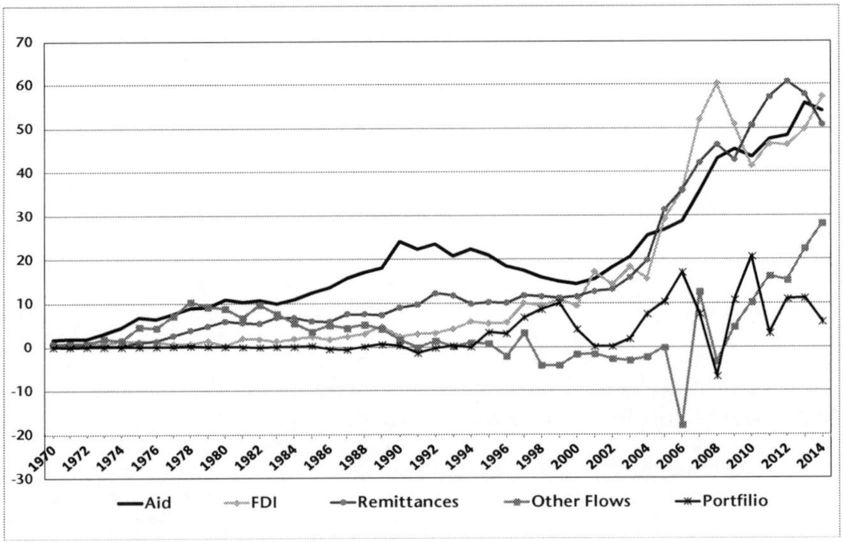

Figure 24.1 Financial flows to Africa

Source: International debt statistics (World Bank)

Table 24.1 Aid relative to income by region

	Africa	*East Asia and the Pacific*	*Latin America and Caribbean*	*Middle East*	*South Asia*
1970–1979	3.12	0.27	0.33	1.87	1.94
1980–1989	3.52	0.19	0.42	1.18	1.64
1990–1999	4.48	0.14	0.34	1.08	1.21
2000–2009	3.46	0.08	0.24	1.28	0.81
2010–2013	2.50	0.05	0.19	0.69	0.67

Source: OECD international development statistics.

Table 24.2 Per capita development assistance by type and by region

		Africa	*East Asia and Oceania*	*Middle East*	*Latin America and Caribbean*	*South Asia*
Total ODA (net of debt relief)	1970–1979	12.79	2.01	29.99	3.96	3.20
	1980–1989	24.32	3.48	37.47	8.48	4.97
	1990–1999	28.44	5.47	25.61	11.20	4.49
	2000–2009	24.84	4.51	45.07	9.53	5.54
	2010–2013	43.62	4.53	68.63	17.28	9.16
Technical assistance	1970–1979	3.11	0.40	1.65	1.17	0.25
	1980–1989	6.57	0.99	3.40	2.67	0.74
	1990–1999	7.59	1.70	5.39	4.60	0.91
	2000–2009	6.04	1.65	5.72	5.03	0.96
	2010–2013	5.33	1.92	4.90	3.50	1.17
Food and humanitarian aid	1970–1979	0.66	0.14	0.41	0.20	0.35
	1980–1989	2.49	0.17	0.37	0.97	0.59
	1990–1999	2.02	0.15	1.84	0.95	0.31
	2000–2009	4.61	0.34	6.04	1.00	0.74
	2010–2013	5.94	0.41	15.15	1.76	1.31
Other ODA flows	1970–1979	9.02	1.47	27.93	2.59	2.59
	1980–1989	15.26	2.32	33.70	4.84	3.63
	1990–1999	18.82	3.62	18.39	5.65	3.26
	2000–2009	14.20	2.51	33.30	3.51	3.83
	2010–2013	32.34	2.20	48.58	12.03	6.69

Source: OECD international development statistics.

Aid receipts have also not been consistent over the years. From a relatively low and steady level of under US$2 billion (in constant US dollars) through the 1960s and early 1970s (but falling in per capita terms), aid levels tripled in nominal terms and more than doubled in real per capita terms between 1973 and 1975 (see Figure 24.2). This was almost certainly a response to the balance of payments and budgetary crises that many African countries faced as a result of a nearly sixfold increase in oil prices during that period. Though fluctuating, real aid flows remained at this comparatively high level until the early 1990s, then dropped sharply and did not regain an upward trajectory until 2001. Since then, aid levels have risen, and since 2005 have been sustained at levels that (in real per capita terms) are not far below the aid levels of the 1980s (see Figure 24.2).

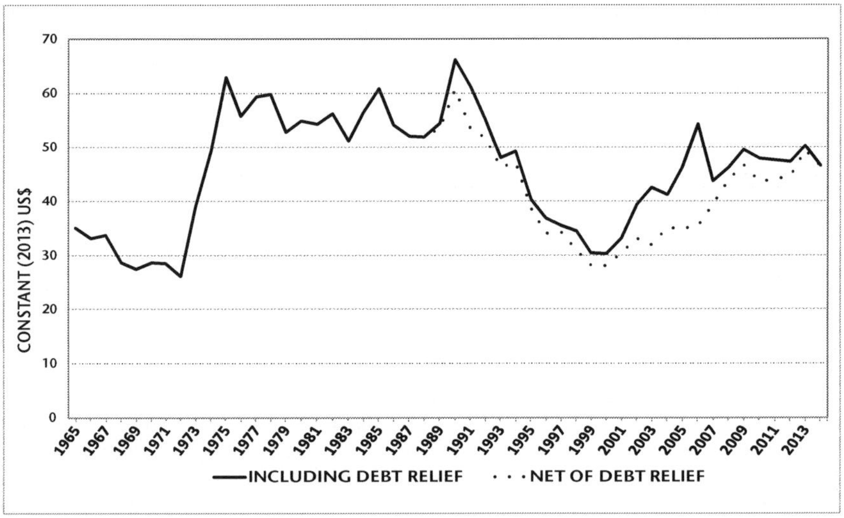

Figure 24.2 Real ODA receipts per capita (all Africa)

Source: OECD international development statistics.

Motives for aid to Africa

Given the fact that the majority of low-income countries are in Africa and that this region had a generally poor growth record during the 1980s and 1990s, the high level of aid receipts for the region would seem to validate the presumption that the giving of aid is prompted mainly by humanitarian motives (such as poverty reduction). Indeed, Bandyopadhyay and Vermann (2013), after examining the aid literature, conclude that the empirical evidence broadly supports the view that poverty reduction is one of the motives of aid giving – but merely one of several motives. Other motives such as strategic interests, market creation and the perpetuation of old colonial ties are also important, and can, over extended periods, dominate the poverty reduction motive. In that regard, an examination of the evolution of aid flows to the region over time strongly suggests the strategic motive was an important (and perhaps even the dominant) impetus for aid flows to Africa until 1990.[6]

As noted earlier, the rapid increase in aid in 1973–1975 was almost certainly a response to the challenges faced by non-oil-producing African countries in the face of the sharp increase in oil prices. When placed alongside the fact that these higher levels of aid disbursements (in real and per capita terms) were maintained through the 1980s – a particularly poor growth period for Africa – the poverty reduction motive might appear to have been the obvious justification for African aid (see Figure 24.2). However, it is notable that most of that aid increase came from increases in bilateral (government-to-government) aid, even though aid meant to address balance of payment and fiscal challenges (the most immediate outcomes of the oil price increases) is primarily the purview of multilateral agencies (the International Monetary Fund (IMF) and the World Bank in particular). More to the point, the real per capita value of aid to Africa began a sharp decline in 1991 that continued through most of the 1990s, even though Africa's growth outcomes were still quite poor. It is not likely a coincidence that this decline was also generated by declines in bilateral aid and that the decline in aid receipts began in the year of the demise of the Soviet Union (1991). Thus, the most cogent explanation for the decline in aid flows to Africa in the 1990s is the reduced need to offer aid as a bulwark against the spread of communism, and this is supported by more direct empirical evidence (Boschini and Olofsgard, 2007). In that light, even though aid may have had poverty reduction implications, the increase in aid in 1973–1975, its sustained levels through to 1990, and its decline thereafter strongly suggest that strategic motives were at the forefront – primarily the need to ensure that these countries did not fall prey to the expansion of communism (in attempts to find answers to very real development challenges).

Poverty reduction better qualifies as the dominant motive for African aid after 2000. Though aid flows to Africa (in real per capita terms) reached a nadir in 2000 (see Figure 24.2), this was also the year that the Millennium Development Goals (MDGs) were agreed upon. This North–South arrangement committed the developing countries (Africa included) to making rapid progress on several fronts, foremost among which was poverty reduction, and obligated the developed world to provide the necessary financing (mostly in the form of aid).[7] Since then, aid amounts have increased, and have again returned to levels not far below those of the 1980s in per capita terms (see Figure 24.2). Though debt relief through the Heavily Indebted Poor Country (HIPC) Initiative and the Multilateral Debt Relief Initiative (MDRI) have also accounted for a substantial portion of aid since 2000, non-debt relief aid has nevertheless made up most of the aid increase (see Figure 24.2).[8] Having said that, the centrality of MDG-related aid for Africa does not altogether preclude strategic motives related to such factors as the 'war on terror' (Bandyopadhyay and Vermann, 2013).

Aid and growth in Africa

Whether aid enhances growth has been a question that has occupied researchers from the earliest years of aid giving, and continues to do so. One of the reasons this question still remains both compelling and unanswered definitively is that it is not a straightforward one. First, we must deal with the issue of intent. If aid has strategic or even humanitarian motives, there need be no necessary relationship between aid and growth (and growth outcomes are not a necessary measure of success for either donor or recipient). Second, we need to consider the means through which aid might enhance growth. Does it do so by increasing investment or another factor of production (the volume route), or does it do so by increasing the efficiency of resource allocation (the efficiency route) (Museru *et al.*, 2014; Arndt *et al.*, 2015)? Third, even if aid can potentially improve growth outcomes, this can happen only if a substantial proportion of aid receipts make their way into the economy. If a sizeable proportion of aid does not make its way into the recipient's economy (because some of that aid is allocated to debt relief, diverted to the accumulation of foreign reserves or used to finance capital flight), any growth effect will be limited, even if aid has substantive growth-enhancing potential (Serieux, 2011).

Donors of bilateral aid, which has always constituted at least two-thirds of all aid given to Africa, have rarely pronounced growth as the objective of aid. More typically, humanitarian objectives (most often poverty reduction) have been the objective attached to aid giving. Therefore, the establishment of a relationship between aid and human development (or poverty reduction) would seem to be the most appropriate way to establish whether bilateral aid is justified. Unfortunately, there have been few empirical studies that have explicitly examined this relationship in general, and fewer still for Africa. Those that have done so have found a negative relationship between aid and human development (Arimah, 2004; Asongu, 2014a, 2014b).

The literature on the relationship between aid and growth (an association more often made by multilateral institutions), by contrast, is much larger for Africa. As noted earlier, aid can affect growth indirectly by influencing the rate of growth of any one of the factors of production – physical capital, human capital or labour (the volume effect). Of course, any short- or even medium-term effect on labour force growth is unlikely,[9] and though influences on the rate of growth of human capital are possible, this angle has not so far been investigated. Several studies have, however, examined the relationship between aid and investment (the rate of growth in physical capital). Most of these studies were specific to sub-Saharan Africa (Levy, 1988; Baldé, 2011; Serieux, 2011; Museru *et al.*, 2014), and only one covered all of Africa (Loxley and Sackey, 2008). Nevertheless, all of these studies found a positive relationship between aid and investment. Ndambendia and Njoupouognigni (2010) replace investment by its determinants (including aid) in a growth equation, and also find a positive relationship, but in this case the volume and efficiency effect cannot be disentangled.

Studies in which aid was included as an additional explanatory variable (along with investment) in growth equations were essentially investigating the efficiency angle – whether aid engendered growth by improving the allocation of resources. There has been less unanimity in the results of these studies. Some studies found a direct and positive relationship between aid and growth (Loxley and Sackey, 2008; Asongu, 2014a; Nwaogu and Ryan, 2015), others found no effect (Levy, 1988; Museru *et al.*, 2014), and others found a negative relationship (Wamboye *et al.*, 2013; Asongu, 2014b). Moreover, Museru *et al.* (2014) found that aid volatility has a negative effect on growth. Suffice it to say, then, that the resource allocation angle finds far from compelling support in the literature, and given the other potential negative (resource allocation) effects that are not captured by the approach described above (aid volatility and donor proliferation – both factors that bedevil aid to Africa) – it seems unlikely that aid has benefited the continent though this pathway.[10]

Given the strong empirical support for a positive relationship between aid and investment in Africa and Africa's high aid-to-income ratio, the region's generally low rate of investment (compared to other regions) would appear to present a conundrum. However, aid can only have a direct effect on investment if it enters the domestic economy. Serieux (2011) showed that, between 1970 and 2006, at least for sub-Saharan Africa (and likely for North Africa as well), on average, only one half of measured aid flows ever made it into the domestic economy. The other half of measured aid flows was diverted into some combination of debt repayment, capital flight and reserve accumulation.[11] Only approximately 13 per cent of any increase in aid flows made its way into investment – meaning that it would take roughly an 8 per cent increase in aid relative to income to increase the investment rate by 1 per cent.

Africa's debt experience

Given that 34 of the 49 formerly heavily indebted countries (HIPCs) were African, it is not surprising that Africa's debt experience has typically been associated with the Enhanced HIPC & MDRI Initiatives (both implemented after 1999), and seen as quite distinct from the debt crisis of the 1980s – which has been more closely associated with Latin American countries. In terms of debt composition and the international response to these countries' debt challenges, that differentiation is valid. However, these crises share the same genesis and early evolution (Serieux, 2001a).

Like aid receipts, debt flows to African countries began a precipitous climb at around the same time that oil prices increased sharply (1973–1975).[12] The increase in debt flows meant that the level of indebtedness of African countries rose steadily from very modest levels in 1972 and continued that upward trend into the 1980s. The same was true for the group of mostly middle-income (and mostly Latin American) countries that would later be referred to as the Brady Plan countries (see Figure 24.3).[13] In fact, in 1982, when Mexico declared its inability to service its debt – precipitating the debt crisis of the 1980s – African countries, on average, had higher (external) debt-to-income ratios than the Brady Plan countries, and the group of African countries that would later become part of the HIPC and Enhanced HIPC Initiatives had even higher (external) debt-to-income ratios than the African average (see Figure 24.3).

This commonality in the initiation and early evolution of these countries' external debt burdens was not accidental. The rapid increases in oil prices during 1973–1975 (and therefore the consequent increases in the value of oil imports) put tremendous pressure on the trade balances of most non-oil-exporting developing countries – generating a need for increased external borrowing to cover balance of payments deficits (and often fiscal deficits as well). Additionally, the low interest rates on foreign debt and the relatively high commodity prices of the late 1970s encouraged continued borrowing by keeping debt service amounts within ostensibly affordable levels.[14]

However, while the debt burdens of these different country groups evolved similarly, the content of their increasing debt burdens was not similar. Middle-income countries, and many of their large firms, sought external financing in bond markets and from international commercial bank loans. The majority of African countries, being, by and large, unable to access these markets, obtained most of their additional external flows from bilateral lending institutions (see Figures 24.4 and 24.5). That difference in debt composition was to have profound implications for the resolutions of the debt crises that would ensue.

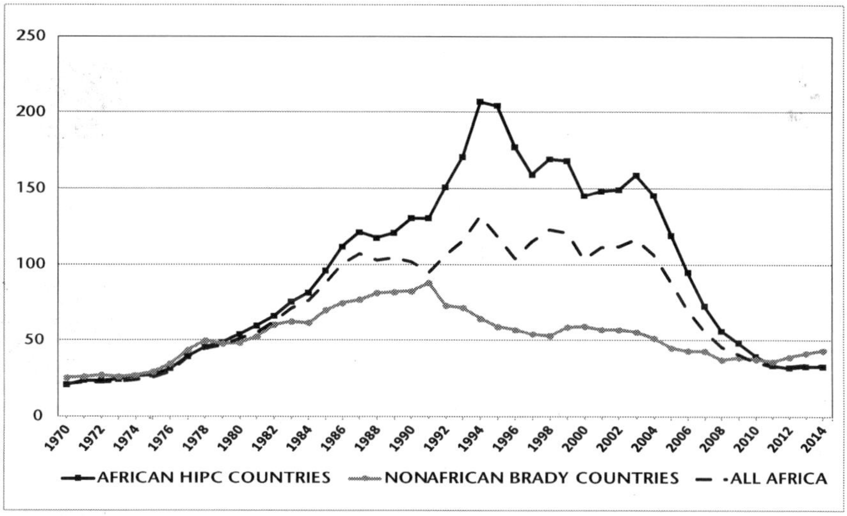

Figure 24.3 The evolution of debt-to-income ratios

Source: International debt statistics (World Bank)

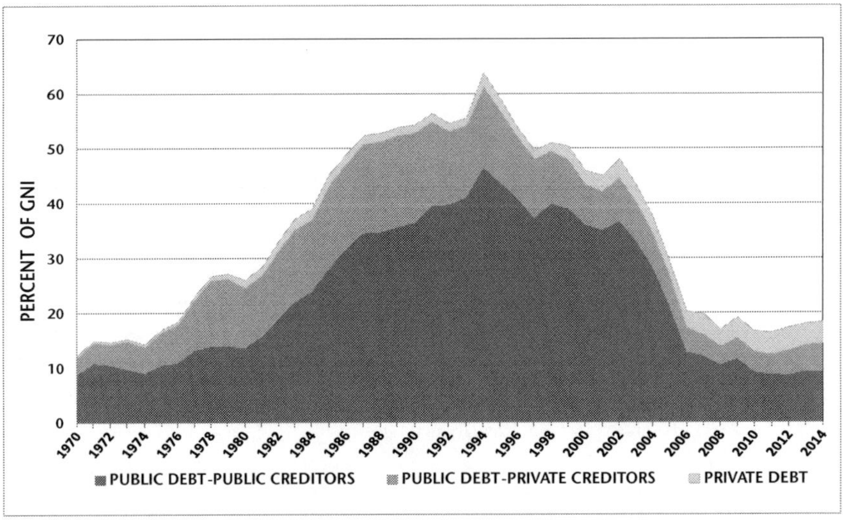

Figure 24.4 Debt composition of African countries

Source: International debt statistics (World Bank)

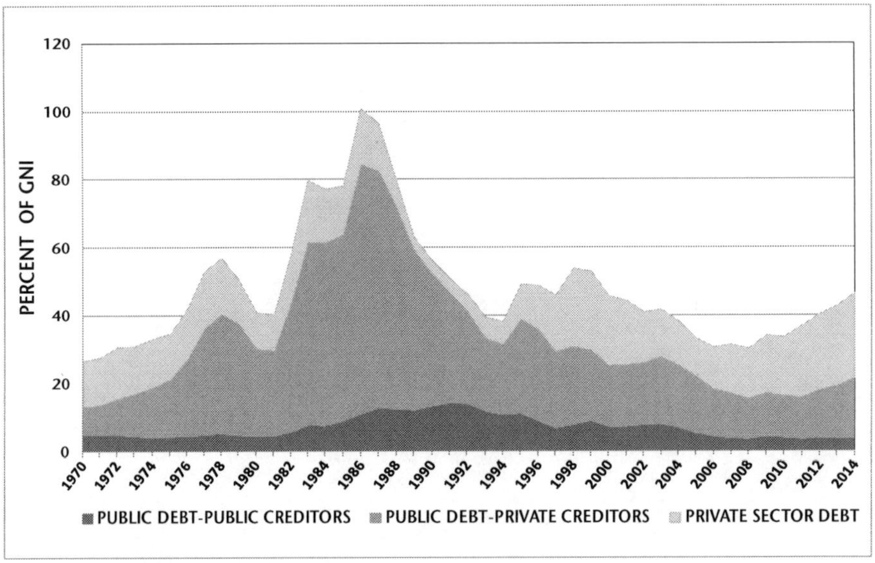

Figure 24.5 Debt composition of Brady Plan countries

Source: International debt statistics (World Bank)

As Figure 24.6 indicates, the accumulated arrears, relative to long-term external debt, of the African HIPC countries was already higher than that of the Brady Plan countries by 1980, and rose through most of the 1980s (a period of high interest rates and low commodity prices) in much the same manner as the arrears of the Brady Plan countries. This suggests that both these groups of countries faced debt overhangs. In fact, given the higher ratios of arrears to

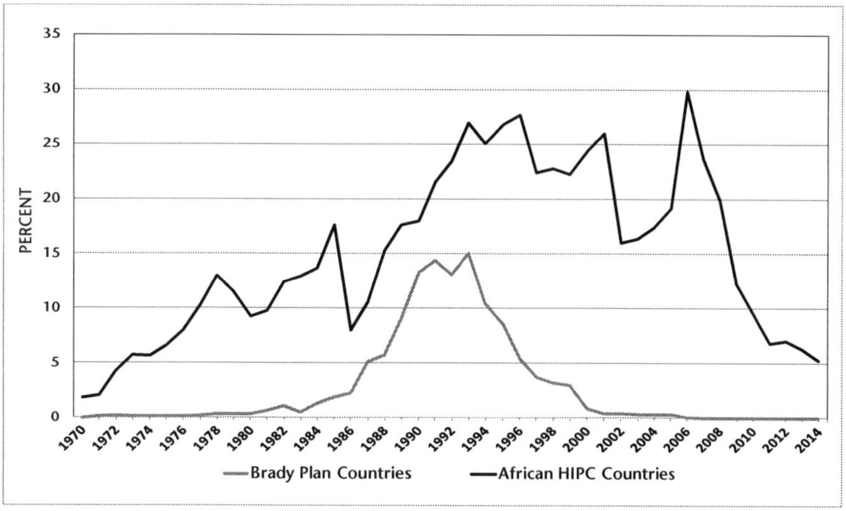

Figure 24.6 Arrears relative to the stock of long-term debt

Source: International debt statistics (World Bank)

debt stock, an even stronger case can be made for the African HIPC countries. However, while the Brady Plan countries were offered debt stock reductions through the Brady Plan of 1989, African countries (except Nigeria – the sole African Brady Plan country) were offered mostly debt rescheduling through the Paris Club in the 1980s and 1990s (see Figure 24.7). As a result, while the Brady Plan countries saw a steady decline in debt-to-income ratios after 1991, the HIPC countries in general (and the African HIPC countries in particular) would have to wait more than a decade longer before a majority of them could obtain substantial debt stock reductions (through the Enhanced HIPC Initiative, and later the MDRI as well). Thus, the debt stocks of African countries (relative to income) remained stubbornly close to or above 100 per cent from 1986 to 2004 and did not match that of former Brady Plan countries until 2011 (see Figure 24.3).

Why was Africa's debt experience so different from that of the Brady Plan countries (all of whom, besides Nigeria, were outside Africa)? The reason lies mostly in the absolute size and the composition of these countries' outstanding debts. The larger, wealthier Brady Plan countries owed large amounts of debt to commercial banks and other private financial institutions of developed countries (see Figure 24.5). A collective default would have seriously threatened the economies and financial systems of many developed countries and the world economy generally (Serieux, 2001a). Though the low-income (mostly African) HIPC countries owed large amounts relative to their economies, they owed a great deal less in absolute value and, perhaps just as importantly, owed most of their debt to public institutions in developed countries (such as export credit agencies) (see Figure 24.4). The threat posed by the potential default of Brady Plan countries compelled an international response, with the IMF as the leading institution. Though the acknowledgement that the Brady Plan countries faced debt overhangs (and not simply liquidity crises) did not come immediately, it did come by the end of the 1980s – as expressed in the prescriptions of the Brady Plan (Vasquez, 1996). The debt of the HIPC countries did not threaten any country's financial system, and definitely not the world economy (Serieux, 2001a). Admission that these counties faced debt overhangs came at first grudgingly and inadequately in

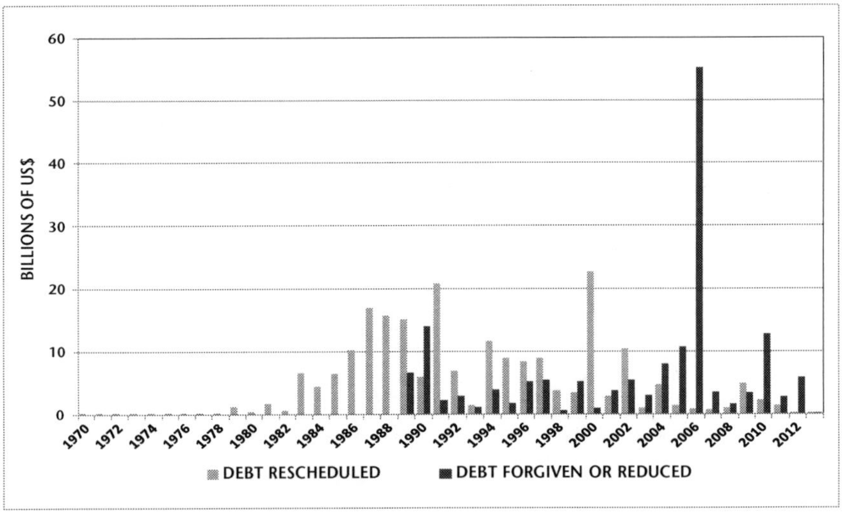

Figure 24.7 Debt rescheduling and forgiveness in Africa

Source: International debt statistics (World Bank)

the HIPC Initiative of 1994, and more realistically in the terms of the Enhanced HIPC Initiative of 1999 – a decade after the Brady Plan was unveiled.

The relationship between aid and debt in Africa

Given the debt challenges faced by so many African countries, it is reasonable to ask whether Africa's high aid receipts had anything to do with the near ubiquity of over-indebtedness across the continent in the 1980s and 1990s. Aid and debt have indeed been related during the past few decades, but in different ways in different time periods.

As noted earlier (see Figure 24.4), the increase in external flows that quickly inflated most African countries' external debt-to-income ratios in the 1970s and 1980s did include debt flows from public agencies, some of which was concessional credit (one of the two types of aid flows).[15] However, it was non-concessional financing that added most to debt burdens during that period as it increased much faster than concessional financing (see Figure 24.8). The fact that grant aid consistently made up the larger proportion of aid flows (over 65 per cent on average) meant that the high aid levels of that period did not necessarily imply rapid debt accumulation through the flows of concessional credit. Thus, though aid-related debt did constitute part of the continent's debt accumulation, it was below (and usually well below) 50 per cent of total external debt before 2000.

On the other hand, the advent of substantive debt relief after 2000 has been one of the factors behind the increase in aid during that period. The mechanics of debt relief (at least under the HIPC Initiative) are such that when donor countries wish to reduce developing country debt, they typically do so by paying off that debt with a 'grant' to the recipient country equivalent to the amount of the debt (and/or arrears) written off. These 'debt relief grants' do not represent actual flows of funds, but rather a way of representing the debt as having been paid off by the creditor country or institution (often to itself) on behalf of the debtor country. Since such a grant is concessional by definition, it is counted as part of official development assistance and

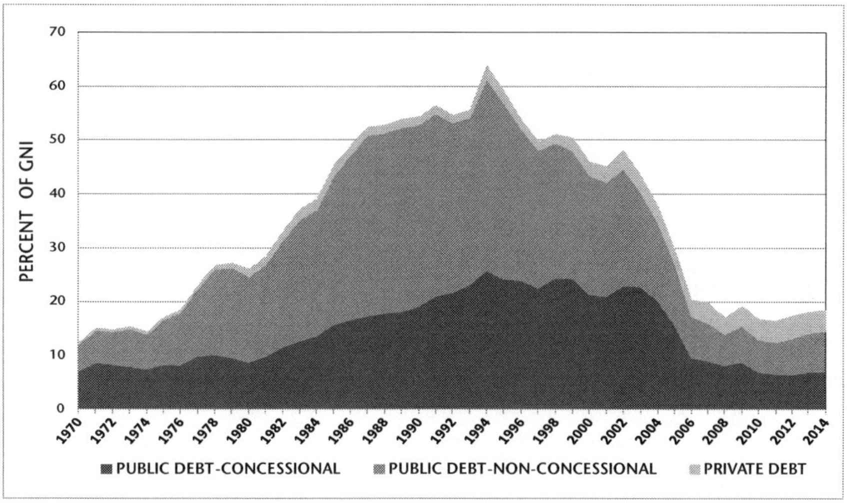

Figure 24.8 Concessional element of external debt (Africa)

Source: International debt statistics (World Bank)

increases measured aid amounts. Thus, because of this accounting convention, debt relief has necessarily implied an increase in measured aid amounts. In that regard, a portion of aid can be said to have been generated by previously accumulated debt (Serieux, 2001b, 2011).

The post-debt relief experience

One of the fears that has been expressed with respect to the debt relief experience is the likelihood that, like several Brady Plan countries after debt reduction, some African countries might soon find themselves requiring another round of debt relief because of the accumulation of new debt that they might find difficult to service if interest rates or the terms of trade move against them (Serieux, 2001b). At face value, though the average debt-to-income ratios of African countries have been rising since 2007, the rate of growth of that ratio does not yet suggest rapid debt accumulation. There are, however, some exceptions to the rule. Gambia, Ghana, Mauritania and Mozambique, who all previously received debt relief under the Enhanced HIPC and MDRI Initiatives, had debt levels over 40 per cent of GNI at the end of 2014 and rising (see Table 24.3). Gambia and Mauritania in particular have debt-to-GNI ratios above 60 per cent, and therefore have reached debt levels that should cause serious concern. Another group of countries that did not receive debt relief (Mauritius and Carbo Verde) have debt levels that are well over 80 per cent of GNI, and also high enough to cause concern (see Table 24.3).

Table 24.3 Total external debt-to-GNI ratios for a select group of countries

Country	*2008*	*2014*	*Country*	*2008*	*2014*
Carbo Verde	35.9	86.4	Mauritania	49.9	73.4
Gambia	39.7	63.9	Mauritius	6.9	90.9
Ghana	20.0	47.7	Mozambique	33.6	47.6

Source: International debt statistics (World Bank)

Perhaps the most remarkable development of the debt relief and post-debt relief period (2000–2014) is the increase in the number of African countries that have begun to raise funds on international bond markets. There has always been a small group of relatively large middle-income countries that could raise funds on international bond markets (South Africa, Côte d'Ivoire, Egypt, Morocco), but a new group of countries, some of them post-HIPC, have also joined that club. These new countries include Angola, DR Congo, Ghana, Guinea, Kenya, Mozambique, Rwanda, Tanzania and Zambia.[16] It is too early to determine whether these countries will be able to navigate what has been for many developing countries very traitorous waters (te Velde, 2014), but it is definitely a development worth paying attention to as it develops.

Conclusion

Africa's status as the highest aid-receiving region, combined with its poor growth record (until recently), has often been treated as a bit of a conundrum. However, that is so only if we assume that growth is a primary objective of aid and that a significant proportion of measured aid flows contribute to domestic investment activity. Neither would appear to be the case. But aid has also not contributed greatly to the debt accumulation of the 1970s and the debt distress of the 1980s and 1990s. But the resolution of the continent's debt problems has much to do with recent increases in aid flow. The reduction in the debt burdens of African HIPCs since 2000 has meant that the continent has entered a new low-debt era accompanied by modest to rapid growth of a majority of economies. One of the features of this new era is the presence of a much larger number of African countries in international bond markets. To the extent that this is an indication that these countries have managed to develop more credible debt management profiles, it is to be applauded. To the extent that it exposes countries that are still quite poor as well as economically fragile to a market known for its unforgiving character, this development is cause for concern.

Notes

1 Thirty-seven of Africa's 54 countries became independent between 1956 and 1968.
2 For Africa, aid and official development assistance (ODA) are equivalent terms. Aid has been used to refer to both 'official development assistance' received to developing countries and 'official aid' received by transition economies. In the case of Africa, only the former definition is relevant since there are no transition economies on the continent. In general, aid covers grants and concessional loans, which are loans with repayment terms that are at least 25 per cent below market terms (measured at net present values).
3 However, if compared in continental terms, Africa has typically received less aid annually than Asia (West, Central, East and South Asia combined), but more than the Americas and Oceania.
4 Since 2000, the Middle East has received the highest amount of food and humanitarian aid per capita.
5 These conclusions are derived from analyses of the aid data from the Organisation for Economic Co-operation and Development (OECD) international development statistics.
6 Strategic motive is here used to cover a range of objectives that fall broadly under the international politics umbrella. This would include attempts to reduce the threat of communism during the Cold War era, desire for support in international forums, attempts to engender peace in the immediate neighbourhood, and more recently the need to encourage and reward support for global anti-terrorist programmes and policies.
7 These commitments are encoded in General Assembly Resolution 55/2 of the UN (2000).

8 Most debt relief was provided in the form of grants given to the debt holder (often by the debt holder) on behalf of the debtor. These grants count as aid even if they do not typically involve any actual movement of funds.
9 There has been no suggestion in the literature (theoretical or empirical) that aid has an effect on migration. It is possible that aid can affect population growth (positively or negatively), but any such effect could only affect labour force growth in the long run.
10 Aid proliferation refers to increases in the number of different donors offering and managing aid programmes in a country. Besides the duplication and conflict that it engenders, the human resource cost and distraction from domestic policy challenges can be enormous.
11 It is still possible for aid diverted to debt servicing and reserve accumulation to have indirect beneficial effects on the economy, but these effects are likely to be of the second order.
12 Some aid flows (concessional loans) are also debt flows, but for African countries, grants, which are non-debt flows, made up one-third of total aid on average.
13 The Brady Plan countries were: Argentina, Brazil, Bulgaria, Costa Rica, the Dominican Republic, Ecuador, Jordan, Mexico, Nigeria, Panama, Peru, the Philippines, Poland, Russia, Uruguay, Venezuela and Vietnam. However, the data provided here do not include Nigeria (because it is African) and Argentina, Poland, Russia, Uruguay and Venezuela because no time series data on these countries' debt stocks could be found.
14 This perception turned out to be fallacious in the end. Commodity prices were above their long-run average and interest rates were below their long-run average.
15 The other is, of course, grants from public agencies.
16 This is indicated by the presence of public and public sector guaranteed bonds in the debt profile of these countries (based on the World Bank's international debt statistics).

References

Arimah, B.C. (2004) 'Poverty reduction and human development in Africa', *Journal of Human Development*, 5(3): 399–415.

Arndt, C., Jones, S. and Tarp, F. (2015) 'Assessing foreign aid's long-run contribution to growth and development', *World Development*, 69: 6–18.

Asongu, S.A. (2014a) 'Development thresholds of foreign aid effectiveness in Africa', *International Journal of Social Economics*, 41(11): 1131–1155.

Asongu, S.A. (2014b) 'The questionable economics of development assistance in Africa: hot-fresh evidence – 1996–2010', *Review of Black Political Economy*, 41(4): 455–480.

Baldé, Y. (2011) 'The impact of remittances and foreign aid on saving/investment in sub-Saharan Africa (SSA)', *African Development Review*, 23(2): 247–262.

Bandyopadhyay, S. and Vermann, E.K. (2013) 'Donor motives for foreign aid', *Federal Reserve Bank of St. Louis Review*, 95(4): 327–336.

Boschini, A. and Olofsgard, A. (2007) 'Foreign aid: an instrument for fighting communism?', *Journal of Development Studies*, 43(4): 622–648.

Kimura, H., Mori, Y. and Sawada, Y. (2012) 'Aid proliferation and economic growth: a cross-country analysis', *World Development*, 40(1): 1–10.

Levy, V. (1988) 'Aid and growth in sub-Saharan Africa: the recent experience', *European Economic Review*, 32(9): 1777–1795.

Loxley, J. and Sackey, H.A. (2008) 'Aid effectiveness in Africa', *African Development Review*, 20(2): 163–199.

Museru, M., Toerien, F. and Gossel, S. (2014) 'The impact of aid and public investment volatility on economic growth in sub-Saharan Africa', *World Development*, 57: 138–147.

Ndambendia, H. and Njoupouognigni, M. (2010) 'Foreign aid, foreign direct investment and economic growth in sub-Saharan Africa: evidence from pooled mean group estimator (PMG)', *International Journal of Economics and Finance*, 2(2): 39–45.

Nwaogu, U.G. and Ryan, M.J. (2015) 'FDI, foreign aid, remittances and economic growth in developing countries,' *Review of Development Economics*, 19(1): 100–115.

Serieux, J. (2001a) 'Debt of the poorest countries: anatomy of a crisis kept on hold', *Canadian Journal of Development Studies*, 21(2): 305–342.

Serieux, J. (2001b) 'The enhanced HIPC Initiative and poor countries: prospects for a permanent exit,' *Canadian Journal of Development Studies*, 21(2): 527–548.
Serieux, J. (2011) 'Aid and resource mobilization in sub-Saharan Africa: the role of reverse flows', *Journal of Development Studies*, 47(7): 1080–1107.
te Velde, D.W. (2014) *Sovereign Bonds in Sub-Saharan Africa: Good for Growth or Ahead of Time?* London: Overseas Development Institute.
UN (United Nations) (2000) *United Nations Millennium Declaration, Res 55/2*, New York: United Nations.
Vasquez, I. (1996) 'The Brady Plan and market-based solutions to the debt crisis', *Cato Journal*, 16(2): 233–243.
Wamboye, E., Adekola, S. and Bruno, B.S. (2013) 'Economic growth and the role of foreign aid in selected African countries', *Development*, 56(2): 155–171.

25

Faith in development in sub-Saharan Africa

Emma Tomalin and Adriaan van Klinken

Introduction

Our aim in this chapter is to explore some of the ways in which religion is relevant for development in sub-Saharan Africa and the extent to which this has been accommodated within international development policy and practice, and with what consequences. As Bompani (2015: 102) writes, 'the great majority of people in sub-Saharan Africa are deeply and actively religious and spirituality is integral in their understanding of the world'. While the majority belong to either Christianity or Islam, many still practise, often in coexistence with Christianity and Islam, forms of African religious beliefs (Pew Forum, 2010). However, not only do people's religious beliefs shape their views about what is important to them and how they conceive of 'development', but faith-based organisations (FBOs) have been central to welfare and service delivery long before the emergence of the contemporary international development project in the post-Second World War era (Tomalin, 2013). This includes traditional mechanisms within different religions in Africa for helping those in need, as well as the European Christian missionaries who accompanied explorers and colonialists, often shaping and assisting their work. Missionaries have a long history of development work, with significant contributions to health, education and infrastructure provision (Gifford, 2009: 46). Indeed, Manji and O'Coill (2002: 568) have argued that the evolution of the role of development non-governmental organisations (NGOs) in Africa 'represents a continuity of the work of their precursors, the missionaries and voluntary organizations that cooperated in Europe's colonization and control of Africa' (see also Deacon and Tomalin, 2015). Thus, Christianity was of great importance for colonial development in Africa (and continues to be so to this day) in a variety of ways, and correspondingly appears to have played a role in creating the foundations for the modern development project – described by Manji and O'Coill (2002: 568) as 'a vast institutional and disciplinary nexus of official agencies, practitioners, consultants, scholars and other miscellaneous "experts" producing and consuming knowledge about the "developing world"'.

However, the apparent religious roots of some aspects of the modern development project have been obscured by a commitment to modernisation theory and secularism, both by international development actors and by the political elites in African settings. As Bompani (2015: 102) writes:

> Dominant postcolonial discourses that Africa would inevitably become less religious and more secular predicated neither the role of religion in navigating and making sense of rapid external change, or its enduring role in individual, institutional or public life across Africa. There was a limited understanding that religious organisations such as missionary or Islamic schools had often been a driver of change and transformation, and as such religious belief would almost inevitably gain traction in Africa, in the face of change, not disappear.

More recently, the inevitability of straightforward processes of secularisation in Africa, and more widely across the globe, has been challenged by firm evidence of 'religious resurgence' (Thomas, 2005) or the 'deprivatisation of religion' (Casanova, 1994). Simplistic theories of modernisation and secularisation have been critiqued and indeed thoroughly debunked. Rather than disappearing or completely diminishing in significance, it is now widely recognised that the significant and continuing role of religion in public life (as well as in people's private lives), whether it is increasing or simply recognised, demands that the relationship between religion and society, even if it is both complex and controversial, needs to be taken seriously. New religious configurations, including the growth of Pentecostalism (van Klinken, 2015) and the rise of political Islam (Soares and Otayek, 2007), also have to be faced by political elites in Africa and the scholars that study these formations. The earlier certainties of the dwindling of religious identities as countries modernised have been replaced by uncertainties over the political, social and economic challenges and opportunities faced by rising religious affiliation.

In response to this 'religious resurgence', within the past decade or so, there has been a noticeable shift within some areas of international development policy, practice and research to include religion as a relevant factor, with a growth in publishing in this area, as well as an increased openness to working with faith actors in pursuit of development goals (Tomalin, 2013, 2015). Despite this shift, however, there have been critiques that the focus of this engagement has been largely 'instrumentalist' (Deneulin and Bano, 2009). This has followed a path where faith actors have been sought, and even encouraged to shape themselves, to reflect approaches to development that resonate with (secular) neo-liberal values and practices, and/or that reflect styles of religiosity that voice their faith position 'passively' and are more or less purged of supernatural elements. This leaves out a whole range of religious positionalities that require quite a different treatment, which may offer alternatives to neo-liberalism and/or be suffused with spirits and other supernatural forces.

In the next section of the chapter, we look more closely at the relationship between religious world views, religious institutions and development, and will examine the implications of the nature of the 'turn to religion' by mainstream development actors. This will be followed by a discussion of three key areas that are particularly relevant for thinking about the religion–development nexus in sub-Saharan Africa: socio-economic change, public health, and gender and sexuality.

Religious world views, religious institutions and development

> We contend that it is largely through religious ideas that Africans think about the world today, and that religious ideas provide them with a means of becoming social and political actors.
>
> *(Ellis and ter Haar, 2004: 2)*

A key argument of Gerrie ter Haar and Stephen Ellis is that the widespread misunderstanding among scholars about what religion actually is causes a major obstacle to understanding the role of religion in development. Because in the West, religion is often equated with its institutional expression, development agencies have indeed worked with religious organisations and their leaders in African contexts. However, 'they have devoted far less attention to the religious ideas that underlie the behaviour of religious believers and communities' (ter Haar and Ellis, 2007: 356) – ideas that relate to 'a belief in the existence of an invisible world, distinct but not separate from the visible one, which is home to spiritual beings that are deemed to have effective powers over the material world', which, according to these authors, is at the heart of religion in African contexts (Ellis and ter Haar, 2004: 14). Taking seriously that for many people in African settings, religion is primarily about interaction with a spirit world, ter Haar and Ellis discuss a number of examples demonstrating how such religious ideas are relevant in fields such as conflict prevention and peace-building, governance, wealth creation and production, health, and education. Hence, they conclude that development scholarship and practice should not just add religious organisations to the existing list of partners useful to implementing development policies, but should fundamentally rethink development policy 'in terms of the world-views of those most immediately concerned, the very people whom development policies seek to assist' (ter Haar and Ellis, 2007: 365). An interesting case study is provided by Smith (2008), whose ethnographic account shows how people in the Taita Hills of Kenya have appropriated development and make sense of the expectations and frustrations it generates, in a moral-spiritual framework of witchcraft. However, rather than being a 'traditional' framework, it appears that understandings of witchcraft are subject to change themselves, connected as they are to the dynamics of development in complex ways.

Reviewing recent literature on religion and development in Africa and other parts of the 'Global South', Bompani and Frahm-Arp (2010: 2) conclude:

> [W]hat is missing, and becoming an obstacle to our investigation and understanding the role of religion in development and politics, is a genuine understanding of the meaning of religion itself and the way that it is embedded in the everyday life of millions of people in the South. There has been an overwhelming focus on the consequences but not on the causes, on what religion does but not what religion is and this implies an a priori exclusion of a deeper understanding of the role of religion in contemporary societies.

In addition to a lack of understanding about what religion is and how it manifests in people's lives, as already noted, engagement has so far tended to be with organisations and individuals that are sympathetically matched to Western neo-liberal models of development, where development policymakers and agencies pick and choose which types of religion to engage with (Deneulin and Bano, 2009).

This selective approach to religious engagement might also be one of the reasons for a lack of attention to Islam in 'faith and development in Africa' discourse:

> Even the idea of 'faith-based development' . . . referring to the importance of religious organizations in bringing about development and which has recently become fashionable, is usually seen as only relating to Christian initiatives. A possible explanation for the inattention to Islamic NGOs is that the concepts of NGOs and 'civil society' are part of the neoliberal project; they therefore tend to be filled in by categories that match the agenda of liberal democracy . . . and Western conceptions of development, in which the very idea of Islamic NGOs appears almost unthinkable.
>
> *(Kaag, 2008: 4)*

However, while it might be more common to see engagement with Christian NGOs in Africa, not all forms of Christianity are necessarily attractive to development actors. For instance, development donors are more likely to seek engagement with organisations that express their faith 'passively' than those that obviously combine their development work with activities that aim to gain converts (Clarke, 2008). Paul Gifford (2015) has recently discussed Catholicism and Pentecostalism as two opposite forms of Christianity in Africa: the latter one being enchanted, that is, viewing the world as pervaded by spiritual forces and helping people to deal with such forces (very much in line with Ellis and ter Haar's definition of religion in Africa quoted above), while the former, standing in the European tradition of enlightenment, generally discounts such spiritual realities and is involved in 'secular' development work in sectors such as education and health, governance, conflict resolution, and human rights – what Gifford (2015: 97) refers to as 'the internal secularization of Catholicism into an NGO'. Gifford's binary scheme can easily be criticised for being rather simplistic and overgeneralising; however, his evaluation of enchanted forms of Christianity in Africa (which he acknowledges himself can also be found in Catholic circles) as an impediment to development – because it would encourage fear and distrust and diminish human accountability and agency – also raises fundamental questions about how we 'read' religious world views and the practices they give rise to, and how we evaluate their significance for development and for our understanding of 'development' and its underlying notion of modernity.

While international development actors might selectively choose to work with religious organisations that passively express their religious identity and that are affiliated with styles of religiosity that are largely purged of supernatural forces, this fails to capture much religious assertion in sub-Saharan Africa. In the following sections, we will briefly explore three key areas that have received attention in the literature and are particularly relevant for thinking about the religion–development nexus in sub-Saharan Africa: socio-economic change, public health, and gender and sexuality.

Faith and socio-economic change

Bompani and Frahm-Arp (2010: 7) suggest that 'The ability of religion to deal with the challenges of the post-colonial period and rapid social changes could be one of the sociological explanations of the rise of Christianity and Islam in Africa in the last few decades'. Many of these challenges and changes relate to socio-economic processes such as urbanisation and the introduction of neoliberal economic policies, effecting changes in family relations, the emergence of a middle class and new opportunities for entrepreneurship but also new economic insecurities. How, then, is religion able not only to make sense of the new situation people find themselves in, but also to help them respond to the new challenges and opportunities? Much of the literature in this field has focused on revivalist movements, in particular Pentecostal-Charismatic forms of Christianity.

Dena Freeman suggests that Pentecostalism offers a third alternative to mainstream development work represented by NGOs, on the one hand, and radical post-development theorists, on the other hand:

> While they [Pentecostals] embrace the mainstream capitalist 'spirit of development' with its desire for wealth and commodity consumption, they maintain a magico-religious worldview in sharp contrast to mainstream development's rational secularism. And while they acknowledge the existence of traditional practices and values, they seek to break away from them, in stark contrast to the post-development theorists who seek to base new models for the future in these traditional pasts.
>
> *(Freeman, 2012: 9)*

The rapid growth of Pentecostalism in Africa in the 1980s is often related to the economic crisis hitting the continent and the subsequent neo-liberal structural adjustment programmes that seriously affected the living standards of many Africans. However interesting this link might be, it is important to note that Pentecostalism has kept growing ever since, also in recent years when many African countries have witnessed significant economic growth. In other words, Pentecostalism appears to be attractive to both the winners and losers of new socio-economic realities.

Pentecostalism's ability to contribute to development and social change is related to a key characteristic that Pentecostal churches and movements generally share: the strong emphasis on personal transformation that comes along with spiritual resources – that is, the power of the Holy Spirit – to achieve such a transformation. Hence, 'Pentecostalism is one of the few modalities – religious or secular – that both demands and legitimises radical behaviour change, including the restructuring of families, communities and social relations' (Freeman, 2012: 13–14). However, since the 1980s, a new feature has emerged, particularly in neo-Pentecostal churches that started preaching the so-called prosperity gospel. Breaking with the ascetic, 'other-worldly' attitude characterising classical forms of Pentecostalism, new forms of Pentecostal-Charismatic Christianity started to express a strong this-worldly concern in which salvation is not just understood in spiritual, but also material, terms. Preached in different tones and nuances, the prosperity gospel basically claims that God wants Christians to live in abundance and to be prosperous, which can be achieved through practices of faith and obedience, such as tithing (donating to the church) and 'positive confession' where the believer claims divine blessings, along with engaging in business and working hard (Sharpe, 2013).

A key question in the study of Pentecostalism has been whether and how Max Weber's thesis of the Protestant ethic and the spirit of capitalism in early modern Europe can be meaningfully employed to analyse the relationship between Pentecostalism and neo-liberal capitalism in contemporary Africa. Freeman (2012) warns of drawing conclusions too quickly here, and she highlights the very real differences between the historical and socio-economic contexts, as well as between the types of capitalism and between Protestantism and Pentecostalism. Rather than expecting a simple repetition of Weber's argument, she points out, we need to ask whether Pentecostalism in Africa (intentionally or unintentionally) plays a similar role as Protestantism in Europe once did, that is, changing people's values and subjectivity and motivating new behaviours, which makes the new economic system seem moral (Freeman, 2012). She concludes that Pentecostalism does indeed play a similar role in Africa today:

> It is a form of Protestantism that not only fits with African sensibilities, but also stimulates a transformation of behaviour that can lead to success, or at least upward mobility, in the contemporary neoliberal economy. It motivates new behaviours and renders them moral.
>
> *(Freeman, 2012: 20)*

In particular, the prosperity gospel makes it moral for Pentecostal believers not only to strive to become richer, but also to keep the gained wealth in the nuclear family and use it for investment in business instead of sharing it with the extended family in line with the traditional redistributive economy.

Several authors have identified reasons why Pentecostal churches might be more successful in contributing to development and bringing about socio-economic change than secular NGOs: first, Pentecostal churches receive most of their funding from their followers, and therefore have to be responsive to their demands; second, they are primarily concerned with transforming individuals and seem to be effective in realising such a personal transformation; third, being embedded in the local community, these churches are good at fostering participation, and are also seen as moral, trustworthy and meaningful institutions; fourth, with its holistic ontology, Pentecostalism both takes seriously traditional world views in which spirits and witches are part of reality, and provides people with the means of dealing with such forces and breaking away from them (Freeman, 2012; Jones, 2012; Piot, 2012; Smith, 2012). Arguably, these factors are also largely applicable to Islamic reformist groups, but their contribution to development has received much less attention so far.

As far as the relation between Islamic groups and socio-development in Africa has been explored, the focus has mainly been on Islamic NGOs who offer material relief and contribute to social welfare, funded by practices of voluntary giving (*sadaqah*) and prescribed alms (*zakat*), as well as by religious endowment (*waqf*) to support larger-scale projects (Weiss, 2002; Skinner, 2010). It is said that 'in a context of structural adjustment, an increased spread of Western consumption ideals through mass communication, and a growing sense of the global context in which one is living, these organizations aim to influence people's material and moral well-being' (Kaag, 2008: 3). Writing about transnational Islamic NGOs in Chad, Mayke Kaag makes the important point that even if the amounts of money involved in these organisations is less than the funds channelled to Africa by Western NGOs, and their work might be more 'traditional' – concerned with concrete relief work rather than with 'capacity-building' – these organisations make an important contribution as they offer 'new ways of connecting to the wider *umma*, the global community of the faithful – with consequences both on the local and the global level', as they contribute to processes of Islamisation and Arabisation (Kaag, 2011: 464). In reformist Islamic movements such as Tablighi Jamaat, there is a strong emphasis on purity and moral transformation of the individual, often resulting in a 'withdrawing from extended family life and its concomitant social and ritual obligations' (Janson, 2014: 131) – a pattern that is reminiscent of the Pentecostal concern with holiness and born-again conversion, and its social consequences.

Faith and public health

> The history of religion can barely be separated from that of health. Most, if not all, religions are bound up with some comprehensive conception of health and well-being, whether cast in cyclical or linear patterns of redemption, salvation or fullness of life.
>
> *(Gunderson and Cochrane, 2012: 21)*

Health and religion have traditionally been closely related, and it is only with the widespread introduction of the Western biomedical model, first in the Global North and then increasingly across the Global South (initially via Christian missionaries), that this relationship became weakened. While the biomedical model perceives 'health' in terms of curing diseases that affect the body, illness in many communities has a religious meaning in terms of understanding its causes and hence appropriate cures (e.g. a punishment from God that requires prayers, or a curse that requires ritual intervention). Further, approaches to healing within religious settings are often more focused on the 'whole person' – in which the body and soul are recognised as responsive to each other – rather than the elimination of physical symptoms alone. As with a biomedical model, there are both strengths and weaknesses to faith-inspired ways of thinking about health, and many are critical of explanations for illness that may discourage people from seeking medical help (e.g. the Pentecostal preacher 'curing' people of HIV and AIDS) (Tomalin *et al.*, 2015).

In African settings, the Christian missionaries first introduced Western medical practices, with the setting up of hospitals and clinics, as one way of justifying the missionary project to an increasingly secularising Europe. In many cases, these missionary set-ups were able to step in to support weak state provision, and this became particularly salient in the post-independence period in many African contexts. According to Gunderson and Cochrane (2012), however, by the post-Second World War period, Christian missions in Africa and other developing settings began to ask why their health services were not having as great an impact as they would have liked. Over the following decade or so, Christian fora, including the World Council of Churches (WCC) and the Lutheran Federation, began to seriously debate these global health issues, and in 1968 the WCC founded the Christian Medical Commission (CMC) to establish links between faith providers and the UN agencies, in particular the WHO. As Olivier (2015: 347) tells us, the CMC was central in the 'shaping of the Alma Ata Declaration of 1975 on primary healthcare and "Health for All", which has remained a core ideal of global health'. A number of religious health associations were also set up in African settings around this time to coordinate faith-based healthcare in newly independent nations, and today these remain 'a highly significant part' of African health systems (Gunderson and Cochrane, 2012: 29).

The HIV and AIDS crisis, beginning in the 1980s, however, brought issues of healthcare in Africa to the attention of a far wider international audience, including the role of religious leaders. On the one hand, Christian healthcare services were in a position to provide much-needed support to people with HIV and AIDS. On the other hand, the moralising discourses – which linked HIV and AIDS to sin – was strongly criticised by international development actors (Olivier and Wodon, 2012). Nonetheless, faith-based actors in Africa were increasingly targeted as recipients of funding to help them fight HIV and AIDS, including from the Global Fund and PEPFAR (see Olivier and Wodon, 2014).[1]

This flow of international funding for religious healthcare soon raised questions about the evidence for the efficacy of faith-based health services in Africa. In response, the WHO funded a significant project in this area, carried out by the African Religious Health Assets Programme (ARHAP), based at the University of Cape Town, that aimed to document the 'contribution made by religion and religious entities to the struggle for health and wellbeing in Zambia and Lesotho, in a context dominated by poverty, stressed public health systems and the HIV/AIDS pandemic' (ARHAP, 2006: 1; Holman, 2015: 135).

This study involved an enquiry 'into the nature and potential contributions of religious entities to the struggle against HIV/AIDS, to universal access to treatment, care and prevention, and to health and wellbeing more broadly' (ARHAP, 2006: 2). It addressed concerns that there have been 'tensions around religion and public health, especially when it comes to HIV/AIDS and matters to do with sexuality, condoms and stigma' (ARHAP, 2006: 131), arguing that this was 'part of the past' and that 'the sheer human impact of the pandemic is drawing religious entities into new and significant contributions to health and wellbeing' (ARHAP, 2006: 131).

Overall, approximately 500 religious and partner organisations working in the area of HIV and AIDS, including about 350 at the local level, were mapped and made visible. The study found that:

> in Lesotho, approximately 40% of national health service is estimated to be provided by 9 Christian hospitals and 75 health centers whose managing churches are members of the Christian Health Association of Lesotho (CHAL); in Zambia, the numbers are similar, with 30% of services provided by 30 hospitals and 66 rural health centers whose managing churches are members of the Churches Health Association of Zambia (CHAZ) . . . [but also that] . . . religious entities have also been shown to have other critical, but less tangible assets, including extensive volunteer networks, expert lay professionals, congregations/churches/mosques/temples in even the most remote locations, and deep systems of community support, 'voice,' advocacy, trust, consolation, and hope.
>
> *(ARHAP, 2006: 7)*

Olivier points out that prior to this project, the global public health community had never undertaken research to find out the nature and extent of religious healthcare services, and she wonders how this had happened, considering 'all the world's main religions have health and healing as central tenets of their faith' (Olivier, 2015: 347). Moreover, as outlined above, 'religious bodies were also strongly involved in early public health advocacy, and in shaping global health policy' (Olivier, 2015: 347). While there has been 'significantly increased attention, dialogue and publications', Olivier argues that:

> this area of inquiry has remained fairly inconclusive . . . [since] . . . there continues to be broad generalizations made within the literature and commentary about the suspected generic strengths and weaknesses of all 'faith-based organisations' – and these continue to be easily challenged from those operating at a policy level.
>
> *(Olivier, 2015: 352; see also Olivier and Wodon, 2012)*

One particularly dominant discourse within the international development literature, which has caused discomfort among religious organisations, is that faith-based organisations, for instance those involved in HIV and AIDS work in Africa, are resources to be exploited 'for the broader goals of development and public health' (Olivier, 2015: 351). Olivier concludes that there needs to be a shift in how we think about and approach FBOs involved in health services where we turn 'away from generalizing statements about "faith-based organisations" and large-scale advocacy about valuing the faith sector', and instead of researching religious institutions as a separate sector, to consider them 'as part of or in comparison to the broader systems in which they operate' (Olivier, 2015: 354–355).

This reinforces the point made earlier in this chapter, where ter Haar and Ellis (2007) argued that misunderstanding among scholars about what religion actually is comprises a major obstacle to understanding its role in development. Instead of viewing religion as an integral part of the sociocultural, economic and political systems in which people live, it tends be viewed as a separate entity (reinforcing the modernist distinction between the secular and the religious) that can be selectively brought into contemporary development activities to serve predetermined goals (Deneulin and Bano, 2009).

Faith, gender and sexuality

> As structures of sociability, learning and mutual financial support, women's 'religious' associations allow them to claim collective relevance at the interstices of domestic, semi-public and public settings, and to initiate new nodes of particular between society and the state. As such, female Muslim activism in Mali can be seen as a particular modality of 'politics from below', one that aims not at political protest but at the transformation of the personal and the social.
>
> *(Schulz, 2010: 89)*

This quotation draws attention to the significant social capital afforded to women via involvement in religious associations. Considering that the majority of women in African settings are likely to be religious, the role that religion plays in women's lives – both positively and negatively – is immense. Yet as with other areas of development policy and practice, 'gender and development' (GAD) approaches until recently have tended to ignore or marginalise religious issues (Tomalin, 2011). The strong secularist leanings of feminist movements across the globe not only tended to see religion as inimical to women's empowerment and development, but this has impacted upon the concepts and strategies employed within gender and development programmes. As already noted above, by the 2000s there had been a shift whereby international donors and NGOs increasingly viewed faith-based organisations (FBOs) as desirable partners and forged links with faith leaders in pursuit of development goals such as the fight against HIV and AIDS in Africa (e.g. UNFPA, 2005, 2008). However, critics have argued that there has been somewhat of a 'rush to find the religious' (Balchin, 2011) in an anxiety to engage with a group of previously neglected stakeholders. Considering the very deep-seated patriarchy within most religious traditions, this runs the risk of undermining the battle to mainstream gender equity into development objectives (Bartelink and Buitelaar, 2006; see also Pearson and Tomalin, 2007; Tadros, 2010, 2011).

Nonetheless, consideration of the intersections between gender, religion and development are ever more important; when religious strategies are adopted by GAD actors, 'key allies' (e.g. religious leaders or faith-based groups) need to be carefully identified to ensure that they are committed to diversity and gender equality. To give just a couple of examples, Adamu (1999), writing about northern Nigeria, has argued that the failure of international donors to incorporate Islam into gender policies and programmes has compounded the view that GAD is a 'tool of the West'. This has meant that some of the principles adopted in international documents, such as CEDAW (the UN's Convention on the Elimination of All Forms of Discrimination Against Women) are perceived to be at odds with Islamic values about the family. Another example of where engaging with religious ideas and actors could have beneficial results for gender equality is regarding the role that certain FBOs can play in mediating between development actors and conservative religious groups. For instance, Marshall and Taylor (2006) discuss the work of the UK-based Christian NGO Tearfund, which works through local evangelical Christian partner organisations in over 70 countries. A project was set up for the purpose of establishing the extent to which evangelical churches in Africa were dealing with issues around sexual activity and behaviour, and whether they were doing so in an informed and progressive manner. One of the findings of the project was that where churches taught that women should serve and be subservient to men, women were less likely to be able to 'challenge the unfaithfulness of their husbands, or negotiate the use of a condom for safer sex' (Marshall and Taylor, 2006: 370). Overall, Tearfund was critical of the evangelical churches – a position that is perhaps easier for it to adopt as an 'insider' to the tradition – and was funding a small pilot initiative aimed at changing attitudes.

For a long time, debates on gender and development in Africa were primarily, if not solely, concerned with the position of women. Only recently, development scholarship and practice has begun to widen its scope to include 'the other half of gender'. In African contexts, this shift was mainly inspired by the HIV epidemic drawing attention to the problematic intersections of popular forms of masculinity with sexuality, gender and family relations, and health, resulting in men being 'targeted for a change' (van Klinken and Chitando, 2015). Religion, in this context, appears to be 'a double-edged sword': on the one hand, it can reinforce widespread notions of male supremacy, power and control through patriarchal ideologies, while on the other hand it can also contribute to constructive notions of masculinity such as responsibility, self-discipline, respect, maturity, and so on. Pentecostal Christianity in particular has been associated with a potential to realise change in male subjectivity and contribute to a transformation of masculinities in the face of challenges such as HIV and AIDS and gender-based violence, among others, through a redefinition of patriarchal notions such as 'male headship' (van Klinken, 2013). However, Islamic reformist groups in Africa also appear to be concerned with the construction of pious forms of masculinity that appropriate and transform local versions of being and becoming a man (Wario, 2012). Going beyond the concern with gender and sexual relations characterising current studies of religion and masculinities in Africa, further research will need to analyse the specific socio-economic impacts of religious masculinity politics in African contexts. We are only at the beginning of understanding how religious organisations might help men who feel economically and socially disempowered to become agents, again, in their own lives and to take a responsible and constructive place in their communities.

Marc Epprecht (2013) has criticised development studies for understating the importance of sexuality in general, and same-sex sexuality in particular, to broad questions of political and economic change. He further argues that homophobia can have 'rippling negative impacts upon health, democratic culture and economic development' in African societies (Epprecht, 2013: 14). Religion being a key factor fuelling homophobia in Africa, there is a need both to unravel the intersections of religion and contemporary politics of homosexuality on the continent and to address the role of faith-based organisations. As Jonathan Connor (2011) points out with reference to the optimistic talk about what religion can contribute to development, basically the question is, 'Development for whom?' Discussing the case of World Vision in particular, he warns that international faith-based organisations involved in development work in Africa 'might, in fact, be strengthening cultures of exclusion', and 'will remain complicit in nurturing bad civil societies' if they do not 'take an explicit stand against homophobia and discrimination' (Connor, 2011: 862). As relevant as Connor's point is, it also raises the more fundamental question of what or whose agenda development work is supposed to promote, and how. The current politics of homosexuality in various parts of Africa – with new anti-gay legislation having been passed in countries such as Uganda, Nigeria and Gambia – suggests that there are different, and indeed conflicting, understandings of what 'development' means and of how 'modernity' is envisioned. As van Klinken and Chitando (forthcoming) point out, it is becoming clear:

> that there is no singular progressive and liberal narrative of 'modernity', neither in the West nor in Africa. Rather African societies navigate their own paths of modernity, and these do not necessarily lead to a greater cultural, social and political acceptance of same-sex relationships or recognition of LGBTI rights.

Thus, sexuality is one of the key fields manifesting the public 'resurgence of religion' and revealing the complexities of thinking about the role of faith in development, and thus it raises fundamental challenges to development studies in a post-secular and postcolonial era.

Conclusion

Obviously, religion needs to be acknowledged as a fundamental force in society, in politics and indeed in development policy and practice in contemporary African contexts. This chapter could only explore some of the ways in which the major religions on the continent, Christianity, Islam and indigenous religions, are relevant for key development issues in sub-Saharan Africa. The argument of this chapter is not simply that religious organisations, religious leaders and religious beliefs and practices do play a role – either positively or negatively – in processes of development, and that therefore religion needs to be taken seriously as a force contributing and/or hindering development. More critically, the significance of religion in shaping people's word view and enabling people's agency also means that development scholars and actors need to engage with religion and understand the meaning and importance of religious ideas in African contexts. In different ways, the sections in this chapter all refer to the same problem of development policymakers and practitioners not understanding what religion is, and instrumentalising it in a way that suits a secular neo-liberal agenda – either in relation to economic development, health, or issues of gender and sexuality. In view of African realities, it might be time to call for a post-secular turn in development studies, which would not just include asking how religion can contribute to (a Western idea of) development, but how development actually is imagined and conceived from religio-cultural perspectives, leading into a self-critical reflection on what this might mean for our development scholarship and practice.

Note

1 AIDS relief funds announced by Bush in his January 2003 State of the Union Address, which promised US$15 billion for the 'President's Emergency Plan for AIDS Relief' (PEPFAR) (www.avert.org/pepfar.htm). By 2005, a quarter of all PEPFAR partners were FBOs, reflecting PEPFAR's explicit position that 'faith-based groups are priority local partners' because of their important role as healthcare providers (PEPFAR, 2005).

References

Adamu, F. (1999) 'A double-edged sword: challenging women's oppression within Muslim society in northern Nigeria', *Gender and Development*, 7(1): 56–61.

ARHAP (African Religious Health Assets Programme) (2006) *Appreciating Assets: The Contribution of Religion to Universal Access in Africa*, available at: www.arhap.uct.ac.za/downloads/ARHAPWHO_entire.pdf (accessed 13 September 2012).

Balchin, C. (2011) 'Religion and development: a practitioner's perspective on instrumentalisation', *IDS Bulletin*, 42(1): 15–21.

Bartelink, B. and Buitelaar, M. (2006) 'The challenges of incorporating Muslim women's views into development policy: analysis of a Dutch action research project in Yemen', *Gender and Development*, 14(3): 351–361.

Bompani, B. (2015) 'Religion and development in sub-Saharan Africa: an overview', in E. Tomalin (ed.), *The Routledge Handbook on Religions and Global Development*, London and New York: Routledge, pp. 101–113.

Bompani, B. and Frahm-Arp, M. (2010) 'Introduction: development and politics from below – new conceptual interpretations', in B. Bompani and M. Frahm-Arp (eds), *Development and Politics from Below: Exploring Religious Spaces in the African State*, New York: Palgrave Macmillan, pp. 1–22.

Casanova, J. (1994) *Public Religions in the Modern World*, Chicago, IL: University of Chicago Press.

Clarke, G. (2008) 'Faith-based organizations and international development: an overview', in G. Clarke and M. Jennings (eds), *Development, Civil Society and Faith-Based Organizations*, London: Palgrave Macmillan, pp. 17–45.

Connor, J. (2011) 'Development for whom? Homosexuality and faith-based development in Zimbabwe', *Development in Practice*, 21(6): 860–869.

Deacon, G. and Tomalin, E. (2015) 'A history of faith-based aid and development', in E. Tomalin (ed.), *Routledge Handbook on Religions and Global Development*, London and New York: Routledge, pp. 68–79.

Deneulin, S. and Bano, M. (2009) *Religion in Development: Rewriting the Secular Script*, London: Zed Books.

Ellis, S. and ter Haar, G. (2004) *Worlds of Power: Religious Thought and Political Practice in Africa*, London: C. Hurst & Co.

Epprecht, M. (2013) *Sexuality and Social Justice in Africa: Rethinking Homophobia and Forging Resistance*, London: Zed Books.

Freeman, D. (2012) 'The Pentecostal ethic and the spirit of development', in D. Freeman (ed.), *Pentecostalism and Development: Churches, NGOs and Social Change in Africa*, New York: Palgrave Macmillan, pp. 1–40.

Gifford, P. (2009) *Christianity, Politics and Public Life in Kenya*, London: C. Hurst & Co.

Gifford, P. (2015) *Christianity, Development and Modernity in Africa*, London: C. Hurst & Co.

Gunderson, G. and Cochrane, J. (2012) *Religion and the Health of the Public: Shifting the Paradigm*, Basingstoke: Palgrave Macmillan.

Holman, S. (2015) *Beholden: Religion, Global Health, and Human Rights*, Oxford and New York: Oxford University Press.

Janson, M. (2014) *Islam, Youth, and Modernity in The Gambia: The Tablighi Jama'at*, Cambridge: Cambridge University Press.

Jones, B. (2012) 'Pentecostalism, development NGOs and meaning in eastern Uganda', in D. Freeman (ed.), *Pentecostalism and Development: Churches, NGOs and Social Change in Africa*, New York: Palgrave Macmillan, pp. 181–202.

Kaag, M. (2008) 'Transnational Islamic NGOs in Chad: Islamic solidarity in the age of neoliberalism', *Africa Today*, 54(3): 3–18.

Kaag, M. (2011) 'Connecting to the Umma through Islamic relief: transnational Islamic NGOs in Chad', *International Development Planning Review*, 33(4): 463–474.

Manji, F. and O'Coill, C. (2002) 'The missionary position: NGOs and development in Africa', *International Affairs*, 78(3): 567–583.

Marshall, M. and Taylor, N. (2006) 'Tackling HIV and AIDS with faith-based communities: learning from attitudes on gender relations and sexual rights within local evangelical churches in Burkina Faso, Zimbabwe, and South Africa', *Gender and Development*, 14(3): 363–374.

Olivier, J. (2015) 'Religion at the intersection of development and public health in development contexts: from advocacy about faith-based organizations to systems thinking', in E. Tomalin (ed.), *The Routledge Handbook on Religions and Global Development*, London and New York: Routledge, pp. 346–358.

Olivier, J. and Wodon, Q. (2012) 'Layers of evidence: discourses and typologies of faith-inspired community responses to HIV/AIDS in Africa', in J. Olivier and Q. Wodon (eds), *Strengthening Faith-Inspired Health Engagement, Vol. 3: Mapping, Cost, and Reach to the Poor of Faith-Inspired Health Care Providers in Sub-Saharan Africa*, HNP Discussion Papers, Washington, DC: World Bank, pp. 25–51.

Olivier, J. and Wodon, Q. (2014) 'Increased funding for AIDS-engaged faith-based organizations in Africa?', *Review of Faith and International Affairs*, 12: 53–71.

Pearson, R. and Tomalin, E. (2008) 'Intelligent design? A gender sensitive interrogation of religion and development', in G. Clarke and M. Jennings (eds), *Development, Civil Society and Faith-Based Organizations: Bridging the Sacred and the Secular*, Basingstoke: Palgrave, pp. 46–71.

PEPFAR (President's Emergency Plan for AIDS Relief) (2005) *Community and Faith-Based Organizations*, Report to Congress, September, available at: www.pepfar.gov/progress/76864.htm (accessed 19 March 2012).

Pew Forum (2010) *Tolerance and Tension: Islam and Christianity in Sub-Saharan Africa, Executive Summary*, available at: http://pewforum.org/executive-summaryislam-and-christianity-in-sub-saharan-africa.aspx#quickdefinition (accessed 12 September 2012).

Piot, C. (2012) 'Pentecostal and development imaginaries in West Africa', in D. Freeman (ed.), *Pentecostalism and Development: Churches, NGOs and Social Change in Africa*, New York: Palgrave Macmillan, pp. 111–133.

Schulz, D.E. (2010) 'Remaking society from within: extraversion and the social forms of female Muslim activism in urban Mali', in B. Bompani and M. Frahm-Arp (eds), *Development and Politics from Below: Exploring Religious Spaces in the African State 2*, New York: Palgrave Macmillan, pp. 74–98.

Sharpe, M. (2013) 'Name it and claim it: prosperity gospel and the global Pentecostal reformation', in M. Clarke (ed.), *Handbook of Research on Development and Religion*, Cheltenham: Edward Elgar, pp. 164–179.

Skinner, D.E. (2010) 'Da'wa and politics in West Africa: Muslim Jama'at and non-governmental organizations in Ghana, Sierre Leone and the Gambia', in B. Bompani and M. Frahm-Arp (eds), *Development and Politics from Below: Exploring Religious Spaces in the African State*, New York: Palgrave Macmillan, pp. 99–130.

Smith, J.H. (2008) *Bewitching Development: Witchcraft and the Reinvention of Development in Neoliberal Kenya*, Chicago, IL: University of Chicago Press.

Smith, J.H. (2012) 'Saving development: secular NGOs, the Pentecostal revolution and the search for a purified political space in the Taita Hills, Kenya', in D. Freeman (ed.), *Pentecostalism and Development: Churches, NGOs and Social Change in Africa*, New York: Palgrave MacMillan, pp. 134–158.

Soares, B. and Otayek, R. (eds) (2007) *Islam and Muslim Politics in Africa*, New York: Palgrave Macmillan.

Tadros, M. (2010) *Faith-Based Organizations and Service Delivery: Some Gender Conundrums*, UNRISD, Gender and Development Programme Paper 11, available at: www.unrisd.org/unrisd/website/document.nsf/8b18431d756b708580256b6400399775/592137c50475f6a8c12577bd004fb5a0/$FILE/Tadros.pdf (accessed 13 September 2012).

Tadros, M. (2011) 'Introduction: gender, rights and religion at the crossroads', *IDS Bulletin*, 42(1): 1–9.

ter Haar, G. and Ellis, S. (2007) 'The role of religion in development: towards a new relationship between the European Union and Africa', *The European Journal of Development Research*, 18(3): 351–367.

Thomas, S.M. (2005) *The Global Resurgence of Religion and the Transformation of International Relations: The Struggle for the Soul of the Twenty-first Century*, New York: Palgrave Macmillan.

Tomalin, E. (2011) *Gender, Faith and Development*, Oxford: Oxfam.

Tomalin, E. (2013) *Religions and Development*, London and New York: Routledge.

Tomalin, E. (2015) *Handbook of Religions and Global Development*, London and New York: Routledge.

Tomalin, E., Sadgrove, J. and Russell, A. (2015) *Places of Worship as Minority Ethnic Public Health Settings in Leeds*, available at: www.leeds.gov.uk/phrc/Pages/Places-of-Worship-as-Public-Health-settings.aspx (accessed 12 October 2015).

UNFPA (United Nations Population Fund) (2005) *Culture Matters: Working with Communities and Faith-Based Organizations: Case Studies from Country Programmes*, New York: UNFPA, available at: www.unfpa.org/public/publications/pid/1430 (accessed 13 September 2012).

UNFPA (United Nations Population Fund) (2008) *Culture Matters: Lessons from a Legacy of Engaging Faith-Based Organizations*, New York: UNFPA, available at: www.unfpa.org/public/pid/1353 (accessed 13 September 2012).

van Klinken, A. (2013) *Transforming Masculinities in African Christianity: Gender Controversies in Times of AIDS*, Farnham: Ashgate.

van Klinken, A. (2015) 'Christianity in Africa: developments and trends', in S. Hunt (ed.), *Handbook of Global Contemporary Christianity: Themes and Developments in Culture, Politics, and Society*, Leiden: Brill, pp. 131–151.

van Klinken, A. and Chitando, E. (2015) 'Masculinities, religion and HIV in Africa', in E. Tomalin (ed.), *The Routledge Handbook of Religions and Global Development*, London and New York: Routledge, pp. 127–137.

van Klinken, A. and Chitando, E. (2016) 'Introduction: public religion, homophobia and politics of homosexuality in Africa', in A. van Klinken and E. Chitando (eds), *Public Religion and the Politics of Homosexuality in Africa*, Farnham: Ashgate, pp. 35–48.

Wario, H.A. (2012) 'Reforming men, refining umma: Tablīghī Jamā'at and novel visions of Islamic masculinity', *Religion and Gender*, 2(2): 231–253.

Weiss, H. (ed.) (2002) *Social Welfare in Muslim Societies in Africa*, Uppsala: Nordic Africa Institute.

Section 7
Rural Africa

26
Land tenure[1]

Jon D. Unruh

Introduction

Land tenure: definitions

Land tenure (rural property rights) can be defined in a variety of ways, but is generally regarded to be 'social relations about land'. One useful way to view land tenure is as a bundle of rights within a society or community. In such a bundle, rights can be added, removed or divided to create a very wide variety of different rights to land and land-based resources. While rights to land are often thought to be produced by titles, deeds, registries and leases, in reality these are artefacts of a system that involves, among other elements, enforcement, dispute resolution, evidence, identity, forms of logic, institutional formation and operation, derivation and maintenance of authority and legitimacy, the role of 'community', and most importantly trust in institutions (Bruce and Migot-Adholla, 1994; Bruce, 2000).

Tenure security plays a primary role in the functioning of land tenure. Essentially, tenure security is a 'feeling' regarding how secure one is with regard to continued access and use of one's land. Such security influences decision-making involving investment in and use of lands. Thus, such decisions, along with transactions (buying, selling, renting, loaning, inheriting), are made based on how secure one's rights to land are. Essentially, the logic is that secure tenure leads to making investments (involving time, effort and money) that are viable over the long term. Thus, there is a connection between tenure security and the rational use and management of resources, and development and improvement in livelihoods. While seemingly straightforward, a great deal of research has been undertaken on the nature of tenure security and how it functions, particularly in the developing world, due to its obvious connections to agricultural productivity, and using land as collateral for loans (e.g. Migot-Adholla and Bruce, 1994; Place *et al.*, 1994). However, tenure security can be difficult to understand, measure and enhance in a consistent and predictable way. This is because tenure security is not very tangible, and a wide variety of variables, processes and contexts are thought to influence tenure security. Generally, however, secure tenure embodies the idea that claims to land are secure because they are 'defendable' against other claimants.

A fundamental aspect of tenure security is that it is the means to prove rights to claims by way of evidence. All claims to land are part of a construction of an evidence-based 'argument for claim'. Even formal title is only an argument based on evidence that can be, and often is, contested – as are claims based on tribal, ethnic, religious and other identities. The legitimacy of evidence depends not only on the interpretation or translation of reality into evidence, but also on the acceptance by 'others' (a community) that the inferences, interpretations and conclusions are logical and legitimate. The 'community' aspect of tenure security intersects with how evidence is treated. In many cases, the 'community' is a local village that has specific internal evidence for a claim that is recognised as legitimate by that community. Evidence at this scale can include community membership, lineage, ethnicity, testimony, land use history, current occupation and physical signs of occupation. However, such a community can exist within a wider national community in which different evidence is legitimate to formal law. This evidence can value deeds and titles and the procedures to obtain and defend them. The different evidence applicable to different scales can be problematic because a local community can be unable to defend their claim (using community evidence) against 'others' in a national community (using documentary evidence).

Tenure types

The two broad primary land tenure types in Africa are formal (or statutory), and informal or unwritten (also known as customary). Formal tenure is largely seen as variations on private or state-owned land. While informal or customary tenure can also be variously labelled 'traditional' or 'indigenous', there are distinctions. Indigenous tenure has to do with particular aboriginal groups, and may or may not be customary or traditional. Traditional tenure is largely historical and can have little to do with present-day customary tenure. Thus, customary tenure is largely 'unofficial' and comprises the many ways that people have responded to the need for rights in land. Customary tenure can be anything from ad hoc to hybridised with statutory tenure, and can include traditional or indigenous elements, but it is dynamic and responsive to present day concerns.

Socialist tenure, in various forms, have been common in the past in Africa. With socialist tenure approaches, collective agricultural programmes hold that property rights and land use decisions are not made by individuals, but by either a group made up of the land resource users connected to the state, or a state-appointed committee. Socialist tenure can take a number of forms, but generally an individual household is assigned a portion of a farm or piece of land, but such an allocation might or might not be their first choice about where to farm. Individuals may have shares in the production (farm yield) or in the revenue from sales of the produce. Or, individuals may work as wage labourers on a collectively owned farm. This form of tenure was common with villagisation programmes in Africa, where people were brought together from a more dispersed farming arrangement into nucleated settlements and given land to farm. The logic here was usually that populations were more easily taxed, educated, provided with healthcare and controlled. Villagisation also freed up vacated land for other uses by the state, such as occurred in Tanzania and Ethiopia's history.

Another form of land tenure currently favoured by a number of African countries is 'state ownership', whereby the national government declares all land to be property of the state. In some cases, the notion of state ownership came with colonialism, but in others it was a reaction against colonialism at independence. Usually, this arrangement allows large areas to be declared 'vacant and ownerless' national property (when in many cases it is not) to be subsequently distributed, according to the state's priorities. In this approach, there can be considerable confusion over exactly what rights are possessed by individuals, communities and the state. Land nationalisation can be a problem when a country changes from other tenure systems to state ownership. The logic behind such nationalisations can include the state's attempts to throw off an exploitive system, thus preventing what is thought to be a form of 'recolonisation' via land purchases by large numbers of foreigners, or preventing the creation of a large landless class.

The land tenure disconnect: the primary land rights problem in Africa

Customary versus statutory tenure systems

There exists a pervasive disconnection in land tenure systems in many areas of the world, and particularly in Africa. This occurs where both customary forms of tenure exist alongside formal tenure systems (such as private property, socialist or state-owned approaches), with little ability for the two to connect as a single system with a compatible logic and compatible rights. The outcome of this incompatibility results in land disputes, evictions and tenure insecurity, usually for those holding land customarily. The underlying problems include the fact that customary systems are already in place, and have often developed over long periods of time. Formal systems, on the other hand, are imposed as peoples, areas, and countries are incorporated into a global system, and feel the pressure to have a land tenure system that can accommodate foreign investment. Then, with decolonisation, as newly independent governments, they want to have a national land tenure system, which they view as fairer than what occurred during the colonial era, as governments express the desire for power over land allocation decisions. As governments seek to regularise land tenure from many different informal systems belonging to indigenous groups within a country, conflict arises. The imposition of formal tenure systems onto customary systems can be problematic, because people are asked to change a fundamental way they do things. Formal tenure systems must be enforced by the state, and this means state intervention in the activities of communities, tribes and religious groups, and this can be seen by these groups as an attempt at control, with potentially negative repercussions.

As the most fundamental rural property rights problem in Africa, the large social, legal, economic and often cultural disconnect between statutory land tenure and customary tenure systems has impacted development, agriculture, nation-building, governance and even wars for decades – with a primary problem being non-recognition of customary systems by formal laws (e.g. Bruce and Migot-Adholla, 1994; de Soto, 2000; Okoth-Ogendo, 2000; McAuslan, 2003). Most African populations conduct their lives with the idea that 'ownership' of land and trees is based on occupancy, use, lineage and other inborn rights. However, African governments often ignore customary tenure systems and regard such areas as part of the public domain, while at the same time lacking the capacity to enforce such a claim or resolve the problems that such a claim produces (Bruce and Fortmann, 1989; Evers et al., 2005; Moyo and Yeros, 2005). While African states can have laws and policies for the servicing of statutory efforts in land tenure, a range of problems, including enforcement, politics, identity, ethnicity, financing, corruption and the potential for conflict, preclude the effective establishment of the state's approach to land tenure over the entire national area or even most rural land in the majority of African nations (Berry, 1990; Bruce and Migot-Adholla, 1994; Quan, 2000). Instituting functioning legal and financial institutions and increasing tenure security for small-scale producers is what international development has been attempting for decades, and the lesson is that the challenges to achieving these are immense (e.g. Peluso, 1992; Creedy and Wurzbacjer, 2001; Saunders *et al.*, 2002).

While there are sound arguments for attempting to bring small-scale African landholders into a formal, uniform, national, enforceable, property rights legal code (de Soto, 2000; Saunders *et al.*, 2002), these are largely unconnected to the realities on the ground. Formalising land rights on the continent has been a long-term effort. But formal law in Africa can have very little to do with what most people are actually doing in their day-to-day lives (e.g. de Soto, 2000). And there can be little opportunity or willingness on the part of the state to formalise customs and norms that reflect ongoing informal rights and obligations about land – such as with Liberia, Sierra Leone and Sudan (Unruh, 2006). Given the pervasiveness of customary law, and that attempting to replace customary law with formal law has already been tried often with problematic outcomes, and does not legislate human behaviour regarding land rights into or out of existence (Griffiths, 1986; Bruce and Migot-Adholla, 1994), the problem is more complex than passing legislation, declaring land to be property of the state, seeking community participation, or providing people with title and assuming that individualised tenure and land markets will follow. This is especially true in a number of African countries where the state will be perceived as weak, having separate interests from those of most of the population, and of questionable legitimacy.

Customary sets of rights and obligations in land tenure that have been created and maintained to facilitate property, land, security, and territorial needs and aspirations will predate, and in many cases be significantly stronger than, laws connected to a state. This is particularly the case as mechanisms for disseminating and fairly enforcing statutory laws (especially with agrarian, semi-literate populations) are also weak, biased or non-existent. Previous analysis of this disconnect has examined its pervasiveness, history, and many problems ranging from agricultural development and investment to nation-building (e.g. Cleaver, 2003; Delville, 2003; McAuslan, 2003).

Tenure paradigms

The failure of many past attempts at replacing customary tenure with formal tenure has led many analysts to conclude that there needs to be movement away from the 'replacement paradigm', whereby customary tenure is replaced with statutory tenure, towards an 'adaptation paradigm'. Such an adaptation approach involves not only the recognition of customary tenure by formal law, but also the evolutionary transformation of customary law as well, in reaction to exposure with formal law (Bruce and Migot-Adholla, 1994).

While the adaptation paradigm highlights the evolutionary nature of change in both customary and formal tenure systems as these adapt to each other, this approach is different from the 'evolutionary theory of land rights'. This evolutionary theory holds that population increase results in land scarcity, change in land values, increased uncertainty and conflict, and as a result a population demands and the state delivers more secure property rights via title. By using population increase and land scarcity to be its primary drivers, the theory assumes that the evolution of customary property rights occurs in isolation from interaction with formal tenure systems, which is the central theme of the 'adaptation paradigm' (Bruce and Migot-Adholla, 1994).

More fundamental theories regarding the nature of property are also relevant to these paradigms. The philosopher John Locke is closely associated with the 'labour theory of property', whereby because a person owns him/herself and their own labour, when that labour is exerted upon natural resources, such as land, there is then justification for claiming such land as private property. And in fact, such property ownership could be considered a 'natural right'. Locke further asserted that such a claim could only come about if there remained enough land left in common for other people to enjoy (Tuckness, 2016). However, there are significant critiques to

Locke's theory of property acquisition, notably that it favours a certain type of labour and land use – that associated with agriculture – over other uses and forms of labour, such as nomadism and other extensive forms of land use and claim. Locke's theory also assumes the presence of a frontier beyond which resides unlimited land for claiming and using (Quiggin, 2016).

Key themes in African land tenure

Tree tenure

The pervasive and powerful notion in Africa that tree planting signifies land claim has thwarted many attempts at development projects (MacDicken, 1990). The literature regarding the strong and pervasive tenure role of trees is significantly large (Otsuka *et al.*, 2001; see also Fortmann and Riddell, 1985 for an annotated bibliography on the topic with 414 entries). Purposefully planted trees are notable for their pervasive role as forms of claim within customary tenure systems. That tree planting serves as land claim is underscored by strong customary restrictions on tree planting for certain groups (such as women, tenants, migrants and other 'outsiders'). In African smallholder tenure systems, trees are planted to delimit clear boundaries and to establish or increase security of tenure. In Africa, purposefully planted trees 'play an integral role in defining local cultures and institutions' (Saunders *et al.*, 2002: 1765), and are indicators of broad societal rights (Saunders *et al.*, 2002). Thus, from a customary perspective, tree planting can be used by 'outsiders' to claim land, effectively transferring that land to the outsider's control. This is an important consideration when a development project, forestry department, NGO, private enterprise, etc. seeks to access land to plant trees, even if trees are given to local communities (Riddell, 1987; Unruh, 1995). An additional issue is that useful trees and forest resources in Africa are commonly viewed as complexes of rights in which different parts of useful trees, at different periods of time, differ in terms of who owns, loans, borrows, inherits, uses or disposes of them (Fortmann, 1987; Saunders *et al.*, 2002).

Tenurial pluralism and international development

The need for a uniform set of statutory laws regarding land tenure is commonly thought to be a prerequisite regarding the prospects for development in Africa (e.g. Smith, 2002; Smith and Sherr, 2002; Antle and Diagana, 2003). In Africa, the locations where formal title and land markets prevail are those of most interest to the state and the commercial sector – densely populated areas, locations of intensive agriculture, fertile lands, areas containing high-value resources, and lands in proximity to infrastructure. But these are not areas where many development projects are often located – unless they will produce a higher income than current or prospective uses in the valuable areas. Instead, projects will be put in areas that are marginal, less valuable, degraded, apparently abandoned, inhabited by the poor, less lucrative for intensive agriculture, with political problems, or are otherwise difficult or neglected (Unruh, 1995). Such areas have much less state influence – even though the state can officially 'own' this land. In these areas, there is often considerable variation in perceptions about what rights the individual, household, communities and the state have, resulting in a situation of pronounced temporal and spatial 'legal pluralism' (e.g. Merry, 1988; von Benda-Beckmann, 1995; Lund, 1998). Such pluralism is different than the simpler customary/statutory divide, in that the interaction over time between many forms of customary tenure (themselves often incompatible, e.g. due to migration, ethnic differences, etc.), as well as changing statutory law, produces a very wide variety of understandings about what law applies to whom, when, and how (e.g. Peters, 1994; Shipton, 1994).

The diversity of customary land tenure systems in any one African country has frequently frustrated African states' attempts to derive and implement a land law that applies to everyone (Guadagni, 2002; Maganga, 2003; McAuslan, 2003). This same diversity also means that formal law will, despite the recommendations of those desiring a straightforward application of development projects, not be able to embrace, and thus make legal, even a subset of this variation in ways that are meaningful to the different customary structures, and still be operable as a formal, widely applied and uniform system applicable to everyone equally (Unruh, 2002, 2006). Globally, the most common examples of legal pluralism regarding land occur in Africa, where, due to the existence of both the wide variety of customary tenure, as well as formal tenure systems, accepting legal pluralism in land is an approach to practical governance (Bruce, 2000). However, the 'improved governance' recommended in approaches to international development as a route to effective implementation of development efforts in Africa assumes that the result will be a single land law for an entire national population (e.g. Niles *et al.*, 2002; Smith, 2002). Laudable as this may be in theory (e.g. de Soto, 2000), and workable as it is in the developed North, in Africa the reality is that the engagement of legal pluralism in land tenure as good governance recognises that: (1) significantly different tenurial 'legalities' are a reality, and that these are not easily reconciled; (2) attempts at forcing one set of tenure rules (e.g. formalised, or one group's customary laws, or a harmonisation of different groups' laws) onto such a pluralistic situation has a record of failure – Sudan, Kenya – and can provoke more problems than it seeks to resolve; and (3) a degree of benign neglect can be an important aspect of this form of practical governance. Recognition of this plurality is important given that a number of potentially serious tenure problems can exist in a latent (but peaceful) state, managed by a purposeful neglect or lack of enforcement of formal law by government. But as the state seeks locations where development efforts can be located, the effort will intersect with the three issues noted above. A focus on and intrusion into latently problematic tenure situations can drive to the fore a large number of serious rights, access and use issues that individual states, NGOs, and private, commercial and international development interests are not in a position to adequately handle.

While the international development literature can recommend that governments guarantee rights to smallholders because it so clearly works in Northern contexts, the codification of property as a way to protect the poor (Saunders *et al.*, 2002) can have the reverse effect. This is because in Africa, the poor often need to be protected from governments, and yet governments will be responsible for lawmaking, guaranteeing rights and titling programmes (Saunders *et al.*, 2002). For many rural small-scale property rights holders in Africa, remaining outside of such a codified system or any government-backed land system is a priority. Such anonymity offers a degree of protection from the predations of government or elements within government (Unruh, 2002). The utility of ambiguity as a form of protection from government is not easily undone in the minds of customary landholders.

Large-scale land acquisitions

The ongoing 'global land grab' has caused significant consternation among African countries and the international development community as government and private interests seek to secure large resource-rich and agriculturally endowed tracts of lands outside their own borders. Large-scale land acquisitions in Africa are fuelled by the belief (supported by various assessments) that the continent has an abundance of available 'reserve' land for food and biofuel production (Borras and Franco, 2010; Deininger and Byerlee, 2011; Lambin and Meyfroidt, 2011).

The statutory/customary disconnect in land rights allows large-scale land acquisitions (LSLAs) to significantly erode local land rights. Fairly common in such acquisitions are examples in which local needs and tenure regimes are ignored, which is unsurprising considering that some types of investors appear to be targeting countries where local land tenure is already weak, and governments have demonstrated few qualms about evicting inhabitants, or not protecting customary rights in the face of encroachment by concessionaires (ABN, 2007; Schoneveld *et al.*, 2010). In Mozambique, investors were able to lease land that had been labelled 'exhausted farmland' from the state for a reforestation initiative; however, residents of the area claimed that the lands had purposefully been left fallow for regeneration within their customary system of shifting agriculture (Overbeek, 2010). The lease issued by the state, with minimal local consultation, effectively eliminated the lands from the community's production cycle, increasing the pressure for them to clear additional forest lands for agriculture. In Ethiopia, the government reportedly leased 5,000 hectares of 'sacred' forests to an India-based concessionaire to transform it into a tea and spice plantation (Afrol News, 2011). These sacred forests, in addition to their symbolic value to local communities, were also used for hunting, minor forest product gathering and beekeeping. The government proceeded with the transaction even after local leaders petitioned Ethiopian policymakers and politicians to support their cause.

While African countries with weak land tenure protections can be targeted for attempts at acquisitions, within such countries investors can target the most productive lands – especially those with access to water sources – and those with the best access to markets (McKenna, 2009; Overbeek, 2010). The acquisition of a country's most fertile lands (either currently or previously in forest) is frequently noted as a driver of social and political conflict (Monachon and Gonda, 2011; Zander and Dürr, 2011). The concentration of highly valued lands in the hands of a few individuals/institutions has been a driver of conflict in some African societies. Such concentration also increases pressure on other available lands, resulting in intensified land use practices and/or fuels migration to urban areas due to the absence of sufficient land to maintain local needs (ABN, 2007; Behrman *et al.*, 2011).

One of the more pervasive impacts of LSLAs is the loss of land, often combined with a loss or weakening of local livelihoods. Land speculation or land-grabbing, along with illicit activities associated with the acquisition of land titles, is common and frequently coupled with the reclassification of local community members as 'squatters' on their own land (due to failure to possess title deeds, despite having customary tenure).

While national governments can assert that all land belongs to the state by way of statutory law, or that the entire national population abides by statutory law with regard to land rights, in reality customary groups have long ignored, resisted and confronted such assertions – Ethiopia, Zambia, Namibia – to the degree that in most cases, the actual on-the-ground penetration of statutory law begins to become patchy a short distance from national capitals and urban centres.

Investors seeking to acquire lands likely do not have a firm grasp of local land dynamics, and can take such assertions of the dominance of statutory law to mean that all members of society understand and abide by such law, including land laws, titles, registries, boundaries administered by the state, and court and ministerial decisions. When conflicts arise, the investor often turns to the state for enforcement of the law, further aggravating the conflict dynamic. This is the case in Ethiopia with large dam projects, and in Tanzania with railway projects. In this regard, the problem becomes both the amount of land acquired by an investment project and the way it was acquired.

In Madagascar, the proposed lease of nearly half the country's arable land to foreign investors led to a breakdown in the enforcement of forestry protections, resulting in rampant poaching, and illegal logging of the country's fragile forests, eventually causing the collapse of the government (Williams, 2009). The subsequent government immediately cancelled the concessions. Apart from this example, the spectre of armed conflict over land rights hangs over many countries being targeted for LSLAs. Both Sudan and Mozambique have been sites of armed conflict and both are some of the top targets for LSLAs by foreign and domestic investors (Deininger and Byerlee, 2011). In Ethiopia, over 8.8 million acres in the Gambella region have been leased to state-owned and foreign firms since 2008, with local communities being driven from their land by state security forces, and the communities resisting violently.

Conclusion

Land tenure in Africa is a diverse array of arrangements that embody the influences of environment, land use, ethnicity, lineage, history, trade, power, politics, mythologies and ideologies. While the different arrangements are not always compatible, they are adaptable to varying degrees in response to changing contexts and interactions with other forms of tenure. It is this adaptability, together with past attempts to replace some forms of tenure with others through law, that has led many prominent thinkers (such as John Bruce, Shem Migot-Adholla and Joan Atherton) to embrace the 'adaptation paradigm' of land tenure change on the continent. This paradigm holds that the different forms of tenure, especially the broad statutory/customary constructs, will co-evolve progressively over time to ultimately include all or most land users in a broad, generally understood tenure system accessible to most or all Africans. While this co-evolution will likely be smooth and relatively trouble-free in some locations, in others it will not. Because the primary form of interaction (hence co-evolution) between the various tenure systems is thought to be through adjudication, conflict and disputes will play a primary role. If forms of adjudication are able to be equitable and accessible to the adherents of different tenure systems, then this will contribute significantly to their peaceful co-evolution.

Land tenure definitions

Customary tenure – The regulation of land rights arising through unwritten, customary or informal practice rather than through written codified law.

Formal tenure – Also known as statutory, codified or state law, forms of formal tenure are created by state law, and are usually based on documentary evidence for claims.

Socialist tenure – Socialist approaches to land rights hold that all land belongs to the state, or to various forms of agricultural cooperatives, as opposed to privately or customarily held land.

Tenure security – The situation in which landholders believe their ongoing occupancy and use of land resources to be guaranteed by formal or informal rights and rules.

Common property – A commons or commonly used land resource from which a specific community can exclude non-members.

Communal land tenure – Tenure forms comprising substantial community control over land resources.

Indigenous tenure – A tenure system attached to specific indigenous groups.

Legal pluralism – The presence of multiple legal systems within a single population, country or other geographic region.

Note

1 Portions of this chapter have previously appeared in Unruh (2003, 2005, 2010, 2013) and Corriveau-Bourque *et al.* (2013).

References

ABN (African Biodiversity Network) (2007) *Agrofuels in Africa: The Impacts on Land, Food and Forest*, Thika, Kenya: African Biodiversity Network.

Afrol News (2011) *Ethiopian 'Sacred Forests' Sold to Indian Tea Producer*, 18 February, available at: http://afrol.com/articles/37365 (accessed 5 January 2018).

Antle, J.M. and Diagana, B. (2003) 'Creating incentives for the adoption of sustainable agricultural practices in developing countries: the role of soil carbon sequestration', *American Journal of Agricultural Economics*, 85: 1178–1184.

Behrman, J., Meinzen-Dick, R. and Quisumbing, A. (2011) *The Gender Implications of Large-Scale Land Deals*, IFPRI Discussion Paper 01056, January.

Berry, S. (1990) *Land Tenure and Agricultural Performance in Africa: Report on a Conference*, paper presented at the Conference on Rural Land Tenure, Credit, Agricultural Investment, and Farm Productivity, 4–8 June, Nairobi.

Borras, S. and Franco, J.C. (2010) 'From threat to opportunity? Problems with the idea of a "code of conduct" for land-grabbing', *Yale Human Rights and Development Law Journal*, 13: 507–523.

Bruce, J.W. (2000) 'African tenure models at the turn of the century: individual property models and common property models', *Land Reform*, 1: 17–27.

Bruce, J. and Fortmann, L. (1989) *Agroforestry: Tenure and Incentives*, Land Tenure Center Paper No. 135, Madison, WI: Land Tenure Center.

Bruce, J. and Migot-Adholla, S. (eds) (1994) *Searching for Land Tenure Security in Africa*, Dubuque, IA: Kendall/Hunt Publishing.

Cleaver, F. (2003) 'Reinventing institutions: bricolage and the social embeddedness of natural resource management', in T.A. Benjaminsen and C. Lund (eds), *Securing Land Rights in Africa*, London: CASS Publishing, pp. 11–30.

Corriveau-Bourque, A., Nelson, I., Pritchard, M., Stanfield, D. and Unruh, J. (2013) 'Impacts of forest related large-scale land acquisitions in the Indian Ocean world', *Journal of the Indian Ocean Region*, 9: 208–226.

Creedy, J. and Wurzbacjer, A.D. (2001) 'The economic value of a forested catchment with timber, water and carbon sequestration benefits', *Ecological Economics*, 38: 71–85.

de Soto, H. (2000) *The Mystery of Capital: Why Capitalism Triumphs in the West and Fails Everywhere Else*, New York: Basic Books.

Delville, P.L. (2003) 'When farmers use "pieces of paper" to record their land transactions in francophone rural Africa: insights into the dynamics of institutional innovation', in T.A. Benjaminsen and C. Lund (eds), *Securing Land Rights in Africa*, London: CASS Publishing, pp. 89–108.

Deininger, K. and Byerlee, D. (2011) *Rising Interest in Farmland: Can It Yield Equitable and Sustainable Results?* Washington, DC: World Bank.

Evers, S., Spierenburg, M. and Wels, H. (2005) *Competing Jurisdictions: Settling Land Claims in Africa*, Leiden: Brill.

Fortmann, L. (1987) 'Tree tenure: an analytical framework for agroforestry projects', in J. Raintree (ed.), *Land Trees and Tenure*, Nairobi and Madison, CT: ICRAF and Land Tenure Center.

Fortmann, L. and Riddell, J. (eds) (1985) *Trees and Tenure: An Annotated Bibliography for Agroforesters and Others*, Nairobi and Madison, CT: ICRAF and Land Tenure Center.

Griffiths, J. (1986) 'What is legal pluralism?', *Journal of Legal Pluralism*, 24: 1–52.

Guadagni, M. (2002) 'Trends in customary land property', in M. Jordan and A. Gambaro (eds), *Land Law in Comparative Perspective*, New York: Kluwer Law International, pp. 7–10.

Lambin, E.F. and Meyfroidt, E. (2011) 'Global land use change, economic globalization, and the looming land scarcity', *Proceedings of the National Academy of Sciences*, 108: 3465–3472.

Lund, C. (1998) 'Struggles for land and political power: on the politicization of land tenure and disputes in Niger', *Journal of Legal Pluralism*, 40: 1–22.

MacDicken, K.G. (1990) 'Agroforestry management in the humid tropics', in K. MacDicken and N. Vergara (eds), *Agroforestry: Classification and Management*, New York: John Wiley & Sons, pp. 1–25.

Maganga, F. (2003) 'The interplay between formal and informal systems of managing resource conflicts: some evidence from south-western Tanzania', in T.A. Benjaminsen and C. Lund (eds), *Securing Land Rights in Africa*, London: Frank Cass Publishing, pp. 51–70.

McAuslan, P. (2003) *Bringing the Law Back In: Essays in Land Law and Development*, Burlington, VT: Ashgate.

McKenna, P. (2009) 'All washed up for Jatropha? The draught-resistant "dream" biofuel is also a water hog', *MIT Technology Review*, 9 June, available at: www.technologyreview.com/search/?s=All%20washed%20up%20for%20Jatropha (accessed 5 January 2018).

Merry, S.E. (1988) 'Legal pluralism', *Law and Society Review*, 22: 869–896.

Migot-Adholla, S. and Bruce, J. (1994) 'Introduction: are indigenous African tenure systems insecure?', in J. Bruce and S. Might-Adholla (eds), *Searching for Land Tenure Security in Africa*, Dubuque, IA: Kendall/Hunt Publishing.

Monachon, D. and Gonda, N. (2011) *Liberalization of Ownership Versus Indigenous Territories in the North of Nicaragua: The Case of the Chorotegas*, Washington, DC: International Land Coalition.

Moyo, S. and Yeros, P. (2005) *Reclaiming the Land: The Resurgence of Rural Movements in Africa, Asia, and Latin America*, London: Zed Books.

Niles, J.O., Brown, S., Pretty, J., Ball, A.S. and Fay, J. (2002) 'Potential carbon mitigation and income in developing countries from changes in use and management of agricultural and forest lands', *Philosophical Transactions of the Royal Society of London*, 360: 1621–1639.

Okoth-Ogendo, H.W.O. (2000) 'Legislative approaches to customary tenure and tenure reform in East Africa', in T. Toulmin and J. Quan (eds), *Evolving Land Rights, Policy and Tenure in Africa*, London: Natural Resources Institute, pp. 123–134.

Otsuka, K., Suyanto, S., Sonobe, T. and Tomich, T.P. (2001) 'Evolution of land tenure institutions and development of agroforestry: evidence from customary land areas of Sumatra', *Agricultural Economics*, 25: 85–101.

Overbeek, W. (2010) *The Expansion of Tree Monocultures: Impacts on Local Peasants Communities in the Province of Niassa (Mozambique)*, WRM Series on Tree Plantations No. 14, Montevideo, Uruguay: World Rainforest Movement.

Peluso, N.L. (1992) *Rich Forest Poor People: Resource Control and Resistance in Java*, Berkeley, CA: University of California Press.

Peters, P.E. (1994) *Dividing the Commons: Politics, Policy, and Culture in Botswana*, Charlottesville, VA: University of Virginia Press.

Place, F., Roth, M. and Hazell, P. (1994) 'Land tenure security and agricultural performance in Africa: overview of research methodology', in J. Bruce and S. Migot-Adholla (eds), *Searching for Land Tenure Security in Africa*, Dubuque, IA: Kendall/Hunt Publishing, pp. 15–40.

Quan, J. (2000) 'Land tenure, economic growth and poverty in sub-Saharan Africa', in T. Toulmin and J. Quan (eds), *Evolving Land Rights, Policy and Tenure in Africa*, London: Natural Resources Institute, pp. 31–50.

Quiggin, J. (2016) 'John Locke against freedom', *Jacobin*, available at: www.jacobinmag.com/author/john-quiggin/ (accessed 7 September 2016).

Riddell, J.C. (1987) 'Land tenure and agroforestry: a regional overview', in P. Schebesta (ed.), *Among the Forest Dwarves of Malaya*, London: Oxford University Press, pp. 1–16.

Saunders, L.S., Hanbury-Tenison, R. and Swingland, I.R. (2002) 'Social capital from carbon property: creating equity for indigenous people', *Philosophical Transactions of the Royal Society London*, 360: 1763–1775.

Schoneveld, G.C., German, L., Andrade, R., Chin, M., Caroko, W. and Romero-Hernández, O. (2010) *The Role of National Governance Systems in Biofuel Development: A Comparative Analysis of Lessons Learned*. Bogor, Indonesia: Center for International Forestry Research.

Shipton, P. (1994) 'Land and culture in tropical Africa: soils, symbols, and the metaphysics of the mundane', *Annual Review of Anthropology*, 23: 347–377.

Smith, J. (2002) 'Afforestation and reforestation in the clean development mechanism of the Kyoto Protocol: implications for forests and forest people', *International Journal of Global Environmental Issues*, 2: 322–343.

Smith, J. and Sherr, S.J. (2002) 'Capturing the value of forest carbon for local livelihoods', *World Development*, 31: 2143–2160.

Tuckness, A. (2016) 'Locke's political philosophy', in E. Zalta (ed.), *The Stanford Encyclopedia of Philosophy*, available at: http://plato.stanford.edu/archives/spr2016/entries/locke-political/ (accessed 21 December 2017).

Unruh, J.D. (1995) 'Post-conflict recovery of African agriculture: the role of "critical resource" tenure', *Ambio*, 24: 343–350.

Unruh, J.D. (2002) 'Poverty and property rights in the developing world: not as simple as we would like', *Land Use Policy*, 19: 275–276.

Unruh, J.D. (2003) 'Land tenure and legal pluralism in the peace process', *Peace and Change: A Journal of Peace Research*, 28: 352–376.

Unruh, J.D. (2006) 'Land tenure and the evidence landscape in the developing world', *Annals of the Association of American Geographers*, 96: 754–772.

Unruh, J.D. (2008) 'Carbon sequestration in Africa: the land tenure problem', *Global Environmental Change: Human and Policy Dimensions*, 18: 700–707.

Unruh, J.D. (2010) 'Land tenure', in B. Warf (ed.), *Encyclopedia of Geography*, New York: Sage, available at: http://sk.sagepub.com/reference/geography/n692.xml (accessed 5 January 2018).

Unruh, J.D., Cligget, L. and Hay, R. (2005) 'Migrant land rights reception and "clearing to claim" in sub-Saharan Africa: a deforestation example from southern Zambia', *Natural Resources Forum*, 29: 190–198.

von Benda-Beckmann, F. (1995) 'Anthropological approaches to property law and economics', *European Journal of Law and Economics*, 2: 309–336.

Williams, N. (2009) 'Alarm bells over Africa land deals', *Current Biology*, 19: R1053–R1054.

Zander, M., and Dürr, J. (2011) *Dynamics in Land Tenure, Local Power and the Peasant Economy: The Case of Petén, Guatemala*, paper presented at the International Conference on Global Land Grabbing, 6–8 April, organised by the Land Deals Politics Initiative (LDPI).

27

Deagrarianisation and depeasantisation in Africa

Deborah Fahy Bryceson

Introduction

Deagrarianisation is essentially a multidimensional process of change involving: (i) livelihood reorientation; (ii) occupational work adjustment; (iii) spatial realignment of residential settlement; and (iv) social reidentification, all entailing movement away from agrarian patterns in local and regional economies (Bryceson, 2002a).

Deagrarianisation is ongoing in Africa, resulting in a process of sectoral transformation, evidenced by a declining proportion of the national population and total labour time engaged in agriculture (Bryceson, 1996, 1997a, 1997b). Over recent decades, statistical indicators of agricultural labour time expenditure, volume and value of product output, as well as survey data and qualitative interviews, have increasingly documented this trend (see Appendix). The net outcome of this process is the declining value of agricultural output relative to other sectors of national production.

Deagrarianisation has been coterminous with urbanisation and industrialisation trends for more than two centuries, but Africa experienced a delayed onset relative to other continents. Europe and North America's deagrarianisation largely took place during the early nineteenth to mid-twentieth centuries as the population gradually shifted from rural to urban settlements and from agrarian to industrial employment. Latin America and Japan followed suit during the post-Second World War period from the mid-1940s to the 1970s, South and South East Asia's transition commenced in the 1960s and escalated until the 1980s, whereas China's deagrarianisation has been very rapid and concentrated during the 1990s to the present.

With the exception of South Africa, Africa's deagrarianisation began in the 1950s, only achieving real impetus during the 1980s, and continuing to unfold to the present, with widely varying rates between countries, as well as sporadic stops and starts usually in the form of expanding subsistence agriculture in periods of economic depression. Nonetheless, it is important to note that subsistence agriculture is likely to continue to serve as a vital fallback for many rural dwellers, in view of the uncertain returns from their engagement in alternative non-agricultural economic pursuits, as will be explained below. South Africa's deagrarianisation began alongside mining industrialisation during the early twentieth century, placing it in the forefront of deagrarianisation within the continent, followed by the rest of southern Africa, North Africa, West Africa and lastly East Africa.

The African agrarian context: making then breaking African peasant commodity production

In eighteenth- and nineteenth-century Europe, deagrarianisation was triggered by push and pull forces, the experience of the rural commons enclosures and later urban industrialisation that both Marx (1967 [1867]) and Chayanov (1966) wrote about at length. The African continent was largely unindustrialised during the twentieth century, whereas urbanisation became a salient force for economic and social change from the mid-twentieth century. Deagrarianisation began to appear in the form of youth migration to the cities, but at the same time African governments and foreign donors devoted considerable effort to raising agricultural productivity and the welfare of the rural population in the first couple of decades after national independence. But thereafter, the rural sector experienced fundamental internal change, going well beyond population outflow. This chapter considers the course of deagrarianisation in conjunction with the changing nature of peasantries in Africa, probing how they evolved and why they began unravelling during the 1980s.

Wolf (1966) and Shanin (1976) delineated peasants' characteristics from other rural producers, be they agricultural wage labourers, sharecroppers, subsistence farmers, etc., based on four main criteria: (1) *farming*, which involves the pursuit of an agricultural livelihood, both subsistence *and* commodity production; (2) *family*, which refers to the centrality of family labour, with the household serving as the main unit of production, consumption, reproduction, socialisation, welfare and risk-spreading; (3) *class*, which entails surplus extraction and class differentiation arising from external subordination to state authorities, as well as regional and international markets; and (4) *community*, which relates to the social cohesion provided by collective rural settlement and local conformist attitudes (Bryceson, 2000a).

African peasantries were first formed during European colonialism from the late nineteenth century based on existing rural producing societies who were co-opted into production for the global market (Bryceson, 2000b). Prior to that time, their production was primarily focused on immediate household and community needs, and to a lesser extent on complementary exchange of local specialised products in wider regional markets. The introduction of export cash crops under European colonialism altered production patterns and made incursions on the self-sufficiency of rural producers' cycle of subsistence production and consumption. Tribal agriculturalists took on the mantle of peasant producers by producing products for the global market in increasingly monetised economies and paying taxes to the colonial state. Participating in a cash economy, they could make purchases of imported goods, usually very simple in nature, be it clothing, soap, candles or kerosene lamps for lighting. Peasants' commodity production was pursued alongside traditional subsistence food production. The distinguishing feature of peasant production relative to agriculturalists per se is their combined agricultural production for subsistence and commodity sale.

While colonial agrarian policies varied from colony to colony, the main objective of colonial governments was to generate export agricultural products upon which state revenue collection could be achieved. Peasant commodity production, settler farms and large-scale plantations constituted the main forms of agricultural production. Depeasantisation is, in effect, a subset and a central component of deagrarianisation within Africa, reflective of the widespread expansion of peasant formation during the colonial and early postcolonial period in most African countries.

However, it must be noted that there were three other forms of rural production that colonial states nurtured as an alternative to peasant formation, namely: (1) large-scale, usually foreign-owned, plantations, such as the Belgian Congo's palm oil and rubber plantations and Gambian and Senegalese groundnut plantations; (2) European settler farm populations that displaced

African rural settlements but often involved sharecropping arrangements with local African farmers, notably in Kenya and Southern Rhodesia; and (3) the growth of male labour migration systems connected with mineral production in southern Africa that extracted labour from peasant settlements. Over time, they evolved complicated and sometimes conflictual relations with peasant settlements. Nonetheless, in the nationalist era, peasant populations had expanded over vast tracts of Africa and become a political force whose vote held sway in the achievement of independence in one country after another.

Peasant commodity production expanded in many postcolonial economies during the 1960s and 1970s, the first decades after the establishment of national independent government. World agricultural product prices were generally high. African governments were interested in harnessing tax revenues from peasant production. The expectation was that agricultural surplus extraction from peasant producers could be invested in national industrialisation to build a modern economy. Many countries initiated import substitution industries to conserve foreign exchange expenditure, but there was negligible labour absorption from these embryonic industries. Compared to other parts of the world, African industrial development achievements were unimpressive. Further industrialisation was pre-empted by severe economic recession that engulfed most of the continent during the 1980s after the international oil crisis of 1979.

African governments had to contend with the undermining effect of skyrocketing international oil prices on peasant agriculture. Peasant agricultural sectors in sub-Saharan oil-producing countries (Nigeria, Angola, Equatorial Guinea, Gabon, the Republic of Congo) and in North Africa (Algeria, Libya, Egypt) generally stagnated due to what is known as 'Dutch disease', whereby the rapid rise in foreign exchange inflows from oil sales caused currency appreciation that inflated the costs of the country's other export products, making them less competitive in the world market.[1] Peasant agricultural exports could not compete, catalysing rural outmigration, particularly of youth to urban areas overheated with oil wealth. More commonly, in the African countries that were not net oil exporters, oil prices crippled the peasant sector. The transport of peasants' crops to national ports from their widely dispersed farm production units incurred exceptionally high costs. Their production could not compete with the crop output prices of farmers in the more densely populated rural areas of Asia and Latin America.

Structural adjustment policies largely dismantled the existing state marketing boards and parastatals that had provided single-channel marketing facilities, crop grading, controlled prices and peasants' crop input requirements. The open market trading facilities that eventually replaced them proved to be of variable utility, and tended to ignore farmers situated off the main roads and in remote areas. Increasing crop input costs led to a reallocation of land and labour away from commercial agriculture. African agricultural exports lost market share as importers came to expect below-standard products. The continent's long-famed beverage exports of coffee, cocoa and tea were increasingly displaced by the output of Asian producers,[2] alongside drastic drops in most other cash crops.

The combined effect of the above world market shifts and policy trends invoked an economic chain reaction that had deep social and political reverberations, striking at the core of customary peasant ways of life. What was most salient was the appearance of a proliferation of income earners emerging from rural households. Male heads of households displaced from cash cropping, as well as their wives, and youth who had hitherto not been engaged in income-generating activities, sought alternative income-earning sources. This marked a notable change from the male patriarchal order in which the male head of household largely monopolised cash earning from export crop production during the colonial and early postcolonial period. The male patriarch's dominance in households was an ideologically ingrained feature of African peasant commodity production (Bryceson, 2002a). Agricultural policies, input distribution and extension services had been targeted at male heads of households. As cash crop production eroded, their social status within the household was challenged.

Both women and the youth were finding ways of earning petty cash through trade, services and craft production. The amounts generally were not large. Women's income earning was more poorly remunerated than men's because women were less literate and numerate, and their income earning tended to be restricted to deployment of their homemaking skills. Nonetheless, the scarcity of cash within the household, and youth and women's discretionary spending of their earnings sanctioned or unsanctioned by the male head of household, enhanced their scope for decision-making.

A large number of women resorted to the sale of prepared snacks, beer and other petty retailing. The services they could offer included hair plaiting, tailoring, knitting, soap making and prostitution. In Malawi, beer brewing and food snack sales accounted for 76 per cent of female heads of households' non-agricultural activities (Tellegen, 1997). Negative male attitudes to female earnings were being readily set aside. Many women had husbands very actively encouraging their income-earning activities given household exigencies. This was especially the case for polygamous men who were finding it difficult to support all of their wives and children.

Male youth, be they teenagers or young unmarried adults, were ubiquitous in petty, medium and long-distance trade, facilitated by being unencumbered by family responsibilities. Many shunned agricultural work in preference for trade activities or non-agricultural wage labour. But there were costs in terms of foregone education and skill training for many youth, who dropped out of school because their parents could not afford to pay the school fees and other ancillary costs.

Returning to Shanin's definition of peasants, we can see that in addition to declining amounts of time spent farming, the strength of the farming household as a coherent production unit was eroding. Family ties declined as more individualised rather than family-based labour production took place (Bryceson, 2002b). The growing tendency to engage in non-agricultural income earning had a dissolving effect on long-standing agrarian divisions of labour, as well as economic responsibilities and rights within the household. The patriarchal nature of the peasant household was challenged. Household members' income pooling weakened as people who formerly had not expected to earn an income asserted a moral right to determine how their income was spent. The next section considers how deagrarianisation confounded economic development thinking, before probing what income diversification activities were pursued in the scramble for non-agricultural diversification.

Deagrarianisation and non-agricultural income diversification in the absence of industrialisation

W.A. Lewis's (1955a, 1955b) early work on sectoral transition from agriculture to industry stressed the impact of labour flows from the low-productivity rural agricultural sector towards the higher-productivity urban industrial sector (Bryceson, 2000a). His work was influential in the new congealing field of development economics, which posited that higher productivity led to proliferating goods and services and enhanced material welfare. In this frame of reasoning, agricultural displacement was a positive sign of development. Lewis (1978) deemed that international trade theory was misguided in according comparative advantage in agriculture to agrarian countries. He argued that agricultural productivity in industrial countries was higher than in the so-called 'agricultural countries' where peasant production prevailed. Therein rested one of the central quandaries of economic development: How could export agriculture in developing countries be the path towards development when they had a comparative disadvantage in agricultural production?

Lewis's insight sheds light on what has underpinned African deagrarianisation since the 1980s. As farmers' lack of a comparative advantage in agriculture sank still further, peasant agricultural export commodity production became largely futile. This was compounded by the structural adjustment policies of the World Bank and IMF that enforced African government removal of agricultural input subsidies. Under these circumstances, growing numbers of rural African farmers experienced a deepening sense of self-defeat in agriculture based on export crop production, and instead they started searching for off-farm income-generating activities. A 'scramble in Africa' for non-agricultural cash earnings ensued (Bryceson, 2002a, 2002b).

Liedholm (1973), in the halcyon days of African export crop viability, had pioneered the study of non-farm income diversification. He argued that non-farm activity afforded a virtuous synergy between farm and non-farm activity, with cross-subsidisation and investment between the two serving to raise overall rural household productivity. Haggblade *et al.* (1989) updated the approach in the troubled years of export crop decline, drawing attention to the significance of highly diverse rural household portfolios of agrarian/non-agrarian activities aimed at ensuring basic subsistence, cash income for additional non-food needs, and minimising risk in a context of rising smallholder uncertainty. Thereafter, Bryceson (1996) drew attention to the growing surge of rural African households' non-agricultural activities, followed by Reardon's (1997) estimates that African non-farm income was between 30 and 50 per cent of rural household income. Having identified the income diversification trend away from agriculture, the key question was: How was this affecting rural household welfare? Was impoverishment being averted? Were some achieving virtuous cross-investment and flourishing? What trends were discernable?

The Livelihoods and Diversification Directions Explored by Research (LADDER) research programme, led by Frank Ellis, constructed a methodology for the study of rural livelihood diversification that drew on Chambers and Conway's (1992) sustainable rural livelihoods approach (Ellis, 2000; Ellis and Freeman, 2005). The LADDER comparative study of 1,295 households in 37 villages across four countries (Kenya, Uganda, Tanzania and Malawi) in 2001–2002 involved village household sample surveys that inventorised household agricultural assets (land and cattle), and collected data on rural households' livelihood portfolios that included subsistence and income-earning activities of all members of the household, encompassing remittances, non-farm wages, farm wages and self-employment. In addition, a community-based wealth-ranking exercise was conducted to discern wealth differences among rural households. The findings revealed that peasant commercial agriculture was relatively negligible. Most household agricultural production was destined for own consumption. This pertained to 80 per cent of the staple maize crop. Other auxiliary staple crops were similarly high. One clear trend was that the share of subsistence consumption in total income decreased with wealth. From lowest to highest income quartiles, the percentages were: bottom (41 per cent), low-middle (35 per cent), high-middle (25 per cent) and top (16 per cent), with the overall average being 21 per cent. Finally, evidence for Tanzania (see Table 27.1), one of the case study countries, confirms the importance of non-agricultural livelihood sources, which had tipped over the 50 per cent mark.

Table 27.1 Tanzanian rural household sources of income (percentage of total)

Maize sale	*Other own farm sales*	*Agriculture subtotal*	*Wages*	*Non-farm*	*Transfers (including remittances)*	*Non-agriculture subtotal*
12.4	37.3	49.7	10.5	36.1	10.5	50.3

Source: Ellis and Freeman (2005), LADDER survey data.

Unwieldy directions: income diversification's various dimensions, forms and pathways

In the economic context of structural adjustment and economic liberalisation of the 1980s and 1990s, most rural households' first reaction was to cut back insofar as possible on expenditure. This included not only avoiding use of agricultural crop inputs, but wide-ranging things such as taking children out of school and reduced gift-giving, which traditionally functioned to bring social cohesion to the community, but there was a limit to cutting back on consumption in households, which were close to basic subsistence. The only option was to search for alternative sources of livelihood.

The rest of this chapter is based on findings from a multinational team of researchers participating in the Deagrarianization and Rural Employment (DARE) research programme.[3] The study, which I coordinated, consisted of a methodological mix of village sample surveys, household work histories and in-depth qualitative interviews with key informants in Ethiopia, Malawi, Nigeria, South Africa, Tanzania and Zimbabwe. The aim was to gain an overview of the economic dynamics, labour patterns, spatial mobility, social identity, gender and generational change embedded in deagrarianisation.

Our findings revealed that non-agricultural activities surged between 1985 and 2000, prompted by the implementation of the SAPs in the 1980s and the economic liberalisation policies of the 1990s (Bryceson, 1999, 2002a, 2002b; Bryceson and Bank, 2001). The proliferating range of sources of income demonstrated a movement from agriculture towards non-agricultural work, from household-based to more individualised labour, and from unpaid towards paid work. However, none of these observed trends were necessarily unilinear, and in specific contexts there were exceptions, deviations, dead ends and sometimes U-turns, as would be expected in an experimental learning-while-doing process.

Nonetheless, there were discernable trends arising from the day-to-day economic, social and political tensions rural households faced. The resolution of these tensions created differing degrees of change and various directions, rather than generating strictly dichotomous either/or outcomes. In their attempts to diversify into non-agricultural activities, involving multiple household members, households tended to be engaged in more than one, often several, different non-agricultural activities simultaneously or at different times throughout the year. Often the activities were highly opportunistic in nature, taking the form of quick responses to market demand and supply. Because the returns from non-agricultural experimentation were uncertain, rural households valued the continuation of their farming activities for home consumption. Own subsistence food production necessarily remained a mainstay of most rural households.

People's construction of different non-agricultural portfolios naturally depended to a large extent on their local resource access, existing skills, time availability with respect to slack time in the agricultural calendar, and the degree to which economic liberalisation policies of the 1990s opened up new avenues for income generation. The range of non-agricultural activities varied a great deal from site to site. The activities can be most easily categorised on the basis of being of high or low earning capacity, formal or informal, or modern versus traditional in nature (Bryceson, 2002b).

Low- as opposed to high-income activities signified differing levels of potential earnings, as well as requisite start-up capital. Input investments and output earnings tended to have corresponding magnitudes. Relatively few people had large amounts of capital to invest, but those that did were likely to invest in transport, products and services with high returns, as opposed to the majority with low starting capital who entered petty trade and beer brewing en masse and faced stiff competition. It was not always the case that the most educated monopolised high-earning niches. In Nigeria, Igbo traders, with relatively low levels of education, started their trading careers early, learning on the job, and through determination many succeeded (Chukwuezi, 1999). On the other hand, basic literacy and numeracy were vital in the pursuit of non-agricultural activities, and essentially functioned as necessary but not sufficient conditions for success.

Turning to practitioners of traditional and modern non-agricultural activities, 'traditional activities' deploy age-old production techniques and local materials to provide services and products that tend to be sold locally or in craft markets. For example, traditional midwives and medical practitioners, who in effect had worked as unpaid community service providers and had been accorded an elevated social status for doing so, were increasingly commodifying their work and charging fees (Mwamfupe, 1998). The DARE survey findings indicated that traditional non-agricultural activities were not as numerous as modern activities.

Craft traditions were not readily commercialised. Ethiopian villages, where craft traditions were well established in pottery, leather tanning, metalworking and weaving, faced an attitudinal barrier. The traditional products were not considered to be as durable or practical as modern substitutes. For example, lightweight plastic containers for water carrying were preferred to the traditional heavy pottery jars (Mulat Demeke, 1997). Another barrier that often cropped up was the existing shortage of local materials such as reeds, clay and metals to make traditional products. Craft commercialisation often exacerbated the environmental stress on such local natural resources.

By contrast, many modern services took off very readily, water and firewood carrying services by male cyclists being a striking example. Women were alleviated the drudgery of heavy head loading over long distances, whereas young men gained a reliable daily income from their investment in a bicycle. The spread of imported bicycles in Tanzania under economic liberalisation facilitated traders' haulage of goods over long distances (Madulu, 1998).

The terms 'formal' and 'informal' sector classifications were first coined during the 1970s, but their meaning and significance has been continuously debated. Generally, a distinction was made between waged, legally taxed employment as opposed to non-taxed self-employment. Or more simply, high-paid waged work was juxtaposed to low-paid self-employment. However, under structural adjustment, the distinction between the two blurred, as urban-based formal sector jobs became unremunerative, and urban employees as well as rural peasants began entering 'informal' activities to diversify their income sources.

Deagrarianisation's awkward balancing act: agrarian subsistence fallback versus market experimentation

African peasant producers adopted a dual strategy involving experimentation with new non-agricultural income-earning activities, while maintaining their subsistence food production as a fallback (Bryceson, 2000c). The two were complementary, decreasing the likelihood of total production failure, but they entailed different skills and ways of thinking. One of the paradoxes of rural dwellers' non-agricultural activity was that it was partially triggered by capital constraints in commercial agriculture, while success in non-agricultural activities was largely determined by access to capital. Income diversifiers contended with many difficulties. They had to source sufficient starting capital while avoiding running down their working capital, using it for business operations rather than necessary consumption.

In most cases, rural households' priority was simply to gain a livelihood to secure their basic needs. These inherent dilemmas made for constant shifting between activities, depending on the imperatives of basic need provisioning and profit rates. In Nigeria's Middle Belt, Yunusa (1999) documented that almost half of his respondents abandoned one or more non-farm activity due to the small profit realised or not having enough capital to continue. However, most of the abandoned activities were not the respondents' main non-agricultural activity, which may suggest business consolidation was underway.

In Malawi, successful diversifying households were involved in more highly rewarding activities, as well as having a greater number of activities over a longer period of time. Those pursuing year-round non-agricultural activities were doing far better than those who sporadically engaged in non-agricultural activities (Tellegen, 1997).

Tellingly, no matter how non-agriculturally engaged respondents were, household dependence on own-farm agricultural produce and the espousal of agrarian moral values was the common denominator across all respondents in case study countries. Often people reported their occupation as 'farmer', whereas their reported economic activity suggested otherwise. Few verbally acknowledged that the structure of the local economy and incomes were drifting towards increasing levels of non-agriculture activity. In north-west Tanzania, Madulu (1998) argues that social status was traditionally derived from having large harvests and large families. Eighty per cent of respondents said that they could not survive without farming. But in a Zimbabwean communal area, some admitted that farming was 'a cultural way of life' and 'a domestic task more than a professional occupation' (Berkvens, 1997).

Conclusion

Deagrarianisation in Africa gripped the continent during the 1980s, propelled by the onset of extremely adverse terms of trade in the world market for smallholder peasant exports. The 'scramble' in Africa for non-agriculture income-generating activities was an experimental process that entailed a great deal of economic stress as well as creative ingenuity on the part of hundreds of thousands of African rural households. Driven by material insecurity, they were engaged in assembling non-agriculture work portfolios alongside subsistence agriculture, amounting to intricate economic balancing and psychological interdependence between agriculture and non-agricultural activities. The experimentation has yielded economic success for some as well as deepening impoverishment for others.

Beyond the structural adjustment and economic liberalisation policies that provided the original impetus for the quickening of the deagrarianisation process, it has gained its own internal momentum as people became increasingly involved in non-agricultural work. There is a generational effect whereby younger people are generally far less inclined to see family farming as their long-term work trajectory. African demographic, economic and social data indicate that much along the lines of other continents, the percentage of the African rural population is decreasing, peasantries are fading, and the commercial family farming sectors of national economies are shrinking in the quest for viable livelihoods.

Notes

1 Dutch disease in Nigeria is described in Meagher (2000).

2 The most salient case is the slump of Tanzanian coffee exports while the output of Vietnamese peasants who were new to coffee production surged.

3 The DARE research network was sponsored by the African Studies Centre at Leiden, the Netherlands, funded by the Netherlands Mininstry of Foreign Affairs. I coordinated the research programme, whose membership included L. Bank, B. Chukwuezi, M.A. Iliya, G. Jambiya, N.F. Madulu, C. Manona, P. McAllister, K. Meagher, M. Demeke, C. Mung'ong'o, R. Mustapha, D. Mwamfupe, N. Tellegen and M. Yunusa.

References

Berkvens, R.J.A. (1997) *Backing Two Horses: Interaction of Agricultural and Non-Agricultural Household Activities in a Zimbabwean Communal Area*, Working Paper No. 24, Leiden: African Studies Centre.

Bryceson, D.F. (1996) 'De-agrarianization and rural employment in sub-Saharan Africa: a sectoral perspective', *World Development*, 24(1): 97–111.

Bryceson, D.F. (1997a) 'De-agrarianisation in sub-Saharan Africa: acknowledging the inevitable', in D.F. Bryceson and V. Jamal (eds), *Farewell to Farms: De-Agrarianisation and Employment in Africa*, Aldershot: Ashgate, pp. 3–20.

Bryceson, D.F. (1997b) 'Disappearing peasantries? Rural labour redundancy in the neo-liberal era and beyond', in D.F. Bryceson and V. Jamal (eds), *Farewell to Farms: De-Agrarianisation and Employment in Africa*, Aldershot: Ashgate, pp. 299–326.

Bryceson, D.F. (1999) 'African rural labour, income diversification and livelihood approaches: a long-term development perspective', *Review of African Political Economy*, 80: 171–189.

Bryceson, D.F. (2000a) 'Peasant theories and smallholder policies: past and present', in D.F. Bryceson, C. Kay and J. Mooij (eds), *Disappearing Peasantries? Rural Labour in Africa, Asia and Latin America*, London: Intermediate Technology Publications, pp. 1–36.

Bryceson, D.F. (2000b) 'African peasants' centrality and marginality: rural labour transformations', in D.F. Bryceson, C. Kay and J. Mooij (eds), *Disappearing Peasantries? Rural Labour in Africa, Asia and Latin America*, London: Intermediate Technology Publications, pp. 37–63.

Bryceson, D.F. (2000c) 'Disappearing peasantries? Rural labour redundancy in the neo-liberal era and beyond', in D.F. Bryceson, C. Kay and J. Mooij (eds), *Disappearing Peasantries? Rural Labour in Africa, Asia and Latin America*, London: Intermediate Technology Publications, pp. 299–326.

Bryceson, D.F. (2002a) 'The scramble in Africa: reorienting rural livelihoods', *World Development*, 30(5): 725–739.

Bryceson, D.F. (2002b) 'Multiplex livelihoods in rural Africa: recasting the terms and conditions of gainful employment', *Journal of Modern African Studies*, 40(1): 1–28.

Bryceson, D.F. and Bank, L. (2001) 'End of an era: the development policy parallax', in L. Bank and D.F. Bryceson (eds), *Livelihoods, Linkages and Policy Paradoxes*, special issue of the *Journal of Contemporary African Studies*, 19(1): 1–19.

Chambers, R. and Conway, R. (1992) *Sustainable Rural Livelihoods: Practical Concepts for the 21st Century*, IDS Discussion Paper No. 296, Brighton: IDS.

Chayanov, A.V. (1966) *The Theory of the Peasant Economy*, Glencoe, IL: Irwin.

Chukwuezi, B. (1999) *Deagrarianization and Rural Employment in Rural Igboland, South-East Nigeria*, Working Paper No. 37, Kano: Centre for Documentation and Research and Leiden: African Studies Centre.

Ellis, F. (2000) *Rural Livelihoods and Diversity in Developing Countries*, Oxford: Oxford University Press.

Ellis, F. and Freeman, A. (2005) 'Comparative evidence from four African countries', in F. Ellis and A. Freeman (eds), *Rural Livelihoods and Poverty Reduction Policies*, Routledge: London, pp. 1–30.

Haggblade, S., Hazell, P. and Brown, J. (1989) 'Farm–nonfarm linkages in rural sub-Saharan Africa', *World Development*, 17(8): 1173–1201.

Lewis, W.A. (1955a) 'Economic development with unlimited supplies of labour', *Manchester School of Economic and Social Studies Bulletin*, 22: 139–192.

Lewis, W.A. (1955b) *The Theory of Economic Growth*, London: Allen & Unwin.

Lewis, W.A. (1978) *The Evolution of the International Economic Order*, Princeton, NJ: Princeton University Press.

Liedholm, C. (1973) *Research on Employment in the Non-Farm Sector in Africa*, East Lansing, MI: Michigan State University.

Madulu, N.F. (1998) *Changing Lifestyles in Farming Societies of Sukumaland: Kwimba District, Tanzania*, Working Paper, Dar es Salaam: Institute of Resource Assessment and Leiden: African Studies Centre.

Marx, K. (1967 [1867]) *Capital Vol. 1*, New York: International Publishers.

Meagher, K. (2000) 'Veiled conflicts: peasant differentiation, gender and structural adjustment in Nigerian Hausaland', in D.F. Bryceson, C. Kay and J. Mooij (eds), *Disappearing Peasantries? Rural Labour in Africa, Asia and Latin America*, London: Intermediate Technology Publications, pp. 81–98.

Mulat Demeke (1997) *Rural Non-Farm Activities in Impoverished Agricultural Communities: The Case of North Shoa, Ethiopia*, Working Paper No. 25, Addis Ababa: Organization for Social Science Research in Eastern Africa and Leiden: African Studies Centre.

Mwamfupe, D. (1998) *Changing Village Land, Labour and Livelihoods: Rungwe and Kyela Districts, Tanzania*, Working Paper No. 42, Dar es Salaam: Institute of Resources Assessment and Leiden: African Studies Centre.
Reardon, T. (1997) 'Using evidence of household income diversification to inform study of the rural nonfarm labor market in Africa', *World Development*, 25(5): 735–747.
Shanin, T. (1976) *Peasants and Peasant Societies*, Harmondsworth: Penguin.
Tellegen, N. (1997) *Rural Enterprises in Malawi: Necessity or Opportunity?* Aldershot: Ashgate.
Wolf, E.R. (1966) *Peasants*, Englewood Cliffs, NJ: Prentice Hall.
Yunusa, M.-B. (1999) *Not Farms Alone: A Study of Rural Livelihoods in the Middle Belt of Nigeria*, Working Paper No. 38, Kano: Centre for Documentation and Research and Leiden: African Studies Centre.

Appendix

Basic indicators of African deagrarianisation, 1960–2010

Year	*% rural population*	*Agriculture as % of GDP*	*Agricultural exports as % of exports total*
1960	85.4		
1965	83.7	21.7	
1970	82.0	19.6	
1975	79.9	19.7	8.3
1980	77.9	18.1	
1985	75.6	20.5	
1990	73.0	21.0	
1995	71.0	19.7	
2000	69.2	17.1	4.9
2005	67.1	17.1	
2010	64.8	15.0	3.4

Source: World Development Indicators, available at: http://databank.worldbank.org (accessed 20 November 2015).

28

Rural development

Simon Milligan and Sam Kareithi

Introduction

While the African economy continues to post impressive growth figures, there is another side to the continent's renaissance: that of unacceptably high incidences of poverty and extreme poverty. Sub-Saharan Africa has the highest incidence of rural poverty anywhere in the world (IFAD, 2010a). In 1988, one in eight of the world's rural people who lived on less than US$1.25 per day resided in sub-Saharan Africa (IFAD, 2010a). By 2008, it was one in three people (IFAD, 2010a). Intriguingly, other regions have fared much better. While the percentage of rural people living in extreme poverty (i.e. on less than US$1.25 per day) in sub-Saharan Africa grew from 51.7 per cent in 1988 to 61.6 per cent in 2008, the incidence of extreme poverty in South East Asia (e.g. Indonesia, Malaysia, the Philippines) fell from more than 50 per cent to about 25 per cent over the same period (IFAD, 2010a).

The coexistence of strong economic growth, coupled with stubborn poverty, vulnerability and inequality, in rural Africa continues to attract the attention of policymakers, researchers, civil society groups and aid organisations (e.g. see Dercon, 2006; IFAD, 2006; Ahmed *et al.*, 2007; IFAD, 2010a; Losch *et al.*, 2012). Consensus about necessary policy responses remains elusive. However, what *is* clear is that as the region changes, so must our understanding of rural Africa, of rural poverty and of rural development.

Contemporary rural Africa

Discussion of contemporary rural livelihoods and rural development is complicated by the ambiguity of the term 'rural' and by the presence of peri-urban areas (see ODI, 2002; IFAD, 2010b). In the absence of an exact definition, rural has become synonymous with the notion of the countryside (Wiggins and Proctor, 2001), areas that are characterised by significant distances between rural settlements, between rural settlements and cities, and where there is an abundance of land and other natural resources, such as pastures, forests, mountains and water (Wiggins and Proctor, 2001; ODI, 2002).

Some 973 million people live in sub-Saharan Africa as of 2014 (World Bank, 2016a). While the proportion of people living in rural areas has steadily declined from 71.9 per cent in 1990 to 63.0 per cent in 2014 (World Bank, 2015a, 2016a), population growth means that the rural population is actually growing (cf. Heinemann, 2014). So, too, are the absolute numbers of rural poor (cf. Radeny *et al.*, 2012). In some countries, four out of five people now live and work in rural areas. In Burundi, for example, nearly 90 per cent of people are based in rural areas as of 2014 (World Bank, 2016b). And Burundi is no exception (see Table 28.1).

What does rural Africa look like? Conventional wisdom about rural Africa points to a number of stylised facts. For example, it is often perceived to be a place: (a) where households typically depend on agriculture or agriculture-related activities; (b) where households, particularly poor households, lack access to functioning markets; (c) where rural labour markets are thin or even absent; and (d) where the non-farm economy is weak and comprises mainly self-employed individuals and family workers (ODI, 2002; IFAD, 2006; Oya, 2010). While there are strands of truth to these 'facts', the reality, as we will see, is frequently far more varied and more nuanced than any 'broad brush picture' might suggest. After all, rural Africa is an expanse of diverse geographical, socio-economic, political and cultural spaces. Rural Africa is not a single region in much the same way as the rural poor are not a single, homogenous group with common interests, capacities, capabilities, opportunities and aspirations.

This chapter seeks to describe and explain the complex, nuanced realities of contemporary rural Africa. It traces the evolution of approaches to rural development such that current framings of problems and solutions are placed in their historical contexts. The chapter does not paint a picture of rural Africa in crisis. But neither does it suggest that households will share the spoils of economic growth. The reality is more complex. It argues that a sustained reduction in rural poverty will require a combination of inclusive economic growth and social protection. For this to happen, people require functioning services and markets that offer (and support) viable livelihood choices, coupled with a macro-policy environment that recognises the causes, nature and pace of change in rural areas.

What is 'rural development'?

The term 'rural development' has been widely used since the mid-1970s (Ellis, 2000). At the time, like now, it was acknowledged that the majority of the world's poor were located in rural

Table 28.1 Rural population as a % of total population, 2014 (selected countries)

Gabon	13
Djibouti	23
Congo Rep	35
São Tomé and Príncipe	35
South Africa	36
Chad	78
Eritrea	78
Swaziland	79
Ethiopia	81
South Sudan	81
Niger	82
Uganda	84
Malawi	84
Burundi	88

Source: World Bank (2016b).

areas (World Bank, 1975; Ellis, 2000). As such, it was, and remains, a 'call to action' rather than a theory of economic or social change (Ellis, 2000). Nevertheless, it still represents a useful organising principle for policies, strategies and investments that seek to enable rural people to achieve sustained improvements to their quality of life (Chambers, 1983; Singh, 1999; Ellis, 2000).

What follows is a brief schematic of the ebb and flow of dominant thinking over recent decades, within which we can see the philosophical battles between central and local planning, the changing outlook on the role of the state versus the market, and the fluctuating emphasis placed on the productive sector set against the social sectors (ODI, 2002; Kraybill, 2013).

Rural development from the 1960s to the 1990s

The dawn of the 1960s saw a major revision in rural development thinking, with 'modernised agriculture' replacing the community development models of the 1950s (Ellis and Biggs, 2001; Byerlee *et al.*, 2005). Attention shifted to the need to mechanise farms and introduce 'improved inputs', investments that were expected to drive productivity gains, stimulate multiplier effects and generate growth in the rural non-farm economy (Ellis and Biggs, 2001; ODI, 2002).

The results were disappointing, and concern mounted about the limited success of the green revolution in Africa (see Dawson *et al.*, 2016). Put simply, smallholders did not seem to benefit from the investments in mechanisation and the development of so-called 'miracle grains'. Poverty rates remained high, not least in rural Africa. In 1973, Robert McNamara, the then President of the World Bank Group, announced the adoption of a new development goal: that of rural development. His seminal speech in Nairobi played a major role in shifting attention in the 1970s to poverty reduction, to meeting people's 'basic needs' and to the possibilities offered by 'rural development' (see McNamara, 1973).

With economic growth and productivity remaining critically important to the World Bank agenda, lending to agriculture and rural development initiatives rose sharply. Aid and development investments often took the form of 'integrated rural development programmes' (or 'area development programmes'), i.e. the simultaneous and often top-down delivery of a range of 'anti-poverty measures', from the delivery of improved village healthcare facilities, to the building of rural roads and rural water supplies, to initiatives that sought to increase the productivity of small-farm agriculture, and to measures that were expected to boost non-farm income generation. With the utilisation of modern agricultural techniques and with improved provision of basic needs, such as food, water and education, poverty was expected to decline, employment was expected to grow, and the reliance on imported foodstuffs was projected to weaken (McNamara, 1973; Chambers, 1997; Ellis and Biggs, 2001).

Between 1973 and 1986, the World Bank lent US$19 billion to support 498 rural development projects of this nature across the world (Chambers, 1997). However, the resource-intensive, state-led rural development approaches pursued from the mid-1970s to the early 1980s did not last. Neither did the focus on agriculture and rural development (Takeuchi, 2000). The debt crisis of the 1980s ushered in an era of structural adjustment policies, budget cuts and a trimming of public sector institutions (ODI, 2002). With structural adjustment came the view that trade and unregulated, free markets would drive growth (Chambers, 1997; Ellis and Biggs, 2001; Potter, 2014). The long arm of government was pulled back as recurrent government expenditure was cut back (Chambers, 1997). The effect was that the state withdrew from its earlier role as the implementing agency for rural and agricultural development initiatives (Ellis and Biggs, 2001) and state marketing boards were disbanded. The (supposedly) integrated rural development programmes became regarded as expensive failures, 'white elephants' characterised by poor design and piloting, and expensive and often corrupt implementation (Chambers, 1997; Mander, 2005). Once again, donor thinking retreated to a narrow focus on financial-economic goals (Gready and Ensor, 2005).

As the 1980s gave way to the 1990s, the social effects of structural adjustment became more apparent and economic growth became regarded as a necessary, but ultimately insufficient, condition for development (Desai and Potter, 2014). With limited enthusiasm for so-called 'big government', attention shifted to issues of good governance, decentralisation and the potential of non-state actors, such as NGOs and the private sector (World Bank, 1997; Ellis, 2000; Ellis and Biggs, 2001). This change in tack coincided with the end of the Cold War and the profound shift in geopolitics from the West and East binary, to the North and South binary (Gready and Ensor, 2005). Poverty reduction and rural development returned to the agenda (Kraybill, 2013). Interest grew in the notions of freedom, agency, capabilities and entitlements (e.g. see Sen, 1981; Leach *et al.*, 1997), ultimately giving rise to the sustainable livelihoods paradigm in the mid/late 1990s (e.g. see Carney, 1998; Ellis, 1998; Scoones, 1998; Ashley and Carney, 1999) and sparking greater recognition of the variety, richness and complexity of rural livelihood strategies (e.g. see Ellis, 2000; Sarch and Birkett, 2000; Shackleton *et al.*, 2000; Allison and Ellis, 2001; Mortimore and Adams, 2001; Scoones and Wolmer, 2002).

At the same time, increasing attention was being paid to human rights and welfare, and the gender-defined division of tasks and opportunities (Ellis, 2000; Ellis and Biggs, 2001; Desai and Potter, 2014). A shift was underway from women in development to gender and development. No longer were women being framed and 'treated' as homogenous and isolated categories of people who would benefit from more self-reliance, greater assertion of their own capabilities and improved income-generating opportunities. Instead, attention was moving to the social relations of gender (rather than women *per se*), empowerment and 'gender mainstreaming' (Okali, 2012).

Rural development in the early twenty-first century: the 2000s and 2010s

By the turn of the century, rural development thinking had gravitated towards a view that the state *and* the market had important roles to play, and that functioning productive *and* social sectors were necessary to deliver socio-economic change and development (Ashley and Maxwell, 2001). The nature of rural development investments was changing accordingly. The extent of public investment in agriculture continued its fall from the highs of the 1970s/1980s, and big donor-financed agriculture projects fell out of favour as attention shifted to broader, strategic investments in the 'enabling environment' such as good governance (Cleaver, 2007).

The sustainable livelihood approach of the late 1990s and the early 2000s offered a counter-narrative to that which regarded the 'rural space' as 'the farm' (Madzivhandila, 2014) and that which therefore focused on how to improve the productivity and efficiency of small farms (Ellis, 2000; Ellis and Biggs, 2001). A wealth of evidence emerged about the existence, nature, drivers and consequences of a range of livelihood strategies. Income and livelihood diversification were seen as a means to grow asset bases and to reduce risk and vulnerability (e.g. see Bryceson, 1996; Ellis, 2000; see also Bryceson, Chapter 27, this volume). The distinction between rural and urban spaces became increasingly blurred. Goods, people and services were regarded as being increasingly interconnected. It became apparent, for example, that urban elites were buying land in the countryside, that rural elites were purchasing land in urban areas as a way to secure improved access to schooling and better livelihoods, and that some people were choosing to officially reside in rural areas yet live day-to-day in urban areas

(Bryceson, 2002; Heinemann, 2014; Zoomers, 2014). There was also recognition that where some households were graduating out of poverty, many remained poor or had fallen back into poverty (e.g. see Bird and Shepherd, 2003; Barrientos *et al.*, 2005; Carter and Barrett, 2006; see also Ellis *et al.*, 2009).

With the growing appreciation of this rural complexity came a challenge. Confidence in the rural development 'project' waned and financing declined (Ashley and Maxwell, 2001; Cleaver, 2007). Rural development was lacking a 'convincing narrative': a narrative that offered manageable solutions to incontestable and perhaps even simple problems (Ashley and Maxwell, 2001; Pain, 2007; Madzivhandila, 2014). In 2003, the head of the Overseas Development Institute in London delivered a conference paper that set out 'how to rescue rural development before it's too late' (Maxwell, 2003). The incidence and unacceptability of rural poverty was not in question. What *was* in question was what rural development should look like in policy and practice.

A potential solution was found in the middle of the first decade of the 2000s. Agriculture returned to centre stage. In 2003, the African Union's New Partnership for Africa's Development established the Comprehensive Africa Agriculture Development Programme: a flagship, Africa-owned initiative that sought to boost agricultural productivity and improve food security and nutrition. In 2005, the UK Department for International Development issued its Agricultural Policy (DFID, 2005). Two years later, the World Bank launched its World Development Report (WDR) for 2008, the first WDR for 25 years to focus on agriculture. Its emergence was welcomed (see Cleaver, 2007; Kay, 2009), not least as a timely response to the food price crisis of 2006–2008 that pushed tens of millions of rural poor across the world back into hunger (IFAD, 2010b). The WDR's key message was simple: that 'agriculture continues to be a fundamental instrument for sustainable development and poverty reduction', and while 'agriculture alone will not be enough to massively reduce poverty . . . it has proven to be uniquely powerful for that task' (World Bank, 2007: 1).

So, what can we take away from the evolution in rural development thinking to date? In truth, the policy space in the 2000s and 2010s has been a little more contested than this crude caricature might suggest. And this warning does not only apply to the last 15 years. Ellis and Biggs' attempt in 2001 to plot the dominant rural development narratives from one decade to the next has been described as an 'impossible and probably fruitless' exercise (Kraybill, 2013). As Ellis and Biggs themselves note, it would be a gross oversimplification to suggest that the 1960s could be characterised as the decade of modernisation, the 1970s as the period of state intervention, the 1980s as the decade of market liberalisation, and the 1990s as the period during which participation and empowerment dominated. As they and others have noted (e.g. Scoones, 2009), some narratives ran in parallel to others, others first emerge in one decade but only take hold many years later, and others remain influential despite having superficially faded to the background. However, what is clearly apparent is that mainstream rural development policy continues to grapple with how best to support people to grasp opportunities and to build resilience against shocks and stresses. In turn, this has meant striking the 'right' balance between productive and social sectors, and between the state and the market. But what might the future look like? And what are the necessary policy responses?

Dynamism and change in rural Africa

Rural spaces are dynamic. That change is never more true than now. As across much of the world, the character of Africa's rural spaces is changing. Market liberalisation, new technologies and services, the remarkable penetration of mobile phones, climate change, and other 'megatrends' are shaping economic, political and social landscapes (Jayne *et al.*, 2014). None of these changes guarantees particular outcomes for particular people. Nevertheless, what *is* certain is that for many poor rural people, the economic opportunities open to them, the challenges they face and the risks they confront are evolving. As such, approaches to rural development need a narrative that recognises the rapid change in rural areas, one that acknowledges processes of demographic change, economic globalisation and regionalisation, and the potential of farming and the non-farm rural economy.

Demographic change

The size and structure of the population of sub-Saharan Africa continues to evolve. Between 1950 and 2000, the population of sub-Saharan Africa more than tripled, from 186 million to 670 million. It is forecast to double again by 2060 (Canning *et al.*, 2015). While the proportion of people living in rural areas is set to fall from 61 per cent in 2010 to 42 per cent by 2050, the total rural population will continue to grow in absolute numbers, from 622 million in 2010 to a projected 927 million in 2050 (Proctor and Lucchesi, 2012). Besides some island states (e.g. Mauritius, Seychelles) and a few countries in southern Africa (e.g. Botswana, Namibia, South Africa and Swaziland), fertility rates and youth dependency rates in sub-Saharan Africa are now among the highest in the world (Canning *et al.*, 2015). The average number of children born per woman has fallen only slightly over the last few decades, from 6.5 in 1950–1955 to 5.4 in 2005–2010 (compared to a fertility decline in East Asia of 5.6 to 1.6 over the same period) (Canning *et al.*, 2015). The result is that sub-Saharan Africa has the highest rural dependency rate in the world, with 44 per cent of the population now being younger than 15 years (Heinemann, 2014).

With the age profile being youthful, the region can anticipate a rapidly growing labour force. Only a proportion of this can be absorbed in the urban or industrial sector (Heinemann, 2014), pointing to the need to expand income-generating opportunities in rural areas. But who will benefit from the so-called 'demographic dividend'? The rapid growth in urban populations, coupled with higher incomes and the emergence of a sizeable middle class, is creating new consumption patterns, patterns that represent market opportunities for those smallholders with the capability to exploit them. However, the great majority of new entrants to the workforce will face challenges in finding gainful employment (Losch *et al.*, 2012; Losch, 2013), and the challenges will be felt most acutely by those with limited skills and means. With youth unemployment and underemployment of interesting concern (see White, 2012), the need for a viable 'rural sector' is readily apparent. For agriculture to be attractive to the youth, it will have to be profitable, competitive and dynamic (Brooks *et al.*, 2013). However, with many people turning away from farming, the future of the rural youth must be high on the policy agenda (Proctor and Lucchesi, 2012). Education, employment, labour and investment policies will have to create opportunities for attractive, decent employment (e.g. in the agri-food sector) and the means by which the rural youth have the capabilities to embrace them. This will necessarily involve improving access to a range of services including, *inter alia*, affordable finance, actionable information and relevant training.

Economic globalisation and regionalisation

Rural transformation is integrally linked with wider processes of structural change (Proctor, 2014). While proportional rates of intra-regional exports in sub-Saharan Africa are the lowest in the world, the value of intra-African trade has grown in nominal terms in recent years, from US$45.9 billion in 1995 to US$130.1 billion in 2011 (and undoubtedly much more if the value of informal cross-border trade were included) (UNCTAD, 2013). Such increases are consistent with the growing – yet sometimes rhetorical – commitment to integrate markets by, for example, harmonising and simplifying customs procedures, improving regional road and rail links, and installing one-stop border posts at key border crossings (e.g. see NEPAD, 2014). With the integration of national and regional economies through trade, capital flows and the migration of people, rural areas are becoming more and more exposed to regional and global forces. But is the increased connectivity of goods, services, ideas and people a key to unlocking the potential of rural areas, or a recipe for disaster for some of those that live there?

It is commonly acknowledged that economic growth and poverty reduction have accelerated in those economies that have successfully increased their trade share in income (Dercon, 2006). Proponents for trade liberalisation and integration argue that improved market access and intra-regional trade promises benefits for producers and consumers, including poor households who are net buyers of food. The increased interconnectivity takes various forms. For example, the widespread uptake of mobile telecommunications and associated services are generating new employment opportunities and the promise of higher household incomes. Equally, for those able to secure benefit from rural-to-urban migration, remittances can improve health and education outcomes, smooth consumption patterns, act as a form of insurance, and ultimately offer different livelihood trajectories (World Bank, 2011; Hazell and Rahman, 2014).

In reality, however, the increasing inter-connectivity of rural Africa creates relative winners and losers (see Walther, Chapter 21, this volume). For some, nothing points more to the recent convergence of powerful economic and political forces than the recent spate of so-called 'land grabs' (see Hall *et al.*, 2015). A combination of demand for 'green energy', rising food and energy prices, and cheap and often malleable water and land rights has led to the acquisition of large tracts of land in parts of rural Africa. While the rate of land acquisitions appears to have slowed from the peak of 2007–2008, and though the total land area taken to date remains small, land-grabbing has the potential to redefine rural spaces. In Ethiopia, Mozambique and Sudan, for example, millions of hectares of land have been transferred to investors in recent years (Deininger *et al.*, 2011). The implications for economic growth, rural livelihoods and food security are only now becoming the subject of scholarly interest (e.g. see Hall *et al.*, 2015).

While evidence suggests that trade openness can be an important element in expanding growth, and that growth in turn is a critically important driver of poverty reduction, there is concern that economically marginal countries with poor geographical endowments and a poor policy environment may become further marginalised (Dercon, 2006). Evidence from Ethiopia suggests that much the same applies at a household level. Those with poor endowments in terms of labour and other assets and who are geographically marginalised in terms of infrastructure, communications and natural geography may be negatively affected (Dercon, 2006). Remoteness continues to stifle the ability of the poor to take advantage of that which growth provides (Ahmed *et al.*, 2007). The reality remains that those living in urban areas typically enjoy better access to financial, physical, human and perhaps even social capital, generating benefits for labour productivity and market access (Wiggins and Proctor, 2001). By contrast, over 70 per cent of those living in rural areas still live more than 2 km from all-weather roads, driving up the cost of transporting goods and people (cf. Kraybill, 2013; see also Porter, Chapter 42, this volume).

The future of farming and the rural non-farm economy

African economies continue to rely heavily on agriculture. The sector accounts for about 32 per cent of GDP in sub-Saharan Africa and employs about 65 per cent of the region's labour force

(AGRA, 2013). Inevitably, the contribution of agriculture to national economies varies widely, typically declining as GDP per capita increases and as economies undergo structural transformations (AGRA, 2013). In Kenya, for example, agriculture's value-added share of GDP has fallen from over 50 per cent in the 1980s to less than 29.3 per cent in 2013 (World Bank, 2015b). By contrast, it still accounts for 48.0 per cent of GDP in Ethiopia, 55.1 per cent in Chad and 53.9 per cent in the Central African Republic (all 2012) (World Bank, 2015b). While agriculture's share of the labour force in sub-Saharan Africa is on the decline (see Bryceson, 1996, 1999, 2002), the actual number of people engaged in agriculture is actually rising (Proctor, 2014). Moreover, for low-income and resource-rich countries, where private sector wage employment remains underdeveloped, agriculture will probably remain the dominant employer well into the 2020s, and possibly beyond (Proctor, 2014), even though young rural people and their parents often hold farming and rural life in very low esteem (Sumberg and Okali, 2013). Consequently, the challenge is how best to support those who can (and those who wish) to 'step up' through intensification and commercialisation of climate-smart, smallholder farming, how to support those who want to 'step out' of agriculture into the rural non-farm economy, and how to support those marginalised by a lack of assets and opportunities to 'hang in' through a combination of productive and social protection measures financed by a mix of state and non-state funding sources (see Barrientos, 2008; Dorward, 2009).

It has often been argued that growth in small farm output is critical for improving rural welfare (Ellis, 2000; see also Wiggins *et al.*, 2010). While smallholders in rural Africa have long been engaged with markets, for some, the opportunity to commercialise – to increase engagement with markets – holds real promise (Wiggins *et al.*, 2011). Where conditions are right – for example, where there is access to improved technologies, physical access to markets, demand for higher value products, access to competent traders, access to sufficient land parcels (e.g. 2–4 ha) and an ability to regularly market a surplus – farming offers households the possibility of escaping poverty and, for the regional economy, the possibility of job creation in the local rural economy (Wiggins *et al.*, 2014; see also Porter, Chapter 29, this volume).

In parts of rural Africa, there has been a rapid growth in the number of medium-scale farmers cultivating 5–20 ha of land. Ghana, Kenya and Zambia are cases in point (see Jayne *et al.*, 2014). However, recent studies have suggested that this phenomenon does not necessarily signify the appearance of so-called 'emergent smallholder farmers' who began farming with less than 5 ha of land, but rather unprecedented rates of investments by relatively wealthy rural people and salaried urbanites, attracted by favourable land administration and agricultural spending policies (e.g. see Jayne *et al.*, 2014; Sitko and Jayne, 2014). Barriers and constraints prevail, even for men (who tend to have greater assets and better links to traders and processors than women) (Wiggins *et al.*, 2011). For aspiring smallholders to capitalise on opportunities, they require reliable, functioning input and output markets and knowledge about their operation. Increasingly, they need access to (often complex) supply chains commanded by supermarkets and exporters (IFAD, 2010b), something that has often proved difficult (see Barrett *et al.*, 1999; Bek *et al.*, 2013).

For those who do not have the ability to 'step up', whose on-farm activities are strained, or whose access to land and water rights are vulnerable, can the rural non-farm economy (RNFE) take up some of the slack? Put simply, how much potential is there for members of rural households to 'step out' of farming into the NFRE or to migrate? At a macro level, the signs are encouraging. The economy of sub-Saharan Africa has grown for a couple of decades, and studies of rural income portfolios indicate that about 50 per cent of rural household incomes in sub-Saharan Africa are generated from non-farm activities and financial transfers from urban areas or abroad (Ellis, 2005). However, the realities are more nuanced. For successful 'exits' to occur, the economy must create sufficient numbers of suitable, attractive jobs for those wishing to leave farming, and those pursuing them must have the right blend of experience and skills to grasp the opportunities (see Ellis, 2005; Nagler and Naudé, 2014). These realities give cause for concern. With the rate of urbanisation having stagnated – or even regressed – in some countries (e.g. Zambia, Côte d'Ivoire, Mali and the Central African Republic) (see Potts, 2012a), there appears to be evidence that high unemployment and the existence of limited economic opportunities are making some towns and cities unattractive to many migrants (see Potts, 2012b, 2015). The poor will be hit hardest. Where the poor (who typically lack skills and capital) do work in the RNFE, they tend to take on low-return activities such as small-scale trading and construction (Wiggins and Keats, 2015). Women typically have the fewest options, which is a consequence of their generally more limited education, fewer marketable skills and more limited social contact with the world beyond the immediate vicinity (Wiggins and Keats, 2015).

For those returning to rural areas, and for those who were unable to step up or step out in the first instance, the priority may be to avoid destitution and extreme poverty, and ultimately to escape extreme poverty by 'hanging in'. For many, this will take the form of a mix of local farm production, labouring and off-farm income generation (see Dorward, 2009). For those who are able to farm, emphasis is often placed on the production of subsistence food crops, a rational choice so long as rural food markets remain risky and poor households remain vulnerable to the effects of shocks such as droughts or bereavement (Leavy and Poulton, 2007). And evidence suggests that the risks are getting greater. Climate change is predicted to have significant biological effects on crop yields. Wheat, sorghum, millet and maize harvests are particularly threatened in Africa, and this will be of particular concern to regions that are already food-insecure or that experience periodic shocks (Knox *et al.*, 2012; Cairns *et al.*, 2013). Declining yields will affect price and consumption patterns, and ultimately per capita calorie consumption and nutrition (Nelson *et al.*, 2009).

Impacts will not be restricted to cropping. While quantitative studies on the potential impact on fisheries, and in turn on livelihoods and well-being, are comparatively uncommon, long-term projections suggest that climate change in West Africa will lead to a 21 per cent drop in annual landed value, a 50 per cent decline in fisheries-related jobs and a total *annual* loss of US$311 million in the regional economy over the first half of the twenty-first century (Lam *et al.*, 2012). As such, the growing interest in fusing social protection, climate change adaptation and disaster risk reduction measures is timely, as is the desire to explore new forms of protective, preventive and promotional social protection measures, including weather-indexed crop insurance (Davies *et al.*, 2009). Time will tell how feasible or scaleable some of the initiatives are. What is certain, however, is that climate-smart approaches to rural development are required.

Policy implications

The need for a strong rural economy that benefits all is clear and surely undisputable. However, supporting people to graduate out of poverty, *and then to remain out of poverty during times of stress*, is a complex process that requires investments and reforms in both productive and social sectors. It also requires an appreciation of the different environments within which people live and work, and therefore of the different opportunities and constraints they face. After all, the consequences of demographic change, the liberalisation of markets, the transformation of production systems and the individualisation of productive resources such as land will be experienced very differently by different people in different locations. So, what is required for sustained improvements to the well-being of those living in rural sub-Saharan Africa?

First, the foundations must be laid and maintained. Rural people, like all citizens, need macroeconomic stability, a predictable policy environment, a predictable, accessible and fair legal system, and a peaceful environment. These foundations are incontestable. Equally, the principles of opportunity, choice, empowerment, participation and security should be nurtured and protected. By extension, rights – to land and other productive resources, for example – must be recognised and systems for grievance and redress must be honoured. As that might suggest, all people, including the young, must be able to actively and meaningfully participate in matters that concern them.

Second, there must be investment in public goods and in individual capabilities. Investment in core rural infrastructure, such as roads, electricity and telecommunications, is essential. Equally, people need access to affordable, high-quality education and health services so that they can make informed choices and exploit opportunities as they emerge. The recent appetite for devolved (e.g. Kenya) and decentralised (e.g. Tanzania and South Sudan) models of governance and fiscal responsibility may be beneficial in that regard. However, if the arrangements are not well managed, the reforms may lead to missed opportunities and greater inequality.

Third, those participating in on-farm, off-farm and non-farm activities require access to functioning markets and services. Improved agricultural production and broad-based smallholder commercialisation more broadly require a range of pre- and post-harvest services, including affordable finance and credit, agricultural extension and research, storage and processing, and timely, reliable market information. Innovative remedies to market failures, such as out-grower schemes, may offer improved access to markets, technology, management skills and information, but they will also require the establishment and enforcement of associated governance and safeguard measures.

Fourth, there is a need to protect and enhance the security of the most vulnerable. Some will remain constrained by limited capacities and capabilities. They may also live in remote areas or in areas with low natural potential. Well-designed social protection programmes can help manage and mitigate risk, without undermining rural markets and imposing unreasonable cost burdens on the state. Equally, attention must be paid to climate-smart agricultural strategies that protect both natural resources and livelihoods.

Finally, it must be recognised that there is no simple solution, no magic wand. The 'rural reality' is both complex and dynamic. Responsibility for shaping the right response cuts across different actors, different locations and different ministries. For some, there are opportunities. For others, there are challenges. For all, there are vested interests and different incentives and motivations to act. Dialogue, policy coherence and coordination are required. That implies that the state must play a strong role. But its role must also be appropriate. It should form policy and coordinate or regulate service provision; it need not necessarily be a service provider. Markets have a role to play, and as such the private sector – from individual farmers, traders and low-cost, private schools to major conglomerates – can all, collectively, contribute to future development.

References

AGRA (Alliance for a Green Revolution in Afrcia) (2013) *Africa Agriculture Status Report: Focus on Staple Crops*, Nairobi: AGRA.

Ahmed, A., Hill, R.V., Smith, L., Wiesmann, D. and Frankenberger, T. (2007) *The World's Most Deprived Characteristics and Causes of Extreme Poverty and Hunger*, 2020 Discussion Paper No. 43, Washington, DC: International Food Policy Research Institute.

Allison, E.H. and Ellis, F. (2001) 'The livelihoods approach and management of small-scale fisheries', *Marine Policy*, 25(5): 377–388.

Ashley, C. and Carney, D. (1999) *Sustainable Livelihoods: Lessons from Early Experience*, London: Department for International Development.

Ashley, C. and Maxwell, S. (2001) 'Rethinking rural development', *Development Policy Review*, 19(4): 395–425.

Barrett, H.R., Ilbery, B.W., Browne, A.W. and Binns, T. (1999) 'Globalization and the changing networks of food supply: the importation of fresh horticultural produce from Kenya to the UK', *Transactions of the Institute of British Geographers*, 24(2): 159–174.

Barrientos, A. (2008) 'Financing social protection', in A. Barrientos and D. Hulme (eds), *Social Protection for the Poor and Poorest Concepts, Policies and Politics*, London: Palgrave Macmillan, pp. 300–312.

Barrientos, A., Hulme, D. and Shepherd, A. (2005) 'Can social protection tackle chronic poverty?', *The European Journal of Development Research*, 17(1): 8–23.

Bek, D., Binns, T. and Nel, E. (2013) 'Wildflower harvesting on South Africa's Agulhas Plain: a mechanism for achieving sustainable local economic development?', *Sustainable Development*, 21(5): 281–293.

Bird, K. and Shepherd, A. (2003) 'Livelihoods and chronic poverty in semi-arid Zimbabwe', *World Development*, 31(3): 591–610.

Brooks, K., Zorya, S., Gautam, A. and Goyal, A. (2013) *Agriculture as a Sector of Opportunity for Young People in Africa*, World Bank Policy Research Working Paper No. 6473, Washington, DC: World Bank.

Bryceson, D. (1996) 'Deagrarianization and rural employment in sub-Saharan Africa: a sectoral perspective', *World Development*, 24(1): 97–111.

Bryceson, D. (1999) 'African rural labor income diversification and livelihood approaches: a long-term development perspective', *Review of African Political Economy*, 80: 171–189.

Bryceson, D. (2002) 'The scramble for Africa: reorienting rural livelihoods', *World Development*, 30(5): 725–739.

Byerlee, D., Diao, X. and Jackson, C. (2005) *Agriculture, Rural Development, and Pro-Poor Growth: Country Experiences in the Post-Reform Era*, Agriculture and Rural Development Discussion Paper 21, Washington, DC: World Bank.

Cairns, J.E., Hellin, J., Sonder, K., Araus, J.L., MacRobert, J.F., Thierfelder, C., *et al.* (2013) 'Adapting maize production to climate change in sub-Saharan Africa', *Food Security*, 5(3): 345–360.

Canning, D., Raja, S. and Yazbeck, A.S. (eds) (2015) *Africa's Demographic Transition: Dividend or Disaster?* Washington, DC: World Bank and Agence Française de Développement.

Carney, D. (ed.) (1998) *Sustainable Rural Livelihoods: What Contribution Can We Make?* London: Department for International Development.

Carter, M.R. and Barrett, C.B. (2006) 'The economics of poverty traps and persistent poverty: an asset-based approach', *Journal of Development Studies*, 42(2): 178–199.

Chambers, R. (1983) *Rural Development: Putting the Last First*, Harlow: Longman.

Chambers, R. (1997) *Whose Reality Counts? Putting the First Last*, London: Intermediate Technology Publications.

Cleaver, K. (2007) *Contemporary Issues of Agriculture and Rural Development in Africa, and IFAD's Approach*, PowerPoint presentation, IFAD.

Davies, M., Guenther, B., Leavy, J., Mitchell, T. and Tanner, T. (2009) *Climate Change Adaptation, Disaster Risk Reduction and Social Protection: Complementary Roles in Agriculture and Rural Growth?* IDS Working Paper 320, Centre for Social Protection and Climate Change and Development Centre.

Dawson, N., Martin, A. and Sikor, T. (2016) 'Green revolution in sub-Saharan Africa: implications of imposed innovation for the wellbeing of rural smallholders', *World Development*, 78: 204–218.

Deininger, K., Byerlee, D., Lindsay, J., Norton, A., Selod, H. and Stickler, M. (2011) *Rising Global Interest in Farmland: Can It Yield Sustainable and Equitable Benefits?* Washington, DC: World Bank.

Dercon, S. (2006) *Globalization and Marginalization in Africa: Poverty, Risk and Vulnerability in Rural Ethiopia*, QEH Working Paper Series, Working Paper No. 147, Department of International Development, University of Oxford.

Desai, V. and Potter, R.B. (2014) 'Part I: the nature of development and development studies', in V. Desai and R.B. Potter (eds), *The Companion to Development Studies*, London: Routledge, pp. 1–77.

DFID (Department for International Development) (2005) *Growth and Poverty Reduction: The Role of Agriculture – A DFID Policy Paper*, London: DFID.

Dorward, A. (2009) 'Integrating contested aspirations, processes and policy: development as hanging in, stepping up and stepping out', *Development Policy Review*, 27(2): 131–146.

Ellis, F. (1998) 'Survey article: household strategies and rural livelihood diversification', *Journal of Development Studies*, 35(1): 1–38.

Ellis, F. (2000) *Rural Livelihoods and Diversity in Developing Countries*, Oxford: Oxford University Press.

Ellis, F. (2005) *Small Farms, Livelihood Diversification and Rural–Urban Transitions: Strategic Issues in Sub-Saharan Africa*, paper prepared for the Research Workshop on the Future of Small Farms organised by

International Food Policy Research Institute (IFPRI), Overseas Development Institute (ODI), Imperial College, London, Withersdane Conference Centre, Wye, Kent, UK, 26–29 June.

Ellis, F. and Biggs, S. (2001) 'Evolving themes in rural development 1950s–2000s', *Development Policy Review*, 19(4): 437–448.

Ellis, F., Devereux, S. and White, P. (2009) *Social Protection in Africa*, London: Edward Elgar.

Gready, P. and Ensor, J. (2005) 'Introduction', in P. Gready and J. Ensor (eds), *Reinventing Development? Translating Rights-Based Approaches from Theory into Practice*, London: Zed Books, pp. 1–45.

Hall, R., Scoones, I. and Tsikata, D. (eds) (2015) *Africa's Land Rush: Rural Livelihoods and Agrarian Change*, London: James Currey.

Hazell, P. and Rahman, A. (2014) 'Concluding chapter: the policy agenda', in P.B.R. Hazell and A. Rahman (eds), *New Directions for Smallholder Agriculture*, Oxford: Oxford University Press, pp. 527–558.

Heinemann, E. (2014) 'The changing rural world and livelihood options for resource-poor rural people', in R.B.R. Hazell and A. Rahman (eds), *New Directions for Smallholder Agriculture*, Oxford: Oxford University Press, pp. 377–399.

IFAD (International Fund for Agricultural Development) (2006) *Targeting Policy: Reaching the Rural Poor*, Rome: IFAD.

IFAD (International Fund for Agricultural Development) (2010a) *Facts & Figures: Rural Poverty Report 2011*, available at: www.ifad.org/rpr2011/media/kit/factsheet_e.pdf (accessed 1 December 2015).

IFAD (International Fund for Agricultural Development) (2010b) *Rural Poverty Report 2011*, Rome: IFAD.

Jayne, T.S., Meyer, F. and Ndibongo Traub, L. (2014) *Africa's Evolving Food Systems: Drivers of Change and the Scope for Influencing Them*, IIED Working Paper, London: International Institute for Environment and Development.

Kay, C. (2009) 'Development strategies and rural development: exploring synergies, eradicating poverty', *The Journal of Peasant Studies*, 36(1): 103–137.

Knox, J., Hess, T., Daccache, A. and Wheeler, T. (2012) 'Climate change impacts on crop productivity in Africa and South Asia', *Environmental Research Letters*, 7(3): 1–8.

Kraybill, D. (2013) 'Rural development in sub-Saharan Africa', in G.P. Green (ed.), *Handbook of Rural Development*, Cheltenham: Edward Elgar, pp. 270–292.

Lam, V.W., Cheung, W.W., Swartz, W. and Sumaila, U.R. (2012) 'Climate change impacts on fisheries in West Africa: implications for economic, food and nutritional security', *African Journal of Marine Science*, 34(1): 103–117.

Leach, M., Mearns, R. and Scoones, I. (1997) *Environmental Entitlements: A Framework for Understanding the Institutional Dynamics of Environmental Change*, IDS Discussion Paper No. 359, Brighton: Institute of Development Studies.

Leavy, J. and Poulton, C. (2007) *Agricultural Commercialisations: A Level Playing Field for Smallholders?* Futures Agriculture Policy Brief 17, Brighton: Institute of Development Studies.

Losch, B. (2013) *Structural Change and Employment in Africa: Situation and Development Issues*, paper presented at the Rural Development Forum in Africa, Cotonou, 24 May.

Losch, B., Fréguin-Gresh, S. and White, E.T. (2012) *Structural Transformation and Rural Change Revisited: Challenges for Late Developing Countries in a Globalizing World*, African Development Forum Series, Washington, DC: World Bank.

Madzivhandila, T.S. (2014) 'The misconception of rural development in South Africa: a recipe for socio-economic stagnation', *Mediterranean Journal of Social Sciences*, 5(8): 89–93.

Mander, H. (2005) 'Rights as struggle: towards a more just humane world', in P. Gready and J. Ensor (eds), *Reinventing Development? Translating Rights-Based Approaches from Theory into Practice*, London: Zed Books, pp. 233–252.

Maxwell, S. (2003) *Six Characters (and a Few More) in Search of an Author: How to Rescue Rural Development Before It's Too Late*, prepared for the 25th International Conference of Agricultural Economists (IAAE), Reshaping Agriculture's Contributions to Society, held in Durban, South Africa, 18–22 August.

McNamara, R.S. (1973) *Address to the Board of Governors, World Bank Group, Nairobi, Kenya, September 24, 1973*, available at: www.juerg-buergi.ch/resources/Archiv/Entwicklungspolitik-Archiv/Dokumente-zur-Entwicklungspolitik/McNamara_Nairobi_speech.pdf (accessed 3 May 2016).

Mortimore, M.J. and Adams, W.M. (2001) 'Farmer adaptation, change and "crisis" in the Sahel', *Global Environmental Change*, 11(1): 49–57.

Nagler, P. and Naudé, W. (2014) *Non-Farm Entrepreneurship in Rural Africa: Patterns and Determinants*, Discussion Paper No. 8008, Bonn: Forschungsinstitut zur Zukunft der Arbeit (IZA) (Institute for the Study of Labour).

Nelson, G.C., Rosegrant, M.W., Koo, J., Robertson, R., Sulser, T., Zhu, T., *et al.* (2009) *Climate Change Impact on Agriculture and Costs of Adaptation.* Washington, DC: International Food Policy Research Institute.

NEPAD (New Partnership for Africa's Development) (2014) *CAADP: Implementation Strategy and Roadmap to Achieve the 2025 Vision on CAADP*, available at: www.nepad.org/system/files/Implementation%20 Strategy%20Report%20English.pdf (accessed 4 December 2015).

ODI (Overseas Development Institute) (2002) *Rethinking Rural Development*, ODI Briefing Paper, London: ODI.

Okali, C. (2012) *Gender Analysis: Engaging with Rural Development and Agricultural Policy Processes*, Future Agricultures Working Paper No. 26, available at: www.future-agricultures.org/publications/research-and-analysis/working-papers/1539-gender-analysis-engaging-with-rural-development-and-agricultural-policy-processes/file (accessed 4 May 2016).

Oya, C. (2010) *Rural Inequality, Wage Employment and Labour Market Formation in Africa*, Working Paper No. 97, Geneva: International Labour Office.

Pain, A. (2007) *Agriculture for Development World Development Report 2008: A Critical Overview – Can Agriculture Drive Development & Reduce Poverty?* World Development Report Seminar Series, Department of Urban & Rural Development, SLU, 26 October.

Potter, R.B. (2014) 'The nature of development studies', in V. Desai and R.B. Potter (eds), *The Companion to Development Studies*, London: Routledge, pp. 25–27.

Potts, D. (2012a) *Whatever Happened to Africa's Rapid Urbanisation?* London: Africa Research.

Potts, D. (2012b) 'What do we know about urbanization in sub-Saharan Africa and does it matter?', *International Development Planning Review*, 34(1): v–xxi.

Potts, D. (2015) 'Debates about African urbanisation, migration and economic growth: what can we learn from Zimbabwe and Zambia?', *The Geographical Journal*, DOI: 10.1111/geoj.12139.

Proctor, F.J. (2014) *Rural Economic Diversification in Sub-Saharan Africa*, IIED Working Paper, London: International Institute for Environment and Development.

Proctor, F.J. and Lucchesi, V. (2012) *Small-Scale Farming and Youth in an Era of Rapid Rural Change*, London: International Institute for Environment and Development.

Radeny, M., van der Berg, M. and Schipper, R. (2012) 'Rural poverty dynamics in Kenya: structural declines and stochastic escapes', *World Development*, 40(8): 1577–1593.

Sarch, M.-T. and Birkett, C. (2000) 'Fishing and farming at Lake Chad: responses to lake-level fluctuations', *The Geographical Journal*, 166(2): 156–172.

Scoones, I. (1998) *Sustainable Rural Livelihoods: A Framework for Analysis*, IDS Working Paper No. 72, Brighton: Institute of Development Studies.

Scoones, I. (2009) 'Livelihoods perspectives and rural development', *Journal of Peasant Studies*, 36(1): 171–196.

Scoones, I. and Wolmer, W. (2002) *Pathways of Change in Africa: Crops, Livestock & Livelihoods in Mali, Ethiopia & Zimbabwe*, London: James Currey.

Sen, A. (1981) *Poverty and Famines: An Essay on Entitlement and Deprivation*, Oxford: Oxford University Press.

Shackleton, S., Shackleton, C. and Cousins, B. (2000) *Re-Valuing the Communual Lands of Southern Africa: New Understandings of Rural Livelihoods*, London: Overseas Development Institute.

Singh, K. (1999) *Rural Development: Principles, Policies and Management.* New Delhi: Sage.

Sitko, N.J. and Jayne, T.S. (2014) 'Structural transformation or elite land capture? The growth of "emergent" farmers in Zambia', *Food Policy*, 48: 194–202.

Sumberg, J. and Okali, C. (2013) 'Young people, agriculture, and transformation in rural Africa: an "opportunity space" approach', *Innovations: Technology, Governance, Globalisation*, 8(1/2): 259–269.

Takeuchi, S. (2000) *African Studies and Rural Development*, paper delivered at a conference, 'African Rural Development Reconsidered', Institute of Developing Economies, 27 January.

UNCTAD (United Nations Conference on Trade and Development) (2013) *Economic Development in Africa. Report 2013. Intra-African Trade: Unlocking Private Sector Dynamism*, Geneva: UNCTAD.

White, B. (2012) 'Agriculture and the generation problem: rural youth, employment and the future of farming', *IDS Bulletin*, 43(6): 9–19.

Wiggins, S. and Keats, S. (2015) *Topic Guide: Stepping out of Agriculture*, London: Overseas Development Institute.

Wiggins, S. and Proctor, S. (2001) 'How special are rural areas? The economic implications of location for rural development', *Development Policy Review*, 19(4): 427–436.

Wiggins, S., Kirsten, J. and Llambi, L. (2010) 'The future of small farms', *World Development*, 38(10): 1341–1348.

Wiggins, S., Argwings-Kodhek, G., Leavy, J. and Poulton, C. (2011) *Small Farm Commercialisation in Africa: Reviewing the Issues*, Futures Agriculture Research Paper No. 23, Brighton: Institute of Development Studies.

Wiggins, S., Argwings-Kodhek, G., Gebreselassie, S., Asuming-Brempong, S., Chirwa, E., Matita, M.M., *et al.* (2014) *Cautious Commercialisation: Findings from Village Studies in Ethiopia, Ghana, Kenya, Malawi & Tanzania*, Future Agricultures Working Paper No. 82, Brighton: Institute of Development Studies.

World Bank (1975) *Rural Development: Sector Policy Paper*, Washington, DC: World Bank.

World Bank (1997) *World Development Report 1997: The State in a Changing World*, Washington, DC: Oxford University Press.

World Bank (2007) *World Development Report 2008: Agriculture for Development*, Washington, DC: World Bank.

World Bank (2011) *Migration and Remittances Factbook 2011*, Washington, DC: World Bank.

World Bank (2015a) *Data*, available at: http://data.worldbank.org/indicator/SP.RUR.TOTL.ZS/countries/1W?display=graph (accessed 8 December 2015).

World Bank (2015b) *Popular Indicators*, available at: http://databank.worldbank.org/data/reports.aspx?Code=NV.AGR.TOTL.ZS&id=af3ce82b&report_name=Popular_indicators&populartype=series&ispopular=y (accessed 3 December 2015).

World Bank (2016a) *Sub-Saharan Africa*, available at: http://data.worldbank.org/region/SSA (accessed 4 May 2016).

World Bank (2016b) *Agriculture and Rural Development*, available at: http://data.worldbank.org/topic/agriculture-and-rural-development (accessed 4 May 2016).

Zoomers, A. (2014) 'Rural livelihoods in a context of new scarcities', in V. Desai and R.B. Potter (eds), *The Companion to Development Studies*. London: Routledge, pp. 230–234.

29

Rural marketing in sub-Saharan Africa

Some current issues

Gina Porter

Introduction

This chapter explores some of the prevailing opportunities and challenges that confront rural Africa in the marketing sector, and how such issues are being addressed, with an emphasis towards the social development and anthropological literature. It focuses principally on marketing of farm produce from the rural smallholder sector (whether the produce is being marketed in rural or urban centres). The chapter is divided into two major sections, the first of which considers some of the broad major changes that have been occurring in patterns of rural marketing in sub-Saharan Africa over the last decade or so. The second focuses entirely on one topic of growing significance: the role of new technologies in the (re)shaping of market access and market practices – from mobile phones to Internet and web-enabled devices.

Trends in smallholder rural marketing in sub-Saharan Africa

Policy efforts in sub-Saharan Africa aimed at moving agriculture away from a subsistence orientation towards higher productivity and market orientation have impacted substantially on rural market structures and actors over recent decades (Dercon and Gollin, 2014); this is evident on the ground, and is reflected in the changing focus of literature concerned with rural marketing in Africa. Thus, while much twentieth-century research up to the late 1980s focused on the operations conducted within periodic rural markets (an area where research interest appears to have declined), research in other areas has gathered pace, not least that with a particular focus on the expansion of global agribusiness interests through contract farming and the impact of supermarket-driven operations on rural marketing systems. In this section, I consider first the continuing importance of traditional periodic market systems and persistent debates regarding the potentially exploitative role of middlemen, before moving on to examine the impact of the development of alternative marketing channels through agribusiness contracts and supermarkets.

Periodic markets

Up to the 1980s, much twentieth-century research on marketing in sub-Saharan Africa was focused specifically on rural periodic marketplaces, on their organisation within wider trading networks, and on the diverse actors – retailers, wholesalers, dealers and suchlike – operating within them (Hodder, 1962; Smith, 1971; Wood, 1978; Clough, 1981). These markets, which have long formed a vital component in the supply chain linking rural producers to urban consumers, especially across much of western and central Africa, are typically in a state of flux:[1] while some are in decline and may ultimately die, others will flourish where conditions prove conducive. Changes occur in the relative importance and functions of such individual market centres in response to diverse influences, from new patterns of trade and commodity demand, to environmental conditions and political factors. In particular, there has been a long history of market reorganisation associated with road transport developments. Where new roads are built or much improved, traders continue to move their focus (as they have done in the past) from remote villages to the roadside, where passing transport commonly provides good business. Meanwhile, urban-based produce dealers and their assistants increasingly concentrate on buying produce in those more accessible markets. Eventually, remoter bush markets often collapse entirely, unless they have a specific role to play (such as bulking particular bush products or handling contraband goods), while larger roadside and peri-urban markets may become so successful that they are transformed from periodic to daily markets. This follows a long-established pattern of evolution. The casualties may include those women resident off-road who have always been expected to headload their family's produce to the market, but now face even longer journeys (Porter, 2011). Not all farmers take their produce for sale at such markets, however: some choose to sell instead at the farm gate to itinerant buyers (whether because of distance from market, lack of transport or other reasons). Nonetheless, rural marketplaces – which may perform bulking, wholesale and retail functions, or just a subset of those functions – continue to play a critical role in interlinking and supporting rural economies across Africa, both within individual countries and, in some cases, internationally (where linkages traverse national boundaries).

The middleman: a persistent bogeyman

The place of informal social institutions versus the state in the shaping of economic practices is a key contextual factor affecting rural marketing in Africa. The potentially exploitative role of middlemen in rural produce marketing has been debated for many years, and negative views about their so-called 'parasitical practices' prevail widely – not least among politicians aiming to justify continued state regulation of agricultural output markets – despite detailed field research over the years that has offered evidence to the contrary. Thus, in West Africa, Lyon and Porter (2007) have argued that the transactions conducted by ostensibly 'powerful', predatory dealers and commission agents are associated with considerable risk-taking on their part, particularly since they must so often play a significant role in informal financing of farm operations, given the reluctance of banks to lend to small-scale traders (or producers). A duality of trust and control underpins such relationships: if the dealers were to engage in massive exploitation, this would militate against the maintenance of the cooperation and trust that lies at the heart of the economically effective social networks that can be widely observed at work across diverse locations.[2]

More recently, Sitko and Jayne (2014) have provided a strong case supporting the positive benefits that rural assemblers bring to the maize trade in eastern and southern Africa, especially for farmers living in remote areas. They argue that states such as Zambia, Malawi and Kenya have been pursuing market policies that limit the capacity of private sector traders, including assembly traders, to participate in output markets, by public spending on parastatal marketing boards, which often procure grains from farmers at above-market, pan-territorial prices. Drawing on a substantial data set for four countries (Kenya, Zambia, Malawi and Mozambique), they demonstrate the highly competitive nature of village-level grain assembly, both in terms of the numbers of buyers in villages and the low price margins they operate on, and emphasise the crucial significance of the assembly trading sector to the vast number of smallholders in the region whose operations are too small to enable them to enjoy the economies of scale needed to directly link to large commercial buyers for their produce. The assembly trading sector brings significant benefits to smallholders in terms of local market access and lowered search and transport costs. By directing public spending in ways that undermine the capacity of assembly traders to participate in output markets, they argue that governments have hindered the ability of poor farmers with small quantities from effectively engaging in markets. Consequently, they assess those continued state interventions that undermine this stage in the value chain as unwise.

The expansion of contract farming

Over the last three decades, an alternative route to achieving improved market access has become increasingly available to many smallholders through contracts to produce crops for large companies – usually processor, exporter or retail chains. This is generally in exchange for inputs and technical assistance on credit that are repaid at harvest when the crop is sold to the company. The reorganisation of supply chains through contract farming and similar practices – unfamiliar territory for most African smallholders in the 1980s – is now common, for instance, in the horticultural sector in Kenya, in tobacco production in Malawi and sugar cane production in Swaziland. Early research into contract farming emphasised actual and likely negative impacts in terms of social tensions and inequality (notably Little and Watts, 1994), but more recent work has observed some more positive impacts for smallholders (e.g. Bolwig, 2012 for Ugandan organic coffee producers, though with fewer benefits for women). Overall, evidence to date suggests that small farmers growing high-value or perishable crops, or crops requiring particular technical expertise (as opposed to non-perishable grains and suchlike, easily sold on spot markets), can achieve higher productivity due to their access to inputs and technical assistance, and may achieve higher prices (due to improved quality and their access to new markets extending across global value chains). However, this has to be balanced against the new vulnerabilities to which farmers may be exposed if they become integrated within such chains, including imbalances of power that give some potential for company manipulation of agreements, for instance by using quality standards to evade agreements. There may also be increased local inequalities where some (usually smaller) farmers are excluded or family household labour loses out when contracts and payments are made to the (usually male) head of household. Contracts are only likely to be available to smallholders when companies judge that the commodity, the market and the policy environment are conducive: there is, unsurprisingly, evidence of a high rate of turnover of schemes in many countries, not least Kenya (Minot, 2011).

It is important to bear in mind that although contract production has grown in visibility across sub-Saharan Africa over the last two decades, and is likely to expand further, it is still a minority element in the rural marketing arena, especially where smallholders are concerned. In the first place, this is because small farmers, for the most part, are focused on supplying food markets for domestic consumption, as they respond to the growth of urban markets, and this is a sector where contract farming is (as yet) of little significance. Meanwhile, agribusiness companies tend to prefer to contract medium-sized or larger farmers (Oya, 2012). This may well apply to many of the large retailers and supermarkets (discussed further in the following section) that are playing an expanded role in the rural market sector in Africa, because there are issues around food safety, 'traceability' and quality standards that arguably make reliance on smallholders risky (Maertens and Swinnen, 2009; Minot, 2009).

Supplying supermarkets: the potential for small farmer involvement

Over a decade ago, Weatherspoon and Reardon (2003) drew attention to the way supermarket chains based in southern Africa were starting to transform the food retail sector across the subcontinent as they moved towards new intermediaries, sidestepping or transforming the traditional wholesale system. Shoprite, which opened in Zambia in 1995, now has 22 supermarkets across the country. In recent years, they and other companies have started to move into West Africa. Pick n Pay, with 11 stores in Zambia and plans to double that number, is also now planning stores in Ghana and Nigeria (England, 2015). While warning about the likely dangers for small producers supplying dynamic urban markets, they also noted the potential for development programmes to help small farmers participate in these new supply chains. There has been much debate subsequently. Neven *et al.* (2006) found that in Kenya, medium-sized fast-growing farms dominated supply as supermarkets selected preferred supplier farmers, and smallholder farmers were largely excluded due to capacity constraints in production and marketing; reliable communication technology was already becoming a factor in supplier selection (see also Barrett and Mutambatsere, 2005). By contrast, in Limpopo province, South Africa, relative remoteness has encouraged small farmers and local supermarkets to work together and develop an effective local vegetable supply chain (Bienabe and Vermeulen, 2007). More recently, Rao and Qaim (2011) found various organisations in Kenya trying to link smallholders to supermarket and export channels. One NGO, for instance, is facilitating farmer collective marketing approaches, and offers training on production techniques and special supermarket requirements and a mechanism to enable even relatively poor households with immediate cash needs to participate in supermarket channels, despite the characteristic delayed payment schedule. Rao and Qaim (2011) conclude, from a survey of over 400 farmers in one district, that supermarkets can contribute to income growth and poverty reduction in the small farm sector. They observe poverty rates among supermarket suppliers as 20 per cent lower than they would be were there no supermarkets, but argue that widespread impact would require broad infrastructure development as well as targeted institutional support. This suggests that while tensions in this sector still prevail regarding the potential of smallholders to participate, with some concerted effort and very careful regulation and monitoring to ensure smallholders are able to obtain a fair deal over the long term, there may be potential to overcome them.

The role of new technologies in the (re)shaping of market access and market practices: from mobile phones to Internet and web-enabled devices

Access to information in diverse areas, from prices and supply conditions to transport availability, is crucial for both traders and for smallholders who wish to market produce to their maximum advantage. Unsurprisingly, then, technological developments in the ICT sector have been seen as a highly promising route to the improved organisation of rural trade in sub-Saharan Africa. The potential for mobile phone communication and associated ICT applications as a leapfrogging device for enabling rural-based farmers and traders to improve their distance management in the marketing arena (extending well beyond transfer of price information) began to emerge around the turn of this century: charting of their impact has become a feature of much rural-focused literature over the last decade. At the same time, of course, the potential for mobile phones to also increasingly expose the 'bottom of the pyramid' in rural African markets to international capital (whether the object is to sell Coca-Cola or fertiliser, or indeed airtime for mobile phones) has also become increasingly evident. It is noteworthy that borderless roaming, an African innovation (first introduced by Celtel in East Africa in 2006), increasingly allows ease of access to markets across national borders (The Economist, 2009).

Over the last decade, a number of researchers have focused on the way the mobile phone can transform the potential of small farmers and rural-based traders to obtain information in a timely manner and at low cost. Reductions in information search and improved two-way communications within supply chains between traders and key customers or suppliers, enabled by the mobile phone, can be extremely important, especially when the difficulties and dangers of African travel are brought into the equation (Porter, 2015). Even a decade ago, when mobile phones were still a relative novelty across much of Ghana, Overå (2006) was able to demonstrate, through a series of cases (including a yam wholesaler, a young onion wholesaler-retailer, and women operating in the wholesale fish trade), how traders were using mobile phones to great advantage in Ghana, enabling them not only to substantially reduce transaction costs (through reduced time and cost of travel) as they substituted calls for much face-to-face communication or messages through intermediaries, but also to build their reputations (as they gave their number to customers), link to new customers, chase debts and facilitate money transactions. However, she stressed (Overå, 2006: 1313), importantly, that communication by mobile phone is not a trust-building mechanism in itself, but rather 'a tool to make *already existing trust-building mechanisms* – exchange of information, observation of behavior and sanctions against dishonest action, and contract fulfillment – more efficient' (emphasis added).

Other work conducted around the same time demonstrated how significant phone usage was emerging elsewhere in Africa, in diverse marketing contexts: among rural and travelling traders in Uganda (found to be increasingly accessing market information through mobile phone SMSs; Ferris *et al.*, 2008); in the Kenyan livestock trade (where traders in the north contacted their Nairobi counterparts, friends or kinsmen about market conditions and alerted them when animals had been dispatched by truck; Mahmoud, 2008); with reference to grain markets in Niger (where a quantitative study showed transaction costs lowered due to a reduction in information search costs; Aker, 2008); in the South African tomato trade (where ownership of a mobile phone is a significant factor in improving price management; Chikazunga *et al.*, 2008); and in the Ugandan banana trade (where mobile phone coverage was associated with a 10 per cent increase in farmers' probability of market participation for (perishable) bananas, but not for (less perishable) maize; Muto and Yamano, 2009). Overall, however, the focus in this literature looks to be more towards potential and actual benefits for rural traders rather than small producers.

Despite much positive reportage regarding the potential of mobile phones to assist with trading across Africa, particularly in organising transport, significant reservations were also being logged. These focused especially around the potential of the mobile phone to improve the prices that farmers receive. In particular, Molony (2008a) observed that accessing a market information system and buying outside established customer relationships could jeopardise farmers' access to credit (which they normally obtained through the dealer to whom they sell regularly), unless an alternative credit source were available: the significance of personalised relationships, cemented by face-to-face contact, cannot be ignored in the marketing business. In Tanzania, Molony thus observed how the tie to traders for credit supply appeared to be limiting the benefits of phone use among potato and tomato farmers: face-to-face interaction continued to be very important at the inception of business relations (even if trusted neighbours had vouched for the middleman), and were required by the farmer at a later stage when looking to obtain credit from their buyer (see also Molony, 2007, 2008b, 2009a, 2009b; Porter *et al.*, 2007).

In recent years, phone network operators have been expanding ever further into rural Africa. At the same time, low-cost handsets and intensive competition between providers have brought communication costs down so low that most rural residents within reach of a network – women as well as men, young and old – now have some access to a mobile phone (with a growing proportion personally owning a phone) and regularly take the opportunity to make contact with others by phone. Even if their own phone is out of action because of lack of battery power or airtime credit, a farmer needing to communicate on an urgent matter may well be able to find a household member, friend or neighbour willing to lend a working phone or airtime; such borrowing is extremely widespread. In tandem with this expansion in usage and ownership of mobile phones across rural Africa, a growing body of research is trying to develop and refine earlier observations regarding their impact, as discussed below.

Perhaps unsurprisingly, so far as prices are concerned, the picture regarding the impact of information now available through phones is turning out to be very complex. In Niger, for instance, Aker and Fafchamps (2013) find differences regarding impact by crop (whether or not it is storable), by time of year and by the type of 'agent' (consumer, trader or producer). Although traders reorient their purchases from high- to low-price markets, hurting some farmers and benefiting others, but producing overall net efficiency gains, they find no evidence that the gains translate into higher average prices for producers. As Zanello *et al.* (2014) emphasise with reference to northern Ghana, much trade in agricultural produce still takes place within social networks and depends on social relationships. Farmers have the option of selling at the farm gate (which they may do, for instance, because of poor transport availability) or taking their produce to a close or (larger) distant market; they found no evidence that ICT (phone and/or radio) encouraged farmers to travel to more distant markets. Tadesse and Bahiigwa (2015), who examined the impact of mobile phones on farmers' marketing decisions and prices they received in rural Ethiopia (based on household- and village-level data collected in 2012), suggest that in that country, the impact of phones is still weak among farmers because of lack of relevant information available for farmers to access by phone. Farmers may have access to a phone but are only likely to use it to search for information if they know *where* to search. This leads Tadesse and Bahiigwa (2015) to suggest that the efficiencies created due to mobile phones may be appropriated by traders more than by farmers.

Hans Peter Hahn (2012: 183), drawing on ethnographic fieldwork conducted in Burkina Faso between 2003 and 2008, takes a different tack in downplaying the economic role of the mobile phone. He proposes that 'mobile phones are acquired mainly for social purposes, like calling one's parents or children in town to ask them for remittances or to organize funerals . . . Economically relevant information does not play a major role in the everyday use of mobiles'. This arguably oversimplifies the complex intermixing of social and economic motivations. While usage is often ostensibly directed at maintaining and building social connections merely as a means of keeping in touch with friends, relations and romantic partners, or a wider desire to 'link up' (Horst and Miller, 2006), rather than the pursuit of specific sources of resource, in-depth interviews in research with young people presents a more complex picture. Evidence suggests that it is actually extremely difficult to measure economic as opposed to social usage, since the mobile phone is widely perceived as a multiple-function device, where economic and social conversations are highly and complexly interconnected (Sey, 2011); a social call is so very often densely interwoven with resource requests, and even ostensibly romantic connections commonly have significant resource implications (Porter *et al.*, 2012, 2015). Hahn's observation of the 'sometimes malicious entanglements' between road networks and mobile networks as criminal gangs undertake roadside hold-ups facilitated by mobile phones, moreover, suggests an arena where the economic value of the phone is clearly paramount! On Nigeria's Jos Plateau, there were certainly many stories among tomato traders about how journeys to southern markets were endangered by highway robbery similarly facilitated by phone (author fieldwork, 2001, 2004).

A focus on social relationships can be extremely enlightening when exploring the impact of the mobile phone in marketing. Burrell (2014), writing about mobile phone usage among Ghanaian market women, reflects on experience in urban and suburban Accra, though some of her findings have wider application (for both urban and rural settings in Africa). In particular, she observes how phones have become valued tools in women's businesses, used to manage transactions (especially important in the case of perishable goods), but rather than leading to a weakening of local networks as a consequence of network extension (a possible impact of phone usage discussed by Aker and Mbiti, 2010), they encourage *deeper* one-to-one relationships with trading partners (including long-distance relationships) and also support women's affiliation in group associations and associated marketplace governance, such as early settlement of incipient market disputes (see also Porter *et al.*, 2010). Use of the phone to obtain price information, Burrell (2014) notes, is often of limited concern to these women traders. This point about the limited significance of price in women's market trade is one that has often been noted in West African market contexts, since secure customer relations are often more crucial to them in the long term for the maintenance of a sound business. That was certainly an observable feature of women's trade in some villages in Gomoa district, Ghana, in the late 1990s (author fieldwork). The fact that women are often more limited in their mobility and access to distant markets often means that keeping loyal customers is especially important for them, since redirecting their trade as a response to variations in price is simply not feasible without creating tensions in the household (Porter, 2011). Being able to exchange information over the phone about road conditions, likely arrival time, etc. is of much value in these long-term relationships between the trader and her 'customers'. Burrell (2014) emphasises business phone use among the older generation of market women (finding youth focused on more social phone interactions), but research with young people (9–25) around their phone use in urban and rural Africa (field research in Ghana, Malawi and South Africa, in progress, see www.dur.ac.uk/child.phones/) suggests there is an emphasis also emerging here among some young traders regarding the value of the phone to their businesses.

Burrell and Oreglia (2015) extend the argument around the myth of marketers using mobile phones to obtain market price information with reference to an ethnographic study with fishermen and fish traders in rural Uganda in 2007–2008. Here, they suggest that a rhetoric of phones as enabling farmers to access prices and increase their revenue as a consequence may 'smooth the road for corporate market expansion' (Burrell and Oreglia, 2015: 276). They offer a counter-narrative, emphasising the material resources that fishermen need to be able to act on better information, and that market prices are often irrelevant or subordinate to other factors when it comes to making trade-related decisions that are embedded in a wider set of relationships, policies and institutions. As in Molony's Tanzania case, fishermen often depend on middlemen to whom they are indebted; thus, they cannot move to alternative traders. Meanwhile, building customer relations is essential for traders who need to maintain a constant supply of produce. As has been noted above in other contexts, they also observe that coordination work – avoiding wasted trips and building reputation through organising transactions (i.e. requiring information about available supplies, the location of vehicles, etc.) – is what brings particular value to their phone communication, as opposed to merely getting price information. They also comment on failures in market information services (MIS) where these depend on depersonalised SMS (text) messages; for smaller-scale market actors with few or no assets, such systems have little attraction if they only give information on prices, given concerns about risk and the costs of travelling to distant markets to benefit from reported higher prices. Rather, they suggest, it will be necessary to incorporate other types of localised information geared to the needs of small actors, if MIS are to work for them. Projects that appear to have been successful in targeting small, less well-off farmers are information-centred services, which integrate mobile phones within a broader set of integrated services. Duncombe (2014) cites the positive example of Collecting and Exchange of Local Agricultural Content (CELAC) in Uganda, which uses SMS to facilitate access to markets, as well as sharing good farming practice (i.e. they themselves can upload agricultural guidance and tips to the database), and the (externally supported) Kenya Agricultural Commodity Exchange (KACE), which is designed to match supply and demand, as well as to learn about current market prices. KACE reportedly provides a comprehensive market information system together with an e-marketplace that individual farmers can access through information centres, mobile phone short message service (SMS), interactive voice response (IVR) service, daily radio bulletins, a live radio auction service and online computer services. This emphasises the potential role of the phone here as an intermediating tool that can bring together and coordinate market and non-market actors and institutions (Duncombe, 2014). KACE estimates that it reaches 1 million farmers a day through radio, SMS or direct contact. Additionally, an estimated 250,000 small-scale traders in agricultural commodities access their services on a daily basis (Karugu, 2010).

It is also important to point out how, in recent years, *mobile money/banking services* have added considerably to the potential value of phones in rural marketing. These services have spread with remarkable rapidity across much of Africa since the appearance of M-PESA less than a decade ago (i.e. launched commercially in 2007). Such systems, which allow individuals to make cheap money transfers quickly and securely to another mobile phone user, without having a bank account (so long as they register for an account with the phone company concerned), appear to be mostly widely used in urban areas where agents tend to be more available (in Tanzania, for instance, they are least likely to be used by poor rural women; Kaffenberger and Butt, 2013). Nonetheless, they are also increasingly appreciated and patronised by rural people operating in locations with limited banking services, unsafe transport and a high incidence of highway robbery (who often also now avail themselves of less formal mobile money systems, simply by sending an airtime minute transaction (see Maurer, 2012). Kikulwe *et al.* (2014), in studies of smallholder farm households in rural Kenya, found a positive impact of mobile money on income, including in terms of marketing a larger proportion of their output. Even so, as with usage of phones for information gathering, the benefits in rural areas are likely to accrue far more frequently to traders who have large sums they need to move around in their businesses – especially larger-scale long-distance traders – than to small farmers (though both may benefit when it facilitates international remittances from family now living in the Global North). Tawah (2013), unsurprisingly, finds that no women in her Camerounian market study had used either mobile banking or transfer services (though mobile phones are widely used by these – mostly illiterate – women to call to organise meetings and facilitate business, as well as for making financial calculations using the calculator function).

Conclusion

This chapter has considered a number of current key debates around rural marketing in sub-Saharan Africa, with specific reference to the smallholder sector. The particular vulnerability of smallholders is frequently associated with their position in rural marketing arenas. However, evidence presented in the chapter suggests that, in some respects, market conditions may not be so disadvantageous for smallholder producers as has sometimes been assessed or anticipated, notably regarding their interactions with middlemen. In other areas, particularly contract farming (including contracts with supermarkets), there also now seems to be some potential for more substantial benefits to accrue to smallholders (in terms of livelihoods and associated household security), though new vulnerabilities may well emerge given continuing prevailing imbalances of power in the sector. Careful regulation and associated monitoring of impacts through detailed field investigation must continue here. In the area of ICT interventions (especially mobile phone access, available applications and associated patterns of usage), meanwhile, change is occurring so rapidly that it is difficult to make any firm assessment of impact in the smallholder or rural trading sectors with confidence. Overall, however, it would seem that major gains are more likely to be experienced by rural long-distance traders and bulking agents who now regularly cut costs by using phones to organise supplies and transport. By contrast, for farmers, the evidence of benefits accruing from mobile phone access to information is much less clear, unless this forms part of a wider integrated service.

Notes

1 In much of southern and eastern Africa, by contrast, indigenous marketing systems were actively, and institutionally, undermined during the colonial period in order to protect large-scale European farmers from competition.

2 See also Rousseau *et al.* (2015) for a recent study of the role of wholesalers in coordinating the shea trade in Burkina Faso, playing an important role in its smooth functioning and in 'profit sharing down the chain to the rural poor'.

References

Aker, J.C. (2008) *Does Digital Divide or Provide? The Impact of Cell Phones on Grain Markets in Niger*, BREAD Working Paper No. 177, Berkeley, CA: University of California.

Aker, J.C. and Fafchamps, M. (2013) *Mobile Phone Coverage and Producer Markets: Evidence from West Africa*, CSAE Working Paper WPS/2013-09.

Aker, J.C. and Mbiti, M. (2010) *Mobile Phones and Economic Development in Africa*, Working Paper 211, Washington, DC: Center for Global Development, June.

Barrett, C. and Mutambatsere, E. (2005) Agricultural Markets in Developing Countries, available at: www.researchgate.net/publication/228246702_Agricultural_Markets_in_Developing_Countries (accessed 22 December 2017).

Bienabe, E. and Vermeulen, H. (2007) *New Trends in Supermarket Procurement System in South Africa: The Case of Local Procurement Schemes from Small-Scale Farmers by Rural-Based Retail Stores*, paper presented at the I. Mediterranean Conference of Agro-Food Social Scientists, Barcelona, April.

Bolwig, S. (2012) *Poverty and Gender Effects of Smallholder Organic Contract Farming in Uganda*, available at: www.fao.org/ag/ags/contract-farming/ags-library/library-document-detail/en/c/206892/ (accessed 22 December 2017).

Burrell, J. (2014) 'Modernity in material form? Mobile phones in the careers of Ghanaian market women', *Review of African Political Economy*, 41(142): 579–593.

Burrell, J. and Oreglia, E. (2015) 'The myth of market price information: mobile phones and the application of economic knowledge in ICTD', *Economy and Society*, 44(2): 271–292.

Chikazunga, D., Bienabe, E. and Louw, A. (2008) *Determinants of Small Scale Farmers' Participation in Restructured Food Markets in South Africa: The Case of the Tomato Sector*, 12th EAAE 2008 Congress Session 2648, August, Ghent, Belgium.

Clough, P. (1981) 'Farmers and traders in Hausaland', *Development and Change*, 12: 273–292.

Dercon, S. and Gollin, D. (2014) *Agriculture in African Development: A Review of Theories and Strategies*, CSAE Working Paper WPS/2014 22, available at: www.csae.ox.ac.uk/workingpapers/pdfs/csae-wps-2014-22.pdf (accessed 22 December 2017).

Duncombe, R. (2014) 'Understanding the impact of mobile phones on livelihoods in developing countries', *Development Policy Review*, 32(5): 567–588.

England, A. (2015) 'Shelf life', *FT.com Magazine*, 25/26 July, pp. 23–25.

Ferris, S., Engoru, P. and Kaganzi, E. (2008) *Making Market Information Services Work Better for the Poor in Uganda*, CAPRi Working Paper 77, May.

Hahn, H.P. (2012) 'Mobile phones and the transformation of society: talking about criminality and the ambivalent perception of new ICT in Burkina Faso', *African Identities*, 10(2): 181–192.

Hodder, B.W. (1962) 'The Yoruba rural market', in P. Bohannan and G. Dalton (eds), *Markets in Africa*, Evanston, IL: Northwestern University Press, pp. 103–117.

Horst, H. and Miller, D. (2006) *The Cell Phone: An Anthropology of Communication*, Oxford: Berg.

Kaffenberger, M. and Butt, S. (2013) *Mobile Money: A Path to Financial Inclusion – Findings from the Tanzania Mobile Money Tracker Study*, InterMedia, April.

Karugu, W. (2010) *Kenya Agricultural Commodity Exchange (KACE): Linking Small-Scale Farmers to National and Regional Markets*, GIM Case Study No. B068, New York: UNDP, available at: www.africa-platform.org/sites/default/files/resources/Kenya_KACE_linking_smallscale_farmers_summary.pdf (accessed 22 December 2017).

Kikulwe, E.M., Fischer, E. and Qaim, M. (2014) 'Mobile money, smallholder farmers and household welfare in Kenya', *Plos One*, 9(10): article e109804, DOI: https://doi.org/10.1371/journal.pone.0109804.

Little, P. and Watts, M.J. (eds) (1994) *Living Under Contract: Contract Farming and Agrarian Transformation in Sub-Saharan Africa*, Madison, WI: University of Wisconsin Press.

Lyon, F. and Porter, G. (2007) 'Market institutions, trust and norms: exploring moral economies in Nigerian food systems', *Cambridge Journal of Economics*, 33(5): 903–920.

Maertens, M. and Swinnen, J. (2009) 'Trade, standards and poverty: evidence from Senegal', *World Development*, 37: 161–178.

Mahmoud, H.A. (2008) 'Risky trade, resilient traders: trust and livestock marketing in northern Kenya', *Africa*, 78: 561–581.

Maurer, B. (2012) 'Mobile money: communication, consumption and change in the payments space', *Journal of Development Studies*, 48(5): 589–604.

Minot, N. (2009) 'Contract farming in developing countries: patterns, impact and policy implications', in P. Pinstrup-Anderson and F. Cheng (eds), *Case Studies in Food Policy for Developing Countries, Vol II*, Ithaca, NY: Cornell University Press, pp. 37–49.

Minot, N. (2011) *Contract Farming in Sub-Saharan Africa: Opportunities and Challenges*, paper prepared for the policy seminar 'Smallholder Led Agricultural Commercialization and Poverty Reduction: How to Achieve It?', 18–22 April, Kigali, Rwanda, available at: http://fsg.afre.msu.edu/aamp/Kigali%20Conference/Minot_Contract_farming_%28AAMP%20Kigali%29.pdf (accessed 22 December 2017).

Molony, T.S.J. (2007) '"I don't trust the phone, it always lies": social capital and information and communication technologies in Tanzanian micro and small enterprises', *Information Technologies and International Development*, 3: 67–83.

Molony, T.S.J. (2008a) 'Running out of credit: the limitations of mobile telephony in a Tanzanian agricultural marketing system', *Journal of Modern African Studies*, 46: 637–658.

Molony, T.S.J. (2008b) 'Non-developmental uses of mobile communication in Tanzania', in J.E. Katz (ed.), *The Handbook of Mobile Communication Studies*, Cambridge, MA: MIT Press, pp. 339–352.

Molony, T.S.J. (2009a) 'Trading places in Tanzania: mobility and marginalization at a time of travel-saving technologies', in M. Bruijn, F. Nyamnjoh and I. Brinkman (eds), *Mobile Phones: The New Talking Drums of Everyday Africa*, Leiden: African Studies Centre and Bamenda, Cameroon: Langaa, pp. 92–109.

Molony, T.S.J. (2009b) 'Carving a niche: ICT, social capital, and trust in the shift from personal to impersonal trading in Tanzania', *Information Technology for Development*, 15: 283–301.

Muto, M. and Yamano, T. (2009) 'The impact of mobile phone coverage expansion on market participation: panel data evidence from Uganda', *World Development*, 37: 1887–1896.

Neven, D., Odera, M. and Reardon, T. (2006) *Horticulture Farmers and Domestic Supermarkets in Kenya*, MSU Dept of Agricultural Economics Staff Paper 2006-06.

Overå, R. (2006) 'Networks, distance and trust: telecommunications development and changing trading practices in Ghana', *World Development*, 34: 1301–1315.

Oya, C. (2012) 'Contract farming in sub-Saharan Africa: a survey of approaches, debates and issues', *Journal of Agrarian Change*, 12(1): 1–33.

Porter, G. (2011) '"I think a woman who travels a lot is befriending other men and that's why she travels": mobility constraints and their implications for rural women and girl children in sub-Saharan Africa', *Gender, Place and Culture*, 18(1): 65–81.

Porter, G. (2015) 'Mobile phones, mobility practices and transport organisation in sub-Saharan Africa', *Mobility in History*, 6: 81–88.

Porter, G., Lyon, F. and Potts, D. (2007) 'Urban food supply in sub-Saharan Africa in the twenty-first century: the role of trading intermediaries and market institutions', *Progress in Development Studies*, 7: 115–134.

Porter, G., Lyon, F. Adamu, F. and Obafemi, L. (2010) 'Conflict and cooperation in market spaces: learning from the operation of local networks of civic engagement in African market trade', *Human Organization*, 69(1): 31–42.

Porter, G. Hampshire, K., Abane, A., Munthali, A., Robson, E., Mashiri, M., *et al.* (2012) 'Youth, mobility and mobile phones in Africa: findings from a three-country study', *Journal of Information Technology for Development*, 18(2): 145–162.

Porter, G., Hampshire, K., Abane, A., Munthali, A., Robson, E., Bango, A., *et al.* (2015) 'Intergenerational relations and the power of the cell phone: perspectives on young people's phone usage in sub-Saharan Africa', *Geoforum*, 64: 37–46.

Rao, E.J.O. and Qaim, M. (2011) 'Supermarkets, farm household income, and poverty: insights from Kenya', *World Development*, 39(5): 784–796.

Rousseau, K., Gautier, D. and Wardell, D.A. (2015) 'Coping with the upheaval of globalization in the shea value chain: the maintenance and relevance of upstream shea nut supply chain organisation in western Burkina Faso', *World Development*, 66: 413–427.

Sey, A. (2011) '"We use it different, different": making sense of trends in mobile phone use in Ghana', *New Media and Society*, 13(3): 375–390.

Sitko, N.J. and Jayne, T.S. (2014) 'Exploitative briefcase businessmen, parasites, and other myths and legends: assembly traders and the performance of maize markets in eastern and southern Africa', *World Development*, 54: 56–67.

Smith, R.H.T. (1971) 'West African market places: temporal periodicity and locational spacing', in C. Meillassoux (ed.), *The Development of Indigenous Trade and Markets in West Africa*, London: Oxford University Press, pp. 319–346.

Tadesse, G. and Bahiigwa, G. (2015) 'Mobile phones and farmers' marketing decisions in Ethiopia', *World Development*, 68: 296–307.

Tawah, S. (2013) 'Market women and mobile phones in the northwest region of Cameroon: managing informal market livelihoods and trade routes through mobile phones', *Suomen Antropologi: Journal of the Finnish Anthropological Society*, 38(1): 59–82.

The Economist (2009) *Mobile Marvels: A Special Report on Telecoms in Emerging Markets*, 26 September, available at: www.economist.com/node/14483896 (accessed 5 January 2018).

Weatherspoon, D.D. and Reardon, T. (2003) 'The rise of supermarkets in Africa: implications for agrifood systems and the rural poor', *Development Policy Review*, 21(3): 333–335.

Wood, L.J. (1978) 'Rural markets in Kenya', in R.H.T. Smith (ed.), *Market Place Trade: Periodic Markets, Hawkers, and Traders in Africa, Asia and Latin America*, Vancouver: Centre for Transportation Studies, pp. 64–80.

Zanello, G., Srinivasan, C. and Shankar, B. (2014) 'Transaction costs, information technologies and the choice of marketplace among farmers in northern Ghana', *Journal of Development Studies*, 50(9): 1226–1239.

30

Rural–urban interaction and development in Africa

Kenneth Lynch

Introduction

This chapter sets out to reunite thinking about urban and rural areas in the study of development across the continent of Africa. It will discuss briefly past approaches to understanding development and rural–urban relations, arguing that there is a need to reconnect these two spaces in a profound way. For example, development studies – both theoretical and empirical – are often based on the premise that there is a clear distinction between the urban and the rural. However, this distinction has been challenged. There is research in Africa on 'rural' activities such as agriculture in 'urban' spaces (Lynch *et al.*, 2013), urban activities such as industry in rural spaces (Bryceson, 2002; see also Bryceson, Chapter 27, this volume), and changes taking place at the interface between urban and rural spaces, and on the increasing interdependence between these two realms (Iaquinta and Drescher, 2002; McGregor *et al.*, 2002; Tiffen, 2003). There is therefore a need to bring these disparate themes together. However, before embarking on a discussion of continent-wide trends, a note of caution: the data demonstrate a disparate set of patterns and trends. While 22.5 per cent of Chad's population is urban, this figure stands at 70.7 per cent in Algeria. While eastern Africa has an urbanisation level of 25.8 per cent, it is 51.6 per cent in northern Africa, 61.6 per cent in southern Africa and 45.1 per cent in western Africa (UN-Habitat, 2016). Similar variation exists in urbanisation rates. However, Potts (2008) cautions that much has been written about trends in sub-Saharan Africa based on questionable data that has been aggregated into received wisdom about trends that closer investigation reveals are questionable or at best may only be partially appropriate. So while this chapter sets out to identify some patterns, it is important to bear in mind the variation across the continent, between regions, and between countries across regions.

Theoretical reflections

Some of the earliest research on development focused on modernisation diffusion that explored the relations between city and countryside in the context of models of core–periphery, in the case of Friedman's (1966) core–periphery model or Vance's (1970) mercantile model. These influential theories are primarily focused on settlement hierarchies rather than the interaction between town and country, suggesting an urban focus; however, they are used to theorise about rural–urban interaction. Although not the originator of this concept, Lipton (1977, 1993) made considerable impact on development studies later, presenting a thesis on the ways in which urban-based industrialisation policies can have an adverse impact on the development of rural areas. After Lipton, a number of studies looked into Lipton's ideas of urban bias, some agreeing that urban bias undermined rural development, some arguing that the distinction between urban and rural areas was rather crude and did not reflect the complex reality (Corbridge and Jones, 2005; Bezemer and Headey, 2007). More recently, however, the question of the distinction between urban and rural development is being questioned. This questioning has been particularly strong in the field of demography and migration studies. The impact of the economic crises of the 1980s has also prompted research on the differentials between cities and the rural areas (Potter *et al.*, 2008). Continued urban growth has prompted concerns about the environmental impacts on the countryside (Potter *et al.*, 2008). This is illustrated – bearing in mind Potts' (2008) reservations about continent-wide data – in the developing trend of growing urban populations depicted in Figure 30.1. This chart illustrates the dramatic growth that appears to be taking place in both urban and rural parts of the continent, but the growth of the urban population is predicted to have grown from 200 million in 1950 to over 500 million by 2000, and will reach more than 900 million and exceed the rural population by 2036. Governments of the continent are aware of this trend, and have been working to tackle the challenges it presents. Examples include regional development initiatives based on small towns (Roberts, 2014; Roberts and Hohmann, 2014) and analysis emphasising the importance of maximising rural–urban interaction for development. These disparate critiques have begun to coalesce into a stronger body of research in recent years, though the gathering transition of populations across the developing world into more urban-based populations has meant a greater focus on livelihoods.

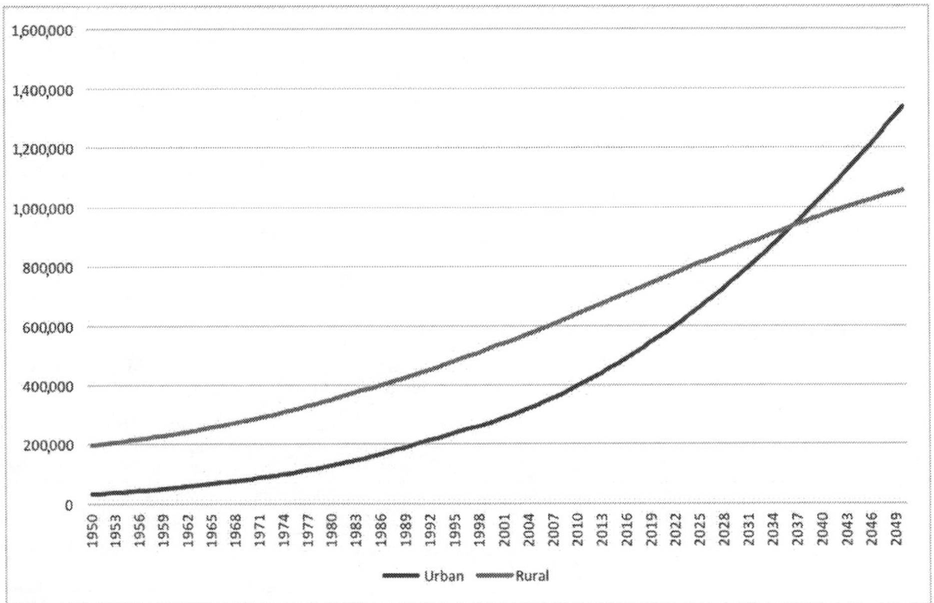

Figure 30.1 Urban and rural population trends in Africa, 1950–2050

Source: Adapted from UN (2014).

This argument does not necessarily break new ground, but proposes the collation of a wide range of research and theory that has relevance for this question of interactions between urban and rural areas. This is proposed within a broader context of research trends that appear to be moving away from the study of the physical interface between urban and rural, to conceiving of the relationships between them as more important. This is even the case in recent approaches to the peri-urban interface where rural and urban come into contact (Tacoli, 1998). Even in the physical space where the urban and the rural meet, there is an emerging consensus that the physical location of these linkages is less important than the way they are constructed and structured (Stephens, 2000; McGranahan and Satterthwaite, 2002). There is evidence that major international oganisations, as well as local NGOs, are moving in a similar way in a realisation that practically, as well as theoretically, the urban and the rural need to be considered alongside each other (see Table 30.1). This chapter therefore proposes a focus on flows between urban and rural areas. This encourages a consideration of goods, people and ideas across the interface between cities and the countryside.

In spite of the challenge of fluidity and fragmented identities that play a role in the lives of the people who communicate, exchange and travel the links that bridge the urban–rural divide, a key contention is that this fluidity is often a deliberate strategy by people living in rural and urban areas in order to maximise their livelihood opportunities. This focus on flows and linkages is about the links between the areas rather than the structures or processes that separate them. This provides more useful and more powerful analysis of the relationships between city and countryside. Many of these linkages are developed in more detail elsewhere (Lynch, 2005), but this chapter will focus briefly on some of the overarching themes.

Table 30.1 International activities focusing on rural–urban links

Organisation	*Activity*	*Weblinks*
IIED	International Institute for Environment & Development, urban–rural linkages research	www.iied.org/urbanisation-rural-urban-linkages
UN-Habitat	Promoting rural–urban linkages; Conference in Nairobi, October 04 Bibliography.	www.unhabitat.org/HD/hdv5n1/contents.htm www.unhabitat.org/interlinkages_conf.asp
IFPRI	International Food Policy Research Institute, rural–urban linkages programme	www.ifpri.org/program/rural-urban-linkages
USAID Urban	Promoting linkages between cities, towns and rural areas for development	https://urban-links.org/
RUDI	Rural Urban Development Initiative, Tanzania; private sector initiative focused on connecting small rural enterprises and communities	www.rudi.or.tz/
FAO Food for the Cities	Food and Agriculture Organization; food links research, publications and policy recommendations	www.fao.org/fcit/fcit-home/en/

Recent trends and reflections

The rapid population growth of Third World cities gives rise to concerns about the changing nature of the relationship between urban and rural. The evidence for this is in a recently growing number of publications, research reports and policy documents of international organisations, which emphasise key development concepts such as decentralisation (see also Table 30.1). The UNDP (2017) estimated that the world's urban population had reached 4 billion by 2015, accounting for 54 per cent of the global total. This is an increase on 30 per cent in 1950, and it is likely to go 66 per cent urban by 2050. During this period, Africa's urban population grew by between 3.24 and 3.55 per cent per annum.

One of the key departures from previous approaches is that a focus on flows presents the possibility for each flow to work in either direction. Under certain conditions, one-way flows may dominate or the emphasis may change over time or from one context to the next. The flows of people, food and natural flows are more visible and tangible. Money and ideas are less tangible and involve service sectors in delivery, and so are less obvious. One of the main limitations in discussing rural–urban linkages is that it immediately represents city and countryside as being clearly separate. This ignores more problematic issues such as the definitions of the rural–urban interface, where distinctions between city and countryside can become blurred. Another important theme is the role of intermediate settlements.

Many major world developments have been linked to the relations between urban and rural realms. For example, according to Owens (2010), in Dar es Salaam there is a long history of links between the city and the rural origin areas of the population. He points out that there is growing evidence that over history, there are patterns of living and households that defy conventional views of linear rural-to-urban movement. This evidence finds instances of circular, stepwise and constant back and forth movement – particularly in southern Africa, where historical government controls and migrant labour patterns kept the labour moving in search of employment. Although there is an expectation that gradual emerging permanence develops, there is evidence that migration is sporadic and episodic, subject to the vagaries of the economics and political instabilities of the country and at an international level. Owens (2010) suggests that the evidence from Dar es Salaam shows that this maintenance of identification with the rural origin location serves to resist proletarianisation, and this is reinforced by migrants returning for cultural festivals and key family events, such as births, weddings or funerals. The way in which mobility is utilised and rurally based cultural identities retained in the city also provides cultural and social capital. For example, particular ethnic groups may specialise in transporting and wholesaling particular rurally produced commodities. See, for example, van Donge (1994) on Waluguru vegetable traders in Tanzania, Gervais-Lambony (2012) on Mina traders in Togo, and FAO (2012) on *maguevas*, or female traders who buy crops direct from growers' fields, for markets in Maputo and Matola in Mozambique.

While the Tanzanian example illustrates a deliberate approach of exploiting the relationships between the urban and rural areas, Hodder (2000) argues that inevitably, rural agricultural sectors and urban industrial sectors play strategic roles in each other's development. He identifies six key reasons for these close links between the two sectors:

1 Agriculture depends on manufactured goods for both the transformation of agriculture (such as farm tolls, machinery, inputs) and for consumer goods that are in demand as agricultural incomes rise (such as radios and bicycles).
2 As agriculture incorporates more technology in its activities, labour becomes a less significant factor. More technologically advanced agriculture releases capital and labour that move into the urban industrial sector.

3 Agriculture provides raw materials for some industries, such as tobacco, cotton and sisal.
4 Agriculture for export can earn foreign exchange, which is important for purchasing items that are vital to industrial processes. These include commodities such as petroleum, chemicals and technology that are not produced locally.
5 There is an important balance to be struck in incomes, prices and taxation between the urban and the rural areas. For example, high food prices provide rewards to farmers and incentives to increase production, but may mean high prices in urban areas, and can lead to poverty and unrest. Taxation in the agricultural sector may be necessary to raise revenues to finance public expenditure, but this may act as a disincentive to farmers, particularly if much of the expenditure is focused on cities or industries.
6 In rapidly urbanising countries, agriculture produces strategically important food for the growing number of urban residents in order to ensure food security at prices that are affordable.

(after Hodder, 2000: 80–82)

These six reasons explain the key interdependencies between rural and urban areas. While the discussion has so far concentrated on the flows between the urban and the rural, one of the main reasons for the separate approaches is the attempt by many to identify the defining characteristics of what is 'urban' and what is 'rural'. However, Rigg (1998), in an analysis of South East Asia, cautions against 'pigeonholing' when it comes to such definitions. For example, he argues that to separate people into urban or rural categories is problematic. He outlines three main difficulties associated with such categorisation:

1 Registration records often do not detect changes in residence. For various reasons, it may be undesirable for recent in-migrants to be registered as urban dwellers. Rigg gives examples of under-reporting of urban residence, particularly in relation to the controversies this can pose during elections, when the definition of the area can have implications for the number of political representatives or the authority into which the representatives are elected (for further discussion of this, see Lynch, 2005).
2 Allocating people to discrete categories such as 'urban' or 'rural' assumes that these categories accurately reflect their realities. Rigg's own empirical research in Thailand, among others, has demonstrated the significance of fluid, fragmented and multilocation households to survival strategies. In Africa, there is also evidence of this, and that this results in households straddling and moving across the rural–urban interface (Smit, 2016) or at least maintaining social and cultural links with their rural origins (Owens, 2010). Thus, categorisation of them as one or the other makes no sense. In fact, boundaries between urban and rural can offer opportunities to trade across (Leybourne and Grant, 1999).
3 Rigg argues that many Asian urban residents do not consider the cities and towns they live in as 'home'. This is because they ultimately intend to return to their rural origins. The same is found in Africa, where many African urban residents speak of their villages. Rigg argues that this brings the issue of identities of the individuals into focus. 'Home' and 'place' are therefore ambiguous and shifting notions, involving multiple identities that can be embodied simultaneously.

A final concern that could be added to Rigg's three points above relates to the blurring of the actual geographical definition of the rural–urban divide. This is particularly the case where cities are physically expanding rapidly and extending their physical limits and their influence outwards into the rural areas. Adding to the difficulty of definition, we find that different sizes of settlement are defined as 'urban' by different countries. For example, Tati (2012) points out that in the DR Congo, an urban area is defined as a *municipalité* made up of a minimum population of 10,000, together with certain service delivery functions. Relatively small settlements are not counted as urban. However, the cities include extended metropolitan areas in their definitions of the city – or city-region. While such data allow for monitoring demographic changes in countries across time – for example, comparing rates of urban growth from one census to the next – this argument about the differences of definition suggests that comparisons between countries are problematic to say the least.

For example, Tsuruta (2003) reports that from its early establishment, Dar es Salaam in present-day Tanzania grew initially from rural in-migrants who maintained their links with their rural origins through cultural and ethnic-based associations focused around music, dance and other cultural activities. The UNHCR (2016) reports that Nigeria, the continent's largest country by population, experienced urban population growth of 6 per cent per annum between 1970 and 1996, and the proportion of the population living in urban areas grew from 7.4 per cent to 20 per cent of the country. The result is that the progression from rural to urban, or vice versa, is very unclear and varies from one country to another. Hardoy *et al.* (2001) suggest that it is best not to consider the proportion of urban and rural populations in terms of precise percentages because of the difficulty of enumerating, but rather in terms of a range of broad trends.

Researchers have also presented arguments for the separate definition of high-density rural areas that can occur under a particular set of characteristics. For example, McGee (1991) describes a kind of high-density rural area that he calls *desakota*, arguing that it is peculiar to Asia and suggesting it forms as a result of the metropolitan urban area extending its economy and influence into the surrounding rural areas, resulting in an intense mixture of agricultural and non-agricultural activities (see later discussion). Examples of *desakota* include areas around Jakarta, Manila and Bangkok. Mortimore (1998), however, carried out research on the environmental implications of what he called the 'close settled zone' around Kano in northern Nigeria, which displayed similar characteristics. This is a densely settled area that is extensive, but has maintained its rural nature. More recently, Cloete (2011) argues that the emergence of *desakota* regions in sub-Saharan Africa are in evidence. Examining population density in some parts of coastal West Africa, areas around Lake Victoria and along parts of the North African coast, there appears to be the emergence of *desakota* on the continent. However, Cloete argues that most urban and rural areas are extending their influence into the peri-urban zone and beyond, and there is evidence that this is an important and neglected feature of African cities. Cloete (2011) argues that a better understanding of and response to urbanisation can be developed in which the borders between urban and rural are broken down to provide new solutions to old urban problems.

Potter *et al.* (2008) argue that geographies of development are about relationships between people, environment and places in different locations and at a variety of scales. The nature of the interaction between city and countryside is at the heart of this focus. The relationships, by implication, involve geographies of groups and individuals. The location of a household, whether city or countryside, has a bearing on its livelihood assets and opportunities. The interaction between urban and rural is initiated in an attempt to take advantage of the differentials or complementarities between the two areas. The differential may be in the form of income and cost of living, supply and demand, or security and hazard. Potter *et al.* (2008: 319) go on to argue that the 'nature and relative significance of these relationships are changing constantly, both through time and space, and are themselves determined to a large extent by complex movements and flows of people, commodities, finance, ideas and information'.

Much of the basis of research that examines population in relation to the rural–urban interface relates to the migration of people across the rural–urban interface. In turn, much of this research focuses on rural-to-urban migration. To some extent, this is understandable, since net migration flows tend to be in this direction. However, there are questions about some of the research that has been carried out. For example, Feldman (1999) argued that previous approaches to the study of rural–urban migration were flawed. Based on his work in South Asia, but also resonating with the work of researchers working in Africa (Potts, 2008; Bennett *et al.*, 2015), he argued that problems can be summarised as:

- They focused on seasonal or temporary migration rather than on permanent migrants.
- They tended to focus on individuals rather than on households, thus focusing on migration as an outcome resulting from rural or urban change rather than as a process.
- There was a lack of research on the impacts of out-migration, such as the loss of agricultural labour capacity, the impact on rural infrastructural needs, and the impacts of the increased incomes on consumption or investment.

The result is an assumption that the remaining rural population is primarily dependent on remittances from urban migrants. This assumption has meant that non-farm employment and micro-industries in rural areas have received little research attention until relatively recently (e.g. see Bryceson, Chapter 27, this volume). Only recently has it been possible to examine the highly varied demographic profiles of migrants and their complex decision-making processes (e.g. see Bennett *et al.*, 2015a, 2015b on migration, gender and children; Potts, 1997, 2008 on migration and livelihoods). In particular, research that focuses on the social networks of migrants has provided powerful analyses of migration processes that are linked to more complex understandings of informal sector activities, slum and squatter communities, rural communities and the flows of people, capital and ideas through the social networks: 'What kinds of continuity and discontinuity shape information and resource flows between (migration) sites? What kind of kin and familial networks facilitate and hinder mobility?' (Feldman, 1999; see also Lynch, 2005; Bennett *et al.*, 2015a, 2015b).

The nature of the relation between town and countryside in the region could be where:

> a large number of urban centers that will have strong economic ties based on exchanges of labor, goods and services with their surrounding rural areas. A network of local economic areas will emerge – unheard of thirty years ago – and just beginning in countries like Nigeria. These 'market watersheds' will cut across national borders in ways that will encourage economic growth.
>
> *(Club du Sahel, 1994: 54)*

Potts (2008) argues that because the implications of migration can be negative as well as positive, the outcome depends on the social relations between the migrants and those affected. This may explain why many households practise livelihood strategies that straddle the urban and rural locations and income opportunities. It is important therefore that governments are sensitive to the informal institutions that structure and facilitate such migration and livelihood processes. It is these institutions that are the key to how migrants' decisions can support livelihoods. This may well be a part explanation for the incredible growth of cellular telephones in sub-Saharan Africa. The importance of social networks provides a demand for technology to ensure those networks are connected (Poole and Lynch, 2003).

Conclusion

This chapter has set out to identify some of the key trends in rural–urban linkages and their significance for thinking and development policy or strategy interventions. This chapter has demonstrated that there is evidence in the literature of urban–rural relationships taking place on a number of levels. Whether it is credit provision or food supply, pollution or water supply, the urban–rural relationship has an impact on both the urban and the rural areas. Until the last 20 years, research has not given significant coverage to rural–urban relationships in Africa, with the exception of research on migration. Recent developments among INGOs strategies suggest a growing interest in the potential of promoting rural–urban interaction. There is now a need to better understand the relations between the cities and the countryside in order that those that use the interstices between the city and the countryside are better able to use such spaces to make a livelihood, produce or trade. Rural–urban relations can be exploitative, but they also provide potential for mutually beneficial links.

As Ellis and Harris (2004: iii) argue, 'poverty reduction is perhaps not so much to do with supporting spatially and temporally static and stagnating sub-sectors of national economies. Rather it is about running with growth where and when it occurs'. This suggests that a focus on the linkages between urban and rural areas may be as important as the focus on the areas themselves.

References

Bennett, R., Hosegood, V., Newell, M.-L. and McGrath, N. (2015a) 'Understanding family migration in rural South Africa: exploring children's inclusion in the destination household of migrant parents', *Population, Space and Place*, 21(4): 310–321.

Bennett, R., Hosegood, V., Newell, M.-L. and McGrath, N. (2015b) 'An approach to measuring dispersed families with a particular focus on children "left behind" by migrant parents: findings from rural South Africa', *Population, Space and Place*, 21(4): 322–334.

Bezemer, D.J. and Headey, D. (2007) *Agriculture, Development and Urban Bias: Research Papers in Economics*, available at: https://ideas.repec.org/p/pra/mprapa/7026.html (accessed 7 November 2016).

Bryceson, D.F. (2002) 'The scramble for Africa: reorienting rural livelihoods', *World Development*, 30(5): 725–739.

Club du Sahel (1994) *West Africa Long Term Perspective Survey*, Paris: Club du Sahel, OECD.

Cloete, J. (2011) *Water-Based Ecosystems, Poverty Alleviation and the Desakota Phenomenon in Sub-Saharan Africa: A Regional Literature Review*, Grahamstown, South Africa: Consortium for Ecosystem Services and Poverty Alleviation in Sub-Saharan Africa, with contributions from Sheona Shackleton, Dylan McGarry and Mike Mortimore.

Corbridge, D. and Jones, G. (2005) *The Continuing Debate About Urban Bias: The Thesis, Its Critics, Its Influence, and Implications for Poverty Reduction*, Research Papers in Environmental and Spatial Analysis No. 99, Department of Geography and Environment, London School of Economics and Political Science, available at: www.lse.ac.uk/collections/geographyAndEnvironment/research/Researchpapers/99%20corbridge%20jones.pdf (accessed 7 March 2007).

Ellis, F. and Harris, N. (2004) *Development Patterns, Mobility and Livelihood Diversification*, keynote paper delivered at DFID Sustainable Development Retreat, University of Surrey, Guildford, 13 July.

FAO (Food and Agriculture Organization) (2012) *Growing Greener Cities in Africa: First Status Report on Urban and Peri-Urban Horticulture in Africa*, Rome: FAO.

Feldman, S. (1999) *Rural–Urban Linkages in South Asia: Contemporary Themes and Policy Directions*, paper presented at Workshop in Poverty Reduction and Social Progress: New Trends and Emerging Lessons, A Regional Dialogue and Consultation on World Development Report 2001 for South Asia, Rajendrapu, Bangladesh, 4–6 April.

Friedman, J. (1966) *Regional Development Policy: A Case Study of Venezuela*, Cambridge, MA: MIT Press.

Gervais-Lambony, P. (2012) 'Lomé', in S. Bekker and G. Thorborn (eds), *Power and Powerlessness: Capital Cities in Africa*, Cape Town: Human Sciences Research Council and CODESRIA, pp. 45–63.

Hardoy, J., Mitlin, D. and Satterthwaite, D. (2001) *Environmental Problems in an Urbanizing World: Finding Solutions for Cities in Africa, Asia and Latin America*, London: Earthscan.

Hodder, R. (2000) *Development Geography*, London: Routledge.

Iaquinta, D. and Drescher, A. (2002) *Defining Periurban: Understanding Rural–Urban Linkages and Their Connection to Institutional Contexts*, paper presented at the Tenth World Congress, IRSA, Rio, 1 August 2000.

Leybourne, S.L. and Grant, M. (1999) 'Bottlenecks in the informal food-transportation network of Harare, Zimbabwe', in M. Koc, R. MacRae, L.J.A. Mougeot and J. Welsh (eds), *For Hunger-Proof Cities: Sustainable Urban Food Systems*, Ottawa: International Development Research Centre, pp. 110–114.

Lipton, M. (1977) *Why Poor People Stay Poor: A Study of Urban Bias in World Development*, London: Temple Smith.

Lipton, M. (1993) 'Urban bias – of consequences, classes and causality', *Journal of Development Studies*, 29(4): 229–258.

Lynch, K. (2005) *Rural Urban Interaction in the Developing World*, London: Routledge.

Lynch, K., Maconachie, R., Binns, J.A., Tengbe, P. and Bangura, K. (2013) 'Meeting the urban challenge? Urban agriculture and food security in post-conflict Freetown, Sierra Leone', *Applied Geography*, 36: 31–39.

McGee, T. G. (1991) 'The emergence of desakota regions in Asia: expanding a hypothesis', in N. Ginsberg, B. Poppel and T.G. McGee (eds), *The Extended Metropolis*, Honolulu, HI: University of Hawaii Press, pp. 3–25.

McGranahan, G. and Satterthwaite, D. (2002) 'The environmental dimensions of sustainable development for cities', *Geography*, 87(3): 213–226.

McGregor, D., Simon, D. and Thompson, D. (2002) 'Contemporary perspectives on the peri-urban zones of cities in developing countries', in D. McGregor, D. Simon and D. Thompson (eds), *The Peri-Urban Interface: Approaches to Sustainable Natural and Human Resources Use*, London: Earthscan, pp. 3–17.

Mortimore, M. (1998) *Roots in the African Dust: Sustaining the Drylands*, Cambridge: Cambridge University Press.

Owens, G.R. (2010) 'Post-colonial migration: virtual culture, urban farming and new peri-urban growth in Dar es Salaam, Tanzania, 1975–2000', *Africa*, 80(2): 249–274.

Poole, N. and Lynch, K. (2003) 'Agricultural market knowledge: appropriate delivery systems for a private and public good?', *Journal of Agricultural Extension and Education*, 9(3): 117–126.

Potter, R.B., Binns, J.A., Elliott, J.A. and Smith, D. (2008) *Geographies of Development*, 3rd edition, Harlow: Addison Wesley Longman.

Potts, D. (1997) 'Urban lives: adopting new strategies and adapting rural links', in C. Rakodi (ed.), *The Urban Challenge in Africa: Growth and Management of Its Large Cities*, Tokyo: United Nations University, pp. 447–494.

Potts, D. (2008) *Rural–Urban and Urban–Rural Migration Flows as Indicators of Economic Opportunity in Sub-Saharan Africa: What Do the Data Tell Us?* Migrating Out of Poverty Working Paper 9.

Rigg, J. (1998) 'Rural–urban interactions, agriculture and wealth: a South East Asian perspective', *Progress in Human Geography*, 22(4): 497–522.

Roberts, B. (2014) *Managing Systems of Secondary Cities: Policy Responses in International Development*, Brussels: Cities Alliance.

Roberts, B. and Hohmann, R. (2014) *The Systems of Secondary Cities: The Neglected Drivers of Urbanising Economies*, CIVIS #7, Brussels: Cities Alliance, available at: http://citiesalliance.org/sites/citiesalliance.org/files/CIVIS%20SECONDARY%20CITIES_Final.pdf (accessed 5 January 2018).

Smit, W. (2016) 'Urban governance and urban food systems in Africa: examining the linkages', *Cities*, 58: 80–86.

Stephens, C. (2000) 'Inequalities in urban environments, health and power: reflections on theory and practice', in C. Pugh (ed.), *Sustainable Cities in Developing Countries*, London: Earthscan, pp. 91–114.

Tacoli, C. (1998) *Bridging the Divide: Rural–Urban Interaction and Livelihood Strategies*, London: International Institute for Environment and Development.

Tati, G. (2012) 'Brazzaville', in S. Bekker and G. Thorborn (eds), *Power and Powerlessness: Capital Cities in Africa*, Cape Town: Human Sciences Research Council and CODESRIA, pp. 104–118.

Tiffen, M. (2003) 'Transition in sub-Saharan Africa: agriculture, urbanization and income growth', *World Development*, 31(8): 1343–1366.

Tsuruta, T. (2003) 'Urban–rural relationships in colonial Dar es Salaam: some notes on ethnic associations and recreations, 1930s–1950s', *Memoirs of the Faculty of Agriculture of Kinki University*, 36: 50–72.

UN (United Nations) (2014) *2014 Revision of World Urbanization Prospects*, New York: UN Department for Economic and Social Affairs, Population Division.
UNDP (United Nations Development Programme) (2017) *Human Development Report, 2016*, New York: UNDP.
UN-Habitat (United Nations Human Settlements Programme) (2016) *Urbanisation and Development, Emerging Futures: World Cities Report 2016*, Nairobi: UN-Habitat.
UNHCR (United Nations High Commission for Refugees) (2016) *Global Report 2016*, Geneva: UNHCR.
Vance, J. (1970) *The Merchant's World: The Geography of Wholesaling*, Englewood Cliffs, NJ: Prentice Hall.
van Donge, J.K. (1994) 'Waluguru traders in Dar es Salaam: an analysis of the social construction of economic life', *African Affairs*, 91: 181–205.

Section 8
Urban Africa

31

The politics of urban management and planning in African cities

Andrea Rigon, Joseph M. Macarthy, Braima Koroma and Alexandre Apsan Frediani

Introduction

Half of Africa's population is expected to live in a city by 2035, up from 40 per cent today. This is a testament to the fact that a quarter of the world's fastest-growing cities are in Africa and 52 African cities already have more than 1 million inhabitants each. But these cities are only projected to absorb a quarter of the growth in urban populations, meaning that small and medium cities will host the majority of new urban dwellers (UN-Habitat, 2014: 23–25). African cities are the most unequal in the world, posing a major challenge to their future (UN-Habitat, 2010: 2).

The starting point of this chapter is that urban management and planning are political issues. This chapter addresses key issues in urban management and planning that pertain to many African cities, including exploring what is managed and planned, by whom, and for whom. Complicating this task is the great diversity of urban realities that, as Myers (Chapter 35, this volume) argues in this book, are impossible to generalise. The diversity of histories and cultures, during both the precolonial and colonial periods, have shaped today's management and planning practice. Different colonial histories brought in a range of different planning and management systems and these intertwined with customary practices. These fusions have generated a variety of postcolonial hybrid models in which the modern state bureaucracy and traditional authorities coexist, particularly in the area of land administration.

The complexity of current arrangements requires specific responses to the distinctiveness of existing management and planning practices. This implies a focus on understanding existing political settlements in cities, which are 'the formal and informal processes, agreements, and practices that help consolidate politics, rather than violence, as a means for dealing with disagreements about interests, ideas and the distribution and use of power' (Laws and Leftwich, 2014: 1). New policies and practices may disrupt existing settlements, so identifying management and planning responses requires exercising political sensitivity and negotiating with all of the local actors involved.

Three fundamental and connected issues across African cities are city-level governance, power decentralisation, and municipal finance. The first looks at governance structures at the city scale; power decentralisation has to do with how city-level government is selected and the powers that it has; and municipal finance is about the capacity of cities to achieve financial autonomy and generate revenues. These three issues are fundamental for examining the legacy of colonial planning, managing and taxing urban land, the informal city, the process for 'planning from below', and urban infrastructure.

City-level and metropolitan governance

Democratic city-level government with sufficient powers and financial autonomy are critical factors for achieving or continuing success in many cities. The devolution of powers to cities has contributed to the transformation of several Latin American cities, which are considered global examples to learn from. These have become central issues in the New Urban Agenda, and African mayors at last year's Habitat III[1] were very vocal in advocating for them.

Many African cities, however, do not have administrative and governance structures that align with city boundaries. For instance, in Nigeria, there are three tiers of government: federal, state and local government areas (LGAs). As a result, there is no democratically elected city-level government. Small towns may more or less fit within the boundaries of LGAs, but LGAs lack power and autonomy. The only exception is the city state of Lagos, where the city has expanded to cover most of the territory of the state, effectively rendering the Governor of Lagos State an elected mayor of the megacity.[2] For all other cities, the number of uncoordinated local government bodies makes consistent city-level planning difficult, particularly in cities that are not state capitals. For example, the city of Ibadan is spread out over five LGAs and Kaduna is spread over four (although two also cover some areas outside the city). LGAs often have limited power and report directly to state governors without an intermediary authority at city level. The lack of unified governance can increase divisions in cities. In Kaduna, the administrative separation of the city reflects the religious divide, complicating efforts to overcome sectarian conflicts between Christians and Muslims. Different LGAs within one city may also be run by different political parties, inhibiting coordination. Moreover, the areas of jurisdiction of LGAs may also include rural or peri-urban areas, meaning that local government departments must plan rural as well as urban interventions (Rigon *et al.*, 2015). In Kampala and other Ugandan cities, decentralisation reforms divided several districts into smaller administrative units. Creating new units by fragmenting existing districts tilted the balance of power between levels of government further towards the central government because subnational units lack bargaining power and administrative capacity, and are now more dependent on the central government for resources (Lewis, 2014).

Even when there is city-level governance, existing boundaries are often inadequate considering the spatial expansion of the city. Much of the urbanisation in African cities is linked to urban and peri-urban areas beyond city boundaries becoming part of the city. However, the expansion of administrative city boundaries can be politically complex because it challenges existing political arrangements and requires acquiescence from the central government. While potentially difficult, such expansion can increase the tax base of the city, especially through taxes levied on land, and give city authorities the power to plan and deliver infrastructure to what are in fact parts of the city.

Area-wide forms of governance and planning are becoming increasingly important, and can help strengthen a city's relationship with surrounding rural areas. Contrary to the views of many urban managers in the continent, area-wide governance should not aim to increase investment in rural areas as a way to slow city growth by staunching migration. Instead, area-wide governance is important because peri-urban areas are increasingly becoming part of the urban system. Food security is one issue in which understanding these rural–urban linkages is critical (see Thornton, Chapter 34, this volume). Moreover, economic activities and other functions of small towns depend on and are integrated with those of the nearby city. Therefore, metropolitan forms of governance may facilitate service and infrastructure planning beyond the city scale. It can also provide a flexible approach to expanding a city when renegotiating borders is difficult and city borders are constantly changing. Emerging urban corridors, where a number of cities are located in the same region, often need linking infrastructure. These corridors may cut across national borders, thus requiring international governance and cooperation.

Devolution of powers and fiscal autonomy

In the previous section, we emphasised the importance of having a form of city-level governance. It is not sufficient to have authorities at city level, however, if they do not have powers or resources to take autonomous management and planning decisions.

Devolution, or the transfer of some powers and functions from higher tiers of government to lower ones, is a major political issue in Africa, and of particular relevance for cities. UN-Habitat calls for a 'radical decentralisation of powers' in African cities, emphasising the devolution of controls over revenue collection (UN-Habitat, 2014: 7). Effective devolution implies applying the subsidiary principle to divide powers between central governments and cities, decentralising powers and functions that are better delivered by government actors closer to citizens. In particular, key issues for cities are: (1) financial autonomy, largely achieved through fiscal autonomy, i.e. the capacity to generate revenue; (2) planning powers; and (3) control over service provision. The first two are key for the latter.

There is a growing consensus and evidence that devolution processes have been critical to the success of many cities, particularly in Latin America. Here, cities control the provision of key services, can raise their own revenues, and offer competitive salaries to highly skilled professionals. In the African contexts, these processes are more difficult, and have to take into account a number of issues. Historically, African countries inherited highly centralised systems of governance from colonial rule that the new political elite did not challenge (UN-Habitat, 2010: 32). This was a way to maintain control but also to prevent the disintegration of many countries whose borders were arbitrarily drawn by colonial rulers, and in which various ethnic identities were stronger than national ones.

African cities also produce a significant share of many countries' economic wealth and they are the places with the greatest potential for future growth. As a result, central governments guard their control over them, which often leads to fragmented urban governance. As mentioned above, Kaduna is divided in different administrative areas under the control of state governors. Cities, and especially capital cities, are central sites of African politics – where coups take place and national parliaments and presidential residences are located – making it more politically problematic for central authorities to relinquish power.

Freetown, Sierra Leone's capital, is another example of local governments that are unable to fully govern. The city council has very limited powers, particularly regarding land and taxation, which remain a prerogative of the central government. This means that the central and local governments have to jointly agree upon any urban project. While the Freetown City Council is responsible for improving the welfare of the people, including through promoting economic development in the city, it does not have the power to register and supervise the use of land. There is a huge potential for the council to generate revenue from taxation on land, especially taxes on property and land value, but it is not permitted to do so. Moreover, because the act establishing local councils is not entrenched in the constitution, the extent to which local councils are allowed to exercise their powers depends greatly on what the central government allows them to do. Local councils in Sierra Leone are under the control of the

Ministry of Local Government and subject to the ministry's political and administrative oversight. They also rely upon the central government for two-thirds of their budgets. Indeed, while the Local Government Act (2004) empowers local councils to create area development plans and to manage and coordinate projects for improving their localities, several councils rarely have the capacity and resources to deliver such plans. As a result, the central government has assumed these functions over time.

Democratic processes brought increased pressure for devolution and opened the debate in a number of countries. Over the last 20 years, approved constitutions – e.g. South Africa (1996), Nigeria (1999) and Kenya (2010) – acknowledge the importance of decentralised governance, albeit not always focusing on the city level. In Kenya, the 20-year struggle for constitutional reform has largely focused on devolution (Rigon, 2010). The principle of devolution from Kenya and Uganda's constitutions appeared in Zimbabwe's 2013 Constitution, which states that while the country remains unitary, government power and functions are devolved through a three-tier cooperative governance system (Moyo and Ncube, 2014).

While not sufficient in itself, since the 1990s, several African countries have introduced major legislation promoting decentralisation. These include Tanzania, Malawi, Ethiopia, Uganda, South Africa, Ghana and Nigeria. In Tanzania, the Local Government Reform Programme worked towards changing local government laws and increasing resources available to local government authorities in the late 1990s, and in Malawi the 1998 Local Government Act created a national decentralisation framework based on democratic principles, accountability, transparency, and public participation in decision-making and development processes. In many cases, however, including Algeria and Morocco, the central government substantially restricts the autonomy of local governments, despite the existence of legislation supporting decentralisation.

In Uganda, the 1995 Constitution and subsequent legislation allowed the transfer of a number of political, administrative and fiscal powers to local governments. To fund these functions, local governments can seek revenue from a variety of sources. Nonetheless, transfers from the central government, particularly in terms of donor funds, remain crucial and are conditional on the local governments meeting specific conditions, thereby limiting their autonomy. These funds largely comprised grants from the Poverty Action Fund, and can only be spent on activities that the central government deems priorities. The central government further retains responsibility for all national projects.

Overall, while elections of local authorities and citizens' democratic demands at the city level have increased, these authorities often have minimal power and insufficient resources, particularly in terms of their ability to provide more and better services. Despite this, there is little incentive to decentralise and what is there decreases further when national and local governments are run by parties that oppose each other.

A key aspect of decentralisation concerns the financial autonomy of city authorities. A central demand of a number of African mayors at Habitat III was to be able to generate their own revenues autonomously in order to reduce dependency on transfers from central governments, which are often unreliable and used to achieve political aims. For example, Nigerian LGAs receive a federal funding allocation that is managed through an account shared with the state government. The state government often uses this arrangement for patronage (NBS, 2012), compromising the financial and political autonomy of LGAs and their capacity to plan and respond to citizens' needs. Civil servants working in LGAs in Kaduna State argued that such governance arrangements and related bureaucratic practices inhibit the capacity of LGAs to implement plans. Political interference, shifting priorities and governors put pressure on the chairs of LGAs to shift their agendas. Indeed, governors often take over LGA functions to build political support (Rigon *et al.*, 2015: 14). In Nigeria, militaries in power have used local institutions for patronage. More recently, state governors still exercise enormous power over local government areas by maintaining control over resources flows (Rigon *et al.*, 2015).

Financial autonomy is also important for cities to access markets to fund infrastructure projects. Large municipalities in South Africa increasingly borrow from local commercial banks or bonds, which can help bridge the gap between revenues and the funding needed to invest in infrastructure. Smaller cities, however, still encounter obstacles accessing credit (Brown *et al.*, 2013).

Devolution is also important in light of the changing urban forms of many African countries. While many countries are still characterised by a major dominant city, there is an increasing number of smaller secondary cities with growing populations. These cities offer opportunities for new management models and for translating residents' democratic claims into practice. Because these cities will experience the bulk of African urban growth in the coming years, and have less institutional capacity and infrastructure than larger cities to accommodate those changes, they have the greatest need for urban management and institution-building (UN-Habitat, 2014: 23).

In conclusion, decentralisation must involve shifting power and resources towards lower tiers of government rather than creating local institutions under central control, as is often the case. Moreover, decentralisation has to take into consideration the 'diversity of traditional practices and the complex local politics of ethnic relations' (UN-Habitat, 2014: 32).

Legacy of colonial planning

Many African cities are still overcoming the legacy of colonial spatial structure and planning regulations (Watson, 2014a). For example, the British colonial state implemented a policy of racial segregation. In Nairobi, this policy was formalised in the 1948 Master Plan, which divided the city into different racial residential areas for Africans, Asians and Europeans. After independence, in Kenya, Namibia, Zimbabwe and many others, the racial spatial segregation of the colonial state became socio-economic residential segregation (K'Akumu and Olima, 2007; UN-Habitat, 2010: 26). The wave of post-independence urbanisation took place on the basis of, and often exacerbated, pre-existing patterns of exclusion and segregation. The spatial inequality and segregation of African cities is an urban form that hides the poor, who often live on small marginal lands. In Nairobi, 55 per cent of the population resides in less than 5 per cent of the land, where they do not have secure tenure (Syagga, 2011: 105). Such spatial inequality and segregation transform cities into archipelagos of intertwined unequal islands where different groups of residents do not meet.

Another major problem is the building codes and regulations inherited from colonial times. For example, Kenyan construction by-laws come from the British, who exported their regulations without adapting them to Kenya's culture, climate, resources and level of economic development (Tuts, 1996: 608). The inherited building codes therefore obligated social housing projects to conform with middle-class standards and prices rather than being tailored specifically to the low-income residents and their ability to pay. Urban planning and regulations should be adapted to the level of development and institutional capacities of a country, and construction standards should be set 'more realistically in order to facilitate rather than restrict the creation of housing and livelihoods' (UN-Habitat, 2010: 2).

The planning systems of many African cities, which are often permeated by colonial regulations, are inadequate for the reality and pace of urban transformations. Despite this, some planners are committed to these systems, and refuse to see urbanisation as a positive process to be managed differently.

Managing and taxing urban land

The importance of land planning powers for cities cannot be overstated. These powers can decide the future direction of city development, and it is here that the potential for taxation and capturing land value reside.

Infrastructure, change of use, or permission to build more or higher enhance the value of land or property. Sometimes, even a planning decision or prospective infrastructure investment can trigger an increase in value such that private owners find the value of their assets exponentially higher without having made any productive contribution. Specific regulatory instruments called land-based financing or land value capture can be employed in these situations to ensure a more equitable distribution of wealth, securing part of the value produced by a public decision for public interests. The money generated from these instruments can help finance the infrastructure that services taxed properties, but can also cross-subsidise infrastructure for poorer areas of the city. In many prosperous Latin American cities, various tools have a long history and have provided significant funding to municipalities. This funding may involve one-off development charges, the (sometimes competitive) sale of development rights, betterment levies, or property tax increments.

Palmer and Berrisford (2015) found that, with the exception of Ethiopia and South Africa, there is no comprehensive land-based financing at the city or national levels anywhere in sub-Saharan Africa. To the contrary, the authors found many situations of reverse value capture in which the cities subsidise internal infrastructure in high-income developments. The study points out how difficult it would be to implement land-based financing in the short term. They identify strong and capable urban governance structures as key enablers of land-based financing, demonstrating the importance of what we discussed above.

Despite their importance, land and property tax are often inadequately implemented. While changes are taking places in a number of African countries, land cadastres are far from being fully functional. They are often used for patronage politics and under the control of the central government, producing significant conflicts between central and local government authorities. Moreover, the implementation of a modern cadastre system often has to adapt and cohabit with the reality of urban land subjected to customary administration practices by traditional authorities. In many cities, there is the development of intertwined formal and informal land markets. In Tanzania, for example, the ineffectiveness of the formal land system gave rise to a large informal land market.

As discussed, property taxes are underutilised, even in those municipalities where they generate a substantial share of the revenues. Revenues for local government need the implementation of new revenue mechanisms such as property tax. Interestingly, cities with high property rates are also those with strong democratic local government traditions such as Harare and Kariba (Zimbabwe), Cape Town and Durban (South Africa), and Kano and Lagos (Nigeria) (Eyoh and Stren, 2007). This demonstrates a link between fiscal autonomy and democratic governance.

In the Western Area of Sierra Leone, which includes Freetown, the Ministry of Lands, Country Planning, and the Environment shares the responsibility for land management with the Land Registry, which is based in the Ministry of Justice. Local councils also claim power over land registration, a function supposedly devolved to them by the 2004 Local Government Act. As a result, land administration in Freetown is plagued by problems of land encroachment, falsification of documents, multiple sales and registrations, unauthorised developments, and improper land demarcation, largely due to the indiscriminate application of land laws, low institutional capacity and a disorganised land market. A project funded by the Investment Climate Facility for Africa attempted to reduce the time and cost of land registration by reviewing and streamlining the registration procedures, and creating an electronic database of all land registration records. However, the lack of comprehensive legislation defining the respective functions of the ministry and of the local councils makes it difficult to set up a modern land management system.

The informal city

The concept of informality is used to describe entire areas and a large sector of economic activities in African cities. The areas are often informal settlements that fall into the problematic UN-Habitat definition of slums, which looks at the characteristics of households in the area. Slum households, according to UN-Habitat (2003), lack at least one of the following: improved water, improved sanitation, sufficient living area, durable housing, and secure tenure. In terms of economic activities, the informal sector is usually defined as employment and production that take place in unincorporated, unregistered or small enterprises, and ILO estimates that this sector accounts for two-thirds of non-agricultural employment in sub-Saharan Africa and just below half in northern Africa (ILO, 2013: 42).

The prevalence of urban slums is a consequence of high inequality in African cities, which UN-Habitat (2010: 2) considers to be 'one of the major threats to African urban stability and, by extension, to overall political stability'. Slums and their residents are often seen as a source of criminality and moral vice to be eradicated. Existing planning approaches criminalise most urban residents living in informal settlements and too often support policies leading to evictions. These may take the form of forced evictions, but may also take more subtle forms of market-led displacements, resulting in human rights violations of the urban poor. While this type of informality of the poor is criminalised, informality of the rich may be ignored or even incentivised. Some high-income developments are also informal from the point of view of planning regulations in that they are built without permits in areas that are not zoned for residential development, against the Master Plan, and often with infrastructure that does not align with city plans. Despite these infractions, however, these developments are not criminalised. This double standard is based on dangerous discourses around the urban poor and linked to the fact that there are profitable informal arrangements between property developers and state-politicised bureaucracy.

In terms of the economy, informality is seen as a separate, self-standing sector that must be formalised in order to extend taxation and regulation or, for those with good intentions, social protection. The way of conceptualising informality through a strict demarcation with the 'formal' fails to consider the 'pervasiveness of informality within formal ways of doing things' (Myers, 2010: 9). In African cities, informality is the norm, with urban value chains and services containing formal and informal elements that are interdependent. Products and services may be delivered through chains with informal and formal stages, for example, and 'formal' enterprises have informal practices, including informal employment of unskilled labour. In addition, employees in the formal sector rely on informal services to deliver productive work, such as transport and food stalls. Understanding the hybridity of formality as part of urban practices across all city actors is a necessary step in the construction of an African urbanism that can underpin the development of planning and management strategies that start with valuing and acknowledging existing urban realities.

Planning from below

As we have just discussed, current government formal planning practices are unable to address the challenges of African cities and provide for the growing demand of housing and services. Many civil servants and local authorities still view planning in terms of restoring a social order lost during the rapid growth of cities. This technocratic discourse emphasises that planning is an exclusively technical and neutral process that professionals have to do *for* people, rather than *with* people, in the context of a colonial planning legacy (Rigon *et al.*, 2015).

Nonetheless, the reality on the ground in many African cities is characterised by 'agency from below', in which urban residents build their houses and provide their own services through various forms of individual and collective action. This self-help approach to housing and services is a major force shaping and making African cities. A significant part of urban management and planning takes place outside the control of city authorities. Under the umbrella of Slum Dwellers International (SDI), national federations of the urban poor organise themselves in saving groups and generate their own censuses and data to negotiate with policymakers. Residents in informal settlements also initiate their own slum upgrading processes (Mitlin and Satterthwaite, 2004). These new emerging citizenship practices push demands on city authorities. Local authorities can respond to citizens' demands if they have the power and capacity to generate their own revenues. Success in these areas could lead to a deepening of democratisation in African cities.

The political power of these forms of collective action is still limited compared to forms of urban collective actions in other places such as India or Brazil, but governments, UN institutions and global fora such as Habitat III, where SDI and their work in African cities had a major presence in high-level dialogues with institutions, are increasingly recognising their importance. In some contexts, these forms of collective actions among the urban poor are slowly changing power relations. This may destabilise existing political settlements, causing conflict. However, it is a risk worth taking because the current situation is also unstable. As UN-Habitat (2010) pointed out, cities that do not work for the majority of urban residents may undermine urban as well as national stability. Under various names (e.g. alliances, coalitions, strategies), a number of citywide initiatives are moving beyond engaging with citizen participation on a project-by-project basis to build citywide participatory governance frameworks. However, their success depends on the capacity to change power relations and address inequalities by allowing a meaningful participation from the urban poor.

The centrality of 'agency from below' in planning and managing cities can contribute to the emergence of a hybrid urbanism that can find new models and concepts appropriate for the specific needs of diverse African cities, leaving behind European models. The value of citizen participation in local governance and planning is increasingly recognised in different African countries, opening opportunities. For instance, the 2010 Constitution of Kenya put a strong emphasis on citizen participation and explicitly mentioned their participation in at least 16 articles. It is considered a national value and principle of governance, key to the achievement of an effective devolution. It also explicitly refers to participation in urban areas: 'National legislation shall provide for the governance and management of urban areas and cities and shall, in particular [. . .] (c) provide for participation by residents in the governance of urban areas and cities' (Republic of Kenya, 2010: Article 184). South Africa's 1996 Constitution states that local authorities should 'work with citizens and groups within the community to find sustainable ways to meet their social, economic and material needs and improve the quality of their lives' (Department of Constitutional Development, South Africa, 1998).

Urban infrastructure

The presence of infrastructure has historically played a key role in the origin and growth of many major African cities, often developed around the main port or railway. However, today African cities suffer from an urban and national infrastructure gap. Poor transport infrastructure accounts for 40 per cent of logistics costs in coastal countries and 60 per cent in landlocked countries (UN-Habitat, 2014: 20). Public transport is costly and lacking, presenting a major challenge to labour mobility. Traffic congestion affects many residents and decreases productivity, with people spending over four hours commuting in some major cities such as Lagos. Over 30 countries experience regular power shortages (UN-Habitat, 2014: 20), so residents and companies that can afford it have to rely on private diesel generators at a great cost, which is also a health hazard. The infrastructure gap is significantly reducing the competitiveness of the African manufacturing sector. It decreases profit margins of smaller businesses and increases consumer prices for goods and services.

Infrastructure is a key challenge for the managers of African cities. For it to be addressed, the issues discussed in this chapter around governance, devolution and finance are crucial. The way in which infrastructure is provided can also contribute to addressing the other challenges of inequality and poverty. Infrastructure can be pro-poor, prioritising the needs of the urban poor and developing mechanisms for cross-subsidising low-income residents, or it can increase inequality and poverty by serving high-income developments and ignoring informal settlements. The choice will depend on the outcome of the politics of urban management and planning, and the capacity of the poor to voice their collective concerns and expand the participatory mechanisms of urban governance.

Conclusions

Reforms of urban management and planning, involving decentralisation of power and fiscal autonomy, are fundamental to prepare African cities to absorb the urban growth forecasted for the next decades. Such reforms are very sensitive because they involve shifts in power relations that may reopen complex processes over other ethnic divisions and unsettle the political settlements at the national and city levels that have allowed cities to function so far.

For example, residents of informal settlements often have to enter political patronage relations or pay a range of brokers to protect their precarious tenure security and receive services. These brokers may include local politicians, traditional rulers, community leaders, police and other civil servants (UN-Habitat, 2014: 36). Therefore, many people thrive in an environment of 'under-regulation' and poor services and infrastructure, which offer profitable opportunities. Any reforms can threaten the status of some groups and may meet resistance and failure if they do not take into account the complexity of existing political arrangements. Therefore, urban management and planning reforms are not managerial changes to make management and services more effective, but complex and negotiated changes in politics.

This need for a political negotiated process and acknowledgement of the city diversity clash with a reality of technocratic models. A number of cities are overcoming colonial master plans and regulations and preparing new plans. However, this often happens through a process of privatisation of planning in which master plans are contracted out to a few multinational consultancy companies that develop standardised plans that reproduce dominant planning principles often exported from Europe. These are plans prepared with little participation and democratic control through processes that do not recognise the role of people's agency in making the city. These plans are disconnected from the reality of most urban residents, and often contain unrealistic 'urban fantasies' based on the model of Dubai, Shanghai or Singapore (Watson, 2014b).

This process is repeating the colonial approach of enforcing universal models, for example conceiving densification and verticalisation as always necessary while denying the emergence of new urbanisms. Can African cities develop indigenous urban models? The New Urban Agenda adopted by the UN at Habitat III acknowledges the right to the city, ensuring all inhabitants can participate in the production and use of their cities. We hope city managers will open spaces to the plurality of forms of agency from below that are already shaping the future of African cities. In a context in which the continent will be shaped by the way in which societies think about cities (Parnell and Pieterse, 2014), we have seen the emergence of a strong body of work exploring the specificities of African urbanism grounded in the lived dynamics of everyday life (Pieterse and Simone, 2013). The capacity of this work to influence urban practitioners will be critical for influencing the management and planning of African cities.

Notes

1 Habitat III is the third bi-decennial United Nations Conference on Housing and Sustainable Urban Development that took place in Quito, Ecuador, in October 2016. The conference produced a New Urban Agenda, which sets a new global standard for sustainable urban development, and will guide the efforts around urbanisation for a wide range of actors for the next 20 years.
2 Another particular case is Nigeria's capital Abuja and its Federal Capital Territory Administration, which does provide some city-level governance, but is effectively a ministry of the federal government rather than a body that democratically represents the capital's residents.

References

Brown, K., Motsoane, T. and Liu, L. (2013) 'South Africa: leveraging private financing for infrastructure', in O. Canuto and L. Liu (eds), *Until Debt Do Us Part: Subnational Debt, Insolvency and Markets*, Washington, DC: World Bank, pp. 495–538.

Department of Constitutional Development, South Africa (1998) *White Paper on Local Government*, available at: www.cogta.gov.za/cgta_2016/wp-content/uploads/2016/06/whitepaper_on_Local-Gov_1998.pdf (accessed 1 May 2017).

Eyoh, D. and Stren, R. (eds) (2007) *Decentralisation and the Politics of Urban Development in West Africa*, Washington, DC: Woodrow Wilson International Center for Scholars.

ILO (International Labour Organization) (2013) *Measuring Informality: A Statistical Manual on the Informal Sector and Informal Employment*, Geneva: ILO.

Laws, E. and Leftwich, A. (2014) *Political Settlements*, Concept Brief 1, Birmingham: DLP.

Lewis, J.I. (2014) 'When decentralization leads to recentralization: subnational state transformation in Uganda', *Regional & Federal Studies*, 24(5): 571–588.

K'Akumu, O.A. and Olima, W.H.A. (2007) 'The dynamics and implications of residential segregation in Nairobi', *Habitat International*, 31(1): 87–99.

Mitlin, D. and Satterthwaite, D. (eds) (2004) *Empowering Squatter Citizen: Local Government, Civil Society, and Urban Poverty Reduction*, London: Earthscan.

Moyo, P. and Ncube, C. (2014) 'Devolution of power in Zimbabwe's new constitutional order: opportunities and potential constraints', *Law, Democracy and Development*, 18: 289–304.

Myers, G.A. (2010) *Seven Themes in African Urban Dynamics*, Discussion Paper 50, Uppsala: Nordiska Afrikainstitutet.

NBS (National Bureau of Statistics) (2012) *Assessment of Good Urban Governance in Nigeria*, Abuja: NBS.

Palmer, I. and Berrisford, S. (2015) *Urban Infrastructure in Sub-Saharan Africa: Harnessing Land Values, Housing and Transport – Final Report on Land-Based Financing for Urban Infrastructure in Sub-Saharan African Cities*, African Centre for Cities & UK Aid, available at: https://assets.publishing.service.gov.uk/media/57a08978ed915d622c000227/61319C_Full-DFID-Report_Web.1.1.pdf (accessed 1 May 2017).

Parnell, S. and Pieterse, E.A. (2014) *Africa's Urban Revolution*, London: Zed Books.

Pieterse, E.A. and Simone, A.M. (2013) *Rogue Urbanism: Emergent African Cities*, Johannesburg: Jacana Media.

Republic of Kenya (2010) *The Constitution of Kenya (Revised 2010)*, available at: www.kenyaembassy.com/pdfs/the%20constitution%20of%20kenya.pdf (accessed 3 January 2018).

Rigon, A. (2010) *20 Years of 'Sweat and Blood': Lessons from the Constitutional Review Process in Kenya*, ISPI Policy Brief 188, available at: www.ispionline.it/it/documents/PB_188_2010.pdf (accessed 1 May 2017).

Rigon, A., Abah, O.S., Dangoji, S., Walker, J., Frediani, A.A., Ogunleye, O., *et al.* (2015) *Well-Being and Citizenship in Urban Nigeria*, London: ICF International and UK Aid.

Syagga, P. (2011) 'Land tenure in slum upgrading projects', in IFRA (ed.), *Slum Upgrading Programmes in Nairobi: Challenges in Implementation*, Nairobi: IFRA, pp. 103–113.

Tuts, R. (1996) 'Cost modelling for appropriate building and planning standards in Kenya', *Habitat International*, 20(4): 607–623.

UN-Habitat (United Nations Human Settlements Programme) (2003) *The Challenge of Slums: Global Report on Human Settlements 2003*, London: Earthscan.

UN-Habitat (United Nations Human Settlements Programme) (2010) *The State of African Cities 2010: Governance, Inequality and Urban Land Markets*, Nairobi: UN-Habitat.

UN-Habitat (United Nations Human Settlements Programme) (2014) *The State of African Cities 2014: Re-Imagining Sustainable Urban Transitions*, Nairobi: UN-Habitat.

Watson, V. (2014a) 'Learning planning from the South: ideas from new urban frontiers', in S. Parnell and S. Oldfield (eds), *The Routledge Handbook on Cities of the Global South*, London: Routledge, pp. 98–108.

Watson, V. (2014b) 'African urban fantasies: dreams or nightmares?', *Environment and Urbanization*, 26(1): 215–231.

32
Housing in Africa

Kamna Patel

Introduction: what is housing?

'Housing' is a concept, and so means much more than a structure that we call a house, flat, apartment, bungalow, condominium or room. Housing is closely related to ideas on basic standards of living (such as the services that ought to accompany a structure, including water, electricity and security of tenure); quality of life (i.e. does housing enable access to amenities, including schools, parks, places of employment and transport services?); and a family's future prospects (i.e. can property be bequeathed and how do property markets affect the value of homes as assets?). The concept of housing, reflected in housing policy, includes some or all of these related ideas. Correspondingly, policies on infrastructure, transport, public services, education, economic growth, and rural and urban planning are relevant to housing. Therefore, housing is best seen as a cross-cutting issue that affects improvements to people's health, safety, livelihoods, wealth, assets, and overall sense of well-being (Patel, 2013). In the African context, housing is typically approached through a development lens and framed as a means to alleviate poor living conditions and improve the current and future economic prospects of individuals and households living and working in African towns and cities. This particular conceptualisation of housing is apparent in the dominant discourses in national housing policies and the policies of international development agencies. Although, by focusing on individuals and households, such policies risk overlooking the ways in which housing helps to construct urban space and life for all residents. Where housing is located, the form it takes, the materials it is made from, and who is able to live in it affects land use (amid competing commercial and public interests for urban land), environmental sustainability, and spatial and economic equality (as who lives where is heavily influenced by a person's income, wealth, class, and in some countries their race and ethnic identity). This makes housing a political issue of great social importance across the continent.

Housing is traditionally regarded as an urban issue, and there is a strong urban orientation to housing debates in global scholarship (see Turner, 1976; Clapham *et al.*, 2012), which is continued in this chapter largely because contemporary housing challenges are intertwined with challenges raised by urbanisation, a growing phenomenon across the continent. Following a brief overview of the history of urban housing policy in Africa directed towards low-income dwellers, where generalisations are unfortunately unavoidable (regional and national histories of housing are referenced in the text), the current relationship between urbanisation and housing is examined, with three key emerging debates relevant to low-income housing in Africa discussed in detail. This discussion engages with the politics of housing over technical aspects such as mechanisms of housing finance and housing design. The chapter concludes with pertinent challenges for research and action on housing.

A brief history of housing policy in urban Africa

Rarely is a single housing policy and its implementation able to respond well to every development issue related to housing. An emphasis on some aspects over others is influenced by the political will and ideology of leaders, the colonial apparatus inherited by African states (embedded in land law, building regulations and urban planning), available resources, and the wider political-economic environment forged by influential actors, including the World Bank and the International Monetary Fund (IMF). The postcolonial history of housing in urban Africa is replete with examples of how these influences play out in the outcomes of national housing policy. Starting with the immediate era leading up to and following decolonisation in much of the continent in the 1950s and 1960s (southern Africa being an exception), large-scale public investment in housing was the norm (as it was in post-war Europe). Much of this housing was built to replace dilapidated 'slum' or squatter housing in core urban areas and to build new public housing for lower-income workers (Stren, 1990). African urban areas were changing in this period. Colonial-era restrictions on African movement to urban areas were relaxed or lifted with the handover of power from European administrations to national ones, and a relative boom in economic growth following the stagnant war years meant more people moved to urban areas for work or to rejoin families. Yet the colonial cities built by Europeans were designed to house predominantly Europeans; black Africans were not well accommodated in formal spaces, and so were often pushed to urban margins in housing of poor quality or crowded into dense urban centres (Chipungu and Adebayo, 2012 discuss this in Zimbabwe during its decolonisation in the 1980s). The housing policies of postcolonial nationalist governments, many influenced by socialist ideas of housing as a social good, sought to address this need for more better-quality housing for low- and middle-income Africans (Tipple, 1994). For example, in Kenya, by the early 1970s, the National Housing Corporation directly oversaw the construction of 2,000 new housing units every year, most of it in central Nairobi. In Tanzania, at around the same time and under the socialist vision of Julius Nyerere, the National Housing Corporation directly built over 5,000 new units in Dar es Salaam (Stren, 1990: 37). However, the pace of building could not keep up with the demand for decent-quality housing. This, accompanied by the very high costs of public housing projects, led to new approaches to housing delivery and shifted the role played by the state.

By the 1970s, under the strong influence of the World Bank, housing challenges were understood by national governments as really infrastructure challenges. So, if people had access to a serviced site with a water and sewerage connection, good access to transport networks, and security of tenure, then with minimal state assistance they could build their own homes incrementally (i.e. building room by room as they could afford to do so, which suits those on a low or irregular income). This era of housing policy was heavily influenced by the ideas of John Turner and his research in Peru, particularly the idea of 'assisted self-help', which maintained that poor people are highly resourceful and capable of meeting their own needs, but require a little assistance with services they cannot provide for themselves, such as road-building (Turner, 1976). Early site and services projects, as these infrastructure-based housing projects were known, were ambitious and expensive, which often hampered the cost recovery principle these schemes were based on. One of the earliest projects was supported by a World Bank loan in Senegal, and planned serviced plots for 152,000 people in Dakar at a cost of US$14.2 million over nine years (Gulyani and Bassett, 2007: 489). As infrastructure projects depended upon access to land, which was scarce and very expensive in urban centres, many site and service schemes ended up on urban peripheries isolated from adequate transport links. This increased transport costs for those already on a low income, and the costs associated with maintaining sewerage and piped water meant many poor people were priced out of site and service schemes, which tended to be occupied by the middle classes (Pugh, 2001), thus maintaining a deficit of low-income housing in African towns and cities. Following this experience, the World Bank moved away from its support for site and service schemes. However, the principles of its agenda for housing in Africa have not significantly changed from the 1970s to now. The World Bank's emphasis remains on supporting cost recovery in low-income housing programmes, thus demonstrating the profitability of the sector to private investors. Additionally, it has encouraged the widespread provision of housing finance (e.g. mortgages, building loans and low-level credit) to support individual self-build and property developers, with the aim of stimulating private sector involvement in housing supply to complement a reduced role for the state (see Van Waeyenberge, 2015 for a thorough discussion of the World Bank and housing finance).

The trend of the state pulling back in housing provision, which started in the 1970s, continued more explicitly from the mid-1980s onwards, initially within the context of structural adjustment policies that affected most African countries beholden to IMF restrictions on public spending. Market-based approaches to housing supply took hold (e.g. Kombe, 2000 discusses this in detail with respect to Tanzanian housing policy; Soliman, 2014 discusses it with respect to Egypt). The role of the state as a housing developer (directly building housing) shifted, as it had globally, to an 'enabling role' (Gulyani and Bassett, 2007; Agunbiade *et al.*, 2013). This is where governments create a positive environment for private sector development (including individual or small developers who may be building their own homes) by providing direct subsidies and/or creating favourable legal and economic frameworks (Pugh, 2001). One area that has proved to be contentious in creating favourable frameworks for building is housing standards, and whether or not a government should lower standards for low-income housing so they are less costly to build, and thus more affordable to buyers. Gulyani and Bassett (2007) point out that in many African countries, building standards were inherited from colonial regimes and were based on European standards of living (e.g. large land parcels for housing developments or a high grade of construction materials), and so may not be appropriate to the varied climate and cultural norms of African countries. However, national governments, they argue, often considered lowering standards for predominately black African housing to be socially unacceptable.

By the 2000s, a range of different approaches to low-income housing provision were in place involving the state, private sector developers and private individuals, and years of investment had been made in the sector by national governments and multilateral development agencies. Yet the housing conditions for people living in African towns and cities, or the housing prospects for those wanting to live there, were poor. There was insufficient formal housing stock, existing formal stock was of a low and deteriorating quality, and where new housing was built by private individuals (either to live in themselves or to rent to others), it was typically without services or secure tenure. These informal settlements became the primary means through which large numbers of low-income people could afford a place to live in urban areas. The severity of the situation was acknowledged by a single target within the Millennium Development Goals to 'improve the lives of at least 100 million slum dwellers by the year 2020' (UNDP, 2000). On the back of this target, UN-Habitat (2003) launched its seminal report 'The Challenge of Slums' to build safer and more inclusive cities. In it, UN-Habitat makes a number of recommendations to national governments to reshape their urban landscape with better planned public spaces, including upgrading informal settlements. This challenge, in policy at least, has been taken up by governments in South Africa (Patel, 2015), Mozambique (Earle, 2014) and Kenya (Otiso, 2003).

As the Millennium Development Goals have given way to the Sustainable Development Goals, within Goal 11 to 'make cities and human settlements inclusive, safe, resilient and sustainable' is the target to 'by 2030, ensure access for all to adequate, safe and affordable housing and basic services and upgrade slums' (UN, 2015). Current scholarship on African housing is heavily focused on examining the challenges and opportunities for the large-scale provision of affordable, safe and adequate housing. These challenges and opportunities are often rooted in the relatively rapid urbanisation of the continent (see Myers, Chapter 35, this volume).

Housing in the context of urbanisation

At its most basic definition, urbanisation is the concentration of a population in an urban area in relation to the total population of a place. It refers to the phenomenon of more people living in greater densities in urban areas (although there is variation in what density classifies an area as 'urban'). This means that urbanisation has a major impact on housing – particularly what type of housing is required, where, and for what type of resident – but not all types of urbanisation will engender the same housing impacts, largely because the housing effects of urbanisation differ depending on the speed and rate at which population changes happen, and what is driving urbanisation.

The oft-quoted statistic is that globally, 50 per cent of the world's population live in an urban area. This figure, however, masks high regional disparity. In countries that are part of the Organisation for Economic Co-operation and Development (OECD), urbanisation levels of 70 per cent or higher are typical, whereas in Africa some of the lowest levels of urbanisation overall are found (e.g. less than 20 per cent in Burundi, Ethiopia and Niger) (Freire *et al.*, 2014: 5). Similarly, within the continent, there is great disparity between heavily urbanised north African countries and middle-income countries such as South Africa, and East African countries, which are the least urbanised but fastest urbanising countries on the continent (it is estimated that by 2040, the number of urban dwellers in East Africa will be five times higher than in 2010) (UN-Habitat, 2014: 11). Globally, this suggests a correlation between economic growth and levels of urbanisation (Knox, 2009). However, there is continuing debate over the relationship between economic growth and urbanisation in African towns and cities. The inverse relationship between growth and urbanisation, where poverty drives urbanisation, which was said to typify African cities (e.g. Buckley and Kalarickal, 2005; Knox, 2009), has been challenged recently on the back of UN-Habitat's 2010 and 2014 *The State of African Cities* reports (UN-Habitat, 2010a, 2014). These reports acknowledge that the nature of urbanisation varies considerably between African cities. In some cases, urban growth drives the development of mega-regions of more than 100 million people, sometimes extending across international boundaries, as in the case of the West African urban region linking Abidjan (Côte d'Ivoire), Accra (Ghana) and Lagos (Nigeria). Urban corridors are another feature, such as the 900 km of transport routes linking Kampala (Uganda), Nairobi (Kenya) and Mombasa (Kenya) in East Africa (Obeng-Odoom, 2013).

The population densities that result from urbanisation, which arise from both rural-to-urban migration and a natural increase in urban populations, and a widely expected increase in urban dwellers in countries that do not yet have a stable urban population present particular challenges for housing. One of the most important concerns competition for land and changing land use in and around towns and cities. For example, with growing numbers of urban dwellers, municipal borders (officially or unofficially) may be forced to expand, changing previously agricultural land bordering a city to residential use (Owusu-Ansah and O'Connor, 2010 discuss this in Kumasi, Ghana), potentially fostering issues for urban food supply. In general, urbanisation increases competition for land from all types of land user and developer, which in turn raises land prices, with effects on affordable housing and the vulnerability of people living in informal or formal low-income settlements where the land on which they dwell is far more valuable without them on it (as LeVan and Olubowale, 2014 discuss with reference to Abuja, Nigeria). Land pressures in urban areas, especially in centres, may also affect the form of urban housing as high-rise residential blocks can be more efficient uses of land than single-storey units, for example. Related to an increased competition for land is an increased competition for housing itself. In a context where existing housing stock is insufficient or of a poor quality, and rising land prices preclude the development of low-income housing (especially when the private sector is active as housing developer), then poorer urban dwellers are increasingly pushed to the low-quality cheap housing typically found in informal settlements, where they are likely to live with tenure insecurity and few services. Thus, if suitable infrastructure (including housing) and livelihood opportunities do not exist for all current and future urban residents, then urbanisation can exacerbate housing inequality and experiences of poverty in towns and cities.

Key housing challenges intensified by urbanisation

The need for affordable, safe and adequate housing engages with three key debates in housing literature, which are discussed in this section. The debates have a clear political perspective; this does not mean these issues are dominant in discussions of housing among politicians or policymakers, but that they concern the role of power in decision-making that determine the distribution of resources in urban areas.

Tenure and housing typology

In every country, many different types of housing and tenure options exist. Some, however, are more politically and culturally favoured than others. In the housing landscape across the Global South, there is a hierarchy of different modes of housing provision. Keivani and Werna (2001) describe two main types: conventional and unconventional housing (also referred to as 'formal' and 'informal'). Conventional housing is provided by three sectors: the public sector, which includes informal settlement upgrading and low-income housing projects; private developers, who include large-scale developers of multiple housing units, or petty developers, who self-build often through self-finance (UN-Habitat, 2011); and cooperatives, which include NGOs and community groups involved in housing provision. Unconventional housing is also provided by private developers, who may illegally subdivide both land and formal housing to accommodate more people. Such developers may build cheap substandard housing, either at a large scale for commercial purposes, or at a much smaller household scale where a subdivided home can be used to supplement basic income. Additionally, private individuals may occupy land that they do not legally own and start construction. Unconventional housing is the quickest response to the rising demand for housing from urbanisation. Within each category of conventional and unconventional housing is a complex network of social relationships, actors and financial flows that enable housing to be built in these ways. Despite this known complexity, housing provision is typically presented in a dichotomous way where conventional housing is preferred in discourse and public policy over unconventional housing; within the latter, the language of 'slums' can play a crucial role to stigmatise and legitimate action against unconventional housing (Gilbert, 2007).

Cutting across these housing types are a myriad of ways housing is occupied (i.e. there are multiple tenure types such as owner-occupier, renter, landlord and squatter). These are also presented in a hierarchy where owner-occupiers are preferred over renters on the under-examined basis that homeowners are fundamental to political, economic and social stability in a country, and as Kumar (2011: 670) argues, 'ownership' is more electorally attractive. The effect of this hierarchy can be seen in state-led approaches to housing provision where criteria to access new housing in upgrade and settlement schemes is restricted to those who own the structure, not renters. Also, there is an explicit encouragement of homeownership through the award of land title in low-income housing schemes, and historically in site and service schemes that provide a plot for (future) homeowners to self-build (Gulyani and Bassett, 2007 provide examples from across Africa). A preference for home ownership means a lack of attention is given to rental housing. There are known difficulties with national data on rental housing; often it is simply not collected or there are difficulties identifying rent-based relationships in unconventional/informal housing in particular. What is well known is that the rental sector is very active in African cities and provides affordable accommodation on a long- or short-term basis that meets the needs of a range of low-income dwellers who might require flexible living arrangements, e.g. a labourer who moves from site to site for employment may prefer to rent flexibly than own a home in the city (Cadstedt, 2010 discusses the low income rental sector in Tanzania; Huchzermeyer, 2007 discusses it in Kenya; Gilbert, 2014 discusses it as a global phenonmenon). One of the consequences of a lack of attention to low-income rental housing in public policy is that provision becomes the preserve of the unregulated informal sector. Tutu (2014) writes that in Accra, Ghana, urbanisation pressures arising from the migration of young men to the city in search of jobs is driving the illegal subdivision of property and shack-building for rental purposes, which exacerbates already poor living conditions and heavy overcrowding.

Given the size of the housing challenge in African towns and cities, which is set to increase with urbanisation, for the past decade scholars have been calling for public policy that supports plurality in housing provision to suit the financial means and cultural and political preferences of a range of urban dwellers (e.g. Keivani and Werna, 2001; Yeboah, 2005). While some governments have responded in policy (e.g. the National Shelter Strategy in Ghana), in practice the strong preference for conventional owner-occupied housing still dominates (see Yeboah, 2005). The power of the discourse of 'informality' and negative traits associated with it may be a key reason for preferences for certain types of housing and tenure arrangements.

Managing informality

'Informality' is a legally, politically, socially and culturally constructed concept. The meaning of informality and its application to spaces, buildings and practices tends to be based on observed differences to formal spaces, buildings and practices, that is, definitions of what constitutes informality in an African town or city is set in legal, political, social and cultural opposition to other ways of living. In the housing sector, the concept of informality is applied to types of settlement, buildings and the people who live there: informal settlements, shacks and 'slum' dwellers, and represents spaces that are unplanned by the state and people who cannot be controlled by the state (through a property tax regime, for example). The common approach across African governments to informality in housing is to eliminate it either through recurring evictions of residents and the razing of buildings, relocation of residents to formal sites and subsequent destruction of previous residence, or through in situ upgrading that allows residents to stay on site, although due to de-densification from overcrowded settlements to well plotted formal buildings, some relocation and eviction is inevitable (Huchzermeyer, 2011). The elimination and eradication of informal housing has been the cornerstone of municipal and national governments' approach to managing urban areas across the continent, at times linked with a 'city beautification' agenda to promote economic investment. This process is underwritten by a particular interpretation of MDG target 11 and UN-Habitat's (2003) *The Challenge of Slums*, which encourage formal housing, service provision and tenure security for low-income dwellers, promoting a discourse of 'slum-free cities' as a form of positive and progressive development (Meth, 2013).

Yet informal housing is the long-standing norm in most African towns and cities. Marie Huchzermeyer (2011: 71) writes:

> In the African context, the transition from pre-modern yet sophisticated and structured forms of habitation, production, trade and governance to the modern western equivalent was rapid and largely imposed. Therefore many expressions of informality could be seen as something the modern state, with its particular approach to urban planning and governing, simply never succeeded in registering, taxing, controlling or supressing.

Through a social historic lens, the presence of informal housing in Africa complicates dichotomous ideas of formal/conventional housing as inherently superior to informal/unconventional housing. This is not to romanticise informal settlements and dwellers, but to shed light on the politics of the language of 'informality' and the positioning of this language in housing discourse as a problem demanding a particular solution: the elimination of informal settlements. This point is underscored by AbdouMaliq Simone (2004a, 2004b), who writes that so much of what makes African urban life innovative, entrepreneurial, communal and successful by a subjective measure exists in the informal economies and social networks found in and around African cities, and that urbanisation increases the importance of the 'informal' for everyday life. His persuasive arguments should lead us to question the association between formal housing and a correct way to live.

Right to the city

The third debate discussed here – the right to the city – is related to the idea of urban equality, which gains poignancy in light of the relationship between urbanisation and inequality. The 'right to the city' is a heavily debated concept, and one enjoying a resurgence in thinking about the effects of urbanisation, particularly the fate of poor urban dwellers in the face of competition for land and resources in urban areas. Its origins date to French urban theorist Henri Lefebvre (1996). At its most basic, the right to the city concerned people's rights to participate in the use and production of urban space, and to be fully involved in urban life, on the basis that through their residence, labour and social contributions, they make urban life in the first place. However, its usefulness as an idea and its intellectual depth as a concept exceeds Lefebvre's original conceptualisation. To different scholars and disciplines, Lefebvre's ideas have prompted questions about different 'rights' that do and ought to exist in cities. For example, political rights and the collective right of all types of people to influence urban political processes by participating in them; socio-economic rights realised through access to resources such as housing; and moral rights that concern the inclusion of people who are disabled, or of a particular race or ethnicity, into urban life (Attoh, 2011; see also Jones, Chapter 47, this volume). In sum, there is no single way to understand a 'right to the city'.

Within scholar activism on low-income housing, the 'right to the city' is a galvanising discourse and a useful way to frame urban struggles among low-income, marginalised and vulnerable urban dwellers who may be excluded or threatened with exclusion from urban spaces (see Huchzermeyer, 2011: Chapter 10). This includes living in informal settlements with a lingering threat of eviction, eradication and elimination, and being excluded from decision-making processes such as where new low-income housing developments are built and what they look like, despite living in urban areas and contributing through their labour and social relations to urban life. Chitekwe-Biti *et al.* (2014) implicitly discuss the 'right to the city' with reference to shack dweller movements in Harare, and how the Zimbabwe Homeless People's Federation has managed to organise and assert its right to make decisions on building affordable housing through negotiations with state actors. Contrast this to the South African shack dweller movement Abahlali baseMjondolo, which has adopted 'the right to the city' as an important slogan and political statement defining a position that the right for shack dwellers to use urban spaces exists on an equal basis to other types of urban dwellers and other (more commercial) interests (Abahlali baseMjondolo, n.d.). Within scholar activism, there are subtle distinctions between how 'the right to the city' is expressed, which does not detract from the broad-based appeal and power of the sentiment that marginalised urban dwellers have a moral, social and, recently in Brazil, legal claim to urban spaces and urban life.

Yet the 'right to the city' as a politically powerful discourse holds a mixed position among international agencies currently working on urbanisation, poverty and housing in Africa. For example, UN-Habitat's *State of the World's Cities 2010/2011* expressly advocates the 'right to the city' as a conceptual framework to understand and therefore better respond to urban exclusion (UN-Habitat, 2010b). By contrast, in UN-Habitat's 2011 *Housing the Poor in African Cities* report, the official discourse on informal settlements acknowledges the distress caused by evictions and living in poor conditions, and even advocates actions to ameliorate this. However, Franklin Obeng-Odoom (2013: 428) notes that this discourse does not frame the struggle for low-income housing in Africa as a political struggle over basic rights (i.e. as an issue that concerns the 'right to the city'). Furthermore, Uchenna Emelonye, a Senior Human Rights Advisor for the Office of the High Commissioner for Human Rights (OHCHR), explained in an interview with UN-Habitat that:

> OHCHR is cognizant of the concept of the 'Right to the City'. We do not, however, promote or affiliate ourselves with the concept for several reasons. Firstly, the concept of 'Right to the City' connotes a certain possible status in international human rights law which it has not attained. Secondly, the right to the city is an academic concept that has been taken on by a variety of civil society organizations to different and sometimes contradicting ends. While some cities have adopted or are in support of the concept, many remain sceptical of the legislative implications at city and national level. That is why UN-Habitat works through mainstreaming human rights in cities for all, focusing on urbanization as the process and the city as the outcome of this process, where the human rights-based approach methodology ensures that no one is left behind.
>
> *(Emelonye, 2015)*

The conceptual ambiguity around the 'right to the city' (exactly which rights are being advocated, for whom and how) means it lends itself to motivating a diverse range of civil society actors, but also can prove difficult to operationalise and implement in state bureaucracy and legislation. Debates on the 'right to the city' look set to continue for many years, and are likely to intensify in parts of the African continent where urbanisation rates are the highest and competition between different actors for urban land and housing is the greatest.

Conclusion: the challenge that lies ahead

At the beginning of this chapter, housing was described as a political issue of great social importance in Africa. It is worth retaining this insight so that housing policy and action plans to provide affordable, safe and adequate housing are not solely occupied by technical challenges around cost recovery, building materials and standards, and the 'how-to' of large-scale housing provision. Instead, the political element of making decisions around the use and distribution of resources in urban areas needs to be brought to the fore and debated in research and in action on housing. The three challenges focused upon here – thinking about a range of housing and tenure types and contesting the basis on which some are preferred over others, questioning the frame in which informality is positioned as a problem, and the debates around 'the right to the city', are likely to become even more important as Africa's urbanisation rates increase and competition for urban space intensifies. This in turn raises a fundamental question: Which people are welcome to enjoy living in an urban area and its associated benefits (access to cultural activities, sites of knowledge production in schools and universities, a range of job opportunities), and which are not? The provision of housing can be an effective way to enable everyone's social and moral claim to a right to the city, or to deny it to certain groups, which makes housing an intently political activity.

Glossary of key terms

Informality – In economic terms, the 'informal' refers to the unregulated or untaxed. It has come to mean much more as a discourse. The informal does not necessarily mean illegal. It may simply refer to practices outside of those the state chooses to recognise.

Slums – UN-Habitat uses 'slum' as a technical category of housing (see UN-Habitat, 2003), though it is widely considered a pejorative term that can denote undesirable and deficient modes of living, and alongside it undesirable people. For example, the language of 'slums' has been used by many governments to associate a place with high crime and unsanitary conditions, which can serve as a useful reason for slum demolition and the eviction of residents (see Gilbert, 2007 on why the language of 'slums' matters).

Tenure security – This refers to how dependable people's claims are to enjoy or own land or a residence without fear of forced eviction, harassment or other threats. Claims can be grounded in laws, cultural norms or socially accepted rules (OHCHR, 2009).

References

Abahlali baseMjondolo (n.d.) *Abahlali baseMjondolo*, available at: http://abahlali.org/ (accessed 1 November 2015).

Agunbiade, M., Rajabifard, A. and Bennett, R. (2013) 'Modes of housing production in developing countries: the contemporary role of land, labour, and capital in Lagos, Nigeria', *Journal of Housing and the Built Environment*, 28(2): 363–379, DOI: 10.1007/s10901-012-9303-8.

Attoh, K.A. (2011) 'What kind of right is the right to the city?', *Progress in Human Geography*, 35(5): 669–685, doi:10.1177/0309132510394706.

Buckley, R.M. and Kalarickal, J. (2005) 'Housing policy in developing countries: conjectures and refutations', *The World Bank Research Observer*, 20(2): 233–257, DOI: 10.1093/wbro/lki007.

Cadstedt, J. (2010) 'Private rental housing in Tanzania: a private matter?', *Habitat International*, 34(1): 46–52, DOI: 10.1016/j.habitatint.2009.05.001.

Chipungu, L. and Adebayo, A. (2012) 'The policy–planning divide: an evaluation of housing production in the aftermath of operation Murambatsvina in Zimbabwe', *Journal of Housing and the Built Environment*, 28(2): 381–396, DOI: 10.1007/s10901-012-9311-8.

Chitekwe-Biti, B., Patel, S. and Mitlin, D. (2014) 'The transnational experience of community-led development: the affordable shelter challenge', in J. Bredenoord, P. van Lindert and P. Smets (eds), *Affordable Housing in the Urban Global South*, London and New York: Routledge, pp. 117–132.

Clapham, D.F., Clark, W.A.V. and Gibb, K. (eds) (2012) *The Sage Handbook of Housing Studies*, London: Sage.

Earle, L. (2014) 'Stepping out of the twilight? Assessing the governance implications of land titling and regularization programmes', *International Journal of Urban and Regional Research*, 38: 628–645, DOI: 10.1111/1468-2427.12112.

Emelonye, U. (2015) *OHCHR on Urban Development and Human Rights in Cities*, 29 April, available at: http://unhabitat.org/ohchr-on-urban-development-and-human-rights-in-cities/ (accessed 23 November 2015).

Freire, M.E., Lall, S. and Leipziger, D. (2014) *Africa's Urbanization: Challenges and Opportunities*, Working Paper No. 7, Washington, DC: Growth Dialogue.

Gilbert, A. (2007) 'The return of the slum: does language matter?', *International Journal of Urban and Regional Research*, 31(4): 697–713, DOI: 10.1111/j.1468-2427.2007.00754.x.

Gilbert, A. (2014) 'Renting a home: the need for a policy response', in J. Bredenoord, P. van Lindert and P. Smets (eds), *Affordable Housing in the Urban Global South*, London and New York: Routledge, pp. 87–101.

Gulyani, S. and Bassett, E.M. (2007) 'Retrieving the baby from the bathwater: slum upgrading in sub-Saharan Africa', *Environment and Planning C: Government and Policy*, 25(4): 486–515, DOI: 10.1068/c4p.

Huchzermeyer, M. (2007) 'Tenement city: the emergence of multi-storey districts through large-scale private landlordism in Nairobi', *International Journal of Urban and Regional Research*, 31(4): 714–732.

Huchzermeyer, M. (2011) *Cities with Slums*, Cape Town: University of Cape Town.

Keivani, R. and Werna, E. (2001) 'Modes of housing provision in developing countries', *Progress in Planning*, 55(2): 65–118, DOI: 10.1016/S0305-9006(00)00022-2.

Knox, P. (2009) 'Urbanization', in S.J. Wood (ed.), *International Encyclopaedia of Human Geography*, London: Elsevier, pp. 112–118.

Kombe, W.J. (2000) 'Regularizing housing land development during the transition to market-led supply in Tanzania', *Habitat International*, 24(2): 167–184, DOI: 10.1016/S0197-3975(99)00036-3.

Kumar, S. (2011) 'The research–policy dialectic: a critical reflection on the virility of landlord–tenant research and the impotence of rental housing policy formulation in the urban Global South', *City*, 15(6): 662–673.

Lefebvre, H. (1996) *Writings on Cities*, translated and edited by E. Kofman and E. Lebas, Oxford: Blackwell.

LeVan, A.C. and Olubowale, J. (2014) 'I am here until development comes: displacement, demolitions, and property rights in urbanizing Nigeria', *African Affairs*, 113(452): 387–408, DOI: 10.1093/afraf/adu030.

Meth, P. (2013) 'Viewpoint: Millennium Development Goals and urban informal settlements – unintended consequences', *International Development Planning Review*, 35(1): v–xiii.

Obeng-Odoom, F. (2013) 'The state of African cities 2010: governance, inequality and urban land markets', *Cities*, 31: 425–429, DOI: 10.1016/j.cities.2012.07.007.

OHCHR (Office of the United Nations High Commissioner for Human Rights) (2009) *The Right to Adequate Housing*, Factsheet 21, Geneva: United Nations.

Otiso, K.M. (2003) 'State, voluntary and private sector partnerships for slum upgrading and basic service delivery in Nairobi City, Kenya', *Cities*, 20(4): 221–229, DOI: 10.1016/S0264-2751(03)00035-0.

Owusu-Ansah, J.K. and O'Connor, K.B. (2010) 'Housing demand in the urban fringe around Kumasi, Ghana', *Journal of Housing and the Built Environment*, 25: 1–17. DOI: 10.1007/s10901-009-9173-x.

Patel, K. (2013) *Topic Guide: Provision and Improvement of Housing for the Poor*, London: Evidence on Demand.

Patel, K. (2015) 'Sowing the seeds of conflict? Low income housing delivery, community participation and inclusive citizenship in South Africa', *Urban Studies*, 53(13): 2738–2757, DOI: 10.1177/0042098015572090.

Pugh, C. (2001) 'The theory and practice of housing sector development for developing countries, 1950', *Housing Studies* 16(4): 339–423, DOI: 10.1080/0267303012006652.

Simone, A. (2004a) 'Critical dimensions of urban life in Africa', in T. Falola and S. Salm (eds), *Globalisation and Urbanisation in Africa*, Asmara: Africa World Press, pp. 11–42.

Simone, A. (2004b) *For the City Yet to Come: Changing African Life in Four Cities*, London: Duke University Press.

Soliman, A. (2014) 'Pathways towards self-help housing innovations in Egypt', in J. Bredenoord, P. van Lindert and P. Smets (eds), *Affordable Housing in the Urban Global South*, London and New York: Routledge, pp. 321–335.

Stren, R. (1990) 'Urban housing in Africa: the changing role of government policy', in P. Amis and P. Lloyd (eds), *Housing Africa's Urban Poor*, Manchester: Manchester University Press, pp. 35–53.

Tipple, G. (1994) 'The need for new urban housing in sub-Saharan Africa: problem or opportunity?', *African Affairs*, 93(373): 587–608.

Turner, J. (1976) *Housing By People: Towards Autonomy in Built Environments*, London: Marion Boyars.

Tutu, R.A. (2014) 'Dilemmatic experiences of young migrants in Accra, Ghana: the merciless hands of pirate urbanization and landlordism', *Journal of Housing and the Built Environment*, 29(4): 637–656, DOI: 10.1007/s10901-013-9371-4.

UN (United Nations) (2015) *Sustainable Development Goals*, available at: https://sustainabledevelopment.un.org/topics (accessed 15 November 2015).

UNDP (United Nations Development Programme) (2000) *Millennium Development Goals (MDGs)*, New York: UNDP, available at: www.undp.org/content/undp/en/home/sdgoverview/mdg_goals.html (accessed 3 January 2018).

UN-Habitat (United Nations Human Settlements Programme) (2003) *The Challenge of Slums*, London: Earthscan.

UN-Habitat (United Nations Human Settlements Programme) (2010a) *The State of African Cities 2010: Governance, Inequality and Urban Land Markets*, Nairobi: UN-Habitat, DOI: 10.1163/156853010X510807.

UN-Habitat (United Nations Human Settlements Programme) (2010b) *State of the World's Cities 2010/2011: Cities for All – Bridging the Urban Divide*, Nairobi: UN-Habitat.

UN-Habitat (United Nations Human Settlements Programme) (2011) *Housing the Poor in African Cities: Housing Finance*, Nairobi: UN-Habitat and Cities Alliance.

UN-Habitat (United Nations Human Settlements Programme) (2014) *The State of African Cities 2014*, Nairobi: UN-Habitat, DOI: 10.1177/0021909614547604.

Van Waeyenberge, E. (2015) *Crisis? What Crisis? The World Bank and Housing Finance for the Poor*, SOAS Department of Economics Working Paper Series, No. 191, School of Oriental and African Studies.

Yeboah, I. (2005) 'Housing the urban poor in twenty-first century sub-Saharan Africa: policy mismatch and a way forward for Ghana', *Geojournal*, 62: 147–161, DOI: 10.1007/s10708-005-8182-x.

33

Perspectives on urban employment in Africa

The case of informal solid waste management in Kinshasa, Democratic Republic of Congo (DRC)

Danny Simatele

Introduction

Two of the contemporary challenges that sub-Saharan Africa faces today are increased levels of unemployment, coupled with high levels of unmanageable poverty that have taken place amid increased deterioration in the African economy and unstable institutional and policy frameworks (Page and Shimeles, 2015). It has been argued, for example, that a large proportion of the population in Africa today lives in absolute poverty and that a significant number of this population has no stable source of income and livelihoods (Page, 2015). What is even more intriguing is that both unemployment and poverty in Africa are increasingly becoming urban phenomena as those in the rural areas migrate to cities (Shimeles, 2015). Recent analyses have shown that urban unemployment and poverty are not only growing rapidly, but have tended to be underestimated (Page and Soderbom, 2014). Binns *et al.* (2012) and Simatele and Etambakonga (2015) observe that the central characteristic of the urban crisis in Africa is not the scale of population growth, but the weakness of both national and local government institutions in the face of rapid urban change.

It has been observed that sub-Saharan Africa as a whole is deeply affected with a high rate of unemployment, especially among the youth (AERC, 2013). Africa has the most youthful population in the world, with the youth covering close to between 60 and 70 per cent of the continent's population (AERC, 2013). This state of affairs poses great threats to the strength and growth of the African economy as employment opportunities have diminished and lagged behind. It has been observed that unemployment in many parts of Africa is often caused by the defective education system that fails to equip learners with the necessary knowledge and skills to be more innovative and create opportunities that would contribute to solving many of the urban problems (AERC, 2013; Page and Soderbom, 2014). Ondari-Okemwa (2009) and Avoseh (2013), for example, argue that the educational systems in Africa do not correspond directly to the economic realities prevailing outside the schools system. Instead of training professionals and people with skills, theory instead of practical is the order of the day (Wolhuter, 2011a).

Golub and Hayat (2014) further observe that the education system in African countries has failed to respond to the existing intergenerational gap, specifically in terms of equipping the younger generation with the skills necessary for survival (see also Simatele and Simatele, 2015). It simply imparts general and literary education devoid of any practical content, and this has played an important role in propagating the high levels of unemployment and poverty as this form of education has prevented people from solving their own problems and managing change (see Avoseh, 2013). Africa's educational policy merely produces individuals whose services do not reflect the economic trends in the job market (Wolhuter, 2011b). The educational structure, especially the curriculum, does not include industrial skills, and hence produces graduates whose skills are often not transferrable (Ondari-Okemwa, 2009; Avoseh, 2013). The open-door policy at the secondary and university level has increased manifold unemployment among the educated that are fit only for white-collar jobs and not for self-employment (see Cheru, 2002; Binns *et al.*, 2012). The lack of formal employment in urban Africa, especially among the youth, has forced many of them to look for alternative sources of employment and income generation activities (Dlamini and Simatele, 2016). It is in this context that African employment has consequently remained overwhelmingly informal, and as observed by the Cities Alliance (2017: 13), 'informality is the main game in [African] towns'.

This chapter will discuss the state of unemployment and the role played by the informal economy in Africa, before moving on to examine a case study from the Democratic Republic of Congo (DRC) that exemplifies these themes.

Unemployment and the informal economy in sub-Saharan Africa

Sub-Saharan Africa has increasingly become an urban continent, with an average annual growth rate of 3.3 per cent, which is currently the highest in the world (Pieterse, 2010). The explosion and growth in Africa's urban population has persisted at a rate that has significantly exceeded the rate of employment creation, and this has given rise to the burgeoning of the informal sector (Golub and Hayat, 2014). A number of commenters on African affairs, for example, argue that the dominant story of production and economic growth in sub-Saharan Africa is anchored on the informal economy (see Pieterse, 2010; Hove *et al.*, 2013). They observe that the informal economy workforce in sub-Saharan Africa is estimated to account for 78 per cent of non-agricultural employment, 93 per cent of all new jobs created, and 61 per cent of urban employment (Cities Alliance, 2017). While these figures are not unique to Africa, they are relatively lower in Latin America and Asia (Simatele and Simatele, 2015).

It is important to note that the growth and development of the informal or parallel economy in sub-Saharan Africa has become an integral part of the urbanisation process that the continent is going through. Some studies, for example, have estimated that the informal economy in sub-Saharan Africa will continue to contribute 7 per cent per annum to the overall creation of employment and job opportunities in the region (Obeng-Odoom, 2011). Furthermore, it has been observed that the informal economy is likely to increase the generation of new employment opportunities at a rate of 2 to 3 per cent per year (Todaro, 2000; Hove *et al.*, 2013). This development, as observed by Musemwa (2010), will significantly benefit women, who currently are under-represented in the formal economy. For women in Africa, the informal sector is estimated to represent 92 per cent of all job and employment opportunities outside agriculture that are available to them (Simatele *et al.*, 2017). It is worthwhile to note that although the informal economy cuts across all sub-sectors of the economy and spatial locations, its prevalence is and will be in cities and towns because its activities are essentially demand-driven, and it is therefore responsive to large concentrations of people, a scenario that is becoming true for sub-Saharan Africa.

Within the urban centres of Africa, it has been argued by Hove *et al.* (2013) that the informal sector employs 60 per cent, on average, of the urban workforce, but accounts for less than one-quarter of urban economic growth output. This state of affairs, as observed by Golub and Hayat (2014), is due to the fact that the contribution of the informal economy to non-agricultural GDP tends to decline with the level of development. They further observe that in Africa, this share is currently estimated to average about 40 per cent (Hove *et al.*, 2013). However, the failure by most African governments to adequately account for this activity could potentially result in a statistical under-representation and appreciation of this sector's contribution to the overall economic output of urban areas in Africa. The informal economy has often been considered and believed to be a sector that is marginal and unproductive, but there is now growing evidence in many developing countries illustrating its contributions in output and importance in employment and job creation. In countries such as the Democratic Republic of Congo (DRC), the informal sector provides urban employment for as much as 80 per cent of the urban population (Simatele and Etambakonga, 2015). In Zimbabwe, it is estimated that between 3 and 4 million people earn their living through the informal economy, supporting another 5 million people, while the formal sector employs about 1.3 million people (ZCTU, 2005).

Although productivity within the informal economy is considered to be low, a considerable proportion of the urban workforce employed in this sector do so as a form of disguised unemployment. It is this sector of the urban economy that is made up of individual and family-owned enterprises, small-scale operations, labour-intensive and adapted technologies, and skills acquired outside the formal schooling systems, and it presents easy entry to a high volume of unemployed youth. As a sector, it has the potential of contributing significantly to economic growth and national development (Cheru, 2002). It is therefore of utmost importance that as African cities continue to grow under conditions of economic stagnation and institutional collapse, the informal economy is strategically positioned and prioritised in national development plans and policies. Pieterse (2010) argues that although the informal economy is characterised by low productivity and incomes, it is a valuable 'safety net' for the unemployed youth and redundant civil servants as it cushions and softens the impacts of economic shocks on them.

In summary, therefore, it would not be an exaggeration to argue that the realities of the informal economy imply that many local authorities in sub-Saharan Africa are faced with a dilemma that cannot be addressed from one approach, but will require the participation of a wide range of stakeholders, including the private sector. While the informal economy is recognised as a major source of incomes and livelihoods for many of the urban poor, many local governments in sub-Saharan Africa are concerned about the contribution of such activities to other urban challenges and problems. Hove *et al.* (2013) observe that as a result of this perception by local authorities, their response has often been to either ignore the informal economy or resort to periodic campaigns targeting their activities and discouraging the occurrence of these activities within the city parameters.

The Democratic Republic of Congo and the informal economy

Since attaining political independence from Belgium in 1960, the Democratic Republic of Congo (DRC) has experienced one of the longest and bloodiest civil wars in Africa. It is argued that the DRC War has been the world's deadliest conflict since the Chinese Civil War, and has claimed an estimated 5.4 million lives since 1998 (Kassay, 2011). More than 90 per cent of these victims have not died in combat, but from diseases such as malaria, diarrhea, pneumonia and malnutrition, which have been aggravated by political instability, resulting in massive displacements and subjecting many people to poor living conditions: poor or no access to shelter, water, food and medicines (Kassay, 2011). It is argued that 40 per cent of these deaths have mainly affected senior citizens and children under the age of 5 (UNDP, 2011). Furthermore, it is argued that the civil conflict in the DRC has also negatively affected agriculture productivity and triggered high levels of rural–urban population movements (Misilu Mia Nsokimieno *et al.*, 2010).

Misilu Mia Nsokimieno *et al.* (2010), for example, are of the view that urban growth has occurred in the face of poor weak institutional frameworks and poor urban planning, and this has been compounded by the absence of productive employment opportunities. The combination of these challenges has resulted in conditions subjecting many urban households and whole communities to profound poverty (71 per cent of the 75 million Congolese), food and nutrition insecurity, as well as other socio-economic deprivations (Iyenda, 2005). As a way of coping with high poverty levels and food insecurity, many individuals and households in Kinshasa have increasingly started adopting multiple and diverse means of seeking a livelihood through turning to the informal sector. One such livelihood strategy is foraging from solid waste, which is being used as a source of food as well as income generation. It has been observed by Iyenda (2005) that an estimated 45 per cent of the urban residents in Kinshasa have turned to foraging from solid waste as a source of income generation and livelihood.

Despite the importance of solid waste collection in the lives of the urban poor in Kinshasa, the activity is not fully recognised and has not received adequate support from urban authorities in terms of integrating it within the broader urban development and planning policy framework. It is argued that integrating informal solid waste into the urban policy presents a number of logistical challenges, such as developing and establishing a resource and infrastructure base (e.g. financial resources and road networks) through which such activities could be effectively and efficiently pursued (see Simatele and Etambakonga, 2015). As a result of such perceptions, solid waste collection and management is one of the major challenges facing the local government authorities in Kinshasa (Din and Cohen, 2013; Kubanza and Simatele, 2016). Increasing amounts of uncollected solid wastes are progressively threatening the survival of urban residents, especially the poor, who in most cases are resident in locations that do not receive or only receive little support from government. Kassay (2011), for example, observes that Kinshasa produces more than 5,000 m^3 of refuse daily. This observation correlates with that of the Kinshasa Provincial Government (2010), who note that, the annual generation of solid waste in the city is estimated to be about 2 million m^3, out of which only 20 to 30 per cent is collected, and the rest is eliminated in unacceptable conditions. Nkeku (2007) further observes that Kinois[1] households dispose of 29.1 per cent of their domestic refuse by using unorganised disposal systems, while 6.4 per cent use informal and unregulated incineration, 14.6 per cent bury the waste, 6.2 per cent use illegal dumping sites (streets and open fields), 6.6 per cent dump their waste in rivers, 21.6 per cent illegally dump in public places, 11.6 per cent use compost bins, and 3.9 per cent use other disposal mechanisms.

Despite this gloomy situation, there are preliminary initiatives by the technical services division of the National Sanitation Programme (PNA) and the Department of Roads and Drainage (DRD) to provide waste removal facilities in the city of Kinshasa. However, these initiatives are being hampered by the continued political instability and lack of resources, both at the national and municipal levels. Consequently, out of the 5,000 m^3 of solid waste produced every day in Kinshasa, the daily capacity of waste removal/disposal by a combined team of the PNA and the DRD is only 600 m^3, or 12 per cent (Lelo, 1998). In the absence of an effective formal waste management system, the informal solid waste sector has become an important part of the processes of the collection and management of solid waste in the city. Hypothetically speaking, the informal solid waste management system, through scavenging, has filled the gap created by the absence of official institutions in solid waste management (SWM), and without it the situation would be worse. In view of these observations, this chapter is particularly interested in examining the extent to which scavenging for solid waste contributes to the urban food baskets, particularly of poor urban households. It is argued in the chapter that scavenging for solid waste in Kinshasa generates a positive externality, both as a source of income and livelihood for the urban poor, and should therefore be integrated in the urban development and planning policy (Iyenda, 2005; Simatele and Etambakong, 2015).

Research approach

The data on which this chapter is based was collected during the month of July 2013, involving 100 research participants who were drawn from three distinct research sites in the city of Kinshasa (see Figure 33.1): *Somba Zikita*, a marketplace where only fresh meat and foodstuffs such as root and leafy vegetables and fruits are traded; *Kinshasa Great Market*, a marketplace predominantly involved in the sale of a range of merchandise ranging from second-hand clothing

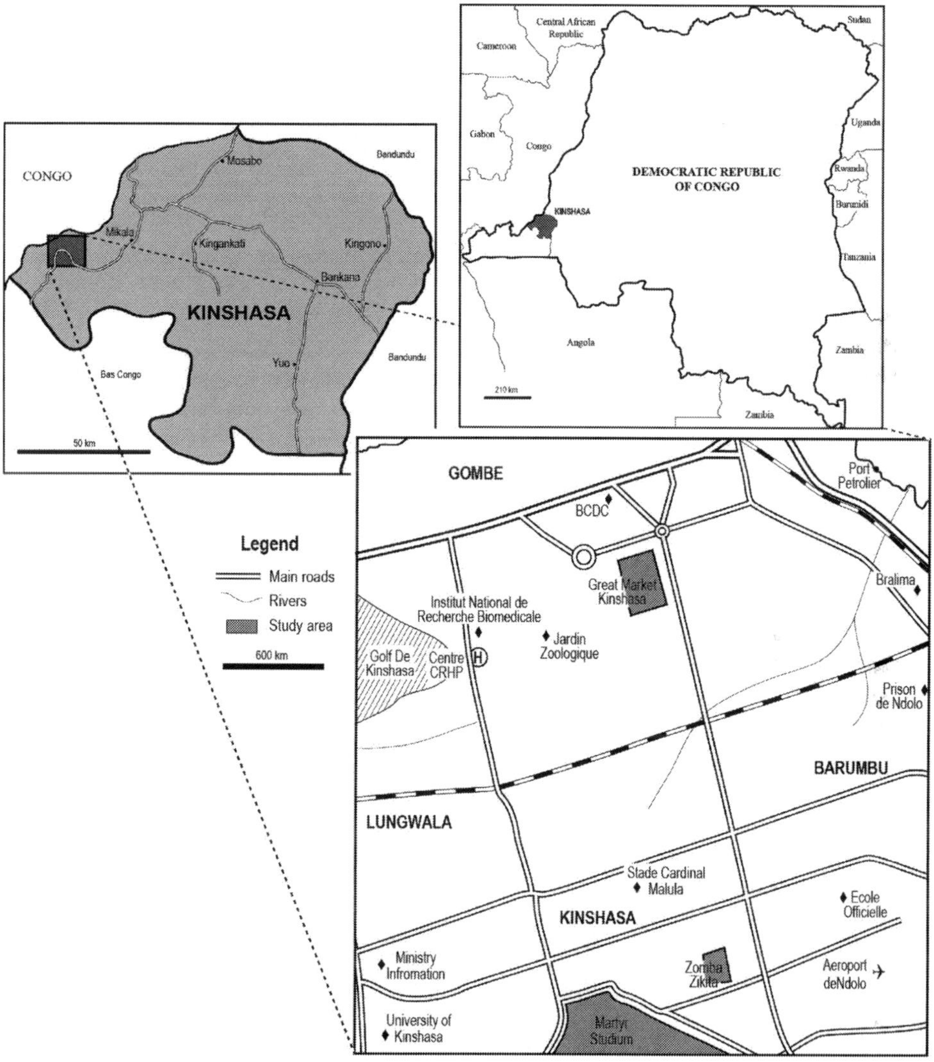

Figure 33.1 Location of the study sites in Kinshasa, DR Congo

Source: Drawn by the Geography Cartographic Unit, University of Witwatersrand.

to electronics and household goods; and finally *Marty Stadium*, which is mainly used as an illegal solid waste dumping site. The research participants were grouped in *four* different categories, namely: informal street and public space cleaners; informal solid waste recyclers; informal solid waste collectors; and itinerant solid waste buyers.

Informal solid waste management and the informal economy: a review of literature

Existing literature on informal solid waste collection predominately took centre stage in the 1970s, with a number of scholars paying attention to the contribution of the informal sector to the urban economy of the newly independent states (Julius, 1978; Furedy, 1984; Cointreau, 1987; de Kock, 1987; Bubel, 1990; Tevera, 1991; Medina, 1997; Rogerson, 2000; Couth and Trois, 2012). Thematically speaking, much of the attention was placed on understanding how colonial institutional structures inherited by the new states would fare in managing high rates of urbanisation that were increasingly resulting in the exponential growth of informal settlements (Beall, 1997; Boyer, 1999; Rogerson, 2000; Binns *et al.*, 2012). It was argued, for example, that after independence, many colonial institutional structures were retained by the new governments without any modification to suit the new political, economic and social reality (Bubel, 1990; Tevera, 1993).

With the lifting and removal of race-based residential and movement restrictions imposed by the colonial systems, the floodgates opened to large numbers of Africans migrating to the capital cities. Cheru (2002) observes that reconciling these two historical phenomena has been a major challenge facing governments across the African continent. As a way of responding to these urban contradictions, many African governments sought to address the situation through the formulation and implementation of 'master planning' projects with large-scale government-sponsored construction of residential dwellings. Cheru (2002) is of the view that such activities merely represented a continuation of colonial practices as the new initiatives proved ineffective in managing urban change. Binns *et al.* (2012), for example, argue that despite the attempt to manage urban development in Africa through 'master planning', urban growth continued unabated. Supporting this assertion, Cheru (2002) observes that the 'master planning' approach was unrealistic, expensive, and devoid of political and economic sense. Policymakers soon began to realise and to articulate an 'anti-urban' position in the hope of discouraging people from migrating to cities. On the contrary, planners saw the growth in urbanisation as inimical to human development. According to Lipton (1977), planners began to argue that urbanisation was a consequence of unbalanced development between rural and urban areas. Subsequently, national governments and aid agencies began to pay greater attention to the development of rural areas in the hope of addressing the rural–urban divide. Cheru (2002), however, argues that in reality, official policy position encouraged both urbanisation and industrialisation, which were considered important symbols of modernisation.

With the oil crisis emerging in the 1970s, coupled with high unemployment resulting from deteriorating economic conditions, there was a significant growth in informal sector activities. The growth in this sector, coupled with the declining economic situation, has coincided with a dramatic deterioration in the supply of basic infrastructure and urban services. Deborah Potts (2007), for example, argues that urban people who have to remain in the cities must work – and if the formal sector cannot provide employment, self-employment in the informal sector becomes the answer. Otang-Ababio (2012) and Otang-Ababio *et al.* (2013) observe that the lack of formal employment opportunities in many sub-Saharan African countries has forced a significant number of individuals and households to turn to scavenging for solid waste as a source of livelihood. They are of the view that due to the lack of formal job opportunities, e-waste scavenging, for example, has emerged as a major livelihood (survival) strategy for a rapidly growing urban population, particularly in West African countries. The failure of the formal sector to generate enough job opportunities to meet the growing numbers of urban jobseekers has compelled many who are qualified but unemployed, and those with low employability, to turn to the informal sector for survival (Simatele and Simatele, 2015).

A major factor in the solid waste management crisis in sub-Saharan Africa is embedded in institutional deficiencies that have resulted in the lack of infrastructural provisions and other logistics that should play a key role in waste management (Taylor, 2014). As a result of this state of affairs, much of the solid waste generated within the urban areas is usually illegally dumped on any available space (Cointreau, 2006). Even in cities where specific areas have been designated as official and legal dumpsites and landfills, there is little control and effective management of the waste (Hove *et al.*, 2013; Otang-Ababio *et al.*, 2013). Instead, this function is usually left to informal solid waste collectors to process the waste (Binns *et al.*, 2012; Couth and Trois, 2012).

Despite the role that the informal sector plays in solid waste management and employment creation in urban areas of sub-Saharan Africa, the International Labour Organization (ILO) observes that 'informal sector activities are largely ignored, rarely supported, often regulated and sometimes actively discouraged by the government' (ILO, 1972: 6; see also Otang-Ababio *et al.*, 2013). In the last two decades, however, several approaches to integrated solid waste management (ISWM) have emerged, with a shift from the victimisation of scavengers to an inclusive discourse focused on achieving effective management of solid waste. This process has included recognising, strengthening and integrating solid waste scavenging activities into solid waste management (SWM) schemes (Bartone, 1986; Tevera, 1991; Boyer, 1999; UNESCO, 2001; Nas and Jaffe, 2003; Medina, 2005; Wilson *et al.*, 2006; Scheinberg, 2012; Dominelli, 2013; Taylor, 2014). In view of this new approach, a number of studies focusing on solid waste scavenging in urban centres have tended to pay particular attention to the operations, social and health circumstances in which individuals and groups of individuals involved in solid waste scavenging operate (Tevera, 1994; Eerd, 1996; Beall, 1997; Thomas-Hope, 1998; Cointreau, 2006). There has been also a considerable literature that pays particular attention to the different approaches, attitudes and perceptions of many governments in developing countries towards solid waste scavenging (Medina, 1997; Nas and Jaffe, 2003; Scheinberg *et al.*, 2010).

Scheinberg *et al.* (2010) argue that the approaches, attitudes and perceptions of municipal authorities towards scavenging activities vary from country to country and sometimes within countries. Medina (1997), for example, categorised the relationship between scavenging schemes and local municipal authorities into four different groups: repression, neglect, collusion and stimulation. The nature of the relationship between the local municipality and scavenging individuals or groups, again, vary from one context to the next depending on the political will of the governors of the local or municipal authority. In Indonesia, for example, scavengers are regarded as a useful part of the waste management system (Scheinberg *et al.*, 2010). The authorities have adopted a more progressive policy, trying to find safer and more equitable win–win situations, rather than trying to either exploit or abolish the informal sector's involvement in solid waste collection and management. In other places, individuals and groups of people involved in scavenging schemes face hostile situations because policymakers consider this informal activity as a blot on the urban landscape and a source of embarrassment to government officials when international visitors come to visit their cities (Rogerson, 2000; Cheru, 2002; Binns *et al.*, 2012). In the city of Kinshasa, for instance, local authorities are indifferent to the informal solid waste sector, and scavengers are often looked at as a nuisance contributing to the difficulties of managing waste effectively (Longondjo, 2010). This situation has been compounded by the fact that solid waste scavenging has been galvanised by the often high levels of involvement of children within the urban economy (Thomas-Hope, 1998; Noel, 2007). In this context, scavenging is often considered as a sign of failure to modernise and transform the urban landscape in the Democratic Republic of Congo, and in the eyes of local government officials it is an uninvited occurrence (Nas and Jaffe, 2003).

It is important to note that informal solid waste collection in sub-Saharan Africa (SSA) is closely connected to issues of unemployment and poverty. A number of scholars, such as de Kock (1987), Tevera (1993), Makovere (1998) and Binns *et al.* (2012), for example, are of the view that scavenging for solid waste in SSA is a response to high levels of poverty and a deterioration in the formal job market. Millions of urban dwellers in these countries often earn a livelihood and income by extracting value from solid waste. Tevera (1993: 95) argues that solid waste scavenging should be viewed 'as representing a survival response by the urban poor, and it is a form of employment, a strategy that is often preferred to begging or stealing'. In the face of rapid urbanisation and high unemployment rates, especially among the youth, and coupled with huge volumes of uncollected waste, the informal sector has increasingly become part of the urban landscape. Despite this situation, scavenging activities still remain outside the urban development and planning policy realms (Simon, 1984; Tevera, 1991, 1994; Makovere, 1998; Nzeadibe, 2009).

Other existing literature has examined informal solid waste collection in Africa from an economic or entrepreneurial perspective (Furedy, 1984; Bartone, 1986; Tevera, 1993; Huysman, 1994; Makovere, 1998; Boyer, 1999; Rogerson, 2000; Scheinberg *et al.*, 2010). From this perspective, it has been argued that informal solid waste activities have provided a significant number of individuals and households with an opportunity to actively engage and participate fully in the formal and urban economy (Potter *et al.*, 2010). Through the trade of collected and recyclable solid waste materials, they implicitly participate in the emerging formal solid waste recycling industry (Otang-Ababio *et al.*, 2013). Thus, there is now an emphasis in some contemporary studies that presents solid waste scavenging as an activity with entrepreneurial characteristics because of its nature of straddling formality and informality (ILO, 2004; Chikarmane and Narayan, 2005; Scheinberg *et al.*, 2010; Wilson *et al.*, 2010; Scheinberg *et al.*, 2011; Scheinberg, 2012). Because of this state of affairs, there is now a movement in some countries towards integrating scavenging activities into the formal economy (e.g. in Indonesia) through improved changes in solid waste management systems (Scheinberg *et al.*, 2010).

Informal solid waste management, employment and income generation in Kinshasa

Solid waste management is one of the issues that presents a real challenge to the urban managers in Kinshasa (Longondjo, 2010; Kubanza and Simatele, 2016). Although solid waste picking in Kinshasa is not in any way different from that in other sub-Saharan African cities, its uniqueness lies in the fact that its incidence has been exacerbated not only by high poverty and unemployment levels, but by increased levels of political instability and civil conflict. This situation has resulted in much of the country's resources being diverted from development projects to either curbing or propagating the civil conflict that the country has experienced over the past six decades (Longondjo, 2010; Kassay, 2011; Simatele and Etambakonga, 2015). Thus, political instability has not only resulted in increased civil conflict, but has also significantly contributed to the weakening of the institutional and policy framework, thereby hampering the effectiveness and efficiency in delivering required social and economic services such as solid waste management. As a result, the solid waste management challenge in

Kinshasa, and the Congo in general, has not only been a result of the lack of resources and knowledge, but has also been largely embedded in the unending civil war, which for the most part has prevented people and institutions from solving urban challenges and the problems that they face. These problems have further been compounded by the fact that the poor and the voiceless have been disenfranchised from participating in urban processes and have been subjected to marginalisation of various degrees.

The challenges in solid waste management in Kinshasa have been further complicated by increased rural–urban migration, and this situation has overwhelmed city authorities, who in most cases are operating under huge budget deficits on one hand, and on the other are implementing urban development approaches that appear to be out of touch with reality on the ground. In the face of increased civil unrest, coupled with a dramatic deterioration in the supply of basic infrastructure and urban services, as well as the declining economic situation, this has left the urban authorities in Kinshasa overwhelmed by the growing demands for solid waste management services (Misilu *et al.*, 2010). With a narrow revenue base, increased civil conflict and limited technical capacities, the municipal authorities in Kinshasa have been unwilling or unable to effectively deliver on their mandates, such as the provision of refuse collection, road maintenance and water supply, to mention but a few. Longondjo (2010) observes that at the city level of Kinshasa, the weak institutional and policy frameworks have struggled to cope and manage the level of urban change.

It has been observed by Din and Cohen (2013), for example, that in Kinshasa, there are a multiplicity of agencies that may deliver urban services, but there is no coordination among them (Kubanza and Simatele, 2016). This argument is supported by Mbuyi (1989), who is of the view that the lack of effective coordination and management systems and the lack of financial resources have led to conditions that are deleterious to the environment and continue to subject the urban poor in Kinshasa to inferior living conditions. The current economic turmoil, coupled with the perpetuation of the civil war, as well as high levels of corruption, have combined to exacerbate problems of solid waste management and push the poor into living miserable lives. Financial resources are central to the effective and efficient management of solid waste management because they determine the level or quality of services that can be provided (Mosha, 1990). Thus, financial availability and sound management systems – or their absence – are key elements to determining the nature of solid waste management in a city, particularly in low-income cities such as Kinshasa.

In the face of institutional and policy challenges in solid waste management and the deterioration in employment opportunities, the informal sector, through activities such as scavenging, is increasingly playing an important role in the management of waste and of employment and livelihood for the urban poor in Kinshasa. The engagement and trade in recycled solid waste with recycling industries, for example, is increasingly proving to be one of the major forms of employment and a source of livelihood for many poor urban households in Kinshasa. An interview discussion with a male research participant aged between 25 and 30 from Zomba Zikita, for example, stated:

> There are very few employment opportunities in Kinshasa these days. I really don't want to join the rebels and fight the government and kill innocent Congolese – I think that won't solve anything. Many of us young people don't want to do that. So I opt to collect used bottles, cupboards and other plastics that I am now selling and getting a bit of money for myself and my family. I don't make enough, but what else can I do?
>
> *(personal communication, 2012)*

Another research respondent aged between 15 and 20 commented:

> I don't have any family. My parents and siblings were killed and I have no one. Picking waste and selling it to the boss at the market is now my life and I make a bit of money for my rent and food.
>
> *(personal communication, 2012)*

The two views expressed by these two participants seem to suggest that scavenging for solid waste has, in a way, been an alternative source of employment and a source of livelihood. Although there's a sense of helplessness from the participants, what is evident in the two views is the fact that solid waste contributes to their welfare and well-being. In spite of the role being played by this informal sector activity, local authorities have an indifferent attitude towards scavengers. Not only do the city officials block scavenging activities, but they are also unwilling to foster any constructive change that would accommodate and recognise the role played by solid waste scavenging. A research participant in his early twenties and from Martyr Stadium, for example, commented:

> Going to the waste dump to some materials is a dangerous business. If council officials catch you there, they will beat the hell out of you and then fine and arrest you. They don't realize that the dump is also my office where I work to make money for my family just as they also make their money from sitting in the offices.
>
> *(personal communication, 2012)*

The negative behaviour towards the solid waste scavengers could be argued as being motivated by a lack of knowledge on the part of urban managers in terms of the contribution that informal solid waste collection makes to solid waste management in the city of Kinshasa. The lack of knowledge, coupled with poor governance practices, has in a large way contributed to the bureaucratic and institutional harassment of the solid waste collectors, who in most cases are less powerful people in society. The current approach to urban planning and management that has been adopted and implemented by city authorities in Kinshasa seems to place very little emphasis on the significance of citizens' involvement in designing and implementing sustainable waste management systems (Longondjo, 2010; Kubanza and Simatele, 2016). Instead, it is an approach that favours and allows the vested interests of powerful and economic elite groups to influence unduly, and even to buy, the policies, regulations and laws of the local authority, making it a difficult task to develop pro-poor policies and integrate solid waste scavenging activities into urban development and planning policies.

Foraging for solid waste in Kinshasa is an activity that seems to be dominated by male actors, and a recent phenomenon is the heavy representation of children. The majority of the solid waste pickers are poor and marginalised groups of people who have some form of education. Seventy per cent of the respondents involved in solid waste collection in the survey were males, almost 42 per cent of all the research participants had attained primary education, 36 per cent had no formal education, and 21 per cent had attained secondary education. Forty-five per cent of all solid waste pickers were aged between 19 and 34, 38 per cent were aged between 10 and 18, and 17 per cent were aged between 35 and 44. These figures suggest that scavenging for solid waste in Kinshasa is an activity mainly dominated by young people who have not been absorbed and integrated into the formal sector.

Solid waste collection in Kinshasa can be classified into four different categories, as shown in Table 33.1. The actors in each of these categories perform different functions, although some

Table 33.1 Activities of solid waste pickers in Kinshasa

Category of solid waste pickers	*Type of activities*	*Reason for involvement in solid waste collection*
Formal cleaners	Cleaning the streets in front of resident homes, cleaning public places such as markets, schools, etc.	Formal employment in order to obtain a salary and obtain food and money for other purposes
Informal collectors	Tour streets pushing trolleys and using pushcarts, tricycles and pickup trucks to collect solids from residents and market for a small fee	Get money for food for the family
Informal recyclers	Pick through garbage bins or dumps to collect recyclable items Collection crews who often sift through garbage bins for sale to scrap dealers Collection of used bottles, cardboards, metal and other items from dumps for recycling	Get money for food and education for the family
Itinerant buyers	Tour the streets, and from house to house searching for recyclable scrap and other materials Retrieve repairable and reusable items that can sell or use themselves	Setting up and running a business to make money and profit for the business to expand

Source: Field-based research materials (2013).

overlap can be observed as they all focus on working with solid waste and are actively involved in collecting solid waste for recycling and trade. Although the activities of the solid waste pickers in Kinshasa do not seem to be organised in any particular manner, the success of each category is dependent on the functionality of other categories. The activities of the itinerant buyers, for example, are linked to those of the formal cleaners, as well as the activities of the informal solid waste collectors and recyclers.

Despite the lack of formal organisation, the waste pickers operate in a rather complex system and contribute significantly to SWM in Kinshasa. Figure 33.2, for example, illustrates the overall picture of solid waste management in Kinshasa. What is particularly interesting in Figure 33.2 is a clear demarcation in the providers that both the formal and informal systems of waste management target. This suggests that the formal system tends to collect waste in the more affluent neighbourhoods, whereas the informal system is focused on the poor neighbourhoods and spaces. One would therefore argue that without the two systems operating side by side, the solid waste situation in the city of Kinshasa would be catastrophic. Although scavenging for solid waste has become an important activity in terms of livelihoods and income generation (see Table 33.2), the generated income is dependent on the category in which one operates (e.g. formal solid cleaner or collector). Table 33.2, for example, suggests that itinerant buyers can earn up to an estimated 32 per cent of their overall monthly total income from buying waste directly from the waste pickers, whereas informal waste recyclers earn up to 28 per cent of their overall monthly income from buying the waste from the itinerant buyers.

The waste pickers, on the other hand, earn an estimated 22 per cent of their income from engaging in waste collection. What is interesting from the information in Table 33.2 is that

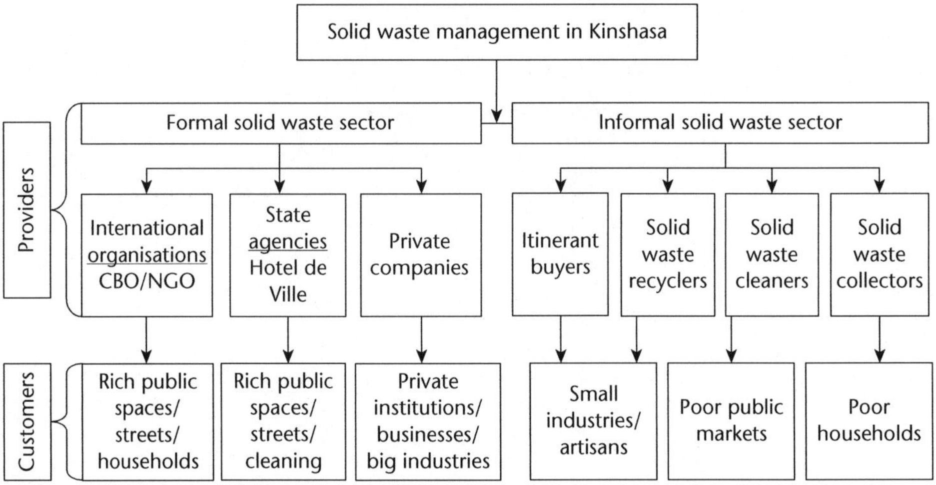

Figure 33.2 Solid waste management system in Kinshasa

Source: Adapted from Noel (2007).

Table 33.2 Average monthly income realised from scavenging

Scavenging category	*Number of respondents*	*Average monthly income (US$)*	*% average monthly income (US$)*
Itinerant buyers	25	65	32
Informal recyclers	25	55	27
Informal waste collectors	25	45	22
Informal cleaners	25	40	20
Total	**100**	**205**	**100**

Source: Field-based research materials (2012).

engaging in solid waste activities does have financial benefits, and it would seem that the groups benefiting more are the itinerant buyers. A probable explanation of this situation may perhaps revolve around the fact that itinerant buyers function as middle agents between the waste collectors and the waste recyclers, and may have more bargaining freedom, putting them in an advantaged position in terms of negotiating the buying and selling prices.

It must also be noted that while scavenging for solid waste contributes to the urban food basket and income generation of poor urban households, these activities also play an important role in environmental management, particularly in reducing greenhouse gas (GHG) from landfill gases and biodegradable solid waste. Couth and Trois (2011), for example, argue that solid waste collection and recycling plays an important role in helping meet GHG emission reduction goals by reducing emissions generated due to waste disposal, extraction of raw materials, and manufacturing of new products. In a city such as Kinshasa, huge amounts of solid waste are disposed of each day (Simatele and Etambakong, 2015), which produces significant amounts of methane (CH_4) emission into the atmosphere (Couth and Trois, 2011).

In a country such as the DRC, where nearly all formal institutions have virtually collapsed, the integration of solid waste scavenging activities must take centre stage in current and future policy initiatives in order to contribute significantly to reducing GHG emissions. Furthermore, given that nothing has been done in the city of Kinshasa to formalise informal waste collection, it is important that local authorities find an efficient solid waste collection strategy or system that will be tailored towards creating job opportunities. Identifying local waste collection solutions that match local needs and possibilities, such as supporting the waste pickers, do not only contribute to improved household well-being, but can also empower communities to actively participate in the formulation of policies and programmes that are critical to the development of sustainable urban environments in the DRC.

Conclusion

The main objective of this chapter was to explore and examine the role that solid waste scavenging plays in contributing to household livelihoods, incomes and environmental management in Kinshasa, DRC. It has been suggested and argued in this chapter that scavenging for solid waste in Kinshasa is a prevalent activity that is predominantly carried out by men, and in some cases children and women. It has also been revealed that there is differentiation between types of scavenging activities that are carried out within the city and by different groups of solid waste collectors. However, it has been argued that of all the actors involved in solid waste scavenging, itinerant buyers seem to benefit the most, while children and other vulnerable groups are overtly exploited. This situation owes much to the lack of existing institutional framework through which the interests of vulnerable solid waste collectors would be protected and enhanced. In view of this observation, it is of uttermost urgency to improve urban governance in Kinshasa and the DRC as a whole. Improving governance may result in a situation where emphasis is placed on the importance of private, individual, community and citizens' group-led initiatives for solid waste management.

It is only through improved governance that the current neglect of the informal sector by local authorities will be addressed and incorporated in the urban development and planning policy. Effective and improved urban governance will, for example, ensure increased local community involvement and participation in decision-making, and this is most likely to generate a greater sense of commitment and responsibility by the community, both in terms of waste generation and management. Furthermore, improved governance will not only facilitate the involvement of local communities, but will also expand the scope for greater participation of both non-governmental and community-based organisations (i.e. NGOs and CBOs). It is important to note that NGOs and CBOs play an important role in urban environments through activities such as citizen empowerment, environmental management, energy, health, water and sanitation, education, and community development.

There is a need by the local authority in Kinshasa to tap into the operations of NGOs and CBOs if an effective solid waste management scheme is to be developed. The local authority must realise that they cannot deliver on all urban promises in the absence of these organisations. There is a need therefore for the local authority to foster partnerships with not only other formal operations, but also with informal actors, which includes those individuals scavenging for solid waste. A more effective way would be moving towards the formation of recognised scavenger cooperatives as a means to promote grassroots development. It is argued here that for an effective solid waste management scheme to be developed in Kinshasa, there is a need to shift the current top-down approach towards recognition, empowerment and incorporation of scavenging activities into the formal waste management plans.

Note

1 The word 'Kinois' is used to describe people who have taken up urban residence in the city of Kinshasa.

References

AERC (African Economic Research Consortium) (2013) *Youth and Unemployment in Sub-Saharan Africa: AERC Senior Policy Seminar XV*, Kigali, Rwanda, 21–22 March, available at: www.aercafrica.org (accessed 6 October 2016).

Avoseh, M.B.M. (2013) 'Proverbs as theoretical frameworks for lifelong learning in indigenous African education', *Adult Education Quarterly: A Journal of Research and Theory*, 63(3): 236–250.

Bartone, C. (1986) 'Recycling waste: the World Bank project on resource recovery', *Development: Seeds of Change*, 4: 35–39.

Beall, J. (1997) 'Thoughts on poverty from a South Asian rubbish dump: gender, inequality and household waste', *Bulletin, Institute of Development Studies*, 28(3): 73–90.

Binns, T., Dixon, A. and Nel, E. (2012) *Africa: Diversity and Development*, London: Routledge.

Boyer, G. (1999) 'Riches in rags', *The Urban Age*, 6(4): 10.

Bubel, N. (1990) 'Waste picking and social waste management', *Environmental Sanitation News*, 30: 53–66.

Cheru, F. (2002) *African Renaissance: Roadmaps to the Challenge of Globalization*, London: Zed Books.

Chikarmane, P. and Laxmi, N. (2005) *Organising the Unorganised: A Case Study of the Kagad Kach Patra Kashtakari Panchayat (Trade Union of Waste-Pickers), KKPKP, Pune, India*, available at: http://wiego.org/wiego/kagad-kach-patra-kashtakari-panchayat-kkpkp (accessed 15 July 2013).

Cities Alliance (2017) *The Context of Urban Development in Sub-Saharan Africa*, available at: www.citiesalliance.org (accessed 20 April 2017).

Cointreau, S. (1987) *Solid Waste Recycling: Case Studies in Developing Countries*, Washington, DC: World Bank.

Cointreau, S. (2006) *Occupational and Environmental Health Issues of Solid Waste Management. Special Emphasis on Middle- and Lower-Income Countries*, World Bank Group, Urban Papers 2, available at: http://siteresources.worldbank.org/INTUSWM/Resources/up-2.pdf (accessed 11 July 2013).

Couth, R. and Trois, C. (2011) *Waste Management Activities and Carbon Emissions in Africa*, Waste Management, University of KwaZulu-Natal, CRECHE, School of Civil Engineering, Survey and Construction, Durban 4041, South Africa.

Couth, R. and Trois, C. (2012) 'Sustainable waste management in Africa through CDM projects', *Journal of Waste Management*, 32(11): 2115–2125.

de Kock, R. (1987) 'The garbage scavengers: picking up the pieces', *Indicator South Africa*, 4: 51–55.

Din, G. and Cohen, E. (2013) 'Modelling municipal solid waste management in Africa: case study of Matadi, the Democratic Republic of Congo', *Journal of Environmental Protection*, 4: 435–445.

Dlamini, S and Simatele, D. (2016) 'Unrecognised informal solid waste recycling in an emerging African megacity: a study of Johannesburg, South Africa', *WIT Transactions on Ecology and The Environment*, 202: 14–25, DOI: 10.2495/WM160021.

Dominelli, L. (2013) 'Environmental justice at the heart of social work practice: greening the profession', *International Journal of Social Welfare*, 22: 431–439.

Eerd, V. (1996) *The Occupational Health Aspects of Waste Collection and Recycling: A Survey of the Literature*, WASTE Working Document 4, part 1, Urban Waste Expertise Program (UWEP), available at: www.waste.nl/page/558 (accessed 11 July 2013).

Furedy, C. (1984) 'Survival strategies of the urban poor: scavenging and recuperation in Calcutta', *GeoJournal*, 8(2): 129–136.

Golub, S. and Hayat, F. (2014) *Employment, Unemployment and Underemployment in Africa*, Wider Working Paper, 2014/014, United Nations University, UNU-WIDER, World Institute for Development Economics Research.

Hove, M., Ngwerume, E. and Muchemwa, C. (2013) 'The urban crisis in sub-Saharan Africa: a threat to human security and sustainable development', *International Journal of Security and Development*, 2(1): 7–18.

Huysman, M. (1994) 'Waste picking as a survival strategy for women in Indian cities', *Environment and Urbanization*, 6(2): 155–174.

ILO (International Labour Organization) (1972) *Employment, Incomes and Equality: A Strategy for Increasing Productive Employment in Kenya*, Geneva: ILO.

ILO (International Labour Organization) (2004) *Addressing the Exploitation of Children in Scavenging: A Thematic Evaluation of Action on Child Labour – A Global Synthesis Report for the ILO*, Geneva: ILO/IPEC, available at: www.ilo.org/iloroot/docstore/ipec/prod/eng/2004_eval_scavenging_en.pdf (accessed 11 July 2013).

Iyenda, G. (2005) 'Street enterprises, urban livelihoods and poverty in Kinshasa', *Environment and Urbanization*, 17: 55–67, available at: http://intl-eau.sagepub.com/content/17/2/55.full.pdf+html (accessed 1 September 2013).

Julius, D. (1978) 'Urban waste as an economic good', in A. Pacey (ed.), *Sanitation in Developing Counties*, Chichester: Wiley, pp. 67–79.

Kassay, N.J. (2011) La gestion des déchets plastiques à Kinshasa: un autre défi environnemental à relever dans la conception des villes durables, *L'Afrique et les defis de XXIeme Siecle*, CODESRIA, 13 eme Assemblee Generale, Rabat, Maroc.

Kinshasa Provincial Government (2010) *The Development Study for Urban Rehabilitation Plan of Kinshasa in the Democratic Republic of the Congo*, available at: http://open_jicareport.jica.go.jp/pdf/12008603_01.pdf (accessed 4 January 2010).

Kubanza, S. and Simatele, D. (2016) 'Social and environmental injustices in solid waste management in sub-Saharan Africa: a study of Kinshasa, the Democratic Republic of Congo', *Local Environment: The International Journal of Justice and Sustainability*, 21(7): 866–882, DOI: 10.1080/13549839.2015.1038985.

Lelo, N.F. (1998) La gestion des déchets domestiques: Bilan annuel d'une expérience pilote de l'Hôtel de ville de Kinshasa, *Actes de 1er colloque sur la problématique des déchets à Kinshasa (Congo)*, AoÛt 12 au 15.

Lipton, M. (1977) *Why Poor People Stay Poor: A Study of Urban Bias in World Development*, London: Temple Smith.

Longondjo, E.C. (2010) 'Challenges facing city of Kinshasa in solid waste management: the case of Kinshasa local municipality', unpublished, MSc thesis, School of Architecture and Planning, Department of Urban and Regional Planning and Development Planning, University of the Witwatersrand, Johannesburg.

Makovere, B. (1998) 'Informal waste collection as livelihood strategy in Johannesburg inner-city', unpublished paper, Department of Geography and Environmental Studies, University of the Witwatersrand, Johannesburg.

Mbuyi, K. (1989) 'Kinshasa: problems of land management, infrastructure and food supply', in R.E. Stren and R.R. White (eds), *African Cities in Crisis: Managing Rapid Urban Growth*, Boulder, CO: Westview Press, pp. 149–175.

Medina, M. (1997) *Informal Recycling and Collection of Solid Wastes in Developing Countries: Issues and Opportunities*, UNU/IAS Working Paper No. 24, July, Tokyo: United Nations University/Institute of Advanced Studies.

Medina, M. (2005) 'Serving the unserved: informal refuse collection in Mexico', *Waste Management and Research*, 23: 390–397.

Misilu, M., Nsokimieno, E., Chen, S. and Zhang, L.-Q. (2010) 'Sustainable urbanization's challenge in Democratic Republic of Congo', *Journal of Sustainable Development*, 3(2): 242–254.

Mosha, A.R. (1990) 'A review of subnational planning experiences in Tanzania', in A.H.J. Helmsing and K.H. Wekwete (eds), *Subnational Planning in Southern and Eastern Africa: Approaches, Finance and Education*. Aldershot: Avebury, pp. 78–101.

Musemwa, M. (2010) 'From "sunshine city" to a landscape of disaster: the politics of water, sanitation and disease in Harare, Zimbabwe, 1980–2009', *Journal of Developing Societies*, 26(2): 165–206.

Nas, P.J.M. and Rivke, J. (2003) *Informal Waste Management: Shifting the Focus from Problem to Potential*, Leiden: CNWS Research School, Leiden University.

Nkeku, E. (2007) 'La gestion des dechets dans la ville de Kinshasa', unpublished MSc in Management and Economics Sciences, Dissertation, University of Kinshasa, DRC.

Noel, C. (2007) 'Solid waste management and urban poverty in Haiti: the livelihood strategies of the urban poor in Greater-Port au-Prince and their contribution to environmental management', unpublished PhD thesis, the University of the West Indies, Institute for Sustainable Development, Environmental Management Unit, Kingston, Jamaica.

Nzeadibe, T.C. (2009) *Solid Waste Reforms and Informal Recycling in Enugu Urban Area, Nigeria*, Habitat International, Department of Geography, Waste Management & Recycling Research Unit, University of Nigeria, 410001 Nsukka, Nigeria.

Obeng-Odoom, F. (2011) 'Informal sector in Ghana under siege', *Journal of Developing Societies*, 27(3–4): 355–392.

Ondari-Okemwa, E. (2009) 'The role of knowledge management in enhancing government service-delivery in Kenya', *South African Journal of Libraries and Information Science*, 75(1): 28–39.

Otang-Ababio, M. (2012) 'When necessity begets ingenuity: e-waste scavenging as a livelihood strategy in Accra, Ghana', *African Studies Quarterly*, 13(1/2): 1–21.

Otang-Ababio, M., Melara Arguello, J.E. and Cabbay, O. (2013) 'Solid waste management in African cities: sorting the facts from the fads in Accra, Ghana', *Habitat International*, 39: 96–104.

Page, J. (2015) 'Structural change and Africa's poverty puzzle', in L. Chandy, H. Kato and H. Kharas (eds), *The Last Mile: In Ending Extreme Poverty*, Washington, DC: Brookings Press, pp. 219–248.

Page J. and Shimeles, A. (2015) 'Aid, employment and poverty reduction in Africa', *African Development Review*, 27(51): 17–30.

Page, J. and Soderbom, M. (2014) *Is Small Beautiful? Small Enterprise, Aid and Employment in Africa*, Helsinki: UN-Wider.

Pieterse, E. (2010) 'Filling the void: towards an agenda for action on African urbanisation', in E. Pieterse (ed.), *African Centres for Cities Urbanisation Imperatives for Africa: Transcending Policy Inertia*, Cape Town: University of Cape Town, pp. 7–45.

Potter, R., Binns, T., Elliot, J. and Smith, D. (2010) *Geographies of Development*, 2nd edition, Harlow: Addison Wesley Longman.

Potts, D. (2007) *The State and the Informal in Sub-Saharan African Urban Economies: Revisiting Debates on Dualism*, Crisis States Working Papers Series No. 2, Destine Development Studies Institute.

Rogerson, C. (2000) *The Waste Sector and Informal Enterpreneurship in Developing World Cities: Working Group Meeting*, Investigation of Appropriate Economic Instruments and Funding Mechanisms for the Implementation of the National Waste Management Strategy, Pretoria, South Africa.

Scheinberg, A. (2012) *Informal Sector Integration and High Performance Recycling: Evidence from 20 Cities*, WIEGO Working Paper (Urban Policies) No. 23, available at: www.cwgnet.net (accessed 11 June 2013).

Scheinberg, A., Simpson, M.H. and Yamini, G. (2010) *Economic Aspects of the Informal Sector in Solid Waste, Final Report and Annexes*, Eschborn, Germany: GIZ (German International Co-Operation), the CWG (Collaborative Working Group on Solid Waste Management in Low- and Middle-income Countries), and the German Ministry of Foreign Affairs, available at www.giz.de (accessed 25 July 2013).

Scheinberg, A., Agathos, N., Gachugi, J.W., Kirai, P., Alumasa, V., Shah, B., Woods, M. and Waarts, Y.R. (2011) *Sustainable Valorisation of Urban Organic Wastes: Insights from African Case Studies*, Wageningen, Netherlands: Wageningen University, available at: www.researchgate.net/publication/239847141_Sustainable_valorisation_of_organic_urban_wastes_insights_from_African_case_studies (accessed 8 January 2018).

Shimeles, A. (2015) *Growth and Poverty in Africa: Shifting Fortunes and New Perspectives*, IZA Discussion Paper No. 8751, 3 January, Göteburg University, available at: https://papers.ssrn.com/sol3/papers.cfm?abstract_id=2544815 (accessed 8 January 2018).

Simatele, D., and Etambakonga, C. (2015) 'Scavenging for solid waste in Kinshasa: a livelihood strategy for the urban poor in the Democratic Republic of Congo', *Habitat International*, 49: 266–274.

Simatele, D. and Simatele, M. (2015) 'Climate variability and urban food security in sub-Saharan Africa: lessons from Zambia using an asset-based adaptation framework', *South African Geographical Journal*, 97(3): 243–263.

Simatele, M.D., Dlamini, S. and Kubanza, N.S. (2017) 'From informality to formality: perspectives on the challenges of integrating solid waste management into the urban development and planning policy in Johannesburg, South Africa', *Habitat International*, 63: 122–130.

Simon, D. (1984) 'Urban poverty, informal sector activity and intersectoral linkages: evidence from Windhoek', *Namibia Development and Change*, 15: 551–576.

Taylor, D. (2014) 'The rise of the environmental justice paradigm: injustice framing and the social construction of environmental discourse', *American Behavioural Scientist*, 43(4): 508–580.

Tevera, D.S. (1991) 'Solid waste disposal in Harare and its effects on the environment: some preliminary observations', *The Zimbabwe Science News*, 25(1/3): 9–13.

Tevera, D.S. (1993) 'Waste-recycling as a livelihood in the informal sector: the case of Teviotdale garbage dumps in Harare', in L. Zinyama, D.S. Tevera and S.D. Cumming (eds), *Harare: The Growth and Problems of the City*, Harare: University of Zimbabwe Press, pp. 83–96.

Tevera, D.S. (1994) 'Dump scavenging in Gaborone, Botswana: anachronism or refugee occupation of the poor?', *Geografiska Annaler*, 76B(1): 21–32.

Thomas-Hope, E. (1998) *Solid Waste Management: Critical Issues for Developing Countries*, Kingston Town: Canoe Press.

Todaro, M. (2000) *Economic Development*, Essex: Pearson Education.

UNDP (United Nations Development Programme) (2011) *Human Development Report Database*, New York: UNDP, available at: http://hdr.undp.org/en/statistics/ (accessed 10 June 2013).

UNESCO (United Nations Educational, Scientific and Cultural Organization) (2001) *National Association of Colombian 'Recicladores Colombian'*, 11 May, available at: www.unesco.org/most/southam4.htm (accessed 1 June 2013).

Wilson, D.C., Rodic, L., Scheinberg, A. and Alabaster, G. (2010) 'Comparative analysis of solid waste management in cities around the world', in *Proceedings Waste 2010: Waste and Resource Management – Putting Strategy into Practice*, Stratford-upon-Avon, Warwickshire, England, 28–29 September.

Wolhuter, C.C. (2011a) 'Comparative education: significance of the field', in G. Likando, C.C. Wolhuter, K. Matengu and J. Mushaandja (eds), *Comparative Education: An Introduction*, Noordbrug: Keurkopie, pp. 35–48.

Wolhuter, C.C. (2011b) 'The spectrum of international educational development', *Journal of Educational Planning and Administration*, 25(3): 235–247.

ZCTU (Zimbabwe Congress of Trade Nations) (2005) *Annual Report*, Harare: ZCTU.

34

Food security in African cities

Linking urban food production with good urban governance

Alec Thornton

> The importance of space – the physical and social environment in which we are born and grow up – is one of the most important determinants of every person's well-being and life chances.
>
> *(South African Government, 2012)*

Introduction

Globally, urban agriculture is increasingly becoming integrated into formal urban planning policy in cities. This is occurring despite increasing urban densities and competing social and economic interests over scarce land and water resources (Kühn, 2003; Hodgson *et al.*, 2011). In surveying the urban food security literature on Africa, the call for tightening the linkages between urban agriculture, social equity, food sovereignty and urban sustainability appears to be growing louder. The quotation above, from South Africa's National Development Plan 2030, suggests a concern regarding the impact of an urbanising population on spatial decision-making, with officials emphasising inclusive, resilient and liveable cities (South African Government, 2012). The development of good urban governance in African cities will play a key role in the actual realisation of inclusive, resilient and liveable cityscapes. Although urban governance is a broad concept with multiple models, this chapter is interested in the roles of institutions and individuals in creating an enabling environment to effectively respond to the needs of all urban residents (Lindell, 2008; UN-Habitat, 2015). Adding to this, an enabling environment can include removing barriers to citizen mobility, as they seek often creative and innovative solutions in meeting a variety of needs, which may lie beyond the service provision capacity of many African cities (Healey, 2004; Resnick, 2014). Many of the urban Africanists discussed in this chapter are optimistic that with supportive urban governance, perhaps urban agriculture can be one driver for enabling 'inclusive, resilient and liveable' spaces in African cities.

Throughout the continent, forms of urban agriculture (UA) can be found in African cities, often appropriating land – public or private – where it is either not wanted, or is tolerated to some degree by city councils in the absence of formal planning or development proposals (McClintock, 2010; Thornton *et al.*, 2010). Accessing land for urban agriculture activities is a real challenge, especially as urban populations expand. Some cities in developing countries, including within Africa, are responding through adopting favourable land tenure and titling policies to allow for various forms of urban agriculture (Orsini *et al.*, 2013). In other cities, municipalities either actively discourage it or acknowledge that the practice is providing some measure of food security, particularly for female-headed households. The fact that people attempt to use vacant land for food production reflects Lefebvre (1991), in his theorising on citizen-driven processes, where people mobilise to adapt their urban environment for the purposes of meeting a variety of needs:

> An existing space may outlive its original purpose and the *raison d'être* which determines its forms, functions, and structures; it may thus in a sense become vacant, and susceptible of being diverted, reappropriated and put to a use quite different from its initial one.
>
> *(Lefebvre, 1991: 94)*

Africa has the highest rate of urbanisation of any continent, and current projections indicate that African cities will host nearly 25 per cent of the global urban population by 2050 (Cobbinah *et al.*, 2015). Rapid urban population growth invites a host of challenges in a developing region that struggled to meet many of the Millennium Development Goals (MDGs), especially the eradication of poverty and hunger (UNECA, 2014). To further emphasise the urban challenge facing this region, the United Nations Human Settlements Programme (UN-Habitat, 2014) projects that nearly 1.3 billion people – more than the current population of China – will be living in cities in Africa in the next 15 years. The post-2015 Sustainable Development Agenda (UNDP, 2015a) emphasises ending hunger, achieving food security for all and promoting sustainable agriculture (SDG2), while creating inclusive, safe, resilient and sustainable cities (SDG11). Although urban food security is, for unknown reasons, not specified in SDG11 (the 'urban SDG'), there is an element of 'co-dependency' in the SDGs, where 'achieving the targets under SDG11 sets the stage for achieving targets in many of the other SDG goals' (UNDP, 2015b). The widely cited definition of food security from the Food and Agriculture Organisation (FAO, 2015) is purposefully inclusive, stating that food security exists when all people, at all times, have physical, social and economic access to sufficient, safe and nutritious food that meets their dietary needs and food preferences for an active and healthy life.

Projected increases in the global urban population, market volatility and a changing climate are challenging the capacities of cities to ensure equitable access and the availability and distribution of fresh, nutritious food (See *et al.*, 2015). Food riots across the African continent during the 2007–2008 global food crisis intensified concerns regarding vulnerabilities in the global food system and sharp increases in the cost of food. Food sovereignty – the right of people to control their own food system – has been taken up by urban food activists in developing countries as a concept that offers more potential than 'food security' to ensure access to nutritious and affordable local food (Wittman *et al.*, 2010). This is particularly the case for cities in sub-Saharan Africa, where urban poor households spend from 50 per cent to over 60 per cent of their income on food (Frayne *et al.*, 2010). This is projected to worsen by 2060, as the ratio of domestic to imported prices for agricultural products will fall in North America and the EU, but rise substantially in developing countries. The steepest price rises will occur in southern Africa, followed by North Africa and the rest of sub-Saharan Africa (SSA) – in that order (AfDB, 2011: 37). Despite this, urban dwellers are proving to be resilient. Although it may seem something of a contradiction, SSA has the world's largest slum population (210 million) while also having the 'greenest' cities, with 40 per cent of urban households comprising urban farmers (FAO, 2012).

In many African cities, urban and peri-urban agriculture (UA or UPA) has a long history of providing fresh food, generating employment, recycling urban wastes, creating greenbelts and strengthening cities' resilience to climate change (FAO, 2012). This chapter will explore the diversity of perceptions and practices of UA as it continues to capture recognition as an urban food security strategy for Africa's expanding cities. This chapter will briefly review UA as an evolving concept and production system, its role in the 'Global South', and the implications of the post-2015 SDGs on UA and the food security agenda.

Urban agriculture

Since the 1980s, UA has undoubtedly generated significant multidisciplinary appeal, contributing to its conceptual development and a critical mass of applied studies in cities in both developed and developing country contexts, concerning production systems, markets, urban farmers, land tenure, climate resiliency, water management and social equity (Sanyal, 1987; Rogerson, 1993; Thornton, 2009; Nel *et al.*, 2009; Barthel and Isendahl, 2013; Cohen and Reynolds, 2015). Various definitions of UA invariably include the cultivation, processing, marketing and distribution of food, forestry, horticultural and aquaculture products that occur in built-up 'intra-urban' areas and the peri-urban 'fringe' (Mougeot, 2000; Thornton, 2008). UA is also a holistic system that goes beyond food production, as it has the potential to contribute to a 'circular economy' through closing food–energy–waste loops in urban and peri-urban areas (Drechsel and Kunze, 2001; Zhang, 2010; Wortman, 2015; Sorkin, 2015). Interest in UA has increased as a result of price fluctuations for basic food staples, inequalities in agro-food networks (Thornton, 2012) and persistent poverty, particularly in sub-Saharan Africa (Kneafsey, 2010; Thornton, 2012; Lynch *et al.*, 2013; Larder *et al.*, 2014). Increasingly, urban and peri-urban horticulture (UPH) is discussed together with UPA (FAO, 2012). In doing so, it joins up the science of plant cultivation (horticulture) with the web of processes – from the fields to the markets – in food and livestock farming (agriculture), in an effort to capture the unique spatial, social and environmental attributes of urban food (and non-food) production in small spaces (e.g. backyards, schools and rooftops) and in larger, open spaces (e.g. parklands and peri-urban corridors).

There is a substantial literature arguing in favour of the need for households to become more proactive in meeting their food needs, as well as to contribute to the overall urban food supply and production chain (Sanyal, 1987; Tinker, 1994; Lynch, 1995; Binns and Fereday, 1996; Foeken and Mwangi, 2000; Rogerson, 2003; Thornton 2008). Some have found that farming near or within the confines of urban areas has historically been a point of contention for urban planners, predating the Industrial Revolution (Mougeot, 2000). Moreover, Lee-Smith and Ali Memon (1994: 3) noted that negative cultural connotations assigned to the notions of the 'city' and 'countryside' date back to the Greco-Roman period. The more recent history of urbanism, associated with the Industrial Revolution, has, for many Western countries, resulted in the separation of 'urban' from 'agriculture' – except for recreational gardening or in times of crisis (Lee-Smith and Ali Memon, 1994). During both World Wars in the twentieth century, urban agriculture was an important part of food production throughout Europe, where backyard 'victory' gardens often meant survival (Lawson, 2005).

In the post-war period, and with the rising dominance of neo-liberal development that emerged in the 1980s, the status of the agricultural sector shifted from concerns over food security to an industrial-style sector based on technological development and commercialisation (Janssen and Braunschweig, 2003; Thornton, 2012). From the 1990s to the present, persistent global hunger, poverty and inequality, and concerns over ecosystem collapse and resource scarcity amid global climate change have led to the promulgation of ideas and approaches concerning sustainable development. In the 2000s, these concerns culminated in a series of global events and processes, culminating in the UN Millennium Declaration and global partnership leading to the creation and adoption of the MDGs, for the period 2002–2015. Building on the momentum of the MDGs, the post-2015 Sustainable Development Goals (SDGs) were promulgated in 2015.

Although many developing regions achieved considerable gains in meeting the MDGs, this was not the case in many parts of Africa, particularly sub-Saharan Africa (Thornton and Rogerson, 2013). Arguably, due to the persistence of poverty and hunger and an increasingly urbanising population, urban Africa may well determine the success, or failure, of the SDGs (Africa Progress Panel, 2015). The relationship between urbanisation, poverty and food insecurity in Africa's cities is persistent (Maxwell, 1999; Maitra and Rao, 2015). Additionally, some African cities have officially prohibited the practice of urban agriculture as a household coping strategy for poverty and food insecurity (Page, 2002; Thornton, 2012). Discussed below, some African cities in the so-called 'Global South' are beginning to recognise the benefits of urban agriculture to urban governance, as part of a local urban food policy planning for food security.

Urban agriculture in the Global South

As a point of contrast to cities in the 'Global North', cities in developing countries appear steadfastly in support of post-Second World War ideas of modernisation through urban industrialisation for economic growth (Thornton *et al.*, 2010). Since the 1990s, and particularly since the World Conference on Human Rights (1993) and World Summit for Social Development (1995), the development paradigm has shifted towards more participatory and inclusive approaches that emphasise social development as central to development and an equal partner to economic concerns (Desai, 2014). This view has contributed to dramatic transitions in many cities in the developed 'North', which emphasise varying levels of commitment to urban social equity in planning decisions, especially for spatially marginalised communities. In these communities, social equity often includes achieving food security and 'food justice', which is expressed through various forms of urban agriculture (Thornton, 2017). By contrast, in cities in the 'South', the scale of inequality, due in large part to differentiated impacts of successive development strategies, do not easily lend themselves to a convergence in ideas and action for re-conceptualising the purpose, potential and promise of inclusive, citizen-led urban spatial (re) production. Although calls for more inclusive urban governance are increasing in parts of Latin America and South Africa (Huchzermeyer, 2012; South African Government, 2012; Brown, 2013; Thornton and Rogerson, 2013; Huchzermeyer, 2014), such calls struggle for recognition elsewhere. As the following example of Lusaka, Zambia, reveals, UA is officially viewed as a practice that is out of place in 'the modern city', despite the fact that it is a fixture in that city, albeit on an informal basis (Rakodi, 1988; Hampwaye *et al.*, 2007; Simatele *et al.*, 2012).

Research has revealed that Lusaka's city planners believe that UA 'is not modern', and urban spaces need to be reserved for investment opportunities (Thornton *et al.*, 2010). This is despite the fact that institutional support for UA in Zambia is not completely unprecedented. From the 1960s to the 1970s, for example, there were national plans to encourage and support urban gardening, including upgrading squatter settlements by increasing plot sizes to allow for food production (Rakodi, 1988; Drescher, 1999). In Zambia's 'Third National Development Plan' (1978–1983), former president Dr Kenneth Kaunda specified the need for making 'self-sustaining cities' by increasing urban food production (Rakodi, 1988; Thornton *et al.*, 2010). Kaunda's humanistic approach of combining self-sufficiency in agriculture with import substitution did not, however, accord with the aspirations of the more neo-liberal-bent members of his government, who eventually accepted the World Bank's deterministic 'modernisation by industrialisation' prescriptions (Quick, 1977). Ultimately, Kaunda's vision for 'self-sustaining cities' in Zambia never materialised, and UA remained a marginalised and informal activity, reflecting the situation in other African cities.

In contrast, the Johannesburg Development Agency (2015) boasts of allocating urban space, previously earmarked for development, for 42 urban gardens, as well as inner-city plots and rooftop gardens in city neighbourhoods, including Hillbrow, Joubert Park, Troyeville, Newtown and Fordsburg. There is also institutional-level interest to include informal peri-urban farmers in the city's formal fresh produce market (Mun Bbun and Thornton, 2013). Specifically, researchers have found that in greater 'Joburg', existing peri-urban land under cultivation by small-scale farmers, for example in Roodepoort and Orange Farm, were being promoted by city officials as ideal sites for local urban food production, to be integrated into the city's fresh produce market initiative, which is part of the South African Government's wider National Fresh Produce scheme and expanding urban agriculture programme (Mun Bbun and Thornton, 2013). South Africa's National Development Plan 2030 views peri-urban spaces as 'vibrant' locations for agricultural production of commodities such as flowers, vegetables, dairy and hydroponics (South African Government, 2012: 9).

The food security agenda

Urban food insecurity is increasingly recognised as a key development challenge (Crush and Frayne 2011a, 2011b; Tawodzera, 2011; Battersby, 2012). Urban food riots in several African countries in 2007–2008 and again in 2010–2011 led to greater awareness of causal factors driving food protests (Sneyd *et al.*, 2013). During this period, Africa experienced more urban food riots than any other global region (Sneyd *et al.*, 2013; Bohstedt, 2014). Studies on urban food riots in this region found that citizens mobilised not so much due to hunger (Sneyd *et al.*, 2013), but to demonstrate a profound dissatisfaction and lack of 'fairness in rights to food, its distribution, and accountability for food supply failures' (Hossain, 2009: 2). Urban agriculture is seen as a stimulus for political and social stability, as it is an activity mobilising the urban poor in a 'moral economy' to ensure food fairness (Prain and Lee-Smith, 2010; Pearson, 2010; FAO, 2014). Urban agriculture has demonstrated that it can play a critical role in varying instances of 'food supply failures' in developing and developed countries, as seen in Cuba, Sierra Leone and Detroit (FAO, 2011).

Although food security in rural areas is experienced at a much higher rate than urban areas (Walsh and van Rooyen, 2015), Crush and Frayne (2011a) stress that the complexities of urban food systems urgently need to be addressed to achieve food security. They critique the 'new international food security agenda' for its 'striking omission' of any consideration of the dimensions and determinants of urban food security (Crush and Frayne, 2011a: 1). They stress the unique spatial and structural qualities of urban poverty and hunger, as pronounced intra-urban disparities in income and living conditions exist within cities. Adding an additional layer of complexity is the space of transition known as the rural–urban interface. Research on the 'rural–urban interface' has resulted in numerous case studies of linked livelihood strategies in 'peri-urban zones' (Lynch, 1995; Simon *et al.*, 2006; Thornton, 2012; Gbanie *et al.*, 2015). As cities expand outwards, they encroach on rural areas, creating peri-urban zones where a variety of exchanges take place, including food, income, ideas and people flows (Simon *et al.*, 2006). This use of peri-urban spaces for intensive large-scale cultivation is gaining traction in some African city councils, contributing further to preserving corridors, or green belts, and controlling urban sprawl on otherwise arable and productive land (Zasada, 2011; Mun Bbun and Thornton, 2013; Gyasi *et al.*, 2014; Gbanie *et al.*, 2015).

African 'food deserts' and good urban governance

A case study of 'food deserts' in 11 southern African towns and cities found that supermarket expansion in low-income urban areas has not improved household access to, and the availability of, fresh food, which is either not stocked or is too costly for poor households (Frayne *et al.*, 2014). Similarly, in a case study of Cape Town, South Africa, Peyton *et al.* (2015) concluded that supermarket expansion has coincided with rapid urbanisation and food insecurity, which has negatively impacted upon informal food economies in low-income areas. Moreover, the African context of food deserts differs from other world regions, as formal and informal markets and rural–urban exchanges were found to form a 'complex nexus of interlinkages' that determine the food security of a household (Battersby and Crush, 2014; Frayne *et al.*, 2014). It is in this context that the notion of expanding profit-driven retail outlets into poor neighbourhoods may be counterproductive as an approach to improving food security in spatially marginalised areas of a city. McClintock (2014) argues that a contradiction exists between alternative food networks, such as urban agriculture, and the formal neo-liberal market. It could be that strengthening food security may depend on reconciling these seemingly irreconcilable differences. Discussed below, such a 'marriage' is arguably more likely to succeed where effective 'good' urban governance can be found, which is politically and socially determined to develop an equitable food policy combined with sustainable urban land use planning.

Although urban agriculture is a feature in many African cities, significantly greater levels of action will be required on the part of councils and communities if urban agriculture is to move beyond its current status as a marginal survival strategy (Thornton *et al.*, 2010). In other developing countries (e.g. Cuba and Brazil), policy support seeking to connect cities to their peri-urban and rural hinterlands is creating employment opportunities for urban poor households, while contributing to improved food and nutrition security in urban (and rural) areas (FAO, 2014). In recognising the importance of urban and peri-urban agriculture to food security, the FAO (2014) is promoting 'city region food systems', defined as:

> [t]he complex relation of actors, relations and processes related to food production, processing, marketing, and consumption in a given geographical region that includes one main or smaller urban centres and surrounding peri-urban and rural areas that exchange people, goods and services across the urban rural continuum.

There are varying perspectives on the role that UA could play in an interconnected 'city food system'. Some would argue that this interconnectivity is unlikely to come to fruition in African cities, where urban agriculture is of 'little significance' to poor households in meeting their food needs and as a source of household income and employment (Frayne *et al.*, 2010; Crush *et al.*, 2011a, 2011b). Moreover, Frayne *et al.* (2014) identified accessibility and affordability of nutritious food as key determinants of urban food security in southern Africa.

Other studies in southern Africa and the wider SSA region (Simatele *et al.*, 2012; Thornton, 2012; Lynch *et al.*, 2013; Mun Bbun and Thornton, 2013; Smart *et al.*, 2015) have suggested that local variables can influence UA significance among various income groups, citing institutional acceptance and beneficial relationships – for example, local responses to economic downturns – as essential determinants to local solutions for urban food security. Key in this regard is the importance of urban governance in playing a facilitating role, or in the least not presenting obstacles to grassroots mobilisation of urban agricultural activities.

Conclusion

This chapter has explored the field of research on urban agriculture, its diversity and complexity as one response to food security challenges facing Africa's cities. In reviewing the literature on food security in African cities, both the potential and constraints of urban agriculture appear as dominant and reoccurring themes. In an era of global urbanisation, urban agriculture is either described as an income-generating, or as a subsistence practice among the urban poor in developing countries. In developed countries, however, the focus is often on the ethical and social responsibility of Westerners in food production and consumption. Increasingly, the literature emphasises issues regarding food access and availability, albeit at different scales, for urban dwellers in both developing and developed countries, where urban agriculture is making positive contributions – particularly when supported by good urban governance.

In reviewing the landscape (or cityscape) of urban food security in Africa, this chapter has revealed an increasing call for more 'inclusive, resilient and liveable cities' through considered integration of urban agriculture and its related activities as part of a broader policy approach to sustainable urban development. This means an approach to good urban governance that is economically and socially responsive and in tune with forms of community-driven adaptation of space for the local production, distribution and consumption of nutritious food. This call is not restricted by boundaries or income levels, and is most critical to the health and well-being of populations in one of the more rapidly urbanising regions – Africa.

References

AfDB (African Development Bank) (2011) *Africa in 50 Years' Time: The Road Towards Inclusive Growth*, Tunis: African Development Bank.

Africa Progress Panel (2015) *Global Goals, African Realities: Building a Sustainable Future for All*, Geneva: Africa Progress Panel, available at: www.africaprogresspanel.org/publications/policy-papers/global-goals-african-realities/ (accessed 27 September 2015).

Barthel, S. and Isendahl, C. (2013) 'Urban gardens, agriculture, and water management: sources of resilience for long-term food security in cities', *Ecological Economics*, 86: 224–234.

Battersby, J. (2012) 'Beyond the food desert: finding ways to speak about urban food security in South Africa', *Geografiska Annaler: Series B, Human Geography*, 94(2): 141–159.

Battersby, J. and Crush, J. (2014) 'Africa's urban food deserts', *Urban Forum*, 25(2): 143–151.

Binns, T. and Fereday, N. (1996) 'Feeding Africa's urban poor: urban and peri-urban horticulture in Kano, Nigeria', *Geography*, 81(4): 380–384.

Bohstedt, J. (2014) 'Food riots and the politics of provisions in world history', *IDS Working Papers*, 444: 1–31.

Brown, A. (2013) 'The right to the city: from Paris 1968 to Rio 2010', *International Journal of Urban and Regional Research*, 37(3): 957–971.

Cobbinah, P., Erdiaw-Kwasie, M. and Amoateng, P. (2015) 'Africa's urbanisation: implications for sustainable development', *Cities*, 47: 62–72.

Cohen, N. and Reynolds, K. (2015) 'Resource needs for a socially just and sustainable urban agriculture system: lessons from New York City', *Renewable Agriculture and Food Systems*, 30(1): 103–114.

Crush, J. and Frayne, B. (2011a) 'Urban food insecurity and the new international food security agenda', *Development Southern Africa*, 28(4): 527–544.

Crush, J. and Frayne, B. (2011b) 'Supermarket expansion and the informal food economy in southern African cities: implications for urban food security', *Journal of Southern African Studies*, 37(4): 781–807.

Crush, J., Drimie, S., Frayne, B. and Caesar, M. (2011a) 'The HIV and urban food security nexus in Africa', *Food Security*, 3: 347–362.

Crush, J., Hovorka, A. and Tevera, D. (2011b) 'Food security in southern African cities: the place of urban agriculture', *Progress in Development Studies*, 11(4): 285–305.

Desai, M. (2014) *The Paradigm of International Social Development: Ideologies, Development Systems and Policy Approaches*, New York: Routledge.

Drechsel, P. and Kunze, D. (eds) (2001) *Waste Composting for Urban and Peri-Urban Agriculture: Closing the Rural–Urban Nutrient Cycle in Sub-Saharan Africa*, Wallingford: CABI.

Drescher, A.W. (1999) 'Urban agriculture in the tropics of central southern Africa: a case study of Lusaka/ Zambia', in M. Koc, R. MacRae, L. Mougeot and J. Welsh (eds), *For Hunger Proof Cities: Sustainable Urban Food Systems*, Ottawa: IDRC, pp. 67–76.

FAO (Food and Agriculture Organization) (2011) *Food, Agriculture and Cities*, available at: www.fao.org/ fileadmin/templates/FCIT/. . ./FoodAgriCities_Oct2011.pdf (accessed 20 October 2015).

FAO (Food and Agriculture Organization) (2012) *Growing Greener Cities in Africa: First Status Report on Urban and Peri-Urban Horticulture in Africa*, Rome: FAO.

FAO (Food and Agriculture Organization) (2014) *Food Security for Sustainable Development and Urbanization: Inputs for FAO's Contribution to the 2014 ECOSOC Integration Segment*, 27–29 May, available at: www. un.org/en/ecosoc/integration/2014/index.shtml (accessed 5 October 2015).

FAO (Food and Agriculture Organization) (2015) *The State of Food Insecurity in the World 2015*, available at: www.fao.org/hunger/glossary/en/ (accessed 28 June 2016).

Foeken, D. and Mwangi, A. (2000) 'Increasing food security through urban agriculture in Nairobi', in N. Bakker, M. Dubbeling, S. Guendel, U. Sabel Koschella and H. De Zeeuw (eds), *Growing Cities, Growing Food: Urban Agriculture on the Policy Agenda*, Feldafing: German Foundation for International Development, pp. 303–327.

Frayne, B., Pendleton, W., Crush, J., Acquah, B., Battersby-Lennard, J., Bras, E., *et al.* (2010) *The State of Urban Food Insecurity in Southern Africa*, Kingston and Cape Town: Queen's University and AFSUN.

Frayne, B., Crush, J. and Lachlan, M. (2014) 'Urbanization, nutrition and development in southern African cities', *Food Security*, 6: 101–122.

Gbanie, S., Thornton, A. and Griffin, A. (2015) '"The diamond of Western Area is land": narratives of land use and land cover change in post-conflict Sierra Leone', *Australasian Review of African Studies*, 36(2): 51–73.

Gyasi, E.A., Kranjac-Berisavljevic, G., Fosu, M., Mensah, A.M., Yiran, G. and Fuseini, I. (2014) 'Managing threats and opportunities of urbanisation for urban and peri-urban agriculture in Tamale, Ghana', in M. Basant, R. Purohit, H. Malano, V.P. Singh and P. Amerasinghe (eds), *The Security of Water, Food, Energy and Liveability of Cities*, Dordrecht: Springer, pp. 87–97.

Hampwaye, G., Nel, E. and Rogerson, C. (2007) 'Urban agriculture as local initiative in Lusaka, Zambia', *Environment and Planning C*, 25(4): 553–572.

Healey, P. (2004) 'Creativity and urban governance', *Policy Studies*, 25(2): 87–102.

Hodgson, K., Campbell, M.C. and Bailkey, M. (2011) *Urban Agriculture: Growing Healthy, Sustainable Places*, Washington, DC: APA Planning Advisory Service.

Hossain, N. (2009) *Towards a New Social Justice Agenda: Understanding Political Responses to Crises*, IDS InFocus Policy Briefing, Issue 11, October 2009, Brighton: Institute for Development Studies.

Huchzermeyer, M. (2012) 'Informal settlements: production and intervention in Brazil and South Africa', *Latin American Perspectives*, 29(1) 83–105.

Huchzermeyer, M. (2014) 'Humanism, creativity and rights: invoking Henri Lefebvre's right to the city in the tension presented by informal settlements in South Africa today', *Transformation: Critical Perspectives on Southern Africa*, 85(1): 64–89.

Janssen, W. and Braunschweig, T. (2003) *Trends in the Organization and Financing of Agricultural Research in Developed Countries: Implications for Developing Countries*, Research Report, 22, The Hague: International Service for National Agricultural Research (ISNAR).

Johannesburg Development Agency (2015) *Rooftop Garden Plants Seeds of Success*, available at: www.jda.org. za/index.php/corporate-social-responsibility/1542-rooftop-garden-plants-seeds-of-success-2 (accessed 20 October 2015).

Kneafsey, M. (2010) 'The region in food: important or irrelevant?', *Cambridge Journal of Regions, Economy and Society*, 3: 177–190.

Kühn, M. (2003) 'Greenbelt and green heart: separating and integrating landscapes in European city regions', *Landscape Urban Planning*, 64(1–2): 19–27.

Larder, N., Lyons, K. and Woolcock, G. (2014) 'Enacting food sovereignty: values and meanings in the act of domestic food production in urban Australia', *Local Environment: The International Journal of Justice and Sustainability*, 19(1): 56–76.

Lawson, L. (2005) *City Bountiful: A Century of Community Gardening in America*, Berkeley, CA: University of California Press.

Lee-Smith, D. and Ali Memon, P. (1994) 'Urban agriculture in Kenya', in A.G. Egziabher, D. Lee-Smith, D.G. Maxwell, P. Ali Memon, L.J.A. Mougeot and C.J. Sawio (eds), Cities Feeding People: An Examination of Urban Agriculture in East Africa, Ottawa: International Development Research Centre (IDRC), pp. 63–78.

Lefebvre, H. (1991) *The Production of Space*, Malden, MA: Blackwell.

Lindell, I. (2008) 'The multiple sites of urban governance: insights from an African city', *Urban Studies*, 45(9): 1879–1901.

Lynch, K. (1995) 'Sustainability and urban food supply in Africa', *Sustainable Development*, 3: 79–88.

Lynch, K., Maconachie, R., Binns, T., Tengbe, B. and Bangura, K. (2013) 'Meeting the urban challenge? Urban agriculture and food security in post-conflict Freetown, Sierra Leone', *Applied Geography*, 36(1): 31–39.

Maitra, C. and Rao, D.P. (2015) 'Poverty–food security nexus: evidence from a survey of urban slum dwellers in Kolkata', *World Development*, 72: 308–325.

Maxwell, D. (1999) 'The political economy of urban food security in sub-Saharan Africa', *World Development*, 27(11): 1939–1953.

McClintock, N. (2010) 'Why farm the city? Theorizing urban agriculture through a lens of metabolic rift', *Regions Economy and Society*, 3(2): 191–207.

McClintock, N. (2014) 'Radical, reformist, and garden-variety neoliberal: coming to terms with urban agriculture's contradictions', *Local Environment: The International Journal of Justice and Sustainability*, 19(2): 147–171.

Mougeot, L. (2000) 'Urban agriculture: definition, presence, potentials and risks', in N. Bakker, M. Dubbeling, S. Guendel, U. Sabel Koschella and H. De Zeeuw (eds), *Growing Cities, Growing Food: Urban Agriculture on the Policy Agenda*, Feldafing: German Foundation for International Development, pp. 1–42.

Mun Bbun, T. and Thornton, A. (2013) 'A level playing field? Improving market availability and access for small-scale producers in Johannesburg, South Africa', *Applied Geography*, 36(1): 40–48.

Nel, E., Hampwaye, G., Thornton, A., Rogerson, C.M. and Marais, L. (2009) *Institutional Responses to Decentralization, Urban Poverty, Food Shortages and Urban Agriculture*, GDN Working Paper Series, Working Paper No. 36, pp. 1–29, Global Development Network (GDN).

Orsini, F., Kahane, R., Nono-Womdim, R. and Gianquinto, G. (2013) 'Urban agriculture in the developing world: a review', *Agronomy for Sustainable Development*, 33(4): 695–720.

Page, B. (2002) 'Urban agriculture in Cameroon: an anti-politics machine in the making?', *Geoforum*, 33(1): 41–54.

Pearson, C. (2010) 'Challenging multidimensional agriculture in cities', *International Journal of Agricultural Sustainability*, 8(1–2): 3–4.

Peyton, S., Moseley, W. and Battersby, J. (2015) 'Implications of supermarket expansion on urban food security in Cape Town, South Africa', *African Geographical Review*, 34(1): 36–54.

Prain, G. and Lee-Smith, D. (2010) 'Urban agriculture in Africa: what has been learned?', in G. Prain, N. Karanja and D. Lee-Smith (eds), *African Urban Harvest: Agriculture in the Cities of Cameroon, Kenya and Uganda*, New York: Springer, pp. 13–35.

Quick, S. (1977) 'Bureaucracy and rural socialism in Zambia', *The Journal of Modern African Studies*, 15(3): 379–400.

Rakodi, C. (1988) 'Urban agriculture: research questions and Zambian evidence', *The Journal of Modern African Studies*, 26(3): 495–515.

Resnick, D. (2014) 'Urban governance and service delivery in African cities: the role of politics and policies', *Development Policy Review*, 32: 3–17.

Rogerson, C.M. (1993) 'Urban agriculture in South Africa: policy issues from the international experience', *Development Southern Africa*, 10(1): 33–44.

Rogerson, C.M. (2003) 'Towards "pro-poor" urban development in South Africa: the case for urban agriculture', *Acta Academica*, 1: 130–158.

Sanyal, B. (1987) 'Urban cultivation amidst modernisation: how should we interpret it?', *Journal of Planning Education and Research*, 6(3): 187–207.

See, L., Fritz, S., You, L., Ramankutty, N., Herrero, M., Justice, C., *et al.* (2015) 'Improved global cropland data as an essential ingredient for food security', *Global Food Security*, 4: 37–45.

Simatele, D., Binns, T. and Simatele, M. (2012) 'Urban livelihoods under a changing climate: perspectives on urban agriculture and planning in Lusaka, Zambia', *Journal of Human Development and Capabilities*, 13(2): 23–46.

Simon, D., McGregor, D. and Thompson, D. (2006) 'Contemporary perspectives on the peri-urban zones of cities in developing areas', in D. Simon, D. McGregor and D. Thompson (eds), *The Peri-Urban Interface: Approaches to Sustainable Natural and Human Resource Use*, London: Earthscan, pp. 1–17.

Smart, J., Nel, E. and Binns, T. (2015) 'Economic crisis and food security in Africa: exploring the significance of urban agriculture in Zambia's Copperbelt province', *Geoforum*, 65: 37–45.

Sneyd, L., Legwegoh, A. and Fraser, E. (2013) 'Food riots: media perspectives on the causes of food protest in Africa', *Food Security*, 5(4): 485–497.

Sorkin, M. (2015) 'City limits: how localizing is the key to our urban future', *New Perspectives Quarterly*, 32(3): 23–27.

South African Government (2012) *National Development Plan: 2030*, available at: www.gov.za/sites/www.gov.za/files/devplan_ch8_0.pdf (accessed 28 September 2015).

Tawodzera, G. (2011) 'Vulnerability in crisis: urban household food insecurity in Epworth, Harare, Zimbabwe', *Food Security*, 3(4): 503–520.

Thornton, A. (2008) 'Beyond the metropolis: small town case studies of urban and peri-urban agriculture in South Africa', *Urban Forum*, 19(3): 243–262.

Thornton, A. (2009) 'Pastures of plenty? Land rights and community-based agriculture in Peddie, a former homeland town in South Africa', *Applied Geography*, 29: 12–20.

Thornton, A. (2012) *Beyond the Metropolis: Urban Agriculture in South Africa*, Lewiston, NY: Edwin Mellen Press.

Thornton, A. (2017) '"The lucky country"? A critical exploration of community gardens and city–community relations in Australian cities', *Local Environment*, 22: 969–985, DOI: http://dx.doi.org/10.1080/13549839.2017.1317726.

Thornton, A. and Rogerson, C. (2013) 'African cities and the Millennium Development Goals: a case for applied geography', *Applied Geography*, 36(1): 1–3.

Thornton, A., Nel, E. and Hampwaye, G. (2010) 'Cultivating Kaunda's plan for self-sufficiency: is urban agriculture finally beginning to receive support in Zambia?', *Development Southern Africa*, 27(4): 613–625.

Thornton, A., Lyons, K. and Sharpe, S. (2017) 'Community gardens for social and food justice: the case of urban agriculture in Australian cities', in L. Shevellar and P. Westoby (eds), *The Routledge Handbook of Community Development Research*, London: Routledge, Chapter 15.

Tinker, I. (1994) 'Foreword: urban agriculture is already feeding cities', in A.G. Egziabher, D. Lee-Smith, D.G. Maxwell, P. Ali Memon, L.J.A. Mougeot and C.J. Sawio (eds), *Cities Feeding People: An Examination of Urban Agriculture in East Africa*, Ottawa: IDRC, p. vii.

UNDP (United Nations Development Programme) (2015a) *The Post-2015 Development Agenda*, available at: www.undp.org/content/undp/en/home/mdgoverview/post-2015-development-agenda.html (accessed 5 September 2015).

UNDP (United Nations Development Programme) (2015b) *SDG Goal 11: Cities Will Play an Important Role in Achieving the SDGs*, available at: www.my.undp.org/content/malaysia/en/home/presscenter/articles/2015/08/04/sdg-goal-11-cities-will-play-an-important-role-in-achieving-the-sdgs.html (accessed 25 September 2015).

UNECA (United Nations Economic Commission for Africa) (2014) *MDG Report 2014: Assessing Progress in Africa Toward the Millennium Development Goals*, Addis Ababa: UNECA/AU/AfDB/UNDP.

UN-Habitat (United Nations Human Settlements Programme) (2014) *The State of African Cities 2014*, Nairobi: UN-Habitat.

UN-Habitat (United Nations Human Settlements Programme) (2015) *Governance*, available at: http://unhabitat.org/urban-themes/governance/ (accessed 20 October 2015).

Walsh, C. and van Rooyen, F. (2015) 'Household food security and hunger in rural and urban communities in the Free State province, South Africa', *Ecology of Food and Nutrition*, 54(2): 118–137.

Wittman, H., Desmarais, A.A. and Wiebe, N. (2010) 'The origins and potential of food sovereignty', in H. Wittman, A.A. Desmarais and N. Wiebe (eds), *Food Sovereignty: Reconnecting Food, Nature and Community*, Oakland, CA: Food First, pp. 1–14.

Wortman, S. (2015) 'Crop physiological response to nutrient solution electrical conductivity and pH in an ebb-and-flow hydroponic system', *Scientia Horticulturae*, 194: 34–42.

Zasada, I. (2011) 'Multifunctional peri-urban agriculture: a review of societal demands and the provision of goods and services by farming', *Land Use Policy*, 28(4): 639–648.

Zhang, T. (2010) 'Urban agricultural development from the perspective of circular economy: an analysis of Sino-Singapore eco-city model in Tianjin', *Asian Agricultural Research*, 2(11): 9–16.

35

The future and Africa's cities

Garth Myers

Introduction

Cities in Africa are facing uncertain and varied prospects. On the one hand, across the continent, cities are worn down by sputtering economic stagnation, failures in basic service provision, political conflict, and daunting environmental crises. On the other hand, there are cities that are brimming with optimism amid growing economies, new inflows of foreign direct investment, consolidating democratic societies, and vibrant cultural productivity. In most cities, one can find manifestations of elements of both of these scenarios to varying degrees, and urbanism straddling optimism and pessimism, dealing with demographic and political change, increasing inequalities and informal settlements, but also increasing wealth and suburbanisation – one sees these diverse elements clearly in cities such as Nairobi, Kenya, Dakar, Senegal, or Lusaka, Zambia, for instance (Myers, 2015). The challenges that Africa's urban areas face now and in the future are myriad and varied, as are the tactics and strategies by which Africans are managing uncertainties and complicated states of precarity.

It is difficult, and in fact really impossible, to generalise about urban futures for a continent as vast, diverse and complex as Africa. According to UN-Habitat's (2014) most recent estimates, there are now 70 cities with more than 750,000 people on the African continent. But these cities manifest diversity in size, shape, biophysical environment, boundaries, climatic features, planning systems, politics and sociocultural dynamics. For now, just consider diversity in population size. Over 50 of these cities contain populations above 1 million, with almost 20 more just below this. Yet these range from the megacities of Lagos, Nigeria, Cairo, Egypt, and Kinshasa, Democratic Republic of Congo (with populations that the UN has conservatively estimated will exceed 15 million, 13 million and 12 million, respectively, by 2020) to cities whose populations will have exceeded 750,000, but not yet reached a million, by 2020. Indeed, most of Africa's major cities will still have less than 2 million people by 2020, according to the UN's data, including nearly half (34) of the 70 that they included in their 2014 statistical annex. It is just as important to an overall understanding of urban environments to appreciate these smaller or medium-sized cities as well as the largest ones. Indeed, many cities with less than 750,000 people matter greatly to a robust understanding of urban futures on the continent, given how rapidly many of these small cities are growing, or how significant they are becoming to the political economies of Africa.

One major challenge of analysing urban futures for the continent lies in the mere facts of these diverse trajectories of growth, and demographic and economic trends. Some patterns and processes of urbanisation in the region are familiar, in that they follow conventional expectations of urban theory, where industrialisation and agglomeration have multiplier effects that draw in rural population, and growth adds to growth. This seems to be the case, for instance, in Accra, Ghana (Quayson, 2014). But there are numerous ways in which urbanisation and urbanism in the region appear to 'upend' the rules, as other cities grow rapidly in population and areal extent without substantial economic growth, formal industry, or classically urban social forms – Douala in Cameroon seems to be a prime example of this trajectory (Simone, 2007). Rather than concluding that African cities need to continue to be thought of through Global North theoretical lenses as cities that do not work or are not cities in the right way, some scholars of urban studies across the region have begun to argue, instead, that new ideas of the urban may be emerging in Africa, which ought to be of relevance to urban studies worldwide (Nuttall and Mbembe, 2008; Myers, 2011; Pieterse and Simone, 2013; Parnell and Pieterse, 2014; Myers, 2016).

Some of these new ideas of the urban are fantastical dreams of what cities can be, while others can be dystopic nightmares. The majority of African ideas of the urban comprise, more commonly, a mixed middle way of getting by in 'muddling through' cities with modest, ordinary futures. In three segments below, I explore the fantasies, nightmares and realities of African urban future-scapes. I highlight examples of the variety within each of the three categories.

Dreams and fantasies

After several decades of retreat, the last decade has brought grand scheme, blueprint master planning bounding forward again on the continent. Between 2007 and 2016, new master plans for Dakar, Nairobi, Lusaka, Addis Ababa and many other cities were published and, at least to some degree, implemented (Watson and Agbola, 2013; Lamson-Hall *et al.*, 2015; Myers, 2015; Bafana, 2016). The return of these grand schemes represents, on one level, the return of the modernist dreams for the continent's cities. Beyond these broad plans, though, the last 10 years have brought numerous efforts to design entirely new cities and towns on the continent. Taken together, the new grand master plans and the new planned cities manifest elite-level urban fantasies (Watson and Agbola, 2013). As Watson and Agbola (2013: 4) put it, 'The fantasy designs for African cities win awards . . . . But the implementation of plans that are unsustainable in the extreme and inappropriate in terms of climate, available infrastructure . . . and affordability, exposes their shortcomings'. They cite the example of the master-planned new city of Kilamba, 20 km south of Luanda, Angola, as a fantastical failure. As Watson (2014: 1) further notes, 'the proposed new urban master plans are now to be found on the websites of international architectural, engineering and property development firms'. But the plans 'appear to disregard the fact that at the moment, the bulk of the population in sub-Saharan Africa cities is extremely poor and living in informal settlements', leaving Watson (2014: 2) to ask whether this new round of urban 'fantasies' manifests the 'dreams' of the continent's urban dwellers, or a return to the worst 'nightmares' of past plans. A cursory examination of some of these plans in their specific contexts and urban realities leaves the answer to Watson's question uncertain.

Dakar, Senegal's capital, as one example, has embraced the renewed fascination for grand schemes – from a 2025 Master Plan for the whole Dakar region to new toll highways, from glitzy new malls to the closest thing to Africa's version of the Statue of Liberty – the African Renaissance Monument (Gridoux, 2012). In the new mall on the oceanside drive in one of Dakar's richest areas, one can shop for elite clothing or appliances, see the latest Hollywood blockbuster at the Cineplex, and stop by the Red Bowl to knock down some bowling pins.

At the same time, in the largely informal city of Pikine on Dakar's outskirts, a city with over a million residents, we find what the deputy mayor called a 'spirit of revolt' that responds to more than 50 per cent unemployment, semi-annual severe flooding, chronic overcrowding, a dire absence of basic services, and serious crime (Diouck, 2013).

After independence for most countries in the early 1960s, there were a handful of notable cases where the new regimes ambitiously went about planning and building entirely new capital cities – Abuja, Yamoussoukro, Lilongwe, Gaborone, Dodoma and Nouakchott, for instance. Most were deemed only marginal success stories that were half-implemented at best (Myers, 2011). The era of structural adjustment largely led to an abandonment of grand urban ideas of that scale. With the rise of neo-liberal models of the local state and urban governance, such fantasies retreated still further. But in recent years, private capital and external donors have spurred on new plans for fanciful new cities. The most well known of these are taking shape outside of Lagos, Nairobi, Luanda, Johannesburg and Kinshasa.

Along the Congo River in Kinshasa, the capital of the DRC, a project to create two artificial islands in the river and to develop them as a new *Cité du Fleuve* was nearing the completion of its first phases by 2015; eventually, the new suburb aims to have 10,000 luxury apartments (for sale at a beginning price of US$175,000 with a minimum deposit of US$50,000) and a few hundred mansions (CORES, 2015). This wealthy neighbourhood sits on the river next to a sprawling mega-cityscape constructed largely by its desperately poor (with a per capita income of US$280 per year) majority; hundreds of fisherfolk and wetlands farmers were forcibly removed to make the new 'city of the river' possible (Kushner, 2013). They join the massive numbers of residents already in Kinshasa's informal settlements (de Boeck, 2012).

As with the new master plans, the new satellite cities thus have a very hazy future. One project in Nairobi's suburbs has been abandoned, another – the Kilamba example mentioned previously, from outside Luanda – received heaps of scorn as a Chinese-built 'ghost city' in Africa, and even those closer to implementation clearly do little to solve the major issues for the continent's largest cities (Watson and Agbola, 2013; Murray, 2016). Instead, these model towns represent an elite or middle-class dream of escape that poses the potential for leaving the millions of urban Africans unable to afford luxury suburban lifestyles, stuck in ever-direr megacity conditions, abandoned by the power structures of their countries.

This hardly stops the urban majorities or the artists who emerge out of these majorities from dreaming grand dreams. Kinshasa's dreamworld, for example, also appears in other forms, such as in the artistic dioramas of Bodys Isek Kingelez. Kingelez, who passed away in 2015, became famous globally for his spectacular models of the future of Kinshasa – composed from trash and recycled materials mounted on cardboard. His 2010 solo exhibition at Paris' Pompidou Center, entitled Dreamlands, featured his masterpieces *Phantom City, Kin* [as in Kinshasa] *of the Third Millennium* and *The City of the Future*. These brilliant, bright, wild dioramas manifested what Kingelez (2008) himself called his 'deep hope for a happy tomorrow'. This deep hope in the artistic dreamland of the future city is far more than the vision of one Congolese artist. Many urban Africans, even in the poorest and most marginalised informal settlements, envision cities much differently from what the visual and physical realities may be. In part because of the importance of religion and spirituality in everyday life, most cities are urban environments profoundly alive to the people who live in them, on multiple dimensions, including the dreamlands. By this, I mean that in many urban cultures in Africa, ordinary people practise religion and live out their spirituality in a manner that makes the distinctions less clear between the physical world and the spiritual or imaginary world.

Let me illustrate this latter point through the case of Zanzibar. I have been researching Zanzibar City in various projects for nearly 30 years, and it is quite evident that the dreamlands (such as the Swahili spirit underworld of Giningi) and fantastical visions of this city are of utmost significance to its residents and their futures. One of the ways that Zanzibaris have talked with me about their city is in relationship to the great Swahili novel of the Tanzanian mainland writer Shaaban Robert (1991 [1951]), *Kusadikika* [*Land of Make-Believe* or *Imaginary Land*]. *Kusadikika* has some narrative parallels with Swift's *Gulliver's Travels*, as Robert tells the satirical tale of Kusadikika, a country that floats in the heavens, and its misrule. Robert first published the novel in 1951, and it is typically taken as a parable about colonial oppression. Many Zanzibaris read it in contemporary times as a fable about the misdeeds of the pirates and tyrants who have ruled the islands since the 1964 socialist revolution. The sense of Zanzibar as an island state unmoored, floating in space, an imaginary, nightmarish land, also hits home for Zanzibaris for reasons beyond the 52 years of revolutionary misdeeds. Zanzibari society, like many societies in urban Africa, is richly and deeply spiritual – and not just in the religious sense, though indeed the overwhelming majority of Zanzibaris are extremely religious. This religiosity intersects every day, as it does in many parts of the continent, with the supernatural, imaginary world. Indeed, in religion and in their daily lives, 'people dwell in proximity with spirits' (Larsen, 2014: 6).

These sorts of dreamscapes can verge into nightmare scenarios. In Zanzibar's urban and suburban residential neighborhoods, perhaps the prevailing social ethos of the dreamworld is best captured in the Swahili word *fitina*, or discord. This *fitina* is most obvious in Zanzibar's electoral politics. In five elections since 1995, the ruling Revolutionary Party has claimed dubious victories over the opposition Civic United Front (CUF), despite CUF's obvious popularity and evidence of them actually winning – which in 2000 and 2015 led to highly problematic rerun elections. Zombies, devils and vampire bats have made appearances in the popular imagination alongside the real ghouls that have stolen these elections. It is hard to envision a bright urban future in conditions of corruption and repression such as this. When asked about how the suburbs of Zanzibar's West District might be reimagined and improved, a local elder said:

> We have a saying in Swahili, 'work the clay while it is wet'. If you want to make a sculpture, if the clay is wet you can make a crocodile, and then change it into a lion, or a cow. If it is already dry, forget it. Do you see what I am saying? The clay is dry here.
>
> *(Anonymous, 2007)*

The interviewee was making a case, allegorically speaking, for how impracticable physical changes to this peri-urban landscape would be, given how intractable and hardened (like that clay) the severe political and ideological differences of Zanzibari society have become.

For many West District residents, the allegory of dry clay stands for the whole landscape of *fitina* that has become the predominant structure of feeling in Zanzibar City and its suburbs. In the realms of *fitina* that predominate in many cities of Africa, the dreamworlds thus edge into dystopia, and the future darkens in the light of the present. At the same time, it is more common to see nightmare scenarios in the scary visions of outsiders, to which I now turn.

Dystopias and nightmares

For many outside observers, including scholars and planners, African cities are a mess, and likely to worsen in the future. Social, political, economic and environmental calamities abound. Environmental dystopias are illustrative, since the tropes of belching, fuming, burning, flooding and foul-smelling cities pervade common imagery of the continent's urban life. The soils are septic from so much overflowing human waste. The surface waters are putrid, left standing because what few drains there are get clogged with solid waste that doesn't get collected. What little water infrastructure that functions brings polluted water to the small percentage of residents with access. Road infrastructures are so poor and traffic so bad that air pollution chokes the many urban pedestrians. Indoor air pollution from charcoal cooking in poorly ventilated small domestic spaces leads to as much toxicity inside as out. Governance structures leave cities highly vulnerable to the negative effects of climate change. Earth, air, water, fire, solid, liquid, gaseous – no matter the element or state, it is in bad shape, according to the nightmare discourse (Myers, 2016).

In urban studies, it remains the case that theorists and scholars think and write across the whole of the continent, and often in a way that is 'obsessed with the less palatable particularities of African [urban] politics and society' (Chabal, 2009: 18). Although there is some evidence here and there of changes, it is still generally found that cities in Africa are ignored, banished to a different, other, lesser category of not-quite cities, or held up as examples of all that can go wrong with urbanism in much of both the mainstream and even critical urban literature.

More often, the broad brush generally has the wider reach, and so generalised dystopias still predominate. For one prominent example, Mike Davis, in *Planet of Slums*, devotes a considerable amount of attention to the continent's cities. Unfortunately, from the book's first sentence, which imagines a woman giving birth 'in the Lagos slum of Ajegunle' as part of a Dickensian vision of megacities in a Third World hell, it is an extreme portrait (Davis, 2005: 1). We read of 'Kinshasa, Luanda, Khartoum, Dar es Salaam' and other African cities as growing 'prodigiously despite ruined import-substitution industries, shrunken public sectors, and downwardly mobile middle classes', and in his crisis-driven narrative he notes that 'the African situation, of course, is more extreme' (Davis, 2005: 16, 18). He uses UN-Habitat data for a chart that shows more than 75 per cent of the urban population of Nigeria (79.2 per cent), Tanzania (92.1 per cent), Ethiopia (99.4 per cent) and Sudan (85.7 per cent) living in 'slums' where nearly all other developing countries listed have percentages below the 50 per cent mark, without any critical discussion of where these data come from or what really constitutes a slum. It is hard to fathom how one would conclude that more than nine out of 10 urban Tanzanians live in 'slums', if slums are equated with the squalor Davis portrays. Davis (2005: 19) is so fixated on exploding slums, with no hope for poverty alleviation, and urbanisms that seem to him comprised mostly of 'pollution, excrement, and decay', that one often loses the valuable insights and broad reading behind the book, such as his dissection of the 'brutal tectonics of neoliberal globalization' (Davis, 2005: 174). He is so driven towards the worst of the worst-case scenarios and 'pathologies' that we, the readers, can only abandon hope, and turn tail heading elsewhere (David, 2005: 128).

Davis's portrayal may have been extreme, but one encounters bleak scenarios from many experts on Africa's cities. One finds a litany of hopelessness, even from African voices, whether in music, literature or popular discourse. In Zanzibar and Dar es Salaam, for example, the graffiti slogan 'hapendwi mtu' ['a person is not loved'] is common on walls around the city, along with 'dunia mti mkavu' ['the world is a bare tree']; Dar's street hip-hop artists confront in song the brutal lack of a future for the many heroin addicts in their midst. Graffiti murals in Dakar replicate these dark sentiments in haunting portraits of those who have died young or of youth in gas masks from the air pollution caused by traffic. Bar art in Lilongwe repeats a common motif in central and East Africa, with an image of a man climbing a tree on a riverbank to escape a lion, not seeing the cobra that is in the tree about to bite his hand, or that the tree is falling into the river and the open mouth of a crocodile: 'death is always before you', the caption reads. In Lusaka, Nairobi, Cape Town and elsewhere, the twenty-first century has been, for many of the urban poor, a moment much like that of the man in the Lilongwe mural. The future is hardly brimming with hope.

Challenges and opportunities

Between the dreams and nightmares, between the fantasies and the dystopias, though, we find the reality of future trajectories for urban Africa. Neither unbridled and universal optimism nor abject despair are warranted. In this final segment, I highlight challenges and opportunities that Africa's cities face in the grounded, in-between realities that neither arise from flights of fancy nor sink into quagmires of dystopia.

Rapid demographic growth without comparable rates of economic growth with broadly spread benefits (and hence rising inequalities) beset many cities on the continent (Fox, 2014). Tied to this, cities frequently find themselves incapacitated by the demands for service provision that come with growing urban populations and limited finances. Many cities are continuing to struggle with the demands for spaces of political freedom, democratisation and progressive openness. And ultimately looming over all other challenges, we see the threats that climate change has in store for the future of cities.

The wide variation in patterns or causes of growth means that it is difficult, again, to create one single playbook for the future of Africa's cities. The closest thing to an attempt to do so lies with the UN-Habitat (2014) *State of African Cities* report, which actually does a decent job of balancing the challenges and opportunities facing the continent's urban areas. The report's vision of what the urban environmental problems are on the continent is familiar: water scarcity, flooding (and sometimes, paradoxically, urban Africa experiences both of these first two at once), water pollution, poor waste management, health and sanitation, and recurring environmental disasters. The main culprits are also fairly familiar: environmental change and climate change, weak governance, and poor management. But from these familiar tropes, the report's authors take off in fairly new and intriguing directions. First, the effort to tie urban and rural dynamics together brings the authors to effectively attend to issues of (urban and rural) food security, energy security and droughts. Second, the report is highly critical of the way in which urban planning has performed across the continent, not for stereotyped reasons ('corruption' in Africa, say), but instead due to the heavy influence of 'the normative orientations of urban planning in the Global North' (UN-Habitat, 2014: 11). These orientations lead planners to obsessions with master planning, building regulation, and control (of buildings and people), and when planners do turn to the environment, 'expensive green technofixes' divert attention 'from needful communities to provide elite green enclaves that entrench inequalities' (UN-Habitat, 2014: 37).

Third, this leads the authors to boldly declare that 'Africa and the world community need to rethink what constitutes a city since the Western concept is no longer the sole legitimate template for its application in Africa' (UN-Habitat, 2014: 37). They thus call for 'a radical re-imagination of African approaches to urbanism' (UN-Habitat, 2014: 7). This call is, at times, rather vague, yet at other times there are detailed or specific recommendations that might be workable, such as the call for a 'radical decentralization of powers' to foster 'community self-organization' that could happen with a 'devolution of controls over revenue collection' in a 'bottom-up system of government' (UN-Habitat, 2014: 11). They are realistic in acknowledging that such radical changes would have to be introduced incrementally, but sharp-elbowed enough to argue that without that radical decentralisation, many urban informal areas 'will detach themselves even further from effective government control' (UN-Habitat, 2014: 11).

Given that the report shows that 'the road to truly democratic governance systems often provides for a rough ride' (UN-Habitat, 2014: 16), it is inspiring that UN-Habitat remains committed to the promotion of 'inclusive, progressive and productive dialogue among urban stakeholders' (UN-Habitat, 2014: 30). Without such dialogue (particularly including African urban youth), 'growing environmental and social strains may exacerbate urban poverty and conflict in the region. The spectre of violence looms large' (UN-Habitat, 2014: 30). The authors see that evident links across scales – global and regional-scale geopolitics and climate change, as well as relations between national and municipal governments and governance networks – all play roles in urban environmental dynamics.

The future brings a set of possibilities for urban Africa that are not as far outside the realm of possibility as Kingelez's vision of Kinshasa 1,000 years from now. One of these is the very real sense, which the UN has picked up on, that Africa's cities are redefining what it means to be a city and reframing the world's idea of what a 'humane urbanism' might be. Africa's cities, even in the worst-case scenarios of political violence and economic decimation, are engines of cultural productivity, inventiveness and ingenuity the likes of which the world has rarely seen. People power has, on occasion, genuinely soared and beaten back the forces of draconian and authoritarian misrule, as in Ouagadougou, the capital of Burkina Faso, in the last two years: an October 2014 popular uprising in the capital overthrew the country's long-time dictator, and the movement, which called itself the 'Citizen's Broom', arose in the streets again a year later to prevent the military from hijacking elections, which produced the return of democratic governance (Farge and Felix, 2014; Tate, 2015).

Conclusion

This is undoubtedly still an era of what the urbanist Edgar Pieterse (2013: 12) calls 'rogue urbanism' in Africa, an era where 'dynamics . . . are so unruly, unpredictable, surprising, confounding and yet, pregnant with possibility'. The power, creativity, dynamism and capacity to shape the cityscape at the grassroots are extraordinary in cities of Africa, but so is the possibility for going 'rogue' in a negative sense, whether led by rogue states, rogue elites, rogue outsiders, or rogue groups at the grassroots, the uncivil society that runs parallel to the civil society. The implications of this 'rogue sensibility' (Pieterse, 2013: 12) for the future are precarious and uncertain. The continent shows us everything from inspiring examples of grassroots organising that leads to sustainable development, to violence-ridden paralysed communities sinking further and further into the mire from incapacitated or fractious grassroots, with most cities' futures lying in between.

For some cities, the twenty-first century has become such a terrain of struggle that productive moments for real change fade into the backdrop. Yet even in the worst scenarios imaginable, there are counterexamples available for the knee-jerk Afro-pessimism that often rules the airwaves of the world. The question seems to be one of how to sustain 'productive moments' of dynamic grassroots engagement that gain state (and sometimes private sector or philanthropic) support – to grow the grassroots, if you will, in a manner that leads to broad environmental justice and long-term political, economic and socio-environmental sustainability. Across the continent, over and over, what we see instead are 'productive moments' that come crashing down far too soon, but the resiliency of urban African communities, then, quite frequently, resurrects the possibility.

A great many well-intended programmes and policies have been developed for Africa's cities. Whether these have the 'aim of ameliorating poverty, rescuing the environment [or] growing local economies', they tend to 'come with built-in assumptions about the nature of the city, its people, cultures and possible futures' (Pieterse and Simone, 2009, cited in Pieterse, 2013: 12). However, there are increasing numbers of institutions, programmes, policies and dialogic processes that mark routes towards what Pieterse (2008) calls 'radical incrementalism,' such as the work of Africulturban and other music and arts institutions in Dakar, the Kibera Public Space Project in Nairobi, and various programmes and activities associated with the African Centre for Cities in Cape Town (Anderson, 2015; Myers, 2016).

What these small examples of positive steps forward teach us, perhaps most of all, is that power is relational. It operates in networks. These networks are constantly shifting and in tension. At the grassroots in marginalised poor urban communities that constitute the majority in most of the continent of Africa's cities, these networks hold less power in the overt world. Yet they are ultimately responsible for a great deal of the production of urban space. Pieterse (2010: 209) writes, 'I have no doubt that the street, the slum, the waste dump, the taxi rank, the mosque and church will become the catalysts of an emancipated African urbanism'. From Pikine to Cairo to Addis to Nairobi to Cape Town, and with many stops in between, it seems that marginal communities do have the potential for transformational energies and dynamism that may produce an 'emancipated African urbanism'.

But their successes are often fleeting, the emancipation transitory or ephemeral. Environmental consciousness varies significantly at the grassroots in Africa's cities. Likewise, the degree and character of the politicisation of that consciousness varies. The rogue sensibilities that predominate in marginalised-majority urban communities are unpredictable and difficult to harness into revolutionary moments that are sustained. The task ahead for the future of Africa's cities is to sustain them.

References

Anderson, P. (2015) 'Reimagining ecology in the city of Cape Town: contemporary ecological research and the role of the African Centre for Cities', in L.A. Sandberg, A. Bardekjian and S. Butt (eds), *Urban Forests, Trees and Greenspace: A Political Ecology Perspective*, New York: Routledge, pp. 261–276.

Bafana, B. (2016) *Africa's Cities of the Future: Proper Planning Key to Sustainable Cities*, available at: www.un.org/africarenewal/magazine/april-2016/africa%E2%80%99s-cities-future (accessed 26 December 2017).

Chabal, P. (2009) *Africa: The Politics of Suffering and Smiling*, London: Zed Books.

CORES (Congo Real Estate Service) (2015) *La Cité du Fleuve and CORES Announce Their New Partnership*, available at: www.cores.cd/en/la-cite-du-fleuve-and-cores-announce-their-new-partnership/ (accessed 16 February 2015).

Davis, M. (2005) *Planet of Slums*, London: Verso.

de Boeck, F. (2012) 'Spectral Kinshasa: building the city through an architecture of words,' in T. Edensor and M. Jayne (eds), *Urban Theory Beyond the West: A World of Cities*, London: Routledge, pp. 311–328.

Diouck, A. (2013) *Author's focus group interview with Alioune Diouck, Mayor of Djidah Thiaroye Kao Commune d'Arrondissement, City of Pikine, Dakar, with Tom Hanlon, Alasse Elhadji, four US undergraduate students and staff members of Partners Senegal*, 4 January, Pikine, Senegal.

Farge, E. and Felix, B. (2014) *Beyond Burkina Faso, Africa's 'Black Spring' Hopes May Be Premature*, 6 November, available at: www.reuters.com/article/2014/11/06/us-burkina-politics-africa-analysis-idUSKBN0IQ1ZE20141106 (accessed 26 December 2017).

Fox, S. (2014) 'Urbanisation as a global historical process: theory and evidence from sub-Saharan Africa', in S. Parnell and E. Pieterse (eds), *Africa's Urban Revolution*, London: Zed Books, pp. 257–283.

Gridoux, M. (2012) 'Trois questions á Khalifa Sall, maire de Dakar', *African Business*, 25, December: 86.

Kingelez, B.I. (2008) *The Art of the Model: An Erudite Art*, available at: www.culturebase.net/artist.php?210 (accessed 13 March 2016).

Kushner, J. (2013) *Income Inequality: In Congo, a Tale of Two Cities*, 28 March, available at: http://pulitzer center.org/reporting/DRC-congo-kinshasa-income-inequality-a-tale-two-cities (accessed 26 December 2017).

Lamson-Hall, P., Degroot, D., Martion, R., Tafesse, T. and Angel, S. (2015) *A New Plan for African Cities: The Ethiopia Urban Expansion Initiative*, New York: Marion Institute of Urban Management, New York University, available from the lead author at: plamsonh@stern.nyu.edu.

Larsen, K. (2014) 'Possessing spirits and bodily transformation in Zanzibar: reflections on ritual, performance, and aesthetics', *Journal of Ritual Studies*, 28(1): 15–29.

Murray, M.J. (2016) *Global Urbanism at the Start of the 21st Century: Emergent Dynamics of Spatial Fragmentation and Social Polarization*, New York and Cambridge: Cambridge University Press.

Myers, G. (2011) *African Cities: Alternative Visions of Urban Theory and Practice*, London: Zed Books.

Myers, G. (2015) 'A world-class city region? Envisioning the Nairobi of 2030', *American Behavioral Scientist*, 59(3): 328–346.

Myers, G. (2016) *Urban Environments in Africa: A Critical Analysis of Environmental Politics*, Bristol: Policy Press.

Nuttall, S. and Mbembe, A. (eds) (2008) *Johannesburg: The Elusive Metropolis*, Durham, NC: Duke University Press.

Parnell, S. and Pieterse, E. (eds) (2014) *Africa's Urban Revolution*, London: Zed Books.

Pieterse, E. (2008) *City Futures: Confronting the Crisis of Urban Development*, London: Zed Books.

Pieterse, E. (2010) 'Cityness and African urban development', *Urban Forum*, 21: 205–219.

Pieterse, E. (2013) 'Introducing rogue urbanism', in E. Pieterse and A.M. Simone (eds), *Rogue Urbanism: Emergent African Cities*, Cape Town: Jacana Media, pp. 12–15.

Pieterse, E. and Simone, A. (eds) (2013) *Rogue Urbanism: Emergent African Cities*, Cape Town: Jacana Media & African Centre for Cities.

Quayson, A. (2014) *Oxford Street, Accra: City Life and the Itineraries of Transnationalism*, Durham, NC: Duke University Press.

Robert, S. (1991 [1951]) *Kusadikika [Land of Make-Believe]*, Dar es Salaam: Mkuki na Nyota.

Simone, A.M. (2007) 'Assembling Douala: imagining forms of urban sociality', in A. Cinar and T. Bender (eds), *Urban Imaginaries: Locating the Modern City*, Minneapolis, MN: University of Minnesota Press, pp. 79–99.

Tate, E. (2015) 'Thomas Sankara and Burkina Faso's "black spring"', *The Bullet*, Socialist Project Bulletin 1078, 9 February, available at: www.socialistproject.ca/bullet/1078.php (accessed 26 December 2017).

UN-Habitat (2014) *State of African Cities 2014: Re-Imagining Sustainable Urban Transitions*, Nairobi: UN-Habitat.

Watson, V. (2014) 'African urban fantasies: dreams or nightmares?', *Environment and Urbanization*, 26: 1–17.

Watson, V. and Agbola, B. (2013) *Who Will Plan Africa's Cities?*, 12 September. Available at: www.africaresearchinstitute.org/publications/who-will-plan-africas-cities/ (accessed 26 December 2017).

Section 9

Economic perspectives

36

Africa's extreme uneven development worsens during global economic turmoil

Patrick Bond

Introduction

A profound myth developed just at the moment Africa's rate of economic growth ceased its rapid 2002–2011 increase: 'Africa Rising'. That myth persists, and as German Central Bank president Jens Weidmann (2017) claimed in mid-2017 at a Berlin G20 conference, without providing any evidence, 'Africa stands ready to benefit from an open world economy. Its economic outlook is positive'. In reality, after the 2011 peak of the commodity super-cycle and subsequent crash, it was simply illogical to proclaim that Africa was prospering, given so many economies' dependence upon primary product exports, especially minerals and oil, and given the ongoing global economic turmoil. A brief commodity price recovery in 2016 and the drop in the value of African currencies did not set the stage for renewed competitiveness, business confidence or investment, but instead another round of fiscal crises, extreme current account deficits and sovereign debt defaults. The continuation of the post-1980 downturn in world fixed capital investment (from the 25.5 per cent of gross domestic product to below 23.5 per cent after the 2008 crash) was also reflected in world profit rate stagnation, in spite of hype over China's 'Belt and Road' mega-infrastructure projects. Moreover, with most major mining houses' value plummeting on the world's stock exchanges from peak 2015 levels by more than 80 per cent that year, not even a more desperate round of exploitation, the entry of the Brazil-Russia-India-China-South Africa (BRICS) bloc (Bond and Garcia, 2015) or the G20's 2017 'Compact with Africa' could disguise the generalised crisis of accumulation in circuits of the world economy most important to Africa.

Even before the 2011 commodity peak and 2015 crash, the neo-liberal export-oriented strategy did enormous damage to genuine popular development, gender equity and Africa's natural environment (Bond, 2006). Below, a series of related aspects of this exploitation are considered: excessive profit-taking through illicit financial flows *and* also licit financial flows; the general outflow of profits and dividends associated with multinational corporate investment; the resulting rise in Africa's foreign debt to unprecedented heights; South African sub-imperial capital accumulation; the subsidisation and financing of new extractive infrastructure and financing; the uncompensated depletion of Africa's 'natural capital'; land grabs; militarisation; and climate change.

Put simply, although there have been slight improvements in poverty levels, mortality and morbidity rates and education (especially after the West's G7 debt relief package in 2005), the conditions for reproduction of daily life in Africa have not improved as a result of the continent's integration into world capitalism since the 1980s. The frenetic globalisation era:

- has for the past third of a century entailed structural adjustment austerity imposed by the Bretton Woods Institutions;
- has been carried out by dictatorships or at best semi-democratic regimes that continue to receive support from the West and the BRICS;
- has had the effect of deepening resource curses due to extractive industry exploitation; and
- has amplified other political, economic and ecological injustices (Bond, 2006).

The 'Great Recession' which the world entered from 2007 exacerbated these problems (Gumede, 2016), ironically just as the mainstream press began to advertise 'Africa Rising' (e.g. Perry, 2012; Robertson, 2013). Africa's per capita GDP levels did indeed rise rapidly, but with most of the gains occurring from 1999 to 2008. There was even a hoax-type claim from the African Development Bank's chief economist Mthuli Ncube that 'one in three Africans is middle class' (Ncube, 2013). In 2017, the bank repeated the claim: 'One of the main drivers of the surge in consumer demand in Africa is the continent's growing population (currently 1 billion) and expanding middle class (estimated at 350 million)' (AfDB *et al.*, 2017: 28–29). But Ncube had defined 'middle class' as those who spend between US$2 and US$20 per day, with 20 per cent in the US$2–4 per day range and 13 per cent from US$4 to US$20. Both ranges are poverty-level in most African cities, whose price levels leave them among the world's most expensive. The share above $20 was less than 5 per cent, and shrinking.

One of the central reasons for the disconnect between 'Africa Rising' and the poverty experienced by the continent's majority is looting: illicit financial flows (IFFs), as well as *legal* financial outflows in the form of profits and dividends sent to transnational corporate (TNC) headquarters, with their inevitable reliance on tax havens. A general case can be made against TNCs based on their excessive profiteering and distortion of African economies. The worst form of foreign direct investment (FDI) tends to come solely in search of raw materials. As a result of the commodity crash, a 15 per cent slowing of annual FDI inflows in the 2008–2016 period meant that the extractive industries' extreme pressures on people and environments became more uneven. In some cases, corporate desperation intensified site-specific extractive industry malpractices, more extreme forms of ecological degradation, social abuse and labour exploitation.

As some firms lost vast amounts of shareholder value, traumatic job losses began. Lonmin's London listing plummeted from a high price of 427,800c per share in 2007 to 41c in 2016. The value of the world's largest commodity firm, Glencore, fell 86 per cent from a 2011 initial listing price of 532c per share to 74c in 2016. The Anglo American Corporation (the largest on the continent over most of the prior century) announced in late 2015, as its London Stock Exchange value shrunk by 93.6 per cent (from 3,540c per share in 2008 to a 2016 low of 227c), that it would scale down mining employment by more than half and sell many African assets to the Indian entrepreneur Anil Agarwal of Vedanta.

In mid-2017, London-based Global Justice Now and several allies released a study by Mark Curtis (2017), *Honest Accounts: How the World Profits from Africa's Wealth*, which estimated that 48 countries in sub-Saharan Africa (i.e. not including those in North Africa) are 'collectively net creditors to the rest of the world, to the tune of $41.3 billion in 2015'. According to Curtis (2017: 1):

- African countries received US$161.6 billion in 2015 – mainly in loans, personal remittances and aid in the form of grants. Yet US$203 billion was taken from Africa, either directly – mainly through corporations repatriating profits and by illegally moving money out of the continent – or by costs imposed by the rest of the world through climate change.
- African countries receive around US$19 billion in aid in the form of grants, but over three times that much (US$68 billion) is taken out in capital flight, mainly by multinational companies deliberately misreporting the value of their imports or exports to reduce tax.
- While Africans receive US$31 billion in personal remittances from overseas, multinational companies operating on the continent repatriate a similar amount (US$32 billion) in profits to their home countries each year.
- African governments received US$32.8 billion in loans in 2015, but paid US$18 billion in debt interest and principal payments, with the overall level of debt rising rapidly.
- An estimated US$29 billion a year is being stolen from Africa in illegal logging, fishing and the trade in wildlife/plants.

Regardless of which BRICS or Western TNC operates in the continent, as shown in the following pages:

- excessive profits are exiting Africa as IFFs;
- *licit* (legal) financial flows are also worthy of more concern;
- FDI continues to leave Africa poorer;
- the need to pay TNCs their profits and dividends in hard currency recently raised Africa's foreign debt to unprecedented heights;
- South African sub-imperial accumulation is worsening;
- new subsidised infrastructure and financing will exacerbate African underdevelopment;
- uncompensated mineral and oil/gas ('natural capital') depletion continues;
- land grabs, militarisation and climate change are all growing threats to the continent; and
- only rising *social* resistance can halt and reverse these trends.

Illicit financial flows

First, the IFFs reflect many of the corrupt ways that wealth is withdrawn from Africa, mostly in the extractives sector. These TNC tactics include mis-invoicing inputs, transfer pricing and other trading scams, tax avoidance and evasion of royalties, bribery, 'round-tripping' investment through tax havens, and simple theft of profits via a myriad of gimmicks aimed at removing resources from Africa. Examples abound:

- In South Africa, Bracking and Sharife (2014) reported that De Beers mis-invoiced US$2.83 billion of diamonds over six years.
- The Alternative Information and Development Centre (2014) showed that Lonmin's platinum operations – notorious at Marikana, not far from Johannesburg, where the firm was complicit in a massacre of 34 of its wildcat-striking mineworkers in 2012 – has also spirited hundreds of millions of dollars offshore to Bermuda since 2000.
- Vedanta's chief executive Agarwal arrogantly bragged at a Bangalore meeting how in 2006 he spent $25 million to buy Zambia's Konkola Copper Mines, which is Africa's largest, and then reaped at least US$500 million profits from it annually, apparently through an accounting scam (Lusaka Times, 2014).

Analysis of IFFs at the continental scale has been carried out by Global Financial Integrity (Kar and Spanjers, 2015) and by political economist Leonce Ndikumana (2015), who showed how Africa is both 'more integrated but more marginalised' in world trade (Ndikumana *et al.*, 2014). Ndikumana subsequently authored a UNCTAD (2016) critique of extractive industries, and his South African and Zambian cases generated intense rebuttals from mining industry representatives about poor statistical quality in data provided by the two governments. While this has required a recalculation, especially of copper and gold exports, the overall critique of IFFs remains intact.

There are also policy-oriented NGOs working against IFF across Africa and the South, including several with Northern roots such as Trust Africa's 'Stop the Bleeding' campaign, Global Financial Integrity, Tax Justice Network, Publish What You Pay and Eurodad. IFFs are a subject of research and economic critique that give hope to many who want Africa's scarce revenues to be recirculated inside poor countries, not siphoned away to offshore financial centres. Nevertheless, the implicit theory of change adopted by the head offices of some such NGOs is dubious; if they argue that because *transparency* is like a harsh light that can *disinfect* corruption, their task is mainly a matter of making capitalism cleaner by bringing problems such as IFFs to light.

To their credit, many NGOs and allied funders and grassroots activists generated sufficient advocacy pressure to compel the African Union and UN Economic Commission on Africa to commission an IFF study led by former South African president Thabo Mbeki (2015). Reporting in mid-2015 and using a conservative methodology, his estimate is that IFFs from Africa exceed US$50 billion a year.

The IFF looting is mostly – but not entirely – related to the extractive industries. In an even more narrow accounting than Mbeki's, the AfDB *et al.* (2013) *African Economic Outlook* estimated that $319 billion was robbed from 2001 to 2010, with the most theft in metals (US$84 billion), oil (US$79 billion), natural gas (US$34 billion), minerals (US$33 billion), petroleum and coal products (US$20 billion), crops (US$17 billion), food products (US$17 billion), machinery (US$17 billion), clothing (US$14 billion), and iron and steel (US$13 billion). The charge that Africa is 'resource-cursed' fits the data well.

From IFFs to LFFs

Even if IFFs were reduced, another reason that FDI leaves Africa much poorer is what can be termed *licit* financial flows (LFFs). These are legal profits and dividends sent home to TNC headquarters after FDI begins to pay off. The payments of such outflows, along with interest and the net trading position, are termed the 'current account'. According to the International Monetary Fund (IMF, 2017) *Regional Economic Outlook*, the last 15 years or so witnessed trade surpluses between sub-Saharan African countries and the rest of the world, reaching 5.6 per cent of GDP in 2011, followed by smaller net surpluses, and then in 2015–2016 deficits of 3.1 and 2.0 per cent of GDP, respectively, with more deficits projected into the future.

The current account measures not only whether imports are greater than exports, but also the flows of profits, dividends and interest. Sub-Saharan Africa had a fair balance (and even in 2004–2008 an average surplus of 2.1 per cent of GDP). But since 2011, it has rapidly fallen into the danger zone, with a current account deficit of 4.0 per cent of GDP in 2016, led by Mozambique (−38 per cent), the Republic of Congo (−29 per cent) and Liberia (−25 per cent). Adding North African countries, the full continent's current account deficit was 6.5 per cent of GDP in 2016 as a result of the fall in oil prices to a low of US$26 per barrel in early 2016. Of 54 African countries, 20 had double-digit current account deficits in 2016. To put that in context, the 1998 crash of leading East Asian economies was catalysed by current account deficits of only 5 per cent.

To cover a current account deficit, flows of external finance are required. Such flows amounted to US$178 billion in 2016, which was US$5 billion less than 2015, largely as a result of a 60 per cent decline in portfolio capital inflows (i.e. purchases of financial shares in debt or stock market investments). Overseas development aid to Africa declined 2 per cent in 2016, and remittances were virtually unchanged. Foreign direct investment is somewhat more complicated, however.

FDI in retreat

Third, partly due to the extended low prices of commodities, the difficulty in raising new hard currency to pay profits and dividends rises as FDI falls. From a US$66 billion peak annual inflow in 2008 to a 2016 level of US$56 billion, FDI remains the second major inflow of hard currency to Africa, trailing only labour remittances. Globally, annual FDI was US$1.56 trillion in 2011, fell to US$1.23 trillion in 2014, rose to US$1.75 trillion in 2015, and then receded to US$1.52 trillion in 2016. As the UNCTAD (2017: x) reported, in 2016, 'FDI flows to Africa continued to slide, reaching $59 billion, down 3 per cent from 2015, mostly reflecting low commodity prices'. The anticipated 2017 uptick to US$65 billion would still be less than 2014's US$71 billion.

The largest 2016 inflows of FDI from and to particular countries were the US with US$385 billion, China (including Hong Kong) with US$231 billion and the UK with US$179 billion; each outstrip Africa's US$59 billion, with Singapore and Brazil following (both at US$50 billion of inflows in 2016). The single largest African FDI project in 2015–2016 was an Egyptian property deal worth US$20 billion by the China Fortune Land Development company, but it will create only 3,000 jobs. A Doha real estate development worth US$8.5 billion, five Italian oil drill projects worth US$8.1 billion and a Chinese oil pipeline costing US$6 billion were the other main mega-project African investment commitments recorded in 2015–2016 (AfDB *et al.*, 2017). Taken together, they suggest an extreme concentration of capital flows that will amplify Africa's uneven development, following a pattern that James Ferguson (2006) describes as 'hopping and skipping' across the continent (instead of smoothly 'flowing').

Curtis (2017) argues that the inflows to sub-Saharan Africa are far lower once both FDI outflows (e.g. US$11 billion from Angola alone in 2016) and corporate lending associated with FDI are subtracted. Hence, in 2015, in sub-Saharan Africa, gross FDI was US$41.2 billion, but outward FDI was US$9.3 billion (leaving a net of US$31.9 billion). Moreover, Curtis (2017: 1) shows, 'Figures from the World Bank suggest that 78 percent of private lending is FDI. This means there was $16.1 billion of loans. Removing this from $31.9 billion leaves $15.8 billion of FDI equity'. To put this in context, the entire set of capital inflows to sub-Saharan Africa that Curtis (2017: 1) measures for 2015 amount to US$162 billion, and the outflows are US$203 billion. Moreover, depletion of natural capital should be included, which the World Bank (2014: vii) estimated amounted to 12 per cent of the subcontinent's US$1.36 trillion GDP in 2010 alone (i.e. another US$163 billion).

The failure to sustain accumulation through FDI is due, in part, to shrinking commodities markets and the slowing of the rapid Chinese gross fixed capital investment wave of 2009–2012. The UNCTAD (2015: 128) also records 'an overall increasing share of regulatory and restrictive policies in total investment policy measures over the last decade' as a result of 'a new realism about the economic and social costs of unregulated market forces', although this may also represent, in part, 'investment protectionism'. This applies less in Africa, although South Africa has become more restrictive about trade as a result of deindustrialisation (e.g. applying steel tariffs against Chinese dumping in 2015–2017), as well as cancelling bilateral investment treaties because they conflict with the country's Black Economic Empowerment policy.

Foreign debt explodes

Fourth, the current account deficit in turn requires that state elites attract yet more new FDI, so as to have hard currency on hand to pay back old FDI (usually as profit and dividend outflows), or if that is less available, as now appears the case, to take on new foreign borrowings. So as to cover the payment deficits and slight trade deficit, Africa's foreign debt is soaring. For sub-Saharan Africa, what was a foreign debt in the $170–210 billion range from 1995 to 2005 (when G7 debt relief lowered it by 10 per cent) rose to nearly $400 billion by 2015 (IMF, 2017). Not only Chinese lending, but also a spate of Eurobonds, became debilitating in several countries, where by 2016 they had become a substantial share of the total public debt stock: Gabon (48 per cent), Namibia (32 per cent), Côte d'Ivoire (26 per cent), Zambia (24 per cent), Ghana (16 per cent), Senegal (15 per cent) and Rwanda (13 per cent).

The 2017 *African Economic Outlook* observed 'tighter financing conditions and increased debt financing have started to worsen debt service burdens, with an upward trend in both the debt-service-to-revenue ratio and the external debt-service-to-exports ratio' (AfDB *et al.*, 2017: 9). For petroleum-based economies, the report continued, there was a 'a seven-fold increase in debt service, from an average of 8 percent of revenues in 2013 to 57 percent in 2016', with Nigeria (66 per cent) and Angola (60 per cent) worst affected. Another fear (AfDB *et al.*, 2017: 10) is domestic debt, as the slowdown has also generated 'a widespread increase in nonperforming loans, triggering higher provisioning, straining banks' profits, and weighing on solvency'.

In the case of the largest African debtor, South Africa, foreign debt rose from $25 billion in 1995 to $35 billion in 2005, and then soared to approximately $150 billion today (i.e. from 20 per cent of GDP in 2005 to more than 40 per cent now). The last time this ratio was reached was in 1985, and the result – thanks also to anti-apartheid activist sanctions pressure against bankers – was that South African president P.W. Botha defaulted on $13 billion of short-term debt coming due and imposed exchange controls. The move signalled to the English-speaking capitalist class that the end of apartheid was near, and thus they should hasten to make favourable post-apartheid arrangements with the African National Congress (ANC), then in exile. Unfortunately, those arrangements entailed drawing South Africa much deeper into the world economy, and as the current account deficit rose, deeper into foreign debt.

Exploitation also comes from within Africa

Fifth, more nuance is important in terms of *which* firms are doing the looting. Western TNCs exploited Africa for centuries, and continue. But the single biggest country-based source of FDI stock in Africa is internal, from South Africa. A dozen companies with Johannesburg Stock Exchange listings draw out very high levels of FDI profits: British American Tobacco, SABMiller breweries (which in 2016 became a subsidiary of ABI based in St Louis), the MTN and Vodacom cell phone networks, Naspers newspapers, four banks (Standard, Barclays, Nedbank and FirstRand), the Sasol oil company, and the local residues of the Anglo American Corporation empire. The South African profit rate for these same firms has been 5 per cent below the rate earned by the same firms in the region (IMF, 2016).

The result is the systematic internal exploitation of the rest of Africa by South African capital, especially as the main retail chains (e.g. Walmart-owned Massmart and its affiliates) use the larger market in the South to achieve production economies of scale production that then swamp and destroy Africa's residual basic needs manufacturing sector. This is a form of looting also based on the IFF strategies used against South Africa by TNCs. Among others, South Africa's MTN cell phone service was reported by the Amabhungane (2015) investigative journalist network to have Mauritian and Dubai financial offices that systematically skim profits for dubious tax avoidance purposes from high-profit operations in Nigeria, Uganda and South Africa (Mauritian company taxes are 3 per cent, with no capital gains). This was a blatant practice when MTN's chairperson was Cyril Ramaphosa, subsequently South Africa's deputy president from 2014. He was also a 9 per cent owner of Lonmin when similar Bermuda platinum 'marketing' operations were a source of tax avoidance payments (Alternative Information and Development Centre, 2014). More than $100 million in a World Bank credit line raised by Lonmin in 2007 was meant to construct more than 5,000 housing units, but just three were built under Ramaphosa's direct responsibility (Bond, 2014). When, in November 2015, MTN was fined $4 billion by Abuja authorities due to its failure to disconnect more than 5 million unregistered Nigerian customers during the state's attempt to crack down on cell phone use by Boko Haram terrorists, there were few defenders of the firm. The fine was reduced to $1 billion only by virtue of Jacob Zuma's personal intervention during a 2016 state visit on MTN's behalf.

Zuma has also intervened on behalf of South African capital militarily. The oil operation of Zuma's nephew Khulubuse in the Democratic Republic of Congo (DRC) was said to be worth $10 billion when the concession was acquired in 2010 (although there were unconfirmed reports he sold the interests). Not far away, 1,350 South African National Defence Force (SANDF) troops were stationed as part of a United Nations ('MONUSCO') peacekeeping force, but in early 2016 it became apparent that these troops were not paying attention as a massacre occurred in their immediate vicinity (Allison, 2016). Instead, as Belgian Royal Museum for Central Africa analyst Theodore Trefon (2016) explained, 'Deployment of South African troops in the Intervention Brigade set up by the United Nations in March 2013 to reinforce MONUSCO in eastern DRC is an indication of President Zuma's motivation to stabilise the region for economic reasons'.

In the neighbouring country to the north, the Central African Republic (CAR), the ANC 'Chancellor House' investment arm sought a diamond monopoly in 2006, codified by Thabo Mbeki and CAR's dictator Francois Bozizé. The latter was rejected by former French sponsors a few years later and, facing a Chadian-backed uprising by the Seleka rebels, won military support from Pretoria. Justifying a five-year commitment to a military presence costing more than $100 million, the deputy foreign minister Ebrahim Ebrahim explained, 'We have assets there that need protection' (Patel, 2013). But on 25 March 2013, more than a dozen corpses of South African soldiers were recovered in Bangui after a two-day battle in which hundreds of local fighters and bystanders were killed. Two hundred SANDF troops were apparently trying to *guard the South African assets*, as Bozizé fled to safety. Seleka invaded his presidential compound, taking state power that day in spite of resistance from the SANDF men that they labelled 'mercenaries'. Two *Sunday Times* reporters (Hosken and Mahlangu 2013: 1) offer quotations from interviews with SANDF troops who made it back alive (emphasis added):

> Our men were deployed to various parts of the city, *protecting belongings of South Africans*. They were the first to be attacked. Everyone thought it was those who were ambushed, but it was the guys outside *the different buildings – the ones which belong to businesses in Jo'burg* . . . We were lied to straight out . . . We were not supposed to be here. We did not come here to do this. We were told we were here to serve and protect, to ensure peace.

Sometimes, when South African capital flows elsewhere in Africa, it carries the baggage of its home base. When xenophobia broke out in 2015, there were many branch plants of Johannesburg firms that became targets of protest by Nigerians, Zimbabweans, Malawians, Mozambicans and Zambians concerned about their relatives' safety. Hostility to Johannesburg capital is also logical because its leadership was named the world's most corrupt according to several crucial indicators compiled by PricewaterhouseCoopers in 2014, as 80 per cent of managers admitted to the firm that they commit economic crimes and South African firms were the 'world leaders in money-laundering, bribery and corruption, procurement fraud, asset misappropriation and cyber crime' (Hosken, 2014: 1).

At the same time, since the late 1990s, South Africa's current account deficit has soared because the country's biggest companies, nearly without exception, relocated to London or New York, and took LFFs with them: Anglo American and its historic partner De Beers, plus SABMiller, Investec bank, Old Mutual insurance, Didata IT, Mondi paper, Liberty Life insurance, Gencor (BHP Billiton) and a few others. As a result, in mid-2015, the South African Reserve Bank (2015) revealed that Johannesburg firms were in 2012–2014 drawing in only half as much in internationally sourced profits ('dividend receipts') as TNCs were taking out of South Africa. But that was an improvement over the 2009–2011 period, when local TNCs pulled in only a third of what foreigners took out. One reason is that Johannesburg firms have been busier in the rest of Africa in the past few years, as mining, cell phones, banking, brewing, construction, tobacco, tourism and other services from South Africa became more available up-continent.

Inclement public subsidisation and private financing of destructive FDI

Sixth, a continual threat to the continent is more frenetic mining and petroleum extraction, notwithstanding falling prices, as a result of state subsidies. In 2017, the G20 proposed a Compact with Africa (CwA) to assure state support for public–private partnerships across the continent and to draw in institutional investors with state guarantees (Weidmann, 2017). As Helmut Reisen (2017: 1) explained, given the anticipated asset base of '$100 trillion by 2020, institutional investors (pension, funds, life insurers and sovereign wealth funds) would need to invest one percent of their annual new inflows to fund Africa's infrastructure gap, estimated at $50 billion per year'. But according to the C20 (2017: 1) group of civil society watchdogs, this strategy will translate into:

> higher costs for the citizens, worse service, secrecy, loss of democratic influence and financial risks for the public and the multinational corporations involved demand that their profits be repatriated in hard currency – even though the typical services contract entails local-currency expenditures and revenues – and that often raises African foreign debt levels, which are now at all-time highs again in many countries. The Compact also is silent regarding problems with (and popular resistance to) investor protection, such as the vague 'fair and equitable treatment' clause in investment agreements and investor-to-state dispute settlement.

Until the G20's focus on just seven African countries, it was generally assumed that the largest donor subsidies would go to the African Union's (2012) Program for Infrastructure Development in Africa (PIDA). The continent-wide trillion-dollar PIDA is mainly aimed at extraction. New roads, railroads, pipelines and bridges are planned, but they largely emanate from mines, oil/gas rigs and plantations, and are mainly directed towards ports. Electricity generation is overwhelmingly biased towards projected mining and smelting needs, although the case of Eskom, in South Africa – perpetually facing load-shedding in early 2015 – is illustrative, as demand for its product fell at least 5 per cent in late 2015 as mine shafts and foundries were shuttered due to adverse economic conditions. In 2015, Eskom suffered regular brownouts, but after winter ended a substantial surplus developed, leading the parastatal to announce major coal-fired power station closures in 2017.

Subsidies of the sort envisaged in the CwA and PIDA could bring back the worst of the FDI, especially from BRICS companies such as Brazil's Vale Mining (in Mozambique), Russia's Rosatom (in a proposed $100 billion nuclear reactor deal with Pretoria, as well as anticipated deals in several other African countries), India's Vedanta (which is extremely exploitative in Zambia), various Chinese parastatals and firms (to illustrate, the Hong Kong-based 88 Queensway Group and the Zimbabwe military looted $13 billion worth of the $15 billion in Marange diamonds, according to 2016 revelations by Robert Mugabe, and the Chinese dam builder Three Gorges Corporation was banned by the World Bank in 2014 due to its African bribery), not to mention the profusion of unethical South African firms (Bond and Garcia, 2015). One route they anticipate receiving indirect financing subsidies in the form of loans at preferential rates is the BRICS New Development Bank (NDB) (Bond, 2016). One NDB director, South Africa's Tito Mboweni, told Bloomberg (2015) that the proposed Russia-South African nuclear deal 'falls squarely within the mandate of the NDB'.

BRICS is already appearing to many in Africa as an intensified version of Western TNCs' exploitative experiences, especially considering the pro-corporate arrangements that exist in their bilateral investment treaties with Africa (Ferrando, 2012; Garcia, 2017). A variety of commentators from the left – including respected intellectuals Walden Bello (2014), Horace Campbell (2014) and Radhika Desai (2013) – have endorsed the new BRICS financial institutions. Yet their arguments have not confronted contradictions, such as the financing of destructive African projects or the upholding of the West's destructive world monetary system and inadequate climate change policy, in which the BRICS are implicated (Bond and Garcia, 2015). The Contingent Reserve Arrangement (BRICS, 2014: 1), for example, requires BRICS countries in financial trouble (such as South Africa will be when short-term foreign debt payments are difficult to fulfil) to go to the IMF for a structural adjustment loan and policy support once they have exhausted 30 per cent of their borrowing quota (Bond, 2016).

Uncompensated natural capital depletion

Seventh, the financing and FDI aimed at extraction are responsible for the depletion of non-renewable resources without the kinds of reinvestment that are more common in sites such as Norway, Australia and Canada, whose economies are also resource-based, but not nearly so resource-cursed as Africa's, in large part because they host headquarters of mining and petroleum TNCs. Many BRICS corporations appear oriented to rapid depletion of Africa's 'natural capital', a term used by economists to describe natural resource endowments. Although the end of the commodity super-cycle will mean a lower rate of extraction, this should not blind Africans to the continent's residual colonial-era bias towards the removal of non-renewable minerals, oil and gas, the exploitation of which leaves Africa far poorer in net terms than anywhere else on earth. That bias towards non-renewable resource depletion without reinvestment has meant that the continent's net wealth fell rapidly after 2001. Even the World Bank (2011) admits that 88 per cent of sub-Saharan African countries suffered net negative wealth accumulation in 2010 (see also World Bank, 2014: vii).

The crisis continues through land grabs, climate change and militarisation

Finally, there are several other devastating features of the contemporary African political economy and political ecology, in the form of land grabs, militarisation and climate change. The most immediate threats face the African peasantry, especially women, and especially those in areas attractive to foreign investors. Already, small farmers are being displaced in sites such as Ethiopia and Mozambique as a result of land grabs by Middle Eastern countries and India, South Africa and China (Ferrando, 2012). The growing role of the US military's Africa Command in dozens of African countries bears testimony to Washington's overlapping desire to maintain control amid rising Islamic fundamentalism from the Sahel to Kenya, which are, coincidentally, theatres of war in the vicinity of large petroleum reserves (Turse, 2014).

Climate change will affect the most vulnerable Africans in the poorest countries, who are already subject to extreme stress as a result of the war-torn socio-economic fabrics in West Africa, the Great Lakes and the Horn of Africa. Although the leader of the Armed Conflict Location Events Data (ACLED) project at Sussex University, Clionadh Raleigh (2017), argues that climate change is not automatically a direct cause of protest, in part because mutual aid systems are activated, there is no doubt that worsening agricultural conditions cause migration to urban areas, which in turn put more stress on the African city's social fabric. As that occurs, further application of neo-liberal state-shrinking public policy is bound to generate yet more social stress, as was witnessed in the case of Syria prior to the 2011 uprising.

Ending the looting of Africa

In large part as a result of worsening socio-economic conditions, African uncivil society activists – those willing to express frustration in means other than what are often termed the 'invited spaces' of official participation – have been protesting at a rising rate. There are various ways to measure this power, including police statistics, journalistic accounts and business executive surveys. According to research carried out at the Universities of Sussex and Texas, protest incidents rose dramatically in 2010–2011 and stayed at remarkably high levels in many African cities (Robert S. Strauss Center for International Security and Law, 2016; ACLED, 2017) (see Figure 36.1).

In 2010, the Armed Conflict Location and Event Data (ACLED, 2017) database recorded scores of protests (especially those that turned violent, typically facing police repression) in Cairo and Alexandria, Mogadishu, Nairobi, the cities and towns on the Gulf of Guinea – especially in Nigeria – and in the four largest South African cities: Johannesburg-Pretoria, Cape Town, Durban and Port Elizabeth. In 2011, dozens of protests in these cities continued. Tunis, Algiers and Cairo were measured as hosting more than 100 protests each.

In the 2015–2016 measurements, the continent witnessed even more intense protests across North Africa, Nigeria and South Africa. In addition, southern Africa witnessed high levels of protests in Harare, Zimbabwe and in Kinshasa and Goma, in the Democratic Republic of Congo, and in Zambia and Madagascar, where the capitals of Lusaka and Antananarivo recorded substantial increases compared to 2011. East Africa and the Horn witnessed scores of protests, in Nairobi, Kenya; Kampala, Uganda; Bujumbura, Burundi; Khartoum, Sudan; and in Ethiopia, Addis Ababa and surrounding towns. West African protests were led by Nigerians, but there were many other scattered sites of social unrest in the Gulf of Guinea. In North Africa in 2016, there were new rounds of protests in the main 2011 sites: Tunisia, Egypt, Libya and Algeria. Although the counter-revolution had prevailed in most of these countries, the activists

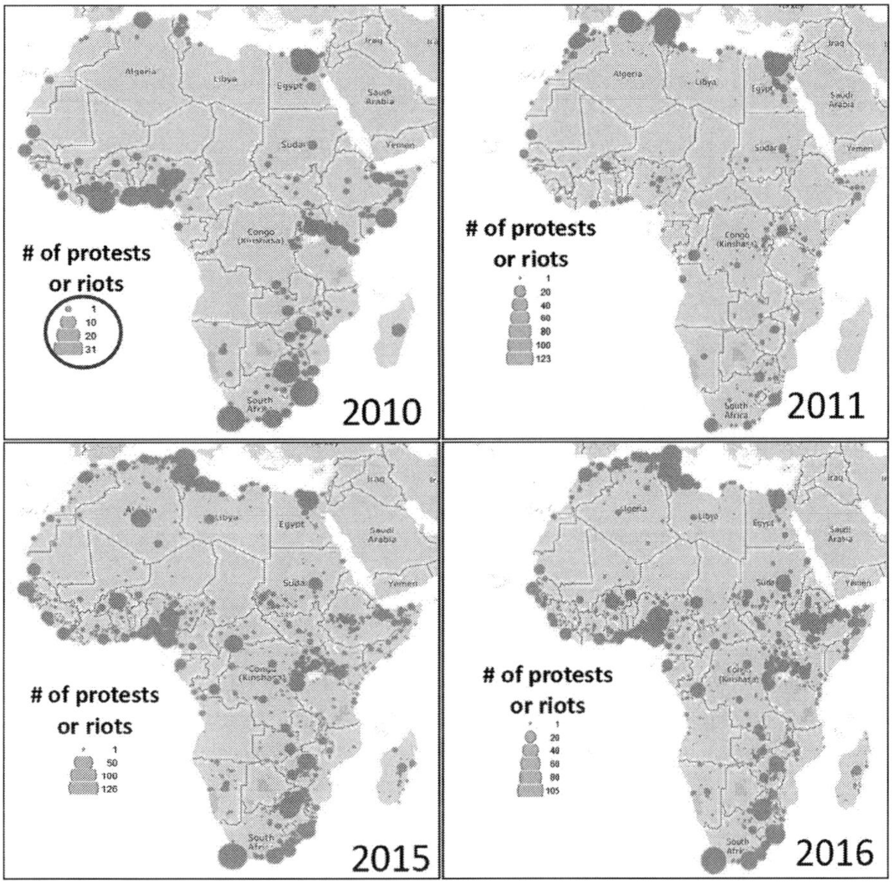

Figure 36.1 Protests and riots in Africa, 2010–2011 and 2015–2016
Source: ACLED (2017).

were not deterred from expressing grievances. State repression intensified in many countries as a response to the protest upsurge (Kode and Ben Garga, 2017).

The African Development Bank, World Bank and Organisation for Economic Co-operation and Development also measure protests, with data based upon Reuters and Agence France Press reports, and in 2017 observed that higher wages and better working conditions consistently ranked as the main reason for protests in recent years (AfDB *et al*., 2017: 129). A good share of the turmoil in Africa prior to the 2011 upsurge took place in the vicinity of mines and mineral wealth (Berman *et al*., 2014). Subsequently, socio-economic protests included the fabled Tunisian Revolution in 2011, catalysed by Mohamed Bouazizi's self-immolation. Both Tunisia and Egypt generated such intense revolutionary bursts of energy because their independent labour movements were also ascendant. Notwithstanding extreme unevenness across and within the continent's trade unions, Africa is ripe for a renewed focus on class struggle.

Indeed, as socio-economic conditions continue to deteriorate, the World Economic Forum's annual *Global Competitiveness Reports* – an annual survey of 14,000 business executives in 138 countries – have ranked the continent's workers as the least cooperative on earth. In 2016, workforces from South Africa (the world's most militant every year since 2012), Chad, Tunisia, Liberia, Mozambique, Morocco, Lesotho, Ethiopia, Tanzania, Algeria and Burundi were in the top 25 most confrontational proletariats (World Economic Forum, 2016), while the most cooperative workers are in Norway, Switzerland, Singapore, Denmark and Sweden.

With GDP growth declining (to just 1.4 per cent in 2016), commodity prices remaining low and declining levels of transnational corporate investment more frantically exploiting the continent, the contradictions may well lead to more sociopolitical explosions. The idea of a 'double-movement' – i.e. social resistance against marketisation, as suggested by Karl Polanyi (1957 [1944]) in *The Great Transformation* – has long applied to Africa. IMF austerity and subsequent 'IMF riots' spread across the continent during the 1980s, and to some extent catalysed democratisation movements during the early 1990s, but mainly failed to establish durable liberal political regimes. With the spoils of exuberant commodity markets going to unaccountable elites, another intense protest wave began in 2011, sparked in North Africa by increasingly urgent socio-economic grievances (Amin, 2011; Arezki and Brückner, 2011; Bond, 2011; Ekine, 2011; Mekay, 2011; Balogun, 2013; Beieler, 2013; Biney, 2013; Mampilly, 2013).

There are so many ways that Africa's entanglement in the world economy leads to underdevelopment that it may be worth heeding the ideas of John Maynard Keynes (1933), the most brilliant economist of the last century when it came to saving capitalism from its worst excesses. As he put it in his 1933 *Yale Review* essay, entitled 'National Self-Sufficiency':

> I sympathise with those who would minimise, rather than with those who would maximise, economic entanglement among nations. Ideas, knowledge, science, hospitality, travel – these are the things which should of their nature be international. But let goods be homespun whenever it is reasonably and conveniently possible and, above all, let finance be primarily national.

Today, we might term this the 'globalisation of people and de-globalisation of capital', and it is a perfect way to sloganise a sound short-term economic strategy appropriate for what we might hope will be a post-FDI world. In Africa, Samir Amin (1990) – the continent's greatest political economist – has argued for this sort of delinking strategy since the 1960s. It is time those arguments are dusted off and put to work, to help Africans continue to uprise against the 'Africa Rising' meme and all that it represents. Those who would dispute this line of argument must confront evidence of the futility of Africa's export-led economic fantasies, whether via the West or BRICS economies: the dramatic downturn in the rate of world trade, the decline in rich country GDP to a 2 per cent annual growth rate, and recessionary conditions in several emerging markets, including South Africa, in early 2017. And as a final clarion call for radical re-envisaging of African political economy, there is also a political-ecological imperative to reboot the fossil fuel-addicted sectors of the economy, as the world necessarily moves to post-carbon systems.

This, then, is the major challenge for Africans who rise up against injustice, especially those forms that can generate solidarity with the rest of the world's progressive people. It is only in sketching out contradictions and opportunities that we can project forward several decades. But at this critical juncture, as the commodity super-cycle's denouement now makes obvious the need for change, at least it is evident that Africans are not lying down.

References

ACLED (Armed Conflict Location and Event Data Project) (2017) *Armed Conflict Location Events Data*, Sussex: ACLED, available at: www.acleddata.com/ (accessed 14 June 2017).

AfDB (African Development Bank), OECD (Organisation for Economic Co-operation and Development), UNDP (United Nations Development Programme) and ECA (Economic Commission for Africa) (2013) *African Economic Outlook: Structural Transformation and Natural Resources*, available at: www.undp.org/content/dam/rba/docs/Reports/African%20Economic%20Outlook%202013%20En.pdf (accessed 5 January 2018).

AfDB (African Development Bank), OECD (Organisation for Economic Co-operation and Development), UNDP (United Nations Development Programme) and ECA (Economic Commission for Africa) (2017) *African Economic Outlook*, Tunis: AfDB, available at: www.afdb.org/en/knowledge/publications/african-economic-outlook/ (accessed 14 June 2017).

African Union (2012) *Programme for Infrastructure Development in Africa*, Addis Ababa: African Union, available at: www.au-pida.org/ (accessed 14 June 2017).

Allison, S. (2016) 'South African peacekeepers accused of failing to prevent DRC massacre', *Daily Maverick*, 21 January, available at: www.dailymaverick.co.za/article/2016-01-21-south-african-peacekeepers-accused-of-failing-to-prevent-drc-massacre/#.WLFrPvl95PY (accessed 14 June 2017).

Alternative Information and Development Centre (2014) *Lonmin, the Marikana Massacre and the Bermuda Connection*, available at: http://aidc.org.za/lonmin-the-marikana-massacre-and-the-bermuda-connection-seminar-and-press-conference/ (accessed 14 June 2017).

Amabhungane (2015) 'Ramaphosa and MTN's offshore stash', *Mail & Guardian*, 8 October, available at: http://amabhungane.co.za/article/2015-10-08-ramaphosa-and-mtns-offshore-stash (accessed 14 June 2017).

Amin, S. (1990) *Delinking*, London: Zed Books.

Amin, S. (2011) 'An Arab springtime?', *Monthly Review*, 2 June, available at: http://monthlyreview.org/commentary/2011-an-arab-springtime (accessed 14 June 2017).

Arezki, R. and Brückner, M. (2011) *Food Prices, Conflict, and Democratic Change*, University of Adelaide School of Economics Research Paper No. 2011-04.

Balogun, J. (2013) 'Africa is rising. Most Africans are not', *Compare Afrique*, 15 February, available at: www.compareafrique.com/africa-is-rising-most-africans-are-not/ (accessed 14 June 2017).

Beieler, J. (2013) *Protest Mapping*, State College, Penn State University, available at: http://johnbeieler.org/protest_mapping/ (accessed 14 June 2017).

Bello, W. (2014) *The BRICS: Challengers to the Global Status Quo*, 29 August, available at: http://fpif.org/brics-challengers-global-status-quo/ (accessed 14 June 2017).

Berman, N., Couttenier, M., Rohner, D. and Thoenig, M. (2014) *This Mine Is Mine! How Minerals Fuel Conflicts in Africa*, Oxford: Oxford Centre for the Analysis of Resource Rich Economies, available at: www.oxcarre.ox.ac.uk/files/OxCarreRP2014141.pdf (accessed 14 June 2017).

Biney, A. (2013) 'Is Africa really rising?', *Pambazuka*, 31 July, available at: http://pambazuka.org/en/category/features/88748 (accessed 14 June 2017).

Bloomberg (2015) *$100 Billion BRICS Lender More Keen on Risk Than World Bank*, 10 July, available at: www.bloomberg.com/news/articles/2015-07-10/brics-100-billion-lender-seeks-riskier-projects-than-world-bank (accessed 14 June 2017).

Bond, P. (2006) *Looting Africa*, London: Zed Books.

Bond, P. (2011) 'Africa's "recovery"', *Africa Insight*, 41(3): 30–45.

Bond, P. (2014) *Elite Transition*, London: Pluto Press.

Bond, P. (2016) 'BRICS banking and the debate over sub-imperialism', *Third World Quarterly*, 37(4): 611–629.

Bond, P. and Garcia, A. (eds) (2015) *BRICS: An Anti-Capitalist Critique*, Johannesburg: Jacana Media.

Bracking, S. and Sharife, K. (2014) *Rough and Polished*, Manchester University Leverhulme Centre for the Study of Value, available at: http://thestudyofvalue.org/wp-content/uploads/2014/05/WP4-Bracking-Sharife-Rough-and-polished-15May.pdf (accessed 14 June 2017).

BRICS (2014) *Treaty for the Establishment of a BRICS Contingent Reserve Arrangement, Fortaleza*, available at: http://brics.itamaraty.gov.br/media2/press-releases/220-treaty-for-the-establishment-of-a-brics-contingent-reserve-arrangement-fortaleza-july-15 (accessed 14 June 2017).

C20 (2017) 'The G20's Compact with Africa', *Pambazuka*, 4 May, available at: www.pambazuka.org/economics/g20%E2%80%99s-compact-africa (accessed 14 June 2017).

Campbell, H. (2014) 'BRICS bank challenges the exorbitant privilege of the US dollar', *TeleSUR*, 24 July, available at: www.telesurtv.net/english/bloggers/BRICS-Bank-Challenge-the-Exorbitant-Privilege-of-the-US-Dollar-20140724-0003.html (accessed 14 June 2017).

Curtis, M. (2017) *Honest Accounts: How the World Profits from Africa's Wealth*, London: Curtis Research, available at: http://curtisresearch.org/publications/honest-accounts-2017-how-the-world-profits-from-africas-wealth/ (accessed 14 June 2017).

Desai, R. (2013) 'The BRICS are building a challenge to Western economic supremacy', *The Guardian*, 2 April, available at: www.theguardian.com/commentisfree/2013/apr/02/brics-challenge-western-supremacy (accessed 14 June 2017).

Ekine, S. (2011) 'Defiant in the face of brutality: uprisings in East and southern Africa', *Pambazuka*, 532, 2 June, available at: http://pambazuka.org/en/category/features/73738 (accessed 14 June 2017).

Ferguson, J. (2006) *Global Shadows: Africa in the Neoliberal World Order*, Durham, NC: Duke University Press.

Ferrando, T. (2012) *BRICS, BITs and Land Grabbing*, Paris: Sciences Po Law School, available at: http://papers.ssrn.com/sol3/papers.cfm?abstract_id=2174455 (accessed 14 June 2017).

Garcia, A. (2017) 'BRICS investment agreements in Africa: more of the same?', *Studies in Political Economy*, 98(1): 24–47.

Gumede, V. (2016) *The Great Recession and its Implications for Human Values: Lessons for Africa*, Johannesburg: Real African Publishers.

Hosken, G. (2014) 'World fraud champs', *Times Live*, 19 February, available at: www.timeslive.co.za/thetimes/article725858.ece (accessed 14 June 2017).

Hosken, G. and Mahlangu, I. (2013) '"We were killing kids"', *Sunday Times*, 31 March, available at: www.timeslive.co.za/local/2013/03/31/We-were-killing-kids1 (accessed 14 June 2017).

IMF (International Monetary Fund) (2016) *2016 Article IV Consultation*, Washington, DC: IMF, available at: www.imf.org/en/Publications/CR/Issues/2016/12/31/South-Africa-2016-Article-IV-Consultation-Press-Release-Staff-Report-and-Statement-by-the-44062 (accessed 14 June 2017).

IMF (International Monetary Fund) (2017) *Regional Economic Outlook: Africa*, Washington, DC: IMF, available at: www.imf.org/en/Publications/REO/SSA/Issues/2017/05/03/sreo0517 (accessed 14 June 2017).

Kar, D. and Spanjers, J. (2015) *Illicit Financial Flows from Developing Countries: 2004–2013*, Washington, DC: Global Financial Integrity.

Keynes, J.M. (1933) 'National self-sufficiency', *Yale Review*, Summer, available at: www.panarchy.org/keynes/national.1933.html (accessed 14 June 2017).

Kode, D. and Ben Garga, M. (2017) 'Activism and the state', *Pambazuka*, 11 May, available at: www.pambazuka.org/democracy-governance/activism-and-state-how-african-civil-society-responds-repression (accessed 14 June 2017).

Lusaka Times (2014) *Video – Vedanta Boss Saying KCM Makes $500 Million Profit per Year*, 13 May, available at: www.lusakatimes.com/2014/05/13/video-vedanta-boss-saying-kcm-makes-500-million-profit-per-year/ (accessed 14 June 2017).

Mampilly, R. (2013) *Urban Protests and Rural Violence in Africa: A Call for an Integrated Approach*, New York: African Futures Forum, Social Science Research Council, 4 February, available at: http://forums.ssrc.org/african-futures/2013/02/04/urban-protests-and-rural-violence-in-africa-a-call-for-an-integrated-approach/#sthash.4KB0d0vO.dpuf (accessed 14 June 2017).

Mbeki, T. (2015) *Track It! Stop It! Get It! Illicit Financial Flow*, Report of the High Level Panel on Illicit Financial Flows from Africa, Addis Ababa: UNECA, available at: www.uneca.org/sites/default/files/PublicationFiles/illicit_financial_flows_why_africa_needs.pdf (accessed 5 January 2018).

Mekay, E. (2011) 'Privatisation aided Egypt revolt, army says', *InterPress Service*, 8 April, available at: www.ips.org/africa/2011/04/privatisation-aided-egypt-revolt-army-says/ (accessed 14 June 2017).

Ncube, M. (2013) *The Middle of the Pyramid: Dynamics of the African Middle Class*, Tunis: African Development Bank, available at: www.afdb.org/fileadmin/uploads/afdb/Documents/Publications/The%20Middle%20of%20the%20Pyramid_The%20Middle%20of%20the%20Pyramid.pdf (accessed 14 June 2017).

Ndikumana, L. (2015) *Integrated Yet Marginalized: Implications of Globalization for African Development*, Amherst, MA: University of Massachusetts/Amherst Political Economy Research Institute, available at: www.peri.umass.edu/236/hash/b3af64ea1d53b2a932a8b6cd57e45e6d/publication/653/ (accessed 14 June 2017).

Ndikumana, L., Boyce, J. and Ndyiaye, A. (2014) *Capital Flight: Measurement and Drivers*, Amherst, MA: University of Massachusetts/Amherst Political Economy Research Institute, available at: www.peri.umass.edu/236/hash/b3af64ea1d53b2a932a8b6cd57e45e6d/publication/653/ (accessed 14 June 2017).

Patel, K. (2013) 'SA troops killed in Central African Republic: why, Mr President?', *Daily Maverick*, 23 March, available at: www.dailymaverick.co.za/article/2013-03-25-sa-troops-killed-in-central-african-republic-why-mr-president/ (accessed 14 June 2017).

Perry, A. (2012) 'Africa Rising', *Time*, 3 December.

Polanyi, K. (1957 [1944]) *The Great Transformation: The Political and Economic Origins of Our Time*, Boston, MA: Beacon Press.

Raleigh, C. (2017) *Climate Violence?* Lecture to the Oxford Martin School, Oxford, available at: www.youtube.com/watch?v=y90WzN5wGq8 (accessed 27 December 2017).

Reisen, H. (2017) *The G20 'Compact with Africa' is Not for Africa's Poor*, Berlin: Shifting Wealth, 8 June, available at: www.tralac.org/news/article/11729-the-g20-compact-with-africa-is-not-for-africa-s-poor-the-finance-framework.html (accessed 14 June 2017).

Robert S. Strauss Center for International Security and Law (2016) *Final Program Report on Climate Change and African Political Stability*, Austin, TX: University of Texas.

Robertson, C. (2013) 'Why Africa will rule the 21st century', *African Business*, 7 January, available at: http://africanbusinessmagazine.com/features/profile/why-africa-will-rule-the-21st-century (accessed 14 June 2017).

South African Reserve Bank (2015) *Quarterly Bulletin*, June, available at: www.resbank.co.za/Publications/Detail-Item-View/Pages/Publications.aspx?sarbweb=3b6aa07d-92ab-441f-b7bf-bb7dfb1bedb4&sarblist=21b5222e-7125-4e55-bb65-56fd3333371e&sarbitem=6776 (accessed 14 June 2017).

Trefon, T. (2016) 'DRC in the Panana Papers', *Congo Masquerade*, 5 April, available at: http://congomasquerade.blogspot.com/2016/04/drc-in-panana-papers.html (accessed 14 June 2017).

Turse, N. (2014) 'Africom becomes a war-fighting combatant command', *TomDispatch*, 13 April, available at: www.tomdispatch.com/blog/175830/tomgram%3A_nick_turse,_africom_becomes_a_%22war-fighting_combatant_command%22/ (accessed 14 June 2017).

UNCTAD (United Nations Conference on Trade and Development) (2015) *World Investment Report 2015*, Geneva: UNCTAD, available at: http://unctad.org/en/PublicationsLibrary/webdiaepcb2013d11_en.pdf (accessed 14 June 2017).

UNCTAD (United Nations Conference on Trade and Development) (2016) *Trade Misinvoicing in Primary Commodities in Developing Countries*, Geneva: UNCTAD, available at: http://unctad.org/en/PublicationsLibrary/suc2016d2_en.pdf (accessed 14 June 2017).

UNCTAD (United Nations Conference on Trade and Development) (2017) *World Investment Report 2017*, Geneva: UNCTAD, available at: http://unctad.org/en/PublicationsLibrary/wir2017_en.pdf (accessed 14 June 2017).

Weidmann, J. (2017) *Improving the Investment Climate in Africa*, Speech to the G20 Africa Summit, 13 June, Frankfurt, Bundesbank Deutschland, available at: www.bundesbank.de/Redaktion/EN/Reden/2017/2017_06_13_weidmann.html (accessed 14 June 2017).

World Bank (2011) *The Changing Wealth of Nations*, Washington, DC: World Bank.

World Bank (2014) *Little Green Data Book*, Washington, DC: World Bank.

World Economic Forum (2016) *Global Competitiveness Report, 2016–17*, Davos: World Economic Forum, available at: www.weforum.org/issues/global-competitiveness (accessed 14 June 2017).

37

Financial sector development, foreign direct investment and economic growth in Africa*

David Fielding

Introduction

One of the main ways in which an economy can grow is through investment, and one of the main forms of investment is investment in physical capital (i.e. machinery, equipment and plants that can help to make work more productive). This can be foreign direct investment (FDI) – that is, capital expenditure by foreign-owned enterprises – or investment by local enterprises.

Past research has shown that a variety of factors are responsible for differences in FDI levels across countries, and for differences between overall levels of FDI in sub-Saharan Africa and FDI elsewhere in the Global South. Not all of these factors are within the control of African governments. For example, international companies seek the best deals for their shareholders, and Cleeve (2008) shows that they are attracted to countries whose governments offer the most generous tax breaks. Competition for FDI between developing countries results in a lower level of tax revenue for all. Solving this problem requires international policy coordination at a level that is probably going to be well beyond the capacity of most current and future African governments.

However, there are other factors that are within the control of governments and that are also important in attracting FDI: macroeconomic stability, the quality of infrastructure and services, and well-defined property rights. Moreover, these factors are also important in encouraging the links between foreign-owned firms and local firms through which international technology and know-how are transferred to the local economy (Amendolagine *et al.*, 2013). Asiedu (2004) shows that in the past, sub-Saharan Africa has lagged behind other parts of the Global South with respect to many of these factors; consequently, FDI in sub-Saharan Africa has been relatively low. Moreover, FDI in sub-Saharan Africa appears to have been less responsive to improvement in some of these factors than FDI in other parts of the world (Asiedu, 2002), and we do not yet fully understand why this is so.

Even in the absence of FDI, a country can achieve high growth through investment by local enterprises. In the short run, this local investment can be funded by borrowing from abroad, but in the long run it must be funded out of local saving, or else foreign debt will grow indefinitely large. In industrialised economies, this funding can take the form of equity, but in Africa, where stock markets are in their infancy, banks and other financial intermediaries perform a crucial role in economic development. By providing finance to the private sector, banks allow firms to take advantage of productive investment opportunities that may not otherwise materialise. There is a large body of evidence indicating that the development of banking systems goes hand in hand with economic growth (e.g. see Levine, 2004). This is a two-way relationship: financial development depends on economic growth (Demetriades and Hussein, 1996), but also a well-functioning banking system promotes growth (Demetriades and Andrianova, 2005).

According to traditional economic theory, banks can promote higher growth either by facilitating higher savings rates, and therefore more rapid capital accumulation (McKinnon, 1976), or by ensuring that resources are allocated to the most productive investment projects (Schumpeter, 1934). This second role is particularly important in emerging markets where information about individual investors is limited. By screening loan applicants, banks can help to address adverse selection in the credit market. By monitoring borrowers, banks can contain moral hazard; for example, they can deter excessively risky investment activity that would undermine a borrower's ability to repay a loan. This process not only helps banks to remain solvent, but also ensures that bank finance is channelled towards productive and sound investments.

There is some evidence that although financial development does lead to higher economic growth, this is not typically through higher savings rates (Beck *et al.*, 2000); such evidence has been cited in favour of a Schumpeterian interpretation of financial development. However, there is reason to believe that in sub-Saharan Africa – which includes countries with the lowest levels of financial development – the lack of a direct link between savings and growth is not simply due to large variations in the efficiency of resource allocation. A study by the World Bank (Honohan and Beck, 2007) has shown that historically, an unusually small fraction of African saving in banks has been channelled into domestic investment projects. This problem is manifest in two ways. First, an unusually large fraction of African bank deposits is held overseas. Honohan and Beck (2007) calculate that the average ratio of offshore deposits to domestic bank deposits in sub-Saharan Africa has been about 0.5, compared with about 0.3 in Latin America and below 0.2 in other parts of the world. Second, an unusually small fraction of African domestic bank deposits has been transformed into domestic bank lending. The average fraction in sub-Saharan Africa has been below 0.3, compared with over 0.7 in high-income countries and over 0.5 in other parts of the developing world. Correspondingly, the fraction of deposits channelled by African banks into liquid assets (0.2) and foreign assets (0.2) has been higher than in other parts of the world. That is to say, savings mobilisation does not appear to have represented a binding constraint on African banks' ability to lend locally. Instead, African banks have complained of a lack of creditworthy borrowers, while at the same time households and firms have complained about lack of credit.

In the rest of this chapter, we first of all review the recent growth performance of sub-Saharan Africa and its levels of FDI compared with other parts of the world. We also look at some of the key financial indicators highlighted by Honohan and Beck (2007), exploring whether they have improved or deteriorated over the last decade. We then investigate some of the underlying causes of changes in these aggregates, summarising recent research published in Andrianova *et al.* (2014). This paper examines the factors driving changes in individual African banks' lending behaviour over time. Finally, we review the implications of these results for public policy in the future.

Recent trends

Figures 37.1 and 37.2 present some recent figures on per capita GDP growth across sub-Saharan Africa over the 2000–2015 period. In Figure 37.1, each of the lines corresponds to an average level of per capita GDP for a different group of countries.[1] Averages are shown for the countries of the UEMOA[2] region (aggregating Benin, Côte d'Ivoire, Guinea-Bissau, Mali, Niger, Senegal and Togo), the countries of the CEMAC[3] region (aggregating Cameroon, the Central African Republic, Chad, Congo Republic, Equatorial Guinea and Gabon), and the rest of sub-Saharan Africa (SSA).[4] The UEMOA and CEMAC regions each have a single currency pegged to and convertible with the euro, and strong financial links with France, the former colonial power. This monetary system is likely to be associated with less national autonomy concerning macroeconomic policy, but greater financial stability. All of the countries of the CEMAC region (with the exception of the Central African Republic) are mineral exporters, while all of the UEMOA countries are mineral importers. Figure 37.2 shows averages for the annual percentage growth rate of per capita GDP, rather than the level, along with a line representing world average per capita GDP growth (figures are taken from World Bank, 2016).

Figure 37.1 illustrates some of the large variation in per capita income levels across Africa, with much higher income in the resource-rich CEMAC region and much lower income in the resource-poor UEMOA region.[5] However, as shown in Figure 37.2, annual growth rates are much more variable in the CEMAC region, where income is closely tied to highly volatile international mineral prices: in the last few years, mineral prices have been falling, and CEMAC income with them. With the exception of the CEMAC group, recent growth rates in sub-Saharan Africa have been slightly higher than the world average: this stands in contrast to previous decades, in which sub-Saharan African growth was consistently lower than the world average. To what extent does this difference correspond to improved investment performance?

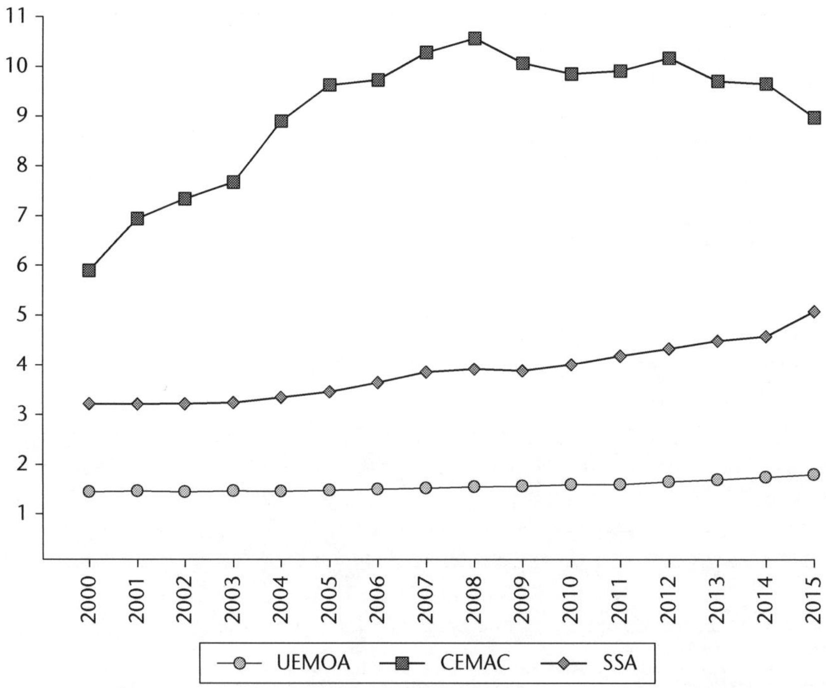

Figure 37.1 Average annual per capita GDP in PPP-adjusted 2011 US dollars for three African regions

Source: World Bank (2016).

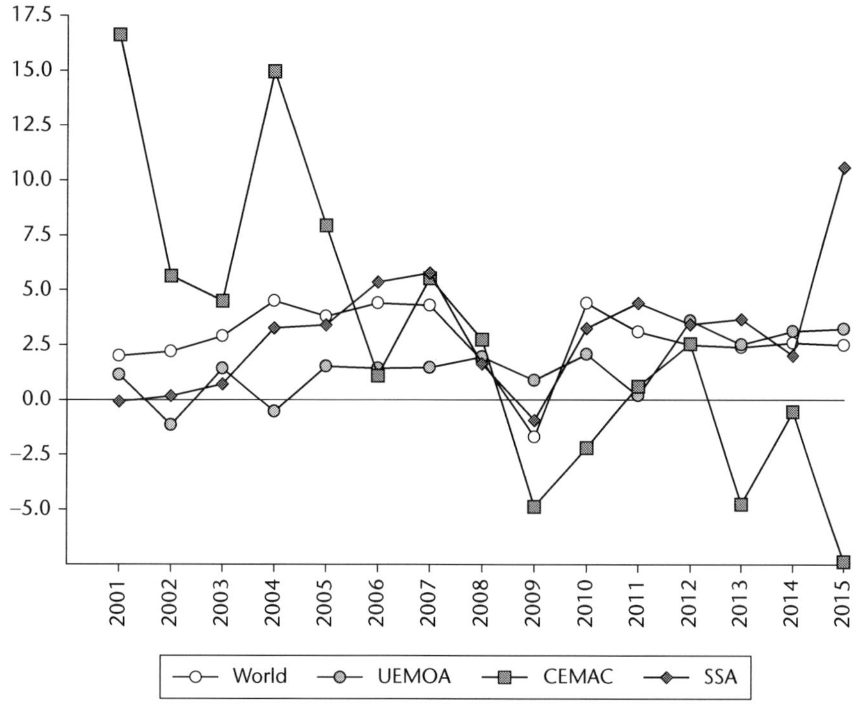

Figure 37.2 Average annual per capita GDP growth rates (in per cent) for three African regions

Source: World Bank (2016).

Figures 37.3–37.6 provide some information regarding investment over 2000–2015. Figure 37.3 illustrates the high cross-country variability of investment measured as gross fixed capital formation as a percentage of GDP, taken from individual countries' national accounts statistics. As the figure shows, some of this variability is attributable to differences in FDI and some is attributable to differences in investment by domestic firms.[6] Moreover, as shown in the scatter-plot in Figure 37.4, there is very little cross-country correlation between FDI and the domestic contribution to investment.[7] Each point in the scatterplot represents a different country, using the data from Figure 37.3, and it can be seen that on average, countries with relatively high levels of FDI have domestic investment levels that are neither relatively high nor relatively low: the least-squares regression line shown in the figure is almost horizontal. This suggests that the factors driving FDI that were discussed in the introduction are not necessarily the same as the factors driving investment by domestic firms. Nevertheless, as shown in Figures 37.5 and 37.6, there has been an upward trend in both FDI and overall gross fixed capital formation over the last 15 years, albeit with high volatility around this trend in the CEMAC region.

The factors driving recent changes in the levels of domestic investment have not yet been fully studied. However, it seems likely that one major factor has been a gradual deepening of banking systems in sub-Saharan Africa, as illustrated in Figures 37.7–37.10, which are based on data in the World Bank Global Financial Development Database (GFDD) (World Bank, 2017) and data published by the Bank of International Settlements (n.d.). First of all, Figure 37.7 shows a general upward trend in the ratio of bank deposits to GDP across Africa over 2000–2015, although the UEMOA and CEMAC countries lag behind the rest of sub-Saharan Africa, despite their financial stability. Broadly speaking, the UEMOA lags behind the sub-Saharan average by about 15 years and the CEMAC lags behind the UEMOA by about 10 years. However, this difference does not necessarily represent underperformance in the mineral-exporting CEMAC countries or a 'resource curse' effect: for a given level of financial development, mineral wealth

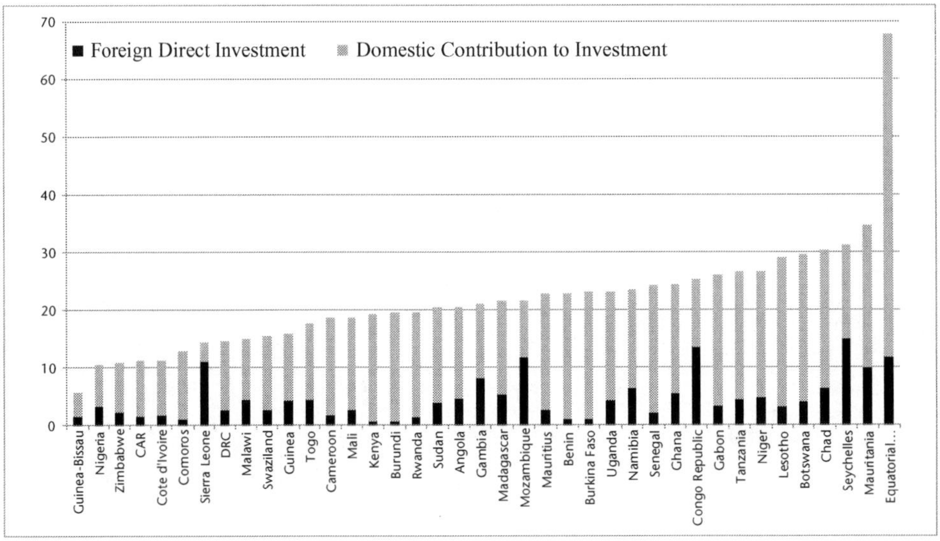

Figure 37.3 Gross fixed capital formation as a percentage of GDP, 2000–2015

Source: World Bank (2016).

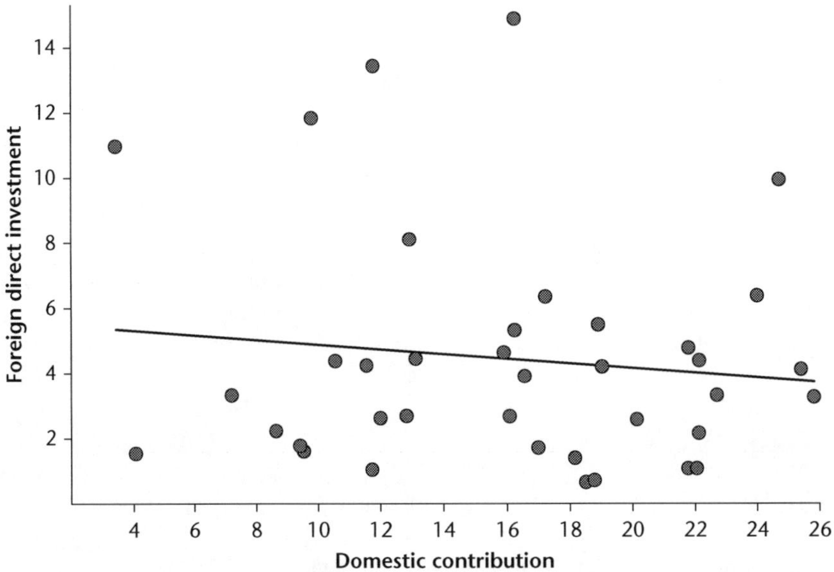

Figure 37.4 Foreign direct investment versus the domestic contribution to investment, 2000–2015

Source: World Bank (2016).

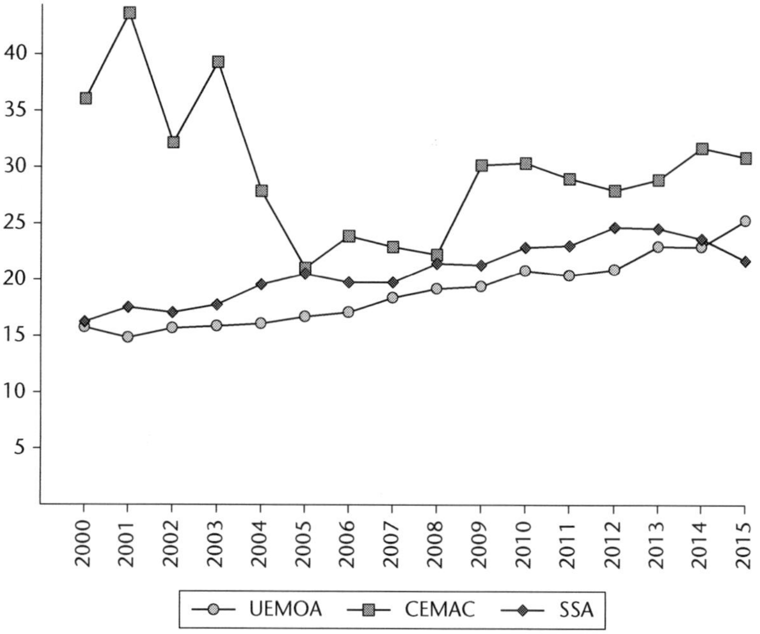

Figure 37.5 Average annual gross fixed capital formation as a percentage of GDP for three African regions

Source: World Bank (2016).

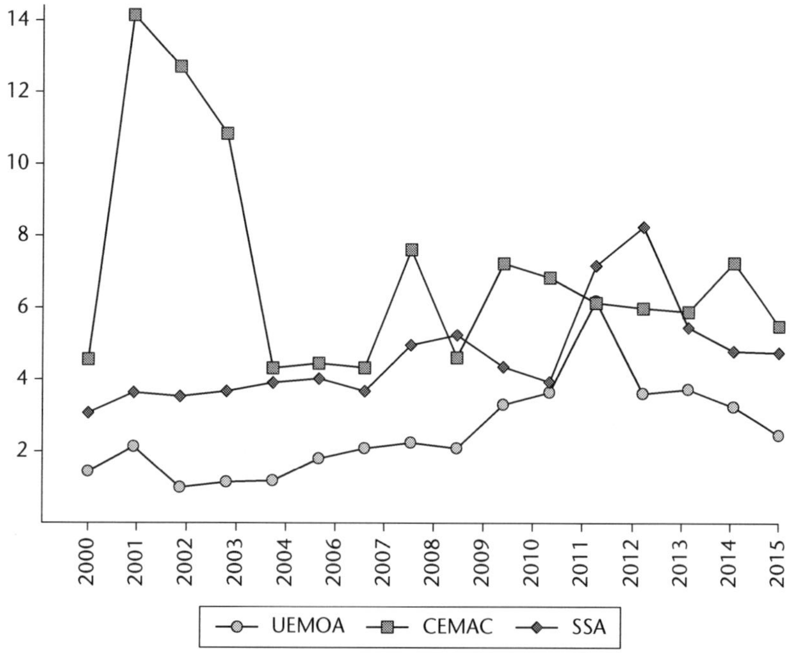

Figure 37.6 Average annual foreign direct investment as a percentage of GDP for three African regions

Source: World Bank (2016).

is likely to be associated with a higher level of per capita GDP, since a large proportion of mining sector capital is funded through FDI.

Overall, Figure 37.7 suggests a healthy propensity to save in Africa – at least on average across all countries. In all three regions shown in the figure, the increase in the ratio of bank deposits to GDP has been about 10 percentage points. However, Honohan and Beck (2007) have pointed out that a large fraction of sub-Saharan African bank deposits are overseas, so Figure 37.8 charts the ratio of overseas deposits to domestic deposits. Here, the trends are at least as strong: in both the UEMOA and CEMAC regions, the ratio has fallen from about 20–25 per cent to about 5 per cent, while in the rest of sub-Saharan Africa the ratio has fallen from about 55 per cent to about 35 per cent. Although this figure must be accompanied by a caveat – investors may have an incentive to hide overseas deposits in order to avoid fiscal scrutiny – it does appear that this obstacle to African financial development has diminished markedly over time.

Figure 37.9 plots the evolution of domestic commercial bank credit as a fraction of bank deposits – that is, the proportion of domestic saving in the banking sector that is channelled into lending. For the UEMOA and CEMAC regions, the picture is much more positive than the one reported by Honohan and Beck (2007), with 85–90 per cent of bank deposits being channelled into domestic credit (there was a dip in the ratio in the CEMAC region during the Global Financial Crisis, but it has since recovered). For the rest of sub-Saharan Africa, there has been a more modest improvement on the Honohan and Beck (2007) figures, with bank credit now at about 70 per cent of deposits, on average. In other words, although the rest of sub-Saharan Africa outperforms the UEMOA and CEMAC regions in terms of saving in the form of bank deposits, these regions outperform the rest of sub-Saharan Africa in terms of transforming deposits into domestic credit.

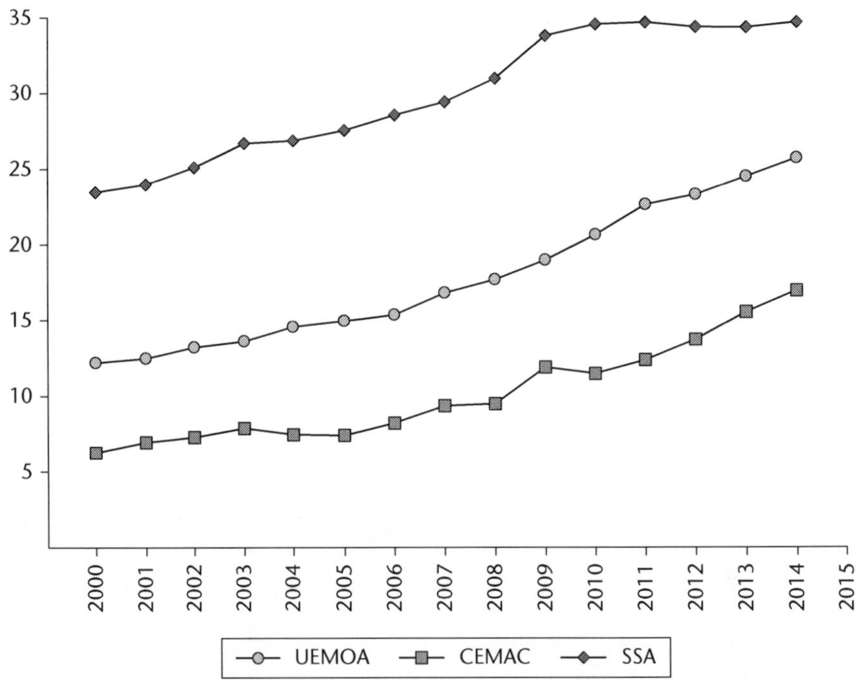

Figure 37.7 Average annual bank deposits as a percentage of GDP for three African regions

Source: World Bank (2017).

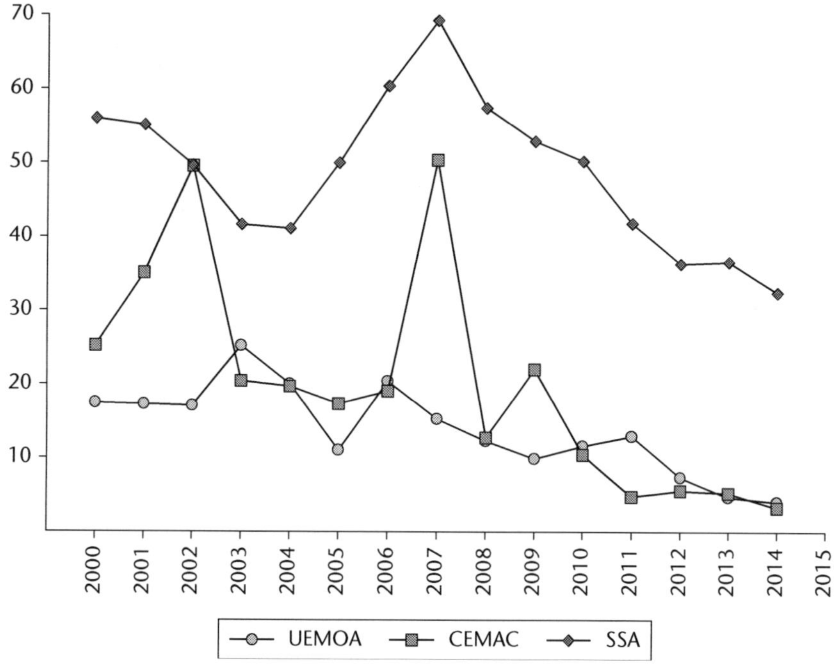

Figure 37.8 Average annual foreign bank deposits as a percentage of domestic bank deposits for three African regions

Source: Bank of International Settlements (n.d.).

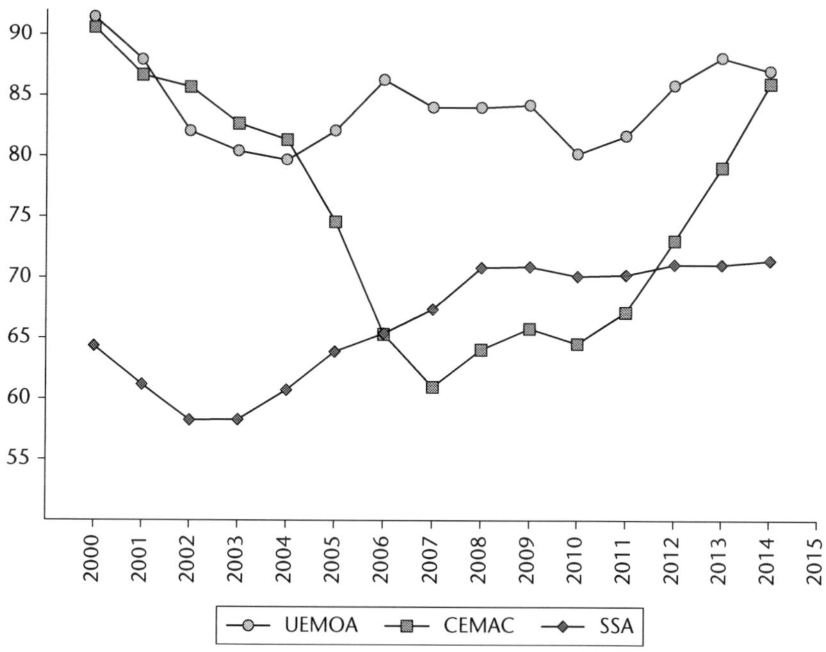

Figure 37.9 Average annual bank credit assets as a percentage of bank deposits for three African regions

Source: World Bank (2017).

These aggregate figures suggest some grounds for optimism: in SSA as a whole, there is evidence of financial sector growth. However, as we will see in the next section, there is a substantial amount of bank- and country-level variation underlying these aggregate figures, and this variation can be used to explore the factors that drive the fraction of domestic saving in bank deposits that are channelled into domestic lending.

Variations in bank asset structure

This section is based on the research presented in Andrianova *et al.* (2014), which proposes a theoretical model of commercial bank behaviour in developing countries, and reports results from a statistical analysis designed to test the main predictions of the theory. In the theoretical model, each bank faces the risk that some of its borrowers could default on their loans. A borrower has an incentive to default on a loan if the cost of repayment exceeds the expected cost of default (that is, the probability of the bank successfully taking action against the borrower multiplied by the penalty imposed in law). Some borrowers may be more likely to exhibit such behaviour than others, because, for example, they value their future reputation more. Banks can screen borrowers to try to determine this likelihood – for example, they can investigate each borrower's credit history – but such screening is likely to be costly. In the absence of low-cost screening technology, as is likely to be the case in a developing country, both the prevalence of loan default among borrowers and the quality of the legal system used to enforce loan contracts will affect the profitability of loans. If a country begins with a very poor contract enforcement system, then (with some variation across banks or across customers in the cost of screening) marginal improvements in the system or in the default rate will increase the fraction of bank assets channelled into business loans. However, as shown in Andrianova *et al.* (2014), there will be a threshold level of contract enforcement quality; above this level, neither further improvements in quality nor improvements in the proportion of opportunistic borrowers will have any effect on loan volumes.

Using data on individual African banks from Bankscope,[8] it is possible to test the prediction that there is such a threshold. The statistical analysis is based on annual observations of the loans-to-assets ratios of 194 banks in 32 countries over the period 1998–2008. It measures the extent to which changes in the following characteristics lead to changes in an individual bank's loans-to-assets ratio:[9]

- the past default rate on the bank's loans;
- bank age and size;
- GDP growth in the country in which the bank operates; and
- the quality of contract enforcement in the country in which the bank operates.

The quality of contract enforcement is measured by the regulatory quality index in the World Bank Worldwide Governance Indicators. Based on a variety of primary sources, this index 'captures perceptions of the ability of the government to formulate and implement sound policies and regulations that permit and promote private sector development' (World Bank, 2018). It is scaled so that across the world it has a mean of zero and a standard deviation of one.

The histograms in Figures 37.10–37.12 illustrate the variation in the three key characteristics in the Andrianova *et al.* (2014) analysis: the loans-to-assets ratio for each bank, the default rate it faces, and regulatory quality in the country in which it operates. It can be seen that there is a great deal of variation in individual banks' ratios: the tails of the distribution lie below 0.1 and above 0.9. The mean figures reported in sources such as Honohan and Beck (2007) are aggregates across banks that vary enormously in their behaviour. The same is true of default rates: while the model default rate is close to zero, a substantial minority of banks face default rates in excess of 50 per cent. Finally, although regulatory quality in Africa is poorer, on average, than in the rest of the world (the histogram in Figure 37.12 lies mostly below the zero line), there is some variation across time and across countries in Africa. At best, African regulatory quality is almost one worldwide standard deviation above the worldwide mean; at worst, it is over two standard deviations below the mean.

The statistical analysis shows that these characteristics are related, but in a way that reveals a threshold effect. First, among banks in countries with a relatively poor level of regulatory quality (below the world mean or worse), marginal improvements in both regulatory quality and the default rate have a significant impact on the loans-to-assets ratio. A one standard deviation increase in regulatory quality can be expected to raise the ratio by seven percentage points in the same year; a 10 percentage point reduction in the default rate can be expected to raise the ratio by two percentage points in the same year. In the long run, as banks adjust to the new conditions, these effects will double in size. Among the other characteristics, bank size and GDP growth also have significant effects. Next, consider banks in countries with a better level of regulatory quality. Among these banks, neither marginal changes in the default rate nor marginal changes in regulatory quality have any significant effect on the loans-to-assets ratio.

On average, banks in a relatively good regulatory environment are performing much better with respect to the loans-to-assets ratio than those in a relatively poor environment. In the relatively good environment (such as Botswana and Ghana, with regulatory quality > 0), there is an average loans-to-assets ratio of 63 per cent, which is within the normal range for other parts of the world. Our analysis suggests that neither further improvements in regulatory quality

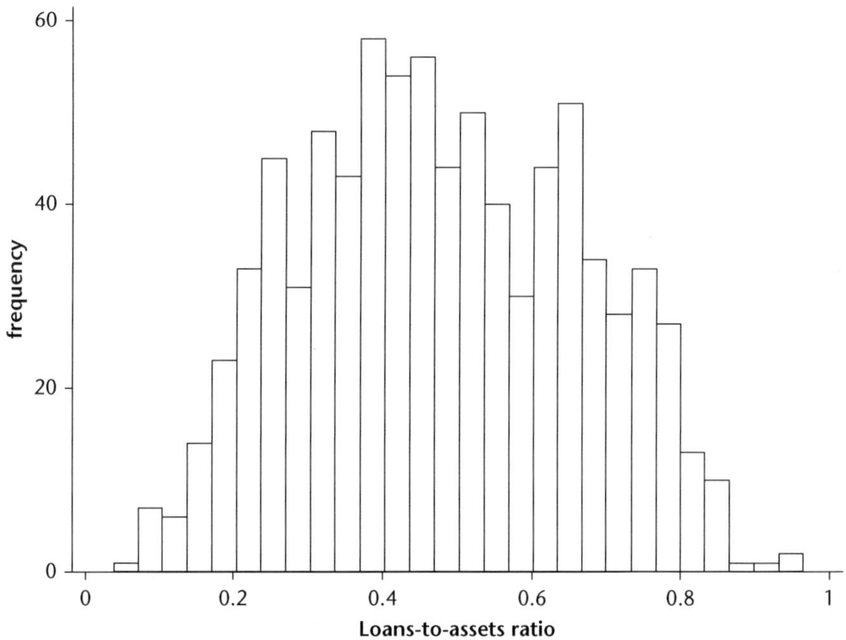

Figure 37.10 Distribution of the loans-to-assets ratio

Source: Andrianova *et al.* (2014).

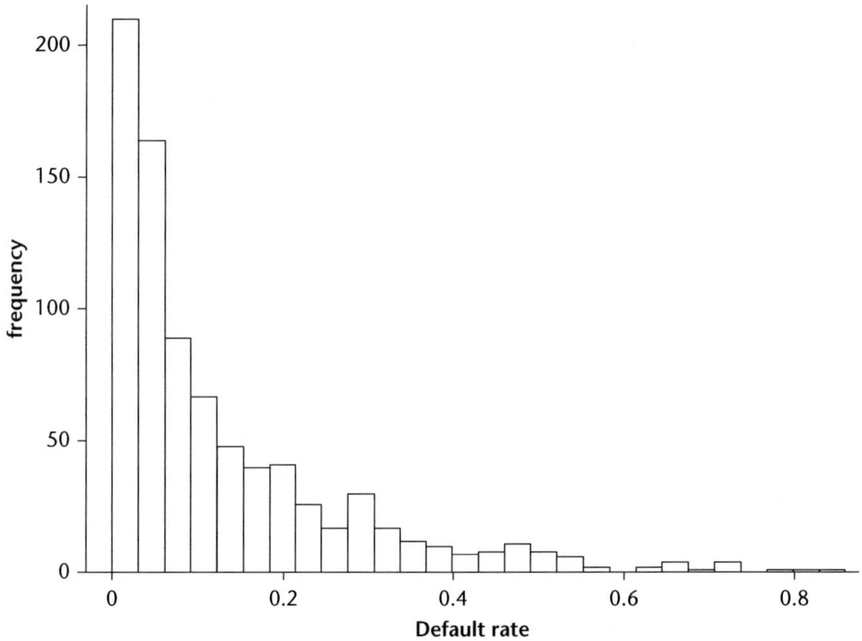

Figure 37.11 Distribution of the default rate

Source: Andrianova *et al.* (2014).

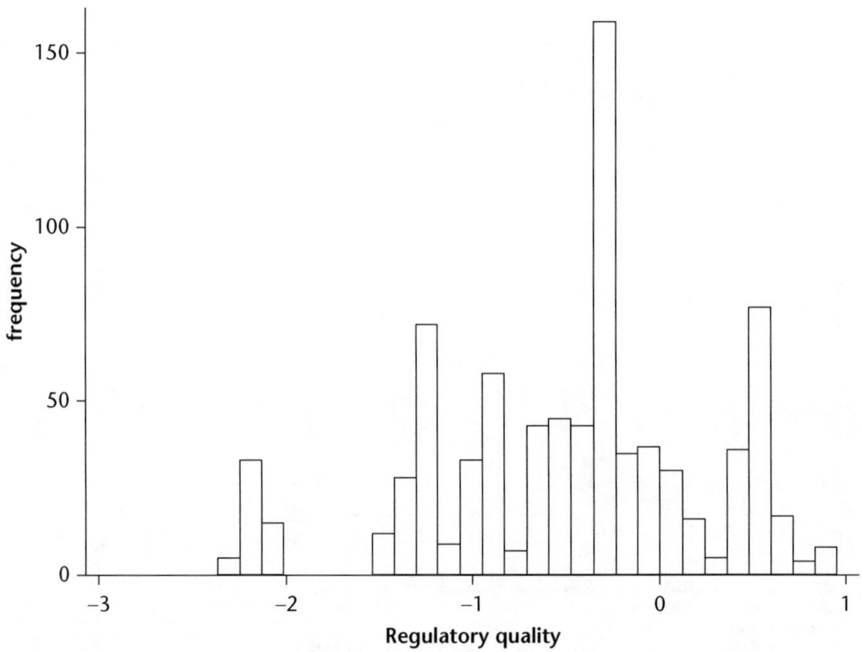

Figure 37.12 Distribution of regulatory quality

Source: Andrianova *et al.* (2014).

nor reductions in default rates will increase the loans-to-assets ratio of banks in this group of countries. This still leaves room for a great deal of variation in economic performance within the group. Botswana, for example, with its long history of channelling mineral revenues into productive investment in human and physical capital, has a per capita GDP of about US$15,000; Ghana, on the other hand, with more limited resources and a more recent history of good economic management, has a per capita GDP of only about US$4,000. But in either country, further economic growth is likely to depend on mobilisation of domestic resources rather than improved financial regulation.

By contrast, in countries with a relatively poor regulatory environment (such as Zambia and Zimbabwe, with regulatory quality < 0), the average loans-to-assets ratio is only 43 per cent. Again, there is substantial variation in economic performance within this group. Zambia has benefited from moderately efficient management of its mineral resources, at least in recent times, and has a per capita GDP of about US$3,600. Zimbabwe's ongoing political and economic woes correspond to a per capita GDP of about US$1,700 and falling. But my analysis suggests that across this group of countries, bank lending is sensitive to marginal changes in regulatory quality and default rates. Here, low savings rates are not necessarily the binding constraint on investment finance, and improvements in the regulatory environment are likely to create a large increase in the supply of loanable funds to the domestic private sector.

Implications for public policy

These results have some implications for policy related to financial development. First, it is not necessary – at least from the point of view of the banking sector – to have an A+ regulatory environment; a B− will work just as well. Very weak legal systems do appear to deter banks from lending locally, but mediocre ones do not. Reaching the historical international average is an appropriate goal for Africa, and incremental steps towards this goal can be beneficial. In anglophone Africa, for example, countries such as Botswana and Ghana, which have reached the historical international mean, are suitable role models for countries such as Zambia and Zimbabwe, which are still several standard deviations below this mean. In the UEMOA, some countries (Burkina Faso and Senegal) now have regulatory environments very close to the worldwide mean, as measured by regulatory quality, and average loans-to-assets ratios in excess of 60 per cent, as recorded in Bankscope. Other UEMOA countries (particularly Côte d'Ivoire, Guinea-Bissau and Togo) have regulatory environments that are somewhat poorer (though not the worst in Africa) and loans-to-assets ratios below 50 per cent. FDI into sub-Saharan Africa still accounts for only 2 per cent of worldwide FDI, and there is scope for African countries with improved regulatory environments to attract much more investment from both the industrialised West and emerging economies in the Far East.

Second, the diminishing rate of return to investment in regulatory quality highlights the potential for institutional investment in other dimensions of financial market quality, for example credit bureaux and other mechanisms that improve information on prospective borrowers (IMF, 2001; Sacerdoti, 2005). Information on credit bureaux in Africa is still too limited for any statistical analysis, but it seems likely that asymmetric information problems in Africa are at least as serious as in other parts of the developing world. There is still ample scope for central banks and other regulatory bodies to improve the conditions under which African banks lend.

Notes

* Part of this paper was presented at the Symposium of the Fiftieth Anniversary of the BCEAO in Dakar, and the author is grateful to symposium participants for their comments. It is partly based on work with Svetlana Andrianova (University of Leicester), Badi Baltagi (Syracuse University) and Panicos Demetriades (University of Leicester) (see Andrianova *et al.*, 2014).

1 GDP is measured in US dollars adjusted for price difference across countries and over time; the base year is 2011 and the base country is the USA.

2 The Union Economique et Monétaire Ouest Africaine.

3 The Communauté Economique et Monétaire de l'Afrique Centrale.

4 The averages are unweighted: each country is counted equally, regardless of its size.

5 There is also considerable variation in the magnitude of income inequality *within* countries, but this is beyond the scope of this chapter.

6 The domestic investment is measured as gross fixed capital formation less FDI.

7 The scatterplot excludes the point for Equatorial Guinea, which is one of the world's least developed countries but has quite recently discovered an unusually high level of mineral wealth, attracting an unusually high level of FDI per capita. The experience of Equatorial Guinea is atypical of the continent.

8 Bankscope was a proprietary database provided by the Bureau van Dyck until the end of 2016. It has now been replaced by Orbis Bank Focus (see Bureau van Dyck, n.d.).

9 The analysis in Andrianova *et al.* (2014) is based on banks' loans-to-assets ratio rather than the loans-to-deposits ratio depicted in Figure 37.9. The former is somewhat smaller than the latter, because total bank liabilities (which must equal total bank assets) include not just deposits, but also banks' borrowing from abroad and liabilities with respect to the government.

References

Amendolagine, V., Boly, A., Coniglio, N.D., Prota, F. and Seric, A. (2013) 'FDI and local linkages in developing countries: evidence from sub-Saharan Africa', *World Development*, 50: 41–56.

Andrianova, S., Baltagi, B.H., Demetriades, P. and Fielding, D. (2014) 'Why do African banks lend so little?', *Oxford Bulletin of Economics and Statistics*, 77: 339–359.

Asiedu, E. (2002) 'On the determinants of foreign direct investment to developing countries: is Africa different?', *World Development*, 30: 107–119.

Asiedu, E. (2004) 'Policy reform and foreign direct investment in Africa: absolute progress but relative decline', *Development Policy Review*, 22: 41–48.

Bank of International Settlements (n.d.) *Bank of International Settlements*, available at: www.bis.org (accessed 5 January 2018).

Beck, T., Levine, R. and Loayza, N. (2000) 'Finance and the sources of growth', *Journal of Financial Economics*, 58: 261–300.

Bureau Van Dijk (n.d.) *Orbis Banks Focus*, available at: www.bvdinfo.com/en-apac/our-products/company-information/international-products/orbis-banks (accessed 5 January 2018).

Cleeve, E. (2008) 'How effective are fiscal incentives to attract FDI to sub-Saharan Africa?', *Journal of Developing Areas*, 42: 135–153.

Demetriades, P. and Andrianova, S. (2005) *Sources and Effectiveness of Financial Development: What We Know and What We Need to Know*, Research Paper RP2005/76, Helsinki: World Institute for Development Economic Research.

Demetriades, P. and Hussein, K. (1996) 'Does financial development cause economic growth? Time-series evidence from 16 countries', *Journal of Development Economics*, 51: 387–411.

Honohan, P. and Beck, T. (2007) *Making Finance Work for Africa*, Washington, DC: World Bank.

IMF (International Monetary Fund) (2001) *Senegal: Financial System Stability Assessment. Country Report 01/189*, Washington, DC: IMF.

Levine, R. (2004) *Finance and Growth: Theory and Evidence*, Working Paper 10766, Cambridge, MA: NBER.

McKinnon, R.I. (ed.) (1976) *Money and Finance in Economic Growth and Development: Essays in Honor of Edward S. Shaw*, New York: Marcel Dekker.

Sacerdoti, E. (2005) *Access to Bank Credit in Sub-Saharan Africa: Key Issues and Reform Strategies*, IMF Working Paper WP/05/166, Washington, DC: IMF.

Schumpeter, J. (1934) *The Theory of Economic Development*, translated by R. Opie, Cambridge, MA: Harvard University Press.

World Bank (2016) *World Development Indicators*, available at: https://data.worldbank.org/data-catalog/world-development-indicators (accessed 5 January 2018).

World Bank (2017) *Global Financial Development Database (GFDD)*, available at: www.worldbank.org/en/publication/gfdr/data/global-financial-development-database (accessed 5 January 2018).

World Bank (2018) *Worldwide Governance Indicators*, available at: http://info.worldbank.org/governance/wgi/#home (accessed 5 January 2018).

38
Agriculture

Roy Maconachie

Introduction

Although in recent decades the African continent has undergone an 'urban transformation' (see Section 8 of this volume), most Africans today continue to live in rural areas. The majority of Africans, in fact, derive their livelihoods as primary producers, often engaging in complex and diverse portfolios of food production activities that include farming, fishing, rearing livestock, and hunting and gathering. Across the continent, rural life continues to centre around agriculture, with some estimates suggesting that the sector employs as much as 65 per cent of Africa's labour force and accounts for 32 per cent of Africa's gross domestic product (GDP) (World Bank, 2013).

Discussions that concern agriculture in Africa are often subdivided into two broad typologies: (1) small-scale indigenous agriculture; and (2) large-scale commercial production. In this chapter, both scales of production will be explored, placing particular emphasis on understanding how agriculture shapes and interacts with different development processes. Indeed, the contribution of the agricultural sector to the GDP of African economies varies greatly from country to country. For example, World Bank (2016) estimates suggest that in 2014, the sector represented 4.8 per cent of GDP in Congo Republic, and less than 3 per cent of GDP in Botswana and South Africa. But alternatively, in Chad, agriculture contributed 52.6 per cent to GDP, and in Sierra Leone it accounted for 54.1 per cent. At the same time, however, in many countries, agriculture is a significant employer, assumes a critical role in the economy, and therefore has the potential to contribute considerably to important development priorities, including food security and the eradication of hunger and poverty, the stimulation of intra-African trade and investment, the development of transportation networks and infrastructure for economic diversification, sustainable resource and environmental management, and employment and wealth creation.

This chapter begins with a critical overview of the small-scale agriculture sector, which provides the main source of food for rural households and often makes up the bulk of the labour force in African countries, but does not usually contribute a large amount of revenue to government coffers. Here, the 'diversity, flexibility and adaptability' of smallholder food production systems, and the resilience of small-scale farmers, are explored in the wake of the significant structural challenges that subsistence farmers must endure. This gives way to a discussion of the changing agrarian dynamics of rural landscapes across the African continent, and more specifically a renewed interest in large-scale commercial agriculture in recent years. While large-scale mechanised agriculture focusing on cash crop production has played a key role in the economies of many African countries since colonial times, some observers have demonstrated recent concern about the way that foreign investment in the sector is reproducing 'neocolonial' processes of accumulation and dispossession in new contexts (see Bernstein, 2010; Li, 2010; Hall *et al.*, 2011). The rapid expansion of the production of bioenergy feedstocks on land that would have formally been used to grow food crops is explored to understand the contradictory impacts of globalisation on food production, and how the sustainability of agriculture in Africa is being shaped and challenged by new global processes and agendas.

Small-scale indigenous agriculture

More so than on any other continent, African food production is dominated by small, family-run farms that are usually dependent on family farm labour. Recent estimates suggest that Africa has 33 million family farms of less than 2 hectares, which account for 80 per cent of all farms (NEPAD, 2013). Such agriculture tends to be labour- rather than capital-intensive, and focuses on subsistence production. At the same time, however, African smallholders are innovators and often highly entrepreneurial, and many are able to intercrop cash crops with staple crops for home consumption. Detailed research carried out in Kono District in Sierra Leone, for example, has demonstrated how farming communities in close proximity to the country's diamondiferous areas produce large quantities of food crops for sale to the mining population, frequently reinvesting their earnings in their homes, families and, most notably, in the expansion of cash crops such as coffee, cocoa, kola nuts and citrus fruits (Maconachie and Binns, 2007; Maconachie, 2011).

Although there have been previous attempts to classify different types of small-scale food production systems, each individual system is varied and complex, and this makes classification challenging (Binns *et al.*, 2012). Farming systems may be extensively cultivated (e.g. shifting cultivation), or intensive, permanently cultivated systems (e.g. rotating fallow systems). The landscapes where farming takes place may also be geographically diverse, with production occurring in a range of ecological settings, including dryland environments (Mortimore, 1989), wetland areas (Dixon, 2003), tropical forests (Tegegne *et al.*, 2016) or woodland savannas (Fairhead and Leach, 1996). African farmers may have no alternative but to adapt to their settings, and various techniques for enhancing yields and safeguarding the environment are widely employed in food production systems, including crop rotation, intercropping, bush fallowing, agroforestry, and the integration of farming and pastoralism.

In some parts of Africa – such as in vast stretches of Ethiopia – mixed farming is practised by smallholders, whereby the integration of different crops and animal husbandry is common. In other regions of the continent where land and agricultural inputs such as chemical fertiliser are exceptionally scarce, farmer–herder linkages have assumed an essential role in maintaining soil fertility and crop output. In the case of northern Nigeria, for example, a symbiotic relationship has long existed between farmers and pastoralists, where cattle are permitted to graze on crop refuse, and in the process deposit manure on farm plots (Mortimore, 1989; Mortimore and Adams, 1999). Here, drought-resistant varieties of sorghum and millet are intercropped with nitrogen-fixing crops, such as groundnuts and cow peas, to further boost soil fertility. One of the most important mixed farming systems in Africa – the 'maize mixed' system – is common in eastern and southern Africa, and is characterised by the intercropping of maize with cash crops such as coffee, cotton or tobacco (Romney *et al.*, 2003). Meanwhile, in North Africa, estimates suggest that some 30 per cent of those involved in agriculture adopt the 'highland mixed' system, where cereal crops (mainly wheat and barley) are intercropped with legumes, fruit and olives (Binns *et al.*, 2012).

While food production systems and the crops produced vary considerably from country to country, it is clear that African smallholders draw upon a variety of farming strategies that directly respond to local social, economic and ecological conditions, while at the same time maximising returns and spreading risk. However, until relatively recently, there has been limited academic interest in understanding the logic and value of African indigenous knowledge in food production systems. Early research on West African rice farming systems carried out in the 1980s, however, suggested that effective agricultural development in resource-poor regions must be based on a respect for indigenous farmers' knowledge and understanding of the environment (Richards, 1985, 1986). Likewise, Dixon's (2003) research on wetland agricultural systems in Ethiopia demonstrated the detailed understanding that local farmers possess of environmental processes, crop selection and sustainable management practices. More recently, researchers have taken an interest in farmers' choice of cropping systems and sowing times in the context of adaptation strategies to climate change (Waha *et al.*, 2013). In each of these cases, such indigenous knowledge is often based on innovation and local experimentation, and is 'embedded' in specific cultural contexts (Richards, 1986). However, traditionally, this type of knowledge has been dismissed as 'primitive, disorganized and unproductive', and inferior to Western scientific knowledge (Binns *et al.*, 2012).

Although farmers must respond and adapt to micro-level conditions, including seasonal patterns of rainfall and temperature, small-scale food production systems in Africa are increasingly being affected by a wider range of factors, including population pressure, proximity to growing urban centres, and more recently global environmental change. Most notably, the role that population plays in shaping smallholder food production systems has long been at the centre of debates that concern the relationship between people and the environment. Gaiser *et al.* (2011) suggest that as rising population densities have put increasing pressure on land under cultivation, some farming systems have become more 'fixed', and fallow periods have become shorter or have disappeared completely. Depending on the availability of land, African smallholders may have to make difficult choices, and face trade-offs between labour-intensive cultivation on small plots or less intensive cultivation on lengthy fallows to regenerate soil fertility (Stock, 2013).

As population densities have continued to increase significantly across the African continent and pressure on land and resources has intensified, much research into African farming systems has been influenced by either the so-called 'optimistic' or the 'pessimistic' school. On the pessimistic side, Malthusian logic has argued that population pressure is detrimental to agriculture and a fundamental cause of environmental degradation. However, on the other side of the debate, following the logic of the 'Cornucopian school', other researchers have convincingly argued that population densities can lead to environmental recovery (Tiffen *et al.*, 1994) and the conditions for 'agricultural growth' (Boserup, 1965).

In the densely populated Kano 'close settled zone' in northern Nigeria, the Boserupian hypothesis has received support from numerous studies at the village and district level, and has greatly influenced much of the research done in the region (Mortimore and Adams, 1999). Many studies that support the 'sustainability' of Kano's intensively cultivated close settled zone derive their foundations from the widely cited work of Boserup (1965), who documented how in response to the growth of rural population densities, farmers shorten their fallow periods, increase the use of technological innovations such as the plough, and adopt multiple cropping cycles to make the land more productive and raise its carrying capacity (Jolly and Torrey, 1993: 3). In short, as Boserup suggests, rather than being a limiting factor, population growth can actually be a prerequisite for agricultural development, recognising that increased pressure on the land can in some cases lead to a reduction in degradation and erosion through various innovations, rather than an increase (see also Tiffen *et al.*, 1994, who came to similar conclusions in their study in Machakos, Kenya).

In most parts of sub-Saharan Africa, there is now also an increasing acknowledgement of the primary role that women play in food production systems, with some estimates suggesting that, on average, they carry out as much as 70 per cent of all agricultural work (Doss and Sofa Team, 2007). For example, statistics from the FAO suggest that in 2010, 55.1 per cent of economically active women in Botswana were involved in agriculture; in Sudan, 65.1 per cent; in Gambia, 86.5 per cent; and in Mozambique, 94 per cent (FAO, 2011). However, in many African countries, patriarchal social norms and the gendered division of labour in agriculture can make the balance between farming activities and off-farm work challenging for women, particularly at certain times of the agricultural calendar. On the domestic front, women bear a disproportionate work burden that can result in 'time poverty'. Women in rural areas in particular are caught up in what has been referred to as a 'triple work burden' across productive, reproductive and social spheres. Not only are rural African women responsible for time- and labour-intensive domestic activities – including collecting and carrying firewood and water (often for long distances), food preparation, cleaning and washing clothes – but they must also simultaneously look after their children and work on their farms. In the poorest households where there is very limited access to assets or additional labour, a woman's time poverty can have a considerable impact on the entire family, having implications for food security, nutrition and health (see also Momsen, 2010).

Large-scale commercial agriculture

Although smallholder farmers constitute the majority of the labour force in rural settings, since the colonial era, large-scale commercial agriculture has also played a vital role in the economies of many African countries. As a consequence of the Berlin Conference of 1884–1885 and the ensuing 'scramble for Africa' at the end of the nineteenth century, European countries established colonies in order to gain access to abundant sources of raw materials and food, as well as pools of cheap, flexible labour (Rodney, 1972). This had a drastic impact on the economies of colonies, which were transformed from being self-sufficient and food-secure to being heavily dependent on the production of commodities for export to foreign markets. Binns *et al.* (2012: 316) go as far as to suggest that during this period, Africa was viewed as a 'giant farm' that was used to produce a host of export crops such as cotton, cocoa, sisal, groundnuts, tea, coffee or tobacco. Expatriate-owned plantations were common in many of the former French colonies, such as Côte d'Ivoire and Guinea, where coffee, cocoa and bananas were grown for export back to France. Meanwhile, in Liberia, rubber was produced on vast concessions owned by the American company Firestone, and the Germans and Belgians established rubber, oil palm and coffee plantations in central Africa (Cameroon and Zaire). White settler farms were established in Kenya, Zimbabwe and South Africa, where a wide range of crops were produced on a commercial scale, including coffee, tea, sugar, cotton, tobacco and pyrethrum (Rodney, 1972).

Today, the legacies of agricultural transformations that took place in many former African colonies are still apparent. Indeed, in some countries in sub-Saharan Africa, more than half of their foreign revenue comes from the export of just a few agricultural commodities. As such, drastic shocks to global commodity prices can have a significant impact on individual incomes, which in turn can affect the levels of poverty and well-being of a country's population (Addison *et al.*, 2016). Diversifying agricultural production and exports, it has been argued, can help to reduce the dependence that single-product economies face, mitigating the social and economic impacts on populations that are caused by commodity price fluctuations (Bowman and Zilberman, 2013; Hazell, 2013).

During the 1970s and 1980s, the development of large-scale capitalist agriculture projects was encouraged in a number of African countries. In some cases, these were established by African governments, and in other cases they were initiated and operated by foreign companies. During the structural adjustment years of the 1980s and 1990s, 'conditionalities' imposed by the World Bank and IMF further stimulated privatisation within the agricultural sector, as well as currency devaluation and the removal of subsidies and tariffs on agricultural products. This had drastic consequences on small-scale agriculture, making it difficult for smallholders to acquire agricultural inputs such as fertiliser and pesticides. Swindell and Illiya (1992) add that in some cases, structural adjustment pressures increased the concentration of land ownership in the hands of wealthy farmers, causing peasant landlessness to 'skyrocket'.

During the years of austerity under structural adjustment, Nigeria, for example, privatised parastatals and encouraged foreign companies to invest in agricultural plantations to produce a wide range of crops such as cotton, oil palm, grain and groundnuts. At the time, the focus was on modernising agricultural production and providing wage labour for local populations. Ironically, however, decades later, these early experiments with large-scale agriculture are now being replicated across Africa in contemporary contexts, as markets have been liberalised and 'opened up' to foreign investment. Most notably, on all corners of the continent, emerging countries (e.g. China, India and South Korea) are engaging in a new 'African land rush' to acquire fertile farmland to produce crops for ensuring food self-sufficiency at home (see also Batterbury and Ndi, Chapter 43, this volume). On a global scale, estimates suggest that 115 million acres of farmland are currently being leased by foreign investors, with the bulk of investment being in Africa (Pena, 2012).

Of course, the marginalisation of smallholder agriculture in favour of large-scale production has not always been a product of external forces from outside Africa. In some cases, it has come from internal factors within the continent. For example, during the apartheid years in South Africa, internal food self-sufficiency was vital to counter the draconian international sanctions placed on the country. In light of this scenario, a well-developed agricultural sector was established in South Africa that was dominated by commercial, mechanised production undertaken by white farmers. Vast stretches of the most fertile agricultural lands were occupied by white farmers, while black Africans who were not providing labour on commercial farms were forced to undertake subsistence production in the overcrowded and less fertile 'homelands' (Bernstein, 2013).

Although South Africa's commercial farms were founded on the back of the inequality associated with apartheid, the country's reputation as a leading agricultural country has continued to this day. South Africa is now internationally recognised for its export of agricultural produce, including high-value-added products such as wines produced in the Cape region or sustainably harvested flowers produced in 'Flower Valley' (Blokker *et al.*, 2015). However, Stock (2013) reports that despite efforts at land reform and redistribution, the structure and ownership of the South African agricultural sector has changed little since the end of apartheid. In their study of the legacies of 'agrarian dualism' in South Africa's Limpopo province, research carried out by Hall *et al.* (2012) focuses on the significant population of black farm labourers who still work and live on white-owned commercial farms in the post-apartheid era. They note that although the employment generated by commercial agriculture remains vital for a considerable proportion of the rural poor, farm labourers living on privately owned farmland often lack the opportunity to use the land independently in order to support their 'precarious' livelihoods.

In an era of rising food prices, the governments of many African countries have made food security through commercial agriculture a top development priority. At the forefront of this agenda has been the promotion of foreign investment, with large land leases for export-oriented plantation agriculture occupying a central place in development strategies. This approach, shaped by Rostovian discourses around 'modernisation' and 'progress', is embedded in the strongly held development narrative that African countries are destined to experience an agrarian transition similar to that of Europe in the eighteenth century (Li, 2009), if they pursue the right path to development.

A central aspect of this trajectory, according to the World Bank (2007), is for governments to stimulate the transition by removing the barriers to agricultural investment. Such a market-led approach for private sector development of commercial agriculture is based on the assumption that the private sector drives the organisation of value chains that bring the market to smallholders and commercial farms, a model that the World Bank has referred to as 'Agriculture for Development' (World Bank, 2007). According to the World Bank (2013), higher agricultural outputs and sustained growth will therefore require attention to five core areas of public action:

- facilitating agricultural markets and trade;
- improving agricultural productivity;
- investing in public infrastructure for agricultural growth;
- reducing rural vulnerability and insecurity; and
- improving agricultural policy and institutions.

In addition to its potential for providing food for home consumption, the commercial agricultural sector also generates employment and can be a catalyst for off-farm activities and local economic development. In recent years, this has been recognised by a wide range of donors and international development organisations, many of whom believe that if the agricultural sector is sufficiently developed, it can serve as an engine of economic growth (Dercon and Gollin, 2014). The New Partnership for African Development (NEPAD) and the African Union (AU), for example, initiated the Comprehensive Africa Agriculture Development Programme (CAADP) in 2003 to stimulate agriculture-driven development (Poulton *et al.*, 2014). CAADP is a pan-African framework that aims to help countries critically review their own agricultural economies and identify future investment opportunities. In trying to stimulate growth and address poverty reduction through agriculture, the framework is primarily concerned with policies that focus on job creation and increasing opportunities, especially for women and youth, food security and improved nutrition, and strengthening resilience. CAADP annual targets focus on achieving 6 per cent growth in agricultural GDP, and an allocation of at least 10 per cent of public expenditures to the agricultural sector. As of March 2015, 41 AU member states have signed CAADP compacts, 33 of which have developed formal national agriculture and food security investment plans (UN, 2016). As a result, UN reports suggest that since the launch of CAADP, public agricultural expenditures have almost doubled, having risen by more than 7 per cent per year across the continent (UN, 2016).

Although the potential of commercial agriculture for driving development in African countries has been widely recognised by donors, NGOs and international development organisations, the sector has also been highly criticised for its uneven impacts. One area where there has been considerable debate over the impacts of globalisation on agriculture in Africa concerns the rapid expansion of agribusiness into biofuel production, with its apparent 'win–win' formula of reducing greenhouse gas emissions without reducing consumption. For some, in the face of high and ever-rising conventional energy costs, biofuel investment represents 'a new profitability frontier' (Borras *et al.*, 2010: 576). And yet it has been at the forefront of enormous controversy that has brought to the fore many of the very issues that have long been at the heart of critiques of industrialised agricultural expansion in Africa. In the final section of the chapter, some of these debates are briefly explored.

The globalisation of agriculture: bioenergy production in Africa

Over the past two decades, there has been a dramatic growth in the production of crops to be used for bioenergy across sub-Saharan Africa, with some estimates suggesting that global commercial biofuel production has increased fivefold (Earth Policy Institute, 2010). Such an expansion of high-intensity industrial agriculture into energy feedstocks, such as sugar cane or oil palm, is the latest development in the ongoing expansion of globalised agricultural production.

At the same time that biofuel production in sub-Saharan Africa has reached unprecedented levels in an era of global economic recession and rising food prices, the sustainability of policies that have supported the growth of the biofuel sector is increasingly being questioned. Although long hailed as a panacea for mitigating climate change and reducing fossil fuel dependence, biofuels' 'green credentials' have come under attack: the land-grabbing associated with their production is now being linked to increased food prices and hunger, incidences of land conflict, and rising CO_2 emissions (see Maconachie and Fortin, 2013).

Proponents of bioenergy investments in Africa have often claimed that they will stimulate rural development by way of direct and indirect 'trickle-down' effects to local communities (de Nie *et al.*, 2009). For example, as summarised by Richardson (2010: 921), who specifically considers the sugar industry in Zambia, 'agro-exports can make an *indirect* contribution to rural development by promoting national economic growth and resilience', and investment in export crops for bioenergy can *directly* support rural development by 'increas[ing] on-farm and off-farm employment and revenue' (original emphasis). This has led to a number of key international financial institutions, including the World Bank, encouraging countries in sub-Saharan Africa to support directly or indirectly the expansion of the production of bioenergy both for domestic consumption and for export (World Bank, 2007). In turn, trade agreements relating to feedstocks for bioenergy have been adopted not only between developed and developing countries, but also between developing and developed countries (Dauvergne and Neville, 2009).

However, the expansion of bioenergy 'for development' in Africa has frequently come under intense criticism (Magdoff, 2008; McMichael, 2009a). The nub of the biofuels controversy for many critics has come down to a 'food versus fuel' debate, characterised by the diversion of food crops from African countries to 'feedstocks' to fuel overconsumption in the developed world. The scale and speed of this conversion has been unprecedented and is often cited as a contributing factor to rocketing food prices across the Global South (Bailey, 2007; McMichael, 2009a). In 2008 and again in 2011, countering their potential contribution to rural development were significant spikes in the prices of food commodities that had a considerable impact on poverty across the African continent. Such spikes have been linked to the dramatic rise in the production of bioenergy feedstocks on land that formerly would have produced food (McMichael, 2010). When the rapid expansion of global agribusiness into biofuels in Africa is underpinned by developed countries' ongoing overconsumption of energy, such a critique of biofuels is thrown into stark perspective.

Resulting from this proliferation of biofuels production has been a dramatic increase in foreign investments in land in developing countries, land that might have been available for growing domestic food crops to ameliorate the situation (Cotula *et al.*, 2008; GRAIN, 2008; Cotula *et al.*, 2009). This is particularly the case in sub-Saharan Africa, which, as McMichael (2009b: 243) notes, has now been dubbed the 'green OPEC' because its extensive land reserves have attracted agrofuel capital from a vast range of investors from Brazil, Saudi Arabia, China, the World Bank, USAID, the European Commission, and various private companies. While a recent World Bank (2011) report entitled *Rising Global Interest in Farmland: Can It Yield Sustainable and Equitable Benefits?* suggests that large-scale acquisitions in land can stimulate the employment of wage workers and contribute to poverty alleviation, such arguments have spawned a series of critiques within the global 'land grab' debate (e.g. see Li, 2011). Many land purchases have been controversial, even in cases where such land has not formerly been used to produce food, but has been defined as 'idle', 'marginal' or 'degraded'.

In the case of the West African country of Sierra Leone, for example, claims by bioenergy companies that their projects are located on marginal land are highly contentious, and do not demonstrate an effective understanding of land use under smallholder farming systems. Upland rice farming – the main food production system in rural areas in Sierra Leone – is based on a system of shifting cultivation, whereby there is a need to frequently relocate to new 'idle' land. This decreases the willingness of landowning farmers to grant land to outsiders on a secure basis, due to the necessity of having those lands on reserve for their own shifting cultivation needs (Unruh and Turray, 2006). What may be perceived to be 'idle' land by the outsider is most often far from the case. Moreover, the local environmental consequences resulting from the continuously cultivated and monocropped plantation agriculture favourable to the commercial production of biofuels are that they adversely affect soil quality and fertility, biodiversity, and water availability and quality (Magdoff, 2008).

In summary, it is apparent that the political-economic drivers that have fuelled global biofuel expansion are complex, contradictory and framed by multiple discourses. Bioenergy feedstocks, such as sugar cane or oil palm, are typical plantation crops of the colonial era, and critics point to the danger of neocolonial processes of accumulation and dispossession simply being reproduced in a new context (Bernstein, 2010; Li, 2010; Hall *et al.*, 2011). Consequently, for some scholars, the global political economy that has emerged around biofuels, and large-scale agricultural deals more broadly, appears set to intensify inequalities in many African countries and 'reinforce and extend previous waves of livelihood displacement' (Dauvergne and Neville, 2010: 632). Other positions, on the other hand, seem more optimistic, pointing out that emerging alliances in support of commercial farming in Africa will allow some actors to reap considerable benefits (e.g. see World Bank, 2011), although these will undoubtedly be unequal and mediated by power relationships. Only time will tell whether or not the latest agricultural transformations taking place across the African continent will leave poor countries vulnerable and exposed to the uneven processes of globalisation, or whether they will serve as a positive driver for change, and a catalyst for growth and socio-economic development in the years to come.

References

Addison, T., Ghoshray, A. and Stamatogiannis, M. (2016) 'Agricultural commodity price shocks and their effect on growth in sub-Saharan Africa', *Journal of Agricultural Economics*, 67(1): 47–61.

Bailey, R. (2007) *Bio-Fuelling Poverty: Why the EU Renewable-Fuel Target May Be Disastrous for Poor People*, Oxfam Briefing Note, Oxford: Oxfam International.

Bernstein, H. (2010) *Class Dynamics and Agrarian Change*, Halifax: Fernwood Publishing.
Bernstein, H. (2013) 'Commercial agriculture in South Africa since 1994: "natural, simply capitalism"', *Journal of Agrarian Change*, 13: 23–46, DOI: 10.1111/joac.12011.
Binns, T., Dixon, A. and Nel, E. (2012) *Africa: Diversity and Development*, London: Routledge.
Blokker, T., Bek, D. and Binns, T. (2015) 'Wildflower harvesting on the Agulhas Plain, South Africa: challenges in a fragmented industry', *South African Journal of Science*, 111(11/12): 63–69, DOI: 10.17159/sajs.2015/20140160.
Borras, J.S.M., McMichael, P. and Scoones, I. (2010) 'The politics of biofuels, land and agrarian change: editors' introduction', *Journal of Peasant Studies*, 37: 575–592.
Boserup, E. (1965) *The Conditions of Agricultural Growth*, London: Allen & Unwin.
Bowman, M. and Zilberman, D. (2013) 'Economic factors affecting diversified farming systems', *Ecology and Society*, 18(1): 33, DOI: http://dx.doi.org/10.5751/ES-05574-180133.
Cotula, L., Dyer, N. and Vermeulen, S. (2008) *Fuelling Exclusion? The Biofuels Boom and Poor People's Access to Land*, London: FAO/IIED.
Cotula, L., Vermeulen, S., Leonard, R. and Keeley, J. (2009) *Land Grab or Development Opportunity? Agricultural Investment and International Land Deals in Africa*, London and Rome: FAO/IIED/IFAD.
Dauvergne, P. and Neville, K.J. (2009) 'The changing North–South and South–South political economy of biofuels', *Third World Quarterly*, 30: 1087–1102.
Dauvergne, P. and Neville, K.J. (2010) 'Forests, food, and fuel in the tropics: the uneven social and ecological consequences of the emerging political economy of biofuels', *Journal of Peasant Studies*, 37(4): 631–660.
de Nie, D., Sayer, J. and McCormick, N. (2009) *Indirect Effects of Bioenergy: Effects on Landscapes and Livelihoods*, Proceedings of the Workshop on Indirect Effects of Bioenergy, Rotterdam, the Netherlands, 9 June, available at: http://cmsdata.iucn.org/downloads/iucn_paper_indirect_effects_of_bioenergy_final_2.pdf (accessed 3 August 2012).
Dercon, S. and Gollin, D. (2014) 'Agriculture in African development: theories and strategies', *Annual Review of Resource Economics*, 6: 471–492.
Dixon, A. (2003) *Indigenous Management of Wetlands: Experiences in Ethiopia*, Aldershot: Ashgate.
Doss and Sofa Team (2007) *The Role of Women in Agriculture*, FAO: Rome, available at: www.fao.org/docrep/013/am307e/am307e00.pdf (accessed 27 December 2017).
Earth Policy Institute (2010) *Biofuels Data from World on the Edge*, available at: www.earthpolicy.org/data_center/C26 (accessed 2 August 2012).
Fairhead, J. and Leach, M. (1996) *Misreading the African Landscape: Society and Ecology in a Forest-Savanna Mosaic*, Cambridge: Cambridge University Press.
FAO (Food and Agriculture Organization) (2011) *The State of Food and Agriculture: Women in Agriculture – Closing the Gender Gap for Development*, Rome: FAO.
Gaiser, T., Michael Judex, M., Mouinou Igué, A., Paeth, H. and Hiepe, C. (2011) 'Future productivity of fallow systems in sub-Saharan Africa: is the effect of demographic pressure and fallow reduction more significant than climate change?', *Agricultural and Forest Meteorology*, 151(8), 15 August: 1120–1130.
GRAIN (2008) *Seized: The 2008 Land Grab for Food and Financial Security*, GRAIN Briefing, available at: www.grain.org/briefings_files/landgrab-2008-en.pdf (accessed 2 August 2012).
Hall, D., Hirsch, P. and Li, T.M. (2011) *Powers of Exclusion: Land Dilemmas in Southeast Asia*, Singapore: NUS Press.
Hall, R., Wisborg, P., Shirinda, S. and Zamchiya, P. (2012) 'Farm workers and farm dwellers in Limpopo province, South Africa', *Journal of Agrarian Change*, 13(1): 47–70.
Hazell, P.B.R. (2013) 'Options for African agriculture in an era of high food and energy prices', *Agricultural Economics*, 44: 19–27, DOI: 10.1111/agec.12047.
Jolly, C. and Torrey, B. (1993) 'Introduction', in C. Jolly and B. Torrey (eds), *Population and Land Use in Developing Countries: Report of a Workshop*, Washington, DC: National Academy Press, pp. 1–14.
Li, T.M. (2009) 'Exit from agriculture: a step forward or a step backward for the rural poor?', *Journal of Peasant Studies*, 36(3): 629–636.
Li, T.M. (2010) 'To make live or let die? Rural dispossession and the protection of surplus populations', *Antipode*, 41: 63–93.
Li, T.M. (2011) 'Centering labor in the land grab debate', *Journal of Peasant Studies*, 38: 281–298.
Maconachie, R. (2011) 'Re-agrarianizing livelihoods in post-conflict Sierra Leone? Mineral wealth and rural change in artisanal and small-scale mining communities', *Journal of International Development*, 23(8): 1054–1067.
Maconachie, R. and Binns, T. (2007) 'Farming miners or mining farmers? Diamond mining and rural development in post-conflict Sierra Leone', *Journal of Rural Studies*, 23(3): 367–380.

Maconachie, R. and Fortin, E. (2013) '"New agriculture" for sustainable development? Biofuels and agrarian change in post-war Sierra Leone', *The Journal of Modern African Studies*, 51(2): 249–277.

Magdoff, F. (2008) 'The political economy and ecology of biofuels', *Monthly Review*, 60: 34–50.

McMichael, P. (2009a) 'A food regime analysis of the "world food crisis"', *Agriculture and Human Values*, 26: 281–295.

McMichael, P. (2009b) 'Banking on agriculture: a review of the World Development Report 2008', *Journal of Agrarian Change*, 9(2): 235–246.

McMichael, P. (2010) 'Agrofuels in the food regime', *The Journal of Peasant Studies*, 37: 609–629.

Momsen, J. (2010) 'Gender in rural areas', in J. Momensen (ed.), *Gender and Development*, London: Routledge, pp. 140–175.

Mortimore, M. (1989) *Adapting to Drought: Farmers, Famines and Desertification in West Africa*, Cambridge: Cambridge University Press.

Mortimore, M. and Adams, W.M. (1999) *Working the Sahel: Environment and Society in Northern Nigeria*, London: Routledge.

NEPAD (2013) *African Agriculture, Transformation and Outlook*, available at: www.un.org/en/africa/osaa/pdf/pubs/2013africanagricultures.pdf (accessed 27 December 2017).

Pena, D. (2012) *Mapping the Land Grab in Africa*, available at: www.goodfoodworld.com/2012/07/mapping-the-land-grab-in-Africa/ (accessed 27 December 2017).

Poulton, C., Berhanu, K., Chinsinga, B., Cooksey, B., Golooba-Mutebi, F. and Loada, A. (2014) *The Comprehensive Africa Agriculture Development Programme (CAADP): Political Incentives, Value Added and Ways Forward*, Future Agricultures Consortium Working Paper 077, University of Sussex, Brighton, February, available at: www.future-agricultures.org/publications/research-and-analysis/working-papers/1833-the-comprehensive-africa-agriculture-development-programme-caadp-political-incentives-value/file (accessed 27 December 2017).

Richards, P. (1985) *Indigenous Agricultural Revolution: Ecology and Food Production in West Africa*, London: Hutchinson.

Richards, P. (1986) *Coping with Hunger: Hazard and Experiment in an African Rice-Farming System*, London: Allen & Unwin.

Richardson, B. (2010) 'Big sugar in southern Africa: rural development and the perverted potential of sugar/ethanol exports', *The Journal of Peasant Studies*, 37: 917–938.

Rodney, W. (1972) *How Europe Underdeveloped Africa*, London: Bogle-L'Ouverture Publications.

Romney, D., Thorne, P., Lukuyu, B. and Thornton, P. (2003) 'Maize as food and feed in intensive smallholder systems: management options for improved integration in mixed farming systems of east and southern Africa', *Field Crops Research*, 84(1–2), October–November: 159–168.

Stock, R. (2013) *Africa South of the Sahara: A Geographical Interpretation*, 3rd edition, New York: Guilford.

Swindell, K. and Iliya, M.A. (1992) *Accumulation, Consolidation and Survival: Non-Farm Incomes and Agrarian Change in Northwest Nigeria*, Project on African Agriculture Working Paper No. 6, New York: Social Science Research Council.

Tegegne, Y., Lindner, M., Fobissie, K. and Kanninen, M. (2016) 'Evolution of drivers of deforestation and forest degradation in the Congo Basin forests: exploring possible policy options to address forest loss', *Land Use Policy*, 51: 312–324.

Tiffen, M., Mortimore, N.J. and Gichuki, F. (1994) *More People, Less Erosion: Environmental Recovery in Kenya*, Chichester: Wiley.

UN (2016) *Office of the Special Adviser on Africa (OSAA). Comprehensive Africa Agriculture Development Programme (CAADP)*, available at www.un.org/en/africa/osaa/peace/caadp.shtml (accessed 27 December 2017).

Unruh, J. and Turray, H. (2006) *Land Tenure, Food Security and Investment in Postwar Sierra Leone*, LSP Working Paper 22, Rome: FAO.

Waha, K., Müller, C., Bondeau, A., Dietrich, J., Kurukulasuriya, P., Heinke, J., *et al.* (2013) 'Adaptation to climate change through the choice of cropping system and sowing date in sub-Saharan Africa', *Global Environmental Change*, 23(1), February: 130–143.

World Bank (2007) *World Development Report, 2008: Agriculture for Development*, Washington, DC: World Bank.

World Bank (2011) *Rising Global Interest in Farmland: Can It Yield Sustainable and Equitable Benefits?* Washington, DC: World Bank.

World Bank (2013) *Fact Sheet: The World Bank and Agriculture in Africa*. Washington, DC: World Bank, available at: http://web.worldbank.org/WBSITE/EXTERNAL/COUNTRIES/AFRICAEXT/0,,contentMDK:21935583~pagePK:146736~piPK:146830~theSitePK:258644,00.html (accessed 6 May 2016).

World Bank (2016) *Databank: Agriculture, Value Added (% of GDP)*, Washington, DC: World Bank, available at: www.data.worldbank.org/indicator/NV.AGR.TOTL.ZS (accessed 13 October 2016).

39
Mining

Roy Maconachie

Introduction

Africa's contribution to the world's mineral economy is significant. The continent is endowed with an abundance of natural resource wealth (see Table 39.1), and in recent years escalating global demands for oil and strategic minerals, particularly from emerging economies, have made African resources increasingly important. Consequently, a number of African countries have become global leaders in the production of key minerals, a development that has generated substantial revenue flows for their economies. Mineral wealth across the continent is diverse and vast: according to World Bank (2012) statistics, in 2010 Guinea alone produced over 8 per cent of the world's bauxite supply, Zambia and the Democratic Republic of Congo jointly produced 6.7 per cent of global copper, and Ghana and Mali accounted for 5.8 per cent of global gold production. South Africa holds the world's largest reserves of gold, and it remains the leading global source of chromium and manganese (Baker *et al.*, 2014). The Democratic Republic of Congo has the world's largest reserves of coltan and significant quantities of the world's cobalt, and collectively African countries produce half or more of the world's diamonds, platinum and cobalt (Stock, 2013). Africa has also increasingly become a region of major oil and gas discoveries, with over 30 countries now involved in the extraction of petroleum products (African Development Bank, 2009).

Although mineral extraction has featured prominently in the economies of many African countries since precolonial times, today most mining countries on the continent do not consume or add value to these raw materials. Contemporary resource-rich Africa remains dependent on exporting primary commodities, and many poor countries are highly susceptible to volatile global mineral markets and the demands of powerful international mining companies. In fact, growing evidence suggests that African countries with significant natural resource wealth have not only reaped limited rewards, but have instead often experienced underdevelopment, corruption, political instability, and in some cases violent conflict (Barma *et al.*, 2012). This has prompted a vociferous debate as to whether or not Africa is characterised by a 'resource curse', where it remains unclear if an abundance of natural resources is a blessing or a hindrance for political and socio-economic development (Ross, 1999).

Table 39.1 Africa's mineral wealth and key natural resources

Country	*Key resources*
Algeria	Petroleum, natural gas, iron ore, phosphates, uranium, lead, zinc
Angola	Petroleum, diamonds, iron ore, phosphates, copper, feldspar, gold, bauxite, uranium
Benin	Small offshore oil deposits, limestone, marble, timber
Botswana	Diamonds, copper, nickel, salt, soda ash, potash, coal, iron ore, silver
Burkina Faso	Manganese, limestone, marble; small deposits of gold, phosphates, pumice, salt
Burundi	Nickel, uranium, rare earth oxides, peat, cobalt, copper, platinum, vanadium, arable land, hydropower, niobium, tantalum, gold, tin, tungsten, kaolin, limestone
Cabo Verde	Salt, basalt rock, limestone, kaolin, fish, clay, gypsum
Cameroon	Petroleum, bauxite, iron ore, timber, hydropower
CAR	Diamonds, uranium, timber, gold, oil, hydropower
Chad	Petroleum, uranium, natron, kaolin, fish (Lake Chad), gold, limestone, sand and gravel, salt
Comoros	Fish
Congo, Democratic Republic of	Cobalt, copper, niobium, tantalum, petroleum, industrial and gem diamonds, gold, silver, zinc, manganese, tin, uranium, coal, hydropower, timber
Congo, Republic of	Petroleum, timber, potash, lead, zinc, uranium, copper, phosphates, gold, magnesium, natural gas, hydropower
Côte d'Ivoire	Petroleum, natural gas, diamonds, manganese, iron ore, cobalt, bauxite, copper, gold, nickel, tantalum, silica sand, clay, cocoa beans, coffee, palm oil, hydropower
Djibouti	Potential geothermal power, gold, clay, granite, limestone, marble, salt, diatomite, gypsum, pumice, petroleum
Egypt	Petroleum, natural gas, iron ore, phosphates, manganese, limestone, gypsum, talc, asbestos, lead, rare earth elements, zinc
Equatorial Guinea	Petroleum, natural gas, timber, gold, bauxite, diamonds, tantalum, sand and gravel, clay
Eritrea	Gold, potash, zinc, copper, salt, possibly oil and natural gas, fish
Ethiopia	Small reserves of gold, platinum, copper, potash, natural gas, hydropower
Gabon	Petroleum, natural gas, diamond, niobium, manganese, uranium, gold, timber, iron ore, hydropower
Gambia	Fish, clay, silica sand, titanium (rutile and ilmenite), tin, zircon
Ghana	Gold, timber, industrial diamonds, bauxite, manganese, fish, rubber, hydropower, petroleum, silver, salt, limestone
Guinea-Bissau	Fish, timber, phosphates, bauxite, clay, granite, limestone, unexploited deposits of petroleum
Guinea	Bauxite, iron ore, diamonds, gold, uranium, hydropower, fish, salt
Kenya	Limestone, soda ash, salt, gemstones, fluorspar, zinc, diatomite, gypsum, wildlife, hydropower
Lesotho	Water, agricultural and grazing land, diamonds, sand, clay, building stone
Liberia	Iron ore, timber, diamonds, gold, hydropower
Libya	Petroleum, natural gas, gypsum
Madagascar	Graphite, chromite, coal, bauxite, rare earth elements, salt, quartz, tar sands, semiprecious stones, mica, fish, hydropower
Malawi	Limestone, arable land, hydropower, unexploited deposits of uranium, coal and bauxite

(continued)

Table 39.1 Continued

Country	*Key resources*
Mali	Gold, phosphates, kaolin, salt, limestone, uranium, gypsum, granite, hydropower (*Note*: bauxite, iron ore, manganese, tin and copper deposits are known but not exploited)
Mauritania	Iron ore, gypsum, copper, phosphate, diamonds, gold, oil, fish
Mauritius	Arable land, fish
Morocco	Phosphates, iron ore, manganese, lead, zinc, fish, salt
Mozambique	Coal, titanium, natural gas, hydropower, tantalum, graphite
Namibia	Diamonds, copper, uranium, gold, silver, lead, tin, lithium, cadmium, tungsten, zinc, salt, hydropower, fish (*Note*: suspected deposits of oil, coal and iron ore)
Niger	Uranium, coal, iron ore, tin, phosphates, gold, molybdenum, gypsum, salt, petroleum
Nigeria	Natural gas, petroleum, tin, iron ore, coal, limestone, niobium, lead, zinc, arable land
Rwanda	Gold, cassiterite (tin ore), wolframite (tungsten ore), methane, hydropower, arable land
São Tomé and Príncipe	Fish, hydropower
Senegal	Fish, phosphates, iron ore
Seychelles	Fish, coconuts (copra), cinnamon trees
Sierra Leone	Diamonds, titanium ore, bauxite, iron ore, gold, chromite
Somalia	Uranium and largely unexploited reserves of iron ore, tin, gypsum, bauxite, copper, salt, natural gas, likely oil reserves
South Africa	Gold, chromium, antimony, coal, iron ore, manganese, nickel, phosphates, tin, rare earth elements, uranium, gem diamonds, platinum, copper, vanadium, salt, natural gas
South Sudan	Hydropower, fertile agricultural land, gold, diamonds, petroleum, hardwoods, limestone, iron ore, copper, chromium ore, zinc, tungsten, mica, silver
Sudan	Petroleum; small reserves of iron ore, copper, chromium ore, zinc, tungsten, mica, silver, gold; hydropower
Swaziland	Asbestos, coal, clay, cassiterite, hydropower, forests, small gold and diamond deposits, quarry stone, talc
Tanzania	Hydropower, tin, phosphates, iron ore, coal, diamonds, gemstones, gold, natural gas, nickel
Togo	Phosphates, limestone, marble, arable land
Tunisia	Petroleum, phosphates, iron ore, lead, zinc, salt
Uganda	Copper, cobalt, hydropower, limestone, salt, arable land, gold
Western Sahara	Phosphates, iron ore
Zambia	Copper, cobalt, zinc, lead, coal, emeralds, gold, silver, uranium, hydropower
Zimbabwe	Coal, chromium ore, asbestos, gold, nickel, copper, iron ore, vanadium, lithium, tin, platinum group metals

Source: CIA (2017).

This chapter critically examines why Africa's mineral economy has so seldom generated meaningful development in its resource-rich countries, despite the importance of African resources in the global economy. Following this introduction, the second section of the chapter provides an historical overview of African mining, exploring mineral extraction in both precolonial times and during the colonial conquest, and reflecting upon how the continent's political economy is deeply ingrained with a long history of natural resource exploitation. This sets the stage for a discussion of Africa's contemporary mining sector in the third and fourth sections, focusing on developments in the post-independence years and the ensuing era of neo-liberal reform, before moving on to the fifth section, where the consequences of extractive-led development trajectories for host communities are explored in greater detail. In the wake of structural adjustment programmes and the 'opening up' of African mineral markets in the 1990s, one significant impact was the rapid expansion of the informal, artisanal mining sector across the continent, which is critically discussed in the sixth section. The seventh section then moves on to consider recent interest from China in Africa's mineral economy, before concluding in the final section with some reflection on the future of mining for African development.

African mining in historical perspective

Precolonial Africa

Mining in Africa has a long, rich history. For thousands of years, the extraction of mineral resources has been an important driver of both development and underdevelopment, with the pursuit of mineral wealth being the cause of the rise and fall of African civilisations. Early evidence suggests that thousands of years ago, the ancient Egyptians mined malachite, turquoise and copper, and extensive amounts of gold were extracted from the mines of Nubia (Stock, 2013). Later, the Romans, who were believed to be the first to develop large-scale mining methods, exploited mineral resources from north-western Africa and sustained their economy with a wide range of mineral resources, including gold, silver, copper, tin and lead (Skaggs *et al.*, 2012).

Minerals were extracted and used by a wide range of precolonial African societies to make both domestic and ceremonial goods, and their exchange was central to interregional and international trade (Stock, 2013). Gold, for example, was an important resource and medium of exchange in both West and southern Africa. In Ghana and Mali, gold was traded extensively in North Africa and Europe in the early eighth century (Hilson, 2002a). Salt was initially brought southwards from the Sahara and exchanged for gold, but later the trade diversified with European and Arab goods being exchanged for commodities such as slaves, skins, gum and spices (UNECA, 2011). Meanwhile in Nigeria, the Nok society was well known for iron making between the seventh and fourth centuries BC, and the Ife and Benin cultures produced fine bronze artwork (UNECA, 2011). Evidence further suggests that African societies have been mining and using copper and its alloys for centuries, with extraction taking place in ancient Egypt through parts of modern Niger, Mauritania, and central and southern Africa (Herbert, 1984).

The colonial era

Although European countries had long been aware that Africa contained rich deposits of valuable minerals, it was not until colonial contact occurred that the true potential of the continent's resource wealth was fully realised by the outside world. Since this time, stretches of resource-rich Africa have served as a resource hinterland for the global capitalist system (Amin, 1972). As far back as the sixteenth and seventeenth centuries, European powers saw African colonies as a source of raw materials to fuel their economies, and natural resource extraction became one of the central pillars of the colonial project (Lanning and Mueller, 1979). Minerals were extracted on an extensive scale, and colonial economic policy involved the introduction of new modes of production and the construction of transportation infrastructure such as roads, railroads and ports to ensure that resources could be shipped back to Europe as quickly and efficiently as possible

(Stock, 2013). As manufacturing became increasingly important to the European economies, especially during the time of the Industrial Revolution, the provision of minerals and other raw materials for industrial development took on greater significance (Willis, 2011).

Perhaps the process of colonial resource exploitation reached its apogee on the African continent during the second half of the nineteenth century. At the time, European interests in Africa increased dramatically due to the continent's wealth of valuable natural resources, including gold, rubber, spices, tea, opium and slaves (Stock, 2013). In 1884–1885, after more than four centuries since initial contact, the European powers formally laid claim to virtually all of Africa when the infamous Berlin Conference was held, thereby initiating the 'scramble for Africa' and an era of new imperialism (Binns *et al.*, 2012). At the Berlin meeting, the imperial chancellor and architect of the German Empire, Otto von Bismarck, convened a conference of 14 states to politically partition Africa. The result was the 'carving up' of the continent into a series of resource colonies that were ruled primarily by the British, French, Germans, Portuguese and Belgians (Adebajo, 2005).

In his now classic book *How Europe Underdeveloped Africa*, Walter Rodney (1972) describes how Africa's natural resources extracted during colonialism were responsible for making European countries tremendously wealthy, while at the same time putting African countries on a path of poverty and despair. In his account, Rodney describes how in South Africa, African mine labourers worked deep underground, enduring abysmal conditions and being paid a poverty wage. While Rodney notes that individuals such as the mining magnate and founder of De Beers, Cecil Rhodes, amassed huge fortunes made from gold and diamonds in southern Africa, it was the shareholders of the mining companies who remained in Europe and North America who benefited most of all. He notes:

> In the mid-1950s, British investments in South Africa were estimated at £860 million and yielded a stable profit of 15% or £129 million every year. Most mining companies had returns well above that average. DeBeers Consolidated Mines made a profit that was both phenomenal and consistently high – between $26 million and $29 million throughout the 1950s.
>
> *(Rodney, 1972: 166)*

In other regions of Africa, countries across the continent also served as resource hinterlands for thriving European economies during colonialism (Rodney, 1972: 167). For example, in North Africa, foreign capital was invested to exploit a wide range of natural resources, including phosphates, oil, lead, zinc, manganese and iron ore. In the West African countries of Guinea, Sierra Leone and Liberia, exploitation was focused on rich deposits of gold, diamonds, iron ore and bauxite. Other key natural resources were channelled back to Europe from Nigeria (tin), Ghana (gold and manganese), Tanganyika (gold and diamonds), Uganda and Congo-Brazzaville (copper).

While many scholars have interpreted colonialism as an economic process driven by the need for territorial expansion, raw materials and labour, it was also a project motivated by the notion of 'progress' and extending 'civilisation'. It was believed that vast 'top-down' mining schemes would transform underdeveloped regions of Africa and contribute to the 'civilising mission'. Subsequently, much post-Second World War development thinking, particularly during the 1950s and 1960s, was highly influenced by the notion of modernisation. Many influential development thinkers such as Gunnar Myrdal and Albert Hirschman developed spatial models of development, believing that economic growth in Africa was unbalanced, did not occur everywhere simultaneously, and should become focused in 'growth poles'. There was also great confidence in the idea that modernisation in the form of urban-based industrial growth should receive priority, and from these areas development would 'trickle down' to 'backwards' sectors and regions of poor countries. The drive for modernisation was therefore frequently associated with 'top-down' development, and the belief that African countries would become better off if they emulated what had happened in developed countries, predominantly by adopting better roads, infrastructure and communication technologies. Progress was linked to industrialisation, a process that demanded the extraction of natural resources to fuel development. To this day, this 'growth pole' narrative has remained highly influential in shaping extractive industry policy across the African continent, playing a defining role in fortifying the large-scale mining bias that still exists in many countries (see Hilson, forthcoming).

Developments since independence

While the highly unequal relationships associated with colonial modes of resource extraction had significant social, political economic and environmental impacts on African populations, with the end of formal colonialism in the 1960s, newly independent countries often found themselves renegotiating relationships with external actors within a new world order. Many of the 'core–periphery' relationships established during the colonial era remained intact, and were in some cases even strengthened, as former colonies became a new untapped source of oil, minerals and cheap raw materials for multinational corporations in the developed world. Moreover, as decolonisation took place against the backdrop of the Cold War, Africa's mineral wealth assumed new geopolitical significance. In return for pledging Western alliances and access to mineral reserves, many Western governments turned a blind eye to repressive African dictatorships. Most notably, France, Belgium and the United States sent troops into Zaire (now the Democratic Republic of Congo) in the 1960s and 1970s to prop up the repressive Mobutu Sese Seko regime and protect their mineral interests. At the same time, more efficient technologies made resource extraction faster and more profitable than the processes that featured during the colonial period, and Rostovian discourses of 'modernisation' and 'progress' continued to drive narratives that justified and legitimised extractive industry investment in corrupt African countries.

During the 1980s, most mineral-producing countries in Africa suffered badly in the global recession. As commodity prices fluctuated drastically and the demand for minerals from industrialised countries plummeted, mineral extraction by transnational companies often became economically unviable, and there were a significant number of mine closures. Few new investments in the extractive industries took place during this period, as there were increasing perceptions that many African nations were difficult places to do business, politically unstable, and characterised by corruption and poor infrastructure. To make matters worse, civil wars also became increasingly common across Africa during the 1980s and 1990s, with the appropriation and mismanagement of high-value natural resources frequently being cited as a key factor in triggering, escalating or prolonging conflicts on all corners of the continent (e.g. see Collier and Hoeffler, 2001; Elbadawi and Sambanis, 2002; Ross, 2003; Bannon and Collier, 2003; Fearon, 2004).

While high-value resources such as gold, diamonds and oil were certainly implicated in a number of African wars in the 1990s, there has been a considerable amount of disagreement among scholars over the causal factors of conflict in resource-rich African countries. Some critics have attributed conflict to resource politics ('grievance') and others have based their analysis on exploitative economic interests ('greed'). In several civil African conflicts, warring factions were able to access 'lootable' resources (i.e. resources such as alluvial diamonds that have a high value-to-weight ratio, and are easily appropriated and transported by unskilled workers) through artisanal extraction, which some scholars have suggested has been particularly amenable to fuelling and prolonging 'greed-based' insurgency. In some cases, such conflicts have also 'spilled over' porous and contested borders into neighbouring African states. For example, West Africa's Mano River Union countries – Liberia, Sierra Leone, Guinea and Côte d'Ivoire – provide a lucid example of how smuggling and the illicit trade of 'lootable' resources can exacerbate regional conflicts. Sierra Leone's protracted civil war of the 1990s is a case in point, where a cross-border exchange of diamonds for weapons took place with Liberian President Charles Taylor, an arrangement that both fuelled and prolonged the decade-long war. Likewise, in Côte d'Ivoire, evidence suggests that rebel groups illegally mined and exported diamonds and gold to finance the conflict that broke out in September 2002 (UN, 2012). As Silberfein and Conteh (2006) note, illicit resource flows across borders are very difficult to regulate or contain, and they usually continue to occur even when sanctions and embargoes are in place because they are often controlled by powerful private interests.

Mining in an era of neo-liberal reform

Although the 1990s were characterised by a number of protracted resource-driven conflicts in Africa, it was apparent that by the end of the decade, the geographies of investment in the international mining industry were beginning to change. In particular, economic liberalisation policies were adopted by many African states, and a broad range of natural resources became drawn into market-led modes of production and governance. These neo-liberal reforms, combined with soaring commodity prices and heightened resource and energy demands from what later came to be known as the so-called BRIC (Brazil, Russia, India and China) countries (and in 2010 they became known as the BRICS when South Africa was added), led to a significant growth in global investment in the extractive industries. Accompanying this increased demand from the world's emerging economies, the globalisation of extractive industries has led to sweeping technological, organisational and regulatory changes in Africa and across the world (Bridge, 2004). Bridge (2004), for example, notes that between 1985 and 2004, more than 90 states adopted new mining laws in an effort to increase foreign direct investment (FDI) in mineral extraction.

In many cases, the reform and deregulation of mining codes in African countries has been adopted as a central pillar of broader neo-liberal agendas, but there remains considerable debate as to whether or not this has had positive impacts on levels of development. For example, Campbell (2004) has documented how the reform of legal and regulatory frameworks in the mining sectors of many African countries has led to a much more favourable environment for FDI. But alternatively, others have noted that accompanying this surge of investment and reform has been an equally remarkable increase in conflict and social mobilisation around the detrimental effects of mining and hydrocarbon projects (Bebbington *et al.*, 2008).

At the forefront of support for the neo-liberal agenda, the World Bank has had a long and controversial history of involvement in supporting extractive industry investments in African countries. Moody (2007) describes how between 1955 and 1990, scores of mining-related public sector grants to developing countries were channelled by the World Bank through its two sub-organisations, the International Bank for Reconstruction and Development (IBRD) and the International Development Association (IDA). These grants were targeted at five main areas: (1) reform and rehabilitation; (2) 'greenfield' mine construction; (3) mineral processing; (4) technical assistance; and (5) engineering work. Moody (2007) reports that during this period of engagement, the World Bank supported nearly 50 mining and mineral processing projects with loans totaling almost US$2 billion.

In response to widespread protests from a range of project-affected communities and civil society organisations, in 2001, the World Bank, under former President James Wolfensohn, launched its first Extractive Industries Review (EIR) to evaluate the role of the bank in the extractive industries. In particular, the review sought to assess whether the World Bank's involvement in the sector was compatible with its mandate of achieving poverty alleviation through sustainable development. Following the inquiry, a series of recommendations were made to advise the World Bank on how to make its involvement in the sector more compatible with its mission, covering a broad range of issues from good governance and increased transparency, to the adoption of more effective environmental and pro-poor policies, to respecting human rights and the protection of indigenous peoples. In 2004, the World Bank released its response to the EIR, and while it did agree with some of the recommendations, very few commitments were actually made. Unsurprisingly, the final submission of the EIR (2003) was very critical, describing the World Bank's actions as 'business as usual with marginal change'.

Mining, community conflict and underdevelopment

While the impacts that transnational mining investments in Africa have had at both national and subnational levels are varied, complex and far-reaching, it goes without saying that mining can, and frequently does, generate serious social, political, and economic problems at the community level. By its very nature, mining is socially and environmentally disruptive, and there are a number of common grievances that are frequently shared by those living in catchment communities across Africa. Many of these issues have been well covered in the literature; for example, the dislocation of livelihoods when land and water sources are appropriated or destroyed through the creation of pollution, or the tension that results through an influx of migrant labourers into resource communities. However, it is also apparent that there are often deep-seated structural dimensions in many African countries that can exacerbate these flashpoints, and can potentially transform resource extraction sites into zones of grievance-driven conflict.

Although the expectations placed upon large-scale mining companies as employment generators are often very high, many assume the form of resource 'enclaves' and are disconnected from wider society (Ferguson, 2006; Ackah-Baidoo, 2012). Any employment gains that do accrue from investments in mining projects can often lead to intensified competition and conflict between local actors, as they vie to capture available job opportunities. Moreover, evidence suggests that because large-scale mining operations are predominantly capital- rather than labour-intensive, the actual number of jobs created is usually miniscule in comparison to the revenue generated. For example, in Mali, estimates suggest that the Sadiola gold mine creates one mining job for every US$700,000 invested, while the Randgold mine directly creates one job for every US$1.23 million invested (Pegg, 2006: 380). As such, some of the strongest community resistance to recent extractive industry expansion in Africa has been in response to the poor track records of mining companies in providing jobs (Pegg, 2006), which is often accompanied by perceptions of poor working conditions and labour exploitation.

Closely related to these observations on employment is evidence to suggest that labour migration into catchment communities can, in itself, also be a considerable source of tension and conflict. One of the most significant impacts of large-scale extraction on local communities is the rapid change in the economic and social fabric of society that takes place when migrant 'strangers' arrive in search of opportunities. Communities' social balance can be disrupted, which impacts resource availability or places further competition on economic activities such as farming and hunting (Hilson, 2002b). The economic shifts caused by in-migration can also lead to increased prices for local goods, which can widen disparities among local populations

(Hilson, 2002b). Compounding this problem, migrants often lack strong social ties, live in disrupted contexts and on borrowed land, and work in harsh conditions for limited incomes. As Weber-Fahr *et al.* (2001: 15) suggest:

> New types of poverty are therefore created, with a mixture of 'original residents' who have been unable to share in employment opportunities, and 'newcomers' who have migrated in with the hope of finding employment, but have been unsuccessful in doing so.

Although extractive industry investments can potentially provide important economic benefits for catchment communities, the significant amount of land and other livelihood resources they appropriate can also cause severe community dislocation and hinder local development. Land use conflicts can become particularly intense when the issue of tenure becomes unclear. In some cases, different systems of tenure may overlap (for example, customary tenure versus state ownership), disputes may arise over disagreements concerning surface versus subsoil ownership, and conflict may be ignited when different claims on valuable minerals are at stake. Indeed, land acquisition associated with mining investments can further obscure tenure rights, particularly when they are insecure or contested (Cotula, 2014). In the case of Liberia, for example, one recent report for the Government Land Commission suggests that the government has issued concessions to commercial entrepreneurs, communities and conservation programmes that exceed 50 per cent of the country's land area. Much of this land, the report adds, has long been utilised by rural populations, opening up the inevitable likelihood for land disputes and conflict to erupt (see EWER, 2012; Land Commission of Liberia, 2012).

As noted by Hilson (2012: 133), '[f]ew industrial activities leave as great an environmental footprint or are as capable of having as much influence on the wellbeing of a society as a large-scale mine or oil and gas project'. Large-scale natural resource investments can have significant environmental costs, including water pollution, deforestation, loss of biodiversity, and soil degradation. Since rural economies in Africa are inextricably linked to the health of natural resources such as forests, soil, rivers, wildlife and fish, a threat to the ecosystem is simultaneously a threat to the subsistence and income of local populations (Horowitz, 2011), placing concerns on 'the security and integrity of livelihoods' (Bebbington *et al.*, 2008: 2890). Indeed, the environmental degradation caused by mining projects can increase the vulnerability of the poor, exacerbate tensions, and trigger conflict. Such sentiments are confirmed in a recent study by Franks *et al.* (2014), who examined publicly available information concerning 50 different cases of prolonged company–community conflict around mining operations, and report that environmental issues were identified as the most common triggers of conflict.

Finally, it is frequently the case that conflict within catchment communities can result from grievances that stem from insufficient compensation for loss of resources, or inadequate consultation with companies and governments. One of the major challenges in managing extractive sector booms and trying to mitigate conflict, therefore, is to ensure that public expectations of the benefits of extractive-led development trajectories are effectively managed and address the needs of a diverse range of stakeholders. Although most African countries have now adopted community development agreements (CDAs) or similar arrangements in their mining laws, companies still often deal directly with central government agencies, bypassing local stakeholders altogether. In some cases where limited community consultation does take place, mining companies may privilege relations with elites or traditional leaders, whose interests can diverge considerably from those of the community. In such situations, elites may capture supplier contracts, employment opportunities and other benefits.

The artisanal and small-scale mining sector

Mining – whether large-scale or small-scale, regulated or informal, capital-intensive or artisanal – can be an important source of income generation in poor, employment-constrained economies, either directly or indirectly, through upstream and downstream multiplier effects. However, over the past two decades, artisanal and small-scale mining (ASM) – low-tech, labour-intensive mineral extraction and processing – has made particular inroads as an employment-generating activity in Africa, serving as a magnet for young, single, unemployed, unskilled labourers in search of job opportunities. Across the continent, numerous minerals are mined artisanally, including diamonds, gemstones, gold, bauxite and coltan. The majority of ASM operators, however, are engaged in the extraction of precious metals and stones, with gold being the most popular mineral mined (Hilson and McQuilken, 2014). According to some sources (see UNECA, 2011), as much as 10 per cent of the world's gold supply originates from the ASM sector.

Much of the recent literature on ASM in Africa (e.g. see Banchirigah, 2006; Spiegel, 2009; Maconachie and Hilson, 2016) suggests that it is a livelihood activity that has been driven by poverty, rather than a desire to 'get rich quickly'. Most notably, the structural adjustment programmes and widespread neo-liberal economic reforms implemented across Africa in the 1980s and 1990s led to a significant downsizing of industries, made life increasingly challenging for smallholder farmers, and were responsible for widespread redundancies in the public sector. Evidence suggests that the ASM economy and the downstream industries it has spawned have absorbed millions of people who were adversely affected by these changes (Hilson and McQuilken, 2014). Although to date, no comprehensive census has been carried out on the extent of ASM across Africa, Table 39.2 lists ASM employment estimates for selected countries across sub-Saharan Africa.

Recent research carried out on ASM in the West African country of Sierra Leone confirms its importance as a key livelihood activity (see Maconachie and Binns, 2007; Maconachie, 2011, 2012). Detailed fieldwork carried out in Kono District in the Eastern province has demonstrated that despite being associated with a host of environmental, health and safety and social concerns, the sector provides a livelihood for hundreds of thousands of poor people in the region, supplying valuable start-up capital for other economic activities, spawning upstream and downstream industries, and helping to nourish smallholder agriculture. But these gains have often come at a cost. Both artisanal gold and diamond production in Sierra Leone are predominantly informal and unregulated activities, and the territories in which they take place are largely under the control of local elite actors. Many new entrants to the sector therefore are forced to forge exploitative relationships with these individuals, including buyers and middlemen, and in the process become trapped in cycles of borrowing and indebtedness.

In Sierra Leone, as elsewhere in Africa, these dynamics have perpetuated the sector's informality, particularly as ASM activities often occur in remote areas where governance is poor, regulatory enforcement is virtually non-existent, and elite capture through 'accumulation by exploitation' is widespread. Porous and unregulated borders between West Africa's four Mano River Union countries have fuelled the emergence of an illicit 'shadow state' economy and fortified the local networks that sustain it. More broadly, across Africa, ASM has subsequently been linked to international criminal networks, providing an effective vehicle for international money laundering and a potential source of resources for diverse 'terrorist' groups (FATF, 2013). Illicit transactions within the ASM economy pose a major challenge to governments, marking a significant loss of revenue that could be used to promote economic growth and development in the countries where minerals are extracted.

Table 39.2 Employment estimates for artisanal and small-scale mining, and minerals extracted, in selected African countries

Country	*Directly working in artisanal and small-scale mining*	*Estimated number of dependents*	*Key minerals mined on a small and artisanal scale*
Angola	150,000	900,000	Diamonds
Burkina Faso	200,000	1,000,000	Gold
CAR	400,000	2,400,000	Gold, diamonds
Chad	100,000	600,000	Gold
Côte d'Ivoire	100,000	600,000	Gold, diamonds
Democratic Republic of Congo	200,000	1,200,000	Diamonds, gold, coltan
Eritrea	400,000	2,400,000	Gold
Ethiopia	500,000	3,000,000	Gold
Ghana	1,100,000	4,400,000	Gold, diamonds, sand
Guinea	300,000	1,500,000	Gold, diamonds
Liberia	100,000	600,000	Gold, diamonds
Madagascar	500,000	2,500,000	Coloured gemstones, gold
Malawi	40,000	240,000	Coloured gemstones, gold
Mali	400,000	2,400,000	Gold
Mozambique	100,000	1,200,000	Coloured gemstones, gold
Niger	450,000	2,700,000	Gold
Nigeria	500,000	2,500,000	Gold
Sierra Leone	300,000	1,800,000	Gold, diamonds, coltan
South Africa	20,000	120,000	Gold
South Sudan	200,000	1,200,000	Gold
Tanzania	1,500,000	9,000,000	Coloured gemstones, gold, diamonds
Uganda	150,000	900,000	Gold
Zimbabwe	500,000	3,000,000	Gold, diamonds, coloured gemstones

Source: Adapted from Hilson and Maconachie (2017). Data extracted from ILO (1999), Dreschler (2001), Mutemeri and Petterson (2002) and Hinton (2005).

Consequently, the push by donors and policymakers to formalise the ASM sector has attracted increasing attention in recent years, with the World Bank and various departments of the United Nations now pressuring African governments to make the sector a centrepiece of their rural development strategies. This stems largely from the growing recognition of ASM's economic impact, the need to gain greater control of its sprawling activities, and the importance of putting regulators in an improved position to tackle the host of environmental problems and social 'ills', including child labour and health and safety concerns, commonly associated with such activities (see Hentschel *et al.*, 2002; UNEP, 2012). While to date there has been minimal effort made to develop the intervention models needed to address these problems, it has been argued that the formalisation of ASM is the key to empowering unregistered miners and making the sector more sustainable.

The new scramble for resources and the rise of China

In addition to the proliferation of ASM activities across the continent, a number of critical scholars have also noted the significant rise in the intensity and geographical extent of large-scale mining investment in Africa over the past two decades (Held *et al.*, 1999). While 'traditional' mining countries such as Canada and Australia have long provided major FDI inflows into the extractive sectors of developing countries, more recently it has been the world's rapidly growing economies, such as Brazil, India and China, that have made significant inroads into the extractive sectors of resource-rich African countries. These new investments, it has been argued by some, have played an important role in reducing the dependence of poor African countries on Western investors or donors such as the World Bank, who have in the past often been criticised for engaging in 'neocolonial plunder'.

China is an interesting case in point, and deserves further discussion. Although at the time of writing, China's economic growth had recently slowed to a rate of 6.5 per cent per annum, for many years the country averaged 10 per cent annually (Binns *et al.*, 2012), becoming the world's second largest oil consumer and the biggest global market for iron, coal, copper, cement, aluminum and nickel (Moody, 2007). While some have argued that China's insatiable demand for raw commodities mirrors the 'core–periphery patterns of exchange' initiated during colonial years and replicated by Western mining companies in the postcolonial years, it seems that the prices paid and extent of investments that China is willing to make are much more significant (Gonzalez-Vicente, 2011). In this vein, China's quest to secure access to natural resource wealth is based on a bold new model of business partnership, with very different characteristics from either the Bretton Woods Institutions or traditional Western business investors. But the country's 'non-intervention' approach to investment has also sparked considerable debate both within scholarly circles and the international community, with critics arguing that Chinese investment in developing countries known for human rights abuses has amounted to little more than an opportunity to provide financial support to otherwise isolated regimes (Gonzalez-Vicente, 2011). To provide one example, Chinese companies have undertaken oil exploration in countries with dubious 'pariah regimes', such as Sudan, to meet the country's vast energy needs (Binns *et al.*, 2012). China reportedly buys 60 per cent of Sudan's oil (Weinstein, 2008), and this, according to a number of critics, played a key role in perpetuating and prolonging the conflict in Darfur (Alden, 2007).

However, the impacts of involvement of Chinese state-owned enterprises in Africa have not been all negative, and proponents have also pointed out a number of positive impacts accruing. For example, with China's resource investments have come additional advances to host countries in the form of infrastructural projects, advances in healthcare, technical and telecommunications support, and educational partnership programmes. Africa has particularly benefited from engagement with China, with trade between Africa and China reaching US$106.84 billion in 2008 (The Guardian, 2009). China's demand for natural resources helped to drive up commodity prices on global markets during the first decade of the new millennium, contributing to the longest period of sustained growth in Africa since the 1970s (Wallis and Dyer, 2007). However, because many of the deals made are not only based on money – but also involve the provisioning of servicing and infrastructure – it has also been argued that there is less potential for all of the investment to be embezzled by corrupt regimes, as has been the case in a number of African countries.

Conclusion

This chapter has provided an overview of contemporary debates that concern mining in Africa, illustrating how the dynamics that shape mineral extraction have evolved over time. It has explored the 'paradox of plenty' that many resource-rich African countries currently face, addressing the question of why so many countries across the continent remain poor despite their mineral wealth. In doing so, it has been argued that Africa has served as a 'resource hinterland' for the global capitalist system since colonial times, and that historical legacies have had great bearing on present-day circumstances.

However, it also appears that much has changed in the last two decades: more efficient technologies and sweeping regulatory changes have made mineral extraction faster and more profitable than ever for a host of powerful new transnational actors. Since the beginning of the new millennium, resource-rich Africa has attracted significant new investments from 'non-traditional' sources, including emerging economies such as China. This has caused global commodity prices to soar in value, and at the same time has made natural resources across Africa increasingly important on global markets. However, host communities at the sites of extraction often continue to be the most marginalised and economically depressed. Alongside large-scale extraction, the development of a buoyant and dynamic artisanal and small-scale mining sector across the continent is an expression of poverty, but has helped millions of poor individuals make ends meet. As a consequence, discussions that concern the formalisation of the artisanal mining sector have entered mainstream policy debates, largely as an attempt to bring illicit mining activities back into the legal domain, while at the same time supporting diggers and creating more sustainable working conditions.

In the case of both the ASM sector and large-scale capital-intensive extraction, it remains an ongoing challenge to convert resource rents into sustainable development trajectories that provide space for meaningful citizen engagement and accommodate the needs of all segments of society. While vast new discoveries of valuable and increasingly accessible resources have propelled many African states to positions of unprecedented economic growth, history has shown that if such resources are badly managed, mounting tension and conflict are inevitable.

Creating more inclusive extractive-led development that provides meaningful benefits to host communities requires a deeper understanding of the politics of participation, and a more nuanced appreciation of how power shapes relationships between different stakeholder groups. But even in situations where citizens are 'invited' to participate in designing more sustainable initiatives in the mining sector (for example, within legal CDAs or voluntary CSR programmes), the inclusion of local actors in decision-making processes cannot be taken as a given (Maconachie, 2010). While the World Bank's Extractive Industries Review (EIR) concluded that the participation of affected communities in processes of free, prior and informed consent (FPIC) is a necessary condition for extractive projects to contribute to poverty alleviation and sustainable development, there has been significant debate and critique over the ways in which consultation actually plays out. Indeed, oppressive histories, unequal power relationships and a lack of capacity can all inhibit citizen participation in extractive industry development in Africa.

References

Ackah-Baidoo, A. (2012) 'Enclave development and "offshore corporate social responsibility": implications for oil-rich sub-Saharan Africa', *Resources Policy*, 37(2): 152–159.

Adebajo, A. (2005) 'The curse of Berlin: Africa's security dilemmas' *Internationale Politik and Gesellschaft*, 4, available at: www.fes.de/ipg/IPG4_2005/07_ADEBAJO.PDF (accessed 27 December 2017).

African Development Bank (2009) *Oil and Gas in Africa*, Oxford: Oxford University Press, available at: www.afdb.org/fileadmin/uploads/afdb/Documents/Publications/Oil%20and%20Gas%20in%20Africa.pdf (accessed 27 December 2017).

Alden, C. (2007) *China in Africa*, London: Zed Books.

Amin, S. (1972) 'Underdevelopment and dependence in black Africa: origins and contemporary forms', *The Journal of Modern African Studies*, 10(4): 503–524.

Baker, L., Newell, P. and Phillips, J. (2014) 'The political economy of energy transitions: the case of South Africa', *New Political Economy*, 19(6): 791–818.

Banchirigah, M.S. (2006) 'How have reforms fuelled the expansion of artisanal mining? Evidence from sub-Saharan Africa', *Resources Policy*, 31(3): 165–171.

Bannon, I. and Collier, P. (eds) (2003) *Natural Resources and Violent Conflict*, Washington, DC: World Bank.

Barma, N., Kaiser, K., Minh Le, T. and Vinuela, L. (2012) *Rents to Riches? The Political Economy of Natural Resource-Led Development*, Washington, DC: World Bank.

Bebbington, A., Bebbington, D., Bury, J., Lingan, J., Muñoz, J. and Scurrah, M. (2008) 'Mining and social movements: struggles over livelihood and rural territorial development in the Andes', *World Development*, 36(12): 2888–2905.

Binns, T., Dixon, A. and Nel, E. (2012) *Africa: Diversity and Development*, London: Routledge.

Bridge, G. (2004) 'Mapping the bonanza: geographies of mining investment in an era of neoliberal reform', *Professional Geographer*, 56(3): 406–421.

Campbell, B. (ed.) (2004) *Regulating Mining in Africa: For Whose Benefit?* Discussion Paper 26, Uppsala: Nordiska Afrikainstitutet.

CIA (Central Intelligence Agency) (2017) *World Fact Book*, Washington, DC: CIA, available at: www.cia.gov/library/publications/the-world-factbook/geos/uk.html (accessed 5 January 2018).

Collier, P. and Hoeffler, A. (2001) *Greed and Grievance in Civil War*, Policy Research Working Paper No. 2355, Washington, DC: World Bank.

Cotula, L. (2014) *Foreign Investment, Law and Sustainable Development: A Handbook on Agriculture and Extractive Industries*, London: International Institute for Environment and Development, available at: http://pubs.iied.org/pdfs/17513IIED.pdf (accessed 28 December 2017).

Dreschler, B. (2001) *Small-Scale Mining and Sustainable Development Within the SADC Region*, Mining, Minerals and Sustainable Development (MMSD), London: International Institute for Environment and Development and World Business Council for Sustainable Development, available at: http://pubs.iied.org/pdfs/G00735.pdf (accessed 8 January 2018).

EIR (Extractive Industries Review) (2003) *Striking a Better Balance: The Final Report of the Extractive Industries Review. Volume 1: The World Bank Group and Extractive Industries*, Jakarta: World Bank.

Elbadawi, I. and Sambanis, N. (2002) 'How much war will we see? Explaining the prevalence of civil war', *Journal of Conflict Resolution*, 46(3): 307–334.

EWER (Early Warning Response Working Group) (2012) *Agricultural Land Concessions and Conflict in Liberia*, Policy Analysis Brief, Monrova, Liberia: EWER, available at: www.lern.ushahidi.com/media/uploads/page/3/EWERPolicyBriefLiberia (accessed 28 December 2017).

FATF (Financial Action Task Force) (2013) *Money Laundering and Terrorist Financing Through Trade in Diamonds*, Paris: FATF.

Fearon, J. (2004) 'Primary commodity exports and civil war', *Journal of Conflict Resolution*, 49(4): 483–507.

Ferguson, J. (2006) *Global Shadows: Africa in the Neoliberal World Order*, Durham, NC: Duke University Press.

Franks, D., Davis, R., Bebbington, A., Ali, S., Kemp, D. and Scurrah, M. (2014) 'Conflict translates environmental and social risk into business costs in the natural resource industries: implications for sustainability science', *Proceedings of the National Academy of Sciences*, 111(21): 7576–7581, DOI: 10.1073/pnas.1405135111.

Gonzalez-Vicente, R. (2011) 'China's engagement in South America and Africa's extractive sectors: new perspectives for resource curse theories', *The Pacific Review*, 24(1): 65–87.

Held, D., McGrew, A., Goldblatt, D. and Perraton, J. (1999) *Global Transformations: Politics, Economics and Culture*, Stanford, CA: Stanford University Press.

Hentschel, T., Hruschka, F. and Priester, M. (2002) *Global Report on Artisanal and Small-Scale Mining. Minerals Mining and Sustainable Development (MMSD) Project*, London: International Institute for Environmental Development.

Herbert, E. (1984) *Red Gold of Africa: Copper in Precolonial History and Culture*, Madison, WI: University of Wisconsin Press.

Hilson, G. (2002a) 'Harvesting mineral riches: 1000 years of gold mining in Ghana', *Resources Policy*, 28(1–2), March–June: 13–26.

Hilson, G. (2002b) 'An overview of land use conflict in mining communities', *Land Use Policy*, 19: 65–73.

Hilson, G. (2012) 'Corporate social responsibility in the extractive industries: experiences from developing countries', *Resources Policy*, 37: 131–137.

Hilson, G. (forthcoming) 'Why is there a large-scale mining "bias" in sub-Saharan Africa?', *Land Use Policy*.

Hilson, G. and Maconachie, R. (2017) 'Formalizing artisanal and small-scale mining: insights, contestations and clarifications', *Area*, 49(4): 443–451.

Hilson, G. and McQuilken, J. (2014) 'Four decades of support for artisanal and small-scale mining in sub-Saharan Africa: a critical review', *The Extractive Industries and Society*, 1(1): 104–118.

Hinton, J.J. (2005) *Communities and Small Scale Mining: An Integrated Review for Development Planning*, Washington, DC: World Bank Mining Group.

Horowitz, L.S. (2011) 'Interpreting industry's impacts: micropolitical ecologies of divergent community responses', *Development and Change*, 42(6): 1379–1391.

ILO (International Labour Organization) (1999) *Social and Labour Issues in Small-Scale Mines*, Report TMSSM/1999, Addis Ababa: ILO.

Land Commission of Liberia (2012) *Land Rights, Private Use Permits and Forest Communities*, Consultant Report by Paul De Wot, Monrovia: Land Commission of Liberia, available at: http://eeas.europa.eu/delegations/liberia/documents/press_corner/20130916_01.pdf (accessed 28 December 2017).

Lanning, G. and Mueller, M. (1979) *Africa Undermined: A History of Mining Companies and Underdevelopment in Africa*, New York: Penguin Books.

Maconachie, R. (2010) '"New spaces" for change? Diamond governance reforms and the micro-politics of participation in post-war Sierra Leone', *Public Administration and Development*, 30(3): 191–202. DOI: 10.1002/pad.558.

Maconachie, R. (2011) 'Re-agrarianising livelihoods in post-conflict Sierra Leone? Mineral wealth and rural change in artisanal and small-scale mining communities', *Journal of International Development*, 23(8): 1054–1067.

Maconachie, R. (2012) 'Diamond mining, urbanisation and social transformation in Sierra Leone', *Journal of Contemporary African Studies*, 30(4): 705–723.

Maconachie, R. and Binns, T. (2007) '"Farming miners" or "mining farmers"? Diamond mining and rural development in post-conflict Sierra Leone', *Journal of Rural Studies*, 23(3): 367–380.

Maconachie, R. and Hilson, G. (2016) 'Re-thinking the child labour "problem" in rural sub-Saharan Africa: the case of Sierra Leone's "half shovels"', *World Development*, 78: 136–147.

Moody, R. (2007) *Rocks and Hard Places: The Globalization of Mining*, London and New York: Zed Books.

Mutemeri, N. and Petersen, F.W. (2002) 'Small-scale mining in South Africa: past, present and future', *Natural Resources Forum*, 26(4): 286–292.

Pegg, S. (2006) 'Mining and poverty reduction: transforming rhetoric into reality', *Journal of Cleaner Production*, 14: 376–387.

Rodney, W. (1972) *How Europe Underdeveloped Africa*, London: Bogle L'Ouverture.

Ross, M. (1999) 'The political economy of the resource curse', *World Politics*, 51(2): 297–322.

Ross, M.L. (2003) 'Oil, drugs and diamonds: the varying roles of natural resources in civil war', in K. Ballentine and J. Sherman (eds), *The Political Economy of Armed Conflict: Beyond Greed and Grievance*, Boulder, CO: Lynne Rienner, pp. 47–70.

Silberfein, M., and Conteh, A.H. (2006) 'Boundaries and conflict in the Mano River region of West Africa', *Conflict Management and Peace Science*, 23: 343–361.

Skaggs, S., Norman, N., Garrison, E., Coleman, D. and Bouhlel, S. (2012) 'Local mining or lead importation in the Roman province of Africa Proconsularis? Lead isotope analysis of curse tablets from Roman Carthage, Tunisia', *Journal of Archaeological Science*, 39: 970–983.

Spiegel, S.J. (2009) 'Resource policies and small-scale gold mining in Zimbabwe', *Resources Policy*, 34: 39–44.

Stock, R. (2013) *Africa South of the Sahara: A Geographical Interpretation*, 3rd edition, New York: The Guildford Press.

The Guardian (2009) *China 'Wants to Set up Factories in Africa'*, available at: www.theguardian.com/world/2009/dec/04/china-manufacturing-factories-africa/print (accessed 23 August 2013).

UN (2012) *Fragile States and Development in West Africa*, Niamey, Niger: Economic Commission for Africa, Sub-Regional Office for West Africa.

UNECA (United Nations Economic Commission for Africa) (2011) *Minerals and Africa's Development: The International Study Group Report on Africa's Mineral Regimes*, Addis Ababa: UNECA.

UNEP (United Nations Environment Programme) (2012) *Analysis of Formalization Approaches in the Artisanal and Small-Scale Gold Mining Sector Based on Experiences in Ecuador, Mongolia, Peru, Tanzania and Uganda*, Geneva: UNEP, Division of Technology, Industry and Economics (DTIE), Chemical Branch.

Wallis, W. and Dyer, G. (2007) 'Wen calls for more access for Africa', *Financial Times*, 16 May, available at www.ft.com/cms/s/34e52c64-0365-11dc-a023-000b5df10621,dwp_uuid=5cdb1d20-feea-11db-aff2-000b5df10621.html (accessed 28 December 2017).

Weber-Fahr, M., Strongman, J., Kunanayagam, R., McMahon, G. and Shelton, C. (2001) *Mining and Poverty Reduction*, Washington, DC: World Bank.

Weinstein, L. (2008) 'The new scramble for Africa', *International Socialist Review*, 60, July–August, available at: http://isreview.org/issues/60/feat-africa.shtml (accessed 23 August 2013).

Willis, K. (2011) *Theories and Practices of Development*, 2nd edition, London: Routledge.

World Bank (2012) *Despite Global Slowdown, African Economies Growing Strongly – New Oil, Gas, and Mineral Wealth an Opportunity for Inclusive Development*, Press Release, 4 October, available at: www.worldbank.org/en/news/press-release/2012/10/04/despite-global-slowdown-african-economies-growing-strongly-world-bank-urges-countries-spend-new-oil-gas-mineral-wealth-wisely (accessed 27 December 2017).

40

Supporting the SMME economy of sub-Saharan Africa

A role for business incubation

Christian M. Rogerson

Introduction

The small, medium and micro-enterprise economy (SMME) is at the heart of contemporary economic progress and employment development across sub-Saharan Africa (SSA). According to several observers, SMMEs account for as much as 95 per cent of all firms in the SSA economy and are responsible for at least 80 per cent of the region's employment (Rogerson, 2001; Fjose *et al.*, 2010; IFC, 2016). By 2020, it is estimated that 122 million new jobs will be required in Africa in order to absorb a growing workforce and to address unemployment. In view of the sclerosis of the large enterprise sector in terms of employment creation, the majority of these jobs are expected to be contributed by SMMEs (McKinsey Global Institute, 2012). The growth potential of Africa's SMME economy is, however, hindered by several factors. Among the key blockages to its expansion are lack of access to finance and new market opportunities, chronic electricity shortages and a weakly developed business environment in which SMMEs are hampered by complex regulations, red tape, corruption, and often the lack of public sector support. It is not surprising therefore that the SMME economy in Africa exhibits high death rates in particular among newly established enterprises in their early and most vulnerable stages of existence (Page and Söderbom, 2015).

The high mortality rate for start-up enterprises is a global phenomenon, albeit it is worst in countries of the Global South (Chandra and He, 2008; InfoDev, 2016). In response to poor survival rates of SMMEs, business incubation has been identified as a tool to help foster start-ups, as well as to support existing SMMEs during their early stages of business development (Rice, 2002; Al-Mubaraki and Busler, 2010; InfoDev, 2010a, 2010b, 2010c, 2010d, 2010e, 2016). Pioneered in Western Europe and North America, the activity of business incubation and the phenomenon of business incubators diffused into developing countries and incorporated as part of frameworks for stimulating innovation, entrepreneurship and business development (Lalkaka, 2003; Akcomak, 2009; Al-Mubaraki and Busler, 2010; InfoDev, 2010d, 2010e; Theodorakopoulos *et al.*, 2014). Thus, business incubators are strategic tools for helping to grow a country's entrepreneurial base while reducing the high mortality of SMMEs (InfoDev, 2016). Business incubators must be seen as a complement to other support structures and initiatives for SMME development. Within a changing policy landscape of interventions that link innovation and entrepreneurship, business incubation is deemed as particularly critical because of its special focus upon strengthening dynamic, growth-oriented early-stage enterprises (Khalil and Olafsen, 2010), and thereby enhancing their prospects of business survival (Henricson Briggs, 2016). The task in this chapter is to analyse the international practice of business incubation and the emerging African policy experience.

The international practice of business incubators

Although the detailed definition of business incubator is contested, there is a general consensus that it provides various targeted business support and technical support services aimed at growing emerging and small start-up business enterprises into financially and operationally independent enterprises (Hackett and Dilts, 2004; Theodorakopoulos *et al.*, 2014; Albort-Morant and Ribiero-Soriano, 2015). For Grimaldi and Grandi (2005: 112), the central push for establishing incubators:

> was to reduce the costs of doing business by offering a set of services ranging from the provision of space, infrastructure and facilities, to more elaborate services, as well as offering access to technical and managerial expertise, assistance in business plan development.

Business incubation has been adopted and supported for varying reasons in several countries (InfoDev, 2014a, 2016). Governments have embraced and supported business incubation within various developmental contexts and in response to different challenges. Although the reasons behind the establishment of business incubators appear similar, often each nation has specific priorities that respond to locally specific challenges (Caiazza, 2014). This said, the most common motives for governments to give support for business incubator programmes relate to issues of employment creation, poverty alleviation, innovation and technology transfer, acceleration of business growth, reduction of the mortality rate of SMMEs, empowerment of specific groups of entrepreneurs, creating value for various stakeholders, urban/rural regeneration, and the revitalisation of local economies (Albort-Morant and Ribiero-Soriano, 2015; Gerlach and Brem, 2015). Beyond the immediate impacts of business incubators upon job creation, enterprise development and local economic revival, it is stressed that business incubators can be 'important change agents in the innovation and entrepreneurship ecosystem' (InfoDev, 2011a: 1). The latest wave of international business incubators are deemed to have 'intellectual capital', which is passed on in terms of intangible assets to the firms that are hosted by the incubator (Calza *et al.*, 2014: 597). From the perspective of one African scholar, business incubators are simply 'one of the most effective ways of promoting entrepreneurship and local economic development' (Adegbite, 2001: 157).

As a concept and practice, business incubation has evolved over time (Grimaldi and Grandi, 2005). It is understood differently by various scholars (Aernoudt, 2004; Akcomak, 2009; Al-Mubaraki and Busler, 2011) leading some commentators to speak of the 'definitional incongruence' surrounding the concept (Theodorakopoulos *et al.*, 2014: 614). According to Hackett and Dilts (2004), most research on business incubation suggests there is uniformity in understanding business incubators as strategic tools to achieve sustainable economic development. In addition, there is consensus among many analysts that a business incubator is an organisation that provides various targeted business support and technical support services aimed at growing emerging and small start-up business enterprises into financially and operationally independent enterprises (Adegbite, 2001; Centre for Strategy & Evaluation

Services, 2002; Adkins, 2009; Akcomak, 2009; InfoDev, 2010a). In some countries, business incubators are an essential driver for making an innovation-based economy (Lalkaka, 2002; InfoDev, 2010d, 2010e). According to the USA's National Business Incubation Association, business incubation is a 'business support process that accelerates the successful development of start-up and fledgling companies by providing entrepreneurs with an array of targeted resources and services' (InfoDev, 2010a: 11). The Information for Development (InfoDev) programme of the World Bank applies the following definition, which guides its international operations:

> Infodev recognizes business incubation as a process aimed at supporting the development and scaling of growth-oriented early-stage enterprises. The process provides entrepreneurs with an enabling environment at the start-up stage of enterprise development. This environment should help reduce the cost of launching the enterprise, increase the confidence and capacity of the entrepreneur and link the entrepreneur to resources required to start and scale a competitive enterprise.
>
> *(InfoDev, 2014a: 8)*

Khalil and Olafsen (2010: 74) assert that from a public policy perspective, the case for public investment in the start-up of business incubators is justified on two grounds. First, it is because effective business incubation yields economic development returns. Second, it is that from a government budget perspective, there is evidence that government contributions towards business incubation can quickly pay for themselves by generating new tax revenues through the jobs and businesses that they germinated. The Brazilian case has been documented and shows that over the past 20 years, an investment of 150 million reals in business incubation has resulted in an estimated 400 million reals in annual tax revenues from graduated enterprises (Khalil and Olafsen, 2010: 75).

According to InfoDev (2016), a typical business incubator consists of a wide variety of business support services that are designed specifically to assist incubatees or clients to start-up and grow small businesses into financially and operationally independent enterprises. Generally speaking, the services offered by business incubators to clients include administrative services, business advice services, training programmes, technical support services, facilitating access to finance; and networking opportunities and marketing (Adegbite, 2001). Three important stages constitute the incubation process, namely pre-incubation, incubation and aftercare. Pre-incubation, stage one, is designed to assist potential incubatees with developing their business ideas into an implementable business before they are admitted as new clients of a business incubator (Bergek and Norrman, 2008). The second stage of incubation involves an intense process of the provision of all essential support services to ensure the graduation of as many small business enterprises as possible into successful firms (InfoDev, 2010a). The final stage is the post-incubation stage, which facilitates the transition and integration of recent graduates into the outside business world for sustained growth and prosperity. Overall, it is stressed as follows:

> Successful incubators accelerate the growth of their clients' businesses as compared to new or small businesses operating independently. They do this by providing advice and guidance and by leveraging linkages to a wide range of innovation and entrepreneurship stakeholders that expand the entrepreneur's access to the knowledge, networks, capital, and markets he or she needs to start and grow the business.
>
> *(InfoDev, 2014a: 12)*

Across the international experience of business incubator development, their core distinguishing characteristics include some of the following features:

- the provision of a comprehensive and integrated business support service under one roof;
- focused mentorship programmes aimed at helping new start-ups get off the ground;
- a management team on site with the necessary expertise to resolve early-stage threats to SMMEs through a network of professionals;
- careful selection of new start-ups through a pre-incubation process before an SMME is admitted into a business incubator;
- provision of a conducive business environment at (often) subsidised cost;
- objective of becoming long-run self-sustaining;
- set time period by which an incubatee must graduate and vacate the incubator; and
- post-incubation support for recent graduates to help integrate them into the unprotected business environment.

Also of note is the increasing trend in many countries for preferring 'incubators without walls' or virtual incubators, as opposed to physical incubators (InfoDev, 2011b; Lewis *et al.*, 2011). It is argued that virtual incubation programmes can be less expensive to operate than residential business incubators that have added expenses surrounding the operation and management of a physical space. This is especially so in rural areas where the client base is spread out over large areas, making commutes difficult such that the option of a virtual incubator might be a good choice (Lewis *et al.*, 2011: 16).

Overall, the incubation concept 'seeks an effective means to link technology, capital and know-how in order to leverage entrepreneurial talent, accelerate the development of new companies, and thus speed the exploitation of technology' (Grimaldi and Grandi, 2005: 111). Nevertheless, business incubators are heterogeneous in nature and can vary dependent upon their main objectives, ownership and management, funding strategies, type of incubatees, facilities, business model or services offered. This said, significant distinctions in operations are forwarded in relation to ownership between the groups of public sector or government-owned and managed incubators, as opposed to those in independent private sector hands. A further option is that of incubators functioning as a form of public–private sector partnership (Albort-Morant and Ribiero-Soriano, 2015).

It is stressed that the business incubator model that evolved in North America and Europe must be adapted to the circumstances and business environment of the Global South (Chandra and Fealey, 2009; Henricson Briggs, 2016). Several analysts concur that Brazil leads one of the most successful business incubation movements in the Global South because of its innovative character and adaptation of incubator models to suit local needs (Chandra, 2007; InfoDev, 2010b; Chandra, 2012; Rogerson, 2016). The landscape of Brazilian business incubators reveals a much broader and diversified range of incubators and operation of different incubator models than is the case for much of the rest of the world (Chandra and Fealey, 2009). It is estimated that 40 per cent of business incubators are technology-based, 18 per cent are for incubation of 'traditional enterprises' (such as clothing and shoes producers), 18 per cent are mixed low-tech and high-tech, 7 per cent are for agri-industrial enterprises, 8 per cent are service enterprises (such as culture and the arts), and the remaining 9 per cent are poverty-focused social business incubators strongly linked to goals of job creation. As compared to other countries, the policy focus in Brazil appears to stress more the provision of softer services such as networking relative to the provision of physical space and hard infrastructure (Chandra, 2012). Above all, the Brazilian experience of business incubation has moved beyond that of much of the rest of the world by its variety and complexity, an innovative adaptation of incubators to specific local needs and circumstances (Rogerson, 2016).

One special form of business incubator that demands a strong focus in the Global South is the agribusiness incubator. The significance of such kinds of business incubator is underscored by the observation in the 2008 World Development Report that the potential of agricultural growth to reduce poverty is four times greater than the potential growth from other sectors (InfoDev, 2011a). A major international review conducted for the World Bank supported the viewpoint that 'agribusiness incubators provide a useful approach toward commercialization and modernization of agriculture and the development of an indigenous agribusiness sector in developing countries' (InfoDev, 2011a: 49). The international record of the practice of agribusiness incubators suggests that they are promising, if not even increasingly essential, components for agricultural progress in the Global South, including Africa (InfoDev, 2014b). In particular, global experience points to the important conclusion that the majority of agribusiness incubators 'have been successful in creating sustainable and competitive enterprises that outweigh the cost, while diffusing a number of technologies as well as product and process innovations' (InfoDev, 2011a: 3).

Business incubation: the African record

Innovative small and medium enterprises are viewed as key drivers for economic and social development across sub-Saharan Africa, and incubators are deemed to be an integral component of an innovation and entrepreneurship ecosystem (D'Costa, 2009; Kiraka, 2009; Henricson Briggs, 2016). It is observed that as compared to other regions of the world, incubation is 'in its infancy on the African continent' (InfoDev, 2006: 3). In a recent international review, Mian *et al.* (2016: 7) isolate 'the absence of research from Africa'. Although a growing number of African countries are experimenting with business incubation – including Ghana, Namibia, Tanzania and Uganda – the countries in sub-Saharan Africa that have instituted the most notable initiatives for business incubation are Kenya, Nigeria and South Africa. It is the diverse and contrasting experiences of these countries that is reviewed in this section.

Much policy interest in business incubators about Africa surrounds the Kenyan experience (Wanyoko, 2013). Three reasons can be advanced. First, in Kenya, 'business incubation is gaining prominence in Government policy' (Ruhiu *et al.*, 2014: 1). Second, Kenya is developing for the first time a national business incubator policy that would place it (along with South Africa and possibly Ghana) at the forefront of African policy debates. Third, Kenya has been highly innovative in terms of using business incubators for promoting goals of industrialisation, as well as for addressing urgent problems of climate change (InfoDev, 2015). Overall, the Information for Development Programme of the World Bank highlights that Kenya has good policy frameworks and support for incubation, and could be a role model for other governments potentially to learn from (InfoDev, 2006: 4).

Kenya is distinguished by an extended history of efforts at business incubation. Some observers locate the origins of business incubation in Kenya as far back as 1967 when the country's Industrial and Commercial Development Corporation established a programme of industrial estates in the country, widely seen as the precursor to business incubator developments (Meru and Struwig, 2011: 113). Indeed, the Kenyan experience is of interest as for some observers it has the status of establishing Africa's first business incubator in 1967 (Henricson Briggs, 2016: 4). A range of different forms of business incubator followed from these early developments. These include programmes operated by the Kenya Industrial Research and Development Institute (KIRDI) to foster technology incubation through both resident and virtual incubation processes, several private sector incubation initiatives (most notably the Kenya Kountry Business Incubator, or KeKoBi) started with support from the World Bank, and the award-winning Kenya Climate Innovation Centre designed to offer targeted services to local climate innovators (InfoDev, 2010f, 2015). In addition, there is a specific business incubator allied to Kenya's export-processing zone programme that seeks to foster export-oriented SMEs (Moyo and Njiraini, 2005). The Export Processing Zones Authority launched this project for nurturing export-oriented Kenyan EPZ small and medium enterprises with a high potential for growth and a demonstrated modern entrepreneurial practice. The landscape of business incubation in Kenya is well organised and includes an association of business incubation stakeholders organised to furnish improved support mechanisms for them (Kiraka, 2009).

Of importance in Kenya is that business incubation is facilitated by the country's innovation-friendly regulatory environment, and importantly by the National Innovation System, defined as a set of institutions whose interactions determine the innovative performance of national firms. The KIRDI stresses industrial R&D is a critical component of national development because of the currently generally low culture of embracing technology. It also supported the establishment of an incubation fund, and the development of an elaborate incubation programme involving key stakeholders in order to address the low survival rate of small enterprises in Kenya (KIRDI, 2011). The organisation, however, faces several challenges in terms of shortages of qualified staff, inadequate funds, the poor dissemination of locally developed innovations, and the importance of ramping up the commercialisation and incubation of local entrepreneurs (KIRDI, 2011: 25). Currently, business incubation in Kenya is viewed as a vehicle to speed up national industrialisation, and in particular through the commercialisation of innovations (Ruhiu *et al.*, 2014: 3).

The most striking new business incubator initiative in Kenya is a special form of business incubator that is targeted to accelerate the development, deployment and transfer of locally relevant climate technologies. Under the Climate Technology Programme (CTP), which links to InfoDev and the World Bank, climate challenges are to be reconfigured into growth opportunities for dynamic small and medium enterprises. With CTP support, small enterprises can assist in emissions reduction and enhance climate resiliency, while at the same time enabling developing countries to capture greater value in the innovation value chain, build competitive sectors and competitive enterprises, and nurture new sources for job creation. In Kenya, 80 per cent of the country's population is currently not served by the electricity grid, and thus constitutes a large potential market opportunity for innovative clean-tech solutions. Local entrepreneurs are being encouraged to respond to these challenges and opportunities by launching new innovative wind and solar technologies that not only foster employment opportunities, but also offer viable off-grid solutions to the country's most marginalised poor communities. Kenya is part of an international network of Climate Innovation Centres (CICs), which are locally owned institutions to drive innovation in clean technology sectors and provide a suite of services to local clean technology small enterprises and climate innovators. Since its foundation in 2012 Kenya's CIC has been described as one of Africa's leading-edge business incubators and aims to catalyse a wave of new clean technology innovation in the country. Further, it seeks to galvanise new private sector-led solutions to clean-tech growth and contribute to sustainable development objectives (World Bank, 2014).

The case of Nigeria offers a contrasting experience in business incubation development. Impetus for the launch of a programme for business incubators in Nigeria was given by the UNDP in 1988. For the past quarter-century, therefore, Nigeria sought to implement a technology incubator programme that exhibits parallels with those in many developed countries (Adegbite, 2001). The first business incubators began to be established in Nigeria in the early

1990s. The objectives of the technology incubator programme in Nigeria parallel those of many other countries, namely to boost the country's industrial base through the commercialisation of R&D and to upgrade and enhance the application of indigenous technologies; to foster the start-up and growth of new innovative businesses engaged in value added and low, medium and high technology activities over time; and to promote functional linkage between research and industry in the country (Adelowo *et al.*, 2012). According to Nigeria's National Board for Technology Incubation, part of the Federal Ministry of Science and Technology, there are 27 technology incubation centres spread across the country seeking to contribute to the mandate of nurturing 'new technology based start-up businesses nationwide' (National Board for Technology Incubation, Nigeria, 2016).

The results of Nigeria's business incubator programme are widely agreed to be disappointing (Obaji *et al.*, 2014; Adelowo *et al.*, 2015; Obaji *et al.*, 2015). Indeed, the Nigerian experience offers a cautionary tale in the reasons for the underperformance of technology business incubation in the African context. Notwithstanding noble programme objectives, several scholars are highly critical of the outputs of the programme over the past 25 years. Adelowo *et al.* (2012: 175) highlight that 'the country has not gained much from the operation of the programme' and write of a record of 'meagre achievement'. This assessment is confirmed by Obaji *et al.* (2015: 440), who state, 'Nigeria's technology incubation programme has not recorded a good success in line with the aim for which it was conceived'. Further, an assessment of the (now) 29 business incubators in the country disclosed that 'most of the centres or the Nigerian incubation programme are not really where it [*sic*] ought to be in comparison with some other countries of equal development' (Obaji *et al.*, 2014: 3).

The reasons for the weak performance of Nigeria's business incubator programme are worth examining as they point to critical issues that can inform policy development in other countries in sub-Saharan Africa. First, there was insufficient adaptation or localisation of the business incubator development model that was imported from developed countries to suit local conditions. Second, it is argued that the Nigerian government adaptation of incubation policies was 'received with open hands by policymakers' who failed to understand what the concept was about, with the consequence that policymakers 'were just groping in the dark as far as the initiative was concerned without having any knowledge related to the "nuts and bolts" of the programme' (Obaji *et al.*, 2015: 441). Third, policy inconsistency rather than stability was a prevalent feature in relation both to the development of small and medium enterprises and the development of science, technology and innovation policy. Fourth, the national innovation system was weak in terms of affording a base for a programme to commercialise technologies, and especially to give impetus to indigenous technologies, as reflected in the shortage of research facilities in Nigeria. Fifth, the programme focus was squarely upon the provision of physical space, with inadequate attention given to the quality and range of business support services that were provided. Sixth, good-sounding policy documents have not been put into practice in terms of implementation, and they are not backed up with appropriate levels of funding. Finally, as with many policies, the incubation policy in Nigeria was manipulated at times for political objectives in terms of selecting certain locations for incubators, irrespective of 'whether it was achievable and viable to have these centres in those regions' (Obaji *et al.*, 2015: 441).

South Africa provides a third case study and offers the most extensive initiatives for using business incubation for development objectives in Africa. In a response to the triple challenge of poverty, unemployment and inequality, the South African government has made substantial commitments to supporting SMMEs as a critical policy vehicle for achieving sustainable economic growth, creating employment opportunities and alleviating poverty (Rogerson, 2004). The importance attached by national government to the role of small businesses is underlined by the establishment during 2014 of a separate dedicated Ministry for Small Business Development. The National Development Plan (NDP), South Africa's blueprint for future development, identifies the SMME sector as central to the realisation of the country's Vision 2030. The NDP Vision 2030 sets forth a target of creating 11 million jobs by 2030, of which 90 per cent is expected to come from SMMEs. Despite the long-standing government commitment and increased support for the sector, South African SMMEs have not reached their intended impact, anticipated or desired by policymakers (Rogerson, 2004). The South African SMME economy continues to experience high mortality rates, with SMMEs failing to grow sufficiently to create the expected employment opportunities, and the country records one of the world's lowest SMME survival rates, with estimates that 80 per cent of South Africa's SMMEs are failed businesses in their first year of existence (Bureau for Economic Research, 2016: 10).

Various forms of business incubation programmes have been applied in South Africa going back to the early 1990s with the establishment of business hives, physical spaces to support new enterprise development mainly situated in the underdeveloped urban townships (Masutha and Rogerson, 2014a, 2014b). During the early 2000s, the programme of business incubation shifted in focus and now was wedded to diffusion of technology through another wave of incubators (SEDA, 2010). The focus was as follows: to increase accessibility to technologies and technical support for small enterprises; to facilitate the acquisition, development and transfer of technology to such enterprises; to facilitate the establishment of women-owned small enterprises; to promote the use of quality systems and standards by small enterprises; to improve small enterprise performance and productivity; to enhance small enterprise profitability and growth; to improve small enterprise competitiveness; and to reduce small enterprise failure rates (InfoDev, 2010c: 18). One of the main achievements of this programme has been to oversee the expansion in the network of business incubators developed by the public sector to reach a total of 42 incubators by 2012 and 49 incubators by early 2016. All these incubators are government-owned and managed through the national Small Enterprise Development Agency (SEDA). All incubators are registered as independent entities, either as a Section 21 or as non-profit companies or trusts. The average incubation process takes approximately three years. Although the three-year period is the norm, it can vary as in certain sectors infant businesses require more time before incubatees are ready to graduate and exit incubators. In terms of the type of incubators favoured, the central focus has been upon supporting sector-specific forms of incubators, mostly linked to manufacturing activities, but with others in ICT, construction and agriculture.

Several performance indicators point to positive impacts. First, and most critically, the network of business incubators reduced the high mortality rate of SMMEs. It is estimated that incubators achieved and maintained a survival rate of eight out of 10 SMMEs as compared to a national failure rate of eight out of 10 SMMEs not receiving support from business incubators. In other words, 80 per cent of SMMEs in these incubators survive their first year of existence. Other positive indicators relate to job creation, with these incubators generating an estimated total of 21,322 jobs between 2004 and 2009 when supporting 1,900 SMMEs (Masutha, 2014). Of this total, the emphasis upon assisting disadvantaged groups is mirrored in statistics that 81 per cent of all businesses supported by the incubators are owned by blacks. Women ownership is at a level of 36 per cent, a finding that suggests that the programme has underperformed in respect of objectives to empower women-owned small businesses (Masutha, 2014).

The most recent phase in the evolution of business incubators in South Africa is a part of national government SMME programmes initiated in 2012. During 2012, the Department of Trade and Industry (DTI) launched the so-called Incubation Support Programme (ISP), which reconfirmed the government's long-term commitment to business incubation as a vehicle to create sustainable enterprises. According to the DTI (2013), the primary objective of the

ISP is to build successful enterprises with the potential to create employment opportunities and revitalise and strengthen both local and national economies. Further, the ISP aims to broaden economic participation by ensuring that incubated SMMEs graduate into the mainstream economy. At the heart of the ISP is the notion of public–private partnerships. Through the ISP, government has called on large private sector businesses to partner and participate meaningfully in national programmes for SMME development through skills transfer, supplier development and marketing assistance. The ISP became effective in September 2012 and is planned to roll out for 10 years. During 2012, the DTI announced the ambitious target of establishing a total of 250 business incubators in South Africa by 2015, a target that, with hindsight, was wildly optimistic. The ISP functions on a grant system. Approved incubators qualify for a maximum of R10 million per year government support for a period of three years. The ISP builds upon and deepens a multi-stakeholder approach to business incubation development that can be traced back to the early 2000s in national policy. During 2015, SEDA released its latest incubation policy, which addresses issues pertaining to the establishment and governance of incubation centres (SEDA, 2015). The policy document restates the critical role of business incubators for provision of 'both financial and non-financial technology and quality support services for small enterprises' (SEDA, 2015: 5). The prime vehicle for delivery of services is via the channel of the suite of technology business incubators that are supported by the programme. Overall, it is made clear that SEDA established incubation centres are dedicated 'to support the transformation of selected early stage businesses with high potential, into self-sufficient, growing and profitable enterprises' (SEDA, 2015: 5).

Although the South African landscape of business incubation remains dominated by SEDA's expanding network of business incubators, recent years have witnessed a welcome broadening in the range of actors engaged in establishing and operating business incubators (Masutha and Rogerson, 2015). Private sector involvement in business incubation is reflected in the establishment of a suite of private sector-operated business incubators by 2016 (Business Partners, 2016). Operational differences exist between the groups of public as opposed to private sector incubators (see Masutha and Rogerson, 2015). It is apparent that whereas public sector incubators place considerable weight on the goals of broadening economic participation, creating employment and transference of skills, the private sector incubators concentrate upon nurturing SMMEs into assets of value and maximising the profitability and turnover of businesses. Recruitment processes into incubators also vary with private sector incubators operating a stricter set of selection criteria than those in the public domain. For private sector business incubators, issues are profitability, growth potential, a bankable business plan, and specific emphasis on characteristics and motivations of the individual entrepreneur. By contrast, the group of public incubators tend to target SMMEs of all sizes and recruit incubatees sensitive to empowerment criteria of ownership by disadvantaged groups. In addition to broadening involvement of different stakeholders engaged in business incubation in South Africa, new policy developments suggest that a widened range of public sector business incubation, including a new Climate Innovation Centre (Climate Innovation Centre, South Africa, 2016) and potential programme extension to such non-traditional sectors as tourism and creative industries (Breytenbach, 2016).

Conclusions

The employment challenges for Africa over the next few decades are immense as the job creation performance of large enterprises is failing to address the needs of the continent's growing labour force (McKinsey Global Institute, 2012). A healthy SMME economy is critical for job creation and the achievement of inclusive economic growth in sub-Saharan Africa (Page and

Söderbom, 2015). As the SMME sector is a logical conduit for governments and international development agencies to support in order to maximise employment expansion, business incubation is a logical approach to adopt. Business incubators are part of a range of policy interventions that can be applied to support the SMME economy. Incubators can be powerful instruments for supporting the healthy development of SMMEs, but they are no substitute for necessary wider changes to the business environment that governments must undertake. Within a changing landscape of interventions that integrate innovation and entrepreneurship, the activity of business incubation is significant because of its special focus upon enhancing the survival of growth-oriented early-stage enterprises. It is argued that Africa can learn from, but should not be bound to replicate, developed country models of business incubation, and instead should seek to chart its own pathway(s) of incubator policy development (Rogerson, 2016). In the Global North it is recognised, however, that no one-size-fits-all incubator practice, policy or service is guaranteed to produce success; instead, there is required a synergy among multiple practices, policies and services in order to generate optimal outcomes.

Critically, the establishment of successful incubation ecosystems demands the adaptation of incubator models to suit local needs (Henricson Briggs, 2016). Business incubators should be conceptualised and planned for as an integral part of a broader national innovation and economic development policy that is geared to strengthening the overall 'ecosystem' of innovation and entrepreneurship. The building and strengthening of an NSI is of paramount importance for the successful development of business incubators. Specific interventions to assist business incubators may be ineffective in the context of a poor or dysfunctional NSI. Incubation should be viewed at one level as a vehicle to develop or create local technologies, and therefore should be firmly aligned to the national innovation framework, with the incubation system representing a channel to the successful commercialisation of local research or innovation. Business incubators should be designed to support and be part of broader national, regional or local strategic frameworks, and are best promoted by an inclusive partnership of public and private sector stakeholders. Moreover, business incubation cannot succeed in a macro-environment of poor governance, lack of transparency, excessive bureaucracy and a culture of corruption.

Although the provision of physical space is central to the incubator model, the value added of incubator operations increasingly hinges upon the type and quality of business support and networking services (soft services) that are furnished. Diversity is to be welcomed in the network of national business incubators, and can be encouraged by experimentation around different formats of business incubator that go beyond the classic technology business incubator. For example, the approach of business incubation can be extended usefully to non-manufacturing or service-based activities to maximise for SMME development the innovation and creativity in tourism and creative industries or the cultural sector. A critical new focus for business incubation, both in urban and rural areas, is upon promoting locally relevant indigenous climate innovations by SMMEs in order to address questions of climate adaptation. The institution of Climate Innovation Centres is a most promising vehicle for such interventions for Africa. In addition, for rural areas, the role of agriculture in economic development and poverty alleviation can also be enhanced through appropriate interventions for agribusiness incubators.

References

Adegbite, O. (2001) 'Business incubators and small enterprise development: the Nigerian experience', *Small Business Economics*, 17(3): 157–166.

Adelowo, C.M., Olaopa, R.O. and Siyanbola, W.O. (2012) 'Technology business incubation as a strategy for SME development: how far, how well in Nigeria', *Science and Technology*, 2(6): 172–181.

Adelowo, C.M., Illori, M.O, Siyanbola, W.O. and Oluwale, B.A. (2015) 'Technological learning mechanisms in Nigeria's technology incubation centre', *African Journal of Economic and Management Sciences*, 6(1): 72–89.

Adkins, D. (2009) 'No policy is a good national policy', *Asia Pacific Journal of Innovation and Entrepreneurship*, 3(3): 5–21.

Aernoudt, R. (2004) 'Incubators: tool for entrepreneurship?', *Small Business Economics*, 23(2): 127–135.

Akcomak, I.S. (2009) *Incubators as Tools for Entrepreneurship Promotion in Developing Countries*. Helsinki: UNU-WIDER Working Papers 2009/52.

Albort-Morant, G. and Ribiero-Soriano, D. (2015) 'A bibliometric analysis of international impact of business incubators', *Journal of Business Research*, 69(5): 1775–1779, DOI: http://dx.doi.org/10.1016/j.jbusres2015.10.054.

Al-Mubaraki, H. and Busler, M. (2010) 'Business incubators: findings from a worldwide survey, and guidance for GCC states', *Global Business Review*, 11(1): 1–20.

Al-Mubaraki, H. and Busler, M. (2011) *Application of Business Incubators in Europe: A SWOT Analysis*, Paper presented at WASD 9th International Conference, Atlantic City, USA, 26–28 October.

Bergek, A. and Norrman, C. (2008) 'Incubator best practice: a framework', *Technovation*, 28(1–2): 20–28.

Breytenbach, M. (2016) 'Incubating enterprises', *Engineering News*, 29 April–5 May: 18–19.

Bureau for Economic Research (2016) *The Small, Medium and Micro Enterprise Sector of South Africa*, Stellenbosch: Bureau for Economic Research, Report Commissioned by SEDA.

Business Partners (2016) *Small to Medium Enterprise (SME) Incubators and Support Programmes*, Johannesburg: Business Partners.

Caiazza, R. (2014) 'Benchmarking of business incubators', *Benchmarking: An International Journal*, 21(6): 1062–1069.

Calza, F., Dezi, L., Schiavone, F. and Simoni, M. (2014) 'The intellectual capital of business incubators', *Journal of Intellectual Capital*, 15(4): 597–610.

Centre for Strategy & Evaluation Services (2002) *Benchmarking of Business Incubators*, Sevenoaks: Centre for Strategy & Evaluation Services for the European Commission.

Chandra, A. (2007) *Business Incubation in Brazil: Creating an Environment for Entrepreneurship*, Indiana State University Networks Financial Institute Working Paper WP 25.

Chandra, A. (2012) 'Business incubation in Brazil: creating an environment for new ventures', *Enterprise Development and Micro Finance*, 20(1): 1–13.

Chandra, A. and Fealey, T. (2009) 'Business incubation in the United States, China and Brazil: a comparison of role of government, incubator funding and financial services', *International Journal of Entrepreneurship*, 13: 67–86.

Chandra, A. and He, W. (2008) 'Government involvement and financial service intensity: a comparative study of three business incubators in China', *International Journal of Innovation and Learning*, 5(3): 266–282.

Climate Innovation Centre, South Africa (2016) *Accelerating the Growth of Innovative Clean Tech Enterprises Towards Low-Carbon, Green Growth Paths*, available at: www.cicsouthafrica.co.za/home (accessed 10 July 2016).

D'Costa, V. (2009) *Harnessing ICT to Reach Africa's Development Goals*, paper prepared for InfoDev, Washington, DC.

DTI (Department of Trade and Industry) (2013) *DTI Launches Incubation Support Programme*, Pretoria: DTI.

Fjose, S., Grunfeld, L.A. and Green, C. (2010) *SMEs and Growth in Sub-Saharan Africa*, Oslo: Menon Business Economics Publication No. 14.

Gerlach, S. and Brem, A. (2015) 'What determines a successful business incubator? Introduction to an incubator guide', *International Journal of Entrepreneurship Venturing*, 7(3): 286–307.

Grimaldi, R. and Grandi, A. (2005) 'Business incubators and new venture creation: an assessment of incubating models', *Technovation*, 25: 111–121.

Hackett, S.M. and Dilts, D.M. (2004) 'A systematic review of business incubation research', *Journal of Technology Transfer*, 29: 55–82.

Henricson Briggs, K.A.K. (2016) *Travels of Business Incubators: Exploring Entrepreneurship from an Embeddedness Perspective in Uganda and Tanzania*, PhD dissertation, Chalmbers University of Technology, Gothenburg, Sweden.

InfoDev (2006) *Promoting Innovation and Entrepreneurship in Africa: Africa Regional Workshop InfoDev Incubator Initiative*, Washington, DC: InfoDev.

InfoDev (2010a) *Global Practice in Incubation: Policy Development and Implementation*, Washington, DC: World Bank.

InfoDev (2010b) *Global Practice in Incubation: Policy Development and Implementation – Brazil Incubation Country Case Study*, Washington, DC: InfoDev.

InfoDev (2010c) *Global Practice in Incubation: Policy Development and Implementation – South Africa Incubation Country Case Study*, Washington, DC: InfoDev.

InfoDev (2010d) *Global Practice in Incubation: Policy Development and Implementation – New Zealand Incubation Country Case Study*, Washington, DC: InfoDev.

InfoDev (2010e) *Global Practice in Incubation: Policy Development and Implementation – Malaysia Incubation Country Case Study*, Washington, DC: InfoDev.

InfoDev (2010f) *Kenya Climate Innovation Centre: CIC – A Business Plan for the Financing and Implementation of a CIC in Kenya*, Report for the Climate Technology Program, Washington, DC: World Bank.

InfoDev (2011a) *Growing Food, Products and Businesses: Applying Business Incubation to Agribusiness SMEs*, Washington, DC: World Bank.

InfoDev (2011b) *Lessons on Virtual Business Incubation Services*, Washington, DC: World Bank.

InfoDev (2014a) *An Evaluation and Impact Assessment of Business Incubation Models in Eastern Europe and Central Asia: An InfoDev Study of Nine Business Incubators in Armenia, Belarus, Kazakhstan, Poland, Romania, the Russian Federation, Serbia and Turkey*, Washington, DC: World Bank.

InfoDev (2014b) *Timbali Technology Incubator: South Africa Case Study*, Washington, DC: World Bank.

InfoDev (2015) *Kenya Climate Innovation Centre Voted 'Most Promising Business Incubator in Africa'*, available at: www.infodev.org/articles/kenya (accessed 28 December 2017).

InfoDev (2016) *Business Incubation Toolkit: iDISC Incubation Good Practice*, available at: www.infodev.org/business-incubation-toolkit (accessed 28 December 2017).

IFC (International Finance Corporation) (2016) *SME Initiatives*, available at: www.ifc.org/wps/wcm/connect/REGION__EXT_Content/Regions/Sub-Saharan+Africa/Advisory+Services/SustainableBusiness/SME_Initiatives/ (accessed 7 August 2016).

Khalil, M.A. and Olafsen, E. (2010) 'Enabling innovative entrepreneurship through business incubation', in A. Lopes-Clares (ed.), *The Innovation for Development Report 2009-2010: Strengthening Innovation for the Prosperity of Nations*, Basingstoke: Palgrave Macmillan, pp. 69–84.

Kiraka, R. (2009) *Innovative Private Sector Development Instruments: An African Perspective*, Vienna: Informationsburo Wirtschaft und Entwicklung Working Paper 1.

KIRDI (Kenya Industrial Research and Development Institute) (2011) *Strategic Plan 2010–2015*, Nairobi: KIRDI.

Lalkaka, R. (2002) 'Technology business incubators to help build an innovation-based economy', *Journal of Change Management*, 3(2): 167–176.

Lalkaka, R. (2003) 'Business incubators in developing countries: characteristics and performance', *International Journal of Entrepreneurship and Innovation Management*, 3(1/2): 31–55.

Lewis, D.A., Harper-Anderson, E. and Molnar, L.A. (2011) *Incubating Success: Incubation Best Practices that Lead to Successful Business Ventures*, Report prepared by the US Department of Commerce Economic Development Administration.

Masutha, M. (2014) 'Small business incubators in South Africa: emergence, geography and local impacts', unpublished MSc dissertation, University of Johannesburg.

Masutha, M. and Rogerson, C.M. (2014a) 'Small enterprise development in South Africa: the role of business incubators', *Bulletin of Geography: Socio-Economic Series*, 26: 141–155.

Masutha, M. and Rogerson, C.M. (2014b) 'Small business incubators: an emerging phenomenon in South Africa's SMME economy', *Urbani Izziv*, 25: S48–S63.

Masutha, M. and Rogerson, C.M. (2015) 'Business incubation for SMME development: South African pathways', *Urban Forum*, 26(2): 223–241.

McKinsey Global Institute (2012) *Africa at Work: Job Creation and Inclusive Growth*, Seoul, San Francisco, CA, London and Washington, DC: McKinsey Global Institute.

Meru, A.K. and Struwig, M. (2011) 'An evaluation of the entrepreneurs' perception of business-incubation services in Kenya', *International Journal of Business Administration*, 2(4): 112–121.

Mian, S., Lamine, W. and Fayolle, A. (2016) 'Technology business incubation: an overview of the state of knowledge', *Technovation*, 50–51: 1–12.

Moyo, E. and Njiraini, P. (2005) *Misallocation of Workspaces for MSEs in Kenya: Some Lessons and Models*, Nairobi: Kenya Institute for Public Policy Research and Analysis Discussion Paper No. 53.

National Board for Technology Incubation, Nigeria (2016) *Our Mandate*, available at: www.nbti.gov.ng/our-mandate (accessed 19 February 2016).

Obaji, N.O., Olugu, M.U. and Balogun, K.O. (2014) 'Innovative policies in technology business incubation: key elements for sustainable entrepreneurship development in Nigeria', *International Journal of Scientific and Research Publications*, 4(11): 1–7.

Obaji, N.O., Olugu, M.U. and Obieke, B.C. (2015) 'Business incubation adaptation and success factors in Nigerian context of a developing country: a literature review', *International Journal of Science, Technology and Management*, 4(1): 436–444.

Page, J. and Söderbom, M. (2015) 'Is small beautiful? Small enterprise, aid and employment in Africa', *African Development Review*, 27: 44–55, DOI: 10.1111/1467-8268.12138.

Rice, M.P. (2002) 'Co-production of business assistance in business incubators: an exploratory study', *Journal of Business Venturing*, 17(2): 163–187.

Rogerson, C.M. (2001) 'In search of the African miracle: successful small enterprise development in Africa', *Habitat International*, 25: 115–142.

Rogerson, C.M. (2004) 'The impact of the South African government's SMME programmes: a ten year review (1994–2003)', *Development Southern Africa*, 21: 765–784.

Rogerson, C.M. (2016) 'Re-energising business incubation policy in South Africa: learning from international experience', in R. Donaldson, G. Visser, J. Kemp and J. de Waal (eds), *#Celebrate Geography: Proceedings of the 11th Biennial Conference of the Society of South African Geographers*, Stellenbosch: Society for South African Geographers, pp. 22–29.

Ruhiu, R.W., Ngugi, P.K. and Waititu, G.A. (2014) 'Effects of managerial skills on the growth of incubated micro and small enterprises in Kenya', *International Journal of Social Sciences and Entrepreneurship*, 1(12): 1–11.

SEDA (Small Enterprise Development Agency) (2010) *Seda Technology Programme Incubators Fast Track SMME Success*, Pretoria: SEDA.

SEDA (Small Enterprise Development Agency) (2015) *Incubation Policy: The Establishment and Governance of Incubation Centres*, Pretoria: SEDA.

Theodorakopoulos, N., Kakabadse, N.K. and McGowan, C. (2014) 'What matters in business incubation? A literature review and suggestion for situated theorizing', *Journal of Small Business and Enterprise Development*, 21: 602–622.

Wanyoko, A.M. (2013) 'Influence of business incubation services on growth of small and medium enterprises in Kenya', *International Journal of Social Sciences and Entrepreneurship*, 1(7): 1–14.

World Bank (2014) *Climate Innovation Center Unlocks Potential for Kenya's Green Entrepreneurs*, available at www.worldbank.org./en/news/feature 2014/10/29 (accessed 11 July 2016).

41

Africa's tourism economy

Uneven progress and challenges

Christian M. Rogerson and Jayne M. Rogerson

Introduction

Conventionally, African debates about economic growth and development are dominated by issues around the potential of agriculture, mining resources and, to a lesser extent, the role of manufacturing. Less acknowledged is the seemingly hidden role and opportunities of Africa's services sector for raising national incomes, employment creation and enhancing growth prospects. Within the service economy, the tourism and hospitality industry is acknowledged as an anchor for sustainable development, and identified also as one of the key sectors with a potential to evolve towards a green low-carbon economy (Scott *et al.*, 2012; UNWTO, 2012; Gössling *et al.*, 2013). At least for the past decade, the tourism sector has been viewed as vital for contributing to achieving several of the 2015 Millennium Development Goals (Saarinen *et al.*, 2013b). Arguably, it will likely remain in the forefront of activities surrounding the 2030 Agenda for Sustainable Development with the United Nations endorsement of 17 Sustainable Development Goals designed to end poverty, fight inequality and injustice, and address global climate change. Especially noteworthy, given the high levels of poverty across the African continent, is the importance attached to tourism as a lever for job creation, poverty reduction, improving livelihoods and catalysing local development prospects (Rogerson and Rogerson, 2010; Snyman, 2012; Spenceley and Meyer, 2012; Okello and Novelli, 2014; Saarinen and Rogerson, 2014; UNWTO, 2015a; de Boer and van Dijk, 2016).

It is argued that while the economic importance of tourism in Africa in the global tourism economy remains modest, as indexed by its 5 per cent share of international arrivals and 3 per cent share of tourism receipts, since 1990 the sector has been steadily rising in terms of its significance for economic development (Rogerson, 2007; UNWTO, 2015b; WTTC, 2015). Despite the global economic turmoil following the 2008 financial crisis, Africa was less hard hit than the world average in terms of international tourism receipts, such that the continent represents one of the key emerging regions of the global tourism economy (Blanke *et al.*, 2011; Cohen and Cohen, 2015). The consensus across several international agencies – including the World Bank, the African Development Bank and the United Nations World Tourism Organization (UNWTO), as well as many academic observers, is that tourism constitutes one of Africa's most promising economic sectors for development in the coming decades (Twining-Ward, 2009;

Rogerson, 2012a; World Bank, 2014; UNWTO, 2015a). Moreover, Blanke *et al.* (2011), the World Bank (2014) and the UNWTO (2015b) concur that Africa's current achievements in the growth of the tourism sector must be set against a backcloth of the continent's still relatively unexploited potential for further tourism development.

The significance of tourism as a potential vehicle for economic and social progress is acknowledged increasingly by African governments in terms of national development plans. Botswana, Egypt, Kenya, Tanzania, Rwanda, Tunisia, Lesotho, Namibia, Malawi and South Africa are among a long list of African nations for which tourism is now an essential element within strategic national development programming. The UNWTO (2015b: 6) underlines that tourism is considered a priority sector for many African countries, and highlights optimism around future tourism growth as a driver for economic and social development, including for poverty alleviation. Set against such a context, in this chapter we seek to provide an overview of the current uneven state of tourism development across Africa, and then isolate several of the most significant challenges that must be addressed in respect of fulfilling the high expectations that surround the tourism sector. Although brief references will be made to tourism issues in North Africa, the central focus is upon the rising economies of tourism development and accompanying challenges across the countries of sub-Saharan Africa.

The state of African tourism

Africa enjoys a broad range of assets that can be the basis for catalysing tourism development (Twining-Ward, 2009; Spenceley, 2010; Novelli, 2015). For long-haul international tourists, the continent's major attractions traditionally are focused on nature tourism and are rooted upon Africa's extraordinary biodiversity. Several countries in East and southern Africa are recognised as world-leading destinations for safari wildlife tourism and viewing the 'big five', namely the African elephant, buffalo, lion, leopard and rhinoceros. Beyond the iconic safari experience, alternative destinations are emerging for other forms of wildlife watching, most notably for gorilla tourism, avitourism, the observation of marine wildlife (especially whales) and viewing agglomerations of wildlife along migration routes (Hottola, 2009; Spenceley *et al.*, 2010). As observed by the UNWTO (2015b: 7), many of these 'wildlife-related tourism products can only be experienced on the African continent and thus represent a unique selling proposition for African tourism'. In the Global North, interest in observing wildlife in their natural habitats has burgeoned as a result of growing media/television coverage, Internet exposure and heightened awareness of conservation issues, because of the risk of species extinction associated with negative impacts of poaching and loss of natural habitats (UNWTO, 2015b).

In addition to nature tourism, other growth points exist in the landscape of African tourism. One of the fastest-expanding segments of the tourism economy is cultural tourism, which several analysts see as offering untapped opportunities for tourism expansion because of Africa's rich traditions of music, art and dance (World Bank, 2014; Novelli, 2015; Manwa *et al.*, 2016; Ngozi and Chinonso, 2016). Twining-Ward (2009: 36) argues that cultural heritage products are 'perhaps the most underdeveloped of Africa's tourism products'. Botswana, Burkina Faso, Ghana, Kenya, Mali, Mozambique, Tanzania, Uganda and Zimbabwe are among a number of African countries using cultural tourism as a niche product in order to diversify their tourism economies (Saarinen *et al.*, 2014; Adiyia *et al.*, 2015; Mgonja *et al.*, 2015; Saarinen and Rogerson, 2015; Manwa *et al.*, 2016). Several African World Heritage Sites, such as the Cradle of Humankind and Mapungubwe National Park in South Africa or Great Zimbabwe in Zimbabwe, are currently being leveraged for local tourism development (Rivett-Carnac, 2011; Sinthumule, 2015; Rogerson and van der Merwe, 2016). For other groups of international leisure tourists, Africa's major attractions are found in its heritage sites, settings of natural beauty, beaches, opportunities for a host of different adventure tourism offerings, and for diaspora tourism (see Twining-Ward, 2009; Spenceley, 2010; McKay, 2013, 2014; Mensah, 2015; McKay, 2016). Volunteer tourism opportunities for groups of mainly youth travellers represent another emerging niche in the rise of African tourism (Stoddart and Rogerson, 2004; Rogerson, 2011; Otoo and Amuquandoh, 2014; Rogerson and Slater, 2014).

Attention must also be drawn to the critical role of business tourism in contemporary Africa. Although the significance of business tourism is uneven across different African countries, it is estimated to represent at least 25 per cent of international arrivals for sub-Saharan Africa as a whole (World Bank, 2014). The majority of these business tourists are intra-African arrivals rather than long-haul international tourists from Europe, North America or Asia. They include both formal sector business travellers and a growing stream of informal business tourists, many of them engaged in cross-border trading and shopping (Rogerson and Letsie, 2013; Rogerson, 2014a, 2015a). Overall, in respect of African tourism, sight must not be lost of the enormous growth of regional travel within the continent and of the extensive flows of domestic tourists, which in numerical terms are far greater than those numbers of international travellers, as shown, for example, in the case of South Africa (Rogerson, 2015b; Harmer and Rogerson, 2016). It is demonstrated in a range of recent research investigations that the growing appetites of Africa's middle classes, as well as its lower-income populations, for domestic and regional travel is driven primarily by motives of leisure, business, and for visits to friends and relatives (VFR) (Rogerson and Hoogendoorn, 2014; World Bank, 2014; Cohen and Cohen, 2015; Rogerson and Harmer, 2015; Rogerson and Mthombeni, 2015).

Arguably, the category of VFR travel is the single largest driver of all tourism flows across Africa, albeit not always captured by official tourism data, which emphasises the collection of statistics relating to international as opposed to domestic tourism. The reliability and quality of official information on tourism in Africa must be seriously questioned (Rogerson, 2007, 2012a; World Bank, 2014). Credible data on domestic tourism flows, the informal sector element of tourism (including VFR travel), and even on the significant sector of business tourism are unavailable for the majority of African countries. Accordingly, only a broad sketch of the state of African tourism can be undertaken, which is based on the available statistics relating to international tourism arrivals published by the UNWTO (2016). For Africa as a whole, the UNWTO (2015a) points out that the leading long-haul source markets are France, the UK, the USA, Germany and Portugal, with China, India and Russia important emerging source markets.

Tables 41.1 and 41.2 present UNWTO (2016) statistics concerning the state of tourism in Africa disaggregated on the basis of UNWTO regional classifications (see Table 41.1) and on the basis of leading individual countries for tourism in sub-Saharan Africa (see Table 41.2).

Table 41.1 Africa: growth of international tourism arrivals and projections, 1980–2030 (million)

UNWTO region	*1980*	*1995*	*2010*	*2020*	*2030*
North Africa	4.0	7.3	18.7	31	46
West and Central Africa	1.0	2.3	6.8	13	22
East Africa	1.2	5.0	12.1	22	37
Southern Africa	1.0	4.3	12.6	20	29
Total: Africa	**7.2**	**18.9**	**50.3**	**85**	**134**

Source: UNWTO (2016: 15).

Note: Egypt and Libya not included in the North Africa region by UNWTO.

Six points can be observed. First is the remarkable progress made in terms of the expansion of international arrivals in Africa as a whole from a 1980 baseline of 7 million to over 50 million based on current (2015) estimates (see Table 41.1). Past images of a 'dark continent' are now being replaced by narratives about Africa awakening and welcoming tourists. Second, at the regional scale of analysis, North African destinations with their mix of sea, sun and sand tourism, as well as heritage assets, capture the majority of international arrivals, in part because of their geographical proximity to European source markets. During the period between 1980 and 2010, the regions of East Africa and southern Africa have accelerated as attractive destinations for international tourism, and in the latter case assisted by the emergence of a democratic South Africa in 1994 and the ending of international sanctions on the country's tourism industry (Rogerson and Visser, 2004). The most undeveloped regions for international tourism continue to be West Africa and central Africa, which are parts of Africa where the growth of tourism demand is constrained by policy challenges surrounding regulatory frameworks, the business environment, tourism infrastructure, and issues of human resources for the tourism industry.

Third, the geographical unevenness of patterns of tourism development across different African countries is evidenced in Table 41.2. Within sub-Saharan Africa, South Africa emerges overwhelmingly as the leading destination for international tourism. This said, it needs to be understood that around 78 per cent of South Africa's international tourists are from other regional source markets in Africa. The majority of South Africa's cohort of international tourist arrivals is constituted by land arrivals, as opposed to air arrivals, with the leading source markets being the proximate countries of Botswana, Lesotho, Mozambique Namibia, Swaziland and Zimbabwe. Although the precise data are unavailable to unpack regional from

Table 41.2 Sub-Saharan Africa: leading individual destinations by international tourism arrivals ('000s)

Country	*2010*	*2013*	*2014*
South Africa	8,074	9,537	9,549
Botswana	1,973	1,544	1,966
Zimbabwe	2,239	1,833	1,880
Mozambique	1,718	1,886	1,661
Namibia	984	1,176	1,320
Uganda	946	1,206	1,266
Kenya	1,470	1,433	1,261
Tanzania	754	1,063	1,113
Ghana	931	994	1,093
Mauritius	935	993	1,039
Zambia	815	915	947
Rwanda	504	864	926
Senegal	900	1,063	836
Swaziland	868	968	N/D
Cameroon	569	912	N/D
Nigeria	1,555	600	N/D
Ethiopia	468	681	770
Angola	425	650	595
Cape Verde	336	503	494
Côte d'Ivoire	252	380	471
Total: Sub-Saharan Africa	**30,743**	**33,971**	**34,877**

Source: Based on UNWTO (2016: 11).

Note: N/D = no data.

long-haul international tourist arrivals in other African countries, it is known that cross-border intra-African travel – often for purposes of business or visiting friends and relatives – represents a substantial segment of total international tourism arrivals across most countries in East and southern Africa (Rogerson, 2015a). Fourth, the lucrative long-haul international leisure market is mainly attracted to the beach and scenic attractions of Mauritius and the 'big five' wildlife tourism destinations of Kenya, Namibia, Botswana, Tanzania and South Africa. What might be called the second wave of growing African destinations for (long-haul) international tourists would include Ethiopia, Uganda, Rwanda, Ghana, Zambia, Zimbabwe and Mozambique. Fifth, tourism in several African destinations has been impacted by health concerns (Ebola outbreak) and terrorism attacks on hotels and beach resorts, with Kenya, Burkina Faso, Sierra Leone and Mali experiencing negative tourism consequences (Onuoha, 2016). Tightened visa regulations can also have deleterious outcomes for international tourism arrivals, as has occurred with South Africa (UNWTO, 2016). Sixth, international tourism is undeveloped and of limited economic significance in a number of African destinations that have experienced or currently are experiencing civil strife and political instability, with South Sudan, Eritrea and Somalia being prime examples.

The above analysis is reinforced by the results of a panel data regression analysis of international tourist arrivals in Africa that pointed to several key determinants of international travel to Africa. The major influential factors were isolated as quality of tourism infrastructure, level of development of the destination, marketing and political stability (Naude and Saayman, 2005).

In an important study, the World Bank (2014) categorises African countries on the basis of income ranking and the scale of tourism development, which is a manifestation of the uneven progress of tourism across the continent. Based upon these criteria, a fourfold typology is proposed of destinations in sub-Saharan Africa (see Table 41.3). First are the *pre-emergent* ones

Table 41.3 World Bank classification of countries in sub-Saharan Africa based on the tourism development level and income ranking

Level of tourism development	*Low income*	*Lower middle income*	*Upper middle income*
Pre-emerging	Central African Republic, Chad, Comoros, Democratic Republic of Congo, Eritrea, Guinea, Guinea-Bissau, Liberia, Niger, Somalia, Togo	Republic of Congo, Equatorial Guinea, Sudan	None
Potential/initiating	Benin, Burundi, Ethiopia, Madagascar, Mali, Mauritania, São Tomé and Príncipe, Sierra Leone	Angola, Cameroon, Côte d'Ívoire, Lesotho, Nigeria, Swaziland	Gabon
Emerging/scaling up	Burkina Faso, Gambia, Malawi, Mozambique, Rwanda, Senegal, Uganda, Zambia, Zimbabwe	None	Seychelles
Consolidating/ maintaining and deepening success	Kenya/Tanzania	Cape Verde, Ghana	Botswana, Mauritius, Namibia, South Africa

Source: World Bank (2014: 6).

that have not yet developed their tourism sector. These are countries that have governance problems, low interest in tourism, and inevitably low tourism growth prospects in the short to medium term. Second are *potential* tourist destinations that are in the process of initiating tourism development. These countries are viewed as having some basic infrastructure for tourism but 'still face market failures, such as with regulation, resources and institutions' (World Bank, 2014: 5). The third group are *emerging* tourism destinations that are scaling up the development of the tourism economy. Although these countries are prioritising tourism development and have solid institutions, several issues remain as blockages to maximising tourism competitiveness. The core issues are considered as shortcomings of government policy and relate to high access costs to the destination, financing, and issues around land access for hotel establishment. The final and highest group of countries in terms of tourism development level are those with established tourism industries that are engaged in the project of *deepening and sustaining* the success of tourism. This entails diversifying tourism products, the building of human resources, and monitoring environmental and social impacts (World Bank, 2014). Across each of these four categories related to the scale of tourism development, it is possible to differentiate African destinations in terms of their income ranking, whether low-, lower-middle or upper-middle income (see Table 41.3).

Africa's tourism development challenges

Over the past two decades, the tourism sector in Africa has been the subject of growing scrutiny by academics and international development agencies. A growing stream of scholarly writings has emerged and consolidated debate about various problems and challenges that confront the tourism sector in maximising opportunities to address Africa's developmental problems, and potentially also to support the United Nations Sustainable Development Agenda goals (e.g. see Ashley and Roe, 2002; Mbaiwa, 2005; Mitchell and Ashley, 2006a; Rogerson, 2006; Mitchell and Faal, 2007; Rogerson and Visser, 2007; Hottola, 2009; Saarinen *et al.*, 2009; Akyeampong, 2011; Magole and Magole, 2011; Manwa, 2011; Rogerson and Rogerson, 2011; Saarinen, 2011; van der Duim *et al.*, 2011; Magombo and Rogerson, 2012; Mayaka and Prasad, 2012; Rogerson, 2012b; Spenceley and Meyer, 2012; Dieke, 2013; Boonabaana, 2014; de Beer *et al.*, 2014; Rogerson and Rogerson, 2014; Rogerson and Visser, 2014; Adiyia *et al.*, 2015; Novelli, 2015; Booyens and Rogerson, 2016a, 2016b; de Boer and van Dijk, 2016; Giampiccoli *et al.*, 2016; Manwa *et al.*, 2016; Nunkoo and Ramkissoon, 2016; Rogerson, 2016a, 2016b). In this analysis three sets of pressing interrelated challenges are isolated for discussion. These relate to: (1) building competitiveness; (2) poverty and inclusion; and (3) addressing climate change.

Building competitiveness

Building the competiveness of Africa's tourism destinations within the global tourism economy is essential for maximising tourism development opportunities (Blanke *et al.*, 2011; Manwa, 2011; Mayaka and Prasad, 2012). Central to enhancing competitiveness is the need for trust as well as innovation in the tourism sector to be led by innovative enterprises and innovative destinations (Booyens and Rogerson, 2016a, 2016b; Nunkoo and Gursoy, 2016). In considering competitiveness, a useful starting point is the concept of a 'tourism destination pyramid', as put forward by the World Bank (2014). By building competitiveness, countries can move up the tourism destination pyramid, and with accompanying expanded tourism flows, more jobs are created and the tourism value chain is strengthened (Magombo and Rogerson, 2012).

In terms of sub-Saharan Africa, several crucial factors are isolated for competitiveness analysis (World Bank, 2014). The limited and costly air transport access to African destinations from major tourism-generating regions in the Global North is of concern, as are the difficulties associated with often infrequent, irregular and inadequate transport access within countries. In addition, tourism is anchored upon the availability of certain infrastructural facilities, not least accommodation services, which may be lacking or of poor quality. Seven essential determining factors are isolated for maintaining and enhancing the competitiveness of African countries as tourism destinations:

- quality of tourism assets;
- quality standards of accommodation services;
- efficiency and safety in transport to and from countries;
- adequacy of infrastructure;
- support from local communities and national government for tourism development;
- safety and security of destinations for tourists, including health issues and terrorism; and
- human resource capacities of officials and employees in the tourism industry that tourists interact with.

(World Bank, 2014)

Another perspective on competitiveness is offered by Blanke *et al.* (2011), who seek to identify the factors and policies that make it attractive to develop tourism in different countries. The analysis highlights three clusters of salient factors that affect competitiveness. First is the regulatory and policy frameworks that impact safety and security, health and hygiene, environmental sustainability, and the overall prioritisation of tourism. Second is a cluster of issues around the business environment and infrastructure, among which are included air transport, ground transport and ICT, representing the essential 'hard' infrastructure for tourism development. The third cluster relates to the 'softer' human, cultural and natural elements of each destination's tourism assets. Using the above indicators, it is revealed that while there are some strong performers in sub-Saharan Africa – notably Mauritius, South Africa, Cape Verde and Botswana – 'most countries receive poor assessments' (Blanke *et al.*, 2011: 91). The majority of countries are considered as having regulatory environments that are not sufficiently supportive of the tourism sector. In addition, shortcomings are flagged in terms of policies and factors that impact environmental sustainability. Once again, problems relating to intra-African air travel and ground transportation, safety and security, health and hygiene, and quality of skills or human resources for tourism development are reiterated (Blanke *et al.*, 2011).

Within the domain of business infrastructural improvement, perhaps the most critical issue for African tourism development surrounds the need for quality upgrading for accommodation services. Tourism growth requires that a range of different forms of accommodation be available to visitors at competitive prices and at acceptable quality standards. The absence or poor quality of such accommodation services is noted in the group of Africa's pre-emerging destinations (World Bank, 2014). By contrast, in those destinations with a mature tourism industry and involved in deepening the success of tourism development, a varied range of forms of accommodation is available. South Africa provides a good case study in the restructuring of the accommodation sector since the country's post-1994 re-entry into the international tourism economy (Rogerson, 2013a, 2013b). As a consequence of the upgrading of accommodation services, tourists in South Africa can now choose from a range of accommodation options, including five-star luxury hotels and safari lodges, boutique hotels, limited service hotels, all-suite hotels, guest houses, bed and breakfasts, self-serviced apartments, backpacker hostels, and, most recently, Airbnb home stays

(Rogerson, 2010, 2011a, 2011b, 2014; Greenberg and Rogerson, 2015). Outside South Africa, there is mounting competition between different international hotel chains – led by Marriott, Radisson and Hilton – in order to access market opportunities for accommodating tourists in other African countries (Rogerson, 2016; Ward, 2016).

Poverty and inclusion

Arguably, as Hall (2007a: 111) observes, 'poverty reduction has become an important item on the tourism agenda'. The capacity of tourism to function as a tool for poverty reduction derives from its potential to distribute incomes to deprived regions or communities on the basis of 'unique qualities of their ecological, historic or cultural heritage' (Akyeampong, 2011: 198). Mitchell and Ashley (2010) argue that such sites become local assets where the discretionary expenditures of visitors yield incomes to residents as well as revenues to wider communities. Overall, it is contended that tourism consumption can be harnessed to serve the poor through unlocking opportunities at all scales of operation in the tourism industry (Spenceley and Meyer, 2012; Saarinen *et al.*, 2013b). Arguably, the importance is clear for localities to evolve appropriate and innovative strategies for place-based economic development in order to maximise opportunities from tourism development (Rogerson, 2014a; Butler and Rogerson, 2016).

Pro-poor tourism (PPT) is an alternative approach to tourism development rather than a particular type of tourism. It represents one of the most important new innovations in tourism scholarship and has generated vibrant debates (Hall, 2007b; Meyer, 2007; Saarinen and Rogerson, 2014). The most distinctive feature is that 'it puts poor people and poverty at the centre' of debates (Ashley *et al.*, 2000: 1). Broadly defined, PPT constitutes an approach that focuses upon accruing net economic benefits to the poor, even if richer groups inevitably are also beneficiaries (Ashley and Roe, 2002). In terms of advancing local development, PPT accords a premium to developing and supporting local small businesses and strengthening linkages with other economic sectors (most notably agriculture and construction), and highlights the necessity of minimising or mitigating negative environmental impacts and the enhancement of local culture (Ashley *et al.*, 2001; Rogerson, 2006; Meyer, 2007; Akyeampong, 2011). It seeks to work with mainstream tourism businesses and recognises that poverty alleviation cannot occur without their involvement. The focus is on maximising, however possible, net benefits to the poor from tourism development projects, including through community participation in decision-making (Mitchell and Ashley, 2010; Akyeampong, 2011).

During the past 15 years, pro-poor tourism has consolidated as one of the central themes in tourism scholarship about the Global South reshaping the research agenda for tourism scholars around the tourism–poverty nexus (Ashley and Roe, 2002; Rogerson, 2006; Hall, 2007a, 2007b; Meyer, 2007; Mitchell and Ashley, 2010; Saarinen *et al.*, 2013b; Rogerson, 2014b; Bakker and Messerli, 2016). Its proponents suggest that it offers qualitative differences with respect to how tourism development relates to the poor (Ashley *et al.*, 2000; Ashley and Roe, 2002); its detractors stress that it is 'another form of neo-liberalism' in that it fails to address the structural reasons for the North–South divide, as well as inequalities within countries (Hall, 2007a; Saarinen *et al.*, 2013b; Saarinen and Rogerson, 2014). Proponents of pro-poor tourism suggest that the benefits of pro-poor tourism interventions are locally significant, and evidence demonstrates that tourism can transfer significant benefits to local economies and communities around tourist destinations (Mitchell and Ashley, 2010; Snyman, 2012). This said, most pro-poor projects have been small-scale and short- to medium-term interventions, such as accommodation establishments seeking to source locally a greater share of required supplies of vegetables (Hunt and Rogerson, 2013; Rogerson, 2014b). In recent debates, the frontiers of discussion are shifting away from pro-poor growth and instead towards focusing on a longer-term approach of 'inclusive growth' and 'inclusive development', which involves a scaling up of the longer-term impacts of tourism development (Bakker and Messerli, 2016).

In the global fight to reduce poverty and attain the goals of the Sustainable Development Agenda, Africa must be at the forefront of attention. Mitchell and Ashley (2006a) argue strongly that tourism can contribute to reducing poverty in Africa for several reasons. As discussed above, tourism is a growing element in several African countries, and the future prospects for African tourism appear promising as the continent's share in global tourism economy expands. In addition, there is a body of evidence from a number of investigations that tourism is having positive impacts (Rogerson, 2006; Meyer, 2007; Mitchell and Ashley, 2010; Saarinen *et al.*, 2013b). Further, the interventions of pro-poor tourism approaches have been shown to work in terms of extending the reach and impact of tourism for poor communities. Mitchell and Ashley (2010) isolate three 'pathways' through which the benefits of tourism can be transferred to poor communities in Africa. First are direct effects, which mainly refer to wage employment and the expansion of work opportunities for the poor. Second are the important secondary effects of tourism for the poor, which can arise from their involvement in the supply chain of tourism businesses in terms of using local construction enterprises, local building material or selling local crafts. By far the most important of these linkages, however, relates to the expanded sale of locally produced agricultural goods to tourism businesses (Hunt *et al.*, 2012; Rogerson, 2012b; Pillay and Rogerson, 2013). Finally, a third way in which poor communities may be beneficiaries of tourism development projects is through 'dynamic effects', which refers to the wider impacts on local communities of tourism development, through, for example, the improvement of local infrastructure in terms of building of roads, increasing entrepreneurial motivations and skills, improved mobile telephone connectivity, and the expansion or enhancement of local education or health facilities (Mitchell and Ashley, 2010).

In terms of policy agendas, the PPT writings emphasise that the challenge in Africa is to evolve strategic interventions to improve local employment, skills/training, local value chains, collective incomes and community involvement in tourism development (Ashley *et al.*, 2000; Ashley and Roe, 2002). Most importantly perhaps is the imperative for policy development to build local economic linkages, and correspondingly to reduce economic leakages (Mitchell and Ashley, 2006b, 2010; Hunt and Rogerson, 2013). It has been documented that where tourism has been shown to have strong pro-poor impacts, the impact is the result of developing strong local linkages and inter-sectoral linkages (Mitchell and Ashley, 2010). The existence of economic leakages is cited as a core reason accounting for tourism's failure to boost the desired or expected level of local development impacts in poor regions and communities (Rogerson, 2014a). This said, the existence of high levels of leakage do not necessarily negate tourism's potential as a vehicle for development in poor rural areas of Africa (Sandbrook, 2010).

Overall, however, the capacity of African destinations to 'capture the gains' from tourism growth depends upon a capacity for strengthening local value chains and the building of new economic relationships between tourism and other sectors in the destination economy (Hunt and Rogerson, 2013; Butler and Rogerson, 2016). Inclusion can be supported by improving the tourism workplace and the environment of work (Maumbe and Van Wyk, 2008; Marobela, 2011; de Beer *et al.*, 2014; Baum *et al.*, 2016). For tourism to drive inclusive growth or development impacts, Bakker and Messerli (2016) stress the need for broader, longer-term interventions that would be part of a country's inclusive growth strategy. It is argued therefore that 'the tourism sector should be developed by the private sector while the government plays a facilitating role offering complementary investments that could benefit other sectors' (Bakker and Messerli, 2016: 6).

Addressing climate change

With less than 3 per cent of the world's total emissions of greenhouse gases, Africa makes no significant contribution to global warming and climate change. Nevertheless, according to the fourth assessment of the Intergovernmental Panel on Climate Change (IPCC, 2007), no continent will be impacted as severely by climate change as Africa. Among the IPCC (2007) projections are: by 2050, an average temperature rise of between 1.5 and 3 °C; major changes in rainfall patterns in terms of annual and seasonal trends, with an increase of extreme events of floods and drought; projected sea level rise in coastal areas; and increased water stress. Further, climate change poses major threats to Africa's critical ecosystems, which contribute significantly to biodiversity; projections are that between 25 and 40 per cent of mammal species in national parks in sub-Saharan Africa will become endangered (IPCC, 2007).

Tourism is viewed as one of the most climate-sensitive sectors according to the IPCC (2007, 2014). Accordingly, it is observed that climate change and its projected 'substantial impacts on tourism worldwide is a pressing issue' (Becken, 2013: 53). During 2007, the Davos Declaration acknowledged climate change as the greatest challenge to sustainable tourism in the twenty-first century (UNWTO, 2007). The tourism sector utilises energy for the transport of tourists to and from, as well as within, destinations, in accommodation establishments, and for a host of tourist activities. With most energy deriving from fossil fuels, the sector contributes to the emission of greenhouse gases, with serious ramifications for climate change (Scott *et al.*, 2010: 395). Becken (2008: 351) contends that climate change cannot be overlooked by tourism policymakers as it 'has the potential to substantially change the way people will travel in the future'.

Africa is the region of the global tourism economy that is anticipated to face moderate to negative impacts of climate change for its tourism sector (Saarinen *et al.*, 2013a; Hoogendoorn and Fitchett, 2016; Rogerson, 2016b). For long-haul international tourism arrivals, African tourism destinations are vulnerable to new legislation that encourages the international aviation sector to curb CO_2 emissions (Gössling *et al.*, 2008). Climate change is viewed as one of the critical potential challenges for sustainable tourism development in Africa, with protected areas often in serious risk (Novelli, 2015; Mkiramweni *et al.*, 2016). By modifying the thermal and moisture balance for vegetation, climate change can significantly alter the character of ecosystems and the behaviour of wildlife in ways that makes them more difficult for tourists to view (UNWTO, 2015a). The subregion of southern Africa is most vulnerable because of the severity of projected physical climate impacts with observed temperature changes for this region, higher than increases indicated for other world regions (Rogerson, 2016b). Environmental modifications linked to climate change threaten to irreparably damage or threaten the foundations upon which tourism development in Africa is based, especially its unique landscapes and wildlife diversity (Hambira *et al.*, 2013). Already there is mounting evidence that climate change is impacting the continent's tourism industry, and in particular represents a severe threat to its iconic nature-based attractions associated with the continent's rich biodiversity (Gössling, 2011; Saarinen *et al.*, 2013a; Hambira and Saarinen, 2015; Hoogendoorn *et al.*, 2016). In Tanzania, for example, climate change is clearly impacting the country's protected areas, most notably the Ngorongoro Conservation Area (Mkiramweni *et al.*, 2016). Njoroge (2015a) and Hoogendoorn *et al.* (2016) provide evidence of negative climate change impacts for Africa's coastal destinations.

How climate change is impacting the African tourism sector is an issue that 'can no longer be ignored' by policymakers (Hambira *et al.*, 2013: 69). The tourism sector necessarily must respond and adapt to climate-induced changes for its long-term economic health and sustainability.

This raises critical policy questions about carbon emission reduction and planning for low-carbon destinations as part of climate compatible development (Gössling, 2011; Njoroge, 2015a). Shifts in global climate change policy, as well as increased fuel costs, heighten the policy significance of the 'greening of tourism' and the making of carbon-neutral destinations (Gössling and Schumacher, 2010). In addition, climate change impacts underscore that environmental management must be an integral component of tourism development planning, especially for sustainable tourism development in sensitive or protected areas (Rogerson, 2012a). South African research points to some limited but innovative greening initiatives starting to take place in accommodation services (Rogerson and Sims, 2012; Booyens and Rogerson, 2016b; Ismail and Rogerson, 2016).

Arguably, based upon international experience, any real, efficient or effective action taken towards addressing the significant threats posed by climate change to tourism in sub-Saharan Africa requires the innovation of mitigation policies and actions, as well as sustainable adaptation strategies (Kajan and Saarinen, 2013; IPCC, 2014; Njoroge, 2015b; Shackleton *et al.*, 2015; Hoogendoorn and Fitchett, 2016). The former seek to reduce the long-term and most severe effects upon the tourism economy that are anticipated from climate change by cutting down on the amounts of anthropogenic pollutants released into the earth's atmosphere. The latter attempt to prepare individual communities and regions for some of the major hazards associated with climate change by seeking infrastructure development and policy actions that attempt to lessen the projected impacts of climate change on those groups considered to be most vulnerable (Kajan and Saarinen, 2013; Njoroge, 2015a). In recent overviews of the nexus of tourism and climate change in Africa, it is stressed that the anticipated severe impacts of climate change on Africa's tourism economy demands an extended knowledge base of research in order to inform locally relevant sustainable adaptation and mitigation strategies (Shackleton *et al.*, 2015; Hoogendoorn and Fitchett, 2016).

Conclusion

Among an increasing number of observers, there is an emerging consensus that the tourism and hospitality sector of Africa is expanding in strength, and arguably that it represents one of the most promising sources for new job creation, livelihood support and inclusive growth (World Bank, 2014; Novelli, 2015). This chapter shows, however, the uneven geographical progress in tourism development across sub-Saharan Africa. For those countries that are pre-emergent or in the process of initiating tourism industries, a degree of caution must be exercised as to its potential for contributing substantively to the Sustainable Development Goals. This said, there are a growing number of African tourism destinations where the role of the tourism sector must be acknowledged as vital for national development programming.

It has been argued here that in leveraging tourism's potential in Africa, a multitude of challenges must be addressed to enhance the prospects of local tourism economies within an increasingly low-growth global economy. Three interrelated sets of challenges were highlighted in this discussion as relating to strengthening the competiveness of African tourism destinations, addressing questions of poverty and inclusion, and grappling with the enormous implications posed by climate change. For successful tourism development across sub-Saharan Africa over the coming decades, there will be required extensive innovation on the behalf of tourism entrepreneurs, policymakers and other stakeholders, including, most notably, the international development agencies. The challenge for Africa's tourism scholars is to advance the research frontier in support of such innovations.

References

Adiyia, B., Stoffelen, A., Jennes, B., Vanneste, D. and Ahebwa, M. (2015) 'Analysing governance in tourism value chains to reshape the tourist bubble in developing countries: the case of cultural tourism in Uganda', *Journal of Ecotourism*, 14(2–3): 113–129.

Akyeampong, O.A. (2011) 'Pro-poor tourism: residents' expectations, experiences and perceptions in the Kakum National Park Area of Ghana', *Journal of Sustainable Tourism*, 19(2): 197–213.

Ashley, C. and Roe, D. (2002) 'Making tourism work for the poor: strategies and challenges in southern Africa', *Development Southern Africa*, 19(1): 61–82.

Ashley, C., Boyd, C. and Goodwin, H. (2000) *Pro-Poor Tourism: Putting Poverty at the Heart of the Tourism Agenda*, Natural Resource Perspectives No. 51, London: Overseas Development Institute, available at: www.odi.org/sites/odi.org.uk/files/odi-assets/publications-opinion-files/2861.pdf (accessed 7 January 2018).

Ashley, C., Roe, D. and Goodwin, H. (2001) *Pro-Poor Tourism Strategies: Making Tourism Work for the Poor*, London: Pro-Poor Tourism Report 1.

Bakker, M. and Messerli, H. (2016) 'Inclusive growth versus pro-poor growth: implications for tourism development', *Tourism and Hospitality Research*, 17(4): 384–391, DOI: 10.1177/1467358416638919.

Baum, T., Cheung, C., Kong, H., Kralj, A., Mooney, S., Thanh, H.N.T., *et al.* (2016) 'Sustainability and the tourism and hospitality workforce: a thematic analysis', *Sustainability*, 8(8): 809–830.

Becken, S. (2008) 'Climate change: beyond the hype', *Tourism Recreation Research*, 33: 351–353.

Becken, S. (2013) 'A review of tourism and climate change as an evolving knowledge domain', *Tourism Management Perspectives*, 6, April: 53–62.

Blanke, J., Browne, C., Garcia, A.F. and Messerli, H.R. (2011) 'Assessing Africa's travel and tourism competitiveness in the wake of the global economic crisis', in World Economic Forum, The World Bank and the African Development Bank (eds), *The Africa Competitiveness Report 2011*, Geneva: World Economic Forum, pp. 89–111.

Boonabaana, B. (2014) 'Negotiating gender and tourism work: women's lived experiences in Uganda', *Tourism and Hospitality Research*, 14(1–2): 27–36.

Booyens, I. and Rogerson, C.M. (2016a) 'Unpacking the geography of tourism innovation in the Western Cape province of South Africa', *Bulletin of Geography: Socio-Economic Series*, 31: 19–36.

Booyens, I. and Rogerson, C.M. (2016b) 'Responsible tourism in the Western Cape, South Africa: an innovation perspective', *Tourism: An International Interdisciplinary Journal*, 64(4): 385–396.

Butler, G. and Rogerson, C.M. (2016) 'Inclusive local tourism development in South Africa: evidence from Dullstroom', *Local Economy*, 31(1–2): 264–281.

Cohen, E. and Cohen, S.A. (2015) 'A mobilities approach to tourism from emerging world regions', *Current Issues in Tourism*, 18(1): 11–43.

de Beer, A., Rogerson, C.M. and Rogerson, J.M. (2014) 'Decent work in the South African tourism industry: evidence from tour guides', *Urban Forum*, 25(1): 89–104.

de Boer, D. and van Dijk, M.P. (2016) 'Can sustainable tourism achieve conservation and local economic development? The experience with nine business-community wildlife-tourism agreements in northern Tanzania', *African Journal of Hospitality Tourism and Leisure*, 5(4), available at: www.ajhtl.com/uploads/7/1/6/3/7163688/article_7_vol_5__4_.pdf (accessed 9 January 2017).

Dieke, P.U.C. (2013) 'Tourism in sub-Saharan Africa: production–consumption nexus', *Current Issues in Tourism*, 16(7–8): 623–626.

Giampiccoli, A., Saayman, M. and Jugmohan, S. (2016) 'Are "Albergo Diffuso" and community-based tourism the answers to community development in South Africa?', *Development Southern Africa*, 33(4): 548–561, DOI: 10.1080/0376835X.2016.1180968.

Gössling, S. (2011) *Carbon Management in Tourism: Mitigating the Impacts of Climate*, London: Routledge.

Gössling, S. and Schumacher, K.P. (2010) 'Implementing carbon neutral policies: issues from the Seychelles', *Journal of Sustainable Tourism*, 18: 377–391.

Gössling, S., Peeters, P. and Scott, D. (2008) 'Consequence of climate policy for international tourism arrivals in developing countries', *Third World Quarterly*, 29: 873–901.

Gössling, S., Scott, D. and Hall, C.M. (2013) 'Challenges of tourism in a low-carbon economy', *WIREs Climate Change*, 4: 525–538.

Greenberg, D. and Rogerson, J.M. (2015) 'The serviced apartment industry of South Africa: a new phenomenon in urban tourism', *Urban Forum*, 26(4): 467–482.

Hall, C.M. (2007a) 'Pro-poor tourism: do tourism exchanges benefit primarily the countries of the South?', *Current Issues in Tourism*, 10(2/3): 111–118.

Hall, C.M. (ed.) (2007b) *Pro-Poor Tourism: Who Benefits?* Clevedon: Cromwell Press.
Hambira, W.L. and Saarinen, J. (2015) 'Policy makers' perceptions of the tourism–climate change nexus: policy needs and constraints in Botswana', *Development Southern Africa*, 32: 350–362.
Hambira, W.L., Manwa, H., Atlhopheng, J. and Saarinen, J. (2013) 'Perceptions of tourism operators towards adaptations to climate change in nature-based tourism: the quest for sustainable tourism in Botswana', *Pula: Botswana Journal of African Studies*, 27(1): 69–85.
Harmer, D. and Rogerson, J.M. (2016) 'Youth tourism festivals in South Africa: the case of Plett Rage', *African Journal of Hospitality, Tourism and Leisure*, 5(4), available at: www.ajhtl.com/uploads/7/1/6/3/7163688/article_20_vol_5__4_.pdf (accessed 7 January 2018).
Hoogendoorn, G. and Fitchett, J.M. (2016) 'Tourism and climate change: a review of threats and adaptation strategies for Africa', *Current Issues in Tourism*, 13: 1–18, DOI: 10.1080/13683500.2016.1188893.
Hoogendoorn, G., Grant, B. and Fitchett, J. (2016) 'Disjunct perceptions? Climate change threats in two low-lying South African coastal towns', *Bulletin of Geography: Socio-Economic Series*, 31: 59–71.
Hottola, P. (ed.) (2009) *Tourism Strategies and Local Responses in Southern Africa*, Wallingford: CABI.
Hunt, H. and Rogerson, C.M. (2013) 'Tourism-led development and backward linkages: evidence from the agriculture–tourism nexus in southern Africa', in G. Visser and S. Ferreira (eds), *Tourism and Crisis*, London: Routledge, pp. 159–179.
Hunt, H., Rogerson, C.M., Rogerson, J.M. and Kotze, N. (2012) 'Agriculture–tourism linkages in Botswana: evidence from the safari lodge accommodation sector', *Africa Insight*, 42(2): 1–17.
IPCC (Intergovernmental Panel on Climate Change) (2007) *Climate Change Impacts, Adaptation and Vulnerability: Contribution of Working Group II to the Fourth Assessment Report of the Intergovernmental Panel on Climate Change*, Geneva: IPCC.
IPCC (Intergovernmental Panel on Climate Change) (2014) *Climate Change 2014: Impacts, Adaptation and Vulnerability*. New York: IPCC.
Ismail, S. and Rogerson, J.M. (2016) 'Retrofitting hotels: evidence from the Protea Hospitality Group of hotels within Gauteng, South Africa', *African Journal of Hospitality, Tourism and Leisure*, 5(3), available at: www.ajhtl.com/uploads/7/1/6/3/7163688/article_10_final_vol_5__3__final.pdf (accessed 7 January 2018).
Kajan, E. and Saarinen, J. (2013) 'Tourism, climate change and adaptation: a review', *Current Issues in Tourism*, 16(2): 167–195.
Magole, L.I. and Magole, L. (2011) 'Revisiting Botswana's high-value, low-volume tourism', *Tourism Analysis*, 16: 203–210.
Magombo, A. and Rogerson, C.M. (2012) 'The evolution of the tourism sector in Malawi', *Africa Insight*, 42(2): 46–65.
Manwa, H. (2011) 'Competitiveness of southern African development community as a tourist destination', *Tourism Analysis*, 16: 77–86.
Manwa, H., Moswete, N. and Saarinen, J. (eds) (2016) *Cultural Tourism in Southern Africa*, Bristol: Channel View.
Marobela, M.N. (2011) 'Tourism and decent work in Botswana: from private sector to a collectivist model', *African Journal of Hospitality, Tourism and Leisure*, 1(3), available at: www.ajhtl.com/uploads/7/1/6/3/7163688/marobela_m.d._article_12_vol.1_31.pdf (accessed 8 January 2018).
Maumbe, K.C. and Van Wyk, L.J. (2008) 'Employment in Cape Town's lodging sector: opportunities, skill requirements, employee aspirations and transformation', *GeoJournal*, 73: 117–132.
Mayaka, M.A. and Prasad, H. (2012) 'Tourism in Kenya: an analysis of strategic issues and challenges', *Tourism Management Perspectives*, 1, January: 48–56.
Mbaiwa, J.E. (2005) 'Enclave tourism and its socio-economic impacts in the Okavango Delta, Botswana', *Tourism Management*, 26: 157–172.
McKay, T. (2013) 'Adventure tourism: opportunities and management challenges for SADC destinations', *Acta Academica*, 45(3): 30–62.
McKay, T. (2014) 'Locating South Africa within the global adventure tourism industry: the case of bungee jumping', *Bulletin of Geography: Socio-Economic Series*, 24: 161–176.
McKay, T. (2016) 'The geography of the South African adventure tourism industry', *African Journal of Hospitality, Tourism and Leisure*, 5(3), available at: www.ajhtl.com/uploads/7/1/6/3/7163688/article_1_vol_5__3_final.pdf (accessed 8 January 2018).
Mensah, I. (2015) 'The roots tourism experience of diaspora Africans: a focus on the Cape Coast and Elmina Castles', *Journal of Heritage Tourism*, 10(3): 213–232.
Meyer, D. (2007) 'Pro-poor tourism: from leakages to linkages – a conceptual framework for creating linkages between the accommodation sector and "poor" neighbouring communities', *Current Issues in Tourism*, 10(6): 558–583.

Mgonja, J.T., Sirima, A., Backman, K.F. and Backman, S.J. (2015) 'Cultural community-based tourism in Tanzania: lessons learned and way forward', *Development Southern Africa*, 32(3): 377–391.

Mitchell, J. and Ashley, C. (2006a) *Can Tourism Help Reduce Poverty in Africa?* London: Overseas Development Institute.

Mitchell, J. and Ashley, C. (2006b) *Tourism Business and the Local Economy: Increasing Impact Through a Linkages Approach*, London: Overseas Development Institute.

Mitchell, J. and Ashley, C. (2010) *Tourism and Poverty Reduction: Pathways to Prosperity*, London: Earthscan.

Mitchell, J. and Faal, J. (2007) 'Holiday package tourism and the poor in the Gambia', *Development Southern Africa*, 24: 445–464.

Mkiramweni, N.P., DeLacy, T., Jiang, M. and Chiwanga, F.E. (2016) 'Climate change risks on protected areas ecotourism: shocks and stressors perspectives in Ngorongoro Conservation Area, Tanzania', *Journal of Ecotourism*, 15(2): 139–157.

Naude, W.A. and Saayman, A. (2005) 'Determinants of tourist arrivals in Africa: a panel data regression analysis', *Tourism Economics*, 11(3): 365–391.

Ngozi, E. and Chinonso, I. (2016) 'The role of cultural resources in tourism development in Awka', *African Journal of Hospitality, Tourism and Leisure*, 5(2), available at: www.ajhtl.com/uploads/7/1/6/3/7163688/2016_article_22_vol_5__2_.pdf (accessed 8 January 2018).

Njoroge, J.M. (2015a) 'Tourism, Millennium Development Goals and climate change in Kenya', *African Journal of Hospitality, Tourism and Leisure*, 4(1), available at: www.ajhtl.com/uploads/7/1/6/3/7163688/article_1_vol_4_1_2015.pdf (accessed 8 January 2018).

Njoroge, J.M. (2015b) 'Climate change and tourism adaptation: literature review', *Tourism and Hospitality Management*, 21(1): 95–108.

Novelli, M. (2015) *Tourism and Development in Sub-Saharan Africa*, London: Routledge.

Nunkoo, R. and Gursoy, D. (2016) 'Political trust and residents' support for alternative and mass tourism: an improved structural model', *Tourism Geographies*, 19(3): 1–22, DOI: 10.1080/14616688.2016.1196239.

Nunkoo, R. and Ramkissoon, H. (2016) 'Stakeholders' views of enclave tourism: a grounded theory approach', *Journal of Hospitality & Tourism Research*, 40(5): 557–588.

Okello, M.M. and Novelli, M. (2014) 'Tourism in the East African community (EAC): challenges, opportunities and way forward', *Tourism and Hospitality Research*, 14(1–2): 53–66.

Onuoha, F.C. (2016) *Hotels in Africa as Terrorist Targets*, Doha: Al Jazeera Centre for Studies.

Otoo, F.E. and Amuquandoh, F.A. (2014) 'An exploration of the motivations for volunteering: a study of international volunteer tourists in Ghana', *Tourism Management Perspectives*, 11: 51–57.

Pillay, M. and Rogerson, C.M. (2013) 'Agriculture–tourism linkages and pro-poor impacts: the accommodation sector of urban coastal KwaZulu-Natal, South Africa', *Applied Geography*, 36: 49–58.

Rivett-Carnac, K. (2011) *Cultural World Heritage Site Scan: Lessons from Four Sites*, Halfway House: Development Bank of Southern Africa.

Rogerson, C.M. (2006) 'Pro-poor local economic development in South Africa: the role of pro-poor tourism', *Local Environment*, 11(1): 37–60.

Rogerson, C.M. (2007) 'Reviewing Africa in the global tourism economy', *Development Southern Africa*, 24: 361–379.

Rogerson, C.M. (2011) 'Youth tourism in Africa: evidence from South Africa', *Tourism Analysis*, 16: 105–120.

Rogerson, C.M. (2012a) 'The tourism–development nexus in sub-Saharan Africa: progress and prospects', *Africa Insight*, 42(2): 28–45.

Rogerson, C.M. (2012b) 'Tourism–agriculture linkages in rural South Africa: evidence from the accommodation sector', *Journal of Sustainable Tourism*, 20: 477–495.

Rogerson, C.M. (2014a) 'Reframing place-based economic development in South Africa: the example of local economic development', *Bulletin of Geography: Socio-Economic Series*, 24: 203–218.

Rogerson, C.M. (2014b) 'Strengthening tourism–poverty linkages', in C.M. Hall, A. Lew and M. Williams (eds), *The Wiley-Blackwell Companion to Tourism*, London: Blackwell, pp. 600–610.

Rogerson, C.M. (2015a) 'Unpacking business tourism mobilities in sub-Saharan Africa', *Current Issues in Tourism*, 18(1): 44–56.

Rogerson, C.M. (2015b) 'Restructuring the geography of domestic tourism in South Africa', *Bulletin of Geography: Socio-Economic Series*, 29: 119–135.

Rogerson, C.M. (2016a) 'Public–private partnerships for tourism infrastructure development: evidence from the Cradle of Humankind, South Africa', *Tourism: An International Interdisciplinary Journal*, 64(4): 419–430.

Rogerson, C.M. (2016b) 'Climate change, tourism and local economic development in South Africa', *Local Economy*, 31(1–2): 322–331.
Rogerson, C.M. and Hoogendoorn, G. (2014) 'VFR travel and second homes tourism: the missing link?', *Tourism Review International*, 18(3): 167–178.
Rogerson, C.M. and Letsie, T. (2013) 'Informal sector business tourism in the Global South: evidence from Maseru, Lesotho', *Urban Forum*, 24(4): 485–502.
Rogerson, C.M. and Mthombeni, T. (2015) 'From slum tourism to slum tourists: township resident mobilities in South Africa', *Nordic Journal of African Studies*, 24(3–4): 319–338.
Rogerson, C.M. and Rogerson, J.M. (2010) 'Local economic development in Africa: global context and research directions', *Development Southern Africa*, 27: 465–480.
Rogerson, C.M. and Rogerson, J.M. (2011) 'Tourism research within the Southern African Development Community: production and consumption in academic journals, 2000–2010', *Tourism Review International*, 15(1–2), 213–222.
Rogerson, C.M. and Rogerson, J.M. (2014) 'Agritourism and local economic development in South Africa', *Bulletin of Geography: Socio-Economic Series*, 26: 93–106.
Rogerson, C.M. and van der Merwe, C.D. (2016) 'Heritage tourism impacts in the Global South: development impacts of the Cradle of Humankind World Heritage Site, South Africa', *Local Economy*, 31(1–2): 234–248.
Rogerson, C.M. and Visser, G. (eds) (2004) *Tourism and Development Issues in Contemporary South Africa*, Pretoria: Africa Institute of South Africa.
Rogerson, C.M. and Visser, G. (eds) (2007) *Urban Tourism in the Developing World: The South African Experience*, New Brunswick, NJ: Transaction Press.
Rogerson, C.M. and Visser, G. (2014) 'A decade of progress in African urban tourism scholarship', *Urban Forum*, 25(4): 407–418.
Rogerson, J.M. (2010) 'The boutique hotel industry in South Africa: definition, scope and organization', *Urban Forum*, 21: 425–439.
Rogerson, J.M. (2011a) 'The limited services hotel in South Africa: the growth of City Lodge', *Urban Forum*, 22: 343–361.
Rogerson, J.M. (2011b) 'The changing all-suite hotel in South Africa: from "extended stay" to African Condo Hotel', *Tourism Review International*, 15(1/2): 107–121.
Rogerson, J.M. (2013a) 'The economic geography of South Africa's hotel industry 1990 to 2010', *Urban Forum*, 24(3): 425–446.
Rogerson, J.M. (2013b) 'Reconfiguring South Africa's hotel industry 1990–2010: structure, segmentation, and spatial transformation', *Applied Geography*, 36: 59–68.
Rogerson, J.M. (2014) 'Hotel location in Africa's world class city: the case of Johannesburg, South Africa', *Bulletin of Geography: Socio-Economic Series*, 25: 181–196.
Rogerson, J.M. (2016) 'Hotel chains of the Global South: the internationalization of South African hotel brands', *Tourism: An International Interdisciplinary Journal*, 64(4): 445–450.
Rogerson, J.M. and Harmer, D. (2015) 'A "rite of passage" youth festival in South Africa: the origins, attendees and organization of Matric Vac', *Nordic Journal of African Studies*, 24(3–4): 221–240.
Rogerson, J.M. and Sims, S.R. (2012) 'The greening of urban hotels in South Africa: evidence from Gauteng', *Urban Forum*, 23(4): 391–407.
Rogerson, J.M. and Slater, D. (2014) 'Urban volunteer tourism: orphanages in Johannesburg', *Urban Forum*, 15(4): 483–499.
Saarinen, J. (2011) 'Tourism, indigenous people, and the challenge of development: The representations of Ovahimbas in tourism promotion and community perceptions towards tourism', *Tourism Analysis*, 16: 31–42.
Saarinen, J. and Rogerson, C.M. (2014) 'Tourism and the Millennium Development Goals: perspectives beyond 2015', *Tourism Geographies*, 16(1): 23–30.
Saarinen, J. and Rogerson, C.M. (2015) 'Setting cultural tourism in southern Africa', *Nordic Journal of African Studies*, 24(3–4): 207–220.
Saarinen, J., Becker, F., Manwa, H. and Wilson, D. (eds) (2009) *Sustainable Tourism in Southern Africa: Local Communities and Natural Resources in Transition*, Bristol: Channel View.
Saarinen, J., Hambira, W.L., Athopeng, J. and Manwa, H. (2013a) 'Tourism and climate change in southern Africa: sustainability and perceived impacts and adaptation strategies of the tourism industry to changing climate and environment in Botswana', in M.V. Reddy and K. Wilkes (eds), *Tourism, Climate Change and Sustainability*, London: Routledge, pp. 243–256.

Saarinen, J., Rogerson, C.M. and Manwa, H. (eds) (2013b) *Tourism and the Millennium Development Goals: Tourism, Local Communities and Development*, London: Routledge.

Saarinen, J., Moswete, N. and Monare, M.J. (2014) 'Cultural tourism: new opportunities for diversifying the tourism industry in Botswana', *Bulletin of Geography: Socio-Economic Series*, 26: 7–18.

Sandbrook, C.G. (2010) 'Putting leakage in its place: the significance of retained tourism revenue in the local context of rural Uganda', *Journal of International Development*, 22(1): 124–136.

Scott, D., Peeters, P. and Gossling, S. (2010) 'Can tourism deliver its "aspirational" greenhouse gas emission reduction targets?', *Journal of Sustainable Tourism*, 18(3): 393–408.

Scott, D., Hall, C.M. and Gössling, S. (2012) *Tourism and Climate Change: Impacts, Mitigation and Adaptation*, London: Routledge.

Shackleton, S., Ziervogel, G., Sallu, S., Gill, T. and Tschakert, P. (2015) 'Why is socially-just climate change adaptation in sub-Saharan Africa so challenging? A review of barriers identified from empirical cases', *Wiley Interdisciplinary Reviews: Climate Change*, 6(3): 321–344.

Sinthumule, N.I. (2015) 'An evaluation of the performance of tourism at Mapungubwe National Park, South Africa', *African Journal of Hospitality, Tourism and Leisure*, 4(1), available at: www.ajhtl.com/uploads/7/1/6/3/7163688/article42vol4(1)2015.pdf (accessed 8 January 2018).

Snyman, S. (2012) 'The role of tourism employment in poverty reduction and community perceptions of conservation and tourism in southern Africa', *Journal of Sustainable Tourism*, 20(3): 395–416.

Spenceley, A. (2010) *Tourism Product Development Interventions and Best Practices in Sub-Saharan Africa: Part 2: Case Studies*, Nelspruit: Report to the World Bank, Tourism Industry: Research and Analysis Phase II.

Spenceley, A. and Meyer, D. (2012) 'Tourism and poverty reduction: theory and practice in less economically developed countries', *Journal of Sustainable Tourism*, 20(3): 297–317.

Spenceley, A., Habyalimana, S., Tusabe, R. and Mariza, D. (2010) 'Benefits to the poor from gorilla tourism in Rwanda', *Development Southern Africa*, 27(5): 647–662.

Stoddart, H. and Rogerson, C.M. (2004) 'Volunteer tourism: the case of Habitat for Humanity South Africa', *GeoJournal*, 60(3): 311–318.

Twining-Ward, L. (2009) *Sub-Saharan Africa Tourism Industry Research*, Report prepared for the Africa Region's Finance and Private Sector Development Department, Washington, DC: World Bank.

UNWTO (United Nations World Tourism Organization) (2007) *Davos Declaration: Climate Change and Tourism – Responding to Global Challenges Davos*, Switzerland, 3 October.

UNWTO (United Nations World Tourism Organization) (2012) *Tourism in the Green Economy*, Madrid: UNWTO.

UNWTO (United Nations World Tourism Organization) (2015a) *Affiliate Members Regional Reports, Volume Four: Tourism in Africa – A Tool for Development*, Madrid: UNWTO.

UNWTO (United Nations World Tourism Organization) (2015b) *Towards Measuring the Economic Value of Wildlife Watching Tourism in Africa*, Madrid: UNWTO.

UNWTO (United Nations World Tourism Organization) (2016) *UNWTO Tourism Highlights: 2016 Edition*, Madrid: UNWTO.

van der Duim, R., Meyer, D., Saarinen, J. and Zellmar, K. (eds) (2011) *New Alliances for Tourism, Conservation and Development in Eastern and Southern Africa*, Delft: Eburon.

Ward, T. (2016) 'Hotel chain development pipelines in Africa: implications for human capital development', *Worldwide Hospitality and Tourism Themes*, 8(2): 134–148.

World Bank (2014) *Tourism in Africa: Harnessing Tourism for Growth and Improved Livelihoods*, Washington, DC: World Bank.

WTTC (World Travel and Tourism Council) (2015) *Travel & Tourism: Economic Impact 2015 Africa*, London: WTTC.

42
Transport[1]

Gina Porter

Introduction: Africa's (high-cost) transport burden

Deficiencies in transport infrastructure and the transport services that operate thereon present massive problems for economic development across sub-Saharan Africa. However, infrastructure per se tends to receive more attention than transport services, a point stressed by Gwilliam (2010), who refers to transport services as 'the forgotten problem'. In this chapter, to help redress that balance, discussion is principally concerned with transport services, after this brief introduction and a review of broad infrastructure issues.

The developmental implications of poor access to transport can be profound. The complex interlinkages between social and economic impacts are emphasised throughout this chapter, as are user needs (as opposed to current transport usage). In a 2008 report, the World Bank estimated that two-thirds of rural people in sub-Saharan Africa lacked reliable access to an all-weather road, and were thus essentially reliant on subsistence living and cut off from healthcare and education, which inevitably impacts on their ability to secure a sound, sustainable livelihood. They noted that since the 1960s, many countries had lost around half their road network (especially rural roads) and that the region currently contributed only 2 per cent of world trade. The vast majority of rural roads were unpaved, and a substantial proportion of the lower trafficked roads un-engineered, with limited or no structures. Unit transport costs were typically three to five times higher than those of developed countries. The report (World Bank, 2008: 64–65) stresses the many factors that push up Africa's transport costs, including some that are mutually reinforcing: thin markets, for example, make it difficult to justify investments to overcome infrastructure deficiencies or to introduce new transport-handling technology.

The scale of the remoteness and accessibility problem in Africa, associated with much lower population densities than those in other developing regions, is one component of the transport problem. Linard *et al.* (2012), who built a high-resolution population distribution data set for Africa and analysed rural accessibility to population centres, estimated the average per person travel time to settlements of more than 50,000 inhabitants at about 3.5 hours, with central and East Africa displaying the longest average travel times. Their analysis highlights the large inequities in access and the isolation of many rural populations: the economic consequences for such populations are massive.

In many urban centres across Africa, meanwhile, traffic congestion is growing at an unprecedented rate, in tandem with burgeoning populations and associated urban sprawl: this also impacts negatively on economic life. Informal motor taxis and motorcycle taxis are increasingly in evidence, despite being expensive, dangerous and polluting, and inefficient in meeting people's everyday mobility needs. Some cities are pressing ahead, however, with mobility solutions transplanted from the developed world, without due consideration of the local context or the mobility needs of local populations: difficulties associated with the development of bus rapid transit in cities such as Dar es Salaam are now starting to generate concern (Rizzo, 2015). The poorest, who tend to live in the least accessible (lowest-cost) peripheral locations, far from major public transport routes, with the least resources to purchase personal transport or to pay transport fares, suffer most, not least in terms of building sustainable livelihood opportunities (Lucas and Porter, 2016).

It is also important to note that although Africa's automobility is far below that of other regions, issues around potential environmental changes associated with growing motorised transport usage are also beginning to emerge, and will have wide-ranging repercussions for economies and societies. With the rapidly increasing motorised means of transport, this sector has become the fastest-growing source of greenhouse emissions on the continent. Motorised vehicles mostly consume non-renewable fuel resources, which in Africa commonly have to be imported, and growing numbers of motorised vehicles, even in rural areas, raises concerns about air pollution and associated environmental damage: there are many (imported, second-hand) old vehicles on the road that are extremely fuel-inefficient and polluting, in addition to their road safety and transport cost implications (Teravaninthorn and Raballand, 2008: 92). Ironically, while the mobility of a large proportion of the population remains highly constrained, the road traffic death rate is the highest in the world (24.1 per 100,000 population; WHO, 2013). Factors such as poor road conditions, inadequate safe spaces for pedestrians and intermediate means of transport, inadequate road safety training, inadequate enforcement of safety measures, old and badly maintained vehicles and transport equipment, and excessive passenger and freight overloading all contribute to this.

Transport infrastructure

Discussion of transport infrastructure brings to the fore a range of complex political economy and environmental problems. On the issue of observed endemic corruption in the sector, Bryceson (2009), for instance, refers to road bias reinforced by the convergence of engineering and road lobbies with African politicians' interests. In recent years, debates about corruption and environmental damage have often implicated the extensive infrastructural investments of (mostly state-owned/sponsored) Chinese companies in roads and railways – their quality standards, use of migrant labour, deficiencies in skills transfer, and concerns over environmental practices (Corkin *et al.*, 2008).

The focus in this chapter is mainly on roads, which provide the key linking transport infrastructure for the movement of goods and passengers across the continent. Rail networks in Africa, for the most part, are a relic of the colonial era: disconnected, reaching inland from the ports, and with little recent investment other than for purely mineral lines, where the Chinese have been particularly active. Foster and Briceno-Garmendia (2010) note that rail investment has usually come from bilateral and multilateral donors: remaining passenger services mostly fail to cover their costs, while freight service tariffs are constrained by road competition. As they point out, South Africa has the most important rail network, dominating in both general rail freight and passenger traffic (mostly urban commuter lines).

Waterways represent another underdeveloped element of the transport sector. This is, in part, a consequence of significant navigation problems on most of Africa's major rivers. Uncoordinated development between different water uses and transport sectors and inadequate funding to develop or make improvements to the important river navigation systems are further impediments. Only a few waterways, most of which are located in the Congo, Nile and Zambezi basins, are internationally navigable. However, African government initiatives of recent years to improve water transport include the 20-year rehabilitation and upgrade plan for ports on Lakes Malawi and Tanganyika being carried out by the Tanzania Ports Authority (TPA) to improve import and export through the seaport of Mtwara in Tanzania and a memorandum of understanding to promote shipping on the Zambezi-Shire water system between the governments of Zambia, Malawi and Mozambique (UNEP, 2010).

So far as sea transport is concerned, matters are little better. Unfortunately, Africa lacks natural ports and its artificial seaports have been poorly developed (AfDB, 2010). Despite the rapid growth of cargo transiting through Africa's ports in recent years, containerisation is still low and the inland transportation linkages remain weak (AfDB, 2010). With the exception of South Africa, ports are mostly small and capacity and port logistics have not been able to keep up with increasing traffic, though major developments are now going ahead in a number of countries, for instance the new container terminal at Pointe Noire (Congo Brazzaville) (AfDB, 2010).

Airport expansion was similarly neglected for many decades, a point well reflected in the fact that Africa still accounts for only 2 per cent of the world's air transport traffic in terms of passenger kilometres flown: African carriers face higher costs than their counterparts elsewhere (Heinz and O'Connell, 2013). Nonetheless, the potential for growth in air transport is widely recognised, given the size and rate of growth of the continent's population, and is reflected in increasing government interest in airport extension programmes (another focus of recent Chinese infrastructure investment).

So far as road transport is concerned, the construction of new roads through 'virgin' territory is particularly contentious, but all-weather tarmac roads still commonly head the wish list of rural populations, and where constructed these tend to promote massive migrations from surrounding bush locations to the new route. Health, social and environmental disbenefits – all with economic implications – may well emerge (from fights around landholding, to increased prostitution, STDs and traffic-related deaths; see Kwamusi, 2002; Mashiri, 2004; Ferguson and Morris, 2006; Feldacker *et al.*, 2011), in tandem with a parallel decline off-road of services and economic viability (trackable notably through assessments of the health of rural markets; see Porter 2002, 2007, 2011).

Weighing the benefits and disbenefits of infrastructure construction (particularly roads) is a difficult and much-debated task, both in urban and rural contexts. Evidence to suggest that all-weather roads often do have a mostly positive impact on poverty alleviation and growth in many regions across rural Africa is presented in various recent publications (Anyanwu and Erhijakpor, 2009; Dercon *et al.*, 2009 for Ethiopia), associated with reduced costs of acquiring inputs, increased output prices, reduced impacts of shocks and permitting entry into new, more profitable activities (Dercon *et al.*, 2009). Taking a broad continental approach, Buys and Uwe (2010) focus on road network upgrading in the context of promoting overland trade expansion, and argue that poor economic integration and isolation from regional and international markets have contributed significantly to poverty in sub-Saharan Africa. Using spatial network analysis techniques and gravity trade model estimations, they quantify the economics of upgrading a primary road network that connects the major urban areas in the region. Their results, they argue, indicate that continental network upgrading is worth serious consideration from an economic perspective: overland trade among sub-Saharan African countries might expand by about US$250 billion over 15 years, with major direct and indirect benefits for the rural poor, but financing the programme would require about US$20 billion for initial upgrading and US$1 billion annually for maintenance.

However, it is important to note that decline in off-road areas that may accompany such road developments tends to be ignored in standard assessments of road impact. No sufficiently sound methodology has been developed, as yet, to adequately assess the complex mix of sociocultural, psychological and economic impacts that may accompany or follow road and other transport infrastructure construction (Porter, 2014).

Transport services

Most transport literature does not adequately take into account the crucial significance of transport services: a major omission, especially where Africa's rural areas are concerned. In most countries, Gwilliam *et al.* (2010: 212) observe that at least 90 per cent of reported traffic on the classified network is carried on the main networks, yet, with the exceptions of Nigeria and South Africa, traffic on the main road network averages only about 500 vehicles per day: absolute volumes of traffic on the rural network average only about 30 vehicles per day, except in Nigeria. Thus, transport remains a major cost and logistical hurdle for many rural people, including smallholder farmers (especially women), yet detailed knowledge of agricultural transport service costs and logistics remains remarkably limited (Naude, 2009; Banjo *et al.*, 2012). The majority of research has focused principally on the impact of road connectivity to markets using data that can be obtained remotely (e.g. Dorosh *et al.*, 2012). For Gwilliam *et al.* (2010: 223), the priority for rural transport services is to maintain basic year-round vehicle access 'for the types of vehicle likely to be operating. The quantity of access is even more important than the quality'. They also observe the potential for better rural telecommunications to more effectively match vehicles to loads and the importance of community organisation for a collective lobby to address the monopoly power of service providers (Gwilliam *et al.*, 2010: 223). This is essential since there is no incentive for road transport providers to improve services or pass on the cost savings that can be obtained (notably from reduced maintenance costs) to users. While competition among transporters tends to rise when roads are improved to all-weather standard, on many earth and gravel roads competition is mostly absent. Roads are not enough, especially where the rural poor are concerned (Dawson and Barwell, 1993; Ellis and Hine, 1998; Bryceson *et al.*, 2008).

The importance of linkages between rural transport systems and agricultural development is well recognised among transport services specialists. Poor rural transport increases the costs of marketing from farms, inhibits use of fertiliser and other inputs, limits the spread of information and increases risk (Banjo *et al.*, 2012: 25). Given the fact that the vast majority of farmers in Africa are smallholders, and about 85 per cent of these are farming under 2 ha, Banjo *et al.* (2012) argue that a stronger focus on smallholder farming transport constraints is crucial and suggests the need to pay particular attention to services along roads and paths at the lower end of the road network. In a similar vein, recent analysis in central Africa suggests that road investments in very isolated rural areas are less attractive than in areas that, albeit some distance from urban markets, are still within 'reasonable reach', because they will not make enough difference in overall travel times to provide adequate market access (Gwilliam *et al.*, 2010: 222). In areas where volumes of production are below the threshold needed to justify use of a truck to collect produce, 'simpler roads targeted more at ensuring accessibility for 2-wheeled vehicles or animal-drawn carts may be more suitable' (Gwilliam *et al.*, 2010: 222). However, an emphasis on intermediate means of transport such as donkeys, bicycles and animal carts is still often rejected by decision-makers as primitive and backwards (Sieber, 1999).

Many agriculture specialists recognise the importance of improving transport services alongside roads as a means to greater accessibility to local output markets and external markets (e.g. Senyolo *et al.*, 2009 for South Africa), but there is a tendency to concentrate on road connectivity to markets rather than the specific role of transport services. This can probably be explained, at least in part, by the relative ease with which broad measures of road connectivity can be obtained and analysed remotely for large regions, compared to the complexities of gathering field data on road transport pricing and other (often hidden) non-economic aspects of transport services operations.

Notable exceptions include work by Jayne *et al.* (2003) with reference to fertiliser farm-gate prices and associated transport access. They note that the sum of importer, wholesaler and retailer profit margins generally account for less than 10 per cent in Kenya, Zambia and Ethiopia, and suggest there are opportunities to reduce domestic marketing costs through a variety of measures including coordinating the timing of fertiliser clearance from the port with up-country transport, reducing transport costs through port, rail and road improvements, and reducing high fuel taxes. Estimated reductions in the farm-gate price of fertiliser from implementing the full range of options identified in each country (not only transport) ranged from 11 to 18 per cent (see also Farrow *et al.*, 2011 on the differential access of households to resources and transport in the case of Malawi fertiliser inputs).

Transport services are also rarely considered in large commercial agriculture contexts in Africa, despite the fact that large firms in agriculture and other industries in Africa are more likely to complain about transport than smaller firms, with substantial losses reported from theft and transport delays for their company-owned transport (Ramachandran, 2008, based on World Bank Enterprise Surveys database). In this respect, work in the South African sugar industry presents a rare exception (Bezuidenhout *et al.*, 2004; Nordengen *et al.*, 2009). There is clearly need for much more empirical research in this arena, not least since many of these large commercial enterprises play an important labour role as employers of the very poor.

Naude (2009) argues more broadly that there is a lack of comparable, cross-country and time series data on transport and logistics costs in African countries, but points out that Africa's relatively high transport costs are an important factor in the continent's slow growth in exports compared to other developing regions. Distance remains one of the most important variables determining transport costs, and distances to international and even local urban markets are commonly high in the region. The dominance and expensiveness of road freight in many countries is a major disadvantage. Being a landlocked country has a particularly cost-inducing effect, through costs of transiting borders and time lost at border delays. The seasonality of crops also raises transport costs.

Transport services may also have a role to play in the expansion of employment opportunities in the non-farm sector. These are especially important because they can increase the sustainability of livelihoods in agriculturally focused communities. Multiplex livelihoods and off-farm income are widely recognised to provide a route out of rural poverty (Bryceson, 1999; Gladwin *et al.*, 2001; Bryceson, 2002). However, such livelihood diversification very often requires travel to the nearest market or service location: a point that has received very little attention. Research in the transport services field has barely started in this area (though see Venter and Cross, 2011 for a valuable South African study). Given recent concerns about high levels of youth unemployment in Africa, further work in this field is urgently needed.

Africa's transport deficiencies from the perspective of some disadvantaged populations

This section examines transport issues from the perspective of less advantaged groups – women, children and older people are each considered in turn. Transport deficiencies play out in a vicious circle of everyday life, where poor access to transport reduces opportunities for education and health-seeking, with wide-ranging repercussions for livelihoods, individual life chances, and the potential for wider community improvement and positive economic change.

Women's and men's access to transport may vary with age, socio-economic status, local culture and other factors, but there are fairly widespread gender distinctions across Africa (particularly among lower-income groups) regarding the use and operation of transport and in broader patterns of mobility. So far as women's access to transport is concerned, it is important to bear in mind that prevailing gender constructs are intimately related to the way people are able to interact between places: just as social processes help shape gendered mobility patterns and access to transport, mobility and transport are strongly implicated in the shaping of social processes (Porter, 1997; Evers and Walters, 2000; Porter, 2002, 2011). The principal operators (and owners) of commercial motorised and non-motorised transport equipment in most rural and urban areas are still mostly male (though Duchene, 2011 provides a few urban examples of women in the commercial transport sector in Africa), while females are commonly the principal pedestrian transporters, usually by head-loading. There are significant educational and associated negative livelihood implications associated with this latter type of transport work, and possibly also health impacts, though we lack firm medical evidence (Porter *et al.*, 2012a, 2013a, 2013b).

Both men and women may be important consumers of transport, but women tend to face more substantial constraints on their mobility and travel mode than men, not least as a result of their relative poverty compared to men (commonly, they have lower incomes and less access to land, labour resources and other productive assets) and the time poverty they face due to heavy work burdens and childcare responsibilities (Blackden and Wodon, 2006). Cultural context can also be highly significant in the shaping of women's mobility and use of transport services, though it is important to bear in mind that restrictions on women's mobility do not necessarily map directly onto religious affiliation: for cases where Islam is not highly restrictive on women's mobility and Christian societies where mobility restrictions prevail, see Mandel (2004) and Porter (2011). Cycling provides some interesting perspectives with reference to cultural constraints since it is a transport mode that is low-cost yet can facilitate considerable movement. The rarity of women cyclists across much of Africa seems – at least in part – to result from male concerns regarding women's mobility beyond the home community (Malmberg-Calvo, 1994; Grieco *et al.*, 1995; Flanary, 2004). Negative views regarding cycling as a pursuit for females remains widespread across much of rural Africa, despite the advantages the associated improved mobility could bring to many women's educational, health and livelihood opportunities (Durham University, 2010).

Young people's spatial mobility and access to transport is shaped by diverse factors, from their age, gender, family socio-economic status and parental status, to disability and the broader socio-cultural and economic context and physical environment, perceptual and cultural factors. It is important to consider the mobility and transport needs of children and young people in the context of their lives within the family and household and in terms of potential life and livelihood trajectories. For instance, access to facilities and services may depend not only directly on transport availability and cost allowing children to travel, but may also be strongly affected by family and household demands for children's work (Porter *et al.*, 2011). Across southern Africa, the high risk to youth of teasing, bullying and rape as they travel, mostly – though not wholly – directed at young girls, has been widely reported (e.g. Leach *et al.*, 2000: 15; Potgieter *et al.*, 2006; Porter *et al.*, 2010; Hampshire *et al.*, 2011).

The significance of the time, effort and/or costs of transport incurred in getting to school is rarely considered in studies of educational access in Africa, yet schools, like health centres, are usually built in central places. The vast majority of children walk to school in many locations: the 2003 South Africa National Household Travel Survey, for instance, found 76 per cent of 'learners' walking to their educational destination, and almost 3 million out of the 16 million total (especially those located in more rural provinces) spend more than an hour a day walking to and from educational institutions (Department of Transport, Republic of South Africa, 2003). When parental contributions and household labour demands are coupled with a long journey to school, these are likely acting together to present a particularly strong deterrent to attendance (Porter *et al.*, 2011, 2012b). However, even children who are able to attend school may be greatly disadvantaged in their school performance by transport constraints in the home environment: at the household level, transport failures may require children to carry water, firewood, etc., as well as to perform other household tasks, both before and after classes. These duties, which tend to fall particularly heavily on girls, in accordance with local cultural norms, delay the time when children leave for school, may cause them to arrive late at school (resulting in punishment from their teachers), and leave them exhausted during lesson time. Transport issues along the route to school (poor roads, unreliable, costly or non-existent transport services, etc.) may add further to their problems. Unsurprisingly, then, in a review of children out of school using DHS surveys in Niger, where there are only 41 girls per 100 boys at school in rural areas (compared to 80:100 in town), distance of home from school was identified as a key factor (DFID, 2001:7).

Difficulties of physical access to health services may impact on young people in particularly life-damaging ways, from failure to obtain immunisation in childhood (Bosu *et al.*, 1997), to high fatality rates among young children who fall ill with cerebral malaria or meningitis, and among girls experiencing complications from pregnancy at an early age. Distance and lack of emergency transport are likely to be critical factors in low and delayed hospital referrals of children from many remote rural areas (Cullinan and Pieterick, 1998; Bossyns *et al.*, 2006).

Older people's mobility and transport needs are only just starting to gain attention, despite the fact that older people form a substantial key component of African populations, not least since in many countries HIV/AIDS has left grandparents supporting and caring for grandchildren. The mobility and mobility constraints older people face, which will impact strongly on their ability to act effectively in this caring role, constitute a major knowledge gap across the continent (Porter *et al.*, 2013b). Only one urban/rural comparative research study (for South Africa), a few urban studies (mostly Nigeria) and one rural study (Tanzania) appear to have been published to date.

Venter (2011), drawing on data from South Africa's 2003 National Household Travel Survey, presents what is probably the only comparative study of urban and rural mobility among older people in the continent. The data, which show that approximately half of older people over 65 ($N = 4{,}522$) do not travel (i.e. did not leave their homes at all on the survey day), give some indication of the limited mobility of this age group compared to the population as a whole (which is approximately twice the figure for that group). The particular disadvantage of rural older people is highlighted: around one-third of older people in urban areas refer to access problems, as opposed to around two-thirds in rural areas, while affordability is a greater issue among the poor rural elderly than the poor urban elderly.

Urban studies encompassing some reflections regarding older people's mobility in Africa extends back to work by Turner and Kwakye (1996; see also Grieco *et al.*, 1996), which emphasises how this often intersects with that of other family members in Accra, with older grandmother traders sometimes taking on the role of stationary domestic anchors for larger households where many adults are engaged in activities across the city. A series of more recent Nigeria-focused transport studies, based on urban household surveys with older people, drew attention to the theme of older people's transport constraints, the inadequacy of current transport provision (not least in the context of continuing travel required for livelihood sustenance in the absence of social security), and the urgent need for government action (Odufuwa, 2006; Ipingbemi, 2010; Olawole and Aloba, 2014). The latter study usefully draws specific attention to age and mobility differentials within the older people sampled, with significantly reduced travel in the 70+ cohort.

The only rural-focused research on older people's mobility published to date seems to be from the study conducted in 10 villages in Kibaha District, Tanzania (Porter *et al.*, 2013b). The new connectivities associated with increasing mobile phone and motorcycle taxi usage among older people here are likely to also be of significance for the future in many parts of rural Africa. Motorcycle taxis (boda boda) are now the main mode of transport in this district, except along the paved road. They have spread rapidly in all the Kibaha study settlements since around 2007–2009, before which transport was limited mainly to bicycle taxis. Their uptake has been facilitated by the availability of cheap imported Chinese motorcycles and the fact that they offer a major employment opportunity for rural youth. We concluded that boda boda have transformed rural lives in this district, even where older people are concerned: 18 per cent of older women and 31 per cent of older men had used one in the week before the survey ($n = 339$) – they are now ubiquitous. While older people would *prefer* other motorised transport (i.e. buses or minibuses, which are cheaper, safer and more comfortable), the only real alternative is usually walking. They are, of course, especially important in health emergencies, but they are also important for livelihoods. The small scale of farming among many older people makes boda boda feasible for transport of farm produce and farm inputs. Moreover, some urban traders are now coming into the district by boda boda to purchase at their farms.

Meanwhile, mobile phones are a complementary new connector in Kibaha District, as in many parts of Africa. Mobile phones were owned by 41 per cent of the older men surveyed and 15 per cent of the older women, but are widely available to many more older people through widespread sharing with relatives and friends. The phones have radically improved older people's access to transport, since mobile phones are now widely used to organise boda boda transport both in emergencies and on a daily basis. Phones have also led to some reported reduction in travel: even remittances from family elsewhere can be accessed by phone mobile money services. However, reduced face-to-face interaction has also reportedly left some older people feeling lonely and isolated.

Conclusion

This chapter has drawn attention to some of the diverse constraints currently facing Africa in the transport sector. The Millennium Development Goals failed to recognise the significance of transport for poverty reduction (Bryceson, 2009), but the new Sustainable Development Goals at least offer the prospect of paying some attention to transport issues in urban Africa. Meanwhile, underlying environmental and social uncertainties associated with global climate change, wherever instigated, are likely to present new challenges.

Despite the evident constraints on improvements of infrastructure and transport services in Africa, it is clear that some positive new developments are emerging, not least the new connectivities associated with the increasing use of phones to aid transport organisation and the benefits that can accrue where mobile phones substitute to some degree for mobility (Porter, 2015, 2016). Of course, physical mobility and transport use per se are not necessarily desirable, especially if they are simply required because of poor access to work and facilities. Indeed, planning for mobility, as opposed to accessibility, may enhance gender biases since the benefits tend to accrue to those already mobile, i.e. male vehicle owners (Masika and Baden, 1997: 9). Non-transport interventions to improve access to resources, and thus substitute for mobility – installation of improved water supplies, community woodlots, more efficient stoves, solar panels and crop-grinding mills – might all bring more significant reductions in transport time than efforts to improve mobility, and have the potential to bring particular benefits to those who would not be able to afford improved transport services. Virtual mobility through use of mobile phones, Internet and other ICTs also has substantial potential to beneficially reduce the transport needs of all sectors of the population.

Note

1 This chapter draws substantially on a literature review prepared by the author for the UK DFID-funded Africa Community Access programme (Porter, 2012).

References

AfDB (African Development Bank) (2010) *Port Development in Africa*, available at: www.afdb.org/fileadmin/uploads/afdb/Documents/Publications/African%20Development%20Report%202010_CH%202.pdf (accessed 23 August 2016).

Anyanwu, J.C. and Erhijakpor, A.E. (2009) 'The impact of road infrastructure on poverty reduction in Africa', *African Political Economic and Security Issues Series*, 1–40.

Banjo, G., Gordon, H. and Riverson, J. (2012) *Rural Transport: Improving Its Contribution to Rural Growth and Poverty Reduction in Sub-Saharan Africa*, Washington, DC: World Bank, SSATP Working Paper 93, November.

Bezuidenhout, C.N., Lusso, C.D., Lyne, P. and Meyer, E. (2004) 'Minimising transport costs in two-phase sugarcane extraction systems through optimal upgrading of roads and transloading zones', *International Sugar Journal*, 106(1269): 466–473.

Blackden, C.M. and Wodon, Q. (eds) (2006) *Gender, Time Use and Poverty in Sub-Saharan Africa*, Washington, DC: World Bank Working Paper No. 73.

Bossyns, P., Abache, R., Abdoulaye, M.S., Miye, H., Depoorter, A.M. and Van Lerberghe, W. (2006) 'Monitoring the referral system through benchmarking in rural Niger: an evaluation of the functional relation between health centres and the district hospital', *BMC Health Services Research*, 6(51), 12 April: 1–7.

Bosu, W.K., Ahelegbe, D., Edum-Fotwe, E., Bainson, K.A. and Turkson, P.K. (1997) 'Factors affecting attendance to immunisation sessions: Ghana', *Acta Tropica*, 68: 259–267.

Bryceson, D. (1999) *Sub-Saharan Africa Betwixt and Between: Rural Livelihood Practices and Policies*, Leiden: Afrika-Studiecentrum, Working Paper 43.

Bryceson, D.F. (2002) 'The scramble in Africa: reorienting rural livelihoods', *World Development*, 30(5): 725–739.

Bryceson, D. (2009) 'Roadmapping development and poverty alleviation: transport and the millennium development goals in Africa', in M. Grieco, M. Ndulo, D. Bryceson, G. Porter and T. McCray (eds), *Africa, Transport and the Millennium Development Goals: Achieving an Internationally Set Agenda*, Newcastle: Cambridge Scholars Publishing, pp. 1–24.

Bryceson, D., Bradbury, A. and Bradbury, T. (2008) 'Roads to poverty reduction? Exploring rural roads' impact on mobility in Africa and Asia', *Development Policy Review*, 26(4): 459–482.

Buys, P.D. and Uwe, W.D. (2010) 'Road network upgrading and overland trade expansion in sub-Saharan Africa', *Journal of African Economies*, 19(3): 399–432.

Corkin, C., Burke, C. and Davies, M. (2008) *China's Role in the Development of Africa's Infrastructure*, Washington, DC: Johns Hopkins University SAIS Working Papers in African Studies.

Cullinan, T.R. and Pieterick, C. (1998) 'Packaged treatment for first line care in cerebral malaria and meningitis', *Bull WHO*, 76(3): 257–264.

Dawson, J. and Barwell, I. (1993) *Roads Are Not Enough: New Perspectives on Rural Transport Planning in Developing Countries*, London: Intermediate Technology Publications.

DFID (Department for International Development) (2001) *Children Out of School*, October, London: DFID.

Department of Transport, Republic of South Africa (2005) *The First South African National Household Travel Survey 2003*, Technical Report.

Dercon, S., Gilligan, D.O., Hoddinott, J. and Woldehanna, T. (2009) 'The impact of agricultural extension and roads on poverty and consumption growth in fifteen Ethiopian villages', *American Journal of Agricultural Economics*, 91(4): 1007–1021.

Dorosh, P., Wang, H.G., You, L. and Schmidt, E. (2012) 'Road connectivity, population, and crop production in sub-Saharan Africa', *Agricultural Economics*, 43(1): 89–103.

Duchene, C. (2011) *Gender and Transport*, International Transport Forum, Discussion Paper 2011.11, available at: www.oecd-ilibrary.org/transport/gender-and-transport_5kg9mq47w59w-en (accessed 29 December 2017).

Durham University (2010) *Children, Transport and Mobility in Sub-Saharan Africa*, Durham: Department of Anthropology, Durham University, available at: www.dur.ac.uk/child.mobility/ (accessed 9 January 2018).

Ellis, S. and Hine, J. (1998) *The Provision of Rural Transport Services: Approach Paper*, SSATP Working Paper No. 37, April.

Evers, B. and Walters, B. (2000) 'Extra-household factors and women farmers' supply response in sub-Saharan Africa', *World Development*, 28(7): 1341–1345.

Farrow, A., Risinamhodzi, K., Zingore, S. and Delve, R.J. (2011) 'Spatially targeting the distribution of agricultural input stockists in Malawi', *Agricultural Systems*, 104(9): 694–702.

Feldacker, C., Ennett, S.T. and Speizer, I. (2011) '"It's not just who you are but where you live": an exploration of community influences on individual HIV status in rural Malawi', *Social Science and Medicine*, 72: 717–725.

Ferguson, A.G. and Morris, C.N. (2006) 'Mapping transactional sex on the Northern Corridor highway in Kenya', *Health and Place*, 13(2): 504–519.

Flanary, R.M. (2004) 'Gender and embodied mobility: learning in Tarsaw, northern Ghana', unpublished PhD thesis, University of Durham.

Foster, V. and Briceno-Garmendia, C. (eds) (2010) *Africa's Infrastructure: A Time for Transformation*, Washington, DC: World Bank.

Gladwin, C.H., Thomson, A.M., Peterson, J.S. and Anderson, A.S. (2001) 'Addressing food security in Africa via multiple livelihood strategies of women farmers', *Food Policy*, 26(2): 177–207.

Grieco, M., Turner, J. and Kwakye, E. (1995) *A Tale of Two Cultures: Ethnicity and Cycling Behaviour in Urban Ghana*, Transport Research Record 1441, Washington, DC.

Grieco, M., Apt, N. and Turner, J. (1996) *At Christmas and on Rainy Days: Transport, Travel and the Female Traders of Accra*, Aldershot: Avebury.

Gwilliam, K. (2010) 'Transport: more than the sum of its parts', in V. Foster and C. Briceno-Garmendia (eds), *Africa's Infrastructure: A Time for Transformation*, Washington, DC: World Bank, pp. 203–210.

Gwilliam, K., Sethi, K., Nogales, A. and Foster, V. (2010) 'Roads: broadening the agenda', in V. Foster and C. Briceno-Garmendia (eds), *Africa's Infrastructure: A Time for Transformation*, Washington, DC: World Bank, pp. 211–228.

Hampshire, K., Porter, G., Mashiri, M., Maponya, M. and Dube, S. (2011) 'Proposing love on the way to school: mobility, sexuality and youth transitions in South Africa', *Culture, Health & Sexuality*, 13(2): 217–231.

Heinz, S. and O'Connel, J.F. (2013) 'Air transport in Africa: toward sustainable business models for African airlines', *Journal of Transport Geography*, 31: 72–83.

Ipingbemi, O. (2010) 'Travel characteristics and mobility constraints of the elderly in Ibadan, Nigeria', *Journal of Transport Geography*, 18(2): 285–291.

Jayne, T.S. Govereh, J. Wanzala, M. and Demeke, M. (2003) Fertilizer market development: a comparative analysis of Ethiopia, Kenya, and Zambia', *Food Policy*, 28(4): 293–316.

Kwamusi, P. (2002) 'Gender and safety in rural transport', in P. Fernando and G. Porter (eds), *Balancing the Load: Women, Gender and Transport*, London: Zed Books, pp. 57–64.

Leach, F., Machakanja, P. and Mandoga, J. (2000) *Preliminary Investigation of the Abuse of Girls in Zimbabwean Junior Secondary Schools*, DFID Education Research Paper, No. 39, available at: www.sussex.ac.uk/webteam/gateway/file.php?name=preliminary-investigation-of-the-abuse-of-girls-in-zimbabwean-schools-paper-39.pdf&site=320 (accessed 8 January 2018).

Linard, C., Gilbert, M., Snow, R.W., Noor, A.M. and Tatem, A.J. (2012) 'Population distribution, settlement patterns and accessibility across Africa in 2010', *PLOS ONE*, 7(2): e31743, DOI: 10.1371/journal.pone.0031743.

Lucas, K. and Porter, G. (2016) 'Mobilities and livelihoods in urban development contexts: introduction', *Journal of Transport Geography*, 55: 129–131.

Malmberg-Calvo, C. (1994) *Case Study on Intermediate Means of Transport: Bicycles and Rural Women in Uganda*, Washington, DC: World Bank, SSATP Working Paper 12.

Mandel, J.L. (2004) 'Mobility matters: women's livelihood strategies in Porto Novo, Benin', *Gender, Place & Culture*, 11(2): 257–287.

Mashiri, M. (2004) *Community Responses to HIV/AIDS Along Transit Corridors and Areas of Intense Transport Operations in Eastern and Southern Africa*, IFRTD/CSIR Transportek, Reports to the UK Department for International Development (Transport Knowledge and Research Programme), April–May.

Masika, R. and Baden, S. (1997) *Infrastructure and Poverty: A Gender Analysis*, Sussex: Institute of Development Studies, BRIDGE Report No. 51.

Naude, W. (2009) 'Geography, transport and Africa's proximity gap', *Journal of Transport Geography*, 17: 1–9.

Nordengen, P.A., Prem, H. and Lyne, P.W.L. (2009) 'Performance-based standards (PBS) vehicles for transport in the agricultural sector', *Sugar Cane International*, 27(2) 56–59.

Odufuwa, O.B. (2006) 'Enhancing mobility of the elderly in sub-Saharan Africa cities through improved public transportation', *IATSS Research*, 30(1): 60–66.

Olawole, M.O. and Aloba, O. (2014) 'Mobility characteristics of the elderly and their associated level of satisfaction with transport services in Osogbo, southwestern Nigeria', *Transport Policy*, 35: 105–116.

Porter, G. (1997) 'Mobility and inequality in rural Nigeria: the case of off-road communities', *Tijdschrift voor Economische en Sociale Geografie*, 88(1): 65–76.

Porter, G. (2002) 'Living in a walking world: rural mobility and social equity issues in sub-Saharan Africa', *World Development*, 30(2): 285–300.

Porter, G. (2007) 'Transport planning in sub-Saharan Africa. Progress report 1: improving access to markets and services', *Progress in Development Studies*, 7(3): 251–257.

Porter, G. (2011) '"I think a woman who travels a lot is befriending other men and that's why she travels": mobility constraints and their implications for rural women and girl children in sub-Saharan Africa', *Gender, Place & Culture*, 18(1): 65–81.

Porter, G. (2012) *Transport Services and Their Impact on Poverty and Growth in Rural Sub-Saharan Africa: Literature Review*, Report to the Africa Community Access Programme, London, Technical Report. London: Department for International Development (DFID), available at: http://dro.dur.ac.uk/14134/1/14134.pdf (accessed 8 Janaury 2018).

Porter, G. (2014) 'Transport services and their impact on poverty and growth in rural sub-Saharan Africa: a review of recent research and future research needs', *Transport Reviews*, 34(1): 25–45.

Porter, G. (2015) 'Mobile phones, mobility practices and transport organisation in sub-Saharan Africa', *Mobility in History*, 6: 81–88.

Porter, G. (2016) 'Mobilities in rural Africa: new connections, new challenges', *Annals of the American Association of Geographers*, 106(2): 434–441.

Porter, G., Hampshire, K., Mashiri, M., Dube, S. and Maponya, G. (2010) '"Youthscapes" and escapes in rural Africa: education, mobility and livelihood trajectories for young people in Eastern Cape, South Africa', *Journal of International Development* [special issue on youth], 22(8): 1090–1101.

Porter, G., Hampshire, K., Abane, A., Tanle, A., Esia-Donkoh, K., Amoako Sekyi, R., *et al.* (2011) 'Mobility, education and livelihood trajectories for young people in rural Ghana: a gender perspective', *Children's Geographies*, 9(3–4): 395–410.

Porter, G., Hampshire, K., Abane, A., Munthali, A., Robson, E., Mashiri, M., *et al.* (2012a) 'Youth, mobility and mobile phones in Africa: findings from a three-country study', *Journal of Information Technology for Development*, 18(2): 145–162.

Porter, G. Hampshire, K., Abane, A., Munthali, A., Robson, E., Mashiri, M., *et al.* (2012b) 'Child porterage and Africa's transport gap: evidence from Ghana, Malawi and South Africa', *World Development*, 40(10): 2136–2154.

Porter, G., Hampshire, K., Dunn, C., Hall, R., Levesley, M., Burton, K., *et al.* (2013a) 'Health impacts of pedestrian head-loading: a review of the evidence with particular reference to women and children in sub-Saharan Africa', *Social Science and Medicine*, 88: 90–97.

Porter, G., Tewodros, A., Bifandimu, F., Gorman, M., Heslop, A., Sibale, E., *et al.* (2013b) 'Transport and mobility constraints in an aging population: health and livelihood implications in rural Tanzania', *Journal of Transport Geography*, 30: 161–169.

Potgieter, C.-A., Pillay, R. and Rama, S. (2006) *Women, Development and Transport in Rural Eastern Cape, South Africa*, Cape Town: HSRC Press.

Ramachandran, N. (2008) *Power and Roads in Africa*, London: Center for Global Development.

Rizzo, M. (2015) 'The political economy of an urban megaproject: the bus rapid transit project in Tanzania', *African Affairs*, 114(455): 249–270.

Senyolo, G.M., Chaminuka, P., Makhura, M.N. and Belete, A. (2009) 'Patterns of access and utilization of output markets by emerging farmers in South Africa: factor analysis approach', *African Journal of Agricultural Research*, 4(3): 208–213.

Sieber, N. (1999) 'Transporting the yield: appropriate transport for agricultural production and marketing in sub-Saharan Africa', *Transport Reviews*, 19(3): 205–220.

Teravaninthorn, S. and Raballand, H. (2008) *Transport Prices and Costs in Africa: A Review of the Main International Corridors*, Washington, DC: World Bank, AICD Working Paper 14.

Turner, J. and Kwakye, E. (1996) 'Transport and survival strategies in a developing economy: case study evidence from Accra, Ghana', *Journal of Transport Geography*, 4(3): 161–168.

UNEP (United Nations Environment Programme) (2010) *Africa Water Atlas*, Nairobi: Division of Early Warning and Assessment (DEWA), available at: www.unep.org/pdf/africa_water_atlas.pdf (accessed 23 August 2016).

Venter, C. (2011) 'Transport expenditure and affordability: the cost of being mobile', *Development Southern Africa*, 28(1): 121–140.

Venter, C. and Cross, C. (2011) 'Location, mobility and access to work: a qualitative exploration in low-income settlements', *Proceedings: Southern African Transport Conference, Pretoria, July 2011*, available at: https://repository.up.ac.za/bitstream/handle/2263/17306/Venter_Location%20%282011%29.pdf?sequence=1 (accessed 8 January 2018).

WHO (World Health Organization) (2013) *Road Safety in the WHO African Region: The Facts*, Geneva: WHO, available at: www.who.int/violence_injury_prevention/road_safety_status/2013/report/factsheet_afro.pdf (accessed 8 January 2018).

World Bank (2008) *Safe, Clean and Affordable: Transport for Development. The World Bank Group's Transport Business Strategy 2008–2012*, Washington, DC: World Bank, available at: http://documents.worldbank.org/curated/en/440361468175472897/Safe-clean-and-affordable-transport-for-development-the-World-Bank-Groups-transport-business-strategy-for-2008-2012 (accessed 8 January 2018).

43
Land-grabbing in Africa

Simon Batterbury and Frankline Ndi

Introduction

Land-grabbing in Africa refers to the purchase or acquisition of use rights to produce food, biofuels or animal feed. Over the last 20 years, private, foreign investors and governments have often secured African land as investments, or to help meet their own national food security and biofuel needs (Daniel and Mittal, 2009). Although widely acknowledged as a global phenomenon, land-grabbing is particularly prevalent in Africa given the continent's favourable biophysical resources and its lack of existing large-scale, industrialised agriculture and plantations compared to other continents (Anseeuw, 2013; Cotula *et al.*, 2014). Many Africa governments, some with weak land tenure regulations and others needing to balance their budgets, breach existing customary and communal land tenure arrangements to reallocate land and forests to firms, foreign governments and speculative investors – particularly for agricultural development.

Land users in regions where commercial agriculture is not yet significant, or where water is abundant, have proven to be particularly vulnerable to land-grabbing (Allan *et al.*, 2013). Firms and other actors promise jobs and technological transfer to local populations in exchange for access to their land through purchase or leases, but as the chapter will show, the gains are often fleeting or absent, and in many cases they do not 'trickle down' to the local communities (O'Brien, 2011; Anseeuw, 2013). The greatest impacts are felt by the poorest of the poor – those forest-dependent, farmer and herder populations who are dispossessed and who lack other options to farm or graze their livestock (Mope Simo, 2011; De Schutter, 2012; Cotula, 2014). They rarely receive adequate compensation for their losses, or any significant opportunities in the new operations that exploit the land. We explore the definitions and workings of land-grabbing in Africa, providing some examples of the process.

Definitions and characteristics

According to Borras and Franco (2012), a 'land grab' is the power to control large quantities of land and landed resources for capital accumulation in response to food security crises, short- or long-term climate change impacts, and financial exigencies. A 'land grab' does not include acquisition for subsurface mining or infrastructure, but is often situated to benefit from adequate water supply (Allan *et al.*, 2013). 'Large-scale land acquisition' or LSLA is a more accurate term, and we will use it here, but it has proven less appealing in media reporting. In 2008, journalist John Vidal identified a related phenomenon: 'green-grabbing'. It involves a similar transfer of rights of ownership, use, and control of resources, but on the basis of 'green' credentials (Fairhead *et al.*, 2012). Thus, environmental projects are also used to justify appropriations or acquisition of land for biodiversity conservation by organisations such as the African Wildlife Foundation. Aside from private and public efforts to preserve habitat, there is strong interest in ecotourism, and in securing forest land for carbon offsetting, energy and sequestration schemes (Borras and Franco, 2010; Borras *et al.*, 2011; Borras and Franco, 2012; Fairhead *et al.*, 2012).

So, in all cases, the term 'grab' signals a loss in access, because it occurs in productive regions of tropical Africa and often where governance is weak. It is generally poorer and marginalised communities that lose land, rather than richer farmers with greater power and influence. The process is therefore a form of privatisation (Green and Adams, 2015) or 'accumulation by dispossession' (Harvey, 2005). It involves the separation of labour from its means of production. Some of the processes Harvey describes include: displacement of peasant farmers in favour of large-scale producers; acquisition of land meant for food crop production to cultivate cash crops; and the privatisation of common resources (such as land and forests). Land deals are therefore part of the 'marginalization of the African peasantry' (Bernstein, 1996). 'Simple capital accumulation' occurs where profits accrued elsewhere are reinvested in land for commercial speculation, because it is a good investment. But from the early stages of African grabbing in the 2000s, 'primitive accumulation' has also occurred – land was not purchased at its true value, but gifted or transferred to new owners, excluding local existing claimants' rights to what they believed to be a common resource. Primitive accumulation harks back to colonial times, when settlers allocated land to themselves at no cost.

The FAO suggests LSLAs involve more than 1,000 hectares of land changing hands in a single deal, the direct involvement of governments or foreign investors, and negative impacts on local food security (Borras and Franco, 2012). Governments or formal state bodies are complicit in justifying and enabling the process of obtaining access to land that is said to be idle or sparsely populated, but the evidence is that most deals involve some form of dispossession (Daniel and Mittal, 2009). This includes land deals that involve national, urban-based elites and firms that also seek to grow food and biofuel for profit – or simply to reserve land for the future – in productive, but sometimes lowly populated, regions (Hall, 2011). Acquisitions can also involve amalgamation of smaller parcels that add up to a significant 'grab', and which may still displace existing land users and land uses (Zoomers and Kaag, 2014). In some cases, transferred land is held as a 'reserve' and its use is unaltered for years or even decades; in others, eviction occurs quickly. In the heavily forested tropics, a long lease to permit oil palm plantations means first cutting and selling the gallery rainforest (a lucrative business) to then plant oil palm.

Data on the scale, geographical distribution, trends, and players (both national and foreign) in large-scale land deals are generally unreliable (Vermeulen and Cotula, 2010). Acquisitions for conservation, tourism and logging are also hard to quantify (Fairhead *et al.*, 2012; see also Zoomers, 2013). Even where data are available, the figures differ. For example, Oxfam (2011) states that Africa's land under acquisition is about 34 million hectares, about the size of Germany (see also Anseeuw *et al.*, 2012). According to the Oakland Institute (2011), in 2009 alone, about 60 million hectares were leased or purchased in Africa. More recently, Aubry *et al.* (2012: 3) claim that between 50 and 80 million hectares were acquired in the 'past few years'. Large land deals are not transparent, and contracts are often kept secret (Holmén, 2015). Information often 'cannot be compared due to variations in methodology, timescale and differing criteria for what makes a land deal' (IIED, 2012: 1), or is simply not available to researchers. Thus, there is great uncertainty about how extensive land-grabbing is (Scoones *et al.*, 2013). Oya (2013) argues that most estimations are based on speculation, coming from media reports and a few more accurate investigations (Holmén, 2015). The most commonly used sources on large land deals are from the Land Matrix (an independent land monitoring initiative) and GRAIN, a small NGO supporting farmers. The Land Matrix cross-checks from a variety of sources. The latter is based on media reports. But this still does not guarantee that a deal is 'real' because 'activists and campaigning civil society organisations circulate a great deal of information among themselves and the same deal can be mentioned many times, thus making the risk of double counting obvious' (Holmén, 2015: 461). Anseeuw *et al.* (2012) discovered that although 134 million hectares had been reported as land-grabbing, only 34 million hectares had been cross-referenced.

Not all of these investments have succeeded in the claims they make about what they will provide. Monoculture crops dominate most land under acquisition, and land most suitable for food crop production has been diverted to plantation tree crop production, including oil palm (Cotula and Vermeulen, 2009). Davis *et al.* (2014: 188) estimated 8.2 million Africans were affected through lost income, and 'poverty reduction is an unlikely result of large-scale land acquisitions'. Job creation is minimal on mechanised plantations with a few wage labourers. In tropical Africa, large-scale land acquisitions survive at the expense of forest ecologies, and to the detriment of forest-dependent communities (De Schutter, 2012), who unwillingly lose all or part of their land to investors.

The global drivers of land acquisition

The driving forces of African land grabs are global. Over the last three decades, rising global demand for food, biofuel and fibre and concerns over water supplies in richer arid and semi-arid nations have combined with the liberalisation of trade and investment regimes (Cotula *et al.*, 2009; Deininger and Byerlee, 2011). Foreign and commercial interests have long held assets in Africa, stretching back centuries, notably to the colonial period, when much land was appropriated by foreign interests. Examples are in southern Nigeria (where palm oil demand helped to fuel Britain's Industrial Revolution), in Algeria (where poor French farmers were resettled without compensating the existing population) and Kenya (where British interests created large tea plantations, farms and ranches on the best land, some still in existence). In Cameroon, for example, the German colonial administration, prior to the First World War, pursued agro-industrial development, growing oil palm, bananas, pepper and other crops in large quantities (Bederman, 1966; see also Mope Simo, 2011). The tradition continued under French and British rule (cf. Konings, 2008). A variety of outgrower and contractor relationships and taxation demands also forced Africans to meet production targets for crops such as cotton and cocoa on their own land (Peluso and Lund, 2011; White *et al.*, 2012).

The politico-economic context of contemporary market-driven valuation of natural resources is different to these earlier colonial ventures (Peluso and Lund, 2011). The sharp increase in world food prices in 2007–2008 and again in 2011 led to bans on food exports by large producer countries to aid their own populations. Increases in the prices of basic commodities such as cereals was an impetus for resource-poor but cash-rich countries to acquire land to secure more reliable food supplies (Lisk, 2013). Agribusiness interests also sought to profit from this same insecurity through African investments. Agrofuel production (biodiesel and bioethanol) received a boost when the European Union required an increase in its use from the early 2000s (however, in 2015, it mandated a reduction of crop-based biofuels). Since the mid-2000s, rich industrialised countries, including their investors from North America, Europe, China, India, the Middle East, Brazil and South Africa, embarked on the purchase and/or lease of millions of hectares of arable land in Madagascar, Ethiopia, Mozambique, Tanzania, Zambia and Ghana, and also in countries emerging from conflict such as the Democratic Republic of Congo, Liberia and Sudan (von Braun and Meinzen-Dick, 2009; Borras *et al.*, 2011; Anseeuw, 2013; Zoomers, 2013).

The United Nations Food and Agriculture Organization (FAO) and the World Bank have expressed concern that food production in developing countries needs to double by 2050 in order to feed the projected world population (Zoomers, 2013). Consequently, they argue that large-scale land deals have the potential to stimulate growth and technological advancement in agriculture, which has been neglected since the 1960s (cf. Anseeuw, 2013), and particularly in sub-Saharan Africa (De Schutter, 2009). They hold that large-scale investments in agriculture will lead to growth in food production, create avenues for the use of new technology, contribute to employment creation, and increase the GNP of most developing countries (De Schutter, 2009; Deininger and Byerlee, 2011).

Research has found the effectiveness of large-scale land acquisition in fostering the livelihood chances of local communities is low (Davis *et al.*, 2014). For instance, the former UN Special Rapporteur on the right to food, Olivier De Schutter, famously argued that investments of this kind threaten the social and economic livelihood of rural agrarian populations, particularly rural farmers without land titles (De Schutter, 2009, 2011, 2012). His reports contend that the process of large-scale land acquisitions will almost inevitably lead to social fragmentation, exclusion, displacement and property enclosure (De Schutter, 2009).

Lorenzo Cotula from the International Institute for Environment and Development (IIED), a UK-based organisation that runs a research programme on land acquisitions, has argued that transnational land deals are part of emerging African 'national strategies' to effect development, and especially in the commercial agricultural sector (Cotula *et al.*, 2014; see also Anseeuw, 2013). Investors have argued that they predominantly acquire marginal and 'unused' land. But 'vacant land', 'empty land' or 'unused land' is a fiction in most parts of Africa (Borras *et al.*, 2011). Depending on the farming system, land may already be cultivated, in fallow to regain soil fertility, grazed, used for traditional rituals, be used for gathering fuelwood, or be part of an important water catchment (Zoomers, 2013).

Dispossessing local groups in favour of large-scale plantation agriculture would appear to be an unsound development pathway for these reasons (Vermeulen and Cotula, 2010; De Schutter, 2012). Foreign investors naturally focus their attention on land with market proximity, higher rainfall, fertile soils, or with access to irrigation (Cotula and Vermeulen, 2009). For example, in Mali, Cotula and Vermeulen (2010) found all recorded land deals are concentrated in the country's highest potential agricultural zones. The land that was most attractive to investors was also claimed and used by local people (Cotula and Vermeulen, 2010). Uncontrolled land leases and purchases alienate small producers from their land with little benefit, despite claims that their labour is required on plantations and arable estates (Davis *et al.*, 2014). Land is turned towards satisfying the demands of international markets (Oakland Institute, 2011; see also De Schutter, 2012). In most instances, local communities' lands are appropriated without alternative sources of livelihoods being available, or at best they are reduced to wage labourers, elsewhere in rural areas, or after migration to cities and towns. Where 'accumulation by dispossession' occurs (Harvey, 2004), compensation packages are rarely delivered; rural people lack power to demand their rights.

Current status

There is some evidence that the pressure from new investment deals has eased in the last few years. The Land Matrix site lists substantial land allocations that have now been cancelled or scaled back. *In 2015*, local farmers in Gassol returned to land seized by Dominion Farms but abandoned in Tabara state, Nigeria (Chow, 2016). Some failures are due to practicalities; a lack of access and infrastructure, or a lack of sufficient financing to prepare and cultivate land. An example is the failed Daewoo project in Madagascar, which lost its licence to operate (Cotula and Vermeulen, 2011). Protests against non-consultative land grabs are also having an impact (Cotula, 2013). Some multinational agencies have begun redirecting their attention to improving the fairness of the agreements and contracts binding global land deals (Kandel, 2015), and in 2016 European MEPs challenged the Group of Eight's New Alliance for Food Security and Nutrition (NAFSN), arguing for a halt to land-grabbing. Others have also pushed for the strengthening of land rights for local communities (Wolford *et al.*, 2013). So too have new government requirements set in place to deny land to undercapitalised and exploitative foreign companies. Land-grabbing is clearly linked to the long-term insecurity of land rights in Africa – indeed, this is the prime reason that many deals have been able to go ahead in the first place, without protection of rights, or social and environmental safeguards. A fairer future for agricultural development would not involve losing land to multinational entities at all, but using it for local and individual benefit. This involves local control of commercial production, while retaining tenure and satisfying local subsistence requirements – the opposite of 'dispossession'.

An example from Cameroon

Despite the slowdown in new deals, certain countries continue to attract commercial interest. Cameroon is among them, with fertile soils and high rainfall in its tropical southern regions. The UN revealed in 2012 that the government had concluded agreements for the transfer of almost 800,000 hectares of arable land, of which almost half was for the development of industrial oil palm plantations; in addition, 300,000 hectares of rainforest was reallocated over a decade (De Schutter, 2012). It is alleged that additional foreign companies are on the way, depending on the fortunes of the oil palm sector (Hoyle and Levang, 2012; see also De Schutter, 2012). The US-based Herakles Farms land acquisition was an early case, and shows how a lack of transparency can lead to accusations of 'grabbing'. Oil palm (*Elaeis guineensis*) is endemic to West Africa. Plantations grew in colonial times, and its oil is widely used in food products, for cooking, and for biofuels. A modern plantation monocrop can yield 3.9 tonnes of crude oil per hectare. The initial request by Herakles for land in the south-west region of the country totalled 73,086 hectares on a 99-year lease; following protests, this was reduced to 19,843 hectares, but still on land that is claimed by and in use by local villagers.

LSLAs in Cameroon have been facilitated almost coincidentally, as a result of changes to the country's land laws that took place in the 1970s in favour of private ownership. From 1974, the government passed decrees that nationalised all land, except private land with valid registration documents. These laws shifted management rights from the hands of traditional leaders to the state – the management went to Land Consultative Boards (LCBs), presided over by the Divisional Officers (DOs). Chiefs and two notables serve as community representatives, identifying boundaries and cultural and sacred sites during social and environmental impact assessments (cf. Fonjong *et al.*, 2016). However, in reality, their roles have been passive, and private individuals or corporate entities have been able to acquire land certificates. The government is happy to promote private sector development and to ensure a constant source of revenue through tax. Nationals can acquire land titles for at least US\$800–1,000 (Fombe *et al.*, 2013), and more depending on the size and location. Applicants pay XAF1,000 per 100 m^2 (about US\$2) for urban land and XAF300 per 100 m^2 (about US\$0.5) in rural areas (XAF is the Central African CFA Franc, the local currency in Cameroon). In addition, they are required to pay other costs, transportation for surveyors, and the cost of placing boundary markers on their land (Firmin-Sellers and Sellers, 1999). These costs are negotiated, in practice, and the process involved is not entirely above board (cf. Fombe *et al.*, 2013). Foreign investors (individual or corporate) pay more, but documents or procedures are rarely made known to the public. It is alleged that the process of land acquisition is corrupt, and requires networking with those occupying higher political positions in the government. The government is seeking both tax income and agricultural modernisation.

The sheer scale of the Herakles proposal cut into land already claimed by communities. Interviews held in Nguti subdivision (in 2015) confirmed that local populations were never consulted; rather, they were merely informed that the national government in Yaounde had already repossessed part of their land for palm plantations, and that they should offer their full collaboration and participation. Apart from knowing the total land allocation, local communities were unaware of what would happen in their community. They were not told the details of the agreement; neither were they informed of any safety nets to provide alternative sources of livelihood or to mitigate risk. The concession boundaries were not entered into the National Land Cadastre (Potter, 2015). Amid this confusion, local communities have argued that all agreements completed without their consent have to be renegotiated.

Unsurprisingly, deals such as this are heavily contested, and investors have been accused of not abiding by national regulations or by international agreements (WWF, 2012; Greenpeace, 2013; Fonjong *et al.*, 2015, 2016). From its inception, the Herakles project was controversial. Local communities, in alliance with international and national NGOs (especially Greenpeace Cameroon and the Oakland Institute) appealed to the government to reverse it, outlining the actual and potential socio-economic and environmental impacts. In December 2013, the concession was downscaled to 19,843 hectares (CED, 2012; WWF, 2012; Greenpeace, 2013). Communities still resist actively and passively through strikes and demonstrations, legal complaints, petition letters, and meetings. For example, in 2014, the people of Babensi II village staged a street protest against the illegal occupation of their ancestral land, demanding the company to quit, but also to compensate those whose farms had been destroyed. Other strikes took place from 2010 to 2013 in the subdivision (e.g. blocking access to the company's oil palm nursery in Nguti). In 2011, a youth organisation, the Bassosi Cultural and Development Association, wrote to the president, refusing a deal to establish an agro-plantation on their ancestral land since it was against local interests. They blamed the government. Also, in 2011, the CEO of Nature Cameroon wrote to the Minister of the Environment and Natural Protection, castigating the results of an environmental and social impact assessment (ESIA) undertaken by the company (Nature Cameroon, 2011). He argued that the lives of 25,000 villagers were being compromised.

Resistance in this case, and in many others across the continent, is spontaneous and uncoordinated (cf. Fonjong *et al.*, 2015). But it led to project downsizing; the company's activities have been slowed, and in practical terms the company is now trying to renegotiate with the communities concerned, especially with women who are most disadvantaged given their strong involvement in farming. Some people want merely to sell palm oil from their own plantations and trees, and without wholesale forest loss resulting from commercial palm plantations. In some concession sites such as Toko and Mundenba, the company has abandoned its operations, although promising to return.

These different forms of resistance are not necessarily against foreign investment in commercial agriculture per se, nor are they intended to challenge the legitimacy of the state or overthrow it, but rather they seek to engage with large-scale land acquisition projects in ways that are appropriate and beneficial to the communities concerned. Nonetheless, local communities whose farmland falls within the reduced concession boundaries still protest. Tropical forests adjacent to the Korup national parks and forest reserves are clearly threatened. Agreements with locals have not been honoured. In general, there is an absence of free, prior and informed consent. Even where plantation jobs are offered, they are low-paid.

Land acquisition is caught up in the fate of rural livelihoods more generally (Ndi, 2017). Where the government exerts legal provision to allocate land to others, customary norms are challenged. Claims to ownership over parcels of land, or even use rights following an LSLA, inevitably provoke complex social changes that result in conflict.

An example of green-grabbing

Hamouchene (2016) argues that green-grabbing in Africa is also beginning to have serious local implications, citing the case of a concentrated solar power (CSP) array installed in Ouarzazate, Morocco. The opening of the first phase of this electricity plant in 2016 was publicised widely in the local and global media for its contribution to renewable energy targets, but the controversy surrounding its construction was not.

Communal land ownership by herders in this arid region did not prevent the sale of 3,000 hectares to a private solar energy company, which went ahead because communities are represented by agents, and they struck a deal in 2010 with the national electricity provider to sell the land they managed at a tenth of its real market value (Rignall, 2012). The Noor-Ouarzazate plant is being built by ACWA Power International, a Saudi Arabian company, in partnership with the government and supported by the Moroccan royal family, and Phase One is now operational. Money from the land sale was not even transferred directly to communities, but was held 'in trust' at the Ministry of the Interior for future development projects. An unrepresentative local 'consultation' and presentation of an environmental impact study was only held after the details of the sale had been agreed. The solar technology uses hundreds of mirrors to concentrate solar radiation, heating salts to produce steam for turbines. Cooling is required, too, but water is scarce in this desert environment. It is sourced from a reservoir also used for irrigation and drinking water. Finance totalling US$9 billion was offered as loans from the European Investment Bank, the World Bank and other international agencies; relatively expensive electricity will be sold back to the national electricity company, and allegedly some will be exported to Europe. The Noor-Ouarzazate project is registered with the UN's Clean Development Mechanism, and thus may receive carbon offsetting payments from developed countries.

As with other land acquisitions, the plateau selected for the plant was deemed to be marginal and 'degraded' by the government, and thus available for other uses. A progressive concentration of wealth in Morocco's rangelands disadvantaging small herd owners in favour of larger ones also made the land sale easier. The form of 'accumulation by dispossession' here was the 'enclosure of public assets by private interests for profit, resulting in greater social inequity' (Hamouchene, 2016).

Conclusion

Too many LSLAs have allowed investors to seek the permission of government to obtain the leases and concessions they need, with little or no effort needed to seek the consent of the local population concerned (Polack *et al.*, 2013). This is now recognised as a major failing of national land laws and regulation (Oakland Institute, 2011; WWF, 2012; Greenpeace, 2013).

One World Bank report argues that corporate investors in land specifically target Third World economies with very weak and porous governance systems (Deininger and Byerlee,

2011; Evers *et al.*, 2013). This explains the negative consequences associated with many land deals in Africa, and indeed some 'green grabs' as well (Deininger and Byerlee, 2011; Anseeuw *et al.*, 2012).

Irrespective of whether large-scale land deals produce positive and/or negative socio-economic outcomes, it is important to note that rural people in Africa are largely unable to participate in the decision-making process for projects that profoundly impact their lives (Cotula *et al.*, 2014; see also von Braun and Meinzen-Dick, 2009). Communities suffer the brunt of large-scale land acquisitions where they are trapped in a web of competing claims, coupled with inadequate access to traditionally accessed land and forest resources, or worse still dispossession – to give way to monoculture plantations, timber extraction and mining (WWF, 2012; Greenpeace, 2014). At least within the context of most rural societies, dispossession or denial of access literally means ruin. Livelihoods, homes and histories are effaced, creating new patterns of migration, resettlement and employment between land users and land occupiers (Lund, 2014).

Fifty years ago, with Africa largely independent from the colonial powers, it would have seemed unlikely that foreign interests would again gain de facto control of productive African land. The plan for Africa's green revolution in the 1960s also missed this possibility: its efforts were directed to enhancing Africa's agricultural output through scientific means, primarily to boost the continent's own food security, and perhaps with the production of some fuel and fibre to meet continental and some international demand. Instead, more recently, poor governance and the might of overseas interests have been diverting the potential of African land and reoriented its productive potential to meet other interests. This is the tragedy of the African land grab. What is needed is greater coherence in national land allocations policies, buttressed by legal controls on accumulation strategies for individuals, states and corporations, and support for African farmers to retain control of productive assets.

References

Allan, J.A., Keulertz, M., Sojamo, S. and Warner, J. (eds) (2013) *Handbook of Land and Water Grabs in Africa: Foreign Direct Investment and Food and Water Security*, London: Routledge.

Anseeuw, W. (2013) 'The rush for land in Africa: resource grabbing or green revolution?', *South African Journal of International Affairs*, 20(1): 159–177.

Anseeuw, W., Wily, L.A., Cotula, L. and Taylor, M. (2012) *Land Rights and the Rush for Land*, London: International Institute for Environment and Development; Rome: International Land Coalition; Montpellier: CIRAD.

Aubry, S., Seufert, P. and Suarez, S.M. (2012) *(Bio) Fuelling Injustice? Europe's Responsibility to Counter Climate Change Without Provoking Land Grabbing and Compounding Food Insecurity in Africa*, Rome: EuropAfrica.

Bederman, S.H. (1966) 'Plantation agriculture in Victoria Division, West Cameroon: an historical introduction', *Geography*, 51(4): 349–360.

Bernstein, H. (1996) 'South Africa's agrarian question: extreme and exceptional?', *The Journal of Peasant Studies*, 23(2–3): 1–52.

Borras Jr, S.M. and Franco, J.C. (2010) 'From threat to opportunity-problems with the idea of a code of conduct for land-grabbing', *Yale Human Rights and Development Journal*, 13(2): 507–523.

Borras Jr, S.M. and Franco, J.C. (2012) 'Global land grabbing and trajectories of agrarian change: a preliminary analysis', *The Journal of Agrarian Change*, 12(1): 34–59.

Borras Jr, S.M., Hall, R., Scoones, I., White, B. and Wolford, W. (2011) 'Towards a better understanding of global land grabbing: an editorial introduction', *The Journal of Peasant Studies*, 38(2): 209–216.

CED (Centre for Environment and Development) (2012) *Herakles' 13th Labour? A Study of SGSOC's Land Concession in South-West Cameroon*, Yaoundé: CED.

Chow, H. (2016) *Farmers Return to Their Land in Nigeria Following Land Grab*, 4 February, London: Global Justice, available at: www.globaljustice.org.uk/blog/2016/feb/4/farmers-return-their-land-nigeria-following-land-grab (accessed 29 December 2017).

Cotula, L. (2013) *African Land Deals: Is a Policy Shift Underway?*, available at: www.iied.org/african-land-deals-policy-shift-underway (accessed 1 September 2016).

Cotula, L. (2014) *Addressing the Human Rights Impacts of 'Land Grabbing'*, Luxembourg: European Parliament, Directorate-General for Internal Policies of the Union Publications Office, DOI: 10.2861/538368.

Cotula, L. and Vermeulen, S. (2009) 'Deal or no deal: the outlook for agricultural land investment in Africa', *International Affairs*, 85(6): 1233–1247.

Cotula, L. and Vermeulen, S. (2010) *Making the Most of Agricultural Investment: A Survey of Business Models That Provide Opportunities for Smallholders*, London: International Institute for Environment and Development.

Cotula, L. and Vermeulen, S. (2011) 'Contexts and procedures for farmland acquisitions in Africa: what outcomes for local people?', *Development*, 54(1): 40–48.

Cotula, L., Vermeulen, S., Leonard, R. and Keeley, J. (2009) *Land Grab or Development Opportunity?* London: IIED, FAO and IFAD.

Cotula, L., Oya, C., Codjoe, E.A., Eid, A., Kakraba-Ampeh, M., Keeley, J., *et al.* (2014) 'Testing claims about large land deals in Africa: findings from a multi-country study', *Journal of Development Studies*, 50(7): 903–925.

Daniel, S. and Mittal, A. (2009) *The Great Land Grab: Rush for World's Farmland Threatens Food Security for the Poor*, Oakland, CA: Oakland Institute, available at: www.oaklandinstitute.org/sites/oaklandinstitute.org/files/LandGrab_final_web.pdf (accessed 29 December 2017).

Davis, K.F., D'Odorico, P. and Rulli, M.C. (2014) 'Land grabbing: a preliminary quantification of economic impacts on rural livelihoods', *Population and Environment*, 36(2), 180–192, DOI: http://doi.org/10.1007/s11111-014-0215-2.

De Schutter, O. (2009) *Large-Scale Land Acquisition and Leases: A Set of Core Principles and Measures to Address the Human Rights Challenge*, Briefing Note, 11 June, Geneva: UN of the High Commissioner for Human Rights.

De Schutter, O. (2011) 'How not to think of land-grabbing: three critiques of large-scale investments in farmland', *The Journal of Peasant Studies*, 38(2): 249–279.

De Schutter, O. (2012) *Report of the Special Rapporteur on the Right to Food: Mission to Cameroon*, Addendum, UN Human Rights Council/22/20/Add2.

Deininger, K. and Byerlee, D. (2011) *Rising Global Interest in Farmland: Can It Yield Sustainable and Equitable Benefits?* Washington, DC: World Bank.

Evers, S.J.T.M., Seagle, C. and Krijtenburg, F. (eds) (2013) *Africa for Sale? Positioning Foreign Large-Scale Land Acquisitions in Africa*, Leiden: Brill.

Fairhead, J., Leach, M. and Scoones, I. (2012) 'Green grabbing: a new appropriation of nature?', *The Journal of Peasant Studies*, 39(2): 237–261.

Firmin-Sellers, K. and Sellers, P. (1999) 'Expected failures and unexpected successes of land titling in Africa', *World Development*, 27(7): 1115–1128.

Fombe, L.F., Sama-Lang, I., Fonjong, L. and Mbah-Fongkimeh, A. (2013) 'Securing tenure for sustainable livelihoods: a case of women land ownership in anglophone Cameroon', *Ethics and Economics*, 10(2): 73–86.

Fonjong, N.L., Sama-Lang, I., Fombe, L. and Abonge, C. (2015) *Disenchanting Voices from Within: Interrogating Women's Resistance to Large-Scale Agro-Investments in Cameroon*, 2015 World Bank Conference on Land and Poverty, 23–27 March, Washington, DC: World Bank.

Fonjong, L., Sama-Lang, I., Fombe, L. and Abonge, C. (2016) 'Land governance and women's rights in large-scale land acquisition in Cameroon', *Development in Practice*, 26(4): 420–430.

Green, E.K. and Adams, W.M. (2015) 'Green grabbing and the dynamics of local-level engagement with neoliberalization in Tanzania's wildlife management areas', *The Journal of Peasant Studies*, 42(1): 97–117.

Greenpeace (2013) *Herakles Farms in Cameroon: A Showcase in Bad Palm Oil Production*, Washington, DC: Greenpeace USA.

Greenpeace (2014) *Forest Echoes*, Issue 2, available at: www.greenpeace.org/africa/Global/africa/2014_Forest%20Echoes%20Issue%202%20Final.pdf (accessed 8 January 2018).

Hall, R. (2011) 'Land grabbing in southern Africa: the many faces of the investor rush', *Review of African Political Economy*, 38(128): 193–214.

Hamouchene, H. (2016) *The Ouarzazate Solar Plant in Morocco: Triumphal 'Green' Capitalism and the Privatization of Nature*, available at: www.jadaliyya.com/pages/index/24124/the-ouarzazate-solar-plant-in-morocco_triumphal-gr (accessed 29 December 2017).

Harvey, D. (2004) 'The "new" imperialism: accumulation by dispossession', *Socialist Register*, 40: 63–87.

Harvey, D. (2005) *A Brief History of Neoliberalism*, Oxford: Oxford University Press.

Holmén, H. (2015) 'Is land grabbing always what it is supposed to be? Large scale land investments in sub-Saharan Africa', *Development Policy Review*, 33(4): 457–478.

Hoyle, D. and Levang, P. (2012) *Oil Palm Development in Cameroon*, Yaoundé: World Wildlife Fund.

IIED (International Institute for Environment and Development) (2012) *The Global Land Rush: What the Evidence Reveals About Scale and Geography*, IIED Briefing. London: IIED.

Kandel, M. (2015) 'Politics from below? Small-, mid- and large-scale land dispossession in Teso, Uganda, and the relevance of scale', *The Journal of Peasant Studies*, 42(3–4): 635–652.

Konings, P. (2008) 'Privatization and labour militancy: the case of Cameroon's tea estate', *Journal of Contemporary African Studies*, 26(1): 51–70.

Lisk, F. (2013) 'Land grabbing or harnessing of development potential in agriculture? East Asia's land-based investments in Africa', *The Pacific Review*, 26(5): 563–587.

Lund, C. (2014) 'Of what is this a case? Analytical movements in qualitative social science research', *Human Organization*, 73(3): 224–234.

Mope Simo, J.A. (2011) *Land Grabbing, Governance and Social Peace-Building Issues in Cameroon: Case Study of the Roles of Elites in Land Deals and Commoditization in North West Region*, paper presented at the International Conference on Global Land Grabbing, 6–8 April.

Nature Cameroon (2011) 'Critical observations of the SG-SOC ESIA in South West: a letter addressed to the Minister of Environment and Nature Protection, 08/09/2011', unpublished manuscript, Nguti: Nature Cameroon.

Ndi, F.A. (2017) 'Land grabbing, local contestation, and the struggle for economic gain: insights from Nguti Village, south west Cameroon', *Sage Open*, 7(1), DOI: https://doi.org/10.1177/2158244016682997.

Oakland Institute (2011) *Understanding Land Investment Deals in Africa: FAQs on Food Security & Western Investors*, Oakland, CA: Oakland Institute.

O'Brien, E. (2011) *Irregular and Illegal Land Acquisition by Kenya's Elites: Trends, Processes and Impacts of Kenya's Land Grabbing phenomenon*, Rome: ILC Collaborative Research Project on Commercial Pressures on Land.

Oxfam (2011) *The Growing Scandal Surrounding the New Wave of Investments in Land*, Briefing Paper 151, Oxford: Oxfam International.

Oya, C. (2013) 'Methodological reflections on "land grab" databases and the "land grab" literature "rush"', *The Journal of Peasant Studies*, 40(3): 503–520.

Peluso, L.N and Lund, C. (2011) 'New frontiers of land control: introduction', *The Journal of Peasant Studies*, 38(4): 667–681.

Polack, E., Cotula, L. and Cote, M. (2013) *Accountability in Africa's Land Rush: What Role for Legal Empowerment?* London: IIED.

Potter, L. (2015) *Managing Oil Palm Landscapes: A Seven-Country Survey of the Modern Palm Oil Industry in South East Asia, Latin America, and West Africa*, Occasional Paper 122, Bogor, Indonesia: CIFOR.

Rignall, K. (2012) *Theorizing Sovereignty in Empty Land: The Land Tenure Implications of Concentrated Solar Power in Pre-Saharan Morocco*, paper presented at the International Conference on Global Land Grabbing II, 17–19 October, organised by Land Deal Politics Initiative (LDPI) and hosted by the Department of Development Sociology at Cornell University, Ithaca, NY.

Scoones, I., Hall, R., Borras Jr, S.M., White, B. and Wolford, W. (2013) 'The politics of evidence: methodologies for understanding the global land rush', *The Journal of Peasant Studies*, 40(3): 469–483.

Vermeulen, S. and Cotula, L. (2010) 'Over the heads of local people: consultation, consent, and recompense in large-scale land deals for biofuels projects in Africa', *The Journal of Peasant Studies*, 37(4): 899–916.

von Braun, J. and Meinzen-Dick, R. (2009) *Land Grabbing by Foreign Investors in Developing Countries: Risks and Opportunities*, IFPRI Policy Brief, 13 April, available at: www.landcoalition.ord/pdf/ifpri_apr_09.pdf (accessed 1 September 2016).

White, B., Borras Jr, S.M., Hall, R. and Scoones, I. (2012) 'The new enclosures: critical perspectives on corporate land deals', *The Journal of Peasant Studies*, 39(3–4): 619–647.

Wolford, W., Borras, S.M., Hall, R., Scoones, I. and White, B. (2013) 'Governing global land deals: the role of the state in the rush for land', *Development and Change*, 44(2): 189–210.

WWF (World Wildlife Fund) (2012) *Emerging Trends in Land-Use Conflicts in Cameroon: Overlapping Natural Resource Permits Threaten Protected Areas and Foreign Direct Investment*, Yaounde, Cameroon: WWF.

Zoomers, A. (2013) 'A critical review of the policy debate on large scale land acquisitions: fighting the symptoms or killing the hearts?', in S.J.T.M. Evers, C. Seagle and F. Krijtenburg (eds), *Africa for Sale? Positioning Foreign Large-Scale Land Acquisitions in Africa*, Leiden: Brill, pp. 55–77.

Zoomers, A. and Kaag, M. (2014) 'Beyond the global land grab hype: ways forward in research and action', in M. Kaag and A. Zoomers (eds), *The Global Land Grab: Beyond the Hype*, London: Zed Books, pp. 201–216.

44

The economics of and prospects for China's Africa return

Lauren A. Johnston

Introduction

China's deepening economic ties with Africa carry 'profound implications for the continent and beyond' – regardless of whether China is 'returning (to Africa) or not' (Wang and Elliot, 2014: 1017). Accordingly, what happens in and between each of China and Africa economically is now at the core of the economic development of both. This chapter comprises an overview of the related economic and political economy trends and transitions.

Economic ties between China and Africa began to grow rapidly in the mid-1990s. Around this time, 'chiefly for profit and resources', China went back to Africa with massive orders, investment and exports (Wang and Elliott, 2014: 1013). More than a decade later, in 2009, China became Africa's largest trade partner. It has remained so since. Africa's trade with China now specifically comprises some quarter of total trade (Pigato and Tang, 2015). China's investments in Africa are similarly increasingly visible, most literally in the construction sector. In recent years, China's FDI flows for sub-Saharan Africa (SSA) have comprised some 10 per cent of total (Pigato and Tang, 2015). Unlike for aggregate foreign trade, these are important within the scale of China's outbound investment portfolio.

Until recently, these trade ties were broadly characterised by a contentious and historical pattern of trade. Large volumes of fuels and minerals flowed from Africa to China while African economies bought value-added manufactured goods from China. As a result, Africa's major commodities exporters not only benefited disproportionately from China–'Africa' trade; they also received the lion's share of China's investment in Africa (see Johnston, 2015a: 391; Chen *et al.*, 2016). Entrenched competition in well-developed energy markets, moreover, led China to frontier energy markets, including in former Sudan and Angola, despite their reputation for poor governance. Thus, where China's interest in resources is generally consistent with Western interests, its share of foreign investment in weakly governed African states is higher (Chen *et al.*, 2016). The propensity for corruption in the management of natural resources, together with China's lack of attention to matters of resource transparency and local mechanisms of oversight, have been a cause for concern (Zafar, 2007).

As the world's second largest economy, home to the world's largest savings pool, an important global trader and mass exporter of manufactured goods, China's influence on Africa is also indirect: through effects on global prices, especially of commodity prices (see Woo, 2016), interest rates, and the terms of trade (Zafar, 2007: 7). Moreover, China has helped to maintain low interest rates and bond yields through its financing of the US deficit, and selective empirical evidence finds that high commodity prices are influenced by low real interest rates (Frankel, 2006, cited in Zafar, 2007: 7).

Such impacts have differing spillover effects on other economies in particular, for example via the terms of trade. China's decades-long low-cost industrial export success offered downward pressure to some industrial prices, including light manufactures such as textiles and clothing, and high-technology products (IMF, 2003, cited in Zafar, 2007: 7). This has contributed to recent terms of trade indices differences between resource-rich and resource-poor economies in Africa (Zafar, 2007). Contemporary structural change in the Chinese economy, however, elaborated in the third section, has substantively shifted the distribution of China-related benefits in Africa in greater favour of commodities importers. An outcome is described in popular media as follows: 'But as China's economy slows and its once seemingly insatiable hunger for Africa's commodities wanes, many African economies are tumbling, quickly' (Onishi, 2016).

In a prescient and pragmatic response to the broader challenges induced by fuel- and mineral-hungry China, in March 2013, former Nigerian Central Bank Governor Sanusi Lamido suggested that Africa should avoid romanticism, and instead must:

> recognize that China – like the US, Russia, Britain, Brazil and the rest – is in Africa not for African interests but its own. The romance must be replaced by hard-nosed economic thinking. Engagement must be on terms that allow the Chinese to make money while developing the continent, such as incentives to set up manufacturing on African soil and policies to ensure employment of Africans.
>
> *(Financial Times, 2013)*

In these footsteps, and visiting Africa in May 2014, Chinese Premier Li Keqiang laid down four principles for the contemporary deepening of China–Africa cooperation: 'sincerity and equality; solidarity and mutual trust; jointly pursuing inclusive development; and innovative pragmatic cooperation' (Johnston, 2014). The latter reflected a new addition to a set of otherwise traditional principles for China's relations with other developing countries. This may infer the direction of ties was pivot differently therein. The concurrent arrival of the new growth model indeed provides this scope. Symbolically, the announcement was made at the Chinese-built African Union (AU) headquarters in Addis Ababa half a century after former Premier Zhou Enlai announced the original principles of China's aid in Africa (see Anshan, 2007). This chapter elaborates the underlying trends of 'innovative pragmatic cooperation' under China's 'new growth model' (Garnaut *et al.*, 2013).

The rest of this chapter is structured as follows. The second section presents the static trends defining the China–Africa economic relationship. The third section presents a chronological discussion of China's economic development, with a focus on seminal turning points that may have most impacted economic ties with Africa. The final section reflects the future of China–Africa ties.

Economic trends

Aid

When China started experimenting with a policy programme of 'openness and reform' in 1979, the national rate of poverty was higher than in Africa as a whole (Ravallion, 2009). For many decades prior, however, this fact did not prevent China from providing then mostly political aid to Africa (see State Council Information Office, 2011). Data on China's now rapidly increasing and more economic-oriented foreign aid are, however, unfortunately scarce. For example, no official numbers for China's foreign aid use Organisation of Economic Co-operation and Development (OECD) aid definitions. This makes reliable level and effectiveness comparisons to the majority of international aid funding impossible. In recent years, however, China has begun to publish a series of foreign aid white papers,[1] of which *China's Foreign Aid 2014* (State Council Information Office, 2014) is the most recent. A handful of aid researchers have, in parallel, painstakingly produced estimates of China's aid. This analysis relies on work by Kitano and Harada (2015), whose research constructs a proxy that attempts an extensive estimation of China's foreign aid using OECD categorisation.

Around half of China's total foreign aid goes to countries in Africa (State Council Information Office, 2011, 2014) (see Figure 44.1). These aid levels are nonetheless small relative both to the aggregate monetary value of China–Africa economic ties and also the total foreign aid volumes going to African countries. China's aid is, however, increasing both absolutely and relatively. China's aid is also increasing net, this being the difference between foreign aid and the repayment of earlier aid-related loans. According to Kitano and Harada (2015), the annual rate of increase since 2001 is greater than 15 per cent. According to China's aid figures, by 2013 this produced total net foreign aid of some US$7 billion, implying some US$3.5 billion of aid to Africa. By comparison, Kitano and Harada (2015) estimate that in 2013, China's OECD-comparable foreign aid figure reached the higher sum of RMB 43,938 million (US$6.6 billion).

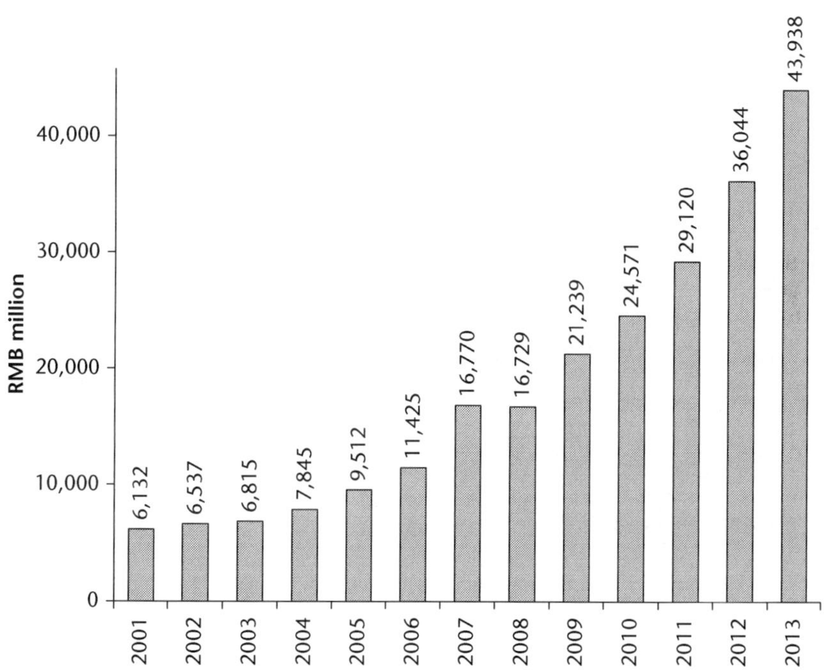

Figure 44.1 Aid flows to Africa

Source: Kitano and Harada (2015).

By sector, transport, storage, energy and communications sectors receive a large share of China's foreign aid. This reflects China's competitiveness, and also that Western aid has recently been directed towards relatively more health and education (Foster, 2009; Amusa *et al.*, 2016). China's concessional loan portfolio is extensive (Kitano and Harada, 2015). In general, however, the lack of systematic and detailed cross-country time series loan data has led to a tendency to overestimate the scale of China's lending to Africa (Benabdallah *et al.*, 2016).

Given these broader data constraints, quantitative research on China's foreign aid and concessional lending to Africa is limited. A recent exception employs a series of fixed-effects regressions to show that current political leaders' birth regions receive substantially larger financial flows than other regions, but no evidence that leaders shift aid towards regions populated by groups who share their ethnicity (Dreher *et al.*, 2016). Qualitative research is more extensive (see Brautigam, 1998; Anshan, 2007; Brautigam, 2010a; Tan-Mullins *et al.*, 2010), but not further herein elaborated.

Trade

Structural change in China and the associated policy shifts in the mid-1990s (see the third section) helped to induce fast-rising trade flows between China and Africa. Momentum in China's industrial development produced a near insatiable demand for raw materials and energy, which Africa has in ample supply. A result was that China traded extensively with African democracies and autocracies alike, making it doubtful that an African country's regime type explains much about its trade with China (Eisenman, 2012: 793).

Figure 44.2 portrays the 20-year Sino-African trade boom. During the boom, China's exports to Africa increased from US$1.5 billion in 1995 to US$57 billion in 2015. China's imports from

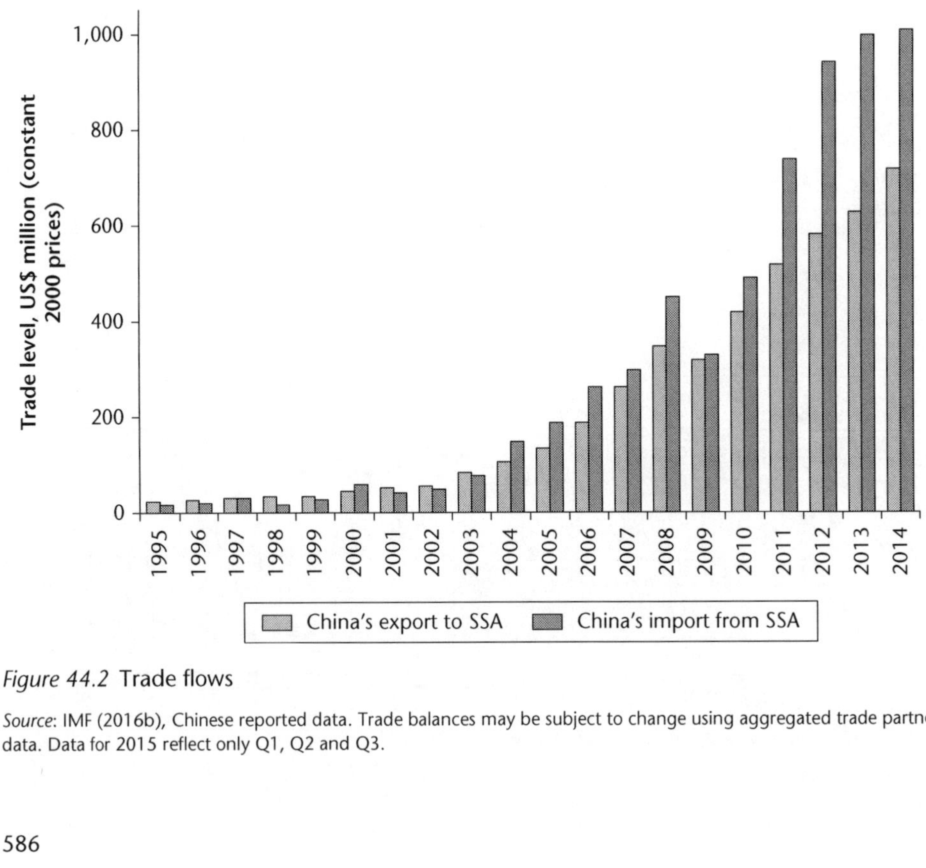

Figure 44.2 Trade flows

Source: IMF (2016b), Chinese reported data. Trade balances may be subject to change using aggregated trade partner data. Data for 2015 reflect only Q1, Q2 and Q3.

Table 44.1 Share of exports to China, selective SSA countries, 2014

Gambia	47.5	Equatorial Guinea	16.5
Congo, DRC	43.1	Gabon	15.4
Angola	42.8	Central African Republic	14.1
Congo, Rep	41.4	Niger	13.8
Sierra Leone	31.1	Ethiopia	11.7
Zimbabwe	27.6	South Africa	11.2
Zambia	25.9	Togo	11.1
Liberia	20.7	Mozambique	10.1
Cameroon	16.7	Rwanda	10.1

Source: IMF (2016b).

Africa rose from US$1 billion to US$110 billion in 2014 over the same period. Where China's low-cost exports were sold relatively consistently across countries in Africa, however, China's imports from Africa were lumpier between countries (see Table 44.1).

Table 44.1 presents the 2014 share of China-bound exports within total national exports where this level reached 10 per cent or more. Top of the list, Gambia's main export to China is timber that is re-exported from neighbouring countries. The majority of that timber is illegally smuggled across the border from Senegal: 'Senegal has lost more than a million trees since 2010 while farmers in Gambia have pocketed $238 million exporting the wood to China where the desire for furniture has exploded in the last few years' (The Herald, 2016). The dominance of fuel and mineral exporters among countries with exports that are relatively dependent on China, together with cases of illegal trade such as a significant share of the timber exported via Gambia, makes Beijing the focal point of new anti-Chinese resistance narratives in Africa (Eisenman, 2012: 793).

Moreover, a long history of work in macroeconomics shows that the presence of natural resources has tended to hinder and not foster long-run development (e.g. Deaton, 1999; Sachs and Warner, 1999, 2001). Raw commodities trade in general does not generate significant value added or enough jobs (UNECA, 2013), and also increases countries' exposure to international exogenous shocks (Gui-Diby and Renard, 2015). Those African countries having recently exported large quantities of high-priced raw materials have in turn broadly failed to utilise the returns to diversify their trade and economies. According to an export diversification indices published by the UN Conference on Trade and Development (UNCTAD), in 2012 the share of African countries for which primary products comprised at least 50 per cent of total exports was three-quarters. One-third of African countries, moreover, had primary products comprising over 90 per cent of total exports (UNECA, 2013).

As implicitly suggested by the fall in trade flows in 2015 (see Figure 44.2), more recently the scale of China's demand for commodities has fallen. Factors driving this structural decline in international demand for China's mainly manufactured products include that industry-induced environmental degradation has reached public tolerance thresholds, the combination of rising domestic wage pressure and a structural decline in demand for China's exports following the Global Financial Crisis. These are elaborated in the third section. The global reach of China's economy means that those changes have also adversely impacted international commodities prices, compounding the fall in value of China's imports from resource-rich Africa. The broader implications for Africa are discussed in the third section.

Investment

Four categories of Chinese investors in Africa have been identified: central-state-owned firms, provincial-state-owned firms, Chinese private firms incorporated in China, and small firms operating in Africa owned by independent Chinese 'migrants' (Kaplinsky and Morris, 2009). Each category has its own distinct characteristics, but state-owned investment is uniquely and often strategically bundled in with aid and trade (Ajakaiye *et al.*, 2009: 481). Private Chinese firms investing in African countries are also typically much smaller than the big SOEs, which dominate the 'mega-deals' (Chen *et al.*, 2016: 6). Where infrastructure and resources dominate the latter, by project the services sector dominates. Services sector projects are nonetheless clustered in countries that are resource-abundant. There are also a significant number of projects in manufacturing (Chen *et al.*, 2016: 6).

China's FDI flows complement China's factor endowment structure, meaning in Africa they flow disproportionately to resource-exporting economies (NBS, 2015). China's FDI flows to Africa reached US$26 billion, compared to US$22 billion going to the US in 2013, and despite the relatively small weight of African economies internationally (Chen *et al.*, 2016: 6). Converse to patterns of trade – which is proportionately more important at the African end – China's investment in Africa is proportionately more important to China than China's investment in Africa is for Africa, and it is growing rapidly (Chen *et al.*, 2016: 6).

In general, however, there are limited if any reliable and accurate Chinese FDI data by investment value. The official data, from the Ministry of Commerce, for example, tend to under-capture the size of outbound FDI (Shen, 2015). Investment projects that are loan-financed also do not appear in China's FDI statistics, and nor do acquisition of African assets that took place in a third jurisdiction, such as the more than US$7 billion purchase of Addax, which was legally transacted in Switzerland (Brautigam, 2010b; SAIS-CARI, 2016). Together with the fact that a not insignificant but unquantified share of China's FDI passes via Hong Kong and other offshore financial centres, China's outbound FDI flows are probably significantly higher than these official statistics suggest (see Table 44.2).

Table 44.2 Selective FDI flows and policies, China and African countries

	FDI inflow, US$ million			*BIT status*	*Double-taxation agreement*
	Y2005	*Y2010*	*Y2014*		
Angola	0.47	101.11	–448.57	n/a	
Benin	1.31	1.76	7.44	Signed	
Botswana	3.69	43.85	52.95	Signed	
Burkina Faso			4.45	n/a	
Burundi			3.45	n/a	
Cameroon	0.19	14.88	29.74	In force	
Cape Verde	0.32	–0.46	0.1	In force	
Central African Republic		25.81	182.24	n/a	
Chad	2.71	2.13	83.12	Signed	
Comoros		–0.01		n/a	
Congo	5.07	34.38	238.6	In force	
Congo DRC	8.11	236.19	157.56	Signed (1997, 2011)	
Côte d'Ivoire	8.74	–5.02	24.26	Signed	
Djibouti		4.23	9.53	Signed	
Equatorial Guinea	6.35	22.08	33.13	In force	
Eritrea		2.94	1.29	n/a	

	FDI inflow, US$ million			*BIT status*	*Double-taxation agreement*
	Y2005	*Y2010*	*Y2014*		
Ethiopia	4.93	58.53	119.59	In force	May 2000
Gabon	2.08	23.44	25.56	In force	
Gambia			0.05	n/a	
Ghana	2.57	55.98	72.9	In force	
Guinea	16.34	9.74	67.7	Signed	
Guinea-Bissau	2.05		1.72	n/a	
Kenya	0.6	101.22	278.39	Signed	
Lesotho		0.56	0.46	n/a	
Liberia	8.65	29.89	40.11	n/a	
Madagascar	0.14	33.58	36.76	In force	
Malawi		9.86	3.4	n/a	
Mali		3.05	23.39	In force	
Mauritania	0.36	5.77	–7.33	n/a	
Mauritius	2.04	22.01	49.43	In force	June 1997
Morocco	0.85	1.75	11.44	In force	November 1999
Mozambique	2.88	0.28	102.51	In force	
Namibia	0.18	5.51	8.02	Signed	
Niger	5.76	196.25	–44.61	Terminated	
Nigeria	53.3	184.89	199.77	In force	February 2010
Rwanda	1.42	12.72	14.94	n/a	
São Tomé and Príncipe		0.02		n/a	
Senegal		18.96	7.06	n/a	
Seychelles	0.05	12.28	7.56	Signed	January 2000
Sierra Leone	0.49		4.92	Signed	
Somalia				n/a	
South Africa	47.47	411.17	42.09	In force	April 1998
South Sudan			–6.82	n/a	
Sudan	91.13	30.96	174.07	In force	July 1998
Swaziland				n/a	
Tanzania	0.96	25.72	166.61	In force	
Togo	0.31	11.77	6.99	n/a	
Tunisia		–0.29	0.71	In force	July 2006
Uganda	0.17	26.5	60.5	Signed	
Zambia	10.09	75.05	424.85	Signed	June 2011
Zimbabwe	1.47	33.8	101.18	In force	

Source: NBS (2008, 2012, 2015).

According to 2014 FDI data, the top 10 recipients of Chinese investment in order were Zambia, Kenya, Congo, Nigeria, the Central African Republic, Sudan, Tanzania, Congo DRC, Ethiopia and Mozambique. East African countries feature prominently in the list for a number of reasons. First, three East African countries were listed in as the preliminary official targets of China's push into labour-intensive manufacturing outside of China – Ethiopia, Kenya and Tanzania. Accordingly, Kenya is the primary African hub of China's flagship Silk Road (outbound investment) initiative. Modern-day Kenya is also home to the landing site of Chinese maritime fleets led by Admiral Zheng He during the Ming Dynasty. Contemporarily, the East African Community grouping of which Kenya is a part also comprises a customs and single market trading union that, with extensive support from Chinese loans and construction companies, is investing in a series of cross-country investment projects. These are intended to unlock the East African market for its own growth and to China's investors and consumer goods.

A second explanation for a relatively prominent role for East African countries, especially coastal East Africa, can be understood through an applied economic geography lens. Specifically, under China's old growth model, the significance of economic geography and growth to potential long-run transformation in Africa was largely ignored (Johnston *et al.*, 2015). China's 'new normal', however, offers Africa's long-run growth strategic coastal and resource-poor economies a triple benefit: (1) a positive shift in the terms of trade; (2) improved prospects for industrial sector labour-intensive foreign investment; and (3) greater investment infrastructure that unlocks ports and intra-country transportation. This in turn may better serve to overcome Africa's unusual economic geography hurdles (e.g. Venables, 2010) and poverty traps (e.g. Collier, 2008). In the process, it is also intended to create new markets for China's goods and financing concurrent with taking advantage of Africa's youthful workforce as populations elsewhere age rapidly. The third section elaborates this point.

In addition to China's capacity and willingness to invest, and recipient country absorptive capacity, the success of China's Africa investments is also dependent on the quality of the respective bilateral commercial relationship (Broadman, 2006: xix). Many policies shape bilateral China–Africa commercial relationships. These are selectively elaborated in State Council Information Office (2013, 2014) and Johnston and Cheng (2015). Table 44.1 presents the cornerstone Bilateral Investment Treaty and double-taxation treaties. Related research suggests some 18 countries have enacted a BIT with China, and a further 13 have signed such an agreement but not yet entered it into law (see Table 44.2). Nine countries in SSA have agreed a double-taxation agreement with China.

China's changing growth model

An important caveat to the next section is that the choice to focus on China should not be taken to mean that macroeconomic change in African economies is unimportant to the China–Africa story. In a handbook that otherwise focuses on Africa, this chapter prioritises elaborating the Chinese side of the China–Africa economic story.

China's old growth model – and Africa

For three decades, until 2011, China's average growth rate was nearly 10 per cent. In the process, millions were lifted from poverty and the global structure of production and distribution of GDP was reshaped. China became the locus of most production of manufactured goods for global markets, and the buyer of most of the growth in supply of energy and metals. A number of related turning points in China in the 1990s transformed ties between China and Africa to those described in the second section.

First, China became a net importer of oil in 1993, and in 1995 its export of machinery and electronics exceeded those of textiles and clothing (Lin and Wang, 2014: 4). To ensure access to needed oil supplies China's largely state-owned oil sector was consequently forced to more deeply integrate with global oil markets. A late entrant among today's major players into the global oil sector, China's state-owned oil giants were forced to enter higher-risk frontier oil markets. This led to Sudan and Angola becoming more prominent among international oil suppliers. Angola became a member of the Organization of the Petroleum Exporting Countries (OPEC) in 2007 thanks to a China-led oil investment boom.

Beyond minerals and fuels, historic forest loss means that China is one of the most forest-deficient countries in the world, with national forest cover of just 22 per cent against a global average of 31 per cent (FAO, 2012, cited in Buckingham, 2016). China is one of the most water-stressed countries in the world, too, with just 6 per cent of the world's fresh water but 21 percent of its population (Wong, 2013, cited in Buckingham, 2016). Advanced competitive industrial manufacturing chains in China nonetheless mean that since 2000, the country has been the main processing hub for the world's forestry sector. These trends have driven deeper – formal and informal – trade ties between China and resource-rich countries. Gabon, Congo Republic, Cameroon and Mozambique are among China's key timber suppliers in Africa. Extensive fishing over decades in Chinese waters and heavy demand for seafood have also pushed Chinese fishermen further afield. Illegal fishing in African waters recently, especially intensively in unpatrolled waters off the coast of poor countries in West Africa, has become a challenge (Greenpeace, 2015; Nelson, 2016).

Concurrent to these developments, China's economic policies evolved to support them. Starting in 1994–1995, China's aid and state financing architecture underwent major reform with the creation of three state-owned policy banks: China Development Bank (see Sanderson and Forsythe, 2012), China Export-Import Bank, and China Agricultural Development Bank. From acquisition of oil to timber, these banks became the principal financiers of China's resources spree. They are also integral to a broader strategy to create export markets concurrent with attempting 'win–win' development outcomes with related partner countries.

A process of SOE reform was also begun in the mid-1990s. Under the reforms, SOEs were stripped of selective government, regulatory and public responsibilities, and at the same time forced to face greater domestic and international competition – such as the oil majors having to compete in global oil markets. An outcome of the Asian Financial Crisis of 1997–1998 supported the process. China, like more crisis-affected countries in the region, had then begun building strategic foreign exchange reserves, which the government was urged to transfer to SOEs in support of acquisition of international assets. Around the same time, a central government mandate directed that China should combine aid, trade and investment together in relations with Africa (Brautigam, 2009: 80). Finally, China began negotiations to join the World Trade Organization in 1994. President Jiang Zemin's visit to Africa in 1996 also marks a shift in the driver of China's relations with Africa, which transitioned from geopolitics to economics (Alden, 2007; Jiang, 2012).

Over these early reform decades, and until the Global Financial Crisis of 2008, the primary drivers of China's decades-long growth miracle were high and rapidly increasing investment levels and export-oriented industrial activity. And these trends were underpinned by a steady flow of workers moving from rural employment to more formal and urban employment. The process of labour transfer from low productivity subsistence rural employment to higher productivity urban employment in economics is captured in the Lewis model. For its duration, the process produces large productivity gains. These dry up when the supply of rural workers dries up, making labour relatively scarce, a point known as the Lewis turning point (LTP). At the LTP, wages rise rapidly and a new less unit of labour-intensive economic driver is required.

Importantly for Africa's industrialisation hopes, China's labour cost advantages are now diminishing. Data in Table 44.3 illustrate the recent demographic trends underlying that bigger economic story. Thanks to the evident consistent falls in the total fertility rate and gains in life expectancy, the share of total population aged over 65 years is increasing – the share has more than doubled since 1980. This means that China's workforce share of population has also been falling since around 2009. The arrival of increasingly rapid population ageing and the LTP in China is also serving to increase labour scarcity and wage pressures. Over the past few decades, China's labour advantage presented a direct and imposing challenge for Africa's basic manufactures exporters since 'Africa has no significant advantage over Asia in terms of labour costs

Table 44.3 Basic ageing population empirics, China

	*Total fertility rate**	*Life expectancy at birth (years)*	*Share of population aged 65 and over (% of total)*	*Number of people aged 65 and over*
1980	2.7	67	4.5	44,155,575
1990	2.5	69.5	5.3	60,164,805
2000	1.5	72.1	6.7	84,597,215
2010	1.7	74.9	8.2	109,691,810
2014	1.6	75.8	9.2	125,512,840

Source: Johnston *et al.* (2016).

Note: * Excludes data for Hong Kong SAR, Macau SAR and Taiwan.

while having large disadvantages in terms of agglomeration economies' (Collier, 2007: 15–16). Now, in contrast, China's economy is moving in a direction that is of a structure that is more complementary with Africa's own development.

China's new growth model – and Africa

Growing labour scarcity and higher labour costs, alongside emerging shifts in priorities for development, now instead have synthesised a 'new normal' growth model since 2011. China's new growth model is characterised by falling investment intensity, less demand for metals and energy, and a shift in emphasis from industrial development to consumption and production of services (see Garnaut *et al.*, 2016). In general, there is a shift away from resource-intensive and export-led growth towards more consumption, services and innovation-led growth. There is also a greater importance for China in utilising its high level of savings to realise growth via outbound investment. The latter can also serve to internationalise selective excess domestic capacities, such as in steel. Growth of some 6 per cent is expected in place of earlier double-digit growth rates (Garnaut *et al.*, 2016). China's economy is, however, much larger now, meaning the volume of growth that China's economy generates will remain very large.

The associated structural decline in China's demand for resources has, however, also induced the end of the decade-long China-led commodities price boom (Garnaut, 2012). This had offered a massive recent growth spike in export revenues in Africa. In that wake, many of the continent's fuel exporters now face macroeconomic distress. In parallel to falling growth in China itself, growth rates in many resource exporters in Africa have fallen, but in the latter case from double-digit rates to somewhere closer to zero (IMF, 2016a). Fiscal limits are consequently being stretched beyond sustainability. On the other hand, for Africa's resource-importing countries, these changes are positive news (Johnston, 2015a). Given the earlier discussed challenges of resource-led growth together with the related adverse consequences of China's old growth model for relatively richly labour-endowed countries, in the medium and long term China's new transition may in fact reflect a 'boom to vusp' turning point that instigates a more diversified integration of African economies into the world economy (Johnston, 2015a). Here, we summarise the related arguments around outbound capital investment, demographic and human capital investment trends.

Outbound investment in physical capital

As Figure 44.3 reveals, China's outbound investment has risen dramatically since the early 1990s. A policy to encourage targeted outbound investment was launched in the late 1990s. The 'Going Out' policy provides incentives to Chinese firms to invest abroad in the acquisition

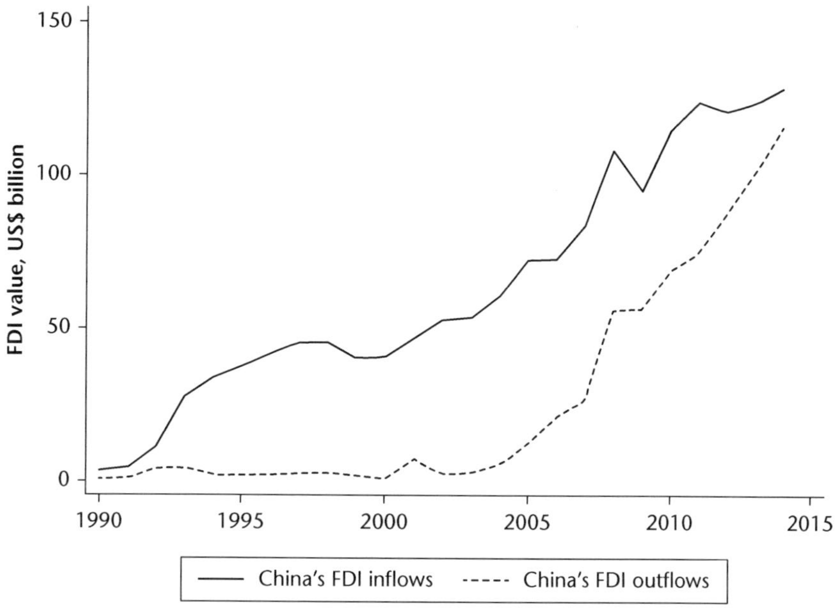

Figure 44.3 Inbound and outbound FDI, China

Source: NBS (2008, 2012, 2015).

of natural resources, to offshore certain industries, and to build global Chinese brand names. Africa's abundant natural resources, youthful population and market development potential make it a target of that investment push (see the second section).

Within investment targeting Africa, infrastructure is presently a big focus (see Foster, 2009). Of the US$60 billion of lending China promised to African countries in late 2015 at the FOCAC Summit in Johannesburg, more than half will be spent on building infrastructure (Awoko, 2016). China has specifically promised to build the 'three major networks' – railway, road and regional aviation. A flagship and Silk Road-connected project among these is the Standard Gauge Rail project, which will initially better connect Kenya's main port at Mombassa with the capital Nairobi. Onward plans will connect by rail a number of landlocked East African Community country members proximate to Kenya, including Uganda, South Sudan, Rwanda and Burundi. In July 2016, Tanzania also agreed a US$7.6 billion loan from China EXIM Bank for the construction of a standard gauge rail corridor that similarly will link Tanzania's economy with neighbours Uganda, Rwanda, Burundi and Congo (Tanzania Invest, 2016). As noted, both Kenya and Tanzania have historical links to China, in Tanzania's case in being home of the Tanzara Railway that is the symbol of China's support for the end of colonialism in Africa. This regional integration approach may also be useful towards overcoming Africa's economic geography hurdles (see Collier, 2006; Johnston *et al.*, 2015). There is an ongoing push within the United Nations to change adoption of the phrase 'landlocked countries' to 'land-linked countries' (Inter Press Service, 2014). Chinese entrepreneurs and policymakers, such as Ethiopian shoe manufacturing frontier investor Helen Hai, use this language already (UN Media, 2014).

Beyond East Africa, China has promised to help Africa as a continent to build the foundations of an integrated transportation network. An estimate of Africa's infrastructure financing requirement is some US$38 billion annually, with a further US$37 billion required annually in operations and maintenance (World Bank, 2013). The sum is the equivalent of 12 per cent of Africa's GDP, the funding gap for which is estimated at US$35 billion (World Bank, 2013). China is also investing in Africa's power network. The African Development Bank has estimated that infrastructure gaps reduce African GDP by some 2 per cent annually (see Brautigam, 2010a). Coming on stream before under construction and planned railways in Kenya, Tanzania and elsewhere, the first standard gauge railway built by China in Africa (and the first outside of China at all) was opened in July 2016 in Nigeria. Chinese firms are, in parallel, investing in iron ore projects and cement and steel plants (see Johnston, 2016). In the power sector, the International Energy Association estimates that between 2010 and 2015, China invested around US$13 billion in power projects in Africa, dwarfing the sums invested by any other foreign investor (Global Construction Review, 2016). A not insignificant share of this is in the renewables sector, opening new markets for China's technologies and helping to relatively preserve the local and international ecosystem.

Such projects are not, however, without their own set of issues. In Botswana in late 2013, the government refused to renew the China National Electric Equipment Company's operation and maintenance contract. In place, a German company, STEAG Energy Services, got the contract on grounds of greater quality assurance. China's loan portfolio monitoring and evaluation is unique, but not lacking stringency (Aguilar and Goldstein, 2009: 1558). Angolan public servants report that Chinese officials are rigorous and demanding when it comes to the use of the credit line and the meeting of its conditions (Aguilar and Goldstein, 2009). Similarly, Farrell (2016) finds that perceptions of the quality of Chinese-constructed roads in Africa are on average worse than the actual quality level delivered. The higher density of Chinese firms active in lower-quality road construction may be one of the causes, in place of Chinese firms failing on average to deliver in higher-grade projects.

Lending institutions reform

To support national intentions for a greater international investor role, in infrastructure markets especially, China has recently led the establishment of new financial institutions. This echoes the establishment of domestic policy banks in the 1990s that similarly supported (domestic) economic reforms. Specifically, China led the establishment of multilateral lending agencies, including the Beijing-based Asian Infrastructure Investment Bank and the Shanghai-based BRICS Development Bank. The latter has among its aims being the world's greenest multilateral lending agency. A $US40 billion Silk Road Fund was established in 2014 to foster development along China's new Silk Road project that broadly sweeps across Asia and the Indian Ocean. A South–South Climate Cooperation Fund, with some $US3 billion in funds, was also announced in 2015 to provide assistance to developing countries on climate issues. The domestic and multilateral financing mechanisms available to Chinese firms for financing their outbound investment plans have, in other words, recently increased dramatically.

Demographic dividend and human talent development

Figure 44.4 illustrates the shifting demographic dividend over the coming decades – in the cases presented, from China to Africa. Recent high fertility rates and falling child mortality rates in Africa are converging to produce a windfall working-age workforce share – akin to what China was able to utilise to transform its economy over recent decades. Now in China, however, family planning-induced falls in fertility rates are in contrast producing a rising old-age dependency ratio – in other words, an increasing population share of dependent old. China, in turn, faces

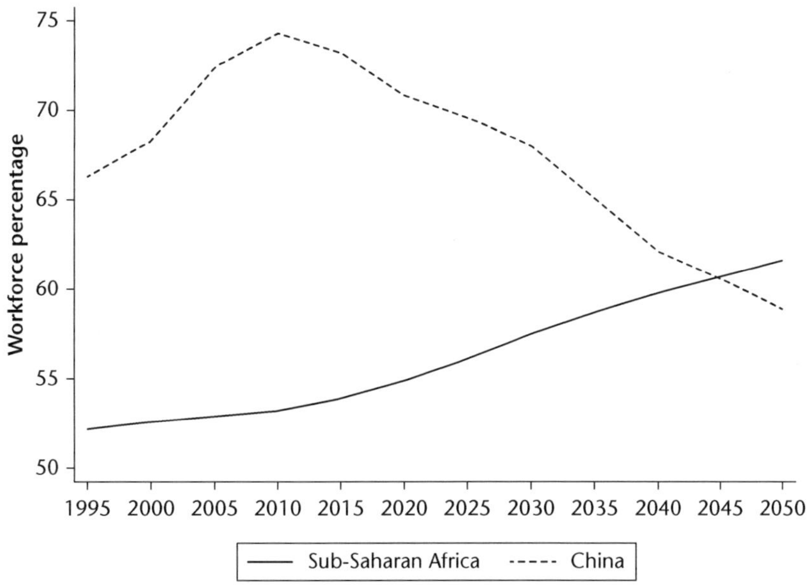

Figure 44.4 Demographic dividend trends, China and Africa

Source: UN DESA (2013).

rising wages and the need to increase productivity per worker in order to maintain total output per person. Outsourcing may help at the margins. Wage comparison between China's historic labour-intensive manufacturing heartland of the southern province of Guangdong against those costs in an industrial park in Ethiopia highlight the potential for low-wage labour-intensive outsourcing from China to Africa. Where the minimum wage in 2016 in Guangdong is some US$300 (China Labour Bulletin, 2015), in Ethiopia's Hawassa Industrial Park, expected to open in October 2016, the average wage will be some US$50 monthly (Addis Fortune, 2016).

Alongside these more commercial and physical capital investments, China is also actively supporting the development of human capital in Africa. At the level of training and education, in late 2015 China's President Xi Jinping announced 40,000 training opportunities for Africans in China and 30,000 government scholarships (MacGregor, 2015). At the education infrastructure level, China has furthermore promised to help build regional vocational education centres and several capacity-building colleges so that African countries can utilise a sufficient number of professional technicians to support their expected medium-term development.

Examples of China's higher education investments in West Africa include the funding and constructing of the Fendell Campus for Engineering at the University of Liberia (Aid Data, 2016) and also the Ghana Medical University, which has since established partnerships with several Chinese medical universities (Yanjian Group, 2015). In mid-2016, the China Shenyang University of Chemical Technology also agreed the establishment of a chemical university of technology in Nigeria. Examples of agreed Chinese investments in East Africa's education sector include an agreement to construct Africa's largest university library at the University of Dar es Salaam in Tanzania (Xinhua, 2016) and an agreement between Mount Kenya University (MKU) and the China University of Petroleum to ensure that MKU can train a sufficient number of high-quality oil and gas experts for Kenya's human resource needs. The Chinese-funded Malawi University of Science and Technology, the country's fourth public university, opened in 2014. These examples, on the other hand, provide no evidence that such projects will be completed and operational without constraints. They do, however, highlight a pattern of investment that is complementary to the recent OECD donor push in the primary education sector (see Johnston, 2015b).

Importantly also, China has since extended its early outbound focus on Kenya, Tanzania and Ethiopia into a broader grouping of countries. In focus now are 'demonstrating and pioneering countries', including Kenya, Tanzania and Ethiopia, plus the Republic of Congo. Priority partners include Angola, Egypt, Guinea, Mozambique, Rwanda, Senegal, Sierra Leone, Togo, Uganda and Zambia. South Africa is considered an engine of China's win–win industrialisation effort with African countries. Such a grouping of countries includes the continent's most advanced economy – South Africa – and some of its poorest, including Guinea and Sierra Leone. The list also includes coastal economies and landlocked economies, as well as countries considered resource-poor and resource-rich. China's investments across them, human and physical, should be expected to increase rapidly in the near future.

Discussion

China, the world's largest developing country, second largest economy, and home to the world's largest pool of savings, is now a major economic partner of Africa. Moreover, in recent years, the structure of China's own development has shifted in ways that are more favourable to Africa's own long-run development than has been the pattern of earlier recent decades. New institutions, including the Silk Road Fund and the Asian Infrastructure Investment Bank, have been established at China's instigation to support China's next economic development phase. These have their eye on green investments in infrastructure in Africa and elsewhere. In Africa, China is similarly investing in human capital, via scholarships, and establishing higher education institutions and educational partnerships and exchange programs between Chinese and African institutions. These improvements in physical and human capital in and between Africa and China should be expected to help both continue to develop their economies while managing respective challenging shifts in demography. A positive outcome to China's Africa return could induce a larger return for Africa, in terms of sustained integration with global value chains.

New opportunity nonetheless will bring its own challenges, and neither offers a solution to all earlier challenges. For example, there are in general fears that China's vision for its own greener environment may end up exporting this type of damage to other countries (e.g. see Sherwin, 2016). In the forestry sector specifically, there are also fears that as Chinese wood product manufacturers increasingly sell to a domestic and not a Western market, pressure to sustainably source timber will diminish, and that this could be catastrophic for forests under vulnerable management in Africa and elsewhere (World of Agroforestry, 2013). China's president has promised that for China's part, it will make great efforts to ensure the preservation of Africa's environment coincides with development of the region's economies. In the face of smuggling and corruption, however, this will take immense political will and persistence. African governments and civil society must invest time and resources to ensure optimal sustainable development outcomes for their continent.

At the macroeconomic level also, just as the difficulty in predicting the duration of price shocks has limited the ability of African policymakers to manage commodity booms and slumps (Cashin and Pattillo, 2000), so may future price shocks, even if less China-led. A greater risk is perhaps also that new opportunity passes without sufficient development gains being realised. African country and regional means of endogenising new opportunity for development will be unique to these environments and those of the world economy over the coming decades. It may thus, for example, be timely for China to invest in policy research and data collection that can more optimally inform and support the immense emerging potential for win–win development between China and African countries. Extensive further research is required both to understand China's economic impact on Africa and also the best means via which African countries, individually, regionally and collectively, can grasp the potential of these new and probably more long-run average favourable international economic conditions for development.

Note

1 Noted by a Chinese government representative as not intended to be a statistical yearbook of China's foreign aid (UN China in the World Seminar, 2014).

References

Addis Fortune (2016) *Bureau Steps up Hawassa IP Recruitment*, available at: http://addisfortune.net/articles/bureau-steps-up-hawassa-ip-recruitment/?platform=hootsuite (accessed 29 July 2016).

Aguilar, R. and Goldstein, A. (2009) 'The Chinisation of Africa: the case of Angola', *The World Economy*, 32(11): 1543–1562.

Aid Data (2016) *Renovate the Fendell Campus of the University of Liberia*, available at: http://china.aiddata.org/projects/1620 (accessed 24 May 2016).

Ajakaiye, O., Kaplinsky, R., Morris, M. and N'Zue, F.F. (2009) *Seizing Opportunities and Confronting the Challenges of China–Africa Investment Relations: Insights from AERC Scoping Studies.*

Alden, C. (2007) *China in Africa*, London: Zed Books.

Amusa, K., Monkam, N. and Viegi, N. (2016) *The Political and Economic Dynamics of Foreign Aid: A Case Study of United States and Chinese Aid to Sub-Sahara Africa*, No. 201628.

Anshan, L. (2007) 'China and Africa: policy and challenges', *China Security*, 3(3): 69–93.

Awoko (2016) *Sierra Leone News: Finance Minister to Make Case for Alone Today in China-Africa Summit*, available at: http://awoko.org/2016/07/28/sierra-leone-newsfinance-minister-to-make-case-for-alone-today-in-china-africa-summit/ (accessed 28 July 2016).

Benabdallah, L., Robertson, W. and Wang, Y. (2016) *China Loans Africa Much Less Than You Think (and Four Other Lessons from a New Database)*, available at: http://africanarguments.org/2016/06/06/china-loans-africa-much-less-than-you-think-and-four-other-lessons-from-a-new-database/ (accessed 8 July 2016).

Brautigam, D. (1998) *Chinese Aid and African Development*, London: Macmillan.

Brautigam, D. (2009) *The Dragon's Gift: The Real Story of China in Africa*, Oxford: Oxford University Press.

Brautigam, D. (2010a) *China, Africa and the International Aid Architecture*, African Development Bank, Working Paper No. 107, available at: www.afdb.org/fileadmin/uploads/afdb/Documents/Publications/WORKING%20107%20%20PDF%20E33.pdf (accessed 8 January 2018).

Brautigam, D. (2010b) 'Looking East: Africa's newest investment partners', *Global Journal of Emerging Market Economies*, 2(2): 173–188.

Broadman, H.G. (2006) *Africa's Silk Road: China and India's New Economic Frontier*, Washington, DC: World Bank.

Buckingham, K. (2016) 'Beyond trees: restoration lessons from China's Loess Plateau', in R. Garnaut, L. Song, F. Cai and L. Johnston (eds), *China's New Sources of Economic Growth: Reform, Resources and Climate Change*, Canberra: ANU Press, pp. 139–158.

Cashin, P. and Pattillo, C. (2000) 'The duration of terms of trade shocks in sub-Saharan Africa', *Finance and Development*, 37(2): 26.

Chen, W., Dollar, D. and Tang, H. (2016) 'Why is China investing in Africa? Evidence from the firm level', *The World Bank Economic Review*, lhw049.

China Labour Bulletin (2015) *Guangdong Finally Announces Plans to Increase Minimum Wage by 19 Percent*, available at: www.clb.org.hk/en/content/guangdong-finally-announces-plans-increase-minimum-wage-19-percent (accessed 29 July 2016).

Collier, P. (2006) *Africa: Geography and Growth*, Oxford: Center for the Study of African Economies, Department of Economics, Oxford University.

Collier, P. (2007) 'Africa's economic growth: opportunities and constraints', *African Development Review*, 19(1): 6–25.

Collier, P. (2008) *The Bottom Billion: Why the Poorest Countries are Failing and What Can Be Done About It*, Oxford: Oxford University Press.

Deaton, A. (1999) 'Commodity prices and growth in Africa', *The Journal of Economic Perspectives*, 13(3): 23–40.

Dreher, A., Fuchs, A., Hodler, R., Parks, B., Raschky, P.A. and Tierney, M.J. (2016) *Aid on Demand: African Leaders and the Geography of China's Foreign Assistance*.

Eisenman, J. (2012) 'China–Africa trade patterns: causes and consequences', *Journal of Contemporary China*, 21(77): 793–810.

Farrell, J. (2016) *How Do Chinese Contractors Perform in Africa? Evidence from World Bank Projects*, SAIS China Africa Research Initiative, Working Paper No. 3, February.

Financial Times (2013) *Africa Must Get Real About Chinese Ties*, available at: www.ft.com/content/562692b0-898c-11e2-ad3f-00144feabdc0 (accessed 10 April 2014).

Foster, V. (2009). *Building Bridges: China's Growing Role as Infrastructure Financier for Sub-Saharan Africa*, Vol. 5, Washington, DC: World Bank.

Garnaut, R. (2012) 'The contemporary China resources boom', *Australian Journal of Agricultural and Resource Economics*, 56(2): 222–243.

Garnaut, R., Song, L. and Fang, C. (2013) *China: A New Model for Growth and Development*, Canberra: ANU Press.

Garnaut, R., Song, L., Fang, C. and Johnston, L. (2016) 'China's new sources of economic growth: a supply-side perspective', in R. Garnaut, L. Song, C. Fang and L. Johnston (eds), *China's New Sources of Economic Growth: Reform, Resources and Climate Change*, Canberra: ANU Press, pp. 1–21.

Global Construction Review (2016) *China Takes the Lion's Share of Electrifying Africa*, available at: www.globalconstructionreview.com/sectors/china-takes-lions-sha7re-electr7ifying-a7frica/ (accessed 26 July 2016).

Greenpeace (2015) *New Evidence Shows Chinese, West African Governments Must Rein in Rogue Fishing Fleet*, available at: www.greenpeace.org/africa/en/campaigns/Defending-Our-Oceans-Hub/Chinese-illegal-fishing-exposed/ (accessed 16 April 2016).

Gui-Diby, S.L. and Renard, M.F. (2015) 'Foreign direct investment inflows and the industrialization of African countries', *World Development*, 74: 43–57.

IMF (International Monetary Fund) (2016a) *Africa Economic Outlook: Time for a Policy Reset*, available at: www.imf.org/external/pubs/ft/reo/2016/afr/eng/sreo0416.htm (accessed 19 May 2016).

IMF (International Monetary Fund) (2016b) *Direction of Trade Statistics*, available at: http://data.imf.org/?sk=9D6028D4-F14A-464C-A2F2-59B2CD424B85 (accessed 8 January 2018).

Inter Press Service (2014) *UN Favours Changeover from Landlocked to 'Land-Linked'*, available at: www.ipsnews.net/2014/11/u-n-favours-changeover-from-landlocked-to-land-linked/ (accessed 28 March 2016).

Jiang, S. (2012) 'China–Africa cooperation, trade and development: history and prospects', *Journal of Practice in Foreign Relations and Trade*, 10, October: 9–13 (in Chinese).

Johnston, L. (2014) 'Premier Li calls for "innovative and pragmatic cooperation" in Africa', *East Asia Forum*, April, available at: www.eastasiaforum.org/2014/06/11/premier-li-calls-for-innovative-and-pragmatic-cooperation-in-africa/ (accessed 8 January 2018).

Johnston, L. (2015a) 'Boom to cusp: prospecting the "new normal" in China and Africa', in R. Garnaut, L. Song, F. Cai and L. Johnston (eds), *China's Domestic Transformation in a Global Context*, Canberra: ANU Press, pp. 383–408.

Johnston, L. (2015b) *China, Australia and the African Frontier*, available at: https://pursuit.unimelb.edu.au/articles/china-australia-and-the-african-frontier (accessed 3 March 2016).

Johnston, L.A. (2017) 'Steel pipe dreams: a China–Guinea and China–Africa lens on prospects for Simandou's iron ore', *The Extractive Industries and Society*, 4(2): 278–289.

Johnston, L.A. and Cheng, Y. (2015) 'China's Africa trade and investment policies: review of a "noodle bowl"', *African East-Asian Affairs*, 4: 6–41.

Johnston, L.A., Morgan, S.L. and Wang, Y. (2015) 'The gravity of China's African export promise', *The World Economy*, 38(6): 913–934.

Johnston, L., Liu, X., Yang, M. and Zhang, X. (2016) 'Getting rich after getting old: China's demographic and economic transition in dynamic international context', in L. Song, R. Garnaut, C. Fang and L. Johnston (eds), *China's New Sources of Economic Growth: Vol. 1 – Reform, Resources and Climate Change*. Canberra: ANU Press, pp. 215–246.

Kaplinsky, R. and Morris, M. (2009) 'Chinese FDI in sub-Saharan Africa: engaging with large dragons', *The European Journal of Development Research*, 21(4): 551–569.

Kitano, N. and Harada, Y. (2015) 'Estimating China's foreign aid 2001–2013', *Journal of International Development*, 28(7): 1050–1074.

Lin, J.Y. and Wang, Y. (2014) *China–Africa Cooperation in Structural Transformation: Ideas, Opportunities, and Finances*, WIDER Working Paper, No. 2014/046.

MacGregor, K. (2015) *China Ups Africa Support with Training, Scholarships*, available at: www.universityworldnews.com/article.php?story=20151211183903298 (accessed 24 February 2016).

NBS (National Bureau of Statistics) (2008) *2007 Statistical Bulletin of China's Outbound Foreign Direct Investment*, Beijing: China Statistical Press.

NBS (National Bureau of Statistics) (2012) *2011 Statistical Bulletin of China's Outbound Foreign Direct Investment*, Beijing: China Statistical Press.

NBS (National Bureau of Statistics) (2015) *2014 Statistical Bulletin of China's Outbound Foreign Direct Investment*, Beijing: China Statistical Press.

Nelson, J. (2016) '24 African countries ban China from fishing illegally in their waters', *Atlanta Black Star*, available at: http://atlantablackstar.com/2016/01/07/24-african-countries-ban-china-from-fishing-illegally-in-their-waters/ (accessed 10 July 2016).

Onishi, N. (2016) 'African economies, and hopes for a new era, shaken by China', *New York Times*, 25 January, available at: www.nytimes.com/2016/01/26/world/africa/african-economies-and-hopes-for-new-era-are-shaken-by-china.html?_r=0 (accessed 2 February 2016).

Pigato, M. and Tang, W. (2015) *China and Africa: Expanding Economic Ties in an Evolving Global Context*, Washington, DC: World Bank.

Ravallion, M. (2009) 'Are there lessons for Africa from China's success against poverty?', *World Development*, 37(2): 303–313.

Sachs, J.D. and Warner, A.M. (1999) 'The big push, natural resource booms and growth', *Journal of Development Economics*, 59(1): 43–76.

Sachs, J.D. and Warner, A.M. (2001) 'The curse of natural resources', *European Economic Review*, 45(4): 827–838.

SAIS-CARI (2016) *Data: Chinese and American FDI to Africa*, available at: www.sais-cari.org/data-chinese-and-american-fdi-to-africa/ (accessed 7 June 2016).

Sanderson, H. and Forsythe, M. (eds) (2012) *China's Superbank: Debt, Oil and Influence – How China Development Bank is Rewriting the Rules of Finance*, Hoboken, NJ: John Wiley & Sons.

Shen, X. (2015) 'Private Chinese investment in Africa: myths and realities', *Development Policy Review*, 33(1): 83–106.

Sherwin, K. (2016). 'China is outsourcing its pollution', *Foreign Policy*, available at: http://foreignpolicy.com/2016/12/07/china-is-outsourcing-its-pollution/ (accessed 8 January 2018).

State Council Information Office (2011) *China's Foreign Aid*, available at: http://english1.english.gov.cn/official/2011-04/21/content_1849913.htm (accessed 10 January 2013).

State Council Information Office (2013) *China–Africa Economic and Trade Cooperation*, available at: http://english.gov.cn/archive/white_paper/2014/08/23/content_281474982986536.htm (accessed 19 September 2015).

State Council Information Office (2014) *China's Foreign Aid 2014*, available at: http://english.gov.cn/archive/white_paper/2014/08/23/content_281474982986592.htm (accessed 15 November 2015).

Tan-Mullins, M., Mohan, G. and Power, M. (2010) 'Redefining "aid" in the China–Africa context', *Development and Change*, 41(5): 857–881.

Tanzania Invest (2016) *China to Lend Tanzania USD7.6bn for Central Corridor Railway*, available at: www.tanzaniainvest.com/transport/china-loan-central-corridor-railway (accessed 28 July 2016).

The Herald (2016) *Senegal's Southern Forests May Disappear by 2018*, available at: www.herald.co.zw/senegals-southern-forests-may-disappear-by-2018/ (accessed 4 June 2016).

UN China in the World Seminar (2014) *UNDP China in the World Series, China's Foreign Aid*, October, Beijing: UNDP.

UN DESA (United Nations Department of Economic and Social Affairs) (2013) *World Population Prospects: The 2012 Revision*. New York: UN DESA.

UNECA (United Nations Economic Commission for Africa) (2013) *Economic Report on Africa 2013: Making the Most of Africa's Commodities Industrializing for Growth, Jobs and Economic Transformation*, Addis Ababa: UNECA.

UN Media (2014) *Goodwill Ambassador Has High Hopes for African Industrialisation*, available at: www.unmultimedia.org/radio/english/2016/01/goodwill-ambassador-has-high-hopes-for-african-industrialization/index.html#.V5yOCZN96CQ (accessed 4 May 2016).

Venables, A.J. (2010) 'Economic geography and African development', *Papers in Regional Science*, 89(3): 469–483.

Wang, F.L. and Elliot, E.A. (2014) 'China in Africa: presence, perceptions and prospects', *Journal of Contemporary China*, 23(90): 1012–1032.

Woo, W.T. (2016) 'The demand-side supplement to supply-side structural reform: termination of the soft-budget constraint', in R. Garnaut, L. Song, F. Cai and L. Johnston (eds), *China's New Sources of Economic Growth: Reform, Resources and Climate Change*, Canberra: ANU Press, pp. 139–158.

World Bank (2013) *Fact Sheet: Infrastructure in Sub-Saharan Africa*, available at: http://web.worldbank.org/WBSITE/EXTERNAL/COUNTRIES/AFRICAEXT/0,,contentMDK:21951811~pagePK:146736~piPK:146830~theSitePK:258644,00.html (accessed 29 December 2017).

World of Agroforestry (2013) *Who Is Importing African Timber to China and What Might This Mean for Sustainability*, available at: http://blog.worldagroforestry.org/index.php/2013/01/14/who-is-importing-african-timber-to-china-and-what-might-this-mean-for-sustainability/ (accessed 29 February 2016).

Xinhua (2016) *China to Build Africa's Biggest University Library in Tanzania*, available at: http://news.xinhuanet.com/english/2016-06/05/c_135414235.htm (accessed 10 July 2016).

Yanjian Group (2015) *Medical University in Volta Region of Ghana Funded by China*, available at: http://english.yanjian.com.cn/Article/ShowArticle.asp?ArticleID=167 (accessed 8 January 2018).

Zafar, A. (2007) 'The growing relationship between China and sub-Saharan Africa: macroeconomic, trade, investment, and aid links', *The World Bank Research Observer*, 22(1): 103–130.

Section 10

Political perspectives on Africa's development

45

Government

African development beyond 'good governance'

David Booth

Introduction

Is good governance good for development? This provocative question was chosen as the title of a stimulating collection of research essays published a few years ago under the auspices of the United Nations Department of Economic and Social Affairs (Sundaram and Chowdhury, 2012). It was an excellent choice. With its double play on the word 'good', it challenged scholars and activists committed to conventional public sector reform and capacity-building efforts in developing countries to reflect on their assumptions. By posing a question rather than providing an answer, it wisely allowed for the fact that many things are still unknown or remain controversial concerning the role of government – and, more broadly, patterns of governance – in the economic and social development of poor countries.

Of course, the topic of governance for development is not only of interest in Africa, but for a number of reasons controversy on the subject has been more prominent there (especially in the sub-Saharan subregion, our main concern here) than elsewhere in the developing world. The challenge issued by Sundaram and Chowdhury (2012) is therefore highly pertinent to the concerns of this handbook. Their question and the range of answers that it has prompted provide a suitable framework for reviewing what has been written, researched and debated about government in Africa over the last 50 years.

Slogans versus evidence

The phrase 'good governance' has been with us for more than half of that 50-year period. It remains an influential shorthand expression not only in the corridors of international development agencies, but among civil society activists, commentators and thought leaders across Africa. While it means different things to different people, it has a common core. The core could be said to be captured in some of the components of the UN's Sustainable Development Goal (SDG) number 16 ('Promote peaceful and inclusive societies for sustainable development, provide access to justice for all and build effective, accountable and inclusive institutions at all levels'). Sub-goals 16.5 to 16.7 commit the international community to substantially reducing corruption and bribery in all their forms, and to developing 'effective, accountable and transparent institutions at all levels' while ensuring 'responsive, inclusive, participatory and representative decision-making', also 'at all levels' (UN, 2015).

Such broadly stated aspirations have undoubted appeal. They accompany some eight other sub-goals of indisputable importance concerning a range of violent abuses, criminalities and injustices on a global scale that call for urgent international action as well as efforts within countries. However, sub-goals 16.5–16.7 are not as helpful as might appear; they serve poorly as guides to action.

By presenting a list of rather abstract governance qualities as self-evidently and universally desirable, at best the SDGs obscure a host of unresolved issues and disagreements about the relationship between forms of government and development outcomes in the real world. They assert, controversially, that there is a single, straightforward set of properties that qualifies a governance system as good. At worst, they give comfort to unthinking application of ideas about 'best practice' governance without due regard to context and feasibility.

The good governance agenda, understood in these terms, has been a persistent influence on thinking about African development, but mainly in policy- and action-oriented contexts. Scholarly analysis and research, in contrast, has made it its business to question the assumptions underlying such seemingly unproblematic declarations of intent. Researchers have insisted on distinguishing between ends and means, between value judgements and testable propositions about causal linkages, and between the ideologies of the powerful and the needs of the rest. What has emerged from this body of work as a whole is a more robust concept of what matters in the governance of development, one that is grounded in historical experience and empirical evidence as well as in common values.

The challenge of economic transformation

It is important that we now have this solid research base. It means we can answer more robustly specific questions about the kind of governance particular countries need. Many countries of sub-Saharan Africa have recently reached a critical juncture in their economic and social development. Steady improvements in key social indicators and regular annual increases in gross domestic product are now the norm. Warfare and civil conflicts are much less widespread than they were. As a result of these changes and realignments in the world economy (especially China), there are real opportunities for these gains now to be turned into processes of economic transformation, where economic growth rests on broad-based improvements in productivity and competitiveness. However, whether or not these opportunities will be taken will depend greatly on what governments are willing and able to do.

This chapter offers a concise summary of literature and evidence in the three areas just touched upon: the 'good governance' agenda as it has affected Africa, the research-based response to this, and the implications for Africa's economic transformation. Thus, the second section discusses the origins of the concern about governance quality in Africa between the 1970s and the 1990s; the third section reviews the rebuttals and reformulations that have led the scholarly debate since the turn of the millennium; the fourth section applies the findings to the challenge of moving Africa on from economic growth to economic transformation; and the fifth section concludes.

(Not) making sense of African governance

Africa is not a country, and there are notorious dangers in generalising across a whole region. Nonetheless, the particularities of countries are best appreciated in the context of some broad trends or cycles of change that have affected nearly all of them in a major way. At the most obvious and general level, most countries were part of the British or French colonial empires until the 1960s or under Portuguese control until the 1970s. Between the 1960s and the 1980s, most of the now independent states abandoned their initial multiparty constitutions and either adopted single-party systems or came to be ruled by the military. Democratisation movements gained ground again in the 1990s. Today, the regional norm is a formally liberal-democratic constitution, under which political practice varies along a spectrum from disguised authoritarianism to imperfect democracy (Crawford and Lynch, 2012; Cheeseman, 2015).

The economic performance of the typical sub-Saharan African country over the equivalent period also featured remarkable ups and downs. As far as we can tell from the available statistics (Jerven, 2013, 2015), national and per capita income growth was impressive in many countries in the first decade or so of independence, but declined sharply towards the end of the 1970s and collapsed in the 1980s. Economies recovered in the 1990s. This happened slowly at first, but since around the middle of that decade, sustained rates of reasonably fast, if poorly distributed, economic growth have become the norm.

Does governance matter?

At this broad level of consideration, there are no obvious or straightforward connections between observed trends in formal political governance and economic performance. That is not surprising, as the general research evidence on the relationship between variables such as the quality of democracy and economic development remains inconclusive, and at best predicts a long-run benefit once democracy is consolidated (Acemoglu *et al.*, 2014; Kelsall, 2014; Masaki and van de Walle, 2014). When the quality of government is conceived in a way that gets closer to the immediate determinants of development outcomes, such as impartiality and low corruption in public administration, the evidence becomes clearer (Rothstein, 2011; Rothstein and Tannenberg, 2015). However, this finding leaves unsettled what are the likely political drivers of this sort of quality.

In spite of these reasons for reserving judgement on the main issues, the dominant perspective on African governance since the early 1990s has been anything but cautious. A central proposition has been that independent Africa has experienced patchy and interrupted economic and social development because, in a comprehensive sense, it has been badly governed. This has been widely taken to mean that the countries of the region need to embrace more wholeheartedly the political institutions and practices adopted by the countries that have been most successful in raising living standards and national power to a high level. In short, they need better governance, and furthermore we know exactly what good governance involves, because it is on display in such places as Stockholm, London and Ottawa.

This perspective came to the fore at a particular historical moment for Africa and the world when perhaps such a simple view of matters seemed consonant with the main facts. That moment, around 1989, featured a confluence of factors and key events that, taken together, pointed many reasonable people to similar conclusions. African economies, or to be exact their formal sectors – the part whose condition is measured – had hit rock bottom and had had started to recover sluggishly at best. Economic stabilisation and structural adjustment programmes funded by the IMF and the World Bank were increasingly the determining factor in national policymaking, and those leading the way in this regard were coming to the conclusion that economic recovery was going to require not just correct policies, but better institutions (World Bank, 1989). The new institutional economics, which argued that markets only function well when supported by sound institutions, was beginning to get the world's attention (North, 1990). The fall of the Soviet Union and the end of the Cold War, meanwhile, was fast removing previous restraints on Western governments' advocacy of democratic politics as the foundation of sound institution-building. Meanwhile, for increasing numbers of urban Africans living under harshly dictatorial regimes, these global events were signalling the possibility of a second liberation. Multi-partyism became the order of the day, even in countries such as Tanzania that had a form of single-party government that remained overwhelmingly popular in rural areas.

Consolidation of an orthodoxy

In this way, an initially uncontested consensus was formed that brought together a broad community of international lenders and donors, on the one hand, and African democrats and reformers, on the other. The concept of good governance was constructed to give coherence to the different elements of the consensus. It came to embrace everything from sound macroeconomic management and civil service efficiency to civic freedoms and broad social participation, including in the design and monitoring of economic adjustment measures. The common strand was that in all these areas, common problems were amenable to evident, experience-based solutions.

In sum, a claim came to be accepted by virtually everybody that Africa: (a) was failing in development for lack of good institutions of governance; and (b) would not succeed until this changed. The tacit assumption was that good institutions were those that been adopted in recent times in the developed West and in Japan: liberal-democratic constitutions and the corresponding arrangements for limiting corruption, permitting transparent decision-making and generating government accountability to citizens. The latest ideas in fashion in the West, including privatisation, 'unbundling' of public sector roles and the creation of semi-autonomous executive agencies to improve civil service efficiency (Bangura and Larbi, 2006), were soon included in the package of remedies.

The immediate manifestations of the new consensus included World Bank programmes that offered concessional loans attached to 'second generation' reforms with a focus on institutions rather than just policies (World Bank, 1994). Apart from the increasing regularity of competitive elections, the innovations included an array of formal arrangements reflecting current conceptions of best practice in public management, including electoral commissions, anti-corruption watchdogs, public defenders, civil service performance enhancement schemes, medium-term expenditure frameworks, participatory planning systems and citizen monitoring committees.

For a long time, the very mixed results obtained with these innovations (Carothers, 1999, 2006; Andrews, 2013) did little to dent the enthusiasm with which they were pursued. Critics of donors and governments typically wanted more of the same, or stronger efforts to make the remedy work better. Even today, the typical traveller on the urban transport system of Nairobi or Lagos wants proper implementation of good governance principles, end of story. However, specialists who moderate their value judgements with the evidence of history and systematic research take a different view.

Bringing back a sense of history

At one level, the good governance consensus about Africa makes perfect sense. In terms of the broad sweep of human history, African economies and social relations are at a relatively early stage of capitalist development. In a historical perspective, therefore, it would be surprising if their political institutions did not have traces of the type associated with pre-capitalism in other parts of the world. But which type exactly? This is one of the questions that preoccupied the first generation of political scientists studying independent Africa, and many of them turned to the writings of Max Weber (1978 [1922]).

Weber distinguished between feudal, patrimonial and modern democratic/bureaucratic forms of political authority, and many Africanists plumped for the idea that contemporary Africa blends modern and patrimonial elements (Eisenstadt, 1972; Médard, 1982). Patrimonial authority is distinguished by centralisation of authority (particularly in comparison with the feudalism of pre-modern Europe and Japan) and by a blurring of the distinction between the wealth of the state and the personal household of the ruler. The term neopatrimonialism was coined to refer to a hybrid form including strong elements of both patrimonial and modern democratic/bureaucratic authority.

The concept of neopatrimonialism

There is little doubt that, like pure patrimonialism, neopatrimonialism has been historically unfavourable to capitalism and economic development. Both centralisation and the use of public resources for the private enrichment of rulers and their families (what today we call grand corruption) work against the enabling conditions for private investment, especially the forms of private investment that help to stimulate further investment – e.g. by financing infrastructure – and those that provide employment on a large scale. Political scientists working in this tradition (Hyden, 1983; Callaghy, 1984; Joseph, 1987; Chabal, 1992) provided ample illustrations of the harm done to African economic development by control orientation and pillage of public resources by the regimes they called neopatrimonial. Later, van de Walle (2001) showed how neopatrimonial drivers of policymaking inhibited and distorted the economic recovery efforts of the 1980s and 1990s. There is a great deal of more recent evidence and theory to support the thesis that under neopatrimonial regimes, as well as under 'limited access orders' (North *et al.*, 2009, 2013) more generally, it is difficult to sustain economic development, and especially real economic transformation.

Around 1989, the political science consensus around the theory of neopatrimonialism did add some academic respectability to the broad good governance position. Partly for that reason, those with reservations about the latter have often concentrated their fire on the former. Some critics claimed, rather unfairly, that the theory implies a belief that the obstacles to African development are bound up with special and unfortunate features of African culture, while others accused the political scientists of misunderstanding Weber (Therkildsen, 2005; Pitcher *et al.*, 2009). They agreed with a third group who found the concept useful in identifying some broad commonalities but mistaken in obscuring the rather substantial differences in the form taken by neopatrimonial leadership across countries and periods, with different implications for possible economic progress (Kelsall, 2013; Mkandawire, 2015).

Africa in comparative perspective

Differences among regimes that share some basic features of neopatrimonial governance become interesting when placed alongside the positions that have been taken by researchers and students of comparative development over the past quarter-century. The crucial arguments are five:

- Even if Africa's inferior development performance is bound up with poor governance rooted in neopatrimonial institutions, there is no sound basis for looking for remedies in the ideologies of institutional best practice that have emerged in the richest countries of the world in the last few decades. If the experience of the developed countries is relevant at all, we should be looking at their earlier history, including as far back as the era of absolute monarchy, when European governance was markedly patrimonial. Otherwise, we may be guilty of what Chang (2002) famously called 'kicking away the ladder', in other words denying to today's developing countries the policy instruments and institutional arrangements that permitted them to emerge from their feudal or patrimonial pasts.

- The relatively recent development successes of East Asian countries, and of China and parts of South East Asia, are arguably a much more relevant source of inspiration in thinking about the relationship between governance and economic performance. Those experiences taken as a whole do not suggest that there is any single overall formula or type of political regime that makes for strong development outcomes. The regimes in question have been very diverse. The South East Asian countries that were most similar to sub-Saharan African countries around 1960 had patterns of governance that were very similar to those observed in Africa then and now. Yet their economic performance has been spectacularly better. The reasons must lie somewhere other than in regime type (Henley, 2015).
- There is hardly any disagreement these days that in a broad sense, the institutions a country has acquired are a fundamental shaper of what is to be expected in terms of development performance. However, there are two important and connected qualifications on this research-based consensus.
- One is that capitalist development does not take hold unless certain basic institutional functions are performed somehow, but the *form* in which they are performed is not universal or pre-determined (Chang, 2007; Rodrik, 2007). To take only the most striking positive example, in China for the last several decades the function of giving private actors confidence to invest has been assured reasonably well, and with spectacular results, without the particular form of well-defined property rights and contract law. The clearest negative example is perhaps Uganda, which is claimed to have the most complete set of transparency and anti-corruption institutions anywhere in the world, but still fails to meet the conditions for dynamic capitalist development (Andrews and Bategeka, 2013).
- The second and related observation about institutions is what can appear to be the exactly the same institutional arrangements function very differently in different contexts where the underlying configurations of political and economic power (what some call the 'political settlement') are different. This point, today associated primarily with the economist Mushtaq Khan (1995, 2010), has also been made repeatedly by the political scientist Robert Bates.

Bates made his reputation with a book inspired by the public choice school of thought (Bates, 1981). This explained why it might be politically rational for African governments to extract 'rents' from their agricultural sectors by means of tax, price and exchange rate policies, and then pour at least some of these resources back into the rural sector in the form of subsidies. According to the book, the discretionary allocation for subsidies was a more effective tool for building and maintaining a political base in the countryside than the simple alternative of providing favourable general conditions for rural development.

State-controlled marketing boards were, and in many places still are, one of the primary instruments of this politically driven repression of African agricultural potential. But when Bates went to Kenya for his next book, he found that Kenyan marketing boards under the presidency of Jomo Kenyatta performed rather well in support of smallholder agriculture, although they ceased to do so under his successor, Daniel arap Moi (Bates, 1989). The reasons had to do with the relationships among the power holders controlling the operations of the boards, and between them and Kenya's main ethno-political constituencies (see also Bates, 2014).

Multiple paths

If we have a proper sense of comparative history, therefore, it is clear that there are routes out of neopatrimonial underdevelopment that do not involve early adoption of the forms of governance that are ideologically favoured in the richer parts of the world today. To begin with, as emphasised by Henley (2015), getting the right policies may be more important than having a sound institutional framework. Second, what counts as a sound institutional framework is context-specific and influenced decisively by the configuration of political and economic power. Lastly, because of all this, we should think in terms of multiple pathways to development success, including some where relevant improvements in patterns of governance are fairly continuous and others where they are not (Levy, 2014). Success in either case is likely to involve a considerable measure of learning by trial and error (Rodrik, 2007).

Governance and economic transformation today

As fast economic growth has become more usual in Africa over recent decades, attention has shifted to the pattern and quality of that growth. Issues of concern include the sectoral and social distribution of the aggregate increases in gross domestic product, the often limited impact of growth on poverty, and the weak capacity of the most dynamic sectors to create jobs. Underlying these issues is that much of the recent growth has been the result of high commodity prices within an economic structure that has changed little since colonial times. Sources of recent growth other than minerals include urbanisation and services for the new upper and middle classes. However, with scant exceptions, both marked improvements in the productivity of agriculture and growth in employment-intensive manufacturing have been lacking.

From growing to transforming economies

One way to express this is to say that countries are getting economic growth, but not economic transformation or 'growth with depth' (ACET, 2014). Economic transformation is understood here as the process of moving labour and other factors of production from lower- to higher-productivity activities, either between sectors or within them (McMillan *et al.*, 2016). While there continue to be strong reasons for emphasising productivity gains in agriculture, especially in the early stages of development (Breisinger and Diao, 2008), convergence in labour productivity between poorer and richer countries has been fastest when investment has moved into manufacturing (Cimoli *et al.*, 2009; Rodrik, 2013; Whitfield *et al.*, 2015). Therefore, the deficit of attention to both agricultural productivity and manufacturing employment is a particularly troubling feature of the political economy of sub-Saharan Africa today.

Until recently, many would have seen this as a challenge too far, arguing that African countries have done too little to improve their governance and should not expect to be able to tackle the tasks of a 'developmental state'. However, there are two sets of reasons for being more optimistic. The first has to do with changes in the world economy, especially in China. The slowing of Chinese economic growth has already meant lower demand for Africa's primary commodities, especially minerals, which has weakened the prospects of African growth on established lines. On the positive side, the current period is presenting a window of opportunity for African manufacturing as wage rises in China impel Chinese manufacturers to move the more labour-intensive parts of global value chains to other world regions (Lin, 2013; Stiglitz and Lin, 2013).

Learning from Asia

The other reasons for being cautiously optimistic stem from the various criticisms of the conventional perspective on African governance problems outlined in the third section. Returning to the experience of successful Asian countries, two observations are particularly relevant.

First, the countries that achieved the fastest industrial progress and/or agricultural transformation were not initially democracies of any kind. During their take-off phase, they lacked many of the qualities associated with 'good governance'. On the other hand, they did benefit from a political leadership that pursued an inclusive form of development, treating economic transformation as a nation-building project. With some exceptions, their regimes constructed quite elaborate systems for ensuring accountability and building consensus. The conventional distinction between democracy and authoritarianism fails to capture some of the most relevant features of Asian development models (MacIntyre, 1994; Campos and Root, 1996; Woo-Cumings, 1999).

Second, the Asian experience shows that it is not necessary for the bureaucracy of the state *as a whole* to become an effective coordinator of policy and driver of private sector investment and productivity growth. Most success stories start in one sector, with a single politically empowered public agency. This was a theme of Johnson's (1982) study of the Ministry of Trade and Investment in Japan. It is part of the story about poverty-reducing agricultural transformation in Indonesia (Henley, 2015).

Building on this, Ansu *et al.* (2016) distinguish four requirements that appear universally relevant to effective policy for economic transformation:

- Constructing a consensus among key actors that establishes *economic transformation as a nation-building project*, with shared commitments extending well beyond a single electoral term.
- Giving at least *one public agency sufficient autonomy, budgetary control and political authorisation* to override interdepartmental coordination problems and engage in a practical way with *credible private sector organisations*.
- Creating institutional arrangements that can coordinate a sufficient set of powerful public and private actors so as to ensure: (1) an appropriate level of technically justified public support to promising sectors or firms; and (2) that this support is conditioned on *mutually enforceable performance standards*.
- Enabling discovery of approaches that work for transformation in the particular country context by means of *explicit experimentation, good feedback and timely correction*.

Getting practical

While several Asian and a few Latin American countries have embraced arrangements of these kinds, examples of their adoption have been quite rare in sub-Saharan Africa. A handful of exceptionally favoured agricultural subsectors (cocoa in Ghana, dairy in Uganda, etc.) provide the principal documented exceptions (Whitfield *et al.*, 2015). It is therefore important to examine the political feasibility and the institutional framing of economic transformation policies in addition to their technical suitability and efficiency. Two broad principles seem generally applicable.

First, reforms and change initiatives need to have modest objectives and be pursued in a politically smart way that takes into account the interests and incentives faced by different economic actors. On being politically smart, there is evidence from countries such as Nigeria, Nepal and the Philippines (Booth, 2014, 2016) that regulatory reforms that improve the long-run conditions for economic transformation can be successful in politically quite unfavourable contexts if their promoters are sufficiently clever about navigating the political-economic obstacles and learn along the way.

Second, policy priorities and initial investment decisions should take into account not just the economic merits of the case, but also the potential to make a difference to the overall political economy and climate of opinion by means of demonstration effects. In Africa, as elsewhere, it is often the case that the barriers to effective policies and institutions for economic transformation are a combination of vested interests and old-fashioned economic theories (Rodrik, 2014; te Velde *et al.*, 2016). Under such circumstances, there is a useful role for initiatives designed specifically to demonstrate what policy for economic transformation looks like. This entails concentrating on one or two promising sectors where the economic benefits and potential profitability of new, high-productivity activities can be revealed in such a striking way that purely intellectual objections begin to fall away. The vested interests then become easier to tackle.

Conclusion

The main implication of these arguments, taken together, is clear. Priorities for African governance have been for too long influenced by perspectives whose starting point is a comparison of where African countries are today with where the capitalist liberal democracies are today (rather than where they used to be when they had economic and social structures more like those of Africa today). There has been far too little attention to the lessons from successive rounds of economic progress in Asia. In this light, the messages about the importance of 'good governance' that formed and became influential between the 1980s and 1990s are misleading at best. It is unfortunate that they continue to be echoed, albeit in a highly abstract way, in the UN declaration on the SDGs. Comprehensive governance improvement is not a precondition for economic progress in developing Africa. Politically smart measures, targeted at specific barriers to transformation and accompanied by rapid feedback and adjustment, are the main requirements, based on experience elsewhere. They should be the main ingredients of future thinking about governance for development in Africa.

References

Acemoglu, D., Naidu, S., Restrepo, P. and Robinson, J.A. (2014) *Democracy Does Cause Growth*, Working Paper 2004, Cambridge, MA: National Bureau of Economic Research.

ACET (African Center for Economic Transformation) (2014) *African Transformation Report 2014: Growth with Depth*, Accra: ACET.

Andrews, M. (2013) *The Limits of Institutional Reform in Development: Changing Rules for Realistic Solutions*, Cambridge: Cambridge University Press.

Andrews, M. and Bategeka, L. (2013) *Overcoming the Limits of Institutional Reform in Uganda*, WIDER Working Paper 2013/111, Helsinki: UNU-WIDER.

Ansu, Y., Booth, D., Kelsall, T. and te Velde, D.W. (2016) *Public and Private Sector Collaboration for Economic Transformation*, paper presented at the African Transformation Forum 2016, Kigali, 14–15 March.

Bangura, Y. and Larbi, G.A. (eds) (2006) *Public Sector Reform in Developing Countries: Capacity Challenges to Improve Services*, Basingstoke: Palgrave Macmillan/UNRISD.

Bates, R.H. (1981) *Markets and States in Tropical Africa: The Political Basis of Agricultural Policies*, Berkeley, CA: University of California Press.

Bates, R.H. (1989) *Beyond the Miracle of the Market: The Political Economy of Agrarian Development in Kenya*, Cambridge: Cambridge University Press.

Bates, R. (2014) 'The new institutionalism', in S. Galiani and I. Sened (eds), *Institutions, Property Rights, and Economic Growth: The Legacy of Douglass North*, Cambridge: Cambridge University Press, pp. 50–65.

Booth, D. (2014) *Aiding Institutional Reform in Developing Countries: Lessons from the Philippines on What Works, What Doesn't and Why*, Working Politically in Practice Working Paper 1, San Francisco, CA and London: Asia Foundation and ODI.

Booth, D. (ed.) (2016) *Politically Smart Support to Economic Development: DFID Experiences*, London: Overseas Development Institute.

Breisinger, C. and Diao, X. (2008) *Economic Transformation in Theory and Practice: What Are the Messages for Africa?* RrSAKSS Working Paper 10, Washington, DC: International Food Policy Research Institute.

Callaghy, T. (1984) *The State–Society Struggle: Zaire in Comparative Perspective*, New York: Columbia University Press.

Campos, J.E. and Root, H.L. (1996) *The Key to the Asian Miracle: Making Shared Growth Credible*, Washington, DC: Brookings Institution.

Carothers, T. (1999) *Aiding Democracy Abroad: The Learning Curve*, Washington, DC: Carnegie Endowment for International Peace.

Carothers, T. (ed.) (2006) *Promoting the Rule of Law Abroad: In Search of Knowledge*, Washington, DC: Carnegie Endowment for International Peace.

Chabal, P. (1992) *Power in Africa: An Essay in Political Interpretation*, London: Macmillan.

Chang, H.-J. (2002) *Kicking Away the Ladder: Development Strategy in Historical Perspective*, London: Anthem Press.

Chang, H.-J. (2007) 'Understanding the relationship between institutions and economic development: some key theoretical issues', in H.-J. Chang (ed.), *Institutional Change and Economic Development*, London: Anthem Press/UN University Press, pp. 17–33.

Cheeseman, N. (2015) *Democracy in Africa: Successes, Failures, and the Struggle for Political Reform*, Cambridge: Cambridge University Press.

Cimoli, M., Dosi, G. and Stiglitz, J.E. (eds) (2009) *Industrial Policy and Development: The Political Economy of Capabilities Accumulation*, Oxford: Oxford University Press.

Crawford, G. and Lynch, G. (eds) (2012) *Democratization in Africa: Challenges and Prospects*, London: Routledge.

Eisenstadt, S.N. (1972) *Traditional Patrimonialism and Modern Neopatrimonialism*, London: Sage.

Henley, D. (2015) *Asia-Africa Development Divergence: A Question of Intent*, London: Zed Books.

Hyden, G. (1983) *No Shortcuts to Progress*, London: Heinemann.

Jerven, M. (2013) *Poor Numbers: How We Are Misled by African Development Statsitics and What to Do About It*, Ithaca, NY: Cornell University Press.

Jerven, M. (2015) *Africa: Why Economists Get It Wrong*, London: Zed Books.

Johnson, C. (1982) *MITI and the Japanese Miracle: The Growth of Industrial Policy, 1925–1975*, Stanford, CA: Stanford University Press.

Joseph, R. (1987) *Democracy and Prebendal Politics in Nigeria: The Rise and Fall of the Second Republic*, Cambridge: Cambridge University Press.

Kelsall, T. (2013) *Business, Politics, and the State in Africa: Challenging the Orthodoxies on Growth and Transformation*, London: Zed Books.

Kelsall, T. (2014) *Authoritarianism, Democracy and Development*, State of the Art 3, Birmingham: Developmental Leadership Program.

Khan, M.H. (1995) 'State failure in weak states: a critique of new institutionalist explanations', in J. Harriss, J. Hunter and C.M. Lewis (eds), *The New Institutional Economics and Third World Development*, London: Routledge, pp. 71–86.

Khan, M.H. (2010) *Political Settlements and the Governance of Growth-Enhancing Institutions*, London: Economics Department, School of Oriental and African Studies, University of London.

Levy, B. (2014) *Working with the Grain: Integrating Governance and Growth in Development Strategies*, New York: Oxford University Press.

Lin, J.Y. (2013) *Against the Consensus: Reflections on the Great Recession*, Cambridge: Cambridge University Press.

MacIntyre, A. (ed.) (1994) *Business and Government in Industrialising Asia*, Ithaca, NY: Cornell University Press.

Masaki, T. and van de Walle, N. (2014) *The Impact of Democracy on Economic Growth in Sub-Saharan Africa, 1982–2012*, Working Paper 2014/057, Helsinki: UNU-WIDER.

McMillan, M., Page, J., Booth, D. and te Velde, D.W. (2016) *Supporting Economic Transformation: An Approach Paper*, London: Overseas Development Institute, SET Programme.

Médard, J.-F. (1982) 'The underdeveloped state in tropical Africa: political clientelism or neo-patrimonialism?', in C. Clapham (ed.), *Private Patronage and Public Power: Political Clientelism in the Modern State*, London: Frances Pinter, pp. 162–192.

Mkandawire, T. (2015) 'Neopatrimonialism and the political economy of economic performance in Africa: critical reflections', *World Politics*, 67(3): 563–612.

North, D.C. (1990) *Institutions, Institutional Change and Economic Performance*, Cambridge: Cambridge University Press.

North, D.C., Wallis, J.J. and Weingast, B.R. (2009) *Violence and Social Orders: A Conceptual Framework for Interpreting Recorded Human History*, Cambridge: Cambridge University Press.

North, D.C., Wallis, J.J., Webb, S.B. and Weingast, B.R. (eds) (2013) *In the Shadow of Violence: Politics, Economics, and the Problems of Development*, Cambridge: Cambridge University Press.

Pitcher, A., Moran, M. and Johnston, M. (2009) 'Rethinking patrimonialism and neopatrimonialism in Africa', *African Studies Review*, 52(1): 125–156.

Rodrik, D. (2007) *One Economics, Many Recipes: Globalization, Institutions, and Economic Growth*, Princeton, NJ: Princeton University Press.

Rodrik, D. (2013) 'Unconditional convergence in manufacturing', *Quarterley Journal of Economics*, 121(1): 165–204.

Rodrik, D. (2014) 'When ideas trump interests: preferences, worldviews, and policy innovations', *Journal of Economic Perspectives*, 28(1): 189–208.

Rothstein, B. (2011) *The Quality of Government: Corruption, Social Trust, and Inequality in International Perspective*, Chicago, IL: University of Chicago Press.

Rothstein, B. and Tannenberg, M. (2015) *Making Development Work: The Quality of Government Approach*, Report 07, Stockholm: Expertgruppen för biståndsanalys (EBA).

Stiglitz, J.E. and Lin, J.Y. (eds) (2013) *The Industrial Policy Revolution: The Role of Government Beyond Ideology*, Basingstoke: Palgrave Macmillan/IEA.

Sundaram, J.K. and Chowdhury, A. (eds) (2012) *Is Good Governance Good for Development?* London: Bloomsbury Academic/United Nations.

te Velde, D.W., Booth, D., Leipziger, D. and Uneze, E. (2016) *Supporting Economic Transformation in Nigeria*, London: Overseas Development Institute, SET Programme.

Therkildsen, O. (2005) 'Understanding public management through neopatrimonialism: a paradigm for all African seasons?', in U. Engel and G.R. Olsen (eds), *The African Exception*, Aldershot: Ashgate, pp. 35–51.

UN (United Nations) (2015) *Transforming Our World: The 2030 Agenda for Sustainable Development*, A/RES/70/1, New York: UN.

van de Walle, N. (2001) *African Economies and the Politics of Permanent Crisis, 1979–1999*, Cambridge: Cambridge University Press.

Weber, M. (1978 [1922]) *Economy and Society*, Berkeley, CA: University of California Press.

Whitfield, L., Therkildsen, O., Buur, L. and Kjær, A.M. (2015) *The Politics of African Industrial Policy: A Comparative Perspective*, Cambridge: Cambridge University Press.

Woo-Cumings, M. (ed.) (1999) *The Developmental State*, Ithaca, NY: Cornell University Press.

World Bank (1989) *Sub-Saharan Africa, from Crisis to Sustainable Growth: A Long-Term Perspective Study*, Washington, DC: World Bank.

World Bank (1994) *Adjustment in Africa: Reforms, Results, and the Road Ahead*, Washington, DC: World Bank.

46

Conflict and post-conflict

Hazel Cameron

[F]or peace and development are interdependent?

(Boutros-Ghali, 1992)

Introduction

Since the late 1990s, there has been a growing awareness of the interrelationship between conflict and development, as well as peace and development, that has led to a greater understanding of correlations between the outbreak of violent conflict and under- or uneven development (Collier, 2003; International Development Committee, 2006). Studies have revealed how developmental factors can trigger violence, but also, on a more positive note, how developmental factors can be instrumental in either preventing or ending violence and initiating post-conflict reconstruction (MacGinty and Williams, 2016). Academics, political leaders and policymakers are now cognisant that development is an important element in conflict prevention. However, conversely, academics, political leaders and policymakers must pay heed to the negative role that development can play in violent conflict. The surge in intra-state conflicts in post-Cold War Africa brought many of its nations' economies to the point of collapse, resulting in humanitarian crisis. For decades, countries such as Liberia, Sierra Leone, Guinea-Bissau and Côte d'Ivoire were crippled by these intra-state conflicts in which violence and relentless killings were rampant. During the first decade of the new millennium, more people in Africa were being killed because of war than in the rest of the world combined. The United Nations Development Programme (UNDP, 2003) described it as a 'decade of despair' for Africa.

This chapter will explore African development through the lens of violent conflict in Africa, highlighting the role of development in conflict as well as the role of conflict on development. The chapter briefly explores non-developmental and developmental explanations for violent conflict in Africa, specifically intra-state wars, before discussing the orthodoxy that development is an essential component of peace-building (Boutros-Ghali, 1992), focusing on the role of development in responding to violent conflict situations through the specific examples of peacekeeping and aid. The chapter thereafter discusses how development can contribute to conflict in Africa using the Rwandan genocide as a case study in which many of the social processes linked with development created conditions in which conflict was less easily restrained, and indeed were casual factors in the escalation of violence.

Grave and ominous challenges

Throughout the past 50 years, Africa has been confronted with a multitude of grave and ominous challenges. Economic growth on the continent is uneven and fragile, and generally poverty is on the rise. According to the International Monetary Fund (IMF, 2016), by the end of 2016, the continent had the lowest GDP per capita, namely US$1,809, compared to the global average of US$10,300. In addition, political instability continues to be persistent and widespread, with at least 80 violent changes of government in Africa since the 1960s. Throughout the period of 1960–1989, coups were the most common means of replacing leaders in Africa; however, the number of coups dropped substantially in the years 1990–2012.

Armed conflict remains a serious problem on the continent, although there is no consensus among analysts on the question of how many armed conflicts are ongoing in Africa at any given time. It is, however, agreed that between 1980 and 1994, almost half of the world's war-affected countries were on the African continent, with Angola, Liberia, Mozambique and Somalia being ranked as the five most severely affected countries in the world (DFID, 2001).

Between 1999 and 2008, the African continent witnessed 13 major armed conflicts, by far the highest total for any region of the world, yet there was cause for optimism when, by 2006, over half of these conflicts had been resolved, and the number of major armed conflicts continued to decline as the end of the decade approached (Harbom and Wallensteen, 2009). Global comparisons provided valuable insights and further cause for optimism. Whereas, as previously noted, by the end of the twentieth century Africa accounted for almost half of the world's major armed conflicts, by the end of the first decade of the new millennium it was home to just one in five (HSRP, 2008). Optimism was to be short-lived; studies show that since 2011, there has been a substantial increase in organised violence across the continent and a related rise in episodes of violence against civilians. This was chiefly due to major new wars in South Sudan, Libya, the Central African Republic and Mali (Harbom and Wallensteen, 2009).

Millions have died because of violent conflict in contemporary Africa, with most deaths a result of disease and malnutrition exacerbated by displacement, rather than death through injuries sustained during violent conflict. However, within the first two months of 2015, approximately 8,300 people died as a direct result of conflict in sub-Saharan Africa, with just five countries, namely Nigeria, Cameroon, Sudan, Niger and Somalia, accounting for approximately 90 per cent of these deaths. The bulk of the violence was concentrated in West Africa, where conflict between Islamist group Boko Haram and armed forces, as well as Boko Haram attacks on civilians, accounted for the vast majority of conflict deaths. The recently published IISS (2017) *Armed Conflict Survey* indicates that currently, four African conflicts, namely Somalia, Sudan, South Sudan and Nigeria, count among the 10 most lethal globally. The report also reveals that although the number of fatalities as a direct result of conflict in sub-Saharan Africa remained high in 2016 at 14,000, it is a substantial reduction from the 24,000 deaths recorded in 2015. Nonetheless, conflicts in South Sudan, Somalia and the Central African Republic (CAR) have worsened since 2015, and displacement rates are at an all-time high. The reduction in deaths in sub-Saharan Africa as a direct result of conflict in 2016 is a result of a 70 per cent decline in fatalities in the Boko Harem conflict, despite the group remaining relatively strong.

It is arguable that Africa is the most conflict-prone region in the world. Furthermore, when peace *is* negotiated that brings an end to violence, unresolved tensions and rivalries generally remain that ensure the peace is often a temporary state of affairs and viewed as unsatisfactory by most participants (Human Security Centre, 2005). The negotiated peace frequently collapses altogether, leading to repetitive cycles of negotiations, signed peace agreements, fragile peace and resumption of violence. This is clearly demonstrated in the conflicts of Somalia, Liberia and South Sudan, where numerous agreements failed to produce any semblance of a stable peace. Armed conflict not only presents a threat to life through direct death and injury, but, as previously noted, death also occurs because of disease and malnutrition through displacement. Violent conflict can and does destroy a country's healthcare and education systems, as well as the infrastructure, and therefore negatively impacts on development. It is estimated that African wars have cost the continent 'well over $700 billion in damages since the year 2000 alone' (Williams, 2016).

Theories of conflict causation are many, and the subject of much debate in both theory and practice. Violent conflicts are triggered by a combination of factors, and as such, analysts who put forward development-related or economic explanations for the outbreak of violent conflicts must also consider non-developmental-related explanations and their relationship with development-related factors. The multiple actors, dynamics and environments involved in conflict has hindered the advancement of a general theory of conflict and conflict management. In Africa and beyond, there are data suggesting multiple causes of conflict that interrelate in decidedly explicit ways according to the prevailing circumstances. Causes of conflict can be either primary or secondary in nature, but the complex nature of societies, economics and human politics ensures that multiple factors are at play when violent conflict is triggered. When examining any violent conflict, analysts are faced with the daunting task of trying to identify all the factors that gave rise to conflict and to establish the relationship between those complex elements. This is a difficult task since, because of the public discourse of conflict, actors may conceal their real motives, or be guarded so to offer little insight into their motivations, as is the case with the state-sponsored militias in Darfur. Added to this, media-led explanations of African wars are frequently overly simplistic, and on occasion outright racist. When it was clear that genocide was taking place in Rwanda, the British media peddled the myth that the genocidal violence was due to ancient tribal hatreds and savagery (Cameron, 2013), and failed to identify that key development issues were a highly significant factor in the violence (as detailed later in this chapter). However, it is not only journalists who fail to make reference to development, economics, poverty or profit as causations of violent conflict.

Non-development causal explanations for conflict in Africa

A significant body of literature on conflict causation depends entirely on theories of history and ancient tribal hatreds (Brown, 1997) and other non-development or non-economic related explanations, and includes theories of conflict causation promoted through the corresponding categories of psychology (Tajfel, 1978), biological disposition (Simmel, 1955), religion (Appleby, 2000), the security dilemma (Posen, 1993), ethnicity (Connor, 1994; Young, 2003), the nature of the state (Tilly, 1985), cultural dysfunction (Kaplan, 1994), identity (Sen, 2006), and incompatible world views (Huntington, 1998). The majority of these theories of conflict causation view economic factors as being secondary to the primary causal explanation.

In terms of theories of conflict causation, a substantial body of literature exists that discusses ethnicity as a non-development causal factor of violent intra-state conflict (Denny and Walter,

2014; Keller, 2014; Cordell and Wolff, 2016). In general, the objective of ethnic conflict is the destruction of specific groups of people, which often also results in the destruction of property and the means of livelihood of the victim groups. Ethnic conflicts also cause significant disruption to social mechanisms. Analysts have focused on two divergent strands, namely 'ethnic hatreds' and 'ethnic fears' (Fujii, 2011: 5). The 'ethnic hatreds' perspective implies ongoing low-level rivalry between ethnic groups that has endured over generations, which escalates when triggered by a change in circumstances, resulting in mass violence directed against the rival group. The 'ethnic fears' perspective highlights the role of political leaders and other powerful elites, who, in order to further their own political agendas, target an ethnic group with arbitrary arrests, organised riots, dissemination of hate propaganda through media, etc., thereby inducing large-scale fear in society of an ethnic 'other'. The resultant insecurity dilemma becomes a trigger for ordinary people to react violently (Fujii, 2011: 5).

Such an explanation is problematic, however, as being ethnically different is in itself no explanation for why groups enter into violent conflict against each other. When ethnic diversity is understood as a fundamental condition for violence, it fails to explain how, since the 1990s, the most peaceful African states have been those that are the most ethnically diverse. Conversely, Rwanda is an African country where ethnic differences were negligible, and yet it witnessed a protracted four-year civil war that culminated in genocide of the Tutsis and the death of around 1 million Rwandans. Ethnic diversity is even less apparent in Somalia, and yet the country continues to suffer from a violent conflict that has been ongoing for over three decades.

Another non-development-related argument is the claim that most conflict environments in contemporary Africa are rooted in European colonialism (see Nasong'o, 2015). The colonial policy of 'divide and rule' intentionally debilitated indigenous institutions and power networks, weakening patrimonial societies and reconfiguring patron–client relationships within artificial boundaries. The artificial borders created by the colonisers across the continent resulted in the abnormal separation of otherwise contiguous groups, and forced non-contiguous groups to exist side by side. These factors undoubtedly caused instability and provided the foundations for violent conflict (Zeleza, 2008). The history of colonialism is an important factor that cannot be ignored by theorists of conflict causation; however, it must be noted that although there are examples of colonisation exacerbating conflict, it is not a trigger or primary cause of contemporary conflicts in Africa. At the end of the nineteenth century, virtually the entire continent was affected equally by the legacies and policies of colonialism. Despite this, many regions of postcolonial Africa have not witnessed violent intra-state conflict. As such, an obvious weakness in the theory of colonialism as an explanation for the outbreak of violent conflict is its failure to recognise African agency, and the responsibility of Africans for wars whose participants are Africans. It also fails to explain why the period between 1980 and the early 2010s were particularly devastating for Africa.

Conversely, a further frequently cited non-developmental explanation for violent conflict in postcolonial Africa is the role of postcolonial ruling elites (e.g. see Bøås and Dunn, 2013). Throughout the past three decades' neopatrimonialism, a legacy of colonialism has been the most recognisable and pervasive form of African governance. Many leading scholars of African studies claim neopatrimonialism encapsulates much political and administrative behaviour in Africa (Medard, 1982; Bratton and van de Walle, 1994; Englebert, 2000; van de Walle, 2001; Erdmann and Engel, 2007). The concept of neopatrimonialism amounts to the rational pursuit of self-interest by a 'big man' and his close cronies, and has been evident in political science and development studies since the 1970s, where it is used to explain political instability and underdevelopment of many African countries (Roth, 1968; Eisenstadt, 1973; Daloz, 2003). Clapham (1985: 48) defines neopatrimonialism as:

> a form of organization in which relationships of a broadly patrimonial type pervade a political and administrative system which is formally constructed on rational-legal lines. Officials hold positions in bureaucratic organizations with powers which are formally defined, but exercise those powers, so far as they can, as a form not of public service but of private property.

Neopatrimonialism facilitates corruption and renders official and formal systems of accountability redundant. Bad governance and corruption have been significant causal factors in violent conflicts in Nigeria, Guinea-Bissau, Sierra Leone, Liberia, Côte d'Ivoire and many other African countries. Sierra Leone's Truth and Reconciliation Commission identified bad governance, corruption and poverty as root causes of the conflict (Fithen, 1999). Since the late 1980s, African development policies, including those of the major international financial institutions such as the World Bank and the IMF, have promoted good governance policies in an effort to challenge corruption and promote economic growth.

Development- and economic-related explanations for conflict causation and maintenance in Africa

However, in efforts to explain conflict causation and maintenance in Africa, some scholars favour economic- and development-related explanations. There are two basic assumptions on the economics of conflict. The first assumption is that economic development offers a route out of conflict at both the international and individual levels. In brief, 'free markets made free men', and free men would not be so reckless as to become involved in war (Mandelbaum, 2004: 265). The alternative view on the economics of conflict highlights the predacious nature of capitalism and how the profound inequalities it causes can exacerbate conflict. This second perspective of economics as a trigger of violent conflict also encompasses the profiteering of warriors. War is a profitable business, whether it be roadside thugs levying a war 'tax', arms manufacturers peddling their goods, or those operating oil-smuggling operations or blood diamond operations in war zones. For warriors profiteering from conflict, peace is a threat to their livelihoods. Economists have developed complex models to indicate how looting and other forms of economic plundering provide powerful incentives for combatants. But such models uncover little of the nature of civil war economies and the challenges faced by non-combatants in times of war. On the other hand, anthropologists have usefully demonstrated the extraordinary measures to which communities, families and individuals will go in an effort to survive violent conflict, as well as the astonishing complexity of civil war economies (Nordstrom, 2004; Cramer, 2006; Butler, 2016).

Over the years, economists have developed a significant body of analysis that presents a correlation between the onset of violent conflict and economic factors; however, the precise nature of the relationship between economics and violent conflict remains a contested issue (Collier and Hoeffler, 2004).

Development discourses on conflict have been influenced by the debates arising from political economy explanations of violent conflict, captured in the 'greed versus grievance' debate. A group of economists affiliated with the World Bank advocated that African violent conflict could be explained by the actions of greedy warlords. These economists argued that civil war in Africa was caused by economic opportunities, disputing that political grievances explained violent conflict since grievances were pervasive across Africa but violent conflict was not. By such rationale, they concluded that the explanation was due to economic and greed factors. As a result, development discourse tended to lay the blame for Africa's wars squarely on the shoulders of greedy warlords (Collier and Hoeffler, 2002).

Collier and Hoeffler's (2002) conclusions resulted in heated debate among scholars who promoted religion or ethnicity as primary causal factors of violent conflict in Africa. They suggested that Collier and Hoeffler's argument was flawed in its assertion that greed and grievance were separate categories, and that explanations of the causes of war were synonymous with explanations of individual motivations, when the reality is that no war is apolitical (Ballentine, 2003; Berdal, 2005; Keen, 2008). However, in recent times, agreement has been reached that both greed *and* grievance are causal factors of violent conflict (Murshed and Tadjoedden, 2009), albeit there is no consensus on the weighting to be afforded to each. Collier later revised his earlier work, stating that 'where rebellion is feasible, it will occur without any special inducements in terms of motivation' (Collier *et al.*, 2009: 25).

It is evident that violent conflicts in Africa are caused by a combination of factors, and academic endeavours to develop theories of conflict causation in Africa must therefore address both development-related or economic explanations, non-development-related explanations, and their interplay (Nhema and Zeleza, 2008). It is important to note here that as well as being a causation of conflict, developmental factors can be instrumental in either preventing or ending violence and initiating post-conflict reconstruction.

African and international efforts to end war in Africa

Development, in the form of peacekeeping and aid, has played a significant role in responding to conflict situations in instances where armed conflict has left regions of Africa in a state of crisis. The complexities of African conflict environments do, however, present significant challenges for global mediators and their peacekeepers, who totalled 95,993 uniformed personnel as of 30 April 2017 (United Nations Peacekeeping, 2017). International attempts to provide humanitarian assistance and foster development in African post-conflict environments have proved onerous. To address these issues, steps have been taken to build peace in Africa and manage conflict and post-conflict zones by reinforcing the continents' own conflict management organisations. The principal African organisation for conflict management between 1990 and 2002 was the Organisation of African Unity (OAU) Mechanism for Conflict Prevention Management and Resolution, in addition to some other initiatives initiated through subregional arrangements on the African continent. In 2002, the African Union (AU) replaced the OAU. The AU Assembly of Heads of States and Government is the foremost organ of the African Union, a continental union that consists of all the 54 fully recognised states on the African continent. The AU Assembly of Heads of States and Government define the objectives and policies of the organisation.

Before the AU was established, continental-level conflict management activities in Africa were generally ad hoc responses to crises, with limited mandates in terms of both military and peace operations, authorised in the name of several subregional organisations. In dismantling the OAU, there was recognition of a need to develop a more effective conflict management mechanism on the continent that would take on a much greater role in terms of military intervention and peace operations, thus moving beyond their usual remit of small observer missions. To this end, Article 4(h) of the AU Constitutive Act gave the assembly the right to intervene in a member state 'in respect of grave circumstances: war crimes, genocide and crimes against humanity' (Kuwali and Viljoen, 2014). It is claimed that the disbanding of the OAU and establishment of the AU was due in no small part to the release of the OAU report into the international failure to prevent Rwanda's genocide. However, the unprecedented right of military intervention in a member state in cases of war crimes, genocide and crimes against humanity in Article 4(h), albeit noble in purpose, has encountered issues in implementation, with a lack of political will to implement Article 4(h). Subsequently, the AU has yet to invoke Article 4(h) to justify military intervention against the will of a fully recognised sovereign government (Franke, 2009: 97).

Despite international debates about the war in Darfur, and overwhelming evidence of 'grave circumstances' as per the AU definition, the Assembly has proved unwilling to mandate the use of military force without the explicit consent of the government of Sudan, who are themselves responsible for the 'grave circumstances'. In December 2015, the Commission took the unprecedented step of invoking Article 4(h) of the Constitutive Act, with a plan to deploy an AU force of 5,000 to Burundi, a country that has been in political turmoil since May 2015. President Nkurunziza was given 96 hours to accept the force. However, the motion was revoked in January 2016 when African leaders refused to endorse the mission. This revealed division between those favouring an interventionist approach within the AU Commission, and member states who overall favoured a less aggressive approach to the crisis in Burundi. It is perhaps unsurprising that those member states with questionable democratic credentials and human rights records are reluctant to support the use of Article 4(h).

International conflict management on the African continent engages in traditional peacekeeping, monitoring of ceasefires, peace enforcement, nation-building, humanitarian aid, police training and peace-building. Undoubtedly, the proliferation of peace operations on the African continent between 1990 and 2015 was the most observable international response to Africa's wars, with over 80 missions, far more than the numbers during the same period on any other continent. This era proved to be the most turbulent in the history of peacekeeping, 'peacekeeping' denoting the traditional United Nations (UN) 'blue helmet' troop deployments, ordinarily to police an agreed separation between antagonists (Bellamy *et al.*, 2010).

These peace operations were authorised and conducted by a number of actors and international organisations, with the vast majority having the support of the United Nations Security Council (UNSC). The UN does not have its own army, and is wholly dependent on member states contributing troops; in the post-Cold War period, there has been a move towards 'outsourcing' or 'subcontracting' peace operations, with the inclusion of regional organisations, INGOs and NGOs to carry out peace support and development activities on behalf of the United Nations (Richmond and Carey, 2005). These missions range from small observer missions to large-scale operations involving many thousands of personnel. The mandates to maintain peace and security in these international operations are varied, and include observing ceasefires, supporting disarmament demobilisation and reintegration programmes (DDR), protecting humanitarian relief supplies, supporting weak governments, strengthening the rule of law and public security, protecting civilians, and counter-insurgency.

The results of these operations in terms of conflict management have been mixed. Over the past six years, there has been a further substantial expansion in UN peacekeeping operations in Africa, although since 2011 the bulk of the UN's peacekeepers on the continent have been deployed in just three countries, namely Sudan, South Sudan and the Democratic Republic of Congo (DRC). Between 2013 and 2014, two international missions were established by the United Nations, with UN peacekeepers replacing African-led forces in Mali (MINUSMA) and the Central African Republic (MINUSCA). As a result of these UN missions, the total number of UN uniformed peacekeepers in Africa numbered more than 80,000, the highest figure on record at that time. By mid-2015, 120,000 uniformed peacekeepers were deployed in nine UN missions, three AU-authorised missions, and various others operated by the EU or France. Undoubtedly, peacekeeping has morphed into a developing world activity, with countries such as Cameroon providing 1,380 troops to the UN, while Senegal has deployed twice this number. Rwanda is Africa's third largest contributor of peacekeepers to UN peace missions, and currently maintains 1,650 troops with the United Nations Mission in South Sudan (UNMISS), and has further personnel deployed in MINUSCA.

Human security and conflict environments

However, significant damage to the reputation of the UN has resulted from situations where UN peacekeepers deployed to conflict environments have failed to stop mass killings, the most notable example being that of the United Nations Assistance Mission in Rwanda (UNAMIR) during the 1994 genocide. Human security is a key development issue, and of all the mandated duties given to peacekeepers, civilian protection is unquestionably the most important (Bellamy and Williams, 2010). Irrespective of the effectiveness of alternative duties being undertaken by the mission, local populations tend to judge peacekeepers on their capacity to protect civilians in danger. Yet until 1999, the UN multidimensional peacekeeping operations in Africa did not actually have an explicit component of civilian protection in their mandates, although that is not to say that peace operations did not attempt to tackle the problem of civilian protection within the limitations of their mandate. In 2003, the inclusion of civilian protection was rolled out to EU peacekeeping missions (notably Operation Artemis and EUFOR missions in Chad and the Central African Republic) and AU missions (notably AMIS, AFISMA and MISCA).

The amendment to mandates was also driven by UN high-level inquiries into the failure to protect victims of mass killings and genocide (see UN docs S/1999/1257, 16 December 1999; A/54/549, 15 November 1999; and A/55/305-S/2000/809). Similarly, the OAU report into the failures to protect civilians during the 1994 genocide in Rwanda (Caplan, 1994) drove changes to mandates of AU peacekeepers and missions. Despite this, even when civilian protection mandates were issued by the UN, they had attached to them geographic, political and operational stipulations, which weakened the ability of UN peacekeepers to respond appropriately to civilians in need of urgent protection. Peacekeepers were required to protect civilians 'without prejudice to responsibility of host nation'. However, it is well documented that the main perpetrators of civilian harm are frequently the armed forces of host governments. Peacekeepers were only supposed to protect civilians within their existing 'capabilities and areas of deployment', while the focus of the peacekeeping mission was limited to protecting only those civilians 'under imminent threat of violence' (Willmot *et al.*, 2016: 262). This ensured that the protection of civilians was doomed to failure. These weaknesses were addressed by the UN, who have since adopted the phrase 'threat of violence' without reference to 'imminent', and stressing that peacekeepers should actively prevent violence against civilians. Media reports and local and international perceptions have generally tended to focus on the human security failures of peacekeepers, but it must be acknowledged that even in the worst-case scenario, the presence of a peacekeeping mission has been beneficial to local civilians rather than exacerbating their situation (Hultman *et al.*, 2013). Unarmed Ghanaian peacekeepers who formed part of the UNAMIR mission in Rwanda during the genocide saved many Rwandan lives through negotiation alone (Dallaire, 2004). UNAMIR peacekeepers assigned to protect government leaders regularly resulted in *génocidaires*, usually in groups of 20 or 30, backing away from the area (Des Forges, 1999). UNAMIR had neither the means nor the mandate to intervene in the killings, yet they are credited with protecting the lives of approximately 30,000 Rwandans (Cameron, 2013).

Peacekeepers have been unable to prevent or stop violent conflict and genocide in Africa, but they can be credited with improving some of the worst conditions, despite a dearth of sufficient resources, personnel and mandates.

Aid and conflict environments

As previously noted, the proliferation of peace operations on the African continent over the past quarter-century is the most observable international response to Africa's wars. Another crucial international response to Africa's wars has been aid, both humanitarian and developmental aid. However, humanitarian assistance and development policies in Africa's conflict zones were rarely grounded in effectual conflict resolution strategies. Whereas the objective of humanitarian assistance was to ease the most dire and immediate effects of armed conflict, international development policies had the objective of minimising the risks of war and formed part of a longer-term project. By the end of the Cold War, international humanitarianism and international development had morphed into a practice that included a greater range of actors at both local and national levels, including regional authorities, transnational firms, NGOs and philanthropists, whereas previously it was the sole remit of governments and international organisations (Riddell, 2007).

Although the primary objectives of those providing aid was officially poverty alleviation and encouraging impartial economic growth, many proponents of development identified aid as a factor in the prevention of violent conflict and a significant component in long-term conflict prevention strategies. Their reasoning was simple: if civil war and violent conflict was 'overwhelmingly a phenomenon of low-income countries', then Africa's civil wars must have occurred in no small part because the continent was poor (Collier and Hoeffler, 1998: 568–571). Their rationale was that if poverty increased the likelihood of war, then development could provide a solution. In some respects, there was a basis for such a claim, since within quantitative literature, it is generally acknowledged that globally high-income states are much less likely to experience violent intra-state conflict than low-income states (Dixon, 2009). Literature with a focus on civil war in Africa between 1965 and 1999 also established a relationship between poverty and civil war (Collier and Hoeffler, 2002). It is therefore of much concern that the numbers of people living with less than US$1.24 per day is rising (DFID, 2009). Such conclusions are corroborated by statistics on African nations compiled by the WB, with figures indicating that development policies are relevant to issues of peace and conflict.

The relationship between development activities and Rwanda's civil war and genocide is a useful case to illuminate the relatedness of development policies and conflict. When the four-year low-level civil war began in October 1990, Rwanda had the eighteenth lowest GNI per capita on the continent. Between 1980 and 1990, Rwandan gross national income (GNI) per capita increased gradually from US$270 to US$350 (World Bank, 2017). However, the country's Human Development Index (HDI) score fell from 0.299 in 1985 to 0.233 in 1990 (UNDP, 2013). Thus, although it is indisputable that Rwanda was one of the world's poorest countries, whether it experienced a clear reduction in prosperity levels in the decade preceding the genocide is debatable. Furthermore, before the onset of civil war, international development actors were already comprehensively invested in Rwanda, with approximately 200 donors already in the country (Uvin, 1998). Within 12 months of the onset of civil war, these international development actors were providing over US$600 million in aid to Rwanda (Uvin, 1998). Such a large international presence indicates that a relationship existed between international development policies and the onset of civil war and genocide in Rwanda. What is not clear is the exact nature of the relationship.

Some commentators have claimed that in the years preceding the civil war and genocide, the institutions of the Rwandan state were destabilised by structural adjustment programmes (SAPs), which destroyed economic activity within the country and led to both unemployment and famine. This then provided conditions in which genocide became possible (Chossudovsky, 2003). However, this is disputed by those who argue that, contrary to claims that Rwanda was a weak state in the years preceding the civil war and the genocide, it was in fact a strong functional state that directed violence towards a subdivision of its own people for political purposes. These analysts claim that the view that the Rwandan state was weak fails to understand the central role of state power in the orchestration of the genocide. What is therefore important to understand is how external agents of development supported 'a repressive and ultimately genocidal state apparatus' (Storey, 2001: 366). It has been noted that those actors of international development in Rwanda sustained the Rwandan government, and thereby allowed 'the government machinery to exist, to expand, to control, to implement' (Uvin, 1998: 226–227). This analysis suggests that violent conflict in Rwanda, culminating in genocide, had its roots in a complex web of exclusion, inequality, marginalisation and racism, facilitated by international development policies that exacerbated these issues rather than abating them (Unvin, 1998: 231). During the 1980s, Rwanda suffered a critical economic crisis that was made worse by the 1986–1987 drop in world coffee prices, which resulted in IFIs, and particularly the World Bank, developing a close relationship with the regime of President Habyarimana. Between 1982 and 1987, World Bank figures indicate that foreign assistance to Rwanda accounted for over 70 per cent of public investment, and this figure continued to rise in the new decade (Uvin, 1998; Storey, 2001). During the early 1990s, international development workers were aware of gross human rights abuses and massacres being orchestrated by the Rwandan state (Adelman and Suhrke, 1996). Despite this, the IFI's policies identify an assumption that the Rwandan government was a benign regime focused on national development, albeit in challenging circumstances, with the IMF subsequently bestowing on Rwanda a US$41 million credit agreement to finance a three-year economic reform programme while the World Bank granted structural adjustment credit (SAC) in June 1991 (Uvin, 1998). That the World Bank and others were of the belief that Rwanda was a concerned state endeavouring to foster national development in difficult conditions allowed the Rwandan government to divert monies awarded by IFIs for commodity imports towards the acquisition of military hardware. The accounting strategies of the IFIs allowed the regime to finance massive arms purchases that would later be used in the 1994 genocide. For example, the Transport Ministry's account was used to purchase trucks imported for the army, while the Health Ministry's account was used to purchase fuel that was in fact intended for use by militia and army vehicles (Toussaint, 2004). Although the IMF and the World Bank put a hold on lending to the government of Rwanda in early 1993, the large sums of money already held in accounts in foreign banks was not frozen, and the Habyarimana regime was able to use these assets to purchase additional arms and weapons that were then distributed during the organisation of the Rwandan genocidal armed forces. It is arguable that the policies and practices of external controlling bodies, but most especially the IMF and the World Bank, facilitated rather than constrained the criminal activities of the government of Rwanda in the preparation and execution of the Rwandan genocide, while the aid industry 'closed its eyes to the racist currents in society' (Uvin, 1998: 8). However, in the context of civil war and the resultant dearth of reliable intelligence on the machinations of the Rwandan state, the IFIs were powerless to block their funds being spent on military hardware. The World Bank was aware of a rise in the government's defence expenditure from 45 per cent to 64 per cent of the budget between 1985 and 1992 (World Bank, 1994), which left them with the dilemma of having to decide whether to continue providing financial support to the government or not. Their decision to continue to provide financial support to the regime of Juvenal Habyarimana was made with the optimism that the health and education sectors would benefit from their investment, even if minimally. The alternative was to suspend the credit agreement programme in its entirety until peace had been negotiated, which would have negated the chance of any financial benefits reaching the health and education sectors.

In sum, since the 1980s, economically weak countries in Africa have had to depend greatly on IFIs for aid and development finance. However, the political economy of development and IFIs 'builds a process in which poverty is, in a counterintuitive sense, not reduced but embedded and reproduced' (Bracking, 2009: xiii). That a relationship exists between violent intra-state conflict and policies of IFIs is clear; matters of structural adjustment and development cannot be divorced from political issues of conflict, and in the case of Rwanda 'the way development was defined, managed and implemented was a crucial element in the creation and evolution of many of the processes that led to genocide' (Uvin, 1998: 3). Albeit a somewhat extreme case, it is unfortunately not exceptional, with much evidence being available of development actors being a significant factor in the escalation of violence across Africa.

Conclusion

Rather than being a ladder out of conflict, this chapter has highlighted how development can contribute to conflict in Africa. Many of the social processes linked with development create conditions in which conflict is less easily restrained or is more easily escalated. Furthermore, there is growing consent among academics, policymakers and others that there is a correlation between development and conflict escalation, maintenance and transformation. The precise nature of these linkages is yet to be determined. Current ongoing violent conflicts in Africa, for example in the Sahel region affecting the West African countries of Niger, Mali, and Mauritania, and low-intensity conflicts surging within notably stable countries such as Nigeria, Ghana and Senegal, are often hinged on several factors, including poverty, human rights violations, bad governance and corruption, ethnic marginalisation, and small arms proliferation. Although many actors, including the AU, the Economic Community of West African States (ECOWAS), civil society and the international community, have been making efforts, conflicts continue to persist. The poor understanding of the complex roots of Africa's violent conflicts ensures that the continent is likely to continue to experience and suffer from these violent intra-state wars.

References

Adelman, H. and Suhrke, A. (1996) 'Early warning and response: why the international community failed to prevent the genocide', *Disasters*, 20(4): 295–304.

Appleby, R.S. (2000) *The Ambivalence of the Sacred: Religion, Violence and Reconciliation*, Lanham, MD: Rowman & Littlefield.

Ballentine, K. (ed.) (2003) *The Political Economy of Armed Conflict: Beyond Greed and Grievance*, Boulder. CO: Lynne Rienner Publishers.

Bellamy, A.J. and Williams, P.D. (2010) 'Protection of civilians in uncivil wars', in S.E. Davies and L. Glanville (eds), *Protecting the Displaced*, Leiden: Brill, pp. 127–162.

Bellamy, A.J., Williams, P.D. and Griffin, S. (2010) *Understanding Peacekeeping*, Cambridge: Polity.

Berdal, M. (2005) 'Beyond greed and grievance – and not too soon', *Review of International Studies*, 31(4): 687–698.

Bøås, M. and Dunn, K.C. (2013) *Politics of Origin in Africa: Autochthony, Citizenship and Conflict*, London: Zed Books.

Boutros-Ghali, B. (1992) *An Agenda for Peace: Preventive Diplomacy, Peace-Making, and Peace-Keeping – Report of the Secretary-General Pursuant to the Statement Adopted by the Summit Meeting of the Security Council on 31 January 1992*, New York: United Nations.

Bracking, S. (2009) *Money and Power*, New York: Pluto Press.

Bratton, M. and van de Walle, N. (1994) 'Neopatrimonial regimes and political transitions in Africa', *World Politics*, 46(4): 453–489.

Brown, M.E. (1997) 'The causes of ethnic conflict: an overview', in M.E. Brown, O.R. Cote, S.M. Lynne-Jones and S.E. Miller (eds), *Nationalism and Ethnic Conflict*, Cambridge, MA: MIT Press, pp. 3–25.

Butler, S.D. (2016) *War Is a Racket: The Antiwar Classic by America's Most Decorated Soldier*, New York: Skyhorse Publishing.

Cameron, H. (2013) *Britain's Hidden Role in the Rwandan Genocide: The Cat's Paw*, London: Routledge.

Caplan, G. (1994) *Rwanda: The Preventable Genocide*, Report of the International Panel of Eminent Personalities to Investigate the 1994 Genocide in Rwanda and the Surrounding Events, Addis Ababa: OAU.

Chossudovsky, M. (2003) *The Globalization of Poverty and the New World Order*, Montreal: Global Research, Centre for Research on Globalization.

Clapham, C. (1985) *Third World Politics: An Introduction*, London: Helm.

Collier, P. (2003) *Breaking the Conflict Trap: Civil War and Development Policy*, Washington, DC: World Bank.

Collier, P. and Hoeffler, A. (1998) 'On economic causes of civil war', *Oxford Economic Papers*, 50(4): 563–573.

Collier, P. and Hoeffler, A. (2002) 'On the incidence of civil war in Africa', *Journal of Conflict Resolution*, 46(1): 13–28.

Collier, P. and Hoeffler, A. (2004) *Greed and Grievance in Civil War*, World Bank Policy Research Working Paper 2355.

Collier, P., Hoeffler, A. and Rohner, D. (2009) 'Beyond greed and grievance: feasibility and civil war', *Oxford Economic Papers*, 61(1): 1–27.

Connor, W. (1994) *Ethnonationalism: The Quest for Understanding*, Princeton, NJ: Princeton University Press.

Cordell, K. and Wolff, S. (eds) (2016) *The Routledge Handbook of Ethnic Conflict*, London: Routledge.

Cramer, C. (2006) *Civil War Is Not a Stupid Thing*, London: Hurst & Company.

Dallaire, R. (2004) *Shake Hands with the Devil: The Failure of Humanity in Rwanda*, Toronto: Random House.

Daloz, J.P. (2003) '"Big Men" in sub-Saharan Africa: how elites accumulate positions and resources', *Comparative Sociology*, 2(1): 271–285.

Denny, E.K. and Walter, B.F. (2014) 'Ethnicity and civil war', *Journal of Peace Research*, 51(2): 199–212.

Des Forges, A. (1999) *Leave None to Tell the Story: Genocide in Rwanda*, New York: Human Rights Watch, available at www.hrw.org/reports/1999/rwanda/Geno15-8-01.htm#P36_10069 (accessed 8 June 2017).

DFID (Department for International Development) (2001) *The Causes of Conflict in Africa: Consultation Document*, London: DFID.

DFID (Department for International Development) (2009) *Eliminating World Poverty: Building Our Common Future*, Cm 7656, London: DFID.

Dixon, J. (2009) 'What causes civil wars? Integrating quantitative research findings', *International Studies Review*, 11(4): 707–735.

Eisenstadt, S.N. (1973) *Traditional Patrimonialism and Modern Neopatrimonialism*, London: Sage.

Englebert, P. (2000) 'Pre-colonial institutions, post-colonial states, and economic development in tropical Africa', *Political Research Quarterly*, 53(1): 7–36.

Erdmann, G. and Engel, U. (2007) 'Neopatrimonialism reconsidered: critical review and elaboration of an elusive concept', *Commonwealth & Comparative Politics*, 45(1): 95–119.

Fithen, C. (1999) *Diamonds and War in Sierra Leone: Cultural Strategies for Commercial Adaptation to Endemic Low-Intensity Conflict*, London: Department of Anthropology, University College London.

Franke, B. (2009) *Security Cooperation in Africa: A Reappraisal*, Boulder, CO: First Forum Press.

Fujii, L.A. (2011) *Killing Neighbors: Webs of Violence in Rwanda*, New York: Cornell University Press.

Harbom, L. and Wallensteen, P. (2009) 'Patterns of major armed conflicts, 1999–2008', in Stockholm International Peace Research Institute (ed.), *The SIPRI Yearbook 2009*, Oxford: Oxford University Press, Appendix 2, available at: www.sipri.org/yearbook/2009/02/appendix2A (accessed 8 January 2018).

HSRP (Human Security Report Project) (2008) *Human Security Brief 2007*, Vancouver: HSRP.

Hultman, L., Kathman, J. and Shannon, M. (2013) 'United Nations peacekeeping and civilian protection in civil war', *American Journal of Political Science*, 57(4): 875–891.

Human Security Centre (2005) *Human Security Report 2005: War and Peace in the 21st Century*, New York: Oxford University Press.

Huntington, S. (1998) *The Clash of Civilizations and the Remaking of World Order*, London: Touchstone.

IISS (International Institute for Strategic Studies) (2017) *Armed Conflict Survey 2017: Chart of Conflict*, London: IISS.

IMF (International Monetary Fund) (2016) *IMF DataMapper*, available at: www.imf.org/external/datamapper/NGDPD@WEO/OEMDC/ADVEC/WEOWORLD (accessed 6 June 2017).

International Development Committee (2006) *Conflict and Development: Peacebuilding and Post-Conflict Reconstruction*, Sixth Report of the Session 2005–2006, Vol. 1, London: HMSO.

Kaplan, R. (1994) 'The coming anarchy', *Atlantic Monthly*, February.

Keen, D. (2008) *Complex Emergencies*, London: Polity.

Keller, E.J. (2014) *Identity, Citizenship, and Political Conflict in Africa*, Bloomington, IN: Indiana University Press.

Kuwali, D. and Viljoen, F. (eds) (2014) *Africa and the Responsibility to Protect; Article 4(h) of the African Union Constitutive Act*, London: Routledge.

MacGinty, R. and Williams, A. (2016) *Conflict and Development*, London: Routledge.

Mandelbaum, M. (2004) *The Ideas That Conquered the World: Peace, Democracy, and Free Markets in the Twenty-First Century*, New York: Public Affairs.

Medard, J.F. (1982) 'The underdeveloped state in Africa: political clientelism or neo-patrimonialism?', in C. Clapham (ed.), *Private Patronage and Public Power: Political Clientelism and the Modern State*, London: Frances Pinter, pp. 162–189.

Murshed, S.M. and Tadjoeddin, M.Z. (2009) 'Revisiting the greed and grievance explanations for violent internal conflict', *Journal of International Development*, 21(1): 87–111.

Nasong'o, W.S. (ed.) (2015) *The Roots of Ethnic Conflict in Africa: From Grievance to Violence*, Basingstoke: Palgrave Macmillan.

Nhema, A.G. and Zeleza, P.T. (2008) *The Roots of African Conflicts: The Causes & Costs*, Oxford: James Currey.

Nordstrom, C. (2004) *Shadows of War: Violence, Power, and International Profiteering in the Twenty-First Century*, Berkeley, CA: University of California Press.

Posen, B. (1993) 'The security dilemma and ethnic conflicts', in M. Brown (ed.), *Ethnic Conflict and International Security*, Princeton, NJ: Princeton University Press, pp. 27–47.

Richmond, O.P. and Carey, H.F. (eds) (2005) *Subcontracting Peace: The Challenges of the NGO Peacebuilding*, Farnham: Ashgate.

Riddell, R.C. (2007) *Does Foreign Aid Really Work?* Oxford: Oxford University Press.

Roth, G. (1968) 'Personal rulership, patrimonialism, and empire-building in the new states', *World Politics*, 20(2): 194–206.

Sen, A. (2006) *Identity and Violence*, New York: Norton.

Simmel, G. (1955) *Conflict: The Web of Group Affiliations*, New York: Free Press.

Storey, A. (2001) 'Structural adjustment, state power & genocide: the World Bank and Rwanda', *Review of African Political Economy*, 28(89): 365–385.

Tajfel, H. (1978) 'Social categorization, social identity and social comparison', in H. Tajfel (ed.), *Differentiation Between Social Groups: Studies in the Social Psychology of Intergroup Relations*, London: Academic Press, pp. 61–76.

Tilly, C. (1985) 'War making and state making as organized crime', in C. Besteman (ed.) (2002), *Violence: A Reader*, New York: NYU Press, pp. 35–60.

Toussaint, E. (2004) *Your Money or Your Life! Tyranny of Global Finance*, London: Pluto Press.

UNDP (United Nations Development Programme) (2003) *Millennium Development Goals: A Compact Among Nations to End Human Poverty*, New York: UNDP.

UNDP (United Nations Development Programme) (2013) *The Rise of the South: Human Progress in a Diverse World*, New York: UNDP.

United Nations Peacekeeping (2017) *Peacekeeping Fact Sheet 31 May 2017*, available at: www.un.org/en/peacekeeping/resources/statistics/factsheet.shtml (accessed 8 June 2017).

Uvin, P. (1998) *Aiding Violence: The Development Enterprise in Rwanda*, Sterling: Kumarian Press.

van de Walle, N. (2001) *African Economies and the Politics of Permanent Crisis, 1979–1999*, New York: Cambridge University Press.

Williams, P.D. (2016) *War and Conflict in Africa*, Cambridge: Polity Press.

Willmot, H., Weller, M., Mamiya, R. and Sheeran, S. (eds) (2016) *Protection of Civilians*, Oxford: Oxford University Press.

World Bank (1994) *Rwanda: Poverty Reduction and Sustainable Growth*, Report No. 12456-RW, Washington, DC: World Bank.

World Bank (2017) *Rwanda*, DataBank Microdata Data Catalog, available at: http://data.worldbank.org/country/rwanda (accessed 8 June 2017).

Young, C. (2003) 'Explaining the conflict potential of ethnicity', in J. Darby and R. MacGinty (eds), *Contemporary Peacemaking: Conflict, Violence and Peace Processes*, Basingstoke: Palgrave Macmillan, pp. 9–18.

Zeleza, P.T. (2008) 'Introduction', in A. Nhema and P.T. Zeleza (eds), *The Roots of African Conflicts*, Oxford: James Currey, pp. 1–35.

47

Human rights in Africa

Impasse, backsliding, or forwards?

Peris Jones

Introduction

Human rights have had impact but still hold enormous potential in providing an additional political avenue down which African – and indeed all – governments and other actors can be held to account. Timely assessments of rights adjudication, for example, tend to show that while compliance is highly variable case to case, never mind country to country, the act of claiming rights both deepens democracy and also meets everyday developmental concerns at the household level (Langford *et al.*, 2017). Thinking about rights as providing tools for accountability provides a useful lens for scrutinising most of the issues discussed in this book. But in order to seize these opportunities, it is argued that shifts in prevalent thinking and focus are required. These changes must see a move away from human rights paradigms that are shorn of political edge and critique. Instead, approaches should be grounded in actually existing struggles, imbrications, practices and uses of rights in vastly different African contexts, which may, or may not, be legally driven. Rather than focus on those specific normative staples of the human rights 'machinery', namely laws, covenants, treaties, reporting mechanisms, etc.,[1] the chapter instead mainly prioritises looking at some of the most pressing challenges currently constraining deeper realisation of human rights discourse and practice experienced across the continent. The chapter discusses the social construction of rights and then some of the critiques of it as a largely state-centric project as played out in key and yet highly flawed institutions such as the International Criminal Court. The final section then suggests how an apparent impasse in human rights may be assisted by supporting more locally grounded approaches.

Challenges: locating and socialising human rights discourse

At the outset of seemingly fragmented views of 'human rights', a core understanding is useful to focus the discussion: What is required for an individual to be a moral agent and to lead a minimally good life? Such a definition inevitably invites various cultural interpretations of 'core', 'individual', 'moral agency' and whatever a 'good life' might entail. But in insisting in acting in a way that protects and enhances an individual's dignity for this 'good' life to be possible, an integral dimension is that rights holders have claims, but also duty bearers have obligations to uphold those claims, or at least to mitigate adverse impact upon them. A considerable array of international and regional human rights law has been developed in the last four decades that gives specific substance to rights in the form of standards that elaborate the rights and map out the corresponding obligations of states. Such measures have provided increasingly concrete coordinates by which to implement human rights. The operationalisation of universal human rights norms, although encompassing administrative and other measures, is often given inordinate focus within the legal sphere and with reference to the compatibility of national laws. The outcome is twofold: first, human rights practice is inherently concerned with *how* authority is exercised, rather than by the strictly democratic sense of *by* whom; second, however, there has been an inherent tendency for human rights to be dominated by legalistic approaches and policy frameworks. For many critical observers, this legal turn has encouraged depoliticisation and the danger of an overtly positivistic, normative and technocratic interpretation of rights to the exclusion of more popular politics. More recent assessments of socio-economic rights adjudication, for example, show, however, there is clearly scope to use litigation as an overtly *political* process (Langford *et al.*, 2017). But outcomes appear uneven. Even in those countries where the legislative and policy framework for the realisation of rights has to a large extent been developed, and there is relatively strong division of powers, such as South Africa, Kenya and others, implementation still requires urgent attention to political obstacles and non-compliance (Langford *et al.*, 2014; Chirwa and Chenwi, 2016).

In the last decade, an important development in human rights thinking has been to place alongside legal positivism and philosophical debates a growing body of multidisciplinary work that does not take rights as given. The anthropologist Richard Wilson (2008) calls this the importance of looking at the 'social life of rights', namely how people speak about them, reinterpret them and actually use them. In this sense, human rights are considered to be a product of social interaction, which is important because alongside identifying obstacles in implementation, there may also be unintended consequences. Without such an awareness of the social context of rights, we risk losing sight of how rights can mobilise not only those who want to enable rights, but also those who may want to constrain them or co-opt them. Appreciating the 'ideological promiscuity' of rights (Wilson, 2008) is critically important for gauging and implementing their practice and impact. Furthermore, human rights research and practice could be greatly enhanced by engaging a geographical imagination (Jones, 2012). Using the analogy of the maiden voyage of a ship, the birth of modern human rights in 1948 covered a vast expanse of an ocean of neatly contained homogenous space. As a universal abstraction, useful therefore in promising a final destination, the critical role of context, however, has been generally bereft in human rights. A geographical imagination is used in the chapter to show a fuller range of processes and complexity encompassing human rights discourse and practice. It highlights the importance of understanding the spatial dimension integral to furthering rights. Awareness across different spaces of rights-based interventions, for example, reveals the great variation that exists in barriers to invoking and enforcing rights. Spatial variation demands, by necessity, a multidisciplinary approach and frees rights from a legal-only conceptualisation, but equally can enable a powerful imbrication of law *and* space. Spatial awareness allows us also to consider the role of barriers and the reinforcing effects of geographies upon human rights. A more open-ended conceptualisation of space therefore allows for the possibility of contestation and struggle, to remake spaces and to show how a spatial imagination is understood, not only as a metaphor for framing understanding of the world, but as a real material entity. The spatial and the socio-legal is a two-way flow: spatiality is constitutive of social and legal processes, as the latter are of spatiality. The spatial construction of rights is also an important reminder that human rights mean different things and do different things at different spatial scales.

State-centric rights

One of the key spatial scales framing the construction of the modern age of human rights has been, particularly since 1948, that they are inextricably part of an international and increasingly regional state system. Historically, they were fashioned from a specific cultural, social and political context of the emergence of secular humanism from which human rights derived its moral power. That particular vision of human rights has tended to reflect what Stephen Hopgood has labelled the 'Human Rights Emporium', namely 'global' human rights as an expression of (Western) geopolitical power: '[w]ho gets to decide global rules and to define legitimate exceptions to them. This is the essence of sovereign power-setting, and breaking the rules. To become the supreme authority' (Hopgood, 2013: x).

So, back in 1948, there were only four independent African countries, three of which ratified the Universal Declaration of Human Rights (Ethiopia, Liberia and Egypt) and one abstaining (South Africa). Subsequently, over time, there are now 53 African country members of the UN, and almost all have ratified the core human rights treaties, reflecting the growing 'global' norm of human rights. There has also been an expansion of additional mechanisms, such as the creation of the International Criminal Court (ICC) in 2002.

The ICC was imbued by new legal precedents such as universal jurisdiction. It was intended as an important step in creating global accountability for crimes against humanity for political and military leaders. Initially welcomed by several African states and gradually ratified by most, many of whom had themselves emerged from gross and systematic human rights violations, it soon became apparent that some significant powers, notably the United States, opted out. The co-option of the global human rights regime by geopolitical power has therefore undermined broader purchase, precisely because of the duplicity of some states. As an example, under international law, the United States and United Kingdom-driven war in Iraq was deemed illegal, yet the ICC is preoccupied to date only with cases involving African conflicts. What should have been an important tool to deal with impunity for atrocities instead has become a political tussle with several African states now seeking to withdraw, purportedly in response to the duplicity. At a recent Assembly of State Parties to the ICC at The Hague, for example, the Kenyan government rallied African countries to seek to withdraw, and it has influenced the African Union to champion this. Others, such as South Africa, are also considering withdrawal. Indeed, the South African government was strongly criticised in June 2015 because it allowed Sudan's President Omar al-Bashir to leave the country despite an ICC arrest warrant for alleged war crimes in Darfur (Munusamy, 2015).

Yet the need for a structure such as the ICC is all too apparent because it is doubtful that a state will in effect submit itself to accountability. Let's consider Kenya, a country that ratified the Rome Statute in 2005. Just two years later, the country witnessed the worst election violence in its history, with the result being contested by the opposition and erupting into widespread violence, which was extensive and visceral: over 1,000 people died and hundreds of thousands had to relocate, often with resulting loss of property. Initially, there were demands within the country to account for the devastation. But parliament and the political establishment, though given ample opportunity to put forward its own investigation, were unable to do so due to political interference and differences. The intervention of Kofi Annan resulted in names of suspects being passed on to the ICC, who opened investigations (Mueller, 2011).

There was an internal process with the long-awaited report of the Kenya Truth, Justice and Reconciliation Commission (the Commission), but that was only delivered to Uhuru Kenyatta in 2013, then newly elected President of Kenya in that same year. The Commission provided a particularly useful window into human rights abuses. The Commission's mandate was to identify causes of violence and human rights abuses stretching from the colonial era and into the post-independence era, and especially up to the post-election violence of 2007–2008. The broad historical perspective in its investigations enabled the Commission to uncover a quite overwhelming picture of gross human rights violations, which they say have been both 'normalised and institutionalised' in Kenya. The sheer scale of violations – associated with every administration from colonialism and since independence – is staggering and provides clues as to the causes, consequences and difficulties in tackling the violence more generally.

What remains to be seen is whether the most pernicious of factors for violence, namely the ties between political economy interests, misuses of the state for economic gain, and use of militia by politicians, can be loosened. An apparent Gordian knot exists tying political and economic interests and the conflict and injustice it spawns. A fundamental dilemma is therefore what to do about the obstacles to alleviating deep-seated problems through many of the eminently sensible suggestions put forward by the TJRC. As the Commission's report shows (TJRC, 2013), arguably the scale of abuses is so vast and implicates so many important people that holding these elites to account looks as likely as sand holding water. The Commission calls boldly for a programme of implementation – including recommendations that 400 people be further investigated, and especially in recovering illegally acquired land. Three years later, none of the recommendations have been implemented. Kenya, however, is in a unique if not unenviable position, with its president, deputy president, several of its MPs and a plethora of high-ranking officials – which include the chairperson of the TJRC himself – implicated in human rights abuses and illegal land acquisition (TJRC, 2013).

No sooner had the report been submitted than had the political elite in quick response already mounted a rearguard action to discredit the findings of the report. Beyond some rather patchy media reporting and a few vocal civil society organisations, it is much harder to see where a more popular impetus for change may come from. It is also understandable that many Kenyans may be cynical about the TJRC when so many previous commissions of inquiry have come and gone, but bereft of any implementation of findings. There are also undoubtedly many who simply wish to move forward rather than dwelling on the uncomfortable past. Indeed, the emphasis upon peace and reconciliation that appeared to be a successful ingredient in the 2013 election campaign for the victors may also now be serving to close down room for those seeking justice and accountability. To compound the difficulties, there is further divisive debate about what remedy is most likely to cut the knot: internal processes such as the TJRC, or ones driven from outside such as the embattled ICC cases against Kenyatta (since dropped) and Deputy Ruto (also since dropped). Kenya agreed to send the cases to The Hague precisely because of the domestic political obstacles. But because the cases moved so slowly, by the time of the hearings in The Hague, two of the accused had been elected to political office, which has presented severe obstacles to the court proceedings. This also enabled the two to mobilise the machinery of their states against the ICC process, lobbying the African Union, and using their parliaments to invoke state sovereignty, whilst also rallying their own ethnic communities against what was constructed as the racist bias of the ICC. More germane to the cases themselves, witnesses were paid off and even disappeared and evidence withheld, which also weakened the cases. Nine years after the outbreak of the violence, nobody has been held accountable. And it is the systematic failure to end impunity that also continues to affect policy-making in other areas, such as extrajudicial executions by police of young men and increasing levels of corruption (Jones *et al.*, 2017).

As the Kenya case demonstrates, the enduring challenge is therefore the paradox that lies at the heart of the human rights system: namely, it is simultaneously dependent upon its implementation by the actions of sovereign states, yet at the same time it is heavily constrained and even held hostage by those same states because the latter are also most likely to be the violators of human rights. Perhaps rights discourse can at times be disempowering because it reifies the duty bearer and de-emphasises the agency of the claimant. It is an imbalance that needs urgently to be rectified.

In another example, during recent negotiations on the resolution entitled 'Recognising the Role of Human Rights Defenders and the Need for Their Protection' at the UN General Assembly, of 14 countries opposing the resolution, it was disconcerting to see, alongside China and Russia, six African countries: Burundi, Nigeria, Kenya, Zimbabwe, Sudan and South Africa (with the latter later changing to support for the resolution). The so-called Africa Group at the General Assembly opposed several provisions, including the release of persons detained or imprisoned for exercising their human rights and fundamental freedoms, and also any measures calling for states to ensure procedural safeguards for human rights defenders. The group also recommended further amendments, such as deleting text condemning use of social media to threaten human rights defenders, their families and representatives. Voting at the UN level of the 'Emporium', of course, tends to reflect the self-interests of state and regional bloc sovereignty. Though hugely disappointing, it is perhaps not surprising that Kenya was particularly influential in offering its support to the group of 14, and to the Africa Group in particular, because many of those, like Kenya, are seeking to withdraw from the ICC. Indeed, for Kenya, whose president and deputy president, as mentioned, were charged at the ICC for crimes against humanity connected to the election violence 2007–2008, labelling the ICC as a tool of foreign (imperialist) powers also becomes a convenient manoeuvre to remove itself from scrutiny (Cheeseman *et al.*, 2014).

As the world moves to a more multipolar geopolitical situation, it is noticeable that the Human Rights Emporium is weakening. There have always been variants of regional particularisms that critique the universalism implicit in human rights (most notably the Asian values debate of the 1990s). But more broadly, there is a pushback on rights in several African countries, with crackdowns on media freedom, civil society organisations and, in addition, with attempts to alter constitutional provisions for the length of term for a serving president. Again, most topically, in the Great Lakes, Burundi's incumbent President Pierre Nkurunziza is pushing for a third term, an insistence that is threatening to unravel the entire country, and which the African Union is seeking to intervene in, though this is still rejected by the president. In Rwanda, Paul Kagame, depending on one's view, is either vilified as increasingly dictatorial or upheld (by donors, at least to date) as a model modern African state leader whose impetus for stability and modernisation seemingly permits overriding the respect of human rights. Either way, Rwanda recently voted, with widespread support, to alter the terms for an incumbent president. Kagame won a third term in August 2017. Similarly, Kabila's recent manoeuvres in the Congo most likely indicate the likelihood of him standing for a third period. In Uganda and Zimbabwe, the issue of length of term in office, however, was resolved long ago, but Zambia is now also moving to a 'third' term. In Ethiopia, the progress in development, including several infrastructural projects that again invokes an old worn image of modernisation, has also seen donors and others remain silent. The regime has displaced pastoralists, and with its plans to increase the administrative sphere of Addis Ababa more recently has fuelled conflict with the Oromia ethnic group, which has resulted in hundreds killed in protests against the government. That a number of states are looking 'East' to encourage Chinese investment in Africa, evident in several countries, such as Kenya, also removes cumbersome conditionalities and tender processes, which further threatens human rights. Notably, leaders such as Kagame are currying favour due to his decision on a third term, in anticipation of Western donors cutting aid. Some countries are even adopting the Chinese currency, such as Zimbabwe. The option of looking

East (including also finance from the Gulf States) is cited as providing more choices to African states, and that it can also assist in deviating from the duplicity of the West. But undermining respect for human rights is arguably undermining democracy even as it may appear to strengthen development, particularly when we look to disaggregate 'national' growth. For all the state-centricity of the human rights 'machinery', it can appear to hold it back as much as to actually protect and promote it. Again, politics – often in the name of development 'modernisation' – rides rough shod over human rights (Likosky, 2006).

Locally grounded rights

An additional but very important consideration, given the dangers of state-centrism, is also therefore the possibility for human rights to mean something else when constructed from more locally driven understanding and uses of human rights, which may serve as an engine of accountability. As Ferguson (2006: 86) notices:

> many of the most important political processes on the continent are occurring . . . at subnational and transnational levels. The local institutions and grassroots social movements . . . must be taken seriously and understood not as regression or throwbacks, but as potentially formidable political responses to contemporary realities.

Contests over the terms, impacts and distribution of the developmental largesse of states across the continent is arguably energising politics *not* within the myopic understanding of civil society as that realm of elite NGO organisations, but rather igniting a broader understanding of 'political society' (Chatterjee, 2004). Though he exaggerates the separation of the 'civil' and 'political' spheres and he underestimates the entanglement between the two, Chatterjee's analysis is extremely useful for showing up the exclusion of grassroots issues and interests from 'the high ground of modernity'. But conversely, to maintain and renew legitimacy, it is not to the NGOs that states must pander, but rather the 'terrain of political society'. An important consideration is therefore the extent to which human rights are embedded in the terrain of the political society.

However, writing in a South Africa context, Neocosmos (2007) maintains that human rights projects can never truly exist in a discrete space outside of the state-centric constraints. He reasons that both human rights and rights-based organisations are institutionalised within states, which renders them deeply ambiguous in relation to power. In effect, social change is channelled through institutional spaces prescribed and largely controlled by the state. These critiques go to the heart of debates over tactics and terrains of struggle across African countries. As such, they also have special salience for NGOs because many of the main preoccupations of rights-based NGOs in South Africa, and to a lesser extent in Zimbabwe, Uganda, Kenya and Nigeria, for example, has been to use institutional arenas such as the legal system to secure rights. In terms of political opportunity and engagement by NGOs, Madlingozi (2014) regards the importance of the scale at which an organisation works as having an important bearing upon rights-based tactics. In other words, he notices a clear distinction existing between national-level NGOs and local grassroots level. The national-level organisations tend to be better resourced and, through elite connections, are able to harness the 'invited' civil and political spaces that exist in a country such as South Africa and, decreasingly, in Kenya. On the other hand, local counter-hegemonic organisations in the political society struggle to mobilise resources and encounter direct oppression and violations of their rights.

But is there any simple binary that exists between national- and local-level civil/political society? First, there is strong empirical evidence of how subaltern communities have worked in alliance with professionals, which may at times serve to sustain smaller organisations and movements (see Madlingozi, 2014). Notable examples also ensure against essentialising NGOs as acting solely through specific modalities or predetermined outcomes, such as their own interests alone. After all, professionals and institutional actors have been involved in various roles but preserve space for the genuine independent self-organisation of people's movements (Madlingozi, 2014). Second perhaps, therefore, the advantages of being located outside institutional spaces are in fact exaggerated and often counterproductive to building robust democracies (see Heywood, 2005). Having said this, the exclusion of the grassroots from the invited political spaces and resources of civil society is noticeable and a considerable hurdle to a more broad-based politics of rights. Third, there is a fine line that exists between autonomous struggles and those that get co-opted into technocratic, legalistic, long-winded and potentially depoliticising litigation cases. Fourth, the danger of a lack of broader popular connectivity is borne out by recent research on the human rights 'movement'. Ron *et al.* (2013), for example, demonstrate empirically, through a multi-country survey, including from Africa, in Morocco, that human rights tend to be heavily monopolised by civil society elites. Furthermore, as an ideology, they continue to be perceived as complicated and/or simply have limited purchase for the majority. Leaving aside important qualifications, such as the specific social understanding of 'human rights' posed in the survey by Ron *et al.* (2013), the fact remains that in many African countries, human rights do not currently represent a sufficiently embedded and popularised ideology. They tend to remain, therefore, somewhat detached and aloof from the real contextual everyday concerns of most poor and marginalised people. The concepts are hollowed out, which is also part of the dynamic that both fuels and is fuelled by the disjuncture between rights-based NGOs and grassroots movements. Human rights discourse can't solely be blamed for a disjuncture, of course, because it might be more to do with funding regimes, class and ethnic divisions within civil society. But it is a reminder that its own unintended outcomes should not, as mentioned earlier, be blinded by good intent. The author's recent work on extrajudicial executions committed by the police in Kenya, for example, reveals how some grassroots activists prefer to identify with a concept of *social justice* rather than rights (see Jones *et al.*, 2017).

Thus, does the essential challenge of human rights discourse and practice also become an opportunity to show how rights-based approaches can become less elitist and more popularly drawn upon? In one sense, the challenge is analytical. There is an exaggerated tendency, in some quarters, to critique rights from a cultural perspective as inappropriate. But this ignores the capacity for rights to become compatible with persistent and multiple modernities. The process of vernacularisation, for example, is useful to understanding the different culturally appropriate and local descriptions and understandings that are not at first glance an obvious rights-based language (Jones and Chingore, 2014), but which are given meaning locally, sometimes in new and surprising ways, to mobilise communities. Indeed, a key strength of human rights interventions has been when it is locally appropriated and given locally relevant meaning that can impact broader (e.g. national) scales for material gains. This approach is apparent in important successes, such as the Treatment Action Campaign (TAC) mobilisation to get access to AIDS medication in the early 2000s in South Africa. The TAC invoked important culturally relevant idioms from the anti-apartheid era through which to provide a mobilising frame. But critically, it encouraged membership engagement through locally driven branches, which gave not only scope, but also much depth of legitimacy, to its work, with a strong participatory element. But equally important were the alliances TAC forged with expert bodies and NGOs, at times using those national and global platforms provided both by the global human rights regime, but also domesticated standards through the constitutional provisions in South Africa (such as the right to health; see Jones, 2009). In this sense, transnational networks and alliances and other imbrications between global and local scales are much more apparent and much less binary than perhaps Hopgood and Chatterjee assume. The literature is increasingly peppered with many examples of local appropriation of rights discourse for progressive agendas (see Jones and Chingore, 2014).

Conclusion

Leading from the TAC example of the success in enabling policy change to broaden access to AIDS drugs, or the plethora of anti-eviction cases, whether in Kenya, Nigeria or South Africa, an emerging and urgent challenge remains as political as perhaps it is empirical: to show the extent to which rights-based approaches can contribute to opening or closing democratic space. Integral to this endeavour is to provide empirical evidence that documents actual outcomes wherever these occur. To avoid premature and myopic views of what rights do and before quick conclusions as to the overall impact they may or may not have in channelling dissent appears, it is thus important to recognise and credit advances. As and when they occur lends an urgency to find an appropriate methodology for capturing human rights impact (Langford *et al.*, 2017). Thus, whether civil society actors become complicit in managing and structuring dissent in a way that does not challenge underlying power relations requires contextual empirical analysis. In addition, theoretical tools should be used that identify when and how claims are presented as rights by political society in ways that challenge the status quo. A key debate becomes one in which the effectiveness of rights needs to be considered and what they further require for broader impact. Significant variables include the type of organisation involved and its capacity to achieve its goals. Another related consideration is that organisations seek alliances with other social forces in order for a rights-based force to be unleashed. The type of right is another key consideration. Ultimately, how rights are used in practice will tell us about how we are to enhance rights-based approaches for ending impunity and encouraging accountability in Africa.

Note

1 There is considerable coverage of this in much mainstream writing on human rights, revolving around important issues such as the impact of reporting. As one example, see the forthcoming study on the impact of reporting on the Convention on the Rights of the Child, and case studies from Africa, which highlight the political and institutionalisation obstacles to implementing child rights (Miamingi, 2016).

References

Chatterjee, P. (2004) *The Politics of the Governed: Reflections on Popular Politics in Most of the World*, New York: Columbia University Press.

Cheeseman, N., Lynch, G. and Wills, J. (2014) 'Democracy and its discontents: understanding Kenya's 2013 elections', *Journal of Eastern African Studies*, 8(1): 2–24.

Chirwa, D. and Chenwi, L. (2016) *The Promotion of Economic, Social and Cultural Rights in Africa*, Cambridge: Cambridge University Press.

Ferguson, J. (2006) *Global Shadows: Africa in a Neo-Liberal World Order*, Durham, NC: Duke University Press.

Heywood, M. (2005) 'Shaping, making and breaking the law in the campaign for a national HIV/AIDS treatment plan', in P.S. Jones and K. Stokke (eds), *Democratising Development: The Politics of Socio-Economic Rights in South Africa*, Leiden: Martinus Njihoff, pp. 181–212.

Hopgood, S. (2013) *The End Times of Human Rights*, Ithaca, NY: Cornell University Press.

Jones, P.S. (2009) *Aids Treatment and Human Rights in Context*, New York: Palgrave Macmillan.

Jones, P.S. (2012) '"Mind the gap": the politics of scale and locality in access to antiretroviral medication in South Africa', *Social Science and Medicine*, 74(1): 28–35.

Jones, P.S. and Chingore, N. (2014) 'Health rights: politics, places and the need for "sites for rights"', in M. Langford, J. Dugard, B. Cousins and T. Madlingozi (eds), *Socio-Economic Rights in South Africa: Symbols or Substance*, Cambridge: Cambridge University Press, pp. 226–252.

Jones, P.S., Kimari, W. and Ramakrishnan, K. (2017) '"Only the people can defend this struggle": the politics of the everyday, extrajudicial executions and civil society in Mathare, Kenya', *Review of African Political Economy*, DOI: 10.1080/03056244.2016.1269000.

Langford, M., Dugard, J., Cousins, B. and Madlingozi, T. (2014) *Socio-Economic Rights in South Africa: Symbols or Substance*, Cambridge: Cambridge University Press.

Langford, M., Rodriguez-Garavito, C. and Rossi, J. (2017) *Social Rights Judgements and the Politics of Compliance*, Cambridge: Cambridge University Press.

Likosky, M. (2006) *Law, Infrastructure and Human Rights*, Cambridge: Cambridge University Press.

Madlingozi, T. (2014) 'Post-apartheid social movements and legal mobilisation', in M. Langford, J. Dugard, B. Cousins and T. Madlingozi (eds), *Socio-Economic Rights in South Africa: Symbols or Substance*, Cambridge: Cambridge University Press, pp. 92–130.

Miamingi, R. (2016) *The Impact of Child Rights Reporting in Africa*, Pretoria: Pretoria University Legal Press.

Mueller, S. (2011) 'Dying to win: elections, political violence, and institutional decay in Kenya', *Journal of Contemporary African Studies*, 29(1): 99–117.

Munusamy, R. (2015) 'ICC plays hardball with South Africa on Al-Bashir', *Daily Maverick*, available at: www.dailymaverick.co.za/article/2015-10-27-icc-plays-hardball-with-south-africa-on-al-bashir/#.WNaRnhIrKAw (accessed 12 March 2016).

Neocosmos, M. (2007) 'Development, social citizenship and human rights: re-thinking the political core of an emancipatory project in Africa', *Africa Development*, 32(4): 35–70.

Ron, J., Crow, D. and Golden, S. (2013) *The Struggle for a Truly Grassroots Human Rights Movement*, available at: www.opendemocracy.net/openglobalrights/james-ron-david-crow-shannon-golden/struggle-for-truly-grassroots-human-rights-move (accessed 2 February 2014).

TJRC (Kenya Truth, Justice and Reconciliation Commission) (2013) *Full Report of the Truth, Justice and Reconciliation Commission*, available at: www.kenyamoja.com/tjrc-report/ (accessed 4 June 2013).

Wilson, R. (2008) 'The social life of human rights', *American Anthropologist*, 108(1): 77–83.

48

The International Criminal Court in Africa

Jonathan Hobson

Introduction

Entering into force on 1 July 2002, the Rome Statue established the International Criminal Court (ICC) with jurisdiction over 'the most serious crimes of concern to the international community': genocide, crimes against humanity and war crimes (ICC, 2002: 3). It is not the purpose of this chapter to explore the differences between these crimes, for such there are many detailed accounts that explain the distinctive and the circumstances in which they occur. For instance, Lemkin's (1944) essay originating the concept of genocide is important for understanding the basis of the act, and works from Schabas (2008), Totten and Bartrop (2009), Jones (2011) and Stanton (2016) contribute important interpretations of the crime. In terms of the links between genocide, war crimes and crimes against humanity, Bauman (2009), Shaw (2003, 2015), and Geras (2015) offer useful insights into broader social concepts of modernity and racism.

This chapter is concerned with the way in which the ICC has been conducting its work in Africa, and it does this in four parts. The first part details key aspects of the legislation underpinning the ICC and outlines some of the criticisms of the ICC's conduct in Africa. The second part explores the difficult relationships the court has with the African Union, discussing the impacts of this for the effective working of the ICC on the continent. The third part examines debates around the ICC's wider impact, particularly on states transitioning from violence. The chapter concludes with a short discussion on the future for the ICC in Africa, outlining some potential solutions to the difficulties it faces.

The International Criminal Court and its conduct in Africa

Although international legislation regarding serious abuses of human rights is relatively recent, the acts that constitute those crimes are not. Human history is full of incidences of widespread violence, destruction and the inflicting of intentional suffering. The ICC was an international response to this: a statement that there are some crimes so significant that, in the words of the Rome Statute, they 'shock the conscience of humanity' (ICC, 2002: 1).

The Rome Statute is a detailed document, establishing the ICC as a global legislative body with an international reach. For the purposes of this chapter, there are four significant clauses that relate to the ICC's work and the criticisms the court faces in Africa. First, the ICC is a court of last resort that works on a principle of complementarity, in other words, that it is first and foremost 'the duty of every State to exercise its criminal jurisdiction over those responsible for international crimes' (ICC, 2002: 1). The ICC will only intervene where 'the State is unwilling or unable genuinely to carry out the investigation or prosecution' (ICC, 2002: 12).

Second, states that have ratified the Rome Statute have a requirement to extradite to the ICC those on trial for crimes under its jurisdiction. Importantly, this requirement applies regardless of a defendant's position, as stated in Article 27: 'Immunities or special procedural rules which may attach to the official capacity of a person, whether under national or international law, shall not bar the Court from exercising its jurisdiction over such a person' (ICC, 2002: 18).

Third, although initially formed by the United Nations, the ICC is intended to be an independent and politically neutral body. It is managed by an Assembly of States Parties comprising representatives of the states that have ratified or acceded to the Rome Statute. It is funded by a contribution model similar to the UN in which contributions are based roughly on member states' GDP. There are 18 judges, elected by the Assembly of States Parties for nine-year terms, and currently those judges are from Argentina, the Dominican Republic, Trinidad and Tobago, Kenya, Botswana, the Democratic Republic of Congo (DRC), Nigeria, Belgium, Italy, the United Kingdom, the Czech Republic, France, Poland, Germany, Hungary, Japan and the Republic of Korea (ICC, 2016a). The ICC initiates investigations in three ways: on the basis of a referral from the UN Security Council; from a petition by a state party or by a state not part of the Rome Statute that accepts the ICC jurisdiction in the case; or through an investigation initiated by the ICC's Office of the Prosecutor (OTP) (ICC, 2016b: 17).

One hundred and twenty-four countries are parties to the Rome Statute, including 34 African states, many of whom were among the first to ratify the legislation. Table 48.1 shows African states that are members, along with their dates of ratification.

Table 48.1 African states party to the Rome Statute of the ICC

State	*Date of ratification*
Senegal, Ghana	1999
Mali, Lesotho, Botswana, Sierra Leone, Gabon, South Africa	2000
Nigeria, Central African Republic	2001
Benin, Mauritius, Democratic Republic of Congo, Niger, Uganda, Namibia, Gambia, United Republic of Tanzania, Malawi, Djibouti, Zambia	2002
Guinea	2003
Burkina Faso, Congo, Burundi, Liberia	2004
Kenya	2005
Comoros	2006
Chad	2007
Madagascar	2008
Seychelles	2010
Tunisia, Cape Verde	2011
Côte d'Ivoire	2013

Source: African Union (2015a).

Although, as Boehme (2017: 56) puts it, 'African countries were deeply involved in creating the Court and all its provisions', it is fair to say that the ICC has a fractious relationship with many of these states. One of the most significant and persistent critiques is that the court 'has preoccupied itself with Africa and failed to investigate equally severe conflicts elsewhere' (Bassiouni and Hansen, 2014). At its most extreme, proponents of this position argue that the ICC is one element of a neocolonial project focused on Africa.

Abdul Tejan-Cole, Executive Director of the Open Society Initiative for West Africa (OSIWA), explains how conflict over the role and perceptions of the ICC in Africa has led some to go so far as 'to accuse the Court of being a neo-colonialist institution peddling a Western agenda that seeks to control African politics through ICC investigations and prosecutions' (Tejan-Cole, 2014). Similarly, in a 2012 opinion piece, Courtenay Griffiths, the lead defense attorney for former Liberian President Charles Taylor, argued that:

> The requirement of international justice is not the *raison d'être* of the International Criminal Court at all. Instead, the court acts as a vehicle for its primarily European funders, of which the UK is one of the largest, to exert their power and influence, particularly in Africa.
>
> *(Griffiths, 2012)*

Associated with arguments of a neocolonialist agenda, although less critical in interpretation, are claims of selectivism when the ICC decides on which cases to prosecute. Knoops (2014: 325), for instance, describes how the 'Achilles heel of the ICC system revolves around the fairness of its selection process of its cases'. There are a number of factors influencing whether or not the ICC pursues a particular case, and the referral processes are clearly established in the Rome Statute. However, it is certainly true that, as with most criminal justice systems, the ICC do not have the resources to prosecute every individual involved in a particular case. In their own words:

> The Court will not be able to bring to justice every person suspected of committing crimes of concern to the international community. The prosecutorial policy of the Office of the Prosecutor is to focus its investigations and prosecutions on those who, having regard to the evidence gathered, bear the greatest responsibility for such crimes.
>
> *(ICC, 2016b: 17)*

The criticism is that selectivism extends beyond decisions *within* cases to decisions about *which* cases the ICC should prosecute. As a consequence, the concern is that geopolitical pressures mean that cases in Africa are more likely to be pursued. For Imoedemhe (2015: 80), it effectively means that 'international crimes are ignored when it is considered politically expedient to do so'.

At the other end of the debate are those such as Keppler (2012: 8), who claims that the 'characterization of the ICC as unfairly targeting Africans is not supported by the facts'. deGuzman (2014) makes a similar case, arguing that 'all of the Court's actions to date have been based on plausible interpretations of the relevant law'. The foundation of this position is the central importance of complementarity in the Rome Statue: in other words, that the ICC is only able to open cases when national governments are failing to investigate crimes in a comprehensive manner. As a consequence of this, any claims that there is an undue focus on one region 'are based on misunderstandings about the extent of the ICC's jurisdiction' (deGuzman, 2014). It is not that those taking this position ignore the distribution of ICC prosecutions in Africa, but they contend that this clustering is a consequence of 'the current situation of the African

Table 48.2 ICC legal action up to April 2017

	Africa	*Asia*	*Europe*	*North America*	*South America*
Investigations not taken to preliminary examination	0	1	0	1	1
Situations currently under investigation	9	0	1	0	0
Investigations taken to preliminary examination	4	2	3 (*) (**)	0	1
In trial or appeal	5	0	0	0	0
Cases closed or defendant found not guilty	5	0	0	0	0
Successful prosecutions (***)	4	0	0	0	0

Source: ICC (2017a).

Notes:

(*) Refers to investigation into alleged war crimes committed by United Kingdom nationals in the context of the Iraq conflict and occupation from 2003 to 2008.
(**) Refers to case opened by Comoros in respect to the 31 May 2010 Israeli raid on the humanitarian aid flotilla bound for the Gaza Strip.
(***) Includes convictions in appeal.

continent, with its wars and poverty, have as their root cause the impunity that thrives in a lack of accountability and the rule of law' (Wilson, 2008: 114).

Whether driven by neocolonialism, politically motivated selectivism or by the demands of procedural justice, as Table 48.2 shows it is clear that the ICC has undertaken the largest portion of its work in situations relating to African states.

The ICC and the African Union

The African Union have been one of the most consistent critics of the ICC's work in Africa. This came to a head in 2009, when the ICC issued a second arrest warrant for Sudanese President Omar al-Bashir, adding genocide to the existing charges of of war crimes and crimes against humanity. The African Union responded at its July 2010 summit, calling on its members not to cooperate with the ICC arrest warrant. Later that year, al-Bashir visited Chad and Kenya, both of which are signatories to the Rome Statute of the ICC, and both refused to arrest and extradite al-Bashir.

The 2010 summit of the African Union was also significant as the African Union rejected a request by the ICC to open a liaison office in Addis Ababa, Ethiopia. This was accompanied by a series of inflammatory comments directed to Luis Moreno-Ocampo, the first ICC Chief Prosecutor, whom they accused of 'making egregiously unacceptable, rude and condescending statement[s]' (Keppler, 2012: 3). These comments reflect a long-standing animosity from the African Union towards Moreno-Ocampo, whom they felt was guilty of 'failing to communicate with his African interlocutors, particularly by failing to partake in discussions with the AU' (du Plessis et al, 2016: 2). Indeed, this relationship was so bad that in 2011, Jean Ping, then chairman of the African Union, said, 'frankly speaking, we are not against the ICC. What we are against is Ocampo's justice' (BBC, 2011).

Regardless of the drivers for the animosity between the African Union and Moreno-Ocampo, it is clear that from the outset, there were significant tensions between the ICC and the African Union. In order to try to overcome these difficulties, between 2012 and 2014, a series of African Union-International Criminal Court (AU-ICC) Joint Seminars were held, the purpose of which was to bring together:

> ICC experts, AU Commission staff, as well as Ambassadors and legal advisors of the Permanent Missions of the African Union Member States in Addis Ababa [in order to] facilitate dialogue and provide an opportunity for frank and open discussions between the ICC and the AU, with a view to fostering understanding of the respective mandates of the two institutions, discussing issues of mutual interest and strengthening the relationship between the two institutions.
>
> *(African Union, 2015b)*

While this series of seminars was taking place, on 12 October 2013 at an extraordinary session in Addis Ababa, Ethiopia, the African Union produced a 'Decisions and Declarations' document on their relationship with the ICC, reiterating the:

> AU's concern on the politicization and misuse of indictments against African leaders by ICC as well as at the unprecedented indictments of and proceedings against the sitting President and Deputy President of Kenya in light of the recent developments in that country.
>
> *(African Union, 2013: 1)*

In particular, the African Union expressed concern over the two ongoing cases regarding President al-Bashir in Sudan, and President Uhuru Kenyatta and Deputy President William Samoei Ruto in Kenya. In both countries, the African Union contested ICC prosecutions on the grounds that they were a misuse of international jurisdiction and that, particularly in the case of Kenya, were a threat to 'the process of addressing the challenges leading to the post-election violence' (African Union, 2013: 1).

As part of the 2013 extraordinary session, the African Union also reaffirmed its support of national and internal law that grants immunity to heads of state and senior officials while in office, stating that 'no charges shall be commenced or continued before any International Court or Tribunal against any serving AU Head of State or Government' (African Union, 2013: 2). Furthermore, a request was made at the extraordinary session that 'any AU Member State that wishes to refer a case to the ICC may inform and seek the advice of the African Union' (African Union, 2013: 3). This move echoed an earlier draft protocol from the African Union that called for the expansion of the African courts to cover human rights abuses, including those crimes currently associated with the ICC (African Union, 2012: 17). For Abbas (2013: 934), such a prospect 'portends some troubling times for the International Criminal Court (ICC), but more so for international criminal justice in Africa'.

The African Union is not the only group to question the mandate of the ICC. The requirement to extradite and the removal of political immunity remains one of the most controversial aspects of the ICC's work. For instance, the United States of America also expresses serious concerns over the potential for politicised decision-making in the ICC: 'The Rome Statute creates a self-initiating prosecutor, answerable to no state or institution other than the Court itself. Without such an external check on the prosecutor, there is insufficient protection against politicized prosecutions or other abuses' (US State Department, 2003). Unlike many members of the African Union, however, the United States has not ratified the Rome Statute: it was the ratification and subsequent refusal to extradite that has had the most significant impact on the authority of the ICC.

In 2015, the African Union's Committee of African Ministers on International Criminal Court reiterated their commitment to suspending or deferring the proceedings against al-Bashir, with a mind to ultimately withdrawing the case. This was accompanied by a request to 'terminate or suspend the proceedings against Deputy President William Samoei Ruto of Kenya until the African concerns and proposals for amendments of the Rome Statute of the ICC are considered' (African Union, 2015c). These continued calls have had a significant impact, as since the initial arrest warrant in 2009, al-Bashir has visited eight African states that are signatories to the Rome Statute of the ICC: Chad, Kenya, Djibouti, Malawi, Nigeria, the DRC, South Africa and Uganda (Boehme, 2017: 51).

Although the anti-ICC rhetoric from the African Union has been persistent, there has not been a universal acquiescence to the African Union's positions. Many states find themselves torn between the two bodies. For instance, Boehme (2017: 52) describes how the African Union's demand for non-compliance in the arrest of Omar al-Bashir 'catapulted the South African executive into a loyalty conflict between its obligation to the African Union and its obligation to the ICC'. This conflict was played out in South Africa's Supreme Court of Appeals, which ruled that the government were in breach of their ICC obligations for failing to extradite al-Bashir during his 2015 visit. As a response to this ruling, in October 2016 the South African government issued an instrument of withdrawal from the ICC, citing that it had 'found its obligations, with respect to the peaceful resolution of conflicts, at times incompatible with the interpretation given by the ICC' (South African Government News Agency, 2016). This decision was again challenged in South Africa's High Court, which in 2017 ruled that without first seeking parliamentary approval, the move was 'unconstitutional and invalid' (BBC, 2017). In March 2017, the South African government formally revoked their withdrawal from the ICC (Mail & Guardian, 2017).

There is an argument to be made around the degree to which the African Union offers a unified voice for African states. For instance, Keppler (2012: 5) claims that the African Union 'does not reflect the range of positions that African governments have regarding the ICC'. For Keppler (2012: 7), a significant part of the pushback to the ICC in Africa comes from a targeted effort by a group of African Union members that have not ratified the Rome Statute and have 'actively sought to exploit unevenness in the application of international justice to present the ICC as a new form of imperialism that should be opposed'. du Plessis *et al.* (2016: 11) take a similar position, claiming that:

> the tendency by some on the AU side to paint the ICC as a tool of Western imperialism and as a neo-colonial project out to get Africa is misplaced and undermines the genuine support and commitment that almost two-thirds of the AU member states have demonstrated by ratifying the Rome Statute.

There is considerable evidence to suggest that the ICC has had, and continues to have, a good level of support across Africa. For instance, the critical messages delivered by the July 2010 African Union summit were by no means universally supported. Several members had 'actively sought a far more favourable text that excluded the call for non-cooperation and criticism of the prosecutor's conduct' (Keppler, 2012: 5). Although this declaration was ultimately included, a clause which called for sanctions on states that continued to cooperate with the ICC was not. Following the release of the declaration, several member states publically distanced themselves from the censure of the ICC, including Botswana and South Africa (although as the above shows, that support was relatively short-lived from the South African government).

Further support for the ICC was voiced in the aftermath of South Africa's 2016 proposed withdrawal, when a number of African Union members publicly backed the court and its work in Africa. This included Nigeria, the DRC, Côte d'Ivoire, Botswana, Tunisia, Ghana, Mali, Burkina Faso, Tanzania, Lesotho and Uganda (Human Rights Watch, 2016a).

The ICC also enjoys significant support from civil society groups and organisations, many of which have 'repeatedly collaborated on letters, analyses and meetings with officials of African ICC states parties to convey the need for strong African government support for the ICC' (Keppler, 2012: 19). In a 2016 open letter to members of the African Union, 200 civil society groups from across Africa wrote to states that are both African Union and ICC members. The open letter urged these states to publicly affirm their support for the court, and to increase their involvement with ICC activities in Africa and beyond (Human Rights Watch, 2016b).

The ICC and transitional states

Another area of the ICC's involvement in Africa that has come under scrutiny is its work in states that are transitioning from periods of significant conflict. These transitional states often face difficult challenges as they seek to rebuild political and social structures while overcoming the psychological and cultural impacts of violence.

For authors such as Okafor and Ngwaba (2015), there has been an overemphasis on the ICC as the route for achieving justice in the face of grave abuses of human rights. This overemphasis is, at least in part, due to an orthodoxy in international approaches that are 'ICC-centric', and which 'consciously or unconsciously advocates the use of the ICC far more robustly in Africa than has ever been done on any other continent' (Okafor and Ngwaba, 2015: 92). Such an approach to managing grave and historic abuses, they argue, fails to take into account the needs of societies transitioning from violence. They have four main criticisms.

First, Okafor and Ngwaba (2015) question the long-term impact of an international jurisdictional approach that bypasses a state's domestic justice system. They highlight the:

> possible negative consequences of the significant displacement or circumvention of local criminal justice institutions that occurs with the utilization of international criminal justice in situations where past crimes fall within the categories of genocide, crimes against humanity and war crimes.
>
> *(Okafor and Ngwaba, 2015: 96)*

Similar criticisms have been levelled at the precursors to the ICC. For instance, Barria and Roper (2005) discuss the efficacy of the international criminal tribunals for the former Yugoslavia and Rwanda, particularly their ability to maintain peace, provide justice and move towards national reconciliation. Indeed, they argue that it was some of these failings in the latter two tribunals that drove the creation of the ICC.

Chief Charles Achaleke Taku, a Lead Counsel for the International Criminal Tribunal for Rwanda, Special Court for Sierra Leone, and for the ICC, echoes criticisms of the ICC's ability to provide longer-term positive impacts in Africa. He argues that there is a risk that the court's approach towards Africa could be 'destabilizing and insensitive to other avenues put in place by the AU to provide African solutions to African problems' (Taku, 2014). To be effective in providing longer-term benefits to transitional states and to develop wider support with bodies such as the African Union, Taku (2014) argues that the 'ICC must be sensitive to the presence of other conflict management and conflict resolution actors in the field in Africa'. In making his case, Taku points to the truth and reconciliation processes that have formed significant parts of transitional contexts in Africa for a number of years. These were arguably most effective in Sierra Leone, but also played significant roles in post-apartheid South Africa and the Gacaca community court system used after the genocide in Rwanda.

Second, Okafor and Ngwaba (2015) identify the potential dangers of politicised justice delivered by international institutions. They argue that the international community is not consistent in its approach to achieving justice, particularly in states transitioning from conflict. To some extent, this should be expected, as the context in which conflicts occur and end are very different, although Okafor and Ngwaba (2015: 97) question whether 'a version of transitional justice tends to be despatched to weaker or less favoured states (many of which are in Africa), as opposed to their stronger or more favoured counterparts (most of which are not African)?' Claims of a politicised and inconsistent approach reflects many of the criticisms levied at the ICC from parts of the African Union. Boehme (2017: 52) argues that, in line with many member states such as South Africa, the African Union has 'a preference for regional solutions to regional conflicts, for quiet over antagonistic diplomacy toward autocratic regimes and for an anti-imperialist world order in which the global south enjoys equal status to the global north'.

Third, Okafor and Ngwaba (2015) problematise an approach sometimes taken by the international community whereby guarantees of impunity are given to certain combatants or groups in order for them to cease fighting. The example they give is of the DRC, in which approximately 330,000 fighters were given guarantees of impunity in order to expedite the end of the conflict (Okafor and Ngwaba, 2015). While this might be a realist approach that takes into account the problems of managing a transition from conflict to peace, it is unlikely, Okafor and Ngwaba (2015: 98) argue, to 'have a deterrent effect against such crimes in the future'.

Although it is true that the international community might negotiate peace deals that involve an element of immunity, it is also important to note that the first case the ICC opened was in the DRC, investigating 'alleged war crimes and crimes against humanity committed mainly in eastern DRC, in the Ituri region and the North and South Kivu Provinces, [committed] since 1 July 2002' (ICC, 2017b). The ICC's first two convictions were also from the DRC: Lubanga Dyilo, found guilty for the war crime of enlisting and using child soldiers and sentenced to 14 years in 2012; and Germain Katanga, found guilty for one count of a crime against humanity (murder) and four counts of war crimes (murder, attacking a civilian population, destruction of property and pillaging) and sentenced to 20 years in 2013 (ICC, 2017b). Other cases in the DRC include the ongoing trial of Bosco Ntaganda, alleged deputy chief of the FPLC, charged with five counts of crimes against humanity and 13 counts of war crimes; the acquittal of Ngudjolo Chui of charges of war crimes and crimes against humanity; and the outstanding arrest warrant for Sylvestre Mudacumura on nine counts of war crimes (ICC, 2017b).

As a counterpoint, Abbas (2013: 946) argues that it is precisely because of the instability in transitional states or states in conflict that the ICC should play a role in offering a path to justice. It is often unrealistic to suppose that those who are 'shielded by the deployment of raw political force' are able to be held to account by national political institutions:

> If there is any serious lesson to be learned from the cat-and-mouse game of the Ugandan regime and the LRA before the referral of the case to the ICC, it is precisely that domestic justice may not be suitable where potential culprits have as much or even more fire power than the state, or where the government is itself morally compromised.
>
> *(Abbas, 2013: 946)*

Finally, Okafor and Ngwaba (2015) argue that it is difficult to apply effective justice in situations where conflict is ongoing, where a fragile post-conflict state remains influenced by an incumbent or outgoing regime, or where a recently militarised group still holds some level of influence and control. In these circumstances, what they describe as the 'international criminal justice-laden and ICC-heavy approaches generally adopted in recent years on the African continent' (Okafor and Ngwaba, 2015: 99) may serve to entrench rather than challenge support for those accused of grave human rights abuses. This might, at least in part, be a driver for some of the anti-ICC sentiment that Keppler (2012) identifies as being present in parts of the African Union.

Although there are some criticisms of the role of the ICC in transitional states, that does not mean there are no benefits to this type of international criminal justice approach. There is a growing body of evidence suggesting that the impact of the ICC goes beyond formal prosecutions. For instance, Bassiouni and Hansen (2014) argue that the 'ICC is an institution with the capacity to change habits and outcomes'. Equally, Hyeran and Simmons (2016: 444) present the work of the ICC as more nuanced than its record of prosecutions, identifying 'multiple mechanisms – legal and social, international and domestic – associated with the ICC's authority that can potentially deter law violation in countries prone to civil violence'. In particular, they identify two potential forms of mutually reinforcing deterrents to which the ICC contributes: prosecution deterrent, both from its own Investigatory powers and its wider impact on domestic laws, and social deterrence from the mobilisation of the international community and domestic civil society (Hyeran and Simmons, 2016: 469).

Broache (2016) supports the idea that interpretations of the ICC's impact should be more circumspect, arguing that the impact of intervention can vary depending on a range of factors, including the nature of conflict at the time of intervention. Through a statistical analysis of ICC action during the conflict in the DRC, he demonstrates 'variation in the effects of prosecutions across stages of the legal process and other relevant categories such as belligerent groups, conflict situations and time' (Broache, 2016: 408). Although careful to point out that this is based on one case study area, the study does demonstrate that the ICC's involvement in Africa is neither wholly positive nor wholly negative.

The future of the ICC in Africa

There are those such as Abbas (2013: 946) who argue that the relationship between the ICC and the African Union 'was a disaster waiting to happen'. This is particularly evident in the case of al-Bashir, around which Boehme (2017: 69) argues that the African Union succeeded in creating a 'non-cooperation norm' in exercising an ICC warrant. Nevertheless, this does not mean that the relationship is irretrievable or that the ICC cannot continue to work in a productive manner across Africa. There are plenty of commentators, such as Imoedemhe (2015: 74), who believe 'the relationship is still salvageable and could be enhanced for the mutual benefit of both institutions with a view to achieving the goal of peace and security'.

One way in which the ICC and African Union might be able to develop a more positive relationship is, somewhat counterintuitively, through the empowerment of the African court to try those crimes associated with the ICC. In support of this, du Plessis *et al.* (2016: 12) argue that 'there could be a valid role for the African Court, given the right political and financial support', particularly if the court were to work closely with the ICC. One route for negotiating this could be through greater dialogue around what constitutes the principle of complementarity and the application of this in practice. For du Plessis *et al.* (2016: 7), properly and clearly articulated complementarity can be a principle through which the ICC and the African Union come to a shared understanding, and thus:

> Rather than perceiving the ICC as an instrument of global or universal (in)justice disrespectful of particularly African states' sovereignty, the very premise of complementarity allows African states to demand that the ICC defers to their competence and right to investigate international crimes.

The ICC faces a difficult role, dealing not only with cases in the aftermath of violence, but often cases occurring as part of ongoing conflict. These conflicts are complicated sociopolitical events, involving competing domestic, regional and at times international parties. It is not the job of the ICC to prevent or interrupt such conflict. While there is some truth in accusations of an overbalance from the ICC to countries in Africa, this does not mean that 'the ICC needs to stop investigating and prosecuting crimes in Africa, but that it needs to also investigate and prosecute crimes elsewhere' (Bassiouni and Hansen, 2014). An undue focus on African cases, perceived or real, does not de facto diminish the impact of the ICC's work. The mandate of the ICC, however, remains tainted in the African context. In part, this is because of a difficult relationship between the ICC and the African Union, in part because of ongoing perceptions of a tendentious focus towards Africa, and in part because of criticisms around the suitability of applying an international justice approach in transitional contexts. For as long as these issues remain unresolved, the ICC will continue to face significant challenges when pursuing its work in Africa.

References

Abbas, A. (2013) 'Prosecuting international crimes in Africa: rationale, prospects and challenge', *The European Journal of International Law*, 24(3): 933–946, DOI: 10.1093/ejil/cht042.

African Union (2012) *Executive Council Twenty-First Ordinary Session, 9–13 July 2012, Addis Ababa, Ethiopia*, available at: https://au.int/en2/sites/default/files/decisions/9652-ex_cl_dec_696-725_xxi_e_final.pdf (accessed 12 April 2017).

African Union (2013) *Extraordinary Session of the Assembly of the African Union, 12 October 2013, Addis Ababa, Ethiopia*, available at: www.iccnow.org/documents/Ext_Assembly_AU_Dec_Decl_12Oct2013.pdf (accessed 12 April 2017).

African Union (2015a) *Meeting of the Open Ended Committee of African Ministers on International Criminal Court (ICC), New York, USA*, available at: www.au.int/web/en/newsevents/13173/meeting-open-ended-committee-african-ministers-international-criminal-court-icc-new (accessed 12 April 2017).

African Union (2015b) *Fourth Annual African Union-International Criminal Court (AU-ICC) Joint Seminar*, available at: www.au.int/web/en/newsevents/13175/fourth-annual-african-union-international-criminal-court-au-icc-joint-seminar (accessed 12 April 2017).

African Union (2015c) *Meeting of the Open Ended Committee of African Ministers on International Criminal Court (ICC), New York, USA*, available at: www.au.int/web/en/newsevents/13173/meeting-open-ended-committee-african-ministers-international-criminal-court-icc-new (accessed 12 April 2017).

Barria, L. and Roper, S. (2005) 'How effective are international criminal tribunals? An analysis of the ICTY and the CTR', *International Journal of Human Rights*, 9(3): 349–368, DOI: 10.1080=13642980500170782.

Bassiouni, M. and Hansen, D. (2014) *The Inevitable Practice of the Office of the Prosecutor*, available at: http://iccforum.com/africa (accessed 12 April 2017).

Bauman, Z. (2009) 'Modernity, racism and extermination', in L. Back and J. Solmos (eds), *Theories of Race and Racism*, London: Routledge.

BBC (2011) *Africa's Fatou Bensouda Is New ICC Chief Prosecutor*, available at: www.bbc.co.uk/news/world-africa-16029121 (accessed 12 April 2017).

BBC (2017) *South Africa's Decision to Leave ICC Ruled 'Invalid'*, available at: www.bbc.co.uk/news/world-africa-39050408 (accessed 16 April 2017).

Boehme, F. (2017) '"We chose Africa": South Africa and the regional politics of cooperation with the International Criminal Court', *International Journal of Transitional Justice*, 11: 50–70, DOI: 10.1093/ijti/ijw024.

Broache, M. (2016) 'Irrelevance, instigation and prevention: the mixed effects of International Criminal Court prosecutions on atrocities in the CNDP/M23 case', *International Journal of Transitional Justice*, 10(3): 388–409.

deGuzman, M. (2014) *Is the ICC Targeting Africa Inappropriately?*, available at: http://iccforum.com/africa (accessed 12 April 2017).

du Plessis, M., Maluwa, T. and O'Reilly, A. (2016) *Africa and the International Criminal Court*, available at: www.chathamhouse.org/sites/files/chathamhouse/public/Research/International%20Law/0713pp_iccafrica.pdf (accessed 12 April 17).

Geras, N. (2015) *Crimes Against Humanity: Birth of a Concept*, Manchester: Manchester University Press.

Griffiths, C. (2012) *The International Criminal Court Is Hurting Africa*, available at: www.telegraph.co.uk/news/worldnews/africaandindianocean/kenya/9373188/The-International-Criminal-Court-is-hurting-Africa.html (accessed 12 April 2017).

Human Rights Watch (2016a) *African Members Reaffirm Support at International Criminal Court Meeting*, available at: www.hrw.org/news/2016/11/17/african-members-reaffirm-support-international-criminal-court-meeting (accessed 12 April 2017).

Human Rights Watch (2016b) *Letter to Presidents of African States Parties to the ICC*, available at: www.hrw.org/sites/default/files/supporting_resources/african_civil_society_group_letter_en.pdf (accessed 12 April 2017).

Hyeran, J. and Simmons, B. (2016) 'Can the International Criminal Court deter atrocity?', *International Organization*, 70, Summer: 443–475, doi:10.1017/S0020818316000114.

ICC (International Criminal Court) (2002) *Rome Statute of the International Criminal Court*, available at: www.icc-cpi.int/nr/rdonlyres/ea9aeff7-5752-4f84-be94-0a655eb30e16/0/rome_statute_english.pdf (accessed 12 April 2017).

ICC (International Criminal Court) (2016a) *The Judges of the Court*, available at: www.icc-cpi.int/iccdocs/pids/publications/judgeseng.pdf (accessed 12 April 2017).

ICC (International Criminal Court) (2016b) *Understanding the International Criminal Court*, available at: www.icc-cpi.int/iccdocs/pids/publications/uicceng.pdf (accessed 12 April 2017).

ICC (International Criminal Court) (2017a) *Al Bashir Case: The Prosecutor v. Omar Hassan Ahmad Al Bashir*, available at: www.icc-cpi.int/darfur/albashir (accessed 12 April 2017).

ICC (International Criminal Court) (2017b) *Situation in the Democratic Republic of the Congo, ICC-01/04*, available at: www.icc-cpi.int/drc/#cases (accessed 12 April 2017).

Imoedemhe, O. (2015) 'Unpacking the tension between the African Union and the international criminal court: the way forward', *African Journal of International and Comparative Law*, 23(1): 74–105, DOI: 10.3366/ajicl.2015.0111.

Jones, A. (2011) *Genocide: A Comprehensive Introduction*, London: Routledge.

Keppler, E. (2012) 'Managing setbacks for the International Criminal Court in Africa', *Journal of African Law*, 56(1): 1–14.

Knoops, G.-J. (2014) *An Introduction to the Law of International Criminal Tribunals: A Comparative Study*, 2nd edition, Boston, MA: Brill Nijhoff.

Lemkin, R. (1944) *Axis Rule in Occupied Europe: Laws of Occupation, Analysis of Government, Proposal for Redress*, Washington, DC: Carnegie Endowment for International Peace.

Mail & Guardian (2017) *South Africa Revokes ICC Withdrawal*, available at: https://mg.co.za/article/2017-03-08-south-africa-revokes-icc-withdrawal (accessed 16 April 2017).

Okafor, O.C. and Ngwaba, U. (2015) 'The International Criminal Court as a "transitional justice" mechanism in Africa: some critical reflections', *International Journal of Transitional Justice*, 9(1): 90–108.

Schabas, W. (2008) *Convention for the Prevention and Punishment of the Crime of Genocide*, available at: http://legal.un.org/avl/pdf/ha/cppcg/cppcg_e.pdf (accessed 12 April 2017).

Shaw, M. (2003) *War and Genocide*, Cambridge: Polity.

Shaw, M. (2015) *What Is Genocide?* 2nd edition, Oxford: Polity.

South African Government News Agency (2016) *SA Formally Withdrawing from ICC*, available at: www.sanews.gov.za/south-africa/sa-formally-withdrawing-icc (accessed 16 April 2017).

Stanton, G. (2016) *The Ten Stages of Genocide*, available at: www.genocidewatch.org/genocide/tenstagesofgenocide.html (accessed 12 April 2017).

Taku, C.A. (2014) *Has the International Criminal Court Inappropriately Targeted Africa?*, available at: http://iccforum.com/africa (accessed 12 April 2017).

Tejan-Cole, A. (2014) *Is the ICC's Exclusively African Case Docket a Legitimate and Appropriate Intervention or an Unfair Targeting of Africans?*, available at: http://iccforum.com/africa (accessed 12 April 2017).

Totten, S. and Bartrop, P. (2009) *The Genocide Studies Reader*, London: Routledge.

US State Department (2003) *Frequently Asked Questions About the U.S. Government's Policy Regarding the International Criminal Court (ICC)*, available at: http://web.archive.org/web/20080109092156/http://www.state.gov/t/pm/rls/fs/23428.htm (accessed 12 April 2017).

Wilson, T. (2008) 'The International Criminal Court: creation, competence, and impact in Africa', *African Journal of Criminology & Justice Studies*, 3(2), July: 85–123.

49 Corruption

Philip Nel

Introduction

Corruption is an umbrella term for activities that are deemed to be inappropriate or improper, not congruent with the expectations applicable to specific roles, contexts or relationships. In Africa, this includes high-level *graft*, that is, the looting of state coffers by 'stationary bandits' (Olson, 1993) such as Omar Bongo Ondimba, who ruled Gabon for 42 years, and is reported to have appropriated 25 per cent of oil-rich Gabon's GDP (Ghosh, 2013). Corruption also includes *cheating* by public office bearers, ranging from the mundane everyday bribes, in which the bribee is compensated for creatively interpreting or breaking a rule, to larger-scale buying and dispensing of political favours. Most discussions of corruption in Africa echo Joseph Nye's widely used definition of corruption as 'behaviour which deviates from the formal duties of a public role because of private-regarding (personal, close family, private clique) pecuniary or status gains; or violates rules against the exercise of certain types of private-regarding influence' (Nye, 1967: 419).

Agency does not lie exclusively with state officials and politicians, though, despite what the proponents of the current good governance programme want us to believe. Bayley's (1966: 720) definition of corruption concerns a larger group of agents. He defines corruption as 'inducement by means of improper considerations to commit a violation of duty'. This conception is echoed by Noonan (1984: xi) in his comprehensive study of one specific form of corruption, the bribe, which he describes as 'an inducement improperly influencing the performance of a public function meant to be gratuitously exercised'. This includes gain-seeking behaviour by domestic and multinational firms who exploit weakly developed governance practices on the continent to grease the wheels of business, avoid taxes, and maximise returns (Bracking, 2015). Enterprise-induced corruption is a major contributor to the large and devastating net resource outflows from the continent, amounting, according to a recent report, to US$800 billion over the period 1980–2012 (Global Financial Integrity, 2015). Bayley and Noonan's definitions also remind us of another set of agents who may have an interest in bribery: the numerous individuals who go about their daily lives and try to eke out a decent existence under conditions of scarcity, both of opportunities for advancement and of the provision of public goods such as infrastructure and the rule of law. Such scarcity is typical of modernising/transitional societies, and inducing corruption is often more an unavoidable means of survival than the result of a reflective choice. The fact that gift-giving and 'showing your gratitude' is deeply embedded in the moral economy of many cultures in Africa, and that the distinction between private gain and civic responsibility is of recent origin and weakly observed, contribute to the normalisation of behaviour that outsiders find perplexing and morally suspect (Ekeh, 1975; Olivier de Sardan, 1999; Smith, 2015). This broader appreciation of the social embeddedness of corruption acts as a useful correction to the narrow public choice approach that reduces the phenomenon to a matter of incentives and duplicating 'right' institutions.

Building up to a discussion of the determinants of corruption in the fourth section, I first discuss sources of cross-national data on corruption, those data that can be used in comparative studies and research aimed at generating substantiated general conclusions in the second section. There are also other very useful sources of information on corruption in Africa, such as country-specific anthropological and sociopolitical studies, both in English and in French (Reno, 1995; Kempe and Chikulo, 2000; Blundo and Olivier de Sardan, 2006, Smith, 2007, 2015; Ellis, 2016). Credit cannot be done to them in a short contribution such as this. Torsello and Venard (2015) provide a very good overview of the anthropological literature, while Blundo's Chapter 2 in Blundo and Olivier de Sardan (2006) is the most comprehensive review of this literature to date. The third section puts some of the sources discussed in the second section to work in reviewing what we know about the developmental consequences of corruption.

Sources

There are, broadly speaking, four types of cross-national data sources on corruption in Africa. The first comprises of surveys among the public or business enterprises of the experiences with, and perceptions of, the extent of corruption. Indices in which expert assessments (local and foreign) are combined with other sources of information (including public surveys) to produce rank-order tables of composite scores for a large number of states populate the second type. The third includes direct observations of corrupt behaviour, and the fourth the indirect statistical estimation of corruption. The latter two are relatively underutilised in the study of Africa, and I look at them very briefly, before focusing on the more prevalent types: surveys and composite indices.

A good example of the direct observation technique is the study by Sequeira and Dankov (2010), who employed researchers to shadow clearing agents who deal with customs and cargo in the ports of Maputo, Mozambique, and Durban, South Africa, both on the Indian Ocean. They recorded incidents of observed bribe payments to port and border officials in a random sample of 1,300 shipments, and found that bribes represent about 14 per cent of shipping costs for a standard container in Maputo, and 4 per cent in the case of Durban. A more indirect method to trace graft was used by Reinikka and Svensson (2004) to compare the amount of a special education block grant funding dispatched by the central government in Uganda in the mid-1990s with the actual amount received by schools. They found that schools received only 13 per cent of what was earmarked for them, with most of the money appropriated by local officials and politicians. The leakage was not uniform, though, suggesting that some schools have more bargaining power than others. Three subsequent public expenditure tracking surveys in Tanzania, Zambia and Ghana produced similar results, the authors reported in the same publication (Reinikka and Svensson, 2004), although the leakage of public funds tended to be lower where the end users were better informed about their entitlements.

An indirect estimation approach can be found in the work of Dreher *et al.* (2007), who treat corruption as a latent (unobserved) variable that can be estimated by means of a structural equation model using variables that are causally associated with corruption and a set of indicators of the extent of corruption in a society. This Multiple Indicators, Multiple Causes (MIMIC) approach has been used successfully to estimate other unobservable economic variables, such as the size of the informal economy (Schneider *et al.*, 2011). Dreher *et al.* (2007) estimate corruption scores for around 100 countries (30 from Africa) over the period 1976–1997. As two of the perception-based indices (WGI and CPI) start only in the mid-1990s, this provides an important complementary, publically accessible and replicable early data source on corruption in Africa. Their figures show that the better performers in Africa, such as Mauritius and Botswana, significantly improved their corruption scores over the 20-year period covered, while the majority of states in the region gradually went from bad to worse. Algeria, the Central African Republic, Côte d'Ivoire, Kenya, Malawi, Nigeria and Togo deteriorated significantly over this period, while scores for Egypt, Morocco and Tunisia stayed very much the same throughout. Overall, Africa's mean score worsened over the 20 years between the mid-1970s and mid-1990s, indicating a general increase in corruption.

Surveys of the public and selected groups, such as firm managers, have become frequent since the 1990s, and those conducted by the Afrobarometer project, the Global Corruption Monitor of Transparency International, and their combined efforts are increasingly being used. Their advantage is that they survey not only perceptions of corruption, but also ask questions about respondents' direct experience with corruption. One recent (2014–2015) continent-wide combined survey (of representative samples in 28 sub-Saharan states) reports that a majority of Africans regard corruption to be on the rise in their states, and they think that their governments are failing to deal with it effectively. Sierra Leone, Nigeria, Liberia and Ghana citizens were the most negative about the scale of corruption in their country (Transparency International and Afrobarometer, 2015). On the basis of answers to the experience-based questions, the report calculates that some 75 million Africans had had at least one experience with corruption in the year preceding the survey, sometimes to escape punishment by police and the courts, but in many cases just to get access to basic services such as primary healthcare, water, sanitation, electricity and education. A similar survey in North Africa and the Middle East (2014–2015) found that two-thirds of respondents blame their governments for not putting a stop to the increasing corruption (both perceived and experienced) in their countries (Transparency International and Afrobarometer, 2016). In Tunisia and Algeria, a large majority responded that corruption is getting worse, while in Egypt and Morocco only about a quarter of all respondents thought so. Studies that have used these and other experience-based results in Afrobarometer surveys conclude that the poor are particularly vulnerable to exploitation in terms of bribery (Justensen and Bjornskov, 2014), and that the prohibitive cost of alternative private sources of basic services (such as schooling, water and electricity) expose them to official whims more than citizens that can afford private alternatives (Peiffer and Rose, 2016).

Surveys of the experiences that firms operating in African states have with corruption add another important dimension to the experience-based data. The World Bank's Enterprise Surveys (WBES) conducted since the mid-2000s cover more than 10,000 firms in 33 African states, and reveal that a quarter of all firms in the region had at least one exposure to bribery involving officials. Just over 30 per cent reported that in their experience, getting a government contract is predicated on 'showing your gratitude' by means of a gift, and 38 per cent perceive corruption to be a major obstacle. Smaller firms seem to rely on bribes more than larger firms, and the more doubt there is about the fairness of the courts, the more inclined firms are to pay bribes. These surveys reveal quite a large variance across the region (see Table 49.1). According to the most recently available surveys, only 4 per cent of firms in Mauritius say that they had to face a bribe request, while the percentages for Liberia (the worst), Guinea and the Democratic Republic of Congo are 70, 60 and 56 per cent, respectively. High as these percentages in individual cases might be, these business surveys show that Africa as a whole is on average not more corrupt (in terms of bribery) than South Asia or Central Asia.

The last of our four types of sources are composite indices. These combine a variety of sources, predominantly the subjective assessments of 'experts' (such as commercial business information providers, non-governmental organisations, multilateral organisations and other public sector bodies) to arrive at country-specific relative scores of perceived corruption.Two such indices have had considerable traction, namely the Corruption Perception Index (CPI) of the global non-governmental organisation Transparency International (Lamsbsdorff, 2003) and the Control of Corruption Index that forms part of the Worldwide Governance Indicators (WGIs) developed under the auspices of the World Bank (Kaufmann *et al.*, 2007, 2010). The methodologies behind both are quite sophisticated, and the care with which a large number of sources are explored and checked do give them credibility (Meon and Weill, 2010). However, concern has been expressed that they do not relate directly to the factors causally related to corruption, and suffer from subjective biases (Dreher *et al.*, 2007; Aidt, 2009; Anderson and Heywood, 2009; Olken, 2009; Charron, 2016). Subjective biases do not render composite indices totally useless, though. It is likely that biases and measurement errors are systematically distributed across these indices, which means that they can be used in cross-country and cross-regional comparisons. They also have the added bonus of covering Africa much better than any other corruption source.

The most recent addition to expert-based perception indices of corruption is constructed by the Varieties of Democracy (VDem) team, overseen by 50 social scientists on six continents, and drawing on the expertise of 2,500 country experts. Although this data set has been used only sparingly in studies of corruption in Africa (see the Vdem website at www.v-dem.net/en/), it holds much promise, and is included in Table 49.1. The project covers 173 states, and assessments date back to 1900. One big advantage of the VDem data is that the academic experts involved distinguish between different types of corruption: judicial, executive, legislative, and corruption in the public sector, thus capturing both graft and cheating, grand and petty corruption. Indices for all four are constructed from various indicators, and all four are combined in a super index called 'political corruption' that runs from 0 to 1, with a higher score indicating more corruption.

Table 49.1 summarises information on Africa from three composite indices, and from the WBES in which firms are asked about their experiences with corruption. Means for the

Table 49.1 Corruption in Africa, 2000s and 2010s

State	*WGI 2000s*	*WGI 2010s*	*CPI 2000s*	*CPI 2010s*	*VDem 2000s*	*VDem 2010s*	*Bribe 2000s*	*Bribe 2010s*
Algeria	0.60	0.51	0.64	0.64	0.72	0.72	19.30	51.30
Angola	1.29	1.35	1.06	1.07	0.82	0.88	n/a	n/a
Benin	0.64	0.79	0.64	0.64	0.65	0.65	n/a	n/a
Botswana	0.95	−0.92	0.69	−0.71	0.23	0.24	5.30	8.40
Burkina Faso	0.21	0.48	0.44	0.49	0.68	0.65	22.70	n/a
Burundi	1.02	1.25	0.96	1.00	0.66	0.78	25.30	30.30
Cameroon	1.00	1.14	0.96	0.96	0.90	0.90	52.15	n/a
Cape Verde	0.53	−0.82	0.34	−0.39	0.16	0.16	7.00	n/a
Central African Republic	1.09	0.95	1.00	0.99	0.87	0.93	n/a	21.00
Chad	1.29	1.27	1.16	1.15	0.93	0.93	19.30	n/a
Comoros	0.78	0.69	0.83	0.87	0.79	0.78	n/a	n/a
Congo	1.01	1.16	0.98	0.98	0.90	0.92	37.50	n/a
Congo, Democratic Republic	1.36	1.34	1.07	1.06	0.89	0.88	68.90	56.80
Côte d'Ivoire	1.08	0.84	0.97	0.94	0.77	0.68	19.30	n/a
Djibouti	0.54	0.39	0.63	0.60	0.72	0.72	n/a	n/a
Egypt	0.53	0.60	0.50	0.52	0.87	0.77	n/a	n/a
Equatorial Guinea	1.54	1.60	1.09	1.09	n/a	n/a	n/a	n/a
Eritrea	0.23	0.67	0.73	0.75	0.61	0.64	n/a	n/a

State	*WGI 2000s*	*WGI 2010s*	*CPI 2000s*	*CPI 2010s*	*VDem 2000s*	*VDem 2010s*	*Bribe 2000s*	*Bribe 2010s*
Ethiopia	0.69	0.58	0.78	0.74	0.65	0.67	6.80	6.50
Gabon	0.77	0.66	0.55	0.57	0.86	0.85	23.00	n/a
Gambia	0.61	0.62	0.81	0.72	0.59	0.59	25.10	n.a.
Ghana	0.14	0.06	0.31	0.29	0.55	0.57	23.00	18.70
Guinea	0.95	1.10	1.13	1.09	0.86	0.74	60.70	n/a
Guinea-Bissau	1.06	1.24	1.04	1.01	0.85	0.86	27.60	n/a
Kenya	0.93	1.00	1.01	1.00	0.78	0.78	42.00	26.40
Lesotho	0.10	−0.17	0.46	0.43	0.44	0.41	14.60	14.60
Liberia	0.85	0.64	0.84	0.72	0.79	0.70	70.50	n/a
Libya	0.94	1.42	0.82	0.86	0.80	0.58	n/a	n/a
Madagascar	0.06	0.53	0.65	0.65	0.80	0.91	11.60	32.90
Malawi	0.65	0.53	0.64	0.57	0.65	0.69	13.70	24.20
Mali	0.48	0.69	0.62	0.64	0.74	0.72	33.30	27.30
Mauritania	0.34	0.73	0.70	0.76	0.76	0.77	52.90	28.90
Mauritius	0.48	−0.50	0.19	−0.23	0.47	0.45	3.90	n/a
Morocco	0.27	0.33	0.42	0.36	0.61	0.60	11.30	37.20
Mozambique	0.52	0.57	0.72	0.74	0.60	0.64	12.40	n/a
Namibia	0.21	−0.28	0.16	−0.17	0.39	0.39	1.20	9.10
Niger	0.83	0.62	0.80	0.79	0.64	0.60	22.00	n/a
Nigeria	1.12	1.15	1.09	1.07	0.87	0.87	40.00	28.90
Rwanda	0.27	−0.60	0.61	0.41	0.65	0.64	7.00	6.90
São Tomé and Príncipe	0.53	0.35	0.71	0.66	0.37	0.37	n/a	n/a
Senegal	0.25	0.35	0.50	0.51	0.60	0.62	21.40	11.10
Seychelles	0.23	−0.33	0.03	−0.08	0.33	0.27	n/a	n/a
Sierra Leone	0.93	0.88	0.96	0.93	0.72	0.71	18.10	n/a
Somalia	1.73	1.66	1.31	1.37	0.91	0.91	n/a	n/a
South Africa	0.34	0.06	0.19	−0.17	0.39	0.40	4.20	n/a
South Sudan	n/a	1.44	n/a	n/a	n/a	0.65	n/a	48.00
Sudan	1.28	1.24	1.07	1.11	0.75	0.77	n/a	17.60
Swaziland	0.35	0.31	0.52	0.52	0.49	0.52	5.70	n/a
Tanzania	0.55	0.72	0.70	0.70	0.56	0.60	21.70	20.80
Togo	0.92	0.99	0.79	0.81	0.78	0.77	19.10	n/a
Tunisia	0.07	0.14	−0.18	−0.15	0.73	0.47	n/a	10.30
Uganda	0.83	0.98	0.84	0.85	0.82	0.85	18.60	22.00
Zambia	0.68	0.44	0.74	0.68	0.46	0.49	7.10	15.80
Zimbabwe	1.29	1.34	0.87	0.85	0.81	0.81	n/a	16.20
Means	***0.72***	***0.64***	***0.72***	***0.65***	***0.68***	***0.67***	***23.56***	***23.65***

Sources and notes:

WGI = Worldwide Governance Indicator's measure of 'control of corruption' reversed. Higher score indicates more (perceived) corruption. This measure is standardised with a mean of zero and a standard deviation of one. A negative score thus means that a state is doing better than the mean for all the states covered by this measure (see http://info.worldbank.org/governance/wgi/index.aspx#home).

CPI = Corruption Perception Index of Transparency International, score reversed and standardised with mean of zero and standard deviation of one. Higher score equals more (perceived) corruption. A negative score thus means that a state is doing better than the mean for the all the states covered by this measure. CPI methodology changed in 2012, and therefore only data up to 2011 is used (see https://transparency.org).

VDem = Political Corruption Index of the Varieties of Democracy project. The index runs from zero to one, with higher score equalling more (perceived) corruption (see www.v-dem.net/en/).

Bribe = Incidence of bribery, expressed as the percentage of firms who experienced at least one bribery payment request. (see www.enterprisesurveys.org/).

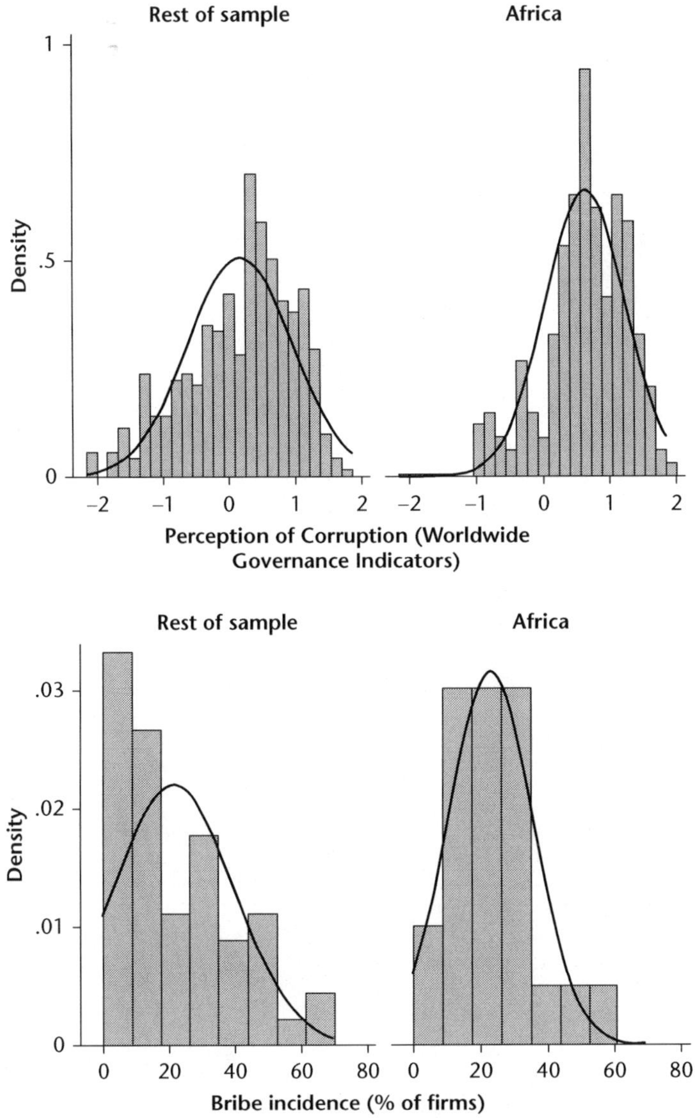

Figure 49.1 Corruption in Africa compared

Sources: Figure 49.1a is based on the 'control of corruption' measure of the Worldwide Governance Indicators, reversed so that a higher score indicates less control (= more perceived corruption), available at: www.govindicators.org/. Figure 49.1b Uses the bribe incidence measure from the World Bank's Enterprise Surveys, available at www.enterprisesurveys.org/.

Note: 'Rest of the sample' refers to 105 (Figure 49.1a) and 88 (Figure 49.1b) states, respectively, excluding Africa, Western Europe, North America, and Australia and New Zealand.

decade of the 2000s are contrasted with means for the 2010s to identify trends. Overall, levels of perceived and reported corruption have stayed high in Africa, but with improvements around the margins. Note that there is a high degree of correlation between the three composite indices (*r* of between 0.87 and 0.97), while the survey results in the last two columns have a correlation of between 0.54 and 0.61 with the composite indices (all pairwise selection).

As a whole, Africa fares poorly when compared to the rest of the developing world and the transition economies, as can be seen from the distribution graphs in Figure 49.1. The measure of (the perception) of corruption used in Figure 49.1a is the standardised version of the WGI's measure of 'control of corruption', but here reversed so that a higher score equals perception of more corruption. The mean for the period 2011–2015 is used. On the left is the distribution for 'rest of the world', that is, all states for which data exists excluding Western Europe, North America, Australia, New Zealand and Africa. In comparison, Africa is perceived to be considerably more corrupt than this larger sample of developing and transition nations. The distribution for Africa also has a relatively short tail to the left, indicating that there are comparatively fewer African states with low levels of perceived corruption (notably Botswana, Mauritius and Tunisia) than the rest of the sample. Twelve of the 20 states with the worst WGI scores in 2015 are in Africa, namely Angola, Burundi, Chad, the Republic of Congo, the Democratic Republic of Congo, Equatorial Guinea, Guinea, Guinea-Bissau, Libya, Nigeria, Somalia and South Sudan. Africa emerges somewhat better from the comparison in Figure 49.1b, which shows the distribution of bribe incidence (percentage of firms required to pay at least one bribe, mean over the period 2011–2015). Africa's distribution is more similar to the rest of the sample compared to Figure 49.1a, but with a larger density in the middle (20–30 per cent), and a thin density to the left compared to the rest. Very few firms operating in Africa report that no bribes were required, while a large number do so in the rest of the sample.

The different sources of corruption reviewed here do not map onto each other one to one, but do provide a rich vein of information that can be exploited systematically. It would be wrong to assume that we will ever have a complete and error-free picture of a murky subject such as corruption, and it is therefore important that we report measurement errors and confidence intervals. The overall picture that emerges from all the sources cited here is not flattering for the continent, although the two mostly used perception-based indices both register a marginal lowering of the average score for Africa since the 2000s. There are pockets of relatively little corruption, such as in Botswana, Cape Verde, Mauritius, Seychelles and Tunisia, but many of the worst are also to be found on the continent.

Consequences

There is a long-standing debate on whether corruption greases the gears of commerce and economic development, or whether it acts as sand that grinds and slows down these same gears. On the side of the 'greasers', Leff (1964: 8) suggests that corruption should be seen as 'an extra-legal institution used by individuals and groups to gain influence over the actions of the bureaucracy'. It provides political access to the politically weak and allows them to invest, produce, barter and trade under conditions of bureaucratic incompetence and/or policy failures. Leff (1964) and Khan (1996) both warn against using idealised benchmarks of economic efficiency against which to compare the presumed inefficiencies induced by corruption. Under certain circumstances, corruption – by getting things done – may compensate for official failings, and this is not only true of developing countries today, but was also true in nineteenth-century USA and eighteenth-century England. The bribe breaks the stranglehold of economic monopolies, cuts red tape (that is, undermines unproductive involvement of the state in micromanaging the economy), facilitates new investment and capital formation, encourages innovation, and assists in national integration (Leff, 1964: 10; Nye, 1967: 419–420). In a recent statistical study focusing on developing and developed countries, Meon and Weill (2010) find that corruption indeed greases the wheels of aggregate economic efficiency when the overall quality of public institutions is suboptimal. In a second-best institutional world, corruption keeps the wheels of commerce spinning by compensating for policy and institutional weaknesses (see also Greenstone, 1966; Choi and Thum, 2005; Hunt and Laszlo, 2012).

Those who believe that corruption acts more like sand in the gears retort that it might be true that corruption corrects for institutional failings, but it is still only a second-best alternative, and it might well contribute to the further weakening of institutions such as the rule of law and public accountability. Overall, corruption creates more inefficiency than it corrects (Aidt, 2009). It encourages the creation of 'extra' levels of regulation that create additional opportunities for rent-seeking (opportunities to maximise monopoly profits), and thus reduces overall productivity. Bayley (1966) – who we saw had some sympathy for the grease argument – also points out that corruption acts as a form of double taxation. This reduces the total amount available for public purposes because the first tax is avoided by paying the second tax (bribe) that ends up in private hands. In addition, public spending priorities might be skewed by ethnic/group favouritism that is encouraged by corruption. By using perception-based measures of corruption, Mauro (1995) reports evidence that corruption undermines growth.

The literature on Africa tends to support the view of the 'sanders' (Mbaku, 2000; Gyimah-Brempong and Munoz de Camacho, 2006). In a study that relies on the CPI for data on corruption, Gyimah-Brempong (2002) finds evidence of a significant and sizeable negative effect on economic growth, attributable directly to its lowering of productive effort and the misuse of resources, and indirectly through reductions in investment in both physical and human capital, as well as degradation of institutions. d'Agostino *et al.* (2016) add a further consideration. They show that corruption encourages public spending in state procurement projects that, by its very nature, allows for *graft* at the highest levels. The secrecy and the size of the contracts involved in military procurement, for instance, create ideal environments for bribers and bribees alike and encourage investment in the defence sector that is suboptimal from an economic growth point of view. The natural resource sector in Africa is also notorious for its rent creation potential, that is, the creation of opportunities to generate monopoly profits. Arezki and Gylfason (2013) find that higher-resource rents lead to more corruption, especially in less democratic states in Africa. They also note that these very states experience fewer internal conflicts, suggesting that the spoils of corruption might well be distributed sufficiently equally to subdue those with the potential of challenging the regime. Over and above this, the resource curse undermines economic efficiency and sustainable development (Taylor, 2016), but does not necessarily endanger political stability in more authoritarian regimes.

Perceptions of and experiences with corruption may also have a negative effect on investment decisions by foreign and domestic investors, and consequently on economic growth. The evidence is mixed, though. Quazi *et al.* (2014) find support for the greasing of the gears argument in that corruption (measured by the CPI) appears to facilitate foreign direct investment more than it discourages it. This effect is conditional on the overall institutional quality of the recipient states, which is exactly what the 'greasers' argue. Abotsi and Iyavarakul (2015) report evidence of a degree of 'corruption tolerance' on the part of foreign investors in Africa, but the 'tolerable level of corruption for investment' is low. International instruments, such as the OECD Convention on Combating Bribery of Foreign Officials and the UN Convention Against Corruption, may well be lowering it even further.

As we all know, development involves more than just economic growth. Whether corruption favours or harms *sustainable development* has not been adequately tested on the African continent. As Aidt (2009) argues, it is only when we look at indicators of sustainable development that we can settle the argument about the developmental effects of corruption conclusively. Using data from enterprise surveys, perception indices and a measure of sustainable development,

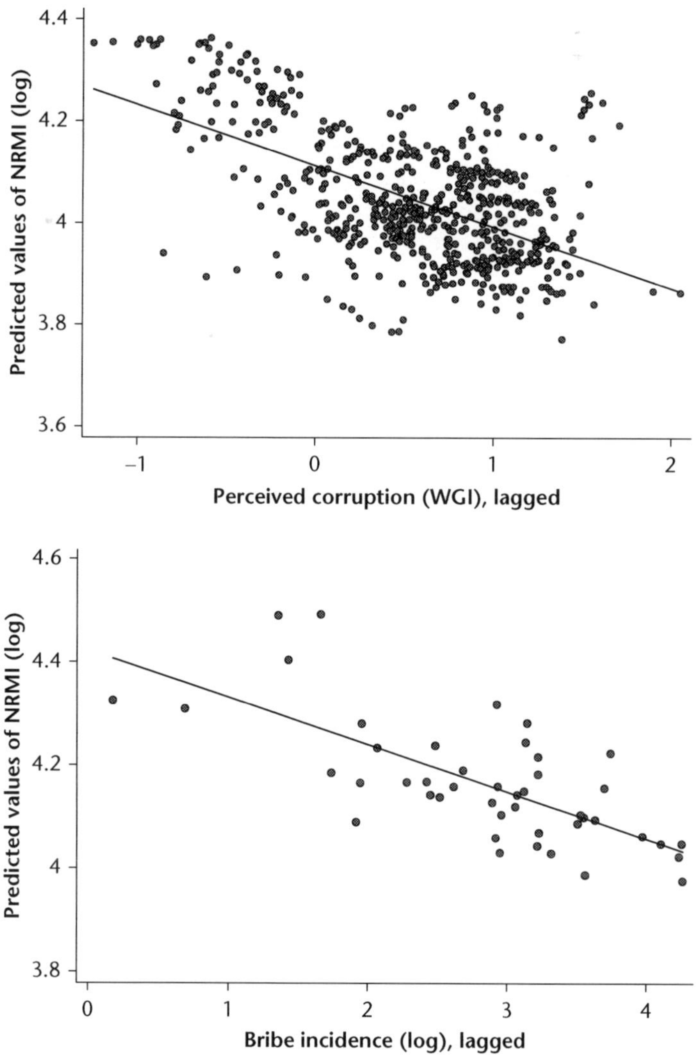

Figure 49.2 Corruption and sustainable development in Africa, 2006–2011

Sources: Figure 49.2a is based on the 'control of corruption' measure of the Worldwide Governance Indicators, reversed so that a higher score indicates less control (= more perceived corruption), available at: www.govindicators.org/. Figure 49.2b uses the bribe incidence measure from the World Bank's Enterprise Surveys, available at: www.enterprisesurveys.org/.

Note: Values of Natural Resource Management Index (NRMI) predicted in panel-clustered, random-effects regression, with GDP per capita and quality of economic institutions as regressors.

Aidt (2009) concludes that while the net effect of actual and perceived corruption on economic growth is close to zero, both are significantly and negatively related to the growth rate of 'genuine wealth', also known as 'net national savings' (Dasgupta, 2001). There are relatively few African states in his sample, though. To see if his general conclusion holds for a larger collection of African states, Figure 49.2 traces the relationship between corruption, measured in two different ways, and an indicator of sustainable development. The latter is the

100-point Natural Resource Management Index (NRMI), a composite index derived from the average of four proximity-to-target indicators for ecoregion protection (weighted average percentage of biomes under protected status), access to improved sanitation, access to improved water and child mortality. The NRMI covers the period 2006–2011, and 53 African states. A higher score records better sustainable management of the human and non-human resources (SEDAC, 2011). The two measures of corruption are the WGI, used above, and the WBES measure of the bribe incidence (the percentage of firms required to pay at least one bribe). Figure 49.2 shows a significant net negative effect of both perceived corruption (Figure 49.2a) and experience of bribery (Figure 49.2b) on predicted values of the NRMI, our measure of sustainable development. These negative effects are *net* of two other determinants of sustainable development, namely GDP per capita (from the World Bank's Development Indicators; World Bank, 2015) and the quality of economic institutions (as measured by Kuncic, 2014). Figure 49.2 implies that a 1 per cent increase in corruption in Africa on average reduces sustainable development by 0.4 per cent. Read in conjunction with Aidt's (2009) findings, Figure 49.2 tells a very cautionary tale. Note, however, that these findings do not necessarily imply that the reduction of corruption is a sufficient determinant of development. It might also well be that the reduction of corruption itself depends on the achievement of a certain standard of development.

Determinants

In the first section, we noted that the majority of Africans (in the 28 states surveyed by Transparency International and Afrobarometer in 2014–2015) identified corruption as a major problem in their states. African leaders appear to agree. In 2003, representatives of 48 member states agreed to the African Union's Convention on Preventing and Combating Corruption. Africa is clearly 'no longer at ease' (as the title of Chinua Achebe's seminal novel on bribery would have it). What can be done to address corruption depends on what one considers to be its main determinants.

There are few works that try to systematically identify determinants of corruption in the region, despite the fact that the general literature has identified a number of candidates in global cross-national studies (Treisman, 2000; Aidt, 2003; Dreher *et al.*, 2007; Treisman, 2007). The exceptions limit the analysis to tracing the effect of a single or a small number of factors (Collier, 2000; Willett, 2009) or focus more on the characteristics of firms and what makes some more inclined to bribe their way to contracts and favours than do others (Pelizzo *et al.*, 2016). Nevertheless, some useful conclusions can be drawn from the general literature. These conclusions warn that there are no quick fixes, though.

The extent of corruption is determined by long-term trends, in particular the legal tradition of a state, level of economic development, and the length of exposure to democracy (Treisman, 2000). The degree of ethnic and linguistic fractionalisation also serves to encourage and perpetuate corruption as rulers favour their own clans/tribes (Treisman, 2007). As we saw in the first section, bribers favour corruption when it is assumed to compensate for the institutional shortcomings and weak general capacity of the state. All of these factors change only very slowly.

Legal institutions based on the common-law tradition are associated with lower levels of corruption than alternative legal traditions, and Africa is no exception. Bribe incidence is on average 5 per cent and perceived corruption 3 per cent lower in African states with a common-law tradition. This tradition entails lawmaking by judges, based on precedents, rather than on civil codes developed by scholars and passed by governments (Treisman, 2000). In Africa, ex-British colonies inherited this tradition, but not all low-corruption states share this tradition (e.g. Mauritius).

Cross-national studies find that states with higher per capita income levels register lower levels of perceived and reported corruption. This effect is also noticeable in Africa, although the association is somewhat weaker. In a panel-data random-effects regression of a sample of 50 developing and transition states and covering the period 1996–2014 (excluding Africa), and controlling for the two other long-term factors discussed here (legal tradition and duration of democracy), a 1 per cent increase in per capita income reduces the incidence of bribery by 0.7 per cent, and significantly so. The corresponding reduction in 42 African states for which we have data is 0.25 per cent, a small but significant reduction. A higher level of GDP per capita accords with larger state capacity, including the ability to tax effectively, securing a larger revenue pool to provide public goods and maintain institutions that can monitor and combat corruption. The distinction between private gain and public resources tend to be more strongly developed in richer states, where there are also higher levels of urbanisation and more diffused educational opportunities. Urbanised, well-educated individuals are less tolerant of corruption. The history of corruption and its eventual decline in the US, the UK and Western Europe underlines that the reduction of corruption is a function of economic and political development, not the other way around (Noonan, 1984; Ryan, 2013). Peck emphasises that the patterns of corruption in early modern England paralleled those in contemporary developing nations, and that these patterns had distinct benefits for the centralisation of state power in tumultuous political times in England (Peck, 1993; see also Huntington, 1968).

In Africa, as in the rest of the developing world, it is the uninterrupted duration of democracy, rather than the mere event of democratisation – the introduction of competitive elections where none existed before – that makes all the difference. Again, controlling for the effects of the other two long-term factors, every additional year of uninterrupted democracy in an African state subsequently reduces the incidence of bribery by 2 per cent. Maturity of democracy brings accountability, more critical media, and restrictions on the terms of office holders – hence, less corruption. It is also accompanied by the development of modern political parties that institutionalise the aggregation of broader societal interests than those of the clan or the family (Huntington, 1968).

Specific policy decisions by state leaders can reduce or exacerbate corruption. The introduction after independence of what Collier (2000: 194) calls 'sweeping economic control regimes which gave officials astonishingly wide powers over private activity' set the scene for the flourishing of corruption (see also Riley, 1998; Mbaku, 2000). The first generation of post-independence leaders appointed large numbers of civil servants to oversee these regulations, and expanded public procurement to unprecedented levels. These leaders favoured light surveillance arrangements. For Collier and Mbaku, this means that corruption was 'made' in Africa, and that it can therefore be 'unmade' once these opportunities for rent-seeking are reduced and better scrutiny of government behaviour is put in place. Collier and Mbaku's prescriptions resonate well with the public choice understanding of corruption, in which corruption is primarily a function of the incentives faced by perpetrators and victims (see Mbaku, 2000, and for a critique, Olken and Pande, 2012). Reduce the incentives – by curtailing state procurement and its size, and streamlining tax and reducing regulatory regimes – and corruption will be tamed. The current 'good governance' consensus driven by donors, international institutions and prominent academic institutions is largely built around this narrow approach. Proponents of this approach single out foreign aid as an additional and unnecessary incentive for corruption, but the evidence is ambiguous. Some authors suggest that aid flows increase corruption (Bräutigam and Knack, 2004). However, Menarda and Weill (2016) find no causal effect in a cross section of developing counties, including African states. As noted in the first section, other transnational factors, such as the availability of means for individuals and firms to drain capital from Africa to tax havens, have a much worse effect than official development assistance could ever have.

There is no easy recipe to tackle corruption, and initiatives launched in Africa have so far met with little success. Better surveillance of national and continent-wide bodies can help, as would measures to curb the illegal drain of resources from the continent. In addition, streamlined state regulation and taxation regimes can also limit opportunities for cheating. Ultimately, though, the reduction of corruption follows from the building of state capacity and economic-political modernisation, and it seems unrealistic to try to bypass these processes. To the extent that the current fashionable good governance agenda ignores or undermines the prerequisites of state developmental capacity (Khan, 2012), it is likely to be counterproductive (Uberti, 2016).

References

Abotsi, A.K. and Iyavarakul, T. (2015) 'Tolerable level of corruption for foreign direct investment in Africa', *Contemporary Economics*, 9(3): 249–270.

Aidt, T.S. (2003) 'Economic analysis of corruption: a survey', *The Economic Journal*, 113: F632–F652.

Aidt, T.S. (2009) 'Corruption, institutions, and economic development', *Oxford Review of Economic Policy*, 25(2): 271–291.

Anderson, S. and Heywood, P.M. (2009) 'The politics of perception: use and abuse of Transparency International's approach to measuring corruption', *Political Studies*, 57: 746–767.

Arezki, R. and Gylfason, T. (2013) 'Resource rents, democracy, corruption and conflict', *Journal of African Economies*, 22(4): 552–569.

Bayley, D. (1966) 'The effects of corruption in a developing nation', *Western Political Quarterly*, 19(4): 719–732.

Blundo, G. and Olivier de Sardan, J.-P. (2006) *Everyday Corruption and the State: Citizens and Public Officials in Africa*, Cape Town: David Philip.

Bracking, S. (2015) 'Corruption and development: the mutable edges of morality in modern markets', in P.M. Heywood (ed.), *Routledge Handbook of Political Corruption*, London: Routledge, pp. 225–241.

Bräutigam, D. and Knack, S. (2004) 'Foreign aid, institutions, and governance in sub-Saharan Africa', *Economic Development and Cultural Change*, 52: 255–285.

Charron, N. (2016) 'Do corruption measures have a perception problem? Assessing the relationship between experiences and perceptions of corruption among citizens and experts', *European Political Science Review*, 8(1): 147–171.

Choi, J.P. and Thum, M. (2005) 'Corruption and the shadow economy', *International Economic Review*, 46(3): 817–836.

Collier, P. (2000) 'How to reduce corruption', *African Development Review*, 12(2): 191–205.

d'Agostino, G., Dunne, J.P. and Pieroni, L. (2016) 'Corruption and growth in Africa', *European Journal of Political Economy*, 43: 1–88.

Dasgupta, P. (2001) *Human Well-Being and the Natural Environment*, Oxford: Oxford University Press.

Dreher, A., Kotsogiannis, C. and McCorriston, S. (2007) 'Corruption around the world: evidence from a structural model', *Journal of Comparative Economics*, 35: 443–466.

Ekeh, P. (1975) 'Colonialism and the two publics in Africa: a theoretical statement', *Comparative Studies in Society and History*, 17(1): 91–112.

Ellis, S. (2016) *This Present Darkness: A History of Nigerian Organized Crime*, Oxford: Oxford University Press.

Ghosh, P. (2013) 'Gabon's Bongo family: living in luxury, paid for by corruption and embezzlement', *International Business Times*, 15 February.

Global Financial Integrity (2015) *Financial Flows and Tax Havens: Combining to Limit the Lives of Billions of People*, available at: https://eiti.org/publishers/global-financial-integrity (accessed 2 January 2018).

Greenstone, J.D. (1966) 'Corruption and self interest in Kampala and Nairobi: a comment on local politics in East Africa', *Comparative Studies in Society and History*, 8(2): 199–210.

Gyimah-Brempong, K. (2002) 'Corruption, economic growth and income inequality in Africa', *Economics of Governance*, 3(3): 183–209.

Gyimah-Brempong, K. and Munoz de Camacho, S. (2006) 'Corruption, growth, and income distribution: are there regional differences?' *Economics of Governance*, 7(3): 245–269.

Hunt, J. and Laszlo, S. (2012) 'Is bribery really regressive? Bribery's costs, benefits, and mechanisms', *World Development*, 40(2): 355–372.

Huntington, S. (1968) *Political Order in Changing Societies*, New Haven, CT: Yale University Press.

Justensen, M. and Bjornskov, C. (2014) 'Exploiting the poor: bureaucratic corruption and poverty in Africa', *World Development*, 58: 106–115.

Kaufmann, D., Kraay, A. and Mastruzzi, M. (2007) *The Worldwide Governance Indicators Project: Answering the Critics*, Washington, DC: World Bank.

Kaufmann, D., Kraay, A. and Mastruzzi, M. (2010) *The Worldwide Governance Indicators: Methodology and Analytical Issues*, Washington, DC: World Bank.

Kempe, R.H. and Chikulo, B.C. (eds) (2000) *Corruption and Development in Africa: Lesson from Country Case-Studies*, Houndmills: Macmillan.

Khan, M. (1996) 'The efficiency implications of corruption', *Journal of International Development*, 8(5): 683–696.

Khan, M. (2012) 'Governance during social transformations: challenges for Africa', *New Political Economy*, 17(5): 667–675.

Kuncic, A. (2014) 'Institutional quality dataset', *Journal of Institutional Economics*, 10(1): 135–161.

Lamsbsdorff, J.G. (2003) *Background Paper to the 2003 Corruption Perceptions Index*, September, Transparency International.

Leff, N.H. (1964) 'Economic development through bureaucratic corruption', *American Behavioral Scientist*, November: 8–14.

Mauro, P. (1995) 'Corruption and growth', *The Quarterly Journal of Economics*, 110(3): 681–712.

Mbaku, J.M. (2000) *Bureaucratic and Political Corruption in Africa: The Public Choice Perspective*, Malabar, FL: Krieger.

Menarda, A.-R. and Weill, L. (2016) 'Understanding the link between aid and corruption: a causality analysis', *Economic Systems*, 40: 260–272.

Meon, P.-G. and Weill, L. (2010) 'Is corruption an efficient grease?', *World Development*, 38(3): 244–259.

Noonan, J.T. (1984) *Bribes*, London: Collier Macmillan.

Nye, J. (1967) 'Corruption and political development: a cost–benefit analysis', *American Political Science Review*, 61(2): 417–427.

Olivier de Sardan, J.-P. (1999) 'A moral economy of corruption in Africa?', *The Journal of Modern African Studies*, 37(1): 25–52.

Olken, B. (2009) 'Corruption perceptions versus corruption reality', *Journal of Public Economics*, 93: 950–964.

Olken, B. and Pande, R. (2012) 'Corruption in developing countries', *Annual Review of Economics*, 4: 479–509.

Olson, M. (1993) 'Dictatorship, democracy and development', *American Political Science Review*, 87: 567–576.

Peck, L.L. (1993) 'Corruption and political development in early modern Britain', in A.J. Heidenheimer, M. Johnston and V.T. LeVine (eds), *Political Corruption: A Handbook*, London: Transaction Publishers, pp. 219–231.

Peiffer, C. and Rose, R. (2016) 'Why are the poor more vulnerable to bribery in Africa? The institutional effect of services', *The Journal of Development Studies*, DOI: http://dx.doi.org/10.1080/00220388.2016.1257121.

Pelizzo, E.A., Pak, A. and Xun, W. (2016) 'Determinants of bribery: theory and evidence from sub-Saharan Africa', *Africa Development Review*, 28(2): 229–240.

Quazi, R., Vemuri, V. and Soliman, M. (2014) 'Impact of corruption on foreign direct investment in Africa', *International Business Research*, 7(4): 1–9.

Reinikka, R. and Svensson, J. (2004) 'Local capture: evidence from a central government transfer program in Uganda', *Quarterly Journal of Economics*, 119(2): 679–705.

Reno, W. (1995) *Corruption and State Politics in Sierra Leone*, Cambridge: Cambridge University Press.

Riley, S.P. (1998) 'The political economy of anti-corruption strategies in Africa', *The European Journal of Development Research*, 10(1): 129–159.

Ryan, A. (2013) 'Conceptions of corruption, its causes, and its cures', *Social Research*, 80(4): 977–992.

Schneider, F., Buehn, A. and Montenegro, C. (2011) 'Shadow economies all over the world: new estimates for 162 countries from 1999 to 2007', in F. Schneider (ed.), *Handbook on the Shadow Economy*, Cheltenham: Edward Elgar, pp. 9–77.

SEDAC (Socioeconomic Data and Applications Center) (2011) *Natural Resource Management Index (NRMI)*, available at: http://sedac.ciesin.columbia.edu/data/collection/nrmi (accessed 2 January 2018).

Sequeira, S. and Djankov, S. (2010) *An Empirical Study of Corruption in Ports*, MPRA Paper, Munich, Germany, available at: https://mpra.ub.uni-muenchen.de/21791/1/MPRA_paper_21791.pdf (accessed 2 January 2018).

Smith, D.J. (2007) *A Culture of Corruption: Everyday Deception and Popular Discontent in Nigeria*, Princeton, NJ: Princeton University Press.

Smith, D.J. (2015) 'The contradictions of corruption in Nigeria', in P.M. Heywood (ed.), *Routledge Handbook of Political Corruption*, London: Routledge, pp. 56–66.

Taylor, I. (2016) 'Dependency redux: why Africa is not rising', *Review of African Political Economy*, 43(147): 8–25.

Torsello, D. and Venard, B. (2015) 'The anthropology of corruption', *Journal of Management Inquiry*, 25(1): 34–54.

Transparency International and Afrobarometer (2015) *People and Corruption: Africa Survey 2015*, available at: www.transparency.org/whatwedo/publication/people_and_corruption_africa_survey_2015 (accessed 2 January 2018).

Transparency International and Afrobarometer (2016), *People and Corruption: Middle East and North Africa*, available at: www.transparency.org/whatwedo/publication/people_and_corruption_mena_survey_2016 (accessed 2 January 2018).

Treisman, D. (2000) 'The causes of corruption: a cross-national study', *Journal of Public Economics*, 76: 399–457.

Treisman, D. (2007) 'What have we learned about the causes of corruption from ten years of cross-national empirical research?', *Annual Review of Political Science*, 10: 211–244.

Uberti, L.J. (2016) 'The "sociological turn" in corruption studies: why fighting graft in the developing world is often unnecessary, and sometimes counterproductive', *Progress in Development Studies*, 16(3): 261–277.

Willett, S. (2009) 'Defence expenditures, arms procurement and corruption in sub-Saharan Africa', *Review of African Political Economy*, 36(121): 335–351.

World Bank (2015) *World Development Indicators 2015*, Washington, DC: World Bank Group, available at: http://data.worldbank.org/data-catalog/world-development-indicators (accessed 2 January 2018).

Section 11
Conclusion

50

Conclusion

Looking forward

Tony Binns, Etienne Nel and Kenneth Lynch

This handbook has considered a wide range of issues that will likely affect the African continent for much of the rest of this century. However, the contributions to the handbook indicate that the Africa that reaches the end of the twenty-first century will be very different from the one that entered it. Much theory about contemporary social, economic and environmental change is based on the analysis of evidence from the past. Africa has been described as 'an experiment' in politics, modernism and economic policy (Bratton and van de Walle, 1997; see also Ndlovu-Gatsheni, Chapter 3, this volume), a part of the world where practitioners have tried out their theories – some of which have been developed within Africa, but many have been introduced or imposed from outside (Brookings Institute, 2017). This concluding chapter sets out to briefly discuss key patterns that appear to be emerging and are likely to influence the future.

Africa is experiencing a number of significant changes. Some of these are common to all or most countries, while some are more pronounced in some countries than others. For example, all countries in the continent are experiencing very high levels of population growth, in particular urban growth. There is a strong trend towards urbanisation and a progressive shift from a predominantly rural population to a predominantly urban population. This demographic change is accompanied by a number of competing cultural and economic trends. For example, the gradual shift to more urban-based population is bringing a large proportion of rural-born people into the city, many of whom retain strong cultural, social and political attachments to their rural origins. As the cities mature, so these cultural attachments to the countryside are beginning to break down, or are at least becoming more complex. A more urbanised population means a population that is more politically active and organised, a population that generally has a greater access to public services, a higher reliance on cash, but this means a reduced direct reliance on ecosystem services.

The year in which this volume is published has marked several significant elections in the continent. Ideally, each democratic election that is successfully completed should bring the continent closer to stable continent-wide governance. However, for each democratic election, there are also political leaders who desire to cling on to power for as long as they possibly can.

It is not unusual in Africa and elsewhere to have some political regimes that retain power at the expense of a democratic expression of the populace. However, African countries are beginning to develop a capacity for democracy, and there are some political leaders who are modelling good practice. African leadership shown by some organisations such as the African Union (AU) and the Economic Community of West African States (ECOWAS) has been important in resolving conflicts and leading peacekeeping interventions. As has been argued in this volume, the emergence of strong democratic governance on the continent is the key to peaceful resolution of conflict, the ending of famine and the sustained reduction of poverty. During the production of this volume, several conflicts came to an end, some were ongoing throughout the writing, while in some cases situations have undoubtedly deteriorated. While Africa is generally experiencing reducing levels of conflict, there are a number of conflicts that appear to be entrenched, or symptomatic of states that are fragile and vulnerable, and which could easily slip back into violent conflict. While there are emerging examples of African solutions for African conflicts, sadly there have been some controversial interventions, often linked to furthering economic interests, and especially focused around acquiring valuable natural resources (see Cameron, Chapter 46, this volume).

A stronger Africa could provide alternative models of development for the rest of the world. If the optimistic examples of African resilience, innovativeness and creativity are allowed to persist, they could be models that may provide positive influence around the world. Some international observers were heartened by such examples as Burkina Faso's 'citizens' broom' (Myers, Chapter 35, this volume) and the Arab Spring uprisings in Tunisia and Egypt. However, the less successful aftermath of the Libyan overthrow of Muamar Gaddafi, and the ongoing problems in the Democratic Republic of Congo, the Central African Republic and South Sudan illustrate the challenges that remain. The evidence of rising urban protests, as discussed by Bond (Chapter 36, this volume), illustrates the uncertain future that lies ahead and the attention that African governments will have to give to their urban populations as they continue to grow as a proportion of national electorates. However, as has been argued by a number of contributors to this volume, the solutions to these challenges have to be African, rather than ones that are imposed from outside (see especially Booth, Chapter 45, this volume).

Future issues for Africa will be difficult to discern, but it seems clear from the process of editing the contributions for this volume that they could well be focused around some of the following questions.

Will Africa achieve gender equality?

There have been significant achievements made under the MDGs, suggesting that conditions have improved for women and girls. However, some of the indicators are still poor by international standards, and on other measures there is still a lot that can be done to improve the conditions and life chances for girls and women, and to improve the opportunities that they have to contribute to the solutions and development across the continent. This is perhaps most evident in the issue of access to education and training (Stephens, Chapter 9, this volume), but also in the framing of men and women's roles in the development project (Barrett, Chapter 7, this volume). The emergence of female heads of state in the twenty-first century is commendable, but women need to join men as politicians, leaders of commerce, industry and public service, and also needed is the enfranchisement of women at a local level in decision-making and ownership of resources. As Barrett (Chapter 7, this volume) has argued, the lack of gender equality is holding back other dimensions of development across the continent.

How can Africa empower its children and youth?

Africa is the world's youngest continent, with under-15-year-olds accounting for 41 per cent of the population (Ansell, Chapter 10, this volume). This leads to a high demand for enfranchisement among those young people, leading particularly to demand for education and employment opportunities. Disaffected youth have been shown to have led to events such as Sierra Leone's civil war and the Arab Spring that affected North Africa. This shows that the future will need plentiful opportunities for youth employment, education and training that can lead to gainful and sustainable employment opportunities. However, as Ansell (Chapter 10, this volume) argues, Africa needs to be able to develop its own construction of youth and childhood in order to provide appropriate theoretical underpinning in response to the challenges of the future.

Will Africa be able to obtain a fair price for its natural resources?

African countries – and before that colonies – have long had their development defined in terms of their relationships with European powers keen to extract natural resources. The twenty-first century has seen the rise of China and India as customers seeking the natural resources that Africa has to offer. African countries should take advantage of the shifting geography of international trade and find ways to obtain a fair price for their natural resources, and also find ways to add value to their raw materials and expand livelihood opportunities (Binns and Nel, Chapter 20, this volume). There are a number of countries that will have to find ways to ensure that they are able to make the most of their wealth in natural resources and avoid the 'resource curse' (Bond, Chapter 36, this volume; Maconachie, Chapter 39, this volume). Examples include Tanzania, Zambia and the Democratic Republic of Congo.

What can Africa do about its 'basket cases'?

Countries such as South Sudan, Somalia, the Central African Republic and the Demographic Republic of Congo suffer from chronic state fragility, frequently falling back into regional civil unrest at best, or outright civil war at worst. These countries remain vulnerable to exploitation of their high-value natural resources, in demand by global transnational corporations (Maconachie, Chapter 39, this volume), and also vulnerable to terrorism, such as Al-Qaeda taking advantage of weaknesses in Mali and Chad, ISIS establishing bases in Libya, and the emergence and persistence of Boko Haram in northern Nigeria.

Will African cooperation help development in the continent?

As Lyons and Jolley (Chapter 5, this volume) point out, many of Africa's major challenges affect all of Africa's countries, so the solutions to many of Africa's challenges could well require the involvement of the African Union (AU). The AU's willingness to become more involved in resolving major international issues on the continent shows that there is some political will among the members to develop the AU in this direction. Greater unification through the AU and other regional groupings, such as the Southern African Development Community (SADC) and Economic Community of West African States (ECOWAS), as well as the re-emergence and expansion of the East African Community (EAC), are all positive signs. These may not be the forms of pan-Africanism in the original vision of Kwame Nkrumah, but they do suggest the emergence of twenty-first-century African international solutions to African international challenges.

How can Africa provide for its growing urban populations?

The current demographic trends and process of urbanisation mean that during the twenty-first century, Africa will both exceed 2 billion people and become predominantly urban-based. These two major trends will shape the future of the continent, but they are uneven in their speed and extent. These will provide both opportunities and threats (Rigon *et al.*, Chapter 31, this volume). The gathering of the majority of the population in urban settlements will hopefully bring opportunities of providing access to public services more easily than to a dispersed rural population, but it is likely that the fastest growth will be in the informal settlements. There are positive models emerging across the continent, including the Kibera Public Space Project and projects led by the Africa Centre for Cities in Cape Town (mentioned by Myers, Chapter 35, this volume). However, there are also examples of informal settlements where living conditions are extremely poor, and in some cases these settlements are at best being neglected and at worst being criminalised or physically cleared. However, as Rigon *et al.* (Chapter 31, this volume) argue, many of the activities in the informal sector are linked to formal sector economic activities, and finding ways to engage and empower the residents of these settlements will be crucial to the success of Africa's growing twenty-first-century cities.

Can Africa develop its own models of development?

The twentieth century saw the development of independent nations after the period of colonisation. African models of development emerged during this period, such as Julius Nyerere's 'ujamaa' development focused on his conceptualisations of African forms of socialism and community cooperation that he developed for Tanzania. Kenneth Kaunda's 'African – or Zambian – humanism' focused on rural development and state control of the economy, while Nkrumah's 'pan-Africanism' focused on African identities and unity while he emerged as the leader of Ghana, the first sub-Saharan state to win independence (Vokes, Chapter 2, this volume). However, the continent is still overwhelmingly dominated by the ramifications of Bretton Woods' models imposed from outside (Booth, Chapter 45, this volume). The twenty-first century is crying out for distinctively African models of development that can respond to the contemporary challenges facing everyday African communities, as well as developing productive rather than exploitative international African relations.

How will external powers impact on the continent?

Bond (Chapter 36, this volume) illustrates the extent of international involvement in the economy and development of Africa, including the traditional economic powers of the USA, the UK and France. Bond also indicates the growth of China and the United Arab Emirates as large investors, both in terms of aggregate foreign direct investment and in developing the largest infrastructural investment projects. One impact of this external involvement has been the growing trend in what has been described as 'land grabs' (Batterbury and Ndi, Chapter 43, this volume). The role of external powers such as China and Brazil has been described as a form of neocolonialism that has emerged as a result of global changes. Johnston (Chapter 44, this volume) points out that Africa is as important to China's twenty-first-century development as China is now to Africa's. The question remains about the extent to which such investments and acquisition of land become a new form of 'resource curse', or whether African governments and communities can develop solutions to ensure that these processes actually benefit the nations and local communities involved.

Can Africa develop continent-wide good governance?

When the international community look at Africa, they tend to readily think of the problem of corruption and how this holds back development and places challenges in the way of engagement. There are, however, two counterarguments to this focus: first, where there is a culture of corruption, it may well have developed during systems of patronage introduced by the European colonialists; and second, where corruption exists in international trade, it is sustained as much by those who pay bribes as it is by those who demand them. There are positive signs emerging in the early period of the twenty-first century in the form of populist protest and elections, for example in the development of urban protests that have taken corruption as a focus. Examples of this include protest in South Africa about recent controversies around the ANC government involvement with international businesses, the 'citizens' broom' ('*le balai citoyen*') sweeping President Blaise Compaoré out of office in Burkina Faso in 2014, or the overthrow of Presidents Zine El Abidine Ben Ali in Tunisia and Hosni Mubarak in Egypt. Social media and access to mobile technology were key components in the emergence of the popular protest movements in these countries. They were also a key element of the populist campaign that emerged in Nigeria and went viral globally around the abduction of the 'Chibok girls' with the #BringBackOurGirls campaign. The future will require effective democratic engagement of such popular campaigns that will lead to positive changes, such as the local and provincial successes of the Democratic Alliance in challenging the post-apartheid dominance of the African National Congress in South Africa, and the successful election of John Pombe Magufuli as President of Tanzania on the back of an anti-corruption campaign and the associated social media campaign that emerged, focused on #WhatWouldMagufuliDo.

What is evident from the discussion of these questions is the overlap between several, and the importance of an integration in developing good governance and achieving youth and gender equity in order to be in a position to tackle the potential challenges presented by the growth in urban populations and the growing international links with emerging economic powers such as China and the Uinted Arab Emirates. Mensa-Bonsu (2016: 12) argues that 'the slogan "African solutions to African problems" will remain hollow unless African researchers step up and produce relevant research to ground proffered solutions in local realities'. This observation underlines the need for researchers working in Africa in the early twenty-first century to give greater focus to analysing the evidence of local realities, in order that their interpretations and conclusions can have impact on the development of African solutions.

References

Bratton, M. and van de Walle, N. (1997) *Democratic Experiments in Africa: Regime Transitions in Comparative Perspective*, Cambridge: Cambridge University Press.

Brookings Institute (2017) *Foresight Africa: Top Priorities for the Continent in 2017*, Washington, DC: Africa Growth Initiative at the Brooking Institution.

Mensa-Bonsu, H. (2016) *'African Solutions for African Problems': Where Is the Research? Director of Legon Center for International Affairs and Diplomacy (LECIAD)*, keynote address delivered at the African Peacebuilding Network (APN) Grantee Training Workshop, LECIAD, University of Ghana, Accra, 27 June.

Index

Note: Indicators for tables are given in **bold**; figures in *italics*.

Made in the USA
Columbia, SC
30 April 2024

35105463R00396